1st International Conference on Civil & Environmental Engineering (CENVIRON 2017)

E3S Web of Conferences Volume 34 (2018)

Penang, Malaysia
28-29 November 2017

Editors:

N. Mohamed Noor **A. W. Azhari**

ISBN: 978-1-5108-5979-1

Printed from e-media with permission by:

Curran Associates, Inc.
57 Morehouse Lane
Red Hook, NY 12571

proceedings.com

Some format issues inherent in the e-media version may also appear in this print version.

Printed by Curran Associates, Inc. (2018)

For additional information, please contact EDP Sciences – Web of Conferences at the address below.

EDP Sciences – Web of Conferences
17, Avenue du Hoggar
Parc d'Activité de Courtabœuf
BP 112
F-91944 Les Ulis Cedex A
France

Phone: +33 (0) 1 69 18 75 75
Fax: +33 (0) 1 69 28 84 91

contact-edps@webofconferences.org

Additional copies of this publication are available from:

Curran Associates, Inc.
57 Morehouse Lane
Red Hook, NY 12571 USA
Phone: 845-758-0400
Fax: 845-758-2633
Email: curran@proceedings.com
Web: www.proceedings.com

TABLE OF CONTENTS

CENVIRON

INTERNATIONAL CONFERENCE ON CIVIL & ENVIRONMENTAL ENGINEERING
PENANG 28 - 29 NOVEMBER 2017

1st International Conference on Civil and Environment (CENVIRON 2017)

28 – 29 November 2017, Penang, Malaysia

Editors:
Fahmi Muhammad Ridwan, Norazian Mohamed Noor, Ayu Wazira Azhari, Sara Yasina Yusof, Farrah Aini Dahalan, Farah Naemah Mohd Saad, Ain Nihla Kamarudzaman, Shamilah Anudai@Anuar

Organizer:
School of Environmental Engineering, Universiti Malaysia Perlis, Malaysia

UniMAP UNIVERSITI MALAYSIA PERLIS

PREFACE

In recent years, increasing number of researches has been conducted in view of achieving sustainable development without compromising the ability of future generations to meet their own needs. In line with this, the School of Environmental Engineering, Universiti Malaysia Perlis (UniMAP) has initiated the very first International Conference on Civil and Environmental Engineering (CENVIRON2017) that will become a perfect platform for academia and the industry to discuss and exchange ideas on the latest development and recent findings on topics related to green and sustainable development.

CENVIRON 2017 is dedicated to cutting edge research that addresses scientific needs and integrates knowledge and progress among academicians, researchers, industries, corporate organizations and interest groups in various civil and environmental related fundamentals and applications. This conference brings together leading researchers, engineers and scientists in the domain of interest from around the world, as it covers the entire range of basic and applied aspects of the current research and development towards achieving green and sustainable environment.

Among the objectives of this conference are; to promote the sharing of knowledge and expertise among academia, researchers and practitioners who work closely in the field of civil and environmental engineering; to provide a platform for discussion and exchange of views on the opportunities that arises in various civil and environmental related fundamentals and application through collaborations between industry and academia; and to support and motivate 'Young Researchers' on establishing their academic and professional relationships. Significantly, to exchange ideas and experiences as well as share their knowledge in understanding and identifying issues related to green and sustainable environment.

The Organizing Committee would like to express the warmest thanks to all Authors of the conference publications for the effort put into their elaboration, and to the Reviewers for the reliable evaluation of the works.

On behalf of the Organizing Committee
Chairman, Ir Dr Mohd Asri Ab Rahim

edp sciences

Web of Conferences

Statement of Peer review

In submitting conference proceedings to *Web of Conferences*, the editors of the proceedings certify to the Publisher that

1. They adhere to its **Policy on Publishing Integrity** in order to safeguard good scientific practice in publishing.
2. All articles have been subjected to peer review administered by the proceedings editors.
3. Reviews have been conducted by expert referees, who have been requested to provide unbiased and constructive comments aimed, whenever possible, at improving the work.
4. Proceedings editors have taken all reasonable steps to ensure the quality of the materials they publish and their decision to accept or reject a paper for publication has been based only on the merits of the work and the relevance to the journal.

Title, date and place of the conference

International Conference on Civil and Environmental Engineering (CENVIRON2017)
28 - 29 November 2017
Penang, Malaysia

Proceedings editor(s):

Norazian Mohamed Noor (PhD)
Ayu Wazira Azhari (PhD)

Date and editor's signature

23 February 2018

The mechanical properties of brick containing recycled concrete aggregate and polyethylene terephthalate waste as sand replacement

Faisal Sheikh Khalid[1*], *Nurul Bazilah* Azmi[1], *Puteri Natasya* Mazenan[1], *Shahiron* Shahidan[1], and *Noorwirdawati* Ali[1]

[1]Jamilus Research Centre for Sustainable Construction (JRC), Faculty of Civil and Environmental Engineering, Universiti Tun Hussein Onn Malaysia, Johor, Malaysia

Abstract. This research focuses on the performance of composite sand cement brick containing recycle concrete aggregate and waste polyethylene terephthalate. This study aims to determine the mechanical properties such as compressive strength and water absorption of composite brick containing recycled concrete aggregate (RCA) and polyethylene terephthalate (PET) waste. The bricks specimens were prepared by using 100% natural sand, they were then replaced by RCA at 25%, 50% and 75% with proportions of PET consists of 0.5%, 1.0% and 1.5% by weight of natural sand. Based on the results of compressive strength, only RCA 25% with 0.5% PET achieve lower strength than normal bricks while others showed a high strength. However, all design mix reaches strength more than $7N/mm^2$ as expected. Besides that, the most favorable mix design that achieves high compressive strength is 75% of RCA with 0.5% PET.

1 Introduction

Cement and sand bricks are a type of bricks that is commonly used in low and medium cost housing development and other commercial constructions in Malaysia since it is easy to produce and cheap [1]. However, there is an issue in producing these materials especially in developing areas where manufactures find it difficult to locate adequate sources of natural cement and aggregate supply [2-3]. Polyethylene Terephthalate (PET) bottles are non-biodegradable substance brings quite problematic because it stays longer in the environment [3]. Bottle waste in food and beverage industries already constitute about 10% of total municipal solid waste [4]. Moreover, PET bottles can produce uncontrolled fire or contaminate the soil and vegetation if the materials be disposed by dumping or burning [5-6].

Earlier research work has shown the potential of using Recycle Concrete Aggregate (RFA) in engineering applications. According to Kumar et al. [7], in order to increase the compressive strength of RFA, these can be improve by the addition of fiber. In addition, aggregates are represented for 70-80% of the concrete components therefore it will be

*Corresponding author: faisalsh@uthm.edu.my

beneficial on preserves a recycle demolished concrete as natural aggregate besides to solve the environmental problems [8-9].

When RCA is demolished by using crusher, it can produce aggregate which the particle of aggregate have variety of sizes and can be categorized by its size either coarse or fine aggregate. Evangelista et al. [9] and Khatib [10] interest in using RCA as fine aggregates in construction materials. They mentioned that fine aggregate would not contribute in the reduction of compressive strength and elastic modulus. Thus, there is a huge potential on substitution of RCA as a fine aggregates in construction materials such as brick. The previous study stated that the value of water absorption of concrete with RCA aggregates is higher compared to concrete with natural aggregate [11]. The durability of RCA seems less durable compared to the NA. It is due to the ratio coarse aggregate replacement and the concrete age itself, make the RA less durable compared to the NA because it has high porosity value. The replacement of natural aggregate from 5 to 100% with the recycled aggregate reduce their compressive strength by 5 to 25%. The water absorbability of RCA is higher, up to 24% compared with the NAC. As stated by Irwan et al. [12] that the compressive strength of concrete containing all types of PET-aggregates are similar to conventional concrete. However, the strength starts decreasing when the volume of PET replacement up to 2.5%. Therefore, the volume replacement is very significant to study to determine the optimum replacement of natural sand.

One of the possibilities is to utilize the RCA and PET bottle wastes as fine aggregate replacement. Therefore, this study aims to investigate the potential of RCA and fine PET wastes as partial replacement fine material compared to conventional sand cement brick. Therefore, the compressive strength and water absorption tests were conducted to meet the objective of the study.

2 Methodology

The design mix for sand cement brick used accordance with BS EN 998-2, where the prescribed mortar for masonry cement sand is 1:3 and it categorized in class M6. The optimum design mix of composite brick was 0.5%, 1.0% and 1.5% of PET and 25%, 50%, 75% of RCA, respectively as partial sand replacement. The brick moulded with a mould size of 215mm in length, 105mm in width, and 65mm in depth. Curing is a process that involves the maintenance of a suitable moisture content and temperature in samples for a suitable period of time immediately after placing and finishing so that desired properties of samples can be achieved. After pressing, the bricks were stacked on timber palettes and marked according to their percentage of aggregate composition.

3 Results and discussions

3.1 Density Test

The results for density against ages for each percentage of RCA and PET in composite sand cement brick were shown in Figure 1. The density steadily decreased with the increasing of RCA and PET content in the mixtures. RCA and PET were known as a material that has low density, which proved the replacement of sand with these materials contributed to the reduction in mass of the brick samples. In addition, the lower value of specific gravity of recycled aggregates and PET shows that these fine aggregates are lighter than that of natural fine aggregates.

 https://doi.org/10.1051/e3sconf/20183401001

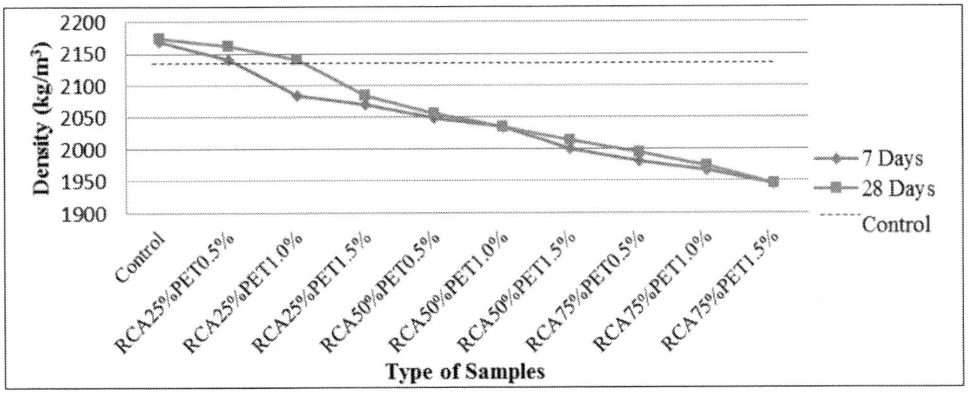

Fig. 1. The average density of composite sand cement bricks with RCA and PET versus.

3.2 Compressive strength

It can be seen clearly that the strength of composite brick on 28 days shows that the strength starts to increase as the replacement of waste is increasing as shown in Fig. 2.

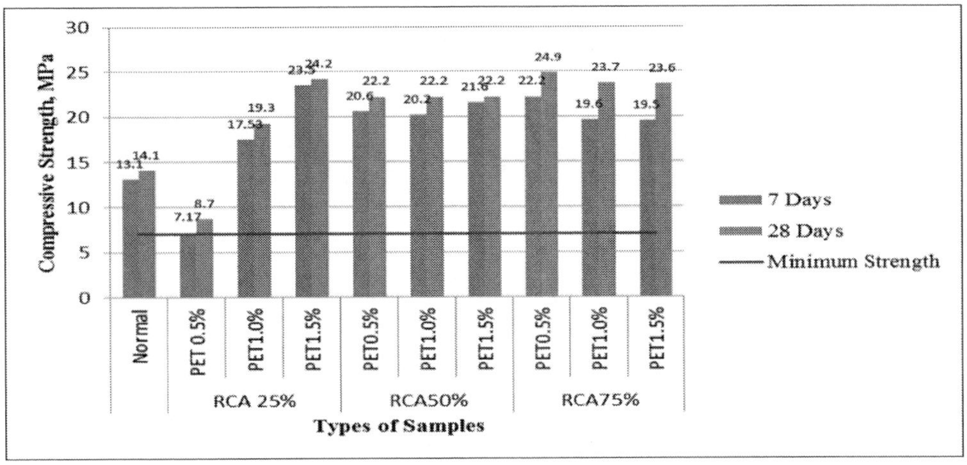

Fig. 2. Compressive Strength at 7 and 28 Days.

All the samples achieve strength more than the minimum strength for both 7 Days and 28 Days as BS 6073-1:1981 stated that the compressive strength of bricks should not be less than 7 N/mm². It is obviously showing that some samples for both curing age achieved high strength from the normal brick. The sample with RCA 25% with PET 1.5% exhibit the second highest strength whereas RCA 25% with PET 0.5% achieve only 7.17 MPa which is more lower strength rather than normal brick. Nevertheless, the specimen with RCA of 25%, it shows the strength increased in line with the increasing of PET volume for 0.5%, 1.0% and 1.5%.

The increasing pattern of compressive strength is seen to be in samples containing RCA 75% for both curing age include 7 and 28 days. The addition of 0.5% PET with 75% of RCA gives the highest strength among the other samples. The reduction of strength is shown when the volume of PET increase from 1.0% to 1.5% for RCA 75%. In addition, the highest percentage of increment strength between 7 days to 28 days among other samples is

17.30% and 17.37% respectively for 75% of RCA with a volume of PET for 1.0% and 1.5%.

Moreover, the results reveal that only certain amount of PET fiber increase the strength with an addition of RCA of 25%, 50% and 75%. The trend of this study having the same pattern with the studied conducted by Irwan et al. [2] and Noorwirdawati et al. [13] that stated the good bonding between aggregate and binder as an increase in synthetic plastic content may result a decreasing in strength of materials.

3.3 Water absorption

Water absorption was carried out to determine the percentage of water absorption of the brick. The details result of water absorption against age for each percentage of RCA and PET were given in Fig. 3. It shows the evolution of water absorption from 7 days to 28 days for all mixtures. The results illustrated that the water absorption increased with the replacement percentage of fine aggregate by the RCA and PET wastes for all mixtures. However, up to RCA50%PET1.0% replacement, a slight decreased of water absorption of composite cement brick for both 7 and 28 days. The results indicate that the sample with RCA50%PET1.0% was less permeable when compared to other samples. Furthermore, the highest water absorption coefficients were obtained in composite sand cement brick with RCA75%PET1.5%. Samples with 75% of RCA absorbed more water than did samples containing 50% and 25% of RCA.

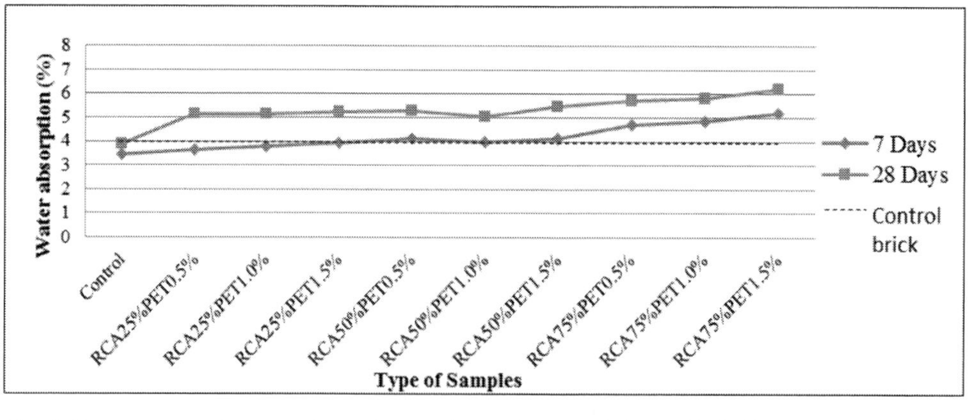

Fig. 3. The average water absorption of composite sand cement bricks with RCA and PET versus ages.

4 Conclusions

As the conclusion, the result can be concluded that the reduction in bricks density was attributed by the lower unit weight of RCA and PET. The sand replacement ratio of RCA and PET at levels 75% and 0.5% has good effects on the compressive strength of the bricks. From the result of water absorption, composite sand cement bricks with recycled fine aggregate and PET significantly affected the increased water absorption characteristic of bricks. However, only bricks with 50% RCA and 1.0% PET show less permeable compared to control brick.

This work was financially supported by University Tun Hussein Onn Malaysia (UTHM) and Minister of Education Malaysia (KPM) through Dana Pembudayaan Penyelidikan (RAGS) Vot. No. R058.

References

1. W. Kubissa, R. Jaskulski, A. Koper, J. Szpetulski, Procedia Engineering. **108**, 30–38 (2015).
2. J.M. Irwan, S.K. Faisal, N. Othman, M.H. Wan Ibrahim, R.M. Asyraf, M.M.K. Annas, Adv. Mater. Res. **795**, 352-355 (2014a).
3. P.N. Mazenan, F.S. Khalid, S. Shahidan, S. Shamsuddin, IOP Conf. Series: Materials Science and Engineering. **271** (2017).
4. O.Y. Marzouk, R.M. Dheilly, M. Queneudec, Waste Management. **27**, 310–318 (2007).
5. J.M. Irwan, M.M.K. Annas, A.K. Aeslina, R.M. Asyraf, S.K. Faisal, Adv. Mater. Res., **911**, 474-478 (2014b).
6. K. Ramadevi, R. Manju, International Journal of Engineering Technology and Advanced Engineering **2**(6), 139-150 (2012).
7. V. Kumar, M. K. Poonia, V. S. Saggu, International Journal for Technological Research In Engineering **2**(11), 2684–2689 (2015).
8. N. B. Azmi, F. S. Khalid, J. M. Irwan, N. Anting, P. N. Mazenan, IOP Conf. Series: Materials Science and Engineering **271** (2017).
9. L. Evangelista, J. D. Brito, Cem. Concr. Res. **29**, 397-401 (2007).
10. J. M. Khatib. Cem. Concr. Res. **35**, 763-769 (2005).
11. S. Yehia, K. Helal, A. Abusharkh, A. Zaher, H. Istaitiyeh, International Journal of Concrete Structures and Materials **9** (2), 219–239 (2015).
12. J.M. Irwan, R.M. Asyraf, S.K. Faisal, N. Othman, Wan Ibrahim M.H., Applied Mechanics And Materials **7**, 911 (2015).
13. N. Ali, N. Din, F. S. Khalid, S. Shahidan, S. R. Abdullah, A. A. Abdul Samad, N. Mohamad, IOP Conf. Series: Materials Science and Engineering **271** (2017).

Assessment of soil compaction properties based on surface wave techniques

Nur Jihan Syamimi Jafri[1,*], *Mohd Asri* Ab Rahim[1], *Mohd Zulham Affandi* Mohd Zahid[2], *Nor Faizah* Bawadi[1], *Muhammad Munsif* Ahmad[2], *Ahmad Faizal* Mansor[2], and *Wan Mohd Sabki* Wan Omar[1]

[1]School of Environmental Engineering, Universiti Malaysia Perlis, Kompleks Pusat Pengajian Jejawi 3, 02600 Arau, Perlis, Malaysia
[2]Department of Civil Engineering Technology, Faculty of Engineering Technology, Universiti Malaysia Perlis, 02100 Padang Besar, Perlis, Malaysia

Abstract. Soil compaction plays an important role in every construction activities to reduce risks of any damage. Traditionally, methods of assessing compaction include field tests and invasive penetration tests for compacted areas have great limitations, which caused time-consuming in evaluating large areas. Thus, this study proposed the possibility of using non-invasive surface wave method like Multi-channel Analysis of Surface Wave (MASW) as a useful tool for assessing soil compaction. The aim of this study was to determine the shear wave velocity profiles and field density of compacted soils under varying compaction efforts by using MASW method. Pre and post compaction of MASW survey were conducted at Pauh Campus, UniMAP after applying rolling compaction with variation of passes (2, 6 and 10). Each seismic data was recorded by GEODE seismograph. Sand replacement test was conducted for each survey line to obtain the field density data. All seismic data were processed using SeisImager/SW software. The results show the shear wave velocity profiles increase with the number of passes from 0 to 6 passes, but decrease after 10 passes. This method could attract the interest of geotechnical community, as it can be an alternative tool to the standard test for assessing of soil compaction in the field operation.

1 Introduction

Soil compaction is an important process in every construction activities in order to increase the bearing capacity of soil. It involved the densification of the ground by applying the mechanical energy such as static compaction, sheepsfoot, vibratory roller and plates, rammers, heavy tamping, vibroflotation and rolling dynamic compaction [1]. Conventionally, the assessment of soil compaction is conducted by soil sampling and laboratory or field testing such as Standard Penetration Test (SPT), Cone Penetration Test (CPT), Dynamic Cone Penetrometer (DCP), in-situ density tests, etc. These techniques are widely accepted within geotechnical projects. However, these techniques are caused time-

* Corresponding author: nurjihansyamimi@gmail.com

consuming to cover large areas. The assessment process must be fast to produce results in the field for immediate assessment and directly evaluate the improvement of properties as an alternative to empirical correlations [2].

Recent years, the surface wave methods have been widely used in many geotechnical engineering projects in terms of cost and time effective compared to the conventional techniques. These applications include assessment of liquefaction potential [3], geotechnical verification of compaction [2, 4, 6], and site characterization [5]. These non-destructive methods have been proved to be efficient compared to the conventional invasive mechanical techniques where the shear wave velocity (V_S) is theoretically related to the shear modulus, G_{max} as shown in the Equation 1 below:

$$G_{max} = \rho \cdot V_S^2 \qquad (1)$$

where, G_{max} is the shear modulus (pa), V_S is the shear wave velocity (m/s) and ρ is the density (kg/m^3).

Surface wave methods have utilised Rayleigh wave for near surface applications, where it travels along the free surface of the earth with elliptical retrograde particle motion. The use of surface waves for determining the dynamic properties of the subsurface are based on their dispersive characteristics, where the velocity propagation of Rayleigh wave is depend on the frequency. Thus, the higher frequency will penetrate the shallow layer while the low frequency will influenced the deeper material.

The first surface wave method was introduced [7], called as Spectral Analysis of Surface Waves (SASW) that used the spectral analysis of ground roll generated by an impulsive source and recorded by two receivers to generate the near-surface of V_S profile. This method is widely used in evaluation of pavement systems [8], evaluation of compacted ground [9, 10] and site characterization [11]. However, a new method called Multichannel Analysis of Surface Waves (MASW) has introduced, to develop an efficient and accurate method to estimate near-surface V_S from ground roll. MASW technique has developed to overcome the weaknesses of the SASW, which used the dispersion characteristics of Rayleigh waves to determine the variation of the V_S in layered systems with depth [12]. MASW technique is an ideal tool to speed up the control process in the evaluation of ground modifications without the requirement to include geogrid reinforcement [2]. MASW technique is a cost-effective tool for obtaining reliable shear wave velocity profiles in estimation the engineering properties of soil [3].

Tokeshi et al. [5] has investigated the use of the non-invasive Multichannel Analysis of Surface Wave (MASW) and Multichannel Simulation with One Receiver (MSOR) techniques to provide an evaluation of compacted ground and a general geotechnical site characterization. The results from both techniques were verified against CPT and SPT data. The correlation observed between the experimental Rayleigh dispersion and the V_S ground model estimated from Rayleigh dispersion inversion has confirmed these techniques are useful for evaluating the Vs ground profile. In agricultural, Donohue et al., [13] also explored the possibility of MASW for assessment of agricultural compaction in a field environment. The results show the significance difference in shear wave velocity between heavily compacted headland and uncompacted soil. The results from this approach has compared with those obtained from measurement of bulk density and penetrometer resistance. However, there is limitation of the surface wave approach for detection compaction, which is the resolution of the technique at very shallow depths.

This paper presented the evaluation of soil compaction after applying the roller compaction using Multichannel Analysis of Surface Wave (MASW) method. The aim of this paper is to determine the shear wave velocity profiles and field density of compacted soils under varying compaction efforts by using MASW method.

 https://doi.org/10.1051/e3sconf/20183401002

2 Methods

2.1 Site location

The test site is shown in Fig. 1(a), which is located near to Litar Pauh Putra Campus, UniMAP. The site has 5 meters depth of soil and was reclaimed using fill material of well graded sandy gravel. The testing was carried out in two stages: before compaction and after compaction. During the first stage, the MASW survey and sand replacement were conducted on the single line survey. The second stage, a 16-tonne roller compactor was used with 2, 6 and 10 passes, continued with the MASW survey and sand replacement test. Fig. 1(b) showed a layout plan of the survey area.

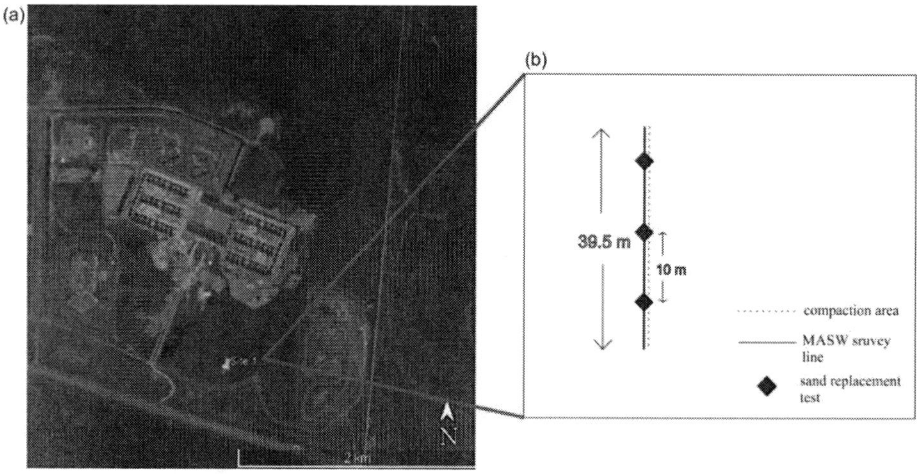

Fig. 1. (a) The site location is shown in Google Earth image and (b) Layout plan of the survey area.

2.2 In-situ density test

Sand replacement method was performed by following the procedure of BS 1377:Part 9:1990 [14] to determine in-situ density of compacted soils. The calibration was conducted before doing the testing to determine the mass of sand in the pouring cylinder. The sampling was performed at three locations along the MASW survey line, as presented in Fig. 1(b) to obtain the mean mass of dry densities of the soil.

Along the survey line, the metal tray was laid and a round hole was excavated approximately 100mm in diameter and maximum of 150mm deep into the ground. All excavated soil from the hole was collected to determine the mass of sand. Then, the pouring cylinder with sand filled was placed on the excavated hole and allowed sand to fill the hole until no movement of the sand. The cylinder was removed. After the mass of poured sand, bulk density of sand and soil, mass of sand filled in the hole and moisture content of the soil was calculated, the dry density of the soil was determined.

2.3 MASW data acquisition and processing

For this testing, 7 kg sledgehammer was used as the seismic source by hitting the striker plate. The seismic source was positioned at 5 m from the first geophone. 24 units of 4.5 Hz vertical geophones were deployed vertically and firmly with equal interval on the soil. The spacing between geophones was kept at 1.5 m, which resulted in a total receiver array length of 34.5 m as shown in Fig. 2. The geophones were connected to the spread cable and recorded using a 24-channel Geode seismograph.

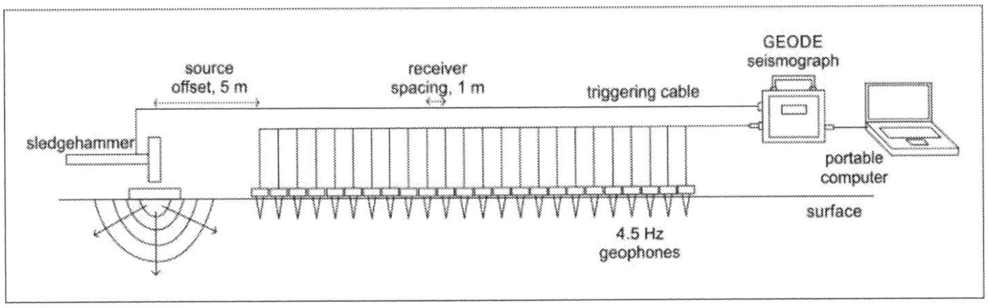

Fig. 2. The field configuration of MASW survey.

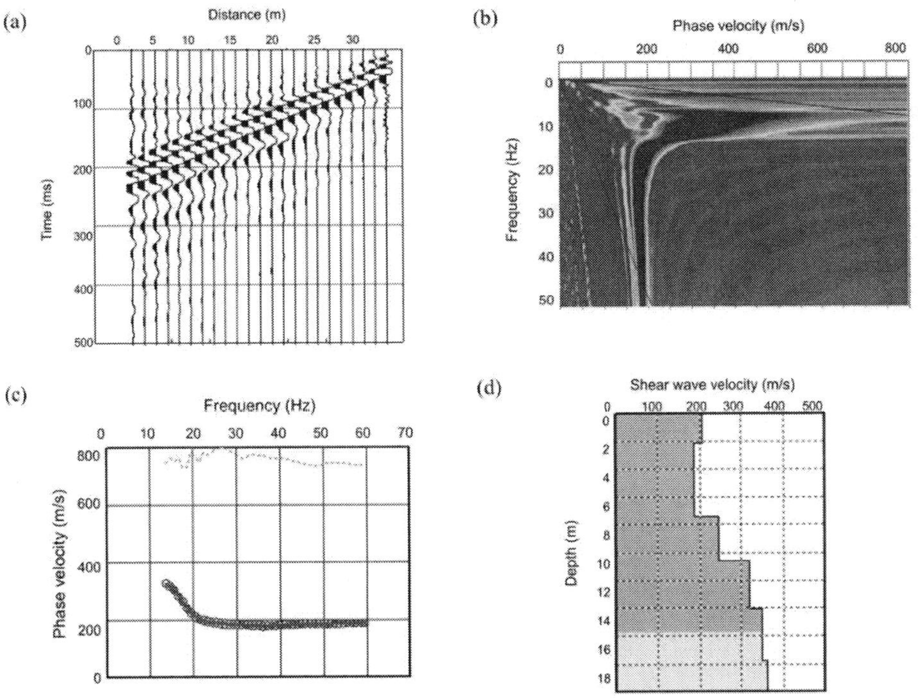

Fig. 3. The procedure of MASW data processing using SeisImager software, a) detection of surface waves, b) image of dispersion curve, c) inversion of dispersion curve, d) 1D shear wave velocity profile.

In order to reduce background noise and prevent interference with resolution and accuracy of the data, the data were acquired by hitting the striker plate during periods of little nearby traffic and/or wind. For data acquisition, the recording time and sampling intervals were set at 1 second and 0.5 milliseconds with three times stacking.

The seismic raw data were processed and analysed using SeisImager/SW software. The procedure of MASW consists of three steps: (1) active data acquisition, (2) developing the dispersion curves and (3) inversion of dispersion data to obtain the shear wave velocity profiles. Fig. 3(a) shows the raw wiggle plot obtained from the field test. The recorded seismic waves were analysed to generate phase velocity images in frequency domain as shown in Fig. 3(b). Then, the dispersion curves were constructed from the phase velocity images using the signal-to-noise ratio (Fig. 3c). The phase velocity image was chosen in the frequency range between 5 to 65 Hz. The inversion of dispersion curve was applied using a non-linear square technique to generate the one-dimensional shear wave velocity profile, as illustrated in Fig. 3(d) [12].

3 Results and discussions

3.1 Dry density of compacted soils

Table 1 shows the results of dry density from MASW survey and sand replacement test. In general, dry density of the ground and the number of passes will increases with depth of soil. However, there were inconsistent results of dry density at the surface of the ground from sand replacement method and MASW as the number of passes increase. This condition explained that the area was reclaimed using fill material of well graded sandy gravel. Thus, it can be due to the reclamation process, the soil particles are in stage to firm.

Table 1. Results of dry density from MASW and sand replacement tests

No. of passes	Shear wave velocity (m/s) below 1 m	Dry density (Mg/m^3) from sand replacement	Moisture content (%)	Dry density (Mg/m^3) from MASW	Percentage difference of dry density (%)
0	192 – 205	1.75	0.76	1.79	2.23
2	170 – 181	1.79	0.72	1.78	0.56
6	178 – 190	1.73	2.05	1.79	3.35
10	175 – 189	1.77	1.34	1.79	1.12

The density from MASW test was calculated using the Equation 2 from Ludwig et al. [15] and the results of soil density with depth from MASW tests is illustrated in Fig. 4.

$$\rho = 1.2475 + 0.399V_p - 0.026V_p^2 \quad (2)$$

3.2 Shear wave velocity profile

The results of the shear wave velocity profile for before and after compaction with 2, 6, and 10 passes has shown in Fig. 4. Generally, the shear wave velocity will increase with depth as the number of passes increases. The ranges of shear wave velocity before compaction (18 m depth from surface) is between 192 m/s to 537 m/s. From ground surface to 3 m depth, the shear wave velocity has decreased after compaction started. This resulted affected due to the process of soil reclamation after applying roller compactor.

Meanwhile, the shear wave velocity for 2 and 6 passes has increased about 7 to 100 m/s at 1 to 18 m depth. This indicates that the soil density increases as the number of passes increases. However, after 10 passes of compaction, shear wave velocity has decreased at 4.5 to 18 m depth. The results in V_S inconsistencies in depth cause due to the low resolution during the process of dispersion curve in picking for low frequencies [3].

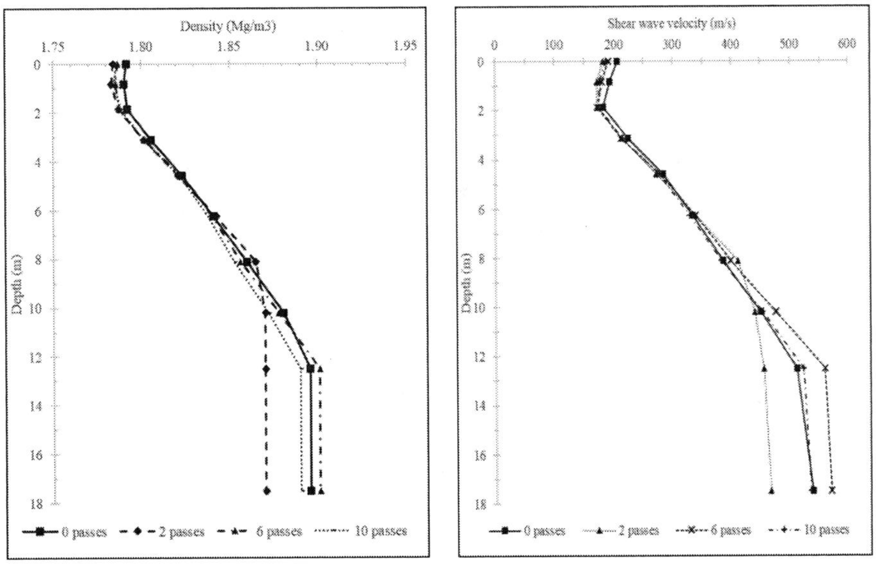

Fig. 4. Soil density and shear wave velocity profile of MASW with different number of passes.

4 Conclusions

The shear wave velocity profiles of before compaction and after compaction were determined using Multi-channel Analysis of Surface Wave (MASW) technique. This research has achieved the objective where the percentage differences of dry density from field density test and MASW survey is less than 10% of standard acceptance, thus achieved a requirement of geotechnical design. The MASW method can be an alternative tool to the standard test for assessing of soil compaction in the field. In addition, MASW method has the advantage of being fast, practical and non-destructive tool for engineers in measuring dry density and shear wave velocity compared to the conventional methods. The procedures of this method need to be explored in depth since it is suitable and more time effective in evaluating the large areas.

The author would like to acknowledge Ministry of Higher Education (MOHE), Malaysia for the sponsor and financial support received from Fundamental Research Grant Scheme (FRGS: 9003-00546).

References

1. M. R. Hausmann, *Engineering Principles of Ground Modification*. New York, McGraw-Hill (1990)
2. A. Bouazza and D. L. Avalle, Australian Geomech, **41**, 101-108 (2006)
3. C. P. Lin, C. C. Chang, T. S. Chang, Soil Dyn Earthq Eng, **24**, 689-698 (2004)
4. R. J. Whiteley, P. Caffi, *Near Surface Geophysics* **12**, 405-414 (2014)
5. K. Tokeshi, P. Harutoonian, C. J. Leo, S. Liyanapathirana, *Adv. Geosci.*, **35**, 37-44 (2013)

6. S. Donohue, and L. Michael, *Near Surface Geophysics* (2008)
7. S. Nazarian, K. H. Stokoe, Research Rep. 256-264 (1983)
8. S. Nazarian, K. H. Stokoe, W. R. Hudson, *Transportation Research Record*, 38-45 (1983)
9. D. S. Kim, M. K. Shin, H. C. Park, Soil Dyn Earthq Eng **21**, 39-46 (2001)
10. A. Bitri, K. Samyn, S. Brûlé, E. H. Javelaud, Geotechnical Assessment and Geoenvironmental Eng., **11**, 683-690 (2013)
11. T. L. Goh, A.R. Samsudin, A. G. Rafek, Application of spectral analysis od surface waves (SASW) method: rock mass characterization, *Sains Malaysiana*, **40**, 425-430 (2011)
12. C. B. Park, R. D. Miller, J. Xia, Multichannel analysis of surface waves. *Geophy.*, **64**, 800-808 (1999)
13. S. Donohue, F. Dermot, A. L. Donohue, Detection of soil compaction using surface waves, *Soil and Tillage Research*, **128**, 54-60 (2013)
14. *Methods for test for soils for civil engineering purposes, In-situ tests*, BS 1377-9 :1990
15. W. J. Ludwig, J. E. Nafe, C. L. Drake, *The Sea*, **4**, 53-84 (1970)

Strength and Microstructure of Concrete with Iron Ore Tailings as Replacement for River Sand

Ali Umara Shettima[1], *Yusof* Ahmad[2*], *Mohd* Warid Hussin[2], *Nasiru* Zakari Muhammad[2], and *Ogunbode* Eziekel Babatude[2]

[1]Department of Civil Engineering Technology, Federal Polytechnic Damaturu, Yobe State. 620001, Nigeria
[2]Faculty of Civil Engineering, Universiti Teknologi Malaysia, Johor Bahru. 81310, Malaysia

Abstract. River Sand is one of the basic ingredients used in the production of concrete. Consequently, continuous consumption of sand in construction industry contributes significantly to depletion of natural resources. To achieve more sustainable construction materials, this paper reports the use of iron ore tailings (IOT) as replacement for river sand in concrete production. IOT is a waste product generated from the production of iron ore and disposed to land fill without any economic value. Concrete mixtures containing different amount of IOT were designed for grade C30 with water to cement ratio of 0.60. The percentage ratios of the river sand replacements by IOT were 25%, 50%, 75% and 100%. Concrete microstructure test namely, XRD and Field Emission Scanned Electron Microscopic/Energy dispersive X-ray Spectroscopy (FESEM/EDX) were conducted for control and IOT concretes in order to determine the interaction and performance of the concrete containing IOT. Test results indicated that the slump values of 130 mm and 80 to 110 mm were recorded for the control and IOT concretes respectively. The concrete sample of 50% IOT recorded the highest compressive strength of 37.7 MPa at 28 days, and the highest flexural strength of 5.5 MPa compared to 4.7 MPa for reference concrete. The texture of the IOT is rough and angular which was able to improve the strength of the concrete.

1 Introduction

The challenge for the civil engineering community with the concept of sustainable development involves the use of waste materials and by-products at reasonable cost with the lowest possible environmental impact. Rapid increase in consumption of river sand due to the increased in construction activity means that sand mining exploration increased in which the river bed is over exploited [1, 2]. Other environmental issues of Sand mining are depletion of virgin deposits, collapsing of river banks, water table lowering and water pollution [3]. The remedies for these impacts are the use of waste material as alternatives to river sand. Increase in the production of iron ore for economic development worldwide has generated massive amount of iron ore tailing, which are frequently being discarded as wastes. This has led to serious environmental deterioration. A statistical survey has shown

* Corresponding author: a-yusof@utm.my

that 5 to 7 billion tonnes of iron ore tailings were produced yearly worldwide [4]. In spite of such huge amount of iron ore tailings stockpiled as waste, its safe disposal or utilization has remained a major unsolved and challenging task for iron ore industries [5]. The objective of this study is to investigate the performance of concrete with IOT as replacement for river sand in terms of strength and microstructure analysis.

2 Materials and mix proportions

Ordinary Portland cement and natural river sand along with crushed gravel of 10 mm coarse aggregates were locally obtained and used in this study. The IOT was obtained from one iron ore mill in southern State of Malaysia. The chemical composition of natural sand and IOT were determined using X-ray fluorescence (XRF) and the results are shown in Table 1.The physical properties of river sand and IOT were also determined and presented in Table 2. The results of particle size distribution of the river sand and IOT are shown in Fig. 1. FESEM/EDS were determined for river sand and IOT. The results of FESEM/EDS for IOT and river sand are presented in Fig. 2 and 3, respectively. The FESEM/EDS result of IOT consists loose particles, porous and highly rough surface with irregular shape that was well dispersed as shown in Fig. 2, while the river sand had a smoother surface as shown in Fig. 3. The EDS consists of silica (Si), which is prominent element present in both materials with a relative percentage by weight of 57.4% and 61.7% for IOT and river sand, respectively. The other elements that were more prominent in IOT were iron (Fe) and alumina (Al), which was about 8.01% and 5.91%, respectively. The alumina element plays a very important role in the formation of calcium alumina silicate hydrate. The high carbon of IOT has resulted high loss of ignition (LOI) value and consequently high water demand. A Polycarboxylic ether based superplasticizer (Rheobuild 1100) complies with ASTM C494 [6] was used to increase the workability of the mix to avoid increase of water. Five mix proportions were prepared as shown in Table 3. The first mix was control, which is a conventional concrete designated as M0 (i.e. without iron ore tailings), and the other four mixes contained IOT. The river sand was replaced with IOT at replacements levels of 25%, 50%, 75% and 100% and designated as M25, M50, M75 and M100 respectively.

Table 1. Chemical composition of river sand and iron ore tailings.

Chemical composition (%)	SiO	Al_2O_3	Fe_2O_3	CaO	MgO	MnO	K_2O	CUO	ZnO	PbO	SO_3	LOI
River Sand	98	0.4	0.1	0.9	0.2	0.02	1.4	-	-	-	-	0.1
IOT	56	10	8.3	4.3	-	1.7	1.5	0.2	0.1	0.4	-	3.3

Table 2. Physical properties of iron ore tailings and river sand.

Type of test	Sand	IOT
Bulk Specific gravity	2.6	2.8
Relative density (g/cm^3)	1.64	1.27
Fineness modulus	2.77	1.05
Water absorption (%)	3	7.0

https://doi.org/10.1051/e3sconf/20183401003

Table 3. Mix proportion of concrete mixes incorporating iron ore tailings.

Description	M0	Proportion of IOT's			
		M25	M50	M75	M100
OPC (kg/m^3)	380	380	380	380	380
Coarse aggregate (kg/m^3)	1009	1009	1009	1009	1009
Natural sand (kg/m^3)	761	571	380.5	190	-
Iron ore tailings (kg/m^3)	-	190	380.5	571	761
Water (kg/m^3)	230	230	230	230	230
Superplastiizer (%)	0.5	0.5	0.5	0.5	0.5
Slump (mm)	130	110	100	90	80
Compaction factor	0.94	0.94	0.91	0.89	0.85
Concrete fresh density (kg/m^3)	2390	2410	2440	2480	2490

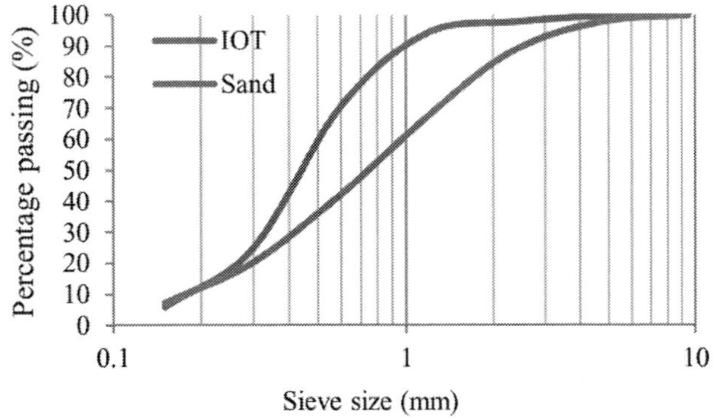

Fig. 1. Grading curve of IOT material and river sand.

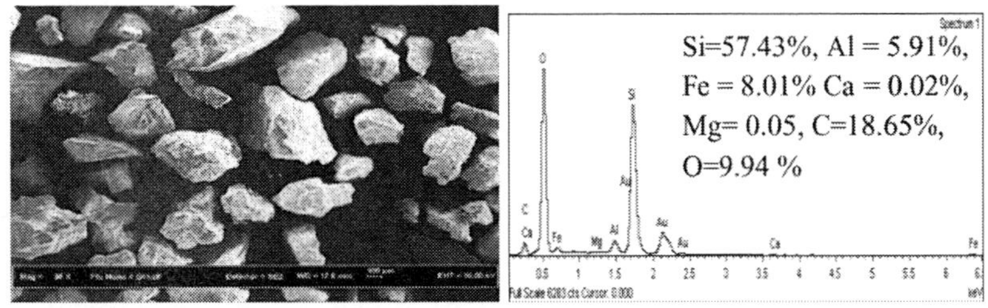

Fig. 2. Energy dispersive X-ray of the spectrum from FESEM of IOT.

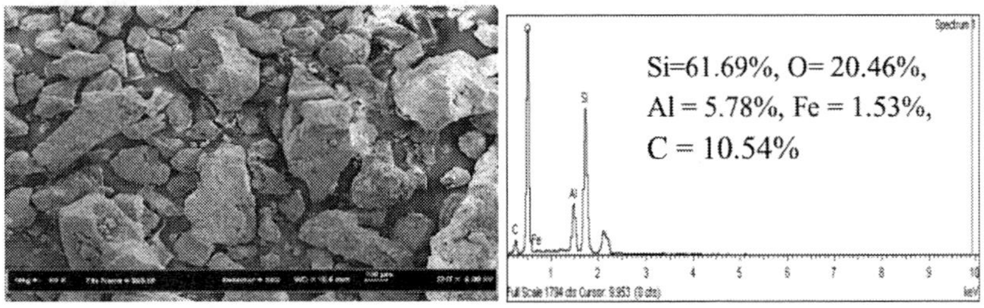

Fig. 3. Energy dispersive X-ray of the spectrum from FESEM of river sand.

https://doi.org/10.1051/e3sconf/20183401003

2.1 Test method

The slump value, compacting factor value and concrete density of the fresh concrete were determined according to standard specification of BS EN 12350-2 [7], BS 1881-103 [8] and BS EN 12350-6 [9], respectively. The results are presented in Table 3. For each of the concrete mixture, specimens of 100mm cubes for compressive strength and 100x100x500mm prisms of specimens for flexural strength were considered. The Microstructure of the concrete at 28 days with various IOT replacements was analyzed using Field Emission Scanning Electron Microscopy (FESEM) Techniques and X-ray diffraction analysis (XRD). The XRD test was conducted at a source of Cu Kα with 40 KV of X-ray acceleration voltage and 30 mA electric current. The scanning rate was 0.05 °/sec with a scanning interval of 2Θ angle from 5to 6 5 ° C.

3 Results and discussions

In this section, properties of concrete in both fresh and hardened form are presented and discussed. Microstructure tests results of XRD and FESEM are also presented and discussed.

3.1 Fresh concrete properties

The results of slump values in Table 3 showed that all the IOT concrete samples reduced the slump value with increase in IOT content due to the high water affinity of the IOT. Although there was decrease in the slump of the IOT concrete samples, but there was improvement in cohesiveness with the addition of superplasticizer. Fresh control concretes (M0) is more cohesive and workable than concrete containing IOT. This might be attributed to the higher specific surface area of IOT, which absorbed more water compared to the river sand. The fineness modulus of 2.07 for IOT; which is less than the minimum value of 2.3 specified for sand by ASTM C33 [10] standard also increased water demand, thus resulting in a decreased in slump workability. The compacting factor test results for all the concrete samples tested, ranges between 0.85-0.94 as shown in Table 3. The British standard BS 1881-103 [8] prescribes for normal weight concrete, compacting factor value range of 0.82 to 0.92.The study of the fresh density of concrete demonstrated in Table 3 shows that fresh density values increase as the IOT content increased in concrete. It can be seen that, although the fresh density was lower in control specimens as compared to those of IOT concrete, all of them are above the minimum limits of 2300 kg/m^3 specified by the BS EN 12350-6 [9]. The higher density values of IOT concrete could be due to the high concentration of iron content in the material, which was heavier than river sand in terms of weight. Another factor could be due to finer particles and angularity which has a better filling effect than the sand, which lead to a higher packing density in the fresh concrete.

3.2 Compressive strength

The compressive strength tests for the control and IOT concrete mixes are shown in Table 4. The results shows that all the IOT concrete samples gave better performance in compressive strength than the control specimen. The incorporation of IOT improves concrete strength up to 50% optimum level and then decreased when the percentage of IOT was more than 50% replacement to sand but still higher than reference concrete (M0). The reasons for this decrease in strength could be attributed to the high water demand for hydration process, which might result to lower compressive strength. The presences of heavy metals in the IOT such as Zn, Cu, Pb and Cd also retard hydration, and lower the strength development. The percentage increase of compressive strength relative to M0 at 28 days were 6.5%, 13.1%, 9.8% and 7% for M25, M50, M75 and M100, respectively. This observed increase in the compressive strength could be partly attributed to the finer particles of IOT, which filled the pore and optimized the pore structure. It is also attributed to the internal curing effect whereby water from the IOT is gradually released into the concrete to further hydrate the matrix. This may be related to chemistry of IOT as observed in the chemical composition. It is equally suspected that the presence of iron in higher concentrations might have positive effect on strength attainment [11]. It was ranged between 10-50% optimum mixtures replacement of natural sand with IOT [12-15].

Table 4. Compressive strength of concrete specimens.

	Compressive strength of concrete mixes (MPa)		
Type of specimen	**7 days**	**14 days**	**28 days**
M0	23.0	28.8	33.3
M25	25.4	30.7	35.8
M50	26.3	32.6	37.7
M75	24.6	29.6	34.5
M100	23.2	28.9	33.8

3.3 Flexural strength

The flexural strength of concrete specimen with various percentages of IOT is shown in Fig.4. It is observed that the flexural strength gain of the concrete containing IOT was superior to control specimens. It could be seen that, the flexural strength at the age of 28 days, concrete containing 50% IOT for M50 was found to developed higher flexural strength of 13% above that of control concrete M0. In general, concrete containing IOT showed higher rates of flexural strength development throughout the curing period. It is interesting to mention that all the flexural strength of control and IOT concretes were higher than 4 MPa as specified in BS 5328-1 [16].

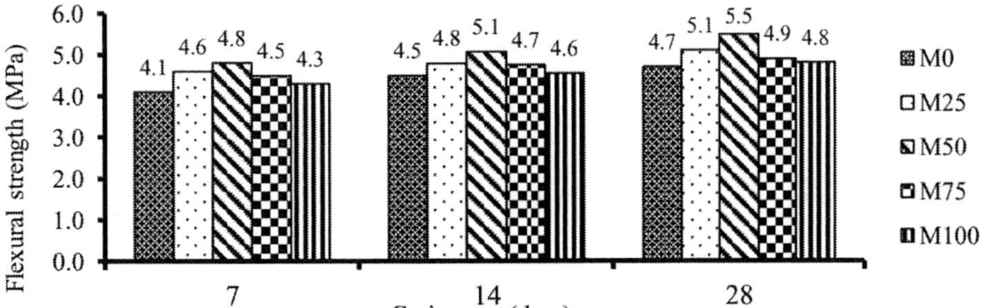

Fig. 4. Flexural strength of respective concrete.

3.3.1 Relationship between flexural strength and compressive strength

The regression line approach was used for the relationship between flexural strength and compressive strength. The coefficient of the linear regression equation with the correlation coefficient for the entire data of the specimens is shown in the Fig.5. Statistical analysis from the equation of the results showed that there is a strong relation between the flexural and compressive strengths. It can be seen from the figure that the correlation coefficient for flexural strength and compressive was 0.99. The figure also shows that when compressive strength is higher, the flexural strengths were also higher. This relationship is linear form and therefore, it is possible to predict the flexural strength of IOT concrete from its compressive strength using the correlation equations. The result of this study shows that IOT concrete has high potentials to be used in construction requiring flexural resistance.

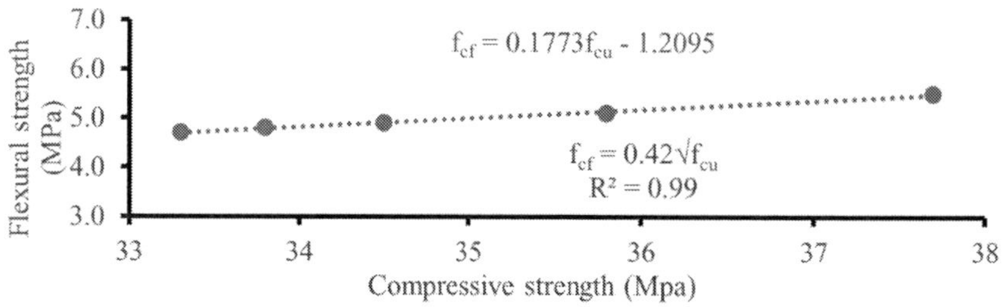

Fig. 5. Relationship between flexural strength and compressive strength.

3.4 X-ray diffraction analysis (XRD)

The XRD microstructural analysis provides useful information on the crystallographic properties of hardened concrete. XRD diffractograms of control specimens and concrete containing IOT at 28 days was presented in Fig. 6. The XRD diffractograms shows the presence of hydrated phases such as quartz (SiO_2), portlandite $Ca(OH)_2$, Dolomite and albite ($NaAlSi_3O_8$) with a peak of calcium silicate hydrate (C-S-H). However, that of IOT concrete mixes contains hematite (Fe_2O_3) in addition to the chemical elements enlisted in the control mixes. The presences of portlandite affected the progress of strength development. The diffraction peak intensity of phase portlandite was more predominant in control concretes as compared to that in IOT concrete samples. The lower portlandite in IOT could be attributed to the reaction between reactive silica in IOT and portlandite which promote the formation of crystallized C–S–H. This reveals that the IOT have the reactivity, as their finer particles are similar to cement or finer than cement.

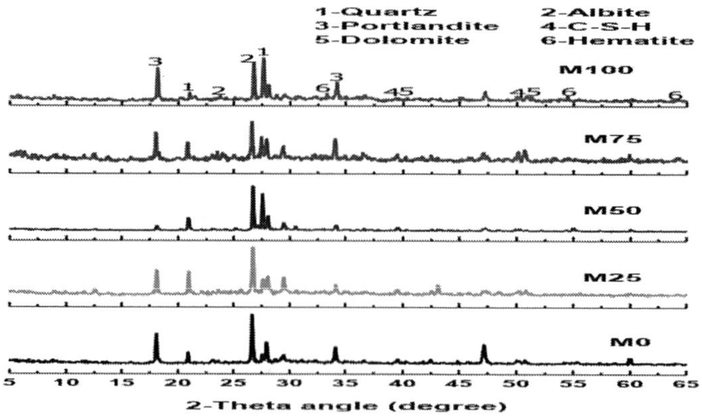

Fig. 6. X-ray diffraction of respective concrete.

3.5 Field Emission Scanning Electron Microscopy

Field-emission scanning electron microscopy (FESEM) provides useful information on the materials and distinguishing morphological features that could be credited to the strength enhancement. Fig. 7 to 11 shows the FESEM morphology of respective concretes.

The morphology of M0 (control specimen) concrete characterised with higher amount of fine needle-like ettringites and porous gaps between the cement and aggregate is shown in Fig. 7. It is also observed that the structures of particles were interconnected and looks like honeycomb with voids. Nonetheless, the microstructure of M25 concrete presented in Figure 8 was relatively less porous with fewer ettringites in the concrete compared with M0. However, there are substantial hexagonal platelets of $Ca(OH)_2$ seen in all the concretes. These $Ca(OH)_2$ crystals were significantly reduced in concrete containing IOT as can be seen from M25 to M100 images. This variation in $Ca(OH)_2$ content can be attributed to the water absorption of IOT materials. It was also noticed that the dark matter in the images stands for IOT. The assumptions regarding the presence of IOT is based on the facts that these medium dark images are seen in almost every sample of IOT except the control specimens. A closer observation indicated that there were no ettringite products in the FESEM of M50, M75 and M100 concretes. Ettringite is generally formed in the early age of hydration and disappeared as the cement hydration continues [17]. This is because after some days depending on the alumina to sulphate ratio of the cement, ettringite become unstable and decomposed to form monosulfoaluminate plate morphology. However, the ettringite in M0 concrete has not been consumed. This could be attributed to dicalcium silicates that do not hydrate at initial hardening were trapped by sulfates to form ettringite after concrete has hardened, due to available moisture in M0 concrete. However, higher alumina and finer particles in IOT help greatly retards the reaction and formation of ettringite. This is an indication that IOT contributed to the densification of the microstructure of the respective concrete due to its micro-filling ability.

4 Conclusions

Due to the high affinity of IOT for water demand, there was significant decrease in the slump of the IOT concrete, which was believed, influenced by the surface area and rough surface of the tailings, but the addition of superplasticiser has improved the cohesiveness. The concrete sample M50 produced the highest value of compressive strength, 37.7 MPa, which gave higher strength than M0 concrete sample, which serves as reference. They all attained the designed strength of grade 30 at 28 days curing.The flexural strength of IOT concrete increased with increase in IOT up to the optimum and then decreased. However, all the values were higher than control specimen. The XRD results revealed the consumption of $Ca(OH)_2$ in IOT concrete during the hydration process and formation of more C-S-H gel in all replacement levels, leading to increase in strength. The morphology of the bulk concrete matrix clearly depicted the influence of the IOT. There was a consistent decrease in calcium hydroxide content with densification of the matrix as a result of increased volume of C-S-H. This has contributed to the achievement of higher compressive strength.

The authors are grateful to Ministry of Education (MOE) and Research Management Centre (RMC), Universiti Tecknologi Malaysia (UTM) for providing the Research University Grant (RUG), Vot. No: Q. J13000. 2509. 06H5

 https://doi.org/10.1051/e3sconf/20183401003

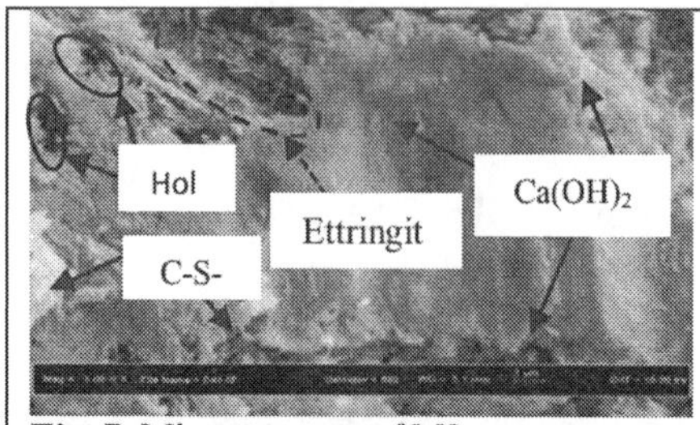

Fig. 7. Microstructure of M0 concrete

Fig. 8. Microstructure of M25 concrete

Fig. 9. Microstructure of M50 concrete

Fig. 10. Microstructure of M75 concrete

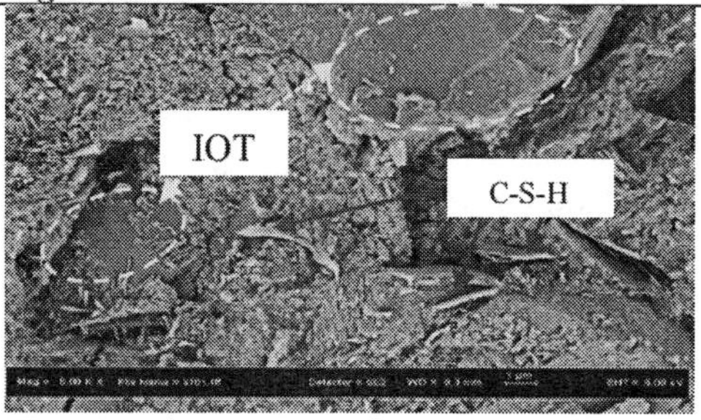

Fig. 11. Microstructure of M100 concrete at 28 days

References

1. Y.F., Hou, X. Jia, Adv Materials Res **671-674**, 1856-1859 (2013)
2. H.Z., Kang, K.W. Jia, L. Yao, Appl Mech Mater, **148-149**, 904-907 (2011)
3. A.T., Ako, S.U. Onoduku, A.S. Oke, I.B. Essien, N.F. Idris, N.A. Umar, A.A. Ahmed, J Geosci Geom, **2** (2), 42-49 (2014)
4. M., Edraki, , T. Baumgartl, E. Manlapig, D. Bradshaw, D.M. Franks and C.J. Moran, J Clean Prod **84**, 411-420 (2014)
5. L., Yu, J.S. Tian, J.X. Zhang and R.J. Yang, Adv Materials Res **250-253**, 1017-1024 (2011)
6. ASTM C494, *American Standard Testing of Materials - Standard Specification for Chemical Admixtures for Concrete*. American society of testing and materials Publication, 10 (2013)
7. BS EN 12350-2, *Testing fresh concrete (Slum test)*. BSI Standards Publication, London (2009)
8. BS 1881-103, *Method for determination of compacting factor*. BSI Standards Publication, London (1993)
9. BS EN 12350-6, *Testing fresh concrete density*. BSI Standards Publication, London (2009)

10. ASTM C33, *American Standard Testing of Materials - Specification for Concrete Aggregates*. American society of testing and materials Publication, 11 (2003)
11. M., Yellishetty, V. Karpe, E.H. Reddy, K.N. Subhash, P.G. Ranjith, Resour Conserv Recy **52**(11), 1283-1289 (2008)
12. G.D., Zhang, X.Z. Zhang, Z.H. Zhou and X. Cheng, Adv Materials Res **838-841**, 152-155 (2013)
13. T.I., Ugama, S.P., Ejeh, International Journal of Advances in Engineering & Technology **7** (4), 1170-1178 (2014)
14. B.N.S., Kumar, R. Suhas, S.U. Shet and J.M. Srishaila, International Journal of Research in Engineering and Technology, **3** (7), 369-376 (2014)
15. M., Manjula, H.S. Sachin, K.M. Shivakumar, D.R. Ranjitha, B.M. Santhosh, International Journal of Informative & Futuristic Research, **2** (10), 3802-3812 (2015)
16. BS 5328-1, *Guide to specifying concrete*. BSI Standards Publication (1997)
17. Neville, A.M., *Properties of concrete*, Pearson Education Limited, London (2011)

Palm Oil Fuel Ash (POFA) and Eggshell Powder (ESP) as Partial Replacement for Cement in Concrete

Mazizah Ezdiani Mohamad [1,*], *Ali A.* Mahmood [1], *Alicia* Yik Yee Min[1], and *Nur Nadhira* A.R.[2]

[1]Dept. of Civil Engineering, University College of Technology Sarawak, 868 Persiaran Brooke, 96000 Sibu, Sarawak, Malaysia
[2]Dept. of Structure and Materials, Faculty of Civil Engineering, Universiti Teknologi Malaysia, 81310 Skudai, Johor Bahru, Malaysia

Abstract. This study is an attempt to partially replace Ordinary Portland cement (OPC) in concrete with palm oil fuel ash (POFA) and eggshell powder (ESP). The mix proportions of POFA and ESP were varied at 10% of cement replacement and compared with OPC concrete as control specimen. The fineness of POFA is characterized by passing through 300 μm sieve and ESP by passing through 75 μm sieve. Compressive strength testing was conducted on concrete specimens to determine the optimum mix proportion of POFA and ESP. Generally the compressive strength of OPC concrete is higher compared to POFA-ESP concrete. Based on the results of POFA-ESP concrete overall, it shows that the optimum mix proportion of concrete is 6%POFA:4% ESP achieved compressive strength of 38.60 N/mm^2 at 28 days. The compressive strength of OPC concrete for the same period was 42.37 N/mm^2. Higher water demand in concrete is needed due to low fineness of POFA that contributing to low compressive strength of POFA-ESP concrete. However, the compressive strength and workability of the POFA-ESP concrete were within the ranges typically encountered in regular concrete mixtures indicating the viability of this replacement procedure for structural and non-structural applications.

1 Introduction

Concrete plays a crucial role in developing the infrastructure of all nations. Nowadays, sustainable construction has gradually become a topic of interest in the engineering as well as other fields. Ordinary Portland cement (OPC) is generally used as a binding material while granite and mining sand are used as aggregate in regular concrete mixes. The production of OPC causes carbon dioxide (CO_2) emissions. As a result, carbon dioxide gas emissions cause harm and pollute the environment. Many research studies have been carried out on utilising the wastes from generated by the industry as cement replacement in the construction industry [1 – 4].

* Corresponding author: mazizah@ucts.edu.my

Palm oil fuel ash (POFA) is one such waste. It is a recycled material that originates from the palm oil industry. POFA is the ash that is produced in the plant boiler by burning the oil palm shell (OPS), and empty fruit bunches (EFB) [5]. According to Pakiam [6], Malaysia is said to be one of the largest palm oil producer in the world with around 41% of the total world supply in the year 2009 to 2010. Most of the POFA produced is typically disposed of in landfills. A few researchers have studied the use of POFA in concrete [7 – 10]. Namini [11] stated that POFA can act as a supplementary cementitious material in concrete. POFA is greyish and almost dark in colour. According to Ekiti et al. [12], POFA can be used as a replacement of cement because it contains a high pozzolanic material, which will help to increase the durability as well as compressive strength of concrete.

Eggshell is a waste produced from the food industry such as restaurants, poultry farms and bakery shops. Re-using eggshells promotes recycling of farm waste and prevent its diversion into landfills. The majority of eggs are utilized in egg breaking plants for mass production of liquid eggs for use in food and non-food related products [13]. Hincke et al. [14] stated that the outer eggshell is made up of calcium carbonate $CaCO_3$. According to Pliya and Cree [13], 1 billion eggs would produce 6600 tonnes of limestone powder and the amounts would not be enough for concrete industry but in total it could be used as partial cement replacement. A limited number of studies have been conducted on the utilization of eggshell waste as an alternative material. Amaral et al. [15] investigated soil bricks containing 0% to 30% of eggshell powder (ESP) as partial replacement of cement. The compressive strength increased by 7% and 12% when cement in bricks was replaced from 10% to 12% compared to bricks without ESP, respectively. However, when 30% ESP was added, the compressive strength decreased by only 2%.

This paper focuses mainly on the effect of different mix proportions of POFA and ESP as partial cement replacement concrete.

2 Experimental Program

2.1 Material and Mix Proportion

Raw POFA was obtained from the Unique Palm Oil Mill, Mukah, Sarawak, Malaysia and oven dried for 24 hours. The ash used in this research was produced by burning the husk fibre and shell of the palm oil fruit. The POFA was then sieved at passing through size of 300 μm sieve. ESP used in this research was blended using a blender and sieved in a 75 μm sieve. The POFA and ESP were varied as 10% cement replacement by weight. Figures 1(a) and 1(b) show ESP and POFA specimens used in this study. The cement used in this study was Ordinary Portland (OPC) Type 1. 10 mm well graded crushed granite and locally available river sand were used as coarse and fine aggregates, respectively. The concrete mixing procedure consisted of various ratios of ESP and POFA at 10% cement replacement, OPC, fine and coarse aggregates. Concrete strength was designed to achieve a compressive strength of 30 N/mm^2 at 28 days. Water to cement ratio was fixed at 0.54. The mix proportions of POFA:ESP at 10% cement replacement were as follows: (i) 0% (control), (ii) 10% POFA:0% ESP, (iii) 8% POFA:2% ESP, (iv) 6% POFA:4% ESP, (v) 4% POFA:6% ESP, (vi) 2% POFA:8% ESP and (vii) 0% POFA:10% ESP. Table 1 shows these mix proportions. Three specimens were casted for each mixture of concrete giving a total of twenty-one (21) specimens. The concrete specimens were cured in a water tank for 28 days before carrying out the compressive strength testing.

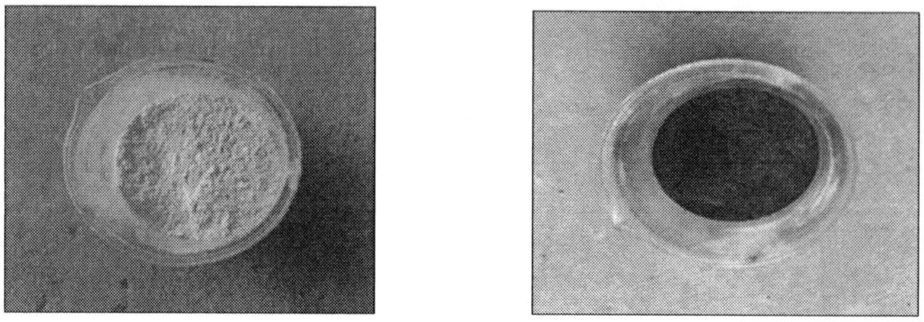

(a) **(b)**

Fig. 1. Materials: a) Eggshell Powder (ESP) and b) Palm Oil Fuel Ash (POFA).

Table 1. Mix Proportions of specimens.

Mix Proportion	Coarse Aggregate (kg)	Fine Aggregate (kg)	Cement (kg)	Eggshell Powder (kg)	POFA (kg)	Water (kg)
Normal mix	3.659	1.885	1.541	-	-	0.769
E10%	3.659	1.885	1.387	0.154	-	0.769
E8%,P2%	3.659	1.885	1.387	0.124	0.031	0.769
E6%,P4%	3.659	1.885	1.387	0.093	0.062	0.769
E4%,P6%	3.659	1.885	1.387	0.062	0.093	0.769
E2%,P8%	3.659	1.885	1.387	0.031	0.124	0.769
P10%	3.659	1.885	1.387	-	0.154	0.769

E – ESP, P – POFA

2.2 Slump Test

The concrete specimens were tested for workability using the slump cone test. The main purpose of the test is to determine the effect of adding ESP and POFA on concrete workability. The testing procedure followed the method given in BS EN 12390-2 [16].

2.3 Density

Density of concrete is the mass to volume ratio. Concrete density depends on the density of the fine and coarse aggregates, the amount of entrapped air, water and cement contents. Normal concrete density is about 2400 kg/m^3. Before conducting the compressive test, the mass of concrete needs to be measured. The mass of the concrete cube can be obtained using a weighing scale. The density can be calculated using Equation 1:

$$\rho = \frac{Mass\ of\ Concrete(kg)}{Volume\ of\ cured\ concrete\ (m^3)} \quad (1)$$

2.4 Compressive Strength Test

Twenty one cubes of (100 × 100 × 100) mm were used to determine the compressive strength. The compressive strength mix was designed for 30 N/mm^2 for each mixture after

28 days of curing. The compression test was carried out using ADR Auto V2 3000 kN compression machine as shown in Fig. 2. The pace rate for the compression test applied was 7.0 kN/s. The loading pace was constantly applied until the specimens visibly failed. The compressive strength for each specimen was determined after 28 days of curing in water. The procedures used in carrying out each test were in accordance with BS EN 12390-3 [17].

Fig. 2. The compression test machine.

3 Results and Discussion

3.1 Workability

Fig. 3 shows the workability of normal mix concrete and POFA-ESP concrete. It is observed that concrete with 10% of ESP increases the workability as the slump height achieved was 98 mm while the normal mix slump was 91.5 mm. When POFA was added to concrete, workability started to decrease. With the gradual addition of more POFA, a corresponding decrease in slump height ensued. At 10% POFA the slump height was 22 mm.

According to Awal [18], the influence of POFA fineness affects both fresh and hardened state properties of concrete. Due to this, a higher water demand was seen in low fineness POFA in concrete compared to high fineness POFA. Hussin et al. [19] showed that unground POFA with low fineness ash particles had larger and higher porosity particles compared to ground POFA with tinier, irregular, and crushed formed particles.

https://doi.org/10.1051/e3sconf/20183401004

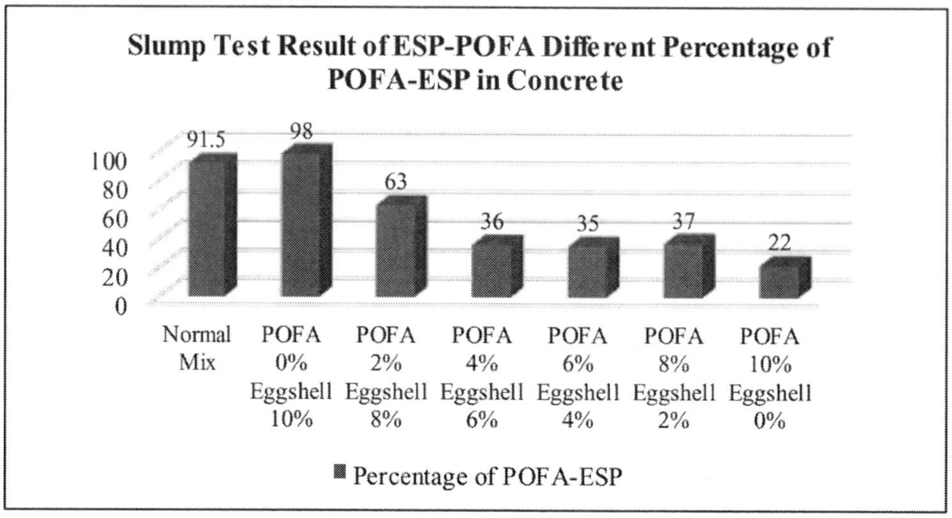

Fig. 3. Density versus percentages of POFA-ESP in concrete.

3.2 Density of Concrete

It is shown in Fig. 4 that the density of normal mix concrete using Ordinary Portland cement is 2415 kg/m^3. The density of concrete was reduced when the eggshell powder and palm oil fuel ash were incorporated in the mix. The highest reduction in density was 2.36% due to the inclusion of 10% ESP in the concrete. The density of the concrete mixtures incorporating 6% POFA:4% ESP and 10% POFA were both reduced by 2.28%. The density of concrete containing 8% POFA:2% ESP was reduced by 1.6% compared to the normal mix concrete. Concrete with 4% POFA:6% ESP had the least significant decrease in density at 0.9% reduction compared to normal mix concrete. The only mixture with an increase in density was the 2% POFA:8% ESP.

The decrease of concrete density when incorporating POFA and ESP might be caused by the low specific gravity of the waste material compared to cement. According to Hussin, et al. [19], cement particles are generally heavier than POFA as the specific gravity of OPC is 3.15 compared to the specific gravity of POFA, which is 2.18. Another study by Yerramala [20] showed that the specific gravity of ESP is 2.37, which is appreciably lower than that of OPC. However, the 2% POFA:8% ESP mixture showed that the density of the POFA-ESP concrete increased by 0.74% compared to OPC concrete. This might be due to weight increase from the aggregates in the POFA-ESP concrete that contributed to this increase in density, in comparison to the OPC concrete density.

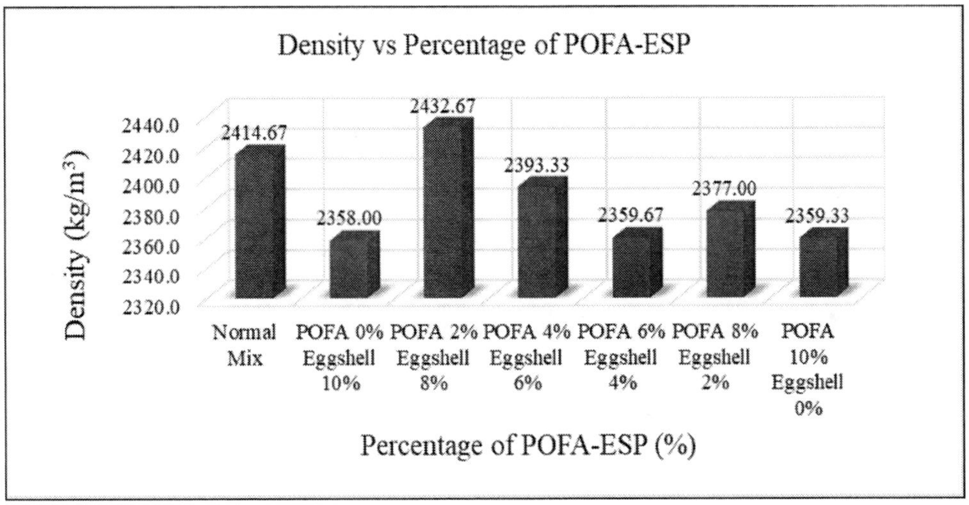

Fig. 4. Density versus percentages of POFA-ESP in concrete.

3.3 Compressive Strength

Fig. 5 shows the relationship between the various mixtures of POFA-ESP and compressive strength. The targets mean strength design for concrete was 30 N/mm^2 at 28 days. The normal mix design of OPC concrete had the highest compressive strength of 42.37 N/mm^2 in comparison to the concrete incorporating POFA and ESP. It is shown in Figure 5 that the optimum strength of ESP-POFA concrete achieved was 38.6 N/mm^2 at 28 days. This was for the concrete mixture with 6% POFA:4% ESP cement. On the other hand, concrete incorporating 4% POFA: 6% ESP had the lowest compressive strength of 34.63 N/mm^2.

A study by Jaturapitakkul et al. [21] found by increasing the ratio of ground POFA fineness increased significantly the rate of pozzolanic activity and compressive strength. Hussin et al. [19] prepared POFA by sieving through a 300 µm sieve and then was ground to obtain its final fineness passing the 45 µm sieve. As a result, partial cement replacement with 45 µm POFA incorporated in aerated concrete had higher compressive strength in comparison to normal aerated concrete [19]. It was shown that the finer particles of POFA contributed to a better pozzalanic material, which can be used as an alternative cement replacement [21].

Moreover, Elaty [22] who did an extensive research study to predict the strength of concrete mixtures, reported values for concrete strength after 28 days of curing from Lea's Chemistry of Cement and Concrete and other references [23]. The values range from 36.1-56.5 N/mm^2. Most of the values found in this research and shown in Figure 5 fall within this range, indicating that using POFA and ESP as partial replacements for cement concrete is viable and strength-wise feasible.

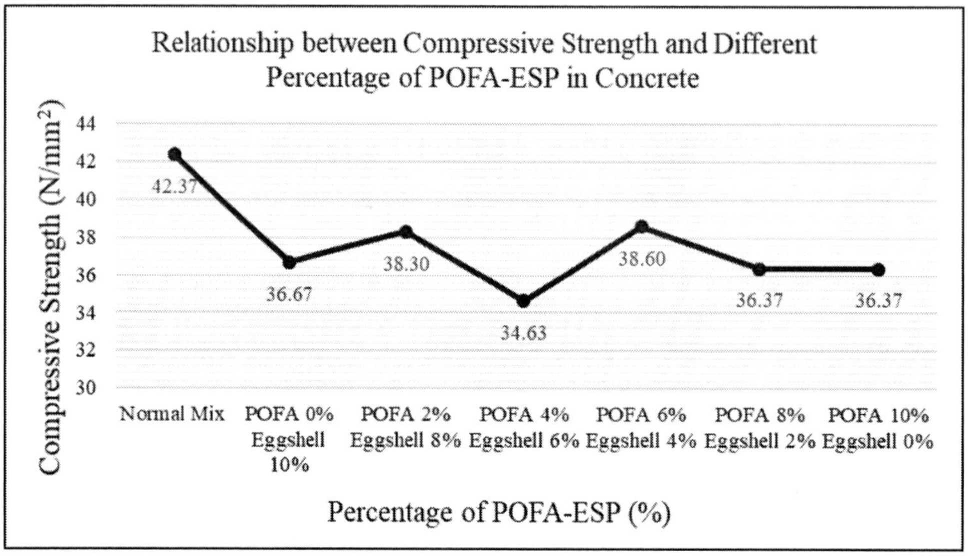

Fig. 5. Relationship between percentages of POFA-ESP and compressive strength.

4 Conclusions

Density, workability and compressive strength testing have been carried out on several POFA-ESP concrete mix proportions. The aim was to replace 10% of cement by weight and investigate experimentally the newly-formed mixtures. Based on the results obtained and above discussion, the following conclusion can be drawn:

POFA-ESP concrete exhibits lower strength development compared to OPC concrete. The results showed that 6% POFA: 4% ESP had the highest strength out of all POFA-ESP concrete mixtures tested. POFA fineness should be increased by grounding and sieving through a 45 μm sieve in order to get higher compressive strength of concrete. The density and workability of POFA-ESP concrete is generally lower compared to normal mix concrete. However, POFA-ESP concrete is viable and strength-wise feasible and can be utilized as partial cement replacement in concrete and used for structural and non-structural purposes. Furthermore, the usage of POFA and ESP will help in reducing environmental problems and landfill disposal problem.

The authors are grateful to Centre for Research and Development, University College of Technology Sarawak (UCTS) for financial support under Grant: UCTS/RESEARCH/1/2016/06. The authors are also thankful to the staff of Civil Engineering Laboratory, Department of Civil Engineering for the facilities and support for experimental work.

References

1. Salam, M. Z. Jumaat, *Utilization of palm oil fuel ash in concrete : A review*, **3730** (2017)
2. R. Karim, M. F. M. Zain, M. Jamil, F. C. Lai, N. Islam, *Energy Procedia*, **12**, 915–919, (2011)
3. V. Sata, C. Jaturapitakkul, K. Kiattikomol, *Constr. Build. Mater.*, **21** (7), 1589–1598 (2007)

4. Jaturapitakkul, J. Tangpagasit, S. Songmue, K. Kiattikomol, Constr. Build. Mater., **25**(11), 4287–4293 (2011)
5. M. Momeen Ul Islam, K. H. Mo, U. J. Alengaram, M. Z. Jumaat, J. Clean. Prod., **115**, 307–314 (2015)
6. R. Pakiam, Palm oil advances as Malaysia's export tax may boost shipments, *http://www.bloomberg.com/news/palm-oil-advances-as-malaysia-s-export-tax-may-boost- shipments.html*. (2013)
7. E. Shahrul Effendie, *Performance of Concrete by Using Palm Oil Fuel Ash (POFA) as A Cement Replacement Material, Bachelor Civ. Eng.*, (2010)
8. T. J. Deepak, A. Elsayed, N. Hassan, C. N., S. Yun Tong, M. B.M., Int. J. Sci. Res. Publ., **3**(1), 138–142 (2014)
9. M. A. A. Rajak, Z. A. Majid, M. Ismail, Procedia Manuf., **2**, 512–518 (2015)
10. S. K. Lim, C. S. Tan, O. Y. Lim, Y. L. Lee, Constr. Build. Material, **46**, 39–47 (2013)
11. H. Golizadeh, S. B. Namini, Journal of Construction in Developing Countries, **20**(1), 85–98 (2015)
12. Ekiti, K. O. Olowe, and V. B. Adebayo, SSRG International Journal of Civil Engineering (SSRG-IJCE), **2**(4), 48–56 (2015)
13. P. Pliya, D. Cree, Constr. Build. Mater., **95**, 1–9 (2015)
14. M.T. Hincke, Y. Nys, J. Gautron, M.D. McKee, Frontiers in Bioscience, **17**(1), 1266–1280 (2012)
15. M. C. Amaral, F. B. Siqueira, A. Z. Destefani, J. N. F. Holanda, Proc. Inst. Civ. Eng. - Waste Resour. Manag., **166**(3), 137–141 (2013)
16. BS 12390-2, *Testing Hardened Concrete. Part 2: C Testing Fresh Concrete Slump test*, (2009)
17. BS EN 12390-3:2009, Testing hardened concrete. Part 3: Compressive strength of test specimens, (2009)
18. M. W. Hussin, Awal ASMA, *Influence of palm oil fuel ash on sulphate resistance of mortar and concrete. Proceedings of the Sixth CANMET / ACI International Conference on Fly Ash, Silica Fume, Slag and Natural Pozzolan in Concrete*, Bangkok, ACI Publication SP-178. **1**, 417-430 (1998)
19. M.W. Hussin, K. Muthusamy, and F. Zakaria, Journal of Materials in Civil Engineering, American Society of Civil Engineers (ASCE), **4**(22) (2010).
20. A. Yerramala, The Indian Concrete Journal, 94–102 (2014)
21. Jaturapitakkul, J. Tangpagasit, S. Songmuea, K. Kiattikomol, Construction and Building Materials, **25**, 4287–4293 (2011)
22. M.A. Elaty, *HBRC J.*, **10**(2), 145–155 (2014)
23. P.C. Hewlett, *New York, Lea's Chemistry of Cement and Concrete*, Elsevier Butterworth-Heinemann, (2005)

Effects of fines content on hydraulic conductivity and morphology of laterite soil as hydraulic barrier

Yamusa Bello Yamusa[1, 2*], *Nor Zurairahetty* Mohd Yunus[1], *Kamarudin* Ahmad[1], *Norhan* Abd Rahman[1], and *Radzuan* Sa'ari[1]

[1]Faculty of Civil Engineering, Universiti Teknologi Malaysia, 81310 Skudai, Johor, Malaysia
[2]Department of Civil Engineering, Nuhu Bamalli Polytechnic, 810001 Zaria, Nigeria

Abstract. Laterite soil was investigated to find out the effects of fines content and to identify the micro-structural and molecular characteristics to evaluate its potentiality as a compacted soil landfill liner material. Tests were carried out on natural soil and reconstituted soil by dry weight of soil samples to determine the physical and engineering properties of the soil. All tests were carried out on the samples by adopting the British Standard 1377:1990. The possible mechanisms that contributed to the clay mineralogy were analyzed using spectroscopic and microscopic techniques such as field emission scanning electron microscopy (FESEM), energy-dispersive X-ray (EDX) and X-ray diffractometry (XRD). The laterite soil was found to contain kaolinite as the major clay minerals. A minimum of 50% fines content of laterite soil met the required result for hydraulic barriers in waste containment facilities.

1 Introduction

The vast increase in population associated with rapid development in industrial and agricultural production, more waste is being produced regularly. This waste is mostly disposed of in landfills. In order to minimize the environmental pollution caused by mankind, numerous scientific studies such as storing, transformation, and annihilation of wastes. But the most economical and the most confident is the impermeable nature of sanitary landfills, with respect to underground and surface waters for solid waste disposal [1]. Hydraulic barriers usually refer to liners and covers used for waste containment structures in engineered sanitary landfill design play the role in preventing/impeding fluid flow and attenuating inorganic contaminants. These hydraulics barrier's structural integrity must be ensured by adopting all the necessary and adequate criteria in their construction [2].

The minimum recommended criteria for sanitary landfill liners varies upon regulations set by different countries. The environmental protection agencies in developed countries, like the USA, UK, and most European countries recommended hydraulic conductivity (k)

* Corresponding author: yamusabello@yahoo.com

 https://doi.org/10.1051/e3sconf/20183401005

of $k \leq 1 \times 10^{-9}$ m/s values for sanitary landfill liners [3]. The criteria set by most researches and regulatory agencies specified a hydraulic conductivity value of $k \leq 1 \times 10^{-9}$ m/s, a volumetric shrinkage value of $\leq 4\%$, and a value of ≥ 200 kN/m^2 shear strength are required for hydraulic barrier systems [4-11]. This paper focused on the hydraulic conductivity criteria.

In constructing compacted soil liners for sanitary landfills, the most common materials used are fine-grained soils because of their relative economy and availability, and good engineering properties when compacted [12]. Though, the morphology, which relates to the structural pattern and mineral formation of soils are controlled by the fines particles. The reason FESEM, EDX and XRD were carried out to examine the minerology of the laterite soil. There is no information in published literature on the minimum percentage of fines required to be adopted in the design of compacted laterite soils as landfill liners and covers in waste containment applications [4]. Therefore, there is need to find the minimum gradation with respect to fines content that can satisfy the hydraulic conductivity criteria used in hydraulic barriers.

2 Materials and methods

2.1 Natural Soil and Reconstituted Soil

In this study, the natural laterite soil sample used was collected from the hilly area near the Faculty of Electrical Engineering at 1 - 1.5 m below the ground. Located on latitude 1°33'39"N and longitude 103°38'44"E, Skudai campus of Universiti Teknologi Malaysia (UTM).

The reconstitution of the laterite soil used in this study was air-dried and then passed through BS 4.75mm aperture sieve to remove oversized gravel. The laterite soil was then sieved into three different grades i.e. fines (<0.063 mm), sand (0.063 mm to 2.00 mm) and gravel (>2.00 mm to ≤4.75 mm). The following gradation of laterite soil specimen were investigated:
Natural laterite soil sample with 30% fines, 40% sand and 30% gravel.
For the reconstituted laterite soil;
(i) 40% fines, 40% sand and 20% gravel.
(ii) 50% fines, 40% sand and 10% gravel.

2.2 Morphological and Mineralogical Composition

Field emission scanning electron microscopy (FESEM) technique was used to examine the morphology and microstructural properties of the laterite soil fabric. Likewise, providing information on the size, shape, and the state of orientation and aggregation of soil particles [13]. This was used in conjunction with Energy Dispersive X-Ray Spectroscopy (EDX), which is a chemical microanalysis method to determine the elements present in the sample. The FESEM and EDX sample preparation is described as follows. Dry sample was prepared and placed on aluminum stub, and to prevent charging effect as well as loss of resolution it was then covered with carbon tape and coated with platinum using a vacuum sputter. Crossbeam 340 machine was used to analyze the sample connected to a computer as shown in Fig. 1. Moreover, X-ray diffraction (XRD) technique is primarily used to identify and characterize compounds based on their diffraction pattern. SmartLab X-ray diffractometer machine was used to analyze the sample connected to a computer as shown in Fig. 2.

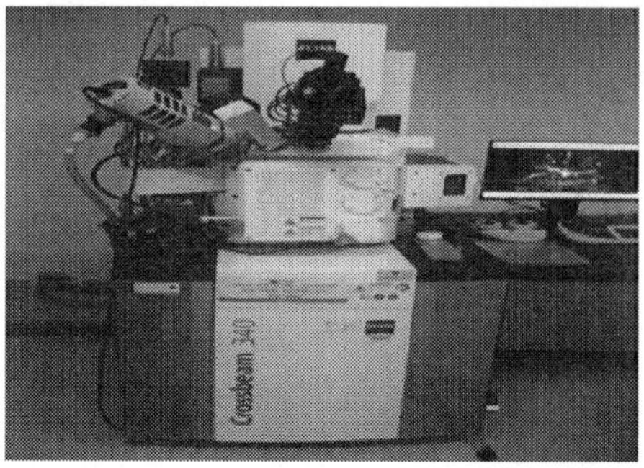

Fig. 1. Crossbeam 340.

Fig. 2. SmartLab X-ray Diffractometer.

2.3 Hydraulic Conductivity

Rigid wall permeameter under falling head condition was used to measure the hydraulic conductivity as recommended by [14]. Soil samples were compacted using the BSL at 30%, 40% and 50% fines contents by varying the moulding water contents at -4%, -2%, 0%, +2% and +4% of the optimum moisture content (OMC). Then the samples were soaked for a minimum period of 48 hours until no air bubbles are obviously observed to allow for full saturation inside a water tank. Then connected to distilled water through a standpipe. Readings of time and distance covered by distilled water in the standpipe were recorded and permeability (k) calculated. Tests were repeated for different gradation at various moisture contents.

3 Results and Discussion

3.1 Index properties

British Standards (BS) [15] was followed in all the laboratory tests carried out to determine the index properties and particle size distribution of the soil as shown in Table 1 and Fig. 3 respectively. From the particle size distribution shown in Fig. 3, the natural laterite soil contains 30% fines, whereas the reconstituted laterite soils contain 40% fines and 50% fines. Furthermore, Atterberg limits results revealed a liquid limit of 76%, plastic limit of 42%, and plasticity index of 34%. Based on these data, the laterite soil is classified according to the BS as sandy silt with gravel of very high plasticity (MV). Generally, low hydraulic conductivity is attained from soils with high liquid limit [16] recommended that the liquid limit of a liner material should be ≥ 20%. Though, clay soils with too high liquid limit (LL) are more susceptible to desiccation cracking [12].

Table 1. Index properties of laterite soil.

Property	Value
Natural Moisture Content, %	34
Specific Gravity	2.7
% Passing BS 63μm sieve	30
OMC, %	30
MDD, Mg/m3	1.35
Liquid Limit, %	76
Plastic Limit, %	42
Plasticity Index, %	34
BS Classification	MV

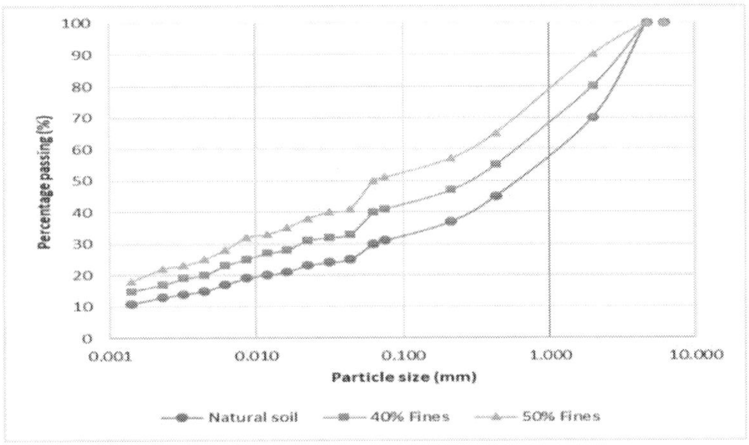

Fig. 3. Particle Size Distribution Curves.

3.2 Morphological and Mineralogical Composition

FESEM technique was used to examine the morphology and microstructural properties of the laterite soil fabric. From the FESEM micrograph shown in Fig. 4 (a & b), there are presence of kaolinite flakes in the soil fabric. This was used in conjunction with EDX which is a chemical microanalysis method to determine the elements present in the sample. Moreover, EDX spectrum in Fig. 5 shows the present of element of oxygen, iron, aluminum, and silica. The oxide composition of the soil as given in Table 2 revealed the sample to be a laterite soil [17-19]. Laterite soil is whose ratio of iron oxide to aluminum oxide is less than 1.33 [20]. As shown in Fig. 4, the neatly arranged book-like kaolinite particles with free oxides present in the soil, and Kaolinitic and Halloysitic Flakes are the major feature of the natural soil in agreement with [21] and [22] respectively. According to [23] the free iron oxides in soils usually have special electrochemical properties, relatively great surface energy, and a strong ability to adsorb, and hence provide structural cementation in the soil. When free iron oxides exist in soil, there is an increase in strength, and a corresponding decrease in swelling-shrinkage capacity. It was observed that the sesquioxide ratio is 1.14 with very high concentrations of Fe_2O_3, Al_2O_3 and SiO_2 of 35.53%, 31.1 and 25.46 respectively, while CO_2 has relatively low concentrations of 7.91%.

X-ray diffraction (XRD) technique is primarily used to identify and characterize compounds based on their diffraction pattern. As a popular fundamental investigation technique, XRD is easy to use and can give a wide range of instantly interpretable data on soil minerals [13, 19]. The XRD quantitative analysis result and pattern that determined the clay mineralogy of the laterite soil was found to contain kaolinite as the major predominant mineral with some mixtures of gibbsite, goethite and cristobalite as presented in Table 3 and Fig. 6 respectively. According to [24] kaolinite clay mineral has the least affinity for water among the clay minerals will usually exhibit moderate shrinkage and swelling on drying and on wetting respectively. It is to mention that the type and quantity of clay minerals present in a soil can make a tremendous influence on the soil behavior when exposed to different environmental conditions [25]. Thus, an investigation method combining macro- and micro-analysis allows one to deeply understand the mechanical properties of clay soils [26]. The quantitative analysis of XRD in Table 3 revealed that the kaolinite clay mineral has the major portion of the laterite soil with 69.5%, 20.4% gibbsite, 6.5% cristobalite and 3.6% goethite.

Table 2. Oxide composition of laterite soil.

Property	**Value (%)**
SiO_2	25.46
Al_2O_3	31.10
Fe_2O_3	35.53
CO_2	7.91

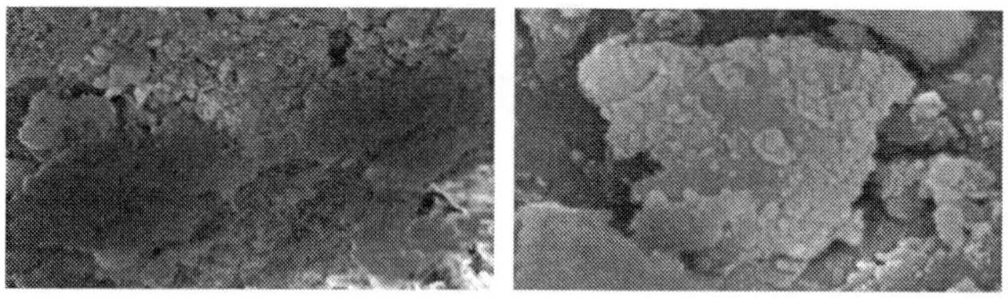

(a) (b)

Fig. 4. (a) and (b) Showing Micrograph using FESEM at: 10,000 and 50,000 Magnification.

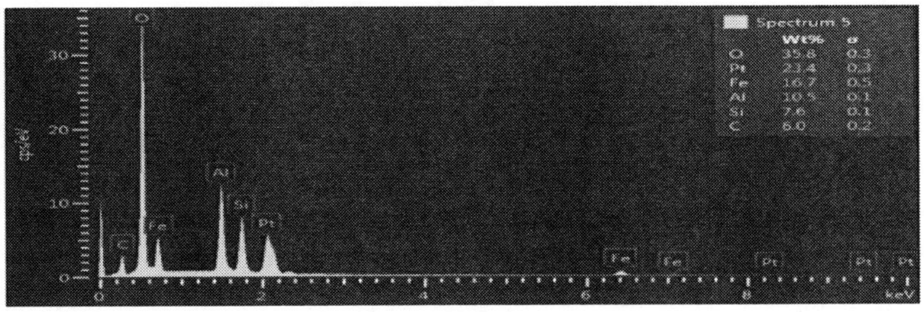

Fig. 5. EDX Spectrum.

Table 3. XRD quantitative analysis.

Property	Value (%)
Kaolinite	69.5
Gibbsite	20.4
Goethite	3.6
Cristobalite	6.5

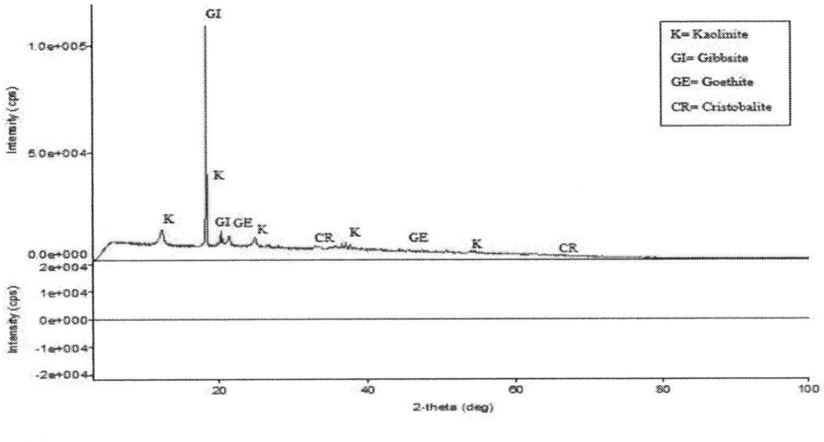

Fig. 6. XRD pattern.

3.3 Hydraulic Conductivity

The effect of fines content on low hydraulic conductivity is expected to be attained when the soil is well graded, and the clay fraction governs the hydraulic behaviour of the soil matrix [27]. The fines in a soil have a higher impermeabilizing effect if they are well distributed so they can most effectively plug voids among the larger particles. Mechanical mixing distributes fines and breaks down some of the soil aggregates thereby supplying fines for void-plugging and destroying large voids [28]. There is a general decrease in hydraulic conductivity with higher fines content as shown in Fig. 7. Hydraulic conductivity of 30%, 40% and 50% fines decreased to 2.5×10^{-6}, 3.78×10^{-8} and 2.44×10^{-9}m/s respectively on OMC. Similarly, hydraulic conductivity of 30%, 40% and 50% fines decrease to 1.55×10^{-7}, 8.13×10^{-9} and 1.02×10^{-9}m/s respectively on the wet sides of optimum. Unlike

OMC and wet of OMC, hydraulic conductivity of 30%, 40% and 50% fines decrease to $9.87x10^{-6}$, $5.08x10^{-8}$ and $9.1x10^{-9}$m/s respectively on the dry side of OMC. The hydraulic conductivity differs with about one order of magnitude with respect to fines content. The maximum regulatory hydraulic conductivity value of 10^{-9}m/s was attained at 50% fines content for the laterite soil investigated. More pronounced result at the 50% fines content is obtained on the wet side of OMC. This indicates that higher moulding moisture contents results to lower hydraulic conductivity.

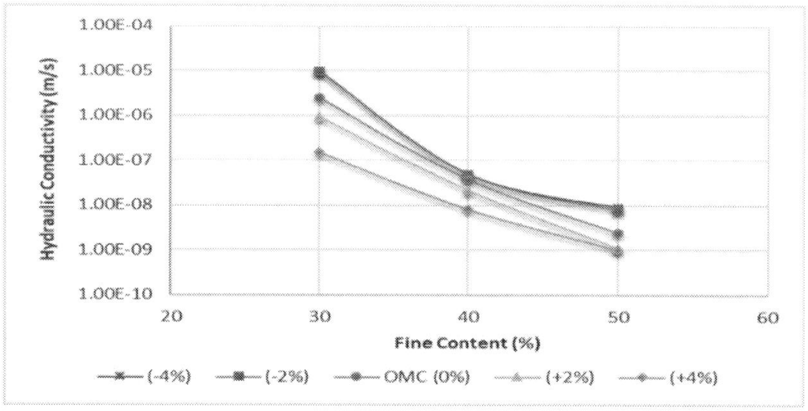

Fig. 7. Hydraulic Conductivity Versus Fines Content.

4 Conclusions

This research technically analyzed the effects of fines content and morphological features of laterite soil as liner in waste containment facilities. The spectroscopic and microscopic techniques revealed the laterite soil to contain kaolinite as the major predominant clay mineral. Hydraulic conductivity decreases with higher fines content. The permeability coefficient of the soil at 50% fines content gives the recommended value of 1×10^{-9} m/s. Percentage of fines content lower than 50% did not meet the hydraulic conductivity requirement. The studied laterite soil is suitable for use as soil liner when contains up to 50% fines content.

The authors thankfully acknowledge the monetary supports provided by the Ministry of Education (MOE) through Universiti Teknologi Malaysia (UTM) under the Fundamental Research Grant Scheme (FRGS – VOT NO 4f885).

References

1. F. Akcanca and M. Aytekin, Env. earth sci **72**, pp. 59-66 (2014)
2. A. Bello, The Pac. J. Sci. Tech **12**, 425-434 (2011).
3. H. M. A. Rashid, J. Wanigarathna, L. Kurukulasuriya, N. Priyankara, A. Alagiyawanna, T. Saito, K. Kawamoto, Env. Earth Sci **76**, 396 (2017).
4. Y. B. Yamusa, K. Ahmad, and N. A. Rahman, Malay. J. Civ. Eng. vol. **29**, 153-164 (2017).
5. A. Bello, The Pac. J. Sci. Tech **16** (2015).
6. K. J. Osinubi and C. M. Nwaiwu, J. geotec. geoenv. eng, **132**, 203-213 (2006).

7. MHLG, Ministry of Housing and Local Government, Criteria for Siting Sanitary Landfills: National Strategic Plan for Solid Waste Management **3**, Appendix 6B, Kuala Lumpur, Malaysia, (2005).
8. D. E. Daniel and Y.-K. Wu, J. Geotec. Eng **119**, 223-237 (1993).
9. USEPA, *Design and construction of RCRA/CERCLA final covers, Soils Used in Cover Systems,* EPA/625/4-91/025. United States Environmental Protection Agency, 1-25 (1991).
10. UKEA. (2014). *Earthworks in landfill engineering: LFE4.*
11. D. E. Daniel, *Geotechnical practice for waste disposal* (Springer Science & Business Media, 2012)
12. K. Osinubi, A. Eberemu, A. Bello, and A. Adzegah, Nig. J. of Tech **31**, 277-287 (2012)
13. N. Latifi, A. S. A. Rashid, A. Marto, and M. M. Tahir, Env. Earth Sci **75**, 1-9 (2016).
14. K. H. Head and R. J. Epps, *Manual of soil laboratory testing* **2**, (Whittles publishing, Dunbeath, Caithness KW6 6EY, Scotland, UK., 2011)
15. BS1377, *Methods of Testing Soil for Civil Engineering Purposes*, (British Standards Institute, London, 1990)
16. C. H. Benson, H. Zhai, and X. Wang, J. Geotec. Eng, **120**, 366-387 (1994)
17. N. Latifi, A. Marto, and A. Eisazadeh, Env. Earth Sci, **73**, 1133-1141 (2015)
18. A. Marto, N. Latifi, and A. Eisazadeh, Arab J. Sci. Eng., **39**, 6949-6958 (2014)
19. N. Latifi, A. S. A. Rashid, N. Ecemis, M. M. Tahir, and A. Marto, Arab J. Geosc, **9**, 1-12 (2016)
20. K. S. Bawa, J. Soil Mec. F. Div, ASCE, **83**, 1 – 15 (1957)
21. A. Eisazadeh, K. A. Kassim, and H. Nur, Eng. geo, **154**, 36-41 (2013)
22. A. A. Bello, KSCE J. Civ. Eng, **17**, 939-948 (2013)
23. X. Zhang, L. Kong, X. Cui, and S. Yin, Eng. Geo. Env, **75**, 1493-1503 (2016)
24. I. Oyediran and C. Iroegbuchu, Ife J. Sci, **15**, 17-30 (2013)
25. A. Eisazadeh and H. Eisazadeh, Env. Earth Sci, **74**, 377-384 (2015)
26. Y. Zhao, Q. Xue, F.-x. Huang, X.-t. Hu, and J.-s. Li, Env. Earth Scie, **75**, 1-9 (2016).
27. D. E. Daniel and C. H. Benson, J. Geotec. Eng, **116**, 1811-1830 (1990)
28. C. H. Benson and J. M. Trast, Clay. clay min., vol. **43**, pp. 669-681 (1995).

The Utilisation of Shredded PET as Aggregate Replacement for Interlocking Concrete Block

M. Mokhtar[1*], *M.* Kaamin[1], *S.* Sahat[1], and *N.B.* Hamid[1]

[1]Department of Civil Engineering, Center for Diploma Studies, Universiti Tun Hussein Onn Malaysia (UTHM), 86400 Parit Raja, Batu Pahat, Johor, Malaysia

Abstract. The consumption of plastic has grown substantially all over the world in recent years and this has created huge quantities of plastic-based waste. Plastic waste is now a serious environmental threat to the modern way of living, although steps were taken to reduce its consumption. This creates substantial garbage every day, which is much unhealthy. Plastic bottles such as Polyethylene terephthalate (PET) was use as the partially component in this making of interlocking blocks concrete. This project investigates the strength and workability of the interlocking block concrete by replacing course aggregate with % PET. The suitability of recycled plastics (PET) as course aggregate in interlocking block concrete and its advantages are discussed here. Moreover, there were more benefits when using interlocking block than using conventional block such as it easy for construction because they are aligning, easy to place, high speed stacking and they offer more resistance to shear and buildings would be even stronger. Based on the test perform, the failure parameter were discussed .From the compressive strength test result, it shows that the strength of concrete block decreased with increased of PET used. From the results, it shows that higher compressive strength was found with 5% natural course aggregate replaced with PET compared to other percentages.

1 Introduction

The interlocking blocks are different from conventional block since they do not require mortar to be laid during bricklaying work. Because of this characteristic, the process of building walls is faster and requires less skilled labour as the blocks are laid dry and lock into place. In addition, the recycled plastics (PET) such as bottles were used to prepare the course aggregates by crushing it to small size thereby providing a sustainable option to deal with the plastic waste.

The normal concrete blocks are inconvenience to use because it takes too long to be stacked, have a low productivity enhancement, high cost and low value of aesthetics. Other than that, plastics are many recycling plants across the world, but as plastics are recycled they lose their strength with the number of recycling. So these plastics will end up as earth fill. In this circumstance instead of recycling it repeatedly, if it is utilized to prepare aggregates for concrete, it will be a boon to the construction industry. Therefore this paper

* Corresponding author: mardiha@uthm.edu.my

aims to study the ability of interlocking block to replace conventional block in terms of reducing the usage of mortar, to study the effect of replacing natural aggregate with plastic aggregate (PET) on workability and compressive strength and to reduce the density of the block by replacing course aggregate with PET.

2 Literature Review

2.1 Interlocking blocks

Interlocking blocks are easy to assemble and disassemble. An interlocking lock system for mortarless wall or other structure assembly in which a plurality of blocks are laid up in courses in a staggered relationship [1]. The blocks are configured so that each end of a block comprises a half interlock and adjacent ends of a pair of blocks in a course together define an interlock portion that interlocks with a mating full interlock portion carried by a block in an adjacent course of blocks [2]. Therefore, the main function of interlocking block is to provide better jointing and has enough strength to support the load bearing walls. The interlocking design techniques are also very effective to solve the workmanship problem to ensure the installation of block works is under good quality.

2.2 Concrete

Concrete is a product consists of cement, fine and course aggregates and water. The materials are mixed in required proportion to form concrete. For normal mix, the amount of mixture for cement, sand and aggregate is measured by ratio of weight or by ratio of volume. However, ratio by weight is preferable. For example, a mixture of 1:2:4 means 1 part of cement, 2 parts of fine aggregate and 4 parts of course aggregate. Other examples of concrete mixture proportion used to produce the required strength of concrete are 1:3:6, 1: 1.5:3 and 1:4:8. Water use in concrete mixture is based on water/cement ratio, for example 0.5 or 0.6. Water/cement ratio is a ratio of weight of water to weight of cement used in concrete mixture. The chemical reaction between cement and water causes concrete to harden. When concrete is wet, it will take up the shaped according to mould [3].

2.2 Polyethylene Terephthalate (PET)

Polyethylene Terephthalate (PET) is a polymer used for former soft drinks that have full durability [4]. Common types of PET plastic bottles used for mineral water bottles, bottles of soda and shampoo. It is made up of an aromatic ester polymerized into large chains. The chemicals that go into making the PET backbone are ethylene glycol and terephthalic acid [5]. Based on ASTM International Resin Identification Coding System, PET is included in Code 1 in the types of identification codes for plastic. Resin Identification Coding (RIC) is a set of symbols appearing on plastic products that identify the plastic resin out of which the product is made. It was developed originally by the Society of the Plastics Industry (now the Plastics Industry Association) in 1988, but has been administered by ASTM International since 2008 [6].

In the recent years, numerous experimental studies were carried out on using waste PET bottles. Fataniya et al. [7] conducted a study on compressive strength of concrete masonry units with plastic bottle cores and PET fibers. They found that the compressive strength increased up to 2% replacement of the fine aggregate with PET bottle fibres and it gradually decreased for 4% and 6% replacements. Hence replacement of fine aggregate with 2% replacement was reasonable. Mokhtar et al. [8] have replacing the use of bricks in building construction by plastic bottles filled with sand. The results showed that as compared to the clay brick, the maximum stress for 250ml and 1.5 Liter bottle bricks is higher than the clay bricks for 4 and 3 times respectively which proved that bottle bricks have the potential to replace standard bricks in building. Arivalagan [9] study the possibility of using plastic aggregate as coarse aggregate in concrete. The author found that the modified concrete mix, with addition of plastic aggregate replacing conventional aggregate up to certain 10% gives strength with in permissible limit.

3 Materials and methods

Material for concrete is Portland cement with the ratio of 1:2:4 (cement: sand: aggregate). The materials used for making formwork are plywood because easy to handle and cost effective. These formworks are prepared well before fill with concrete. The design, size and dimension for standard and interlocking block were as shown in Fig. 1.

Type of block	Description	Diagram (mm)	Size (mm)
A	Standard	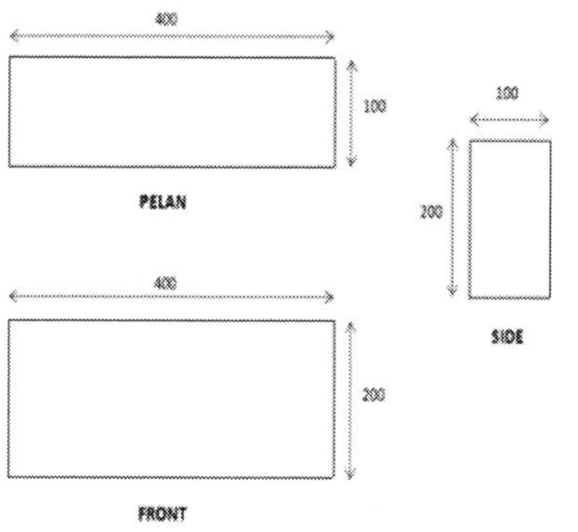	400mm x 200mm x 100mm
B	Interlocking Block	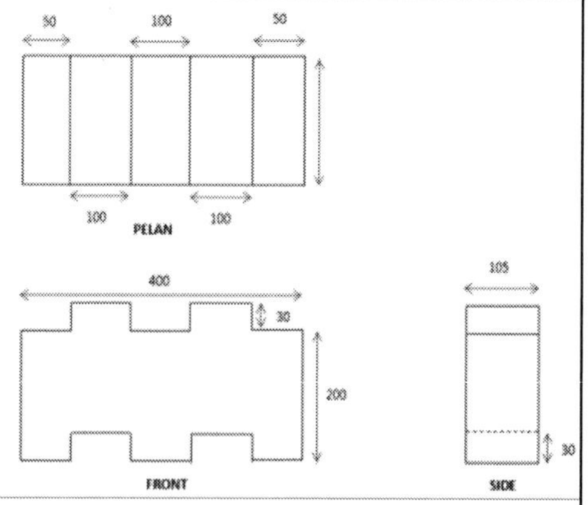	400mm x 200mm x 105mm

Fig. 1. Design, size and dimension for standard and interlocking block.

Slump test was conducted before the concrete work is carried out. It is the most commonly used method of measuring the consistency of concrete [10]. Changes in the reduction shows the changes in the material, the water content or the content of the mix.

 https://doi.org/10.1051/e3sconf/20183401006

Therefore, it is useful for quality control of concrete that will be generated. This test was done to obtain the result whether true slump, shear slump or collapse.

Curing is the process to avoid the water content in the concrete evaporates quickly by hot weather. The concrete was put in a damp at least 7 days after it harden. Rather the purpose is to show that high-performance concrete has to be water cured as early as possible, late curing is of partially no value, but is still better that no curing at all, total shrinkage can be drastically reduced by appropriate early water curing and no water curing at all can be catastrophic [11]. Figure 2 shows concrete cube with 0 – 15% PET.

Type of Cube	Standard	5% PET	10% PET	15% PET
Diagram				

Fig. 2. Concrete cube with 0 – 15% PET.

Compressive strength test was carried out to obtain the strength of the concrete sample. This test was performed by BS 5628: Part 1: 1992. Sample used were concrete sample that had reached the age of 7 days. The main factors affecting the strength of concrete is the number of honeycomb were left in the concrete, the strength will reduced if the concrete more compressible. Therefore, it is imperative that concrete is made as dense as possible. Concrete strength also increases with age, but the rate increases is greatly affected by proper curing methods. Ideally concrete stored in humid conditions to allow the cement hydration occurs properly [12]. Based on the compressive strength results, one of the percentage of PET will be chosen to proceed with construction of standard and interlocking concrete blocks.

4 Results and discussions

Slump test was conducted to measure the workability of concrete. Table 1 shows that measured slump decreased with increase in percentage of PET. This result shows that the workability of mixes increase with increase in the percentage of PET.

Through the process of hardening and curing for 7 days of normal concrete cubes and cubes which replaced with 5%, 10%, 15% PET, therefore compression tests were carry out to obtain the maximum compressive strength of concrete. Based on the results, the value compressive strength normal cubes were 38.7 MPa with density of 2350kg / m^3, while the compressive strength of concrete cube with 5% of PET was 21.5 MPa with a density 2240 kg/m^3. Next, the concrete cube in the mix 10% PET and 15% has a compressive strength of 12.7 Mpa and 5.2 Mpa with densities were 2050 kg/m^3 and 1790 kg/m^3.

Table 1. Slump test results.

Sample	Slump (mm)
Standard	28
5% PET	21
10% PET	17
15% PET	13

From the results it showed that the more the percentage of PET replaced by aggregate, the compressive strength of concrete decreases. This may occurred with the presents of honeycomb. One of the factors that affect the honeycomb is depend on the water cement ratio. The more value of water cement ratio, the less honey comb appears. Fig. 3 and Fig. 4 shows compressive strength against PET replacement and density against PET replacement. From the results it showed that the implementation usage of PET in concrete mixture had the effects on the strength and density. As stated by Pramod et al [13] in their research on innovative techniques of waste plastic used in concrete mixture, the strength of concrete block decreased with increased of PET used. In addition, the observation when stacking the wall blocks, interlocking blocks concrete with containing PET can saves construction time and less amount usage of mortar compared with conventional blocks. Fig. 5 (a and b) shows samples of standard block concrete and interlocking block concrete.

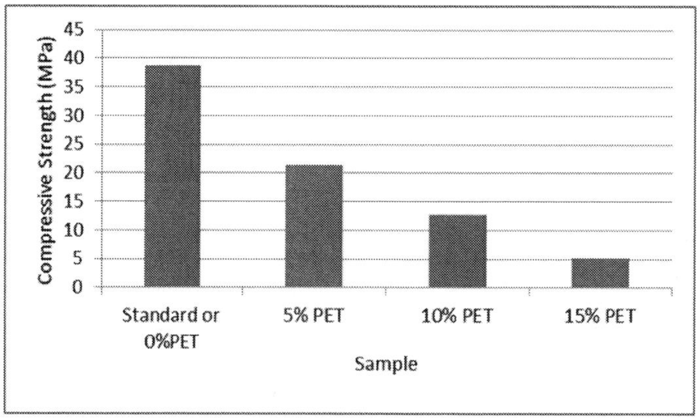

Fig. 3. Compressive strength against PET replacement.

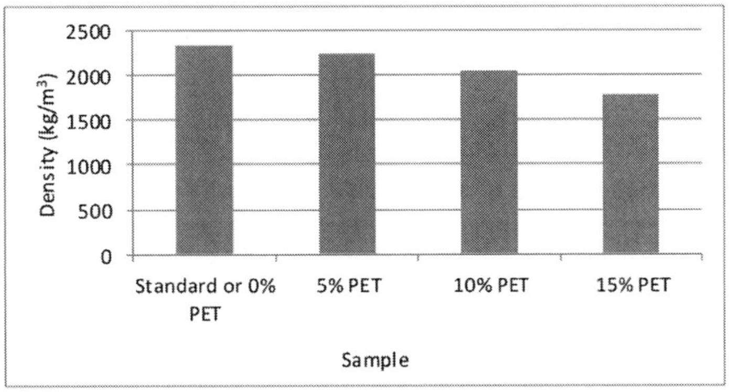

Fig. 4. Density against PET replacement.

Fig. 5 (a). Samples of standard block concrete.

Fig. 5 (b). Samples of interlocking block concrete.

5 Conclusions

From the result of this research, the following conclusion can be drawn:

- The more the percentage of PET replaced by aggregate, the compressive strength of concrete decreases.
- The density of concrete decreased when plastic content increased
- By using recycled waste plastic in concrete can reduce the landfill and environmental issues.
- Compared with conventional blocks, the interlocking blocks concrete with containing PET can saves construction time and less amount usage of mortar.

References

1. Haener Juan, Interlocking Mortarless Building Block System, US 5575128 A. (1996)
2. Timothy A. Bott, Interlcoking Building Block, US6948282 B2 (2005)
3. Yusop bin Paal, Werdah binti Abdul Halim, Sarifah binti Daud, *Pengajian Kejuruteraan Awam Tingkatan 4*, Dewan Bahasa dan Pustaka (2011)
4. Francisco Casonova-del-Angel and Jorge Luis Vazquez-Ruiz, Manufacturing Light Concrete with PET Aggregate, Article ID 287323 (2012)
5. Paul John Dornath, Analysis of Chemical Leaching from Common Consumer Plastic Bottle Under High Stress Conditions (2010)

6. ASTM International, ASTM D7611-13 Standard Practice for Coding Plastic Manufactured Articles for Resin Identification (2013)
7. Rakesh Fataniya, Rihan Maaze, Kalpesh Kapadiya & Vijay F. Pipalia, Experimental IJSRD - International Journal for Scientific Research & Development **3**(4), 2321 (2015)
8. Mokhtar M, Sahat S, Hamid B, Kaamin M, Kesot MJ, Wen LC, Xin L, Ling NP, Lei VS, Journal of Engineering and Applied Sciences **10**, 1-5 (2015)
9. Arivalagan.S, International Journal Of Engineering Sciences & Research Technology **11**, 443-449 (2016)
10. Klieger, Paul and Joseph F, *Significance of tests and properties of concrete and concrete-making materials* ,Philadelphia, PA (1994)
11. Aitcin , *High-Performance concrete*, E.& F.N. Spon. London (1998)
12. Concrete Block Association, *Aggregate Concrete Blocks A Guide to Selection & Specification*, Leicester, United Kingdom, UK (2007)
13. Pramod S. Patil, J.R.Mali, Ganesh V.Tapkire, H. R. Kumavat, *Innovative techniques of waste plastic used in concrete mixture* (2014)

3D resistivity survey for shallow subsurface fault investigations

Kraipat Petrit[1], *Poonnapa* Klamthim[1], and *Helmut* Duerrast[1,2,*]

[1]Geophysics Research Center, Prince of Songkla University, 90112 Hatyai, Songkhla, Thailand
[2]Dept. of Physics, Fac. of Science, Prince of Songkla University, 90112 Hatyai, Songkhla, Thailand

Abstract. The shallow subsurface is subject to various human activities, and the place of occurrence of geohazards, e.g. shallow active faults. The identification of the location and orientation of such faults can be vital for infrastructure development. The aim of this study was to develop a low-cost 3D resistivity survey system, with reasonable survey time for shallow fault investigations. The study area in Songkhla Province, Thailand is located in an old quarry where faults could be identified in outcrops. The study area was designed to cover the expected fault with 100 electrodes arranged in a 10×10 square grid with an electrode spacing of 3 meters in x and y axis. Each electrode in turn was used as a current and potential electrode using a dipole-dipole array. Field data have been processed and interpreted using 3DResINV. Results, presented in horizontal depth slices and vertical xz- and yz-cross sections, revealed through differences in resistivity down to 8 m depths a complex structural setting with two shallow faults and dipping sedimentary rock layers. In conclusion, this study has shown that a 3D resistivity survey can imagine complex tectonic structures, thus providing a far more insight into the shallow subsurface.

1 Introduction

The shallow subsurface, the first few hundred meters below the Earth's surface, is subject to various human activities, e.g., building foundations, groundwater drilling, but also the place of occurrence of potential geohazards, e.g., landslides and shallow active faults. The identification of the location and orientation of shallow faults can be vital for infrastructure development, thus mitigating the impact of potential earthquakes. Various types of geophysical methods can be used to identify and locate faults in the shallow subsurface, e.g. 2D resistivity, seismic reflection, and ground penetrating radar; however these techniques produce only a 2D cross section of the subsurface. In order to get more information about a more complex subsurface either several 2D surveys have to be conducted or real 3D ones. Among the latter ones is the 3D resistivity method. It is a relatively fast and easy method to do in comparison to a real 3D seismic reflection survey, which is commonly applied in petroleum exploration and deeper depths.

[*] Corresponding author: helmut.j@psu.ac.th

The objective of this work is to develop a fully functioning system for 3D resistivity measurements using the available technology at the department and then testing the system and subsequent data processing and interpretation for a study site where the subsurface structures were known from geological field observations.

2 Theories

2.1 Resistivity of rock and material

Resistivity of a material is defined as the resistance in ohms between the opposite face of a unit cube of material. For a cylinder of resistance δR, length δL, and cross-section area δA, the resistivity ρ is given by equation 1 with resistivity is ohm-meter [1].

$$\rho = (\delta R \delta A)/\delta L \qquad (1)$$

Resistivity is one of the physical properties with the largest variation (Fig. 1). Metallic minerals in igneous rocks, for example, conduct electrical currents by their free electrons. For porous sediments and sedimentary rocks, which make up most of the shallow subsurface, the electrical current is carried by ions in the pore fluid; thus the resistivity of shallow subsurface materials depend mainly on the type of pore fluid, saturation, and porosity. This process is called electrolytic conductivity. Resistivity for this kind of material is defined by equation 2, Archie's law [1, 2],

$$\rho = (a\, \rho_w)/(S_w^n\, \emptyset^m) \qquad (2)$$

where a is a constant, ρ_w is the resistivity of the pore fluid, S_w is the water saturation, n is saturation exponent, $\emptyset$ is the porosity, and m is the cementation exponent.

For clay material the situation is different as their surface is electrical charged resulting in colloidal conductivity. Therefore, resistivity values for clay are usually very low (see Fig. 1).

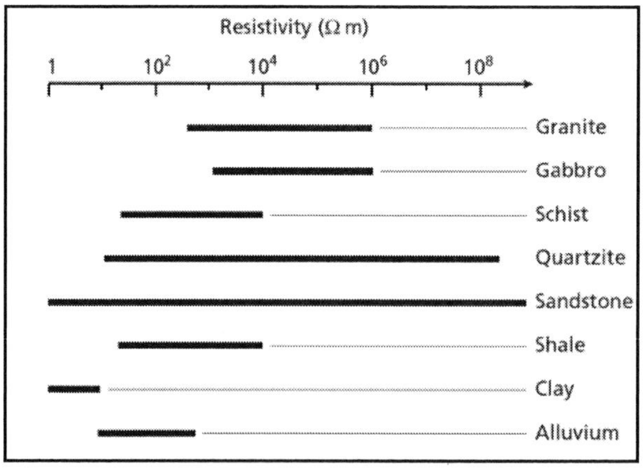

Fig. 1. Resistivity range of some common rock and sediment types, after [2].

Porosity is one of the most important rock properties in describing porous media. It is defined as the ratio of pore volume to bulk volume of a rock sample. This factor is a dimensionless quantity, expressed either in decimal or percentage. Fluid inside the pores of the shallow subsurface is mainly water and air [3]. The permeability of a rock is a measure of the ease with which the rock will permit the passage of fluids. If fluid passes through the rock easily, it has high permeability, and opposite.

2.2 Resistivity survey

Resistivity survey is one type of geophysics method wildly used for shallow subsurface investigations. Here, a source current is injected into the subsurface via two electrodes, with the resulting potential of the subsurface measured by two other electrodes [2]. Measured values of this method are 'apparent resistivity' values, which are calculated from the current, potential, and distance arrangement of the electrodes as shown in equation 3 [2],

$$\rho_a = k(\Delta V/I) = kR \tag{3}$$

where ρ_a is the apparent resistivity, R is the resistance measured with a resistivity meter and k is the geometric factor, which depends on the overall arrangement and distance of the four current and potential electrodes.

A 3D resistivity survey is used to study the resistivity distribution in three dimensions; x, y and z (depth) dimension. Generally, electrodes are set on the ground surface of the study area in a rectangular or square grid. All electrodes are connected to a resistivity meter and also used to inject electrical current as well as to measure the potential [4].

2.3 Inversion method

Inversion method is used to estimate a possible physical model of the subsurface structure from the observed data. In this case, the observed data are apparent resistivity values from field survey and the model parameter are 'true' resistivity values of the subsurface. 3DResINV is available resistivity inversion software, which is used in this work. Inversion routine is based on the smoothness-constrained least-square method shown in equation 4. Procedures of this software make rectangular blocks where the model resistivity values are allowed to vary in all three directions; then try to reduce the difference between calculated data and measured apparent resistivity values by adjusting the resistivity of the model blocks. A measure of this difference is given by the root-mean-square (RMS) error [4, 5].

$$(J^TJ + \lambda F)\, \Delta\, \mathbf{q}_k = J^Tg - \lambda F\, \mathbf{q}_k \tag{4}$$

where J is the Jacobian matrix of the partial derivative, J^T is the transpose of J, λ is a damping factor, q_k is the model perturbation vector, g is a misfit data, and F is a roughness filter.

3 Field survey

3.1 Study area

The study area was in an old quarry located in the northern part of Sadao district, Songkhla province, Southern Thailand (Fig. 2a). The northern part is a flat area because rocks and soils have been taken out. In the southern part still remained the part of a hill and sediment layer can be seen from cliffs marking the quarry boundary. All formations in the quarry are sedimentary rocks with conglomerates, conglomeratic sandstone, and sandstone. In some parts of the area sandstone interbedded with siltstone occurred (Fig. 2b and 3).

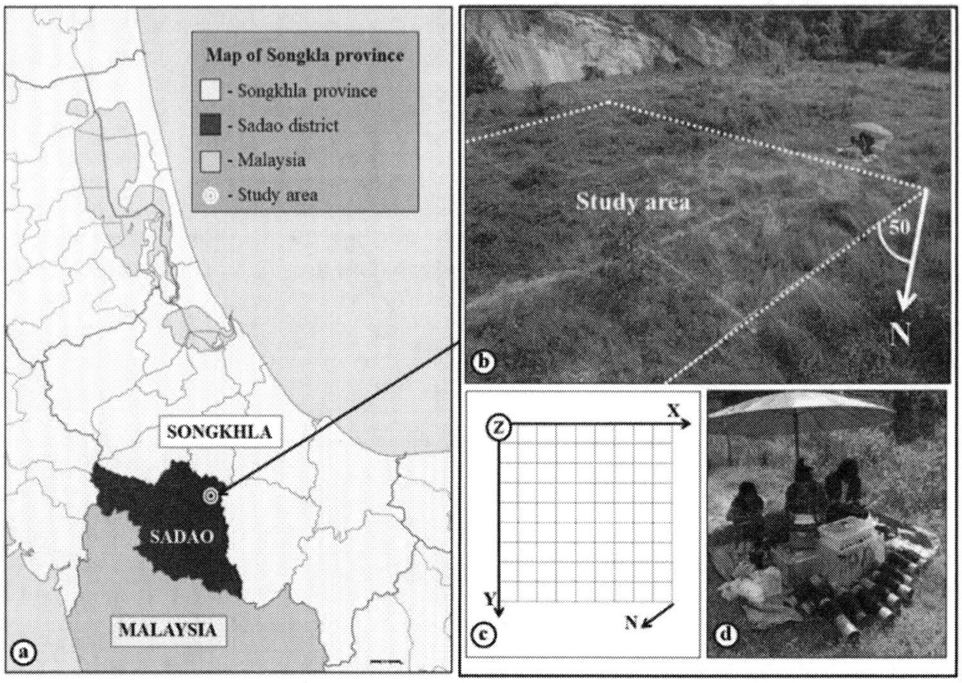

Fig. 2. Study area: (a) Location of study area in northern part of Sadao district, Songkhla province, southern Thailand, (b) Boundaries of the survey area with the direction of the x-axis 50 degrees from north over west, (c) Square grid of 10×10 electrodes with x-, y- and z-direction, and (d) Equipment set up consisting of resistivity meter, 12 V battery, switch box, and cable assembly.

3.2 Structural geology

Structural geological surface investigations in the study area revealed dipping sedimentary layers and faults (Fig. 3). Fig. 3d shows sediment layers dipping into north-west direction (288, dip direction) with dip angles of 15-20 degrees. The upper layer is conglomerate and the lower layer is sandstone and siltstone. Fig. 3a and 3e show a fault that can be found in this area. Measurements of the fault plane show that the fault is dipping into south direction with a dip angle of 64 degrees (dip direction/dip angle, 222/64).

3.3 Data acquisition and processing

To prove that 3D resistivity survey can be used to determine the structure of the subsurface, the study area is designed to cover the expected fault line show in Fig. 3. From Fig. 2b and 2c, the study area is a square grid of x- and y-axis. Direction of x-axis is 50 degree from north over west. Fig. 4 shows that 100 electrodes are set in a square grid with an inline spacing of 3 m for both x- and y-axis. Therefore, the size of the study area is 10×10 electrodes with 27×27 m (Fig. 4c).

Electrode configuration used to determine resistivity values in this work is dipole-dipole array; it consists of one of current electrode pair and one of potential electrode pair, show in Fig. 4b. From the 10×10 square of electrodes, 840 values of apparent resistivity have been measured from all directions; 10 lines in x-direction, 10 lines in y-direction, and 26 lines in both diagonal directions.

https://doi.org/10.1051/e3sconf/20183401007

Fig. 3. Structural geology of study area (a) Fault, looking into NW, dip direction/dip angle of the fault plane is 222/64 and of the sedimentary layering 333/20, (b) Study area, looking into SE, (c) Sedimentary layer in the quarry nearby the study area, looking into SE, and measurement of sedimentary layering is 310/18, (d) Sedimentary layering seen at the cliff locate in the southern part of the study area, (e) Fault, looking into SE, with an orientation of 222/64, and (f) Sedimentary layering near the study area, looking into NW; measurement of layering plane gives 288/15.

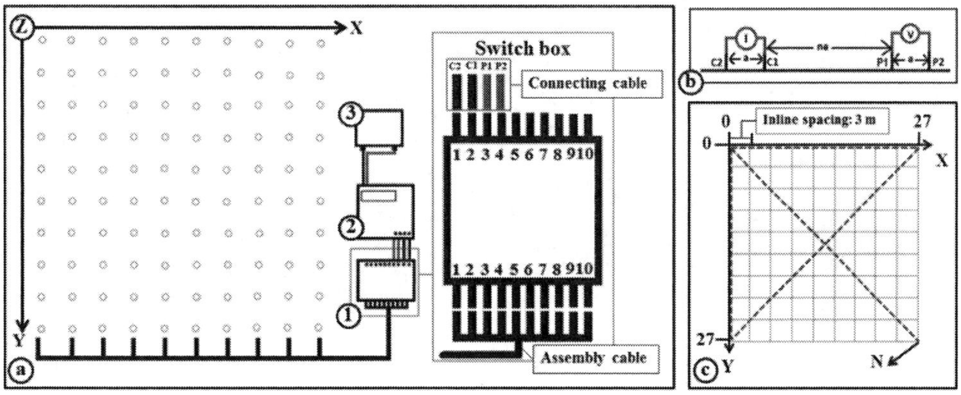

Fig. 4. Survey design and equipment: (a) Diagram of equipment set up consisting of 100 electrodes arranged in a 10×10 square grid, cable, switch box (1), resistivity meter (2) and 12 V battery (3), (b) Arrangement of electrodes in dipole-dipole array, C1-C2 is current electrode pair, P1-P2 is potential electrode pair, a is 3 m inline spacing and n is the ratio varied from 1 to 7 for 10 nodes of electrodes, and (c) Measurement lines, x, y, and both diagonal directions.

Equipment used to measure the apparent resistivity values is a Terrameter SAS 1000, single-channel resistivity meter. The main difference from a multi-channel resistivity meter that is actually used for a 3D resistivity survey is the limitation of a single-channel. Therefore, the resistivity meter cannot measure more than one data point at the same time of injecting current. In this work, a switch box was developed to use with a single-channel resistivity meter. Fig. 4a shows a diagram of the equipment set up with 10 electrodes in one line connected to the switch box by a cable assembly. The switch box is connected to the single-channel resistivity meter by two pairs of cables; a current cable pair and a potential cable pair. By this way, all data points in one line can be measured by moving the current cable pair and the potential cable pair on the switch box.

After data acquisition, the 3DResINV software processes the apparent resistivity values. Results from the software are images of resistivity distribution in horizontal sections (*xy* plane), x-sections (*xz* plane), and y-sections (*yz* plane).

4 Results and discussions

Fig. 5 and 6 present the inversion results of the resistivity model of the study area. From these results resistivity areas can be separated in two zones: A low resistivity (L) area with resistivity values varying from 10 to 2,000 ohm m, indicated by blue and green colors; and a higher resistivity area (H) that has a resistivity range of 2,000 to 5,000 ohm m, indicated by red and purple colors.

Including the geological information, low resistivity area can be considered to be siltstone and sandstone, whereas the high resistivity areas can be considered to be conglomerate. As the porosity and permeability of the conglomerate is lower than siltstone and sandstone, the resistivity of the conglomerate is higher than of the siltstone and sandstone.

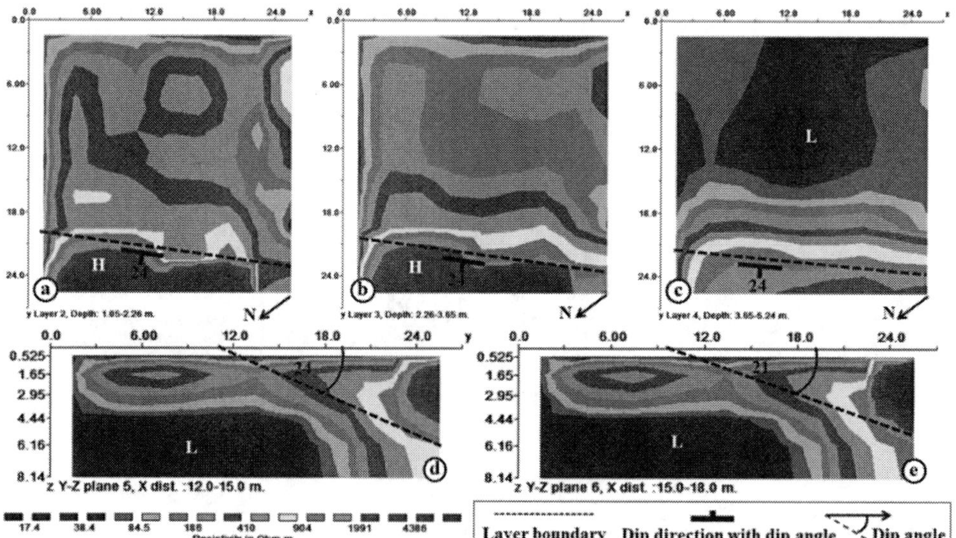

Fig. 5. Structural geology discussion, sedimentary layering: (a), (b), and (c), results of resistivity model in horizontal section of difference depth intervals with an interpretation of dip direction of the sedimentary layers, and (d) and (e) results of resistivity model in y-section with interpretation of dip angle of the layers. L = low resistivity, H = high resistivity.

4.1 Dipping sedimentary layers

Fig. 5a-c presents the resistivity models of the horizontal sections for different depth intervals. Results show that layers of higher resistivity indicated to be conglomerate layers dipping into northwest direction, 225-230 degrees from north. From Fig. 5d and 5e, the dip angle is 20-25 degrees from the surface.

4.2 Faults

The consideration of fault lines in this area is based on discontinuity of resistivity values. As faults are cracks in the subsurface that can separate sedimentary layers they create discontinuities in the resistivity values. Fig. 6 shows that there are two fault lines in the study area; Fault 1 dipping into south direction, with a dip direction of 210 degrees from north and a dip angle of 68 degrees. Fault 2 is dipping in south direction with a dip direction of 215-220 degrees from north and a dip angle of 74 degrees.

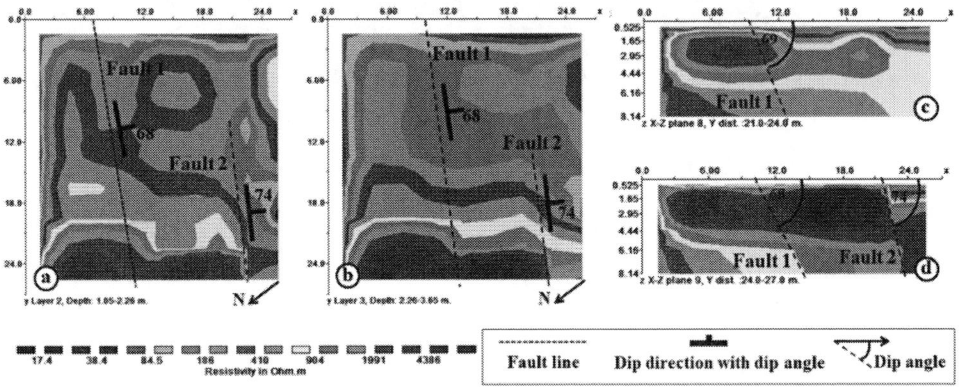

Fig. 6. Structural geology discussion, faults: (a) and (b) results of resistivity model in horizontal section with interpretation of dip direction of faults, and (c) and (d) results of resistivity model in x-section with interpretation of dip angle of the faults. L = low resistivity, H = high resistivity.

5 Conclusions

5.1 Subsurface model

From the resistivity survey, the result shows that the sedimentary layers dipping into northwest direction with a dip angle 20-25 degrees. Information of dipping sedimentary layers correlates with the geology data (see Fig. 3). The upper layer with higher resistivity should be conglomerate and the lower layer with lower resistivity should be siltstone and sandstone, as shown in Fig. 7.

From geological data, Fault 1 correlates with the expected fault line because both have similar dip direction and dip angle. From the results Fault 2 cannot be seen and identified from the surface. Detecting Fault 2 proves that a 3D resistivity survey can be used to determine geological structures in the subsurface that cannot be seen from surface investigations.

https://doi.org/10.1051/e3sconf/20183401007

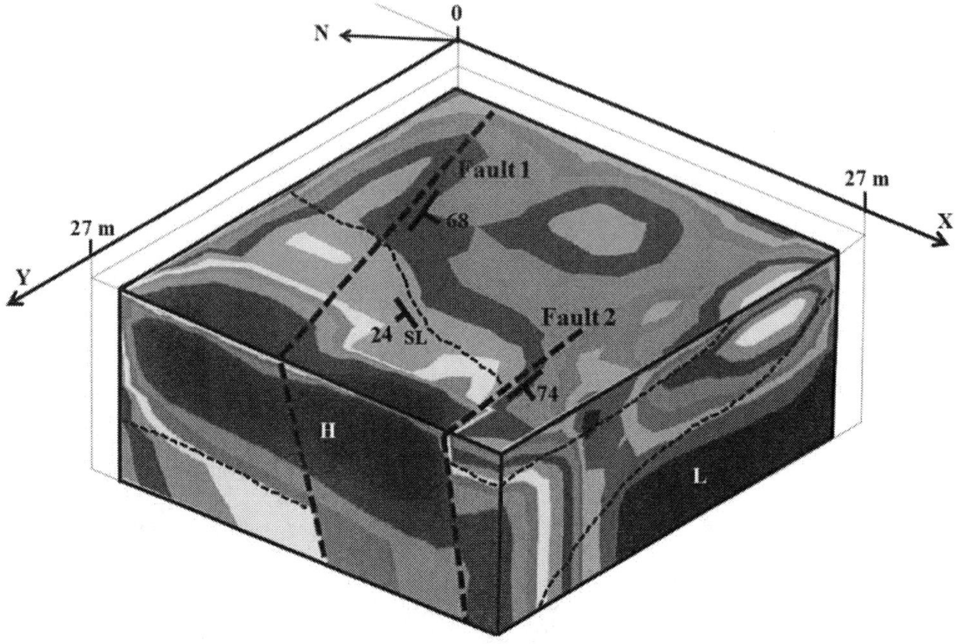

Fig. 7. Subsurface model from 3D resistivity survey: Fault information; Fault 1 is dipping into south direction with 68 degrees dip angle and Fault 2 is dipping into south direction with 74 degrees dip angle. Sedimentary layering (SL) shows dipping into northwest direction with a dip angle of 24 degrees. L = low resistivity, H = high resistivity.

5.2 Acquisition system

Conclusion also can be drawn regarding the 3D resistivity acquisition system. Results presented here show that it is possible to make a 3D resistivity survey by using a single-channel resistivity meter with a switch box and cable assembly instead of a multi-channel resistivity meter that still requires high survey costs nowadays. From this work, the data acquisition system can measure 840 data points in one to two days depending on survey area environment and the survey costs are actually lower than using a multi-channel resistivity meter.

References

1. P. Kearey, M. Brooks, I. Hill, *Introduction to Geophysical Exploration*, 3rd ed. (Blackwell Science Ltd, Oxford, 2002)
2. W.M. Telford, L.P. Geldart, R.E. Sheriff, *Applied Geophysics*, 2nd ed. (Cambridge University Press, New York, 1990)
3. R. Bowen, *Groundwater*, 2nd ed. (Elsevier Applied Science Publisher, Essex, 1986)
4. M.H. Loke, Tutorial: 2-D and 3-D electrical imaging surveys (2015)
5. M.K. Sen, P.L. Stoffa, *Globle Optimization Methods in Geophysics Inversion*, 2nd ed. (Cambridge University Press, New York, 2013)

Drying Shrinkage of Mortar Incorporating High Volume Oil Palm Biomass Waste

Nor Hasanah Abdul Shukor Lim[1]*, *Mostafa* Samadi[1], *Abdul Rahman* Mohd. Sam[2], *Nur Hafizah* Abd Khalid[1], *Noor Nabilah* Sarbini[1], *Nur Farhayu* Ariffin[3], *Mohd Warid* Hussin[1], and *Mohammed* A. Ismail[4]

[1]Department of Structure and Materials, Faculty of Civil Engineering, Universiti Teknologi Malaysia, 81310, Johor, Malaysia.
[2]UTM Construction Research Centre, Institute for Smart Infrastructure and Innovative Construction, Faculty of Civil Engineering, Universiti Teknologi Malaysia,81310, Johor, Malaysia.
[3]Faculty of Civil Engineering and Earth Resources, Universiti Malaysia Pahang, Lebuhraya 8 Tun Razak, 26300 Kuantan, Pahang, Malaysia
[4]Civil and Construction Engineering Department, Faculty of Engineering and Science, Curtin University Sarawak, CDT 250, 98009 Miri, Sarawak, Malaysia

Abstract. This paper studies the drying shrinkage of mortar incorporating oil palm biomass waste including Palm Oil Fuel Ash, Oil Palm Kernel Shell and Oil Palm Fibre. Nano size of palm oil fuel ash was used up to 80 % as cement replacement by weight. The ash has been treated to improve the physical and chemical properties of mortar. The mass ratio of sand to blended ashes was 3:1. The test was carried out using 25 x 25 x 160 mm prism for drying shrinkage tests and 70 x 70 x70 mm for compressive strength test. The results show that the shrinkage value of biomass mortar is reduced by 31% compared with OPC mortar thus, showing better performance in restraining deformation of the mortar while the compressive strength increased by 24% compared with OPC mortar at later age. The study gives a better understanding of how the biomass waste affect on mortar compressive strength and drying shrinkage behaviour. Overall, the oil palm biomass waste can be used to produce a better performance mortar at later age in terms of compressive strength and drying shrinkage.

1 Introduction

Drying shrinkage is strain associated with loss of moisture from the mortar by evaporation of water or hydration of cement. The change in volume of the mortar is related to the volume of water lost. The loss of free water from mortar may induce shrinkage. As the hydration process of the mortar continues, the water in the small capillaries is reduced significantly [1]. The loss of the water may force mortar to shrink causing cracks that can adversely affect the structural performances [2]. If drying shrinkage is not considered

* Corresponding author: norhasanah@utm.my

during the design stage it may have negative effect on the durability and serviceability of mortar.

Few studies have been conducted to examine the effects of oil palm fuel ash (OPFA) on drying shrinkage [3-6]. The results showed that by increasing the unground OPFA percentage as partial cement replacement, the shrinkage increased slightly at later ages. However, the use of 10 to 30% finer POFA (10 μm) as cement replacement reduced the drying shrinkage of concrete. This is due to the pore refinement, the transformation of large pores into fine pores, and reduced the loss of water, thus reducing the drying shrinkage [6]. The transformation of large pores into fine pores decreases the evaporation of water from concrete surface and thus reduces the drying shrinkage [7].

Therefore, exploring OPFA as nano material would create an advanced waste material in improving the performances of mortar. In this paper, high volume of OPFA with nano size was used as cement replacement up to 80% with addition of kernel shell and oil palm fibre. This helps to reduce the waste generation from palm oil plantation.

2 Materials and Methods

2.1 Materials

The cement used in this study is Ordinary Portland cement (OPC) Type I. OPFA was obtained from the burning of palm oil shell and husk (in equal volume). OPFA collected was treated and processed until nanoparticles size [8]. The chemical properties of OPC and nano POFA were shown in Table 1. The chemical composition shows that OPC is rich in Calcium Oxide (CaO) content while nano OPFA is consisting of Silica Oxide (SiO_2). The Oil Palm Kernel Shell was crushed and used as sand replacement. The Oil Palm Fibre added was 7% by the volume of binder. The mortar incorporating the oil palm biomass waste was then known as biomass mortar [9, 10].

Table 1. Chemical composition of OPC and Nano POFA.

Chemical composition (%)	OPC	Nano POFA
SiO_2	16.40	69.30
Al_2O_3	4.24	5.30
Fe_2O_3	3.53	5.10
CaO	68.30	9.15
K_2O	0.22	11.10
MgO	2.39	4.10
CO_2	0.10	0.10
SO_3	4.39	1.59
LOI	2.40	1.30

2.2 Testing Procedures

All mortar specimens were prepared with binder to fine aggregates with the ratio of 1:3, whereby the fine aggregates was prepared into saturated surface dried condition. The mixing was carried out in a room temperature. The mix proportions are based on volume of materials as according to BS EN 998-1:2010 [11]. The test specimens of 70 x 70 x 70 mm cubes were prepared as described in ASTM C109-13 [12]. The specimens were then cured in water until the age of testing. Mix proportion of OPC and biomass mortars were shown in Table 2.

Table 2. Mix proportion of mortars (kg/m^3).

Mortar	Cement	Nano OPFA	Sand	OPKS	OPF	W/C ratio
OPC	525	-	1578	-	-	0.4
Biomass	105	420	789	789	3.7	0.4

Drying shrinkage is strain associated with loss of moisture from the mortar by evaporation of water or hydration of cement. For drying shrinkage, the size of specimens was 25x25x250 mm as shown in Figure 1 and cured in water. The test was conducted according to ASTM C596-14 [13]. After demoulding, mortar specimens were measured and marked to get the centre of the specimens. The marked was necessary as a guide to place the demec disc and demec gauge for shrinkage reading. The demec disc was glued to both sides at a distance of 75 mm from the centre. After that, demec gauge was placed on the demec disc for 5 to 10 minutes and the initial readings were recorded. Shrinkage readings were recorded by using demec gauge at 25±2°C. The drying shrinkage strain of each type of mortar was calculated. Shrinkage was determined as an average of three specimens. It was calculated by the following Equations 1:

$$\text{Percentage of shrinkage} = \frac{Lo-L}{Lo} \times 100 \tag{1}$$

where L_o = original length (length of standard bar) (mm); L = length as measured during or after cure excluding studs (mm).

a) Specimens with demec disc glued.

b) Measuring shrinkage using demec gauge.

Fig. 1. Sample preparation for drying shrinkage test.

3 Results and Discussions

3.1 Compressive Strength

Fig. 2 shows the compressive strength comparison of the OPC and biomass mortars. As the age of water curing increases, the compressive strength of OPC and biomass mortars are increases. This is due to the continuous hydration process and pozzolanic reaction occurred in the mortars. At 28 days curing, the compressive strength of biomass mortar increases by 24% as compared to OPC mortar. This is due to the second production of C-S-H gel from pozzolanic reaction. After 90 days of curing, the OPC and biomass mortars show compressive strength of 32.43 MPa and 42.84 MPa, respectively. Despite the stronger bond produced and filler packing effect, the fibre in biomass mortar help to distribute the load and increase the bond between binder and fine aggregates resulting in an increase in the compressive strength. The compressive strengths obtained at the age of 90 days were approximately 24% higher than the target required design strength of 30 MPa. This continual increase in strength even after 90 days indicates that the biomass mortar keep increase the compressive strength through the gradual hydration process and pozzolanic reaction of the nano size OPFA.

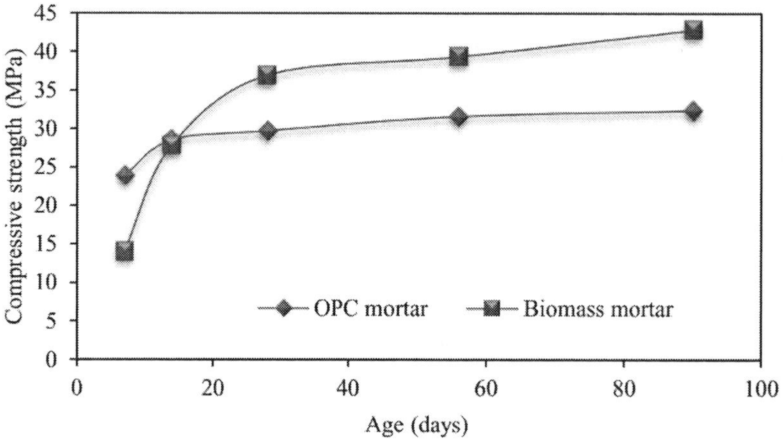

Fig. 2. Strength development of OPC and biomass mortars.

3.2 Ultra Pulse Velocity

Meanwhile, Fig. 3 shows the comparison of biomass and OPC mortars on the UPV values. The biomass mortar also produces lower UPV value at early age and increases as the hydration process increases. Despite the effect of nano OPFA as cement replacement in the biomass mortar, the smaller OPKS size with irregular and crushed particles increases the interlocking friction with the binder. In addition, the fibre content also increased the interlocking between fine aggregates and binder thus makes the biomass mortar denser and less porous. Since the void content in the mortar reduces, the UPV value is increases. Neville, (2011) reported that UPV value more than 4500 m/s generally indicates mortar with excellent quality [14]. In this study, these excellent UPV values could be attributed to the improved pore structure of mortar as a result of optimum particle packing and enhanced filling ability of nano size OPFA. Furthermore, higher UPV values were obtained at later age which could be attributed to the increased densification of the paste resulting from secondary hydration of the reactive nano OPFA. Similar findings were reported by previous research using fly ash as pozzolanic material on the packing effect [15].

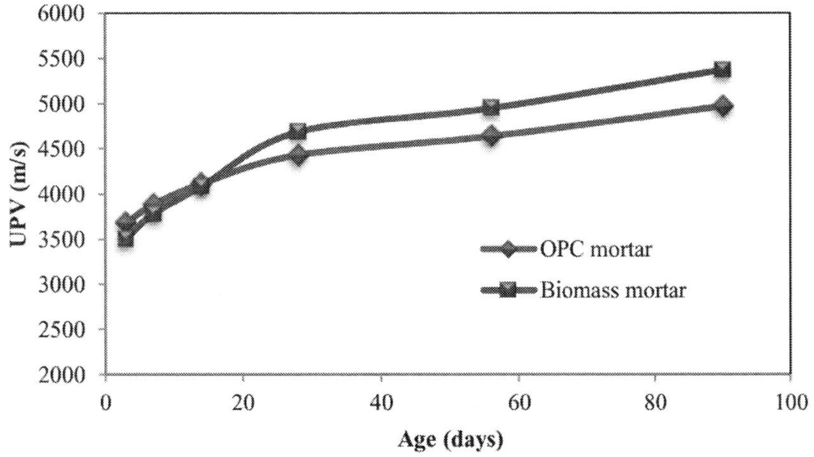

Fig. 3. UPV analysis of OPC and biomass mortars.

The UPV value and the compressive strength of biomass mortar have a strong and positive linear correlation as shown in Fig. 4. The coefficient of determination (R^2) value of 99% showed a good prediction of the relationship. The correlation was determined for the compressive strength values in the range of 13.63 MPa to 47.37 MPa and UPV value in the range of 3503 m/s to 5373 m/s, respectively. Similar correlation was also observed by Demirboga et al., (2004) for the high strength concretes incorporating fly ash and blast furnace slag [16].

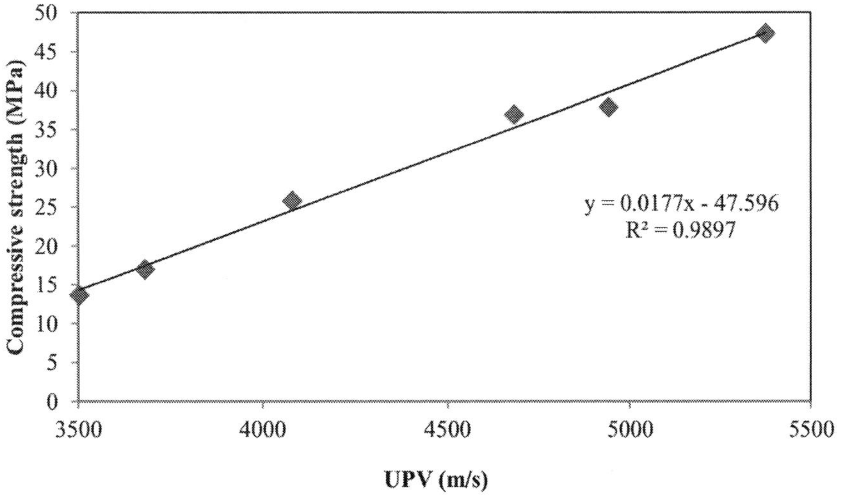

Fig. 4. Relationship between compressive strength and UPV value of biomass mortar.

3.3 Drying Shrinkage

Fig. 5 shows the development of the drying shrinkage OPC and biomass mortars until the age of 90 days. As can be seen in the figure, the drying shrinkage of OPC mortar is higher than that of biomass mortar at all ages. The increasing rate of drying shrinkage of both mortars is very sharp in the first month. At the age of 28 days, the drying shrinkage for OPC and biomass mortars are $853x10^{-6}$ mm/mm and $621x10^{-6}$ mm/mm, respectively. This is due to the dense structure from the gradual hydration process and pozzolanic reaction of nano OPFA as cement replacement [17]. In addition, the fibre content is found to have an effect in reducing the drying shrinkage of mortar. However, the shrinkage cracks were not observed because of the presence of fibres that prevented the occurrence of cracks thus reduces the shrinkage value. Similarly, previous research reported that the fibre content has a direct influence on restrained drying shrinkage in concrete, as well as on the control of cracking [18]. Therefore, by adding the NOPFA, OPKS and OPF, the shrinkage value of biomass mortar reduced by 31% from OPC mortar. Therefore, the results show that biomass mortar has better performance in deformation of mortar.

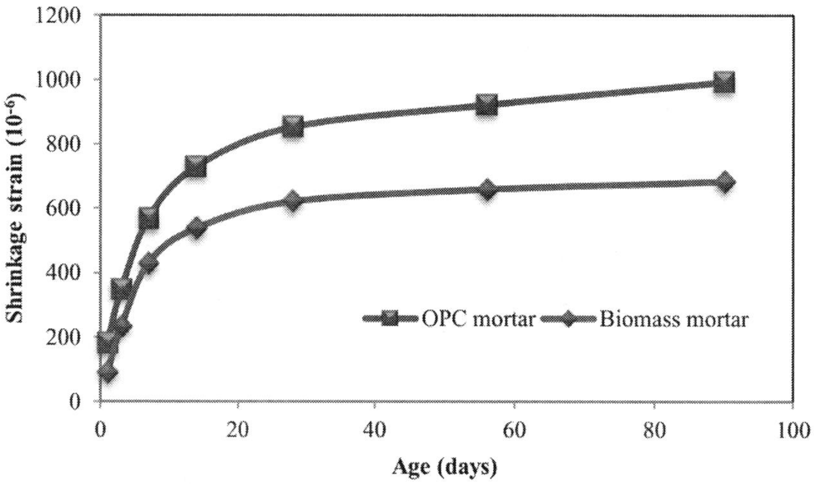

Fig. 5. Drying shrinkage of OPC and biomass mortars.

4 Conclusions

This study generally focuses on the performance of mortar using high volume waste from oil palm biomass. Higher values of ultrasonic pulse velocity were obtained by using finer nano OPFA at all ages due to denser microstructure of the mortar. Similar result was obtained when biomass mortar yielded higher UPV values by 5% compared with OPC mortar after 28 days. The UPV values at 28 days for biomass and OPC mortars are 4683 and 4429 m/s, respectively. Besides, the performance of biomass mortar in drying shrinkage which is $683x10^{-6}$ mm/mm is better than OPC mortar with a value of $993x10^{-6}$ mm/mm at the age of 90 days. With the addition of nano OPFA, OPKS and OPF, the shrinkage value of biomass mortar is reduced by 31% compared with OPC mortar thus, showing better performance in restraining deformation of the mortar.

The authors are grateful to the Ministry of Higher Education, Malaysia (MOHE) and Research Management Centre (RMC), Universiti Teknologi Malaysia (UTM) for financial support under grant Q.J130000.2722.02K88. The authors are also thankful to the staff of Structures & Materials Laboratory, Faculty of Civil Engineering and microstructure lab, Hanyang University Ansan Campus, South Korea for the support throughout the study.

References

1. W. Zhang, M. Zakaria and Y. Hama. Constr Build Mater **49**, 500–510 (2013)
2. B.S. Thomas. Renewable and Sustainable Energy Reviews, **80**, 550-561 (2017)
3. N. Ranjbar, A. Behnia, B. Alsubari, P.M. Birgani and M.Z. Jumaat. Journal of Cleaner Production **112**, 723-730 (2016)
4. A.S.M.A. Awal and M.W. Hussin. Malaysian Journal of Civil Engineering **21**(2),125-134 (2009)
5. W. Tangchirapat and C. Jaturapitakkul. Cement Concrete Comp **32**, 767–774 (2010)
6. W. Tangchirapat, C. Jaturapitakkul and P. Chindaprasirt, Constr Build Mater **23**, 2641–2646 (2009)
7. N.H.A.S. Lim, M. Samadi, A. Shafagat, N.F. Ariffn, A. Keyvanfar and A.R.M. Sam. Journal of Environmental Treatment Techniques, **4**(4), 149-152 (2016)
8. N.H.A.S. Lim, M.A. Ismail, H.S. Lee, M.W. Hussin, A.R.M. Sam and M. Samadi. Constr Build Mater **93**, 29–34 (2015)
9. N.H.A.S. Lim, M.W. Hussin, A.R.M. Sam, M.A.R. Bhutta, N.F. Ariffn, N.H.A. Khalid and M. Samadi, Advanced Materials Research, **1113**, 578-585 (2015)
10. H.L. Muttashar, M.W. Hussin, J. Mirza, N.H.A.S. Lim and G.F. Huseien, Jurnal teknologi **78**(8), 101–108 (2016)
11. British Standard Institution. *Specification for mortar for masonry. Rendering and plastering mortar*. BS EN 998-1:2010 (2010)
12. ASTM C109. Standard test method for compressive strength of hydraulic cement mortars. *Annual Book of ASTM Standards, American Society for Testing and Materials* (2013)
13. ASTM C596. Standard test method for drying shrinkage of mortar containing hydraulic cement. *Annual Book of ASTM Standards, American Society for Testing and Materials* (2009)
14. A.M. Neville. *Properties of Concrete* (Fifth Edition). Harlow, England: Pearson Education Limited (2011)
15. J. Tangpagasit, R. Cheerarot, C. Jaturapitakkul and K. Kiattikomol, Cement and Concrete Research **35**, 1145–1151 (2005)
16. R. Demirboga, I. Turkmen and M.B. Karakoc, . Cement Concrete Comp **34**(12), 2329-2336 (2004)
17. W. Li, Z. Huang, T. Zu, C. Shi, W.H. Dua, S.P. Shah, Journal of Materials in Civil Engineering **28**, 04015068 (2016)
18. Z. Ahmad, A. Ibrahim and P.M.D. Tahir, International Journal of Engineering Science and Technology, **2**(5), 699-708 (2010)

Lateral Displacement and Shear Lag Effect of Combination of Diagrid-Frame

Roslida Abd. Samat [1,*], *Fong Teng* Chua[1], *Nur Akmal Hayati* Mohd Mustakim[1], *Sariffuddin* Saad[1] and *Suhaimi* Abu Bakar[1]

[1]Faculty of Civil Engineering, Universiti Teknologi Malaysia, 81310 Skudai, Johor, Malaysia

Abstract. Diagrid system, which is the portmanteau of diagonal grid member, is an exterior lateral load resisting system for tall building that has gained a wide acceptance in the design of tall buildings. There is abundance of researches that studied the efficiency of diagrid systems, which are constructed from the ground level to the top of the buildings in resisting the lateral load. Nevertheless, no study had been performed on the effectiveness of the diagrid that is constructed above other tall building systems despite the existence of a few buildings in the world that employ such system. The objective of this research is to understand the behavior of the lateral displacement and shear lag effect due to wind load when the diagrid structure is constructed above a frame. Models of 60-story buildings with a footprint of 36m x 36m were analyzed by using Staad.Pro software. The level where the diagrid members started was altered. The lateral displacement was reduced to 60.6 percent and 41 percent of the lateral displacement of a building with full frame system when the combination of frame-diagrid that had the diagrid started at Level 1 and Level 45, respectively were employed. Furthermore, the shear lag ratio was reduced from 1.7 to 1.3 when the level where the diagrid started was increased from Level 1 to Level 45.

1 Introduction

The latest tall building system which now has become favourable among today's designer is diagrid system. Diagrid system is composed of many diagonal structures that connect together to form a triangulated shape or can be seen as grid shape. The term "diagrid" is the combination of "diagonal" and "grid" words, which refers to a structural system that is single-thickness in nature and gains its structural integrity through the use of triangulation [1]. Diagrid structure is different from the braced tube structure as it has no vertical columns present and thus is considered as the evolution of braced tube structure [2]. Diagrid system is favoured due to its structural efficiency and aesthetic potential [3,4]. Diagrid enables the architect to design a building with some unique elements such as irregular angle and nonlinear shape as portrayed in buildings like Swiss Re (30 St. Mary Axe) in London, Capital Gate in Abu Dhabi and CCTV in Beijing. Moon [5] studied the

* Corresponding author: roslida@utm.my

structural performance of complex-shaped tall buildings such as twisted, tilted and freeform towers that were designed with diagrid system.

The aesthetically unique design of skyscraper that diagrid provides is mainly due to its free of exterior columns [6]. Since diagrid system allows the elimination of almost all the vertical columns, this system is able to provide variety of open floor plans. Diagrid building is an efficient system because it is able to achieve the same displacement as the displacement of buildings that have other tall building systems with significant reduction of the steel volume. The implementation of diagrid system to Encana Tower that is located in Calgary in Western Canada had reduced approximately 20 percent of the structural steel weight when compared to a building with a conventional moment frame system [7].

Diagrid, which is an exterior structural system, is designed to control the lateral displacement of tall building. The taller the building is the higher the wind load is as wind speed increases parabolically with height from ground. Damage to non-structural elements such as cracking in reinforced walls, damage to lightweight partitions and impaired operation of windows is common examples of the effect of excessive lateral displacement of tall building [8]. Thus, the control of lateral displacement is a serious issue in the design of tall building. A building should not sway horizontally more than H/500 to consider the design criteria that are strength, serviceability, stability and human comfort [9]. American Society of Civil Engineer [10] recommended that the drift limit to be on the order of 1/600 to 1/400 of the building or story height.

Another important issue in the design of high rise building is shear lag. Shear lag is caused by the lack of shear stiffness that reduces the structural effectiveness of the framed tube. In framed tube building, the tube action behaves like a cantilever box beam that is subjected to overturning moment induced by lateral loading, with a significant contribution from the flange elements (the walls that is normal to the direction of the lateral load). The elementary theory of bending that states a plane section remains plane before and after bending can be achieved only when the shear stiffness of the cross section is infinite. Thus, the linear distribution of bending stress in the cross section of the building that results in the same value of axial force in all columns at the flange of the building cannot be achieved in tubular framed building. The axial forces in the columns at the middle of the flange frames lag behind those near the corner and this is called as shear lag effect. The use of diagrid tube structure minimizes the shear lag effect by reducing the difference in the axial force values in the columns at the middle part of the flange with those in the columns at the corner of the flange of the building, as stated by [11] and [12]. However, the increase of the slope of the braces of diagrid increases the shear lag and becomes more rapid when the angle is higher than 70°. Interestingly, shear lag does not correlate with the lateral displacement of high rise building [12].

Full diagrid structures are buildings that have diagrid system being constructed from the ground floor to the top of the building. Most diagrid buildings in the world are full diagrid structures, for example, Swiss Re (30 St. Mary Axe) in London, CCTV in Beijing, United Steelworkers Building in Pittsburgh, USA and Tornado Tower in Doha, Qatar. There are a few buildings in the world that employ combination of diagrid system and other tall building systems at their perimeter, for example, Lotte World Tower in Seoul, Korea which has its diagrid being constructed only from the 107th floor to the top [13] and Hearst building in New York which has its diagrid system being supported by mega columns and super-diagonals that are constructed from the foundation to the 10th floor [14].

Many studies on the performance of full diagrid structures where the diagrid system begins from the ground floor to the top of building have been conducted. However, there is no systematic research that has been carried out on tall building that has diagrid that are constructed above a frame structure. Thus, the objective of this study is to find the effect of constructing a diagrid system above a frame to the lateral displacement and shear lag of a sixty-storey building due to wind load. The story number of the frame that transmitted the force from the diagrid system to the foundation were varied to investigate how they influenced both the lateral displacement and shear lag.

2 Methodology

The structural analysis software, STAAD.Pro V8i was used to analyse all models statically. All elements in the buildings such as beams, columns and diagrid truss were modelled by using line element where each node has three degrees of freedom, which are independent vertical motion, independent horizontal motion and independent rotational motion.

Generally, each model has 60 stories with 4 m storey height and 36 m x 36 m square floor plan as shown in Fig. 1. All columns were reinforced concrete columns. The size of the internal column is 1200 mm x 1200 mm, 900 mm x 900 mm and 600 mm x 600 mm at Level 1 to 20, Level 21 to 40 and Level 41 to 60 respectively. Full diagrid model did not have any external columns, while the models with the combination of frame and diagrid systems did have external columns in the frame system part.

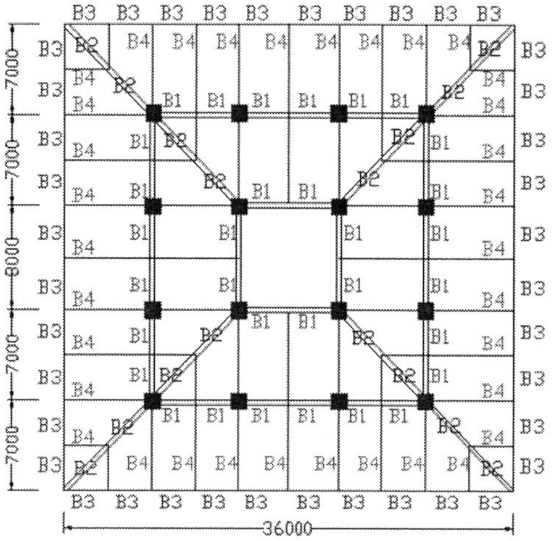

Beam specification :

B1 - 650 mm x 250 mm reinforced concrete beam
B2 - 750 mm x 350 mm reinforced concrete beam
B3 - 700 mm x 700 mm x 25 mm thick steel hollow tube
B4 - UB 406x140x53 steel beam

Fig. 1. Typical plan view of all Building Models.

The angle of the triangulated members was uniform and was approximately about 69 degrees, where one module of diagrid was equalled to two storeys of the building. Moon [15] stated that, 69 degrees is the near optimum angle for 60-story high building with diagrid structure. Furthermore, the intersection between the diagonal members was pin-connection so that axial action merely carries the transverse shear and moment. The diagonal members had pinned supports, while the external and internal columns had fixed-end supports. Models of combination of diagrid and frame systems used reinforced concrete beams of 400mm x 1000 mm as the transfer beams that transmitted the axial forces in the diagrid system to the frame system

Model 1 is a frame model without any diagrid while Model 2 is a full diagrid building. For the study of combination of frame and diagrid system, the frame system was at the lower level while diagrid system was at upper level. The level where the diagrid started was varied to Level 1, 3, 5, 7, 15, 30 and 45 as shown in Fig. 2.

The wind load was calculated based on the procedure provided in the ASCE7-10. Two wind environments that were considered are Malaysian wind speed, which is benign and Hong Kong wind speed, which is more aggressive. Determination of the displacement of a

https://doi.org/10.1051/e3sconf/20183401009

building is a serviceability design while the determination of shear lag is a survivability design. Thus, the 3-second gust wind speed for 10-year return period at 10 m high for Malaysian and Hong Kong wind environment which is 28.14 m/s and 42.48 m/s, respectively were applied during the analysis for displacement. Analysis for axial forces which was used to observe shear lag had the 3-second gust wind speed for 50-year return period at 10 m high for Malaysian and Hong Kong wind environment of 33.5 m/s and 57.4 m/s, respectively. The buildings, further, were assumed to be located in an urban area and thus, exposure B as stated in ASCE 7-10 was used.

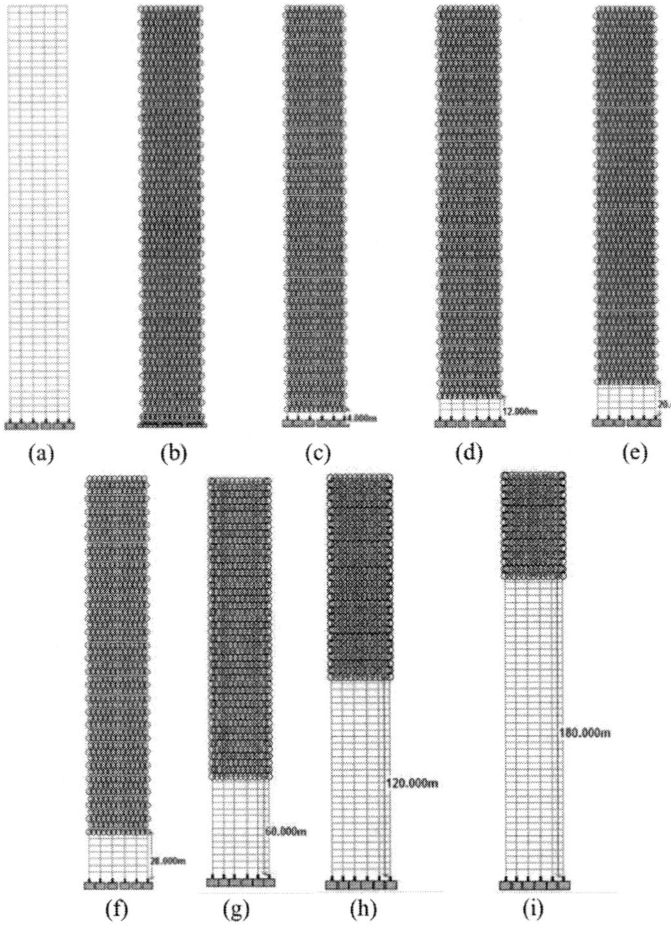

Fig. 2. Side elevation of 60 storey models (a) Model 1: Frame without diagrid, (b) Model 2: Diagrid started at ground level, (c) Model 3: Diagrid started level 1, (d) Model 4: Diagrid started level 3, (e) Model 5: Diagrid started level 5 and (f) Model 6: Diagrid started level 7 (g) Model 7: Diagrid started level 15 (h) Model 8: Diagrid started level 30 (i) Model 9: Diagrid started level 45.

3 Results and Discussions

Results from the dynamic analysis that was performed on all building models show that the lowest natural frequency of all models exceeded 1 Hz. Thus, the buildings are categorized as stiff buildings, and static analysis is sufficient to determine both lateral displacement and stresses of the models, as in accordance with building code in United States of America, ASCE 7-10 [10].

3.1 Lateral Displacement

The graph of lateral displacement of all models at different level is plotted in Fig. 3. Full frame model has the largest lateral displacement under wind load serviceability limit state condition, while full diagrid has the least lateral displacement. The maximum displacement occurred at the top of the building. The maximum displacement of the full frame model was reduced by 66 percent when full diagrid is used. Models with the diagrid constructed at Level 1, 3, 5 and 7 had less percentage of the reduction of the lateral displacement, but the difference is less than 10 percent. Models with the diagrid started at Level 15, reduced the lateral displacement by 50 percent, which is 16 percent less than the full diagrid. The maximum lateral displacement of the models with diagrid started at Level 30 and 45 is about the same where the lateral displacement is reduced by 41 percent when compared to full frame model.

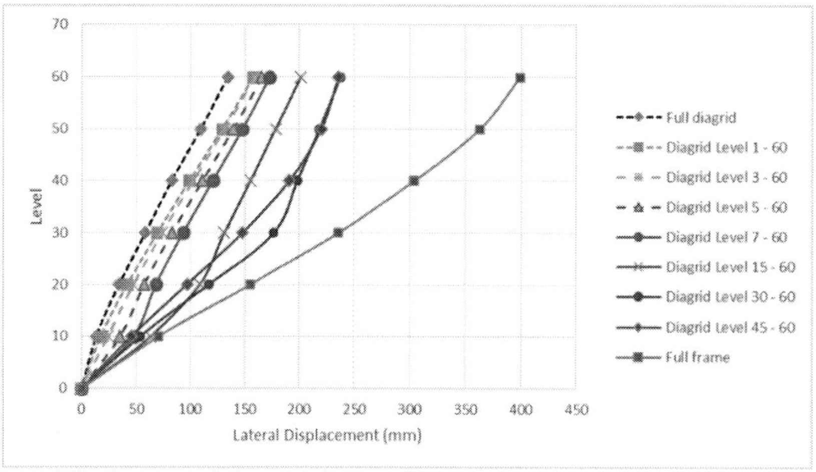

Fig. 3. Lateral displacement of all models in Hong Kong wind environment.

The graph also shows that the interstorey drift became smaller drastically at the level where the diagrid started. Models with diagrid at Level 15-60 has larger interstorey drift from ground level to level 15, but became smaller at level 15 to 60 due to the existence of the diagrid. The same behavior is observed for models with diagrid at Level 30-60 and Level 45-60, where large interstorey drift is observed at lower levels where there was only frame system, but became smaller at the levels where diagrid existed.

3.2 Shear Lag Effect

The axial stress distribution diagram at first storey (z/H = 0.00) of model with full frame system demonstrated the phenomena of positive shear lag where the columns at the edge of flange panel experienced higher axial stresses compared to the middle columns. The lateral load causes compressive axial stresses at the windward face of the building and tensile axial stress at the leeward face of the building. Combining the effect of the weight of the building, results in compressive axial stress at both windward and leeward face of the building where the leeward face has very large compressive axial stress while the windward face has minimal stress. The shear lag effect further, causes the columns at the two corners of the leeward face of the building to have maximum axial stresses while the columns that are located at the middle of the windward face to have minimum axial stresses as depicted in Fig. 4.

Model with full diagrid system had the least value of stresses in both flange and web panel besides being almost uniform (Figure 5). Model 3 which had its diagrid to start at Level 1 had positive shear lag at the first storey (z/H = 0.00), while suffering from the highest stresses at the four sides of the perimeter column, which are 39986 kN and 54240 kN (for Malaysia and Hong Kong wind load case respectively at the building leeward face.

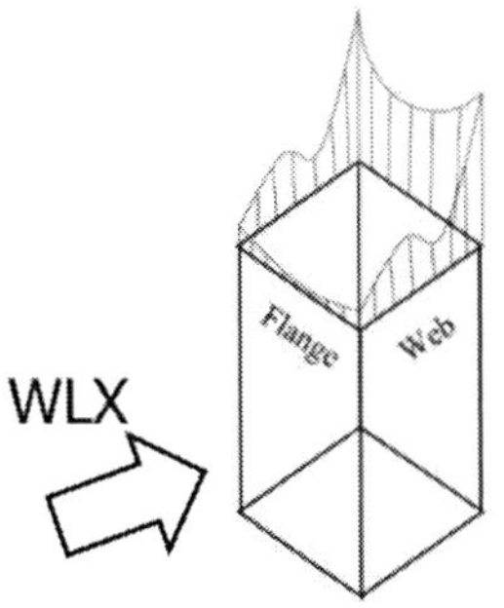

Fig. 4. Axial stress distribution of model with full frame system under wind load.

Furthermore, the axial stress distributions at the ground floor of Model 4, 5, 6, 7, 8 and 9 which are depicted in the graphs in Fig. 5 and Fig. 6, had the same concave up graph shape as Model 3, but lesser value of maximum stresses at the corner columns.

At ground level, the shear lag ratio, which is the ratio of the axial force in corner columns to the axial force in columns at the middle of the panel, increased by 30.8%, which is from 1.3 to 1.7, when one floor high frame is added under the diagrid system. Table 1 shows that the shear lag ratio at the base dropped from 1.57 to 1.26 when the level of where the diagrid started was changed from Level 1 to Level 45. In fact, the shear lag effect is slightly lower than the shear lag of full diagrid, when the diagrid started at Level 45.

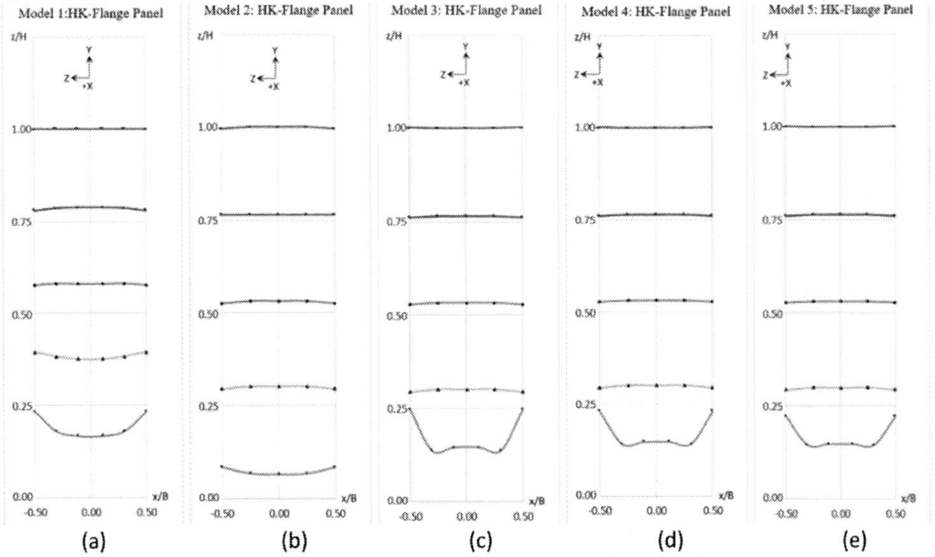

Fig. 5. Axial force distribution at the flange panel of model with (a) full frame (b) full diagrid (c) diagrid at Level 1 – 60 (d) Level 3 – 60 (e) Level 5 – 60.

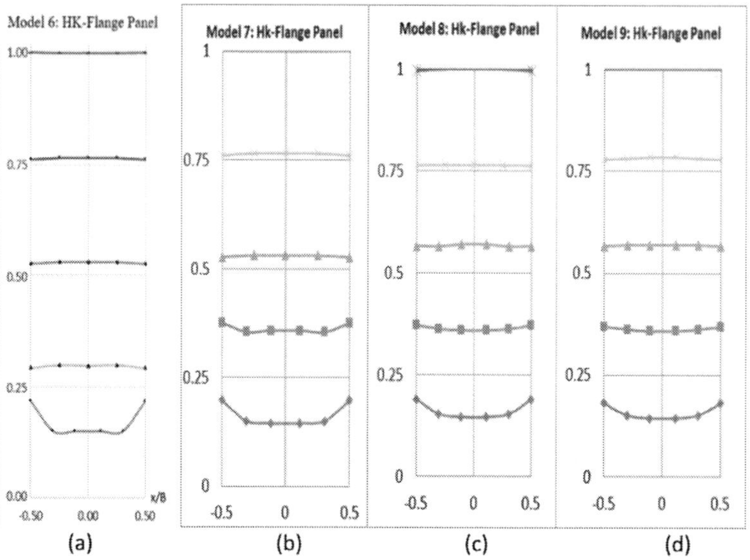

Fig. 6. Axial force distribution at the flange panel of model with diagrid at (a) Level 7 – 60 (b) Level 15 – 60 (c) Level 30 – 60 (d) Level 45 – 60.

In addition, this analysis found that a minor negative shear lag occurred in the region of one-quarter high of the building model from the ground level for full diagrid model and models that have the diagrid started at Level 1, 3, 5 and 7. The models with the diagrid started at Level 15, 30 and 45 did not have negative shear lag at one-quarter high of the model, but had a positive shear lag, instead (Fig. 5 and 6). Negative shear lag phenomena show that the stress at the corners of the flange panel is lower than the stresses at the middle of the flange panel. Furthermore, the stress distribution above one-quarter high of the building model remained the same despite the changes of the level where the diagrid started.

Table 1. Shear lag ratio, *f*, and lateral deflection of all models.

MODEL	HONG KONG WIND LOAD CASE		
	DEFLECTION (mm)	SHEAR LAG RATIO, *f*	
		BASE	MID-HEIGHT
Full Frame	398.696	1.39	0.96
Full diagrid	134.2	1.3	0.79
Diagrid at Level 1-60	157.0	1.7	0.87
Diagrid at Level 3-60	159.0	1.57	0.87
Diagrid at Level 5-60	165.3	1.5	0.87
Diagrid at Level 7-60	173.1	1.46	0.87
Diagrid at Level 15-60	201.329	1.369049	0.88314
Diagrid at Level 30-60	237.092	1.302932	0.946925
Diagrid at Level 45-60	235.266	1.255534	0.964197

4 Conclusions

The study proved that constructing the diagrid above a frame system is still effective in reducing the lateral displacement but not as much as a full diagrid system. The lateral displacement is reduced by 60.6 percent and 41 percent of the lateral displacement of a full frame when the diagrid system is constructed starting from Level 1 and Level 45 of the building respectively.

Shear lag ratio was increased from 1.3 to 1.7 when the diagrid started at Level 1 instead of started at ground level. However, the shear lag ratio was reduced as the higher the level where the diagrid started. As a conclusion, combination of diagrid and frame system is efficient in reducing the lateral displacement but not efficient in reducing the axial force in the external columns.

The study was funded by the Ministry of Education Malaysia and Research Management Centre, Universiti Teknologi Malaysia under Research University Grant No. QJ130000.2622.12J25.

References

1. T.M. Boake, *Diagrid Structures: Systems, Connections, Details* (Birkhauser Verlag GmbH, 2014)
2. S.Yadav, V. Garg, International Journal of Civil and Structural Engineering Research **3**, 394-406 (2015)
3. H. Varsani, N. Pokar, D. Gandhi, International Journal of Advance Research in Engineering, Science & Technology **2**, 1–4 (2015).
4. K.S. Moon, M.M. Ali, Architectural Science Review **50**, 205-223 (2007)
5. K.S. Moon, Procedia Engineering **14**, 1343-1350 (2011)
6. K.S. Moon, Architectural Science Review **51**, 239-251(2008)
7. B.Charnish, T.MCDonnell, *CTBUH 8th World Congress* (2008)
8. L. G. Griffis, *Engineering Journal / American Institute Steel Construction*, First Quarter (1993)
9. P. Jayachandran, *National Workshop on High-rise and Tall Buildings* (2009)
10. American Society of Civil Engineer, *ASCE Standard ASCE/SEI 7-10 Minimum Design Loads for Buildings and Other Structures* (2010)
11. B.S.Taranath, Structural Analysis & Design of Tall Buildings (1988)
12. J. Leonard, *Investigation of Shear Lag Effect in Highrise Buildings with Diagrid System* (2004)
13. http://www.skyscrapercity.com/showthread.php?t=474259&page=293
14. http://www.fosterandpartners.com/projects/hearst-tower/
15. K.S. Moon, International Journal of Sustainable Building Technology and Urban Development **2**, 37-42 (2011)

A Review of Influence of Various Types of Structural Bracing to the Structural Performance of Buildings

S.M. Razak[1,1], *T.C.* Kong[1], *N.Z.* Zainol[1], *A.* Adnan[2], and *M.* Azimi[2]

[1]Universiti Malaysia Perlis, Department of Civil Engineering Technology, 02100 Sungai Chuchoh, Padang Besar, Perlis
[2]Universiti Teknologi Malaysia, Faculty of Civil Engineering, Department of Structural and Material Engineering, Universiti Teknologi Malaysia, 81310 UTM Skudai, Johor

Abstract. Excessive lateral drift can contribute significantly towards crack formation, leading to structural damage. The structural damage will in turn reduce the capacity of the structure and weaken it from the intended design capacity. Generally, lateral drift is more pronounced in higher and longer structure, such as high rise buildings and bridges. A typical method employed to control lateral drift is structural bracing, which works by increasing stiffness and stability of structure. This paper reviews the influence of various types of structural bracing to structural performance of buildings. The history of structural bracing is visited and the differences between numerous structural bracing in term of suitability to different types of buildings and loading, mechanisms, technical details, advantages and limitations, and the overall effect on the structural behaviour and performance are dissected. Proper and efficient structural bracing is pertinent for each high rise building as this will lead towards safer, sustainable and more economical buildings, which are cheaper to maintain throughout the life of the buildings in the future.

1 Introduction

Dynamic loads can impart significantly greater effect towards the structural response of a structure compared to the static loading. As such, the properties of the structure such as lateral stiffness and strength plays pivotal role towards achieving efficient structural performance against dynamic loads which include typhoons, earthquakes, blast and many others. The stiffness of the structure contributes greatly towards the P-delta effect of the structure. Dynamic loading especially those applied laterally such as the base shear due to earthquake ground motion are capable of exerting huge amount of energy on the structure. To avoid structural failure, some of the energy exerted by the dynamic load has to be properly dissipated so that the excessive lateral movement of the structure can be minimized. One of the most efficient methods for lateral movement resistance is structural bracing [1]. Structural bracings work by providing lateral stiffness and stability to the

[1] Corresponding author: skmuiz@unimap.edu.my

structure, especially for the multi-storey and high-rise buildings [2]. This subsequently increases the lateral resistance of the structure and reduces the internal forces through appropriate bracing arrangement. Thus, for economic reason, structural bracing has been widely used worldwide [1].

According to [2], there are three important criteria which govern the stiffness of structure. First, when the path of internal force is more direct, the structure becomes much stiffer. Second, the stiffness of structure increases when the distribution of the structure is uniform. Third, the stiffness of structure increase with the reduction in the internal forces. By incorporating bracings in accordance to these three criteria, a stiffer and more economical structure can be achieved.

2 Early Development of Structural Bracing

One of the earliest buildings to incorporate structural bracing is the Dewitt-Chestnut Building in 1965 in Chicago. The building used the frame tube system which is developed by Fazlur R. Khan. Later in 1970, Fazlur R. Khan developed the braced tube system for the John Hancock Centre, also in Chicago. The student of Khan which is Mikio Sasaki discovered and proposed the first design of diagonally braces tower in 1964 in the seminal thesis writing. In 1968, Robin Hodgkinson developed a brace which uses concrete in outward appearances for the purpose of alternate uses. The concept of exterior brace with concrete is applied on the Ontario Centre in Chicago in 1985. These ideas contribute signicantly to the design of high-rise structure design against the natural forces such as wind load, earthquake and gravity load [3].

Fig. 1. John Hancock Centre.

The first eccentric braced frame design has been introduced to the building in Japan in 1970s [4]. In essence, the concept of Eccentric Braced Frame is derived from the Moment Resisting Frame and Concentrically Braced Frame. Consequently, this means that the advantages provided by both Moment Resisting Frame and Concentrically Braced Frame are integrated into a single structural system for structure which incorporate the use of Eccentric Braced Frame [5]. Chimeh et al [4] noted that there are many varieties in term of the placement of the link beam for the eccentric braced frame. Different placement or location of the link in the structure will lead to different structural response due to impact of lateral load.

3 Concentrically Braced Frames (CBFs)

Moment Resisting Frames (MRF) is traditionally assumed to contribute higher ductility to the structure compared to Concentrically Braced Frames (CBF). The inherent lack of ductility for the concentrically braced frames to be verified by introduced larger lateral strength in the previous seismic studies due to the fact that the specified value of the modification factor for CBF are much smaller compared to the MRF [6]. However, MRF is much more expensive than CBF since simple connection and smaller cross sectional area of beam and column is used in the concentrically braced frame. Besides that, the braces member in the structural frame are able to develop the axial forces and sustain the bending moment demand of the structure [7]. The understanding of the concentrically braced frame was been improved. During 80s, there have a lots of the researcher starting their studies to investigate the features and important of the concentrically braced frame. This means that the realization on strength enhancement of simply structural design will lead to the brittle characteristic and undergoes a serious of damage to the structure under major of seismic event [6]. The example of concentrically braced frames is commonly classified into three major types which is Diagonally Braced Frames, V-Braced Frames and Chevron Braced Frame [8]. Concentrically Braced Frames is effectively in resisting the seismic excitation [7, 8]. This means that they are resisting the lateral seismic acceleration primarily through the axial forces (tension and compression) and deformation of the braces, columns and beams [9].

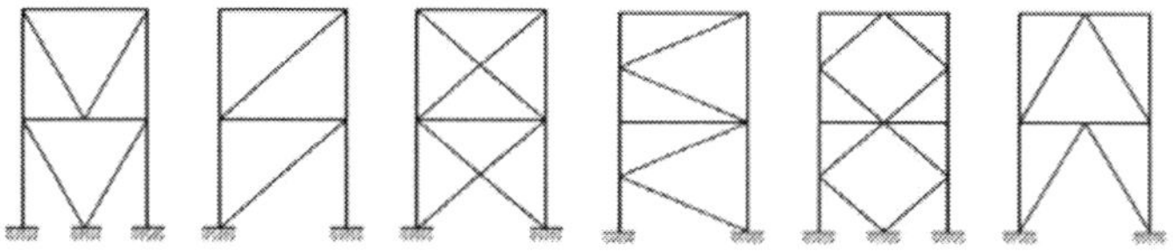

Fig. 2. Different Types of Concentrically Braced Frames.

Besides that, concentrically braced frame has been widely used in the structural frames since the braced frame is practical and cost effective [2, 10]. CBF provide stability and lateral stiffness required by the modern seismic design provision [10]. Concentrically braced frame have a low plastic distribution capacity characteristic properties. This will tends to cause the development of the collapse configuration or called as plastic deformation in a few storeys [11]. Moreover, concentrically braced frames have a desired performance in the lateral stiffness due to the ductility limitation behaviour [12]. According to Eurocode 8 in the design of seismic behaviour of the concentrically braced frame, the elasticity behaviour of the column and beam should be maintained while the energy dissipating behaviour in the braces is required [11].

Assuming the tension is occur at the braces while the others is in compression under post buckling range behaviour, based on the principle, the equally distribution of the internal forces for the column and beam should be obtained. The design bending moment on the beams is due to the unbalances forces should be identified by apply firmly the design moment capacity. The factor that influence the seismic performance of the concentrically braced frame is affected by the slenderness and width-to-thickness ratio of the braces. Besides that, the braces connection is significant influenced by the secondary bracing deformation. Thus, this caused the gusset plate should be large and thick on the current provision. This enable to control the framing members by draw a huge stresses to the gusset plate welds. Unfortunately, the test that has been carried out indicated that gusset plate or at the welds before deformation capacity of the braces can be utilized which caused the fail in such bracing connection. Thus, this leads to recommend the thin and compact gusset plate and which able to improve the drift capacity [10].

4 Diagonal Braced Frames (DBF)

Single diagonal is consider effectively in resisting the lateral load [1]. A study by Maheri and Sahebi (1997) on the in-plane shear resistance among the three steel structural frame that are bare frame structure without brace, diagonally braced frame and double diagonal braced frame (X-bracing) showed that the resistance of diagonally braces and X-bracing are 2.4 and 4.0 times better respectively compared to bare structure [2].

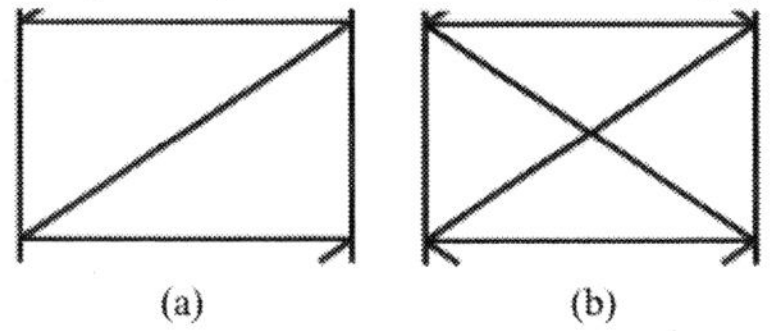

Fig. 3. Diagonal Bracing. (a) Single Diagonal Bracing (b) Double Diagonally Bracing [1].

When lateral load is applied to the braced frame, the diagonal braces are subjected to the compression while the horizontal web acts as the axial tension member in order to maintain the frame structure in equilibrium under lateral load [1]. In comparing the X-bracing and mega X-bracing, the mega X-bracing (diagrids) indicated that reduced in lateral drift with approximate 50% and 25% of material saving in terms of less steel bracing is used [2]. Besides that, diagrids structure provides higher torsion resistance which diagonal braces effective in minimizing shear deformation. In addition, diagrids structure possess higher structural flexibility than other bracing system. This means that diagrids structure is adjustable in the modules and angles(s) of diagonal into optimise structural performance [3].

The code of constraint in least weight optimisation is carried out by Kameshki and Saka (2001) through the genetic algorithms. This show that the weight for the X-bracing, chevron bracing and single diagonal bracing is arranged from the lowest to the highest. This indicated the X-bracing is higher efficiency in contributes the strength control compared to the optimized chevron bracing and single diagonal bracing are contribute to the stiffness control [2]. In term of the effects of brace configuration on the dynamic properties of frame, the V-bracing and chevron bracing had the lowest natural frequency with approximately 75% from X-bracing. The X-bracing had highest natural frequency compared to V-bracing followed by the K-bracing.

On the other hands, a stiffness based design method for the preliminary estimation of the member sizes of steel braced tube structure is purposed by Moon (2010). In the studies, the empirical relationships between the numbers of duplicated modules, the tower aspect ratio and the optimum slope is between 40° to 50° of the X-brace were provided [2]. The slenderness ratio of the braces also influences the capacity of the energy absorption. This means that the smaller slenderness ratio of braces had a high capacity of energy absorption. In addition, the braces with small slenderness ratio decreasing the ductility behaviour but the deterioration of the energy dissipation became faster especially for the X-bracing [7].

The performances of the tension braces member in the design of single diagonal braces and X-bracing depends on the stiffness, resistance and ductility according to the European seismic design standard. This means that the early buckling of compression bracings and their low post buckling strength and stiffness are assumed. In the structural analysis, the compression in the diagonal is reasonably been neglected. The traditional issues in the seismic engineering community generated these code assumption on the improving of structural performance to the maximum state especially in the high seismic situation. However, if these assumption are applied to the low moderate seismicity condition, this could not been become economically feasible. Gioncu and Mazzolani (2014) indicated the most universal typology of lateral excitation resistant steel frame for concentrically braced frames and the compression diagonal in X-bracing is typically to be neglected among the weak points in the design [7].

5 Chevron Braced Frames

A high elastic stiffness and strength are the characteristics which in the chevron braces [13]. Chevron also effectively in the architectural functionality. This is important to allow the placement of the window and doorways at the braces bay [14]. However, the unbalanced force is created on the braces beam This is due to the fact that the braces with compression behaviour are deformed while the tension braces will remain to sustain the tension force during earthquake [15].

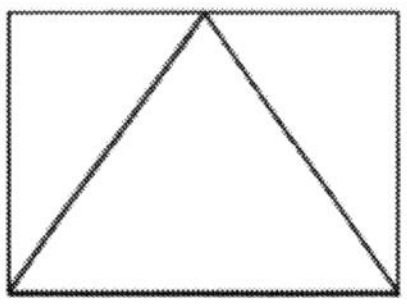

Fig. 4. Chevron Bracing [1]

The unbalanced distributed force is the major influences of the tension and compression braces which can cause the elastic deflection [10]. This causes the poor performances of the overall braces frame due to the deterioration on the braces [14]. This means that one of the braces resists the tension while the other braces are in compression at each storey of braces bay. Both of them distribute the lateral load equally in the elastic range before the buckling point. However, the tension braces remain in tension while the braces in compression lost all the axial load capacity after buckling. This contributes to the unbalanced distributed lateral load and caused large bending moment in the intersection of beam and braces [13]. This leads the mid span beam undergoes the plastic hinge and collapse due to unable to sustain the downward forces [10].

The AICS Seismic Provision have suggested that using a strong beam which is able to provide adequate strength such as deep and heavy beams to prevent the undesirable deterioration of frame [13, 15]. This means that the strong beam mechanism is higher in the ductility and energy dissipation to delaying the braces fracture since the chevron is non-ductile behaviour [12, 15]. Besides that, the outstanding of the hysteretic response can be archived from the strong beam mechanism wherein a reasonable distribution of damage over the height of building can be contributed by the ductile braces [12]. Unfortunately, this unable to provide enough strength during lateral load and caused the soft story behaviour in the lower stories of the structure [15].

Struts are normally added between the story beam in other to replace the use of the deep and heavy beams, which are much more expensive. This arrangement is known as the zipper braces frame (ZBF). The struts mitigate the vertical load induced to the adjacent story braces and keep their axial load capacity after compression braces buckled [13]. Besides, the struts help to redistribute the loads on each storey. Besides that, the strut helps to redistribution the loads on each stories and contribute to the redistribution of deformation demands [12, 13]. This caused to inelastic action along the height of frames and demand of uniform drift. Once the compression braces is buckling at the same time, it lead to the potentially damaging the structure or even possibility collapse is the defect of this configuration. This is because the plastic hinge occur at the beam since the unbalance force required to transfer to the columns by story beams [13]. A hat truss, also known as suspension system, is usually added at the top two stories to maintain the elastic behaviour when the braces and zipper column achieve the maximum capacity [13, 15]. This is means that the zipper column is against the vertical load from framing below since the hat truss acts as a tension member.

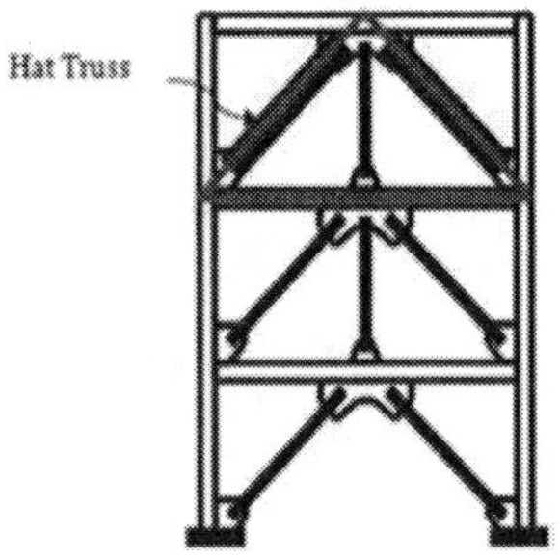

Fig. 5: Hat Truss in Zipper Braced Frames [13].

The elastic features of the top story braces should be maintain after the buckled of the braces below [15]. This is important to conquer the stability and the potential soft story formation problems. In the actual testing that done by previous researcher, it is found that the suspension zipper braces frame able to provide a higher strength and ductility features during seismic excitation [14]. There is more axial compression is developed by the suspended zipper braced frame for the upper stories compared to the chevron braced frame [13]. However, the increases of the required strength for zipper column is directly proportional to the height of the structure. It will caused architecture problems which unable the placement of window and doorway at the braces bay. This is because larger of the top story braces is used [15].

6 Eccentrically Braced Frames

Eccentrically braced frames are made of two short diagonal braces connecting the column to the middle span of beam with short segment of beam, to increase the lateral load resistance in terms of the seismic performance [16]. Besides that, the eccentric braces provides have architecture functionally which similarity with chevron braced frame [1, 5].

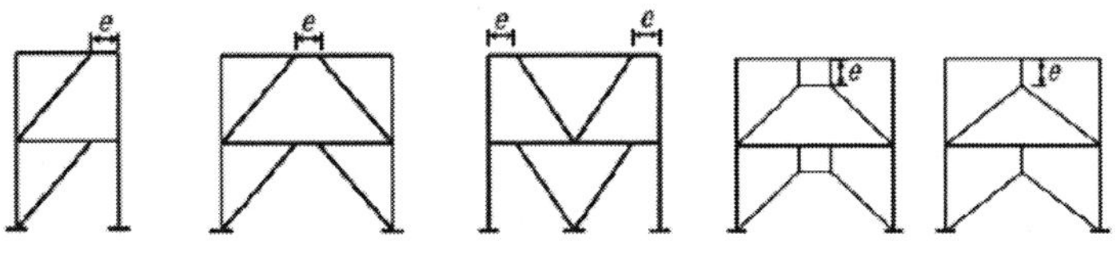

Fig. 6. Types of Eccentrically Braced Frames Depending on the Location of the Link Beam [4].

In the structural frame, the horizontal short segment in between the two braces member connection is called links [5]. The main function of the link beam is used to absorb the energy dissipation resulting from the lateral excitation [4, 12]. This is because the links is used to control plastic deformation and maintain structural stability through dissipates the energy from the seismic excitation [5]. Moreover, the resistance of lateral movement in the eccentric braces is through the bending of the column and beam [1]. Therefore, the links provides the ductility behaviour to counter the impact seismic force from the lateral excitation [17, 18].

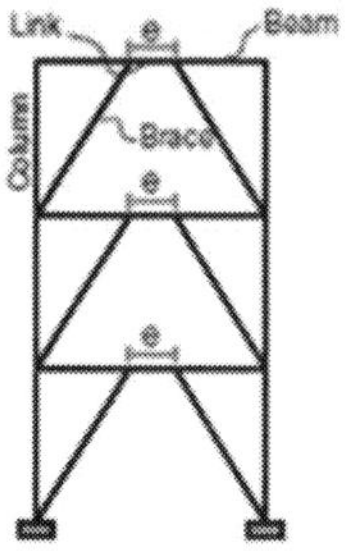

Fig. 7. Horizontal Links on the Eccentrically Braced Frames [5].

The strength, stability and ductility of the eccentric braced frame is influenced by the length of the horizontal links [5]. This is because the links indicated the capacity of energy dissipation inside the system [12]. Shorter horizontal length of links leads to higher shear yielding efficiency [5]. This is because the shear in the storey directly influences the shear in the links especially if the short links are used whereas bending may occur in the longer links [18]. Otherwise, the long links will contribute the flexural yielding compared to shorter links which only in shear yielding. If an intermediate length of the horizontal links to be used in the system, a combination of the shear and flexural yielding would be experienced. However, it has been shown that the short links has better performance in terms of rotation capacities compared to others. The previous researcher indicated that the short links have greater in strength and ductility when it subjected several cyclic load. Even through the longer links is able to provide the architecture freedom especially in the placement of the window and door opening [5].

Nevertheless, the lateral stiffness of eccentric braces is lower than the concentrically braced frames especially the diagonal bracing [1, 4]. This is because the eccentric braces have a more ductile characteristic [4]. In addition, this is time consuming and expensive in the replacement of the horizontal links after it was damaged due to the seismic excitation [12]. However this causes the difficulties of the replacing or repairing the horizontal link beam since it is a primary component in the structure despite the high capacity of energy dissipation [12, 17]. A vertical links is proposal to overcome the above disadvantages. This means that the vertical links is attached to the story beam. The vertical links is similarly with horizontal links which provide the ductility and acts as fuses which able to absorb energy. The vertical links able to deliver and withstand most of the shear force during the dynamic analysis. This means that accumulation of the shear at the links is not desired to be occur. Moreover, the links is deforming plastically. This is due to that links provide the uniform energy dissipation behaviour.

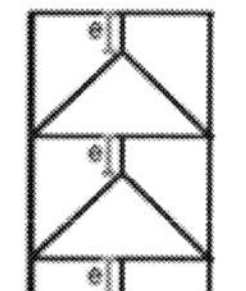

Fig. 8. Vertical Links on the Eccentrically Braced Frames [5].

Also, there is no rotational restraint at the lower end of the vertical links. It does not lead to significant damage to the story beam since the internal energy dissipation and intensive inelastic deformation are accumulated in the links. Hence, the replacement of the vertical links is effective and easier than horizontal links after lateral excitation [17].

Table 1. Comparison of the structural bracing

 https://doi.org/10.1051/e3sconf/20183401010

Diagonal Braced Frames	Chevron Braced Frames	Eccentrically Braced Frames
	Characteristic	
➢ High natural frequency ➢ Smaller slenderness ➢ High Capacity energy absorption ➢ Decreasing the ductility ➢ Deterioration of energy dissipation become faster for X-bracing	➢ Low natural frequency ➢ High elastic stiffness and strength	➢ Absorb energy dissipation ➢ Control plastic deformation ➢ Maintain structural stability ➢ Links provide ductility behaviour ➢ Strength, stability and ductility influenced by length of horizontal links
	Mechanism	
➢ Bracing acts as the compression member while the horizontal web member (beam) acts as tension member in the single diagonal braced frame ➢ Both bracing in the X-bracing acts as the tension and compression member respectively	➢ Compression and tension exert on each side of braces respectively ➢ Unbalances forces created in braces beam ➢ Compression braces are deformed ➢ Tension braces sustain tension ➢ Cause elastic deformation ➢ Poor performance ➢ Both braces distribute lateral load equally before buckling ➢ Compression lost axial load capacity after buckling ➢ Contribute unbalanced distributed load capacity after buckling ➢ Large bending moment on intersection of beam and braces	➢ Resistance of lateral movement through bending of column and beam ➢ Shorter horizontal links lead to high shear yielding efficiency ➢ Longer horizontal links contribute flexural yielding ➢ Intermediate links will leading a combination of shear and flexural yielding experiences ➢ Short links show better performances in terms of rotation capacities ➢ Lateral stiffness lower than CBF especially diagonally bracing due to the eccentrically braces have more ductile ➢ Time consuming and expensive in replacement horizontal links ➢ Difficulties of replacing or repairing since it is primary component of despite high capacity of energy dissipation
	Latest Improvement	
Mega X-Bracing (Diagrid) ➢ Reduced in lateral drift with approximate 50% ➢ 25% of material saving in term of less steel bracing is used ➢ Minimizing shear deformation ➢ Provide maximum resistance against torsion ➢ Possess greater structural flexibility than bracing system ➢ Diagrid can be configured, with some adjustment in modules and angle(s) of diagonals, to meet architectural and structural requirements.	**Strong beam** ➢ Deep and heavy beam to provide adequate strength and prevent undesirable deterioration of frame **Strut** ➢ Replace deep beam and heavy beam due to expensive ➢ Mitigate vertical load induced to adjacent story braces and keep their axial load capacity after compression braces buckled	**Vertical links** ➢ No rotational restraint at lower end ➢ Does not lead to significant damage to beam since internal energy dissipation and intensive elastic deformation accumulate in link ➢ Replacement more efficient and easier

7 Comparison

Table below shows the comparison of the characteristic, mechanism and latest improvement among the structural bracing for diagonal braced frames, chevron braced frames and eccentrically braced frames. An effective structural bracing which able to reduce the structural performance due to lateral excitation by contribute to sustainable development. Eccentric braces able to reduce the structural repairing cost. This means that eccentric braces reduced the potential damage to the structure due to less maintenance activity toward the structure and reduce overall cost of the building. Meanwhile, safety of the occupant in the building is ensured because eccentric braces able to delay the structural response toward the earthquake by prolong the means time escape for the occupant escape from the building and contribute the comfort to the occupant at the same time. This is because the intensity that generated from the ground are reduced by dissipating the energy through the structural bracing. Thus, eccentric braces allow to prolong the half-life of structure.

References

1. Z. Siddiqi, R. Hameed and U. Akmal, Pak. J. Engg. & Appl. Sci., **14**, 17-26, (2014)
2. X. Yu, T. Ji and T. Zheng, Engineering Structures **89**, 147-161, (2015)
3. K. Al-Kodmany and M. M. Ali, International Journal of High Rise Buildings **5**, 271-291, (2016)
4. M. N. Chimeh and P. Homami, 15th World Conferences on Earthquake Engineering (WCEE), Lisbon, (2012)
5. S. K. Azad and C. Topkaya, Journal of Constructional Steel Research **128**, 53-73, (2017).
6. G. Brandonisio, M. Toreno, E. Grande, E. Mele and A. D. Luca, Journal of Construction Steel Research **78**, 22-37, (2012).
7. A. Kanyilmaz, Journal of Construction Steel Research, **133**, 1-18, (2017).
8. R. Sabelli, C. W. Roeder and J. F. Hajjar, *Seismic Design of Steel Special Concentrically Braced Frame Systems: A Guide for Practicing Engineers* (United States of America: U.S. Department of Commerce, 2013)
9. R. Sabelli, *Research on Improving The Design and Analysis of Earthquake-Resistant Steel-Braced Frames* (The 2000 NEHRP Professional Fellowship Report, 2001)
10. T. Okazaki, D. G. Lignos, T. Hikino and K. Kajiwara, Journal of Structural Engineer **139**, 515-525, (2013)
11. M. Bosco, E. Marino and P. Rossi, 15th World Conference on Earthquake Engineering, Lisbon (2012)
12. D. M. Patil and K. K. Sangle, Structures **3**, 282-305, (2015)
13. Y. Ozcelik, A. Saritas and P. M. Clayton, Journal of Construction Steel Research **119**, 169-175, (2016)
14. M. D'Aniello, S. Costanzo and R. Landolfo, Journal of Constructional Steel Research **112**, 305-324, (2015)
15. H. Yu, Y. Guo, Y. Zhang and X. Wang, Journal of Constructional Steel Research **120**, 8-24, (2016)
16. K. Torrie, *Lateral Bracing Systems of Multi-Story Structures*, (College of Engineering and Technology Brigham Young University, Provo, 1994)
17. M. G. Vetr, A. Ghamari and J. Bouwkamp, Journal of Building Engineering **10**, 47-59, (2017)
18. X. M. Han, The 14th World Conference on Earthquake Engineering, Beijing, (2008).

Parametric Study of Fire Performance of Concrete Filled Hollow Steel Section Columns with Circular and Square Cross-Section

Ahmad Nurfaidhi Rizalman[1,*], Ng *Seong Yap*, *Mahmood* Md Tahir[2], and *Shahrin* Mohammad[2]

[1]Civil Engineering Programme, Faculty of Engineering, 88999 Universiti Malaysia Sabah, Kota Kinabalu, Sabah, Malaysia
[2]Faculty of Civil Engineering, Universiti Teknologi Malaysia, 81310, Skudai, Johor, Malaysia

Abstract. Concrete filled hollow steel section column have been widely accepted by structural engineers and designers for high rise construction due to the benefits of combining steel and concrete. The advantages of concrete filled hollow steel section column include higher strength, ductility, energy absorption capacity, and good structural fire resistance. In this paper, comparison on the fire performance between circular and square concrete filled hollow steel section column is established. A three-dimensional finite element package, ABAQUS, was used to develop the numerical model to study the temperature development, critical temperature, and fire resistance time of the selected composite columns. Based on the analysis and comparison of typical parameters, the effect of equal cross-sectional size for both steel and concrete, concrete types, and thickness of external protection on temperature distribution and structural fire behaviour of the columns are discussed. The result showed that concrete filled hollow steel section column with circular cross-section generally has higher fire resistance than the square section.

1 Introduction

Studies have shown that the fire performance of steel hollow section column can be improved by filling the hollow area of the steel section with concrete [1]. Other advantages of concrete filled hollow steel section column include higher strength, ductility, energy absorption capacity, and good structural fire resistance.

Recently, the research methods for fire studies on concrete filled steel hollow section (CFHSS) columns have shifted to the numerical modelling due to the cost and time constraint for conducting the fire testing. Many studies all over the world have been carried out towards the development of the numerical models to predict the fire performance of CFHSS columns [2-4]. However, there has been limited studies done on the numerical models to cater various shapes of steel hollow sections and external fire protection.

* Corresponding author: ahmadnurfaidhi@ums.edu.my

In this paper, an extensive study of the fire performance of slender CFHSS columns at elevated temperature was carried out using the verified 3D numerical models developed by Rizalman and Md Tahir using finite element software, ABAQUS [5,6]. The aim of this research is to establish a fire performance comparison between circular and square CFHSS columns subjected to different variables including cross-sectional shapes, types of concrete, and thickness of fire protection. Thus, the fire resistance and critical temperature of the CFHSS columns for circular and square shapes are discussed in this paper.

2 Methodology

2.1 Material Properties at Elevated Temperature

In this study, the mechanical and thermal properties of the materials employed in the 3D numerical model are temperature-dependant. For the mechanical properties of steel, the yield strength (f_y) was assumed to be 350 MPa and the Elastic Modulus (E_s) was taken as 210000 MPa. The stress-strain relationship of steel material at elevated temperature was defined in accordance to the Eurocode 3 model [7]. The Von Misses yield surface was used to define the plasticity behaviour of the steel elements in the numerical model.

For the mechanical properties of concrete, the compressive strength (f_c) of 30 MPa was adopted and the corresponding strain (ε_s) was assumed as 0.0025. The stress-strain relationship of concrete material at elevated by Lie was defined according to the Lie Model [8]. The linear Drucker-Prager yield surface was used to define the plastic behaviour of the concrete elements in the numerical model. The thermal properties provided in BS EN 1994-1-2 [9] were employed to define the thermal properties for both steel and concrete, correspondingly.

2.2 Description of the Selected CFHSS Columns

In this study, a series of slender CFHSS columns with circular and square cross-sections were used to examine the effect of cross-sectional shapes and sizes, concrete types and the thickness of fire protection, on the structural fire behaviour of CFHSS columns. The selected CFHSS columns were divided into five cases: sectional strength at room temperature, cross-sectional area of steel, cross-sectional area of concrete, types of infill, and fire protection thickness.

All selected slender columns were 3600 mm length with steel tube thickness of 5 mm. The column was subjected to standard fire ISO-834 [10]. Fig. 1(a) and 1(b) show the basic three-dimensional finite element model of slender CFHSS columns for circular and square cross-section, accordingly.

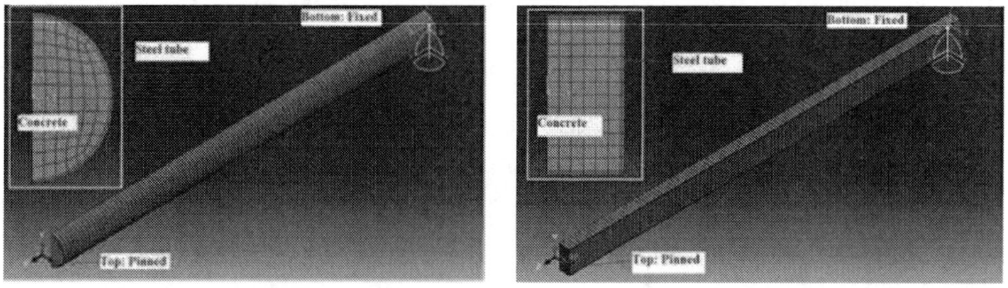

(a) Circular CFHSS column (b) Square CFHSS column

Fig. 1. 3D finite element model for circular and square CFHSS columns.

In this study, there were three different load levels applied to each column specimen calculated as 20%, 40% and 60% of their axial design load N_y at room temperature. The axial load was obtained using the following expression:

$$N_y = A_a f_y + A_c f_c + A_s f_s \tag{1}$$

where A_a, A_c, and A_s are cross-sectional area of steel, concrete and reinforcing bars, respectively. Moreover, f_y, f_c, and f_s are the characteristic material strengths of steel, concrete and reinforcing bars, correspondingly.

As previously mentioned, each column was subjected to three levels of axial loads including $0.2N_y$, $0.4N_y$, and $0.6N_y$. The gravitational load, 9.8 N, was also applied to the numerical model.

The basic geometrical and mechanical characteristics of the selected slender CFHSS columns are presented in the table below.

Table 1. Case A - Columns with Equal Compressive Strengths, N_y at Room Temperature.

Features	Group 1			Group 2		
	CHS1	SHS1	CHS1/SHS1	CHS2	SHS2	CHS2/SHS2
Size(mm)	177 x 5	150 x 150 x 5		234.4 x 5	200 x 200 x 5	
A_s (mm^2)	2702	2900	0.932	3603	3900	0.924
A_c (mm^2)	21904	19600	1.118	39549	36100	1.096
N_y (kN)	1603	1603	1.000	2448	2448	1.000

Table 2. Case B - Columns with Equal Cross-Sectional Area of Steel, A_s.

Features	Group 3			Group 4		
	CHS3	SHS3	CHS3/SHS3	CHS4	SHS4	CHS4/SHS4
Size (mm)	189.6 x 5	150 x 150 x 5		253.3 x 5	200 x 200 x 5	
A_s (mm^2)	2900	2900	1.000	3900	3900	1.000
A_c (mm^2)	25334	19600	1.293	46392	36100	1.288
N_y (kN)	1775	1603	1.107	2760	2448	1.127

Table 3. Case C - Columns with Equal Cross-Sectional Area of Concrete, A_c.

Features	Group 5			Group 6		
	CHS5	SHS5	CHS5/SHS5	CHS6	SHS6	CHS6/SHS6
Size (mm)	189.6 x 5	150 x 150 x 5		253.3 x 5	200 x 200 x 5	
A_s (mm^2)	2560	2900	0.883	3446	3900	0.884
A_c (mm^2)	19607	19600	1.000	36103	36100	1.000
N_y (kN)	1484	1603	0.926	2289	2448	0.935

Table 4. Case D – Columns with Different Types of Infill.

	CHS7	CHS8	CHS9	SHS7	SHS8	SHS9
Size (mm)	234.4 x 5			200 x 200 x 5		
Concrete types	None	Plain concrete	Reinforced concrete	None	Plain concrete	Reinforced concrete
A_s (mm^2)	3603			3900		
A_c (mm^2)	39549			36100		
A_a (mm^2)	298			298		
N_y (kN)	2448			2448		

Table 5. Case E – Columns with Different Thickness of Fire Protection.

	CHS2	CHS10	CHS11	SHS2	SHS10	SHS12
Size (mm)	234.4 x 5			200 x 200 x 5		
Thickness of fire protection	None	10 mm	25 mm	None	10 mm	25 mm
A_s (mm^2)	3603			3900		
A_c (mm^2)	39549			36100		
A_a (mm^2)	298			298		
N_y (kN)	2448			2448		

2.3 Thermal and Structural Analysis Procedure

In this paper, the slender CFHSS columns were exposed to standard fire ISO-834 [10] for 120 minutes. In a fire, the heat was transferred from the fire source to an exposed steel outer surface by convection and radiation. The conductive coefficient of 25 W/m^{2}k proposed by BS EN 1991-1-2 [11] was employed. Meanwhile, the emissivity of the exposed surface and the configuration factor for radiation at the exposed surface was assumed as 0.8 and 1.0, respectively. It was also assumed that gap conductive and gap radiative were formed between the steel tube and concrete core interface during fire. The gap conductive was taken as 200 W/m2K whereas the gap radiative was defined by employing the surface emissivity of 0.8 and the configuration factor of 1.0.

The structural analysis was governed by two steps. First, the axial load was applied to the column via the loading plate at room temperature. Then, the load was kept constant during the second step where the thermal loads was applied by extracting the temperature distribution results from the thermal analysis. The normal and tangential interaction at the steel and concrete interface was defined as "hard" contact and friction coefficient of 0.3, respectively. The accuracy of both thermal and structural analysis adopted in this paper have been validated in the works done by Rizalman and Md. Tahir [5,6]

 https://doi.org/10.1051/e3sconf/20183401011

3 Results and discussions

The following sections discussed the comparison of fire resistance and critical temperature between circular and square CFHSS columns.

3.1 Column with Equal Compressive Strength at Room Temperature

Fig. 2 shows the relative comparison of fire resistance and critical temperature between circular and square CFHSS columns with equal compressive strength at room temperature. The comparison indicates that the circular column has longer fire resistance than the square column as illustrated in Fig. 2(a). The same trend was observed for higher load levels but with lower fire resistance. The figure also shows columns with larger cross-section (CHS2 and SHS2) have higher fire resistance than the smaller ones (CHS1 and SHS1).

A similar pattern was also observed on the critical temperature of the selected CFHSS columns (as shown in Fig. 2(b)) where it decreases with increasing load levels. However, higher critical temperatures were observed in circular columns than the square columns because of longer fire resistance.

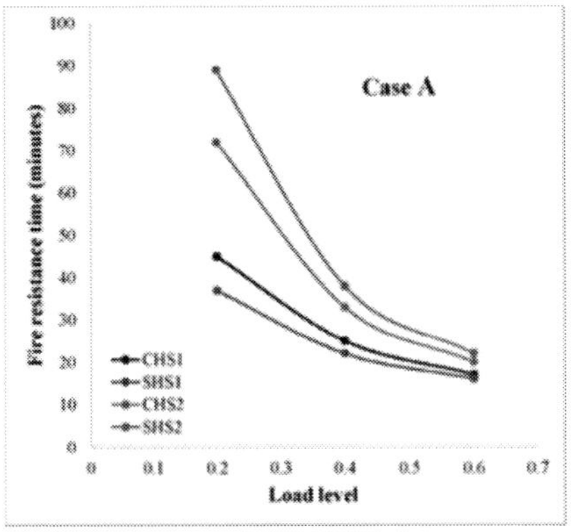

(a) Fire resistance

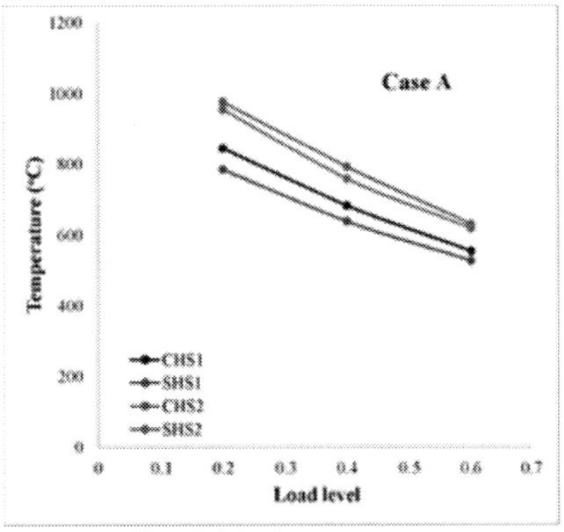

(b) Critical temperature

Fig. 2. Relationship of load ratio to fire resistance and critical temperature for Case A.

3.2 Column with Equal Cross-Sectional Area of Steel

Fig. 3 shows the relative comparison of fire resistance and critical temperature for columns with equal cross-sectional area of steel. As expected, the analysis produced similar results as Case A where circular column has longer fire resistance than the square column (Fig. 3(a)). However, the difference in the fire resistance between circular and square shapes were more noticeable compared to Case A. This is because the difference in the cross-sectional area of concrete in Case B was larger than in Case A. This indicates that the size of concrete has significant role in increasing the fire resistance of the column. A similar trend was observed for the critical temperature for the columns in Case B, as illustrated in Fig. 3(b).

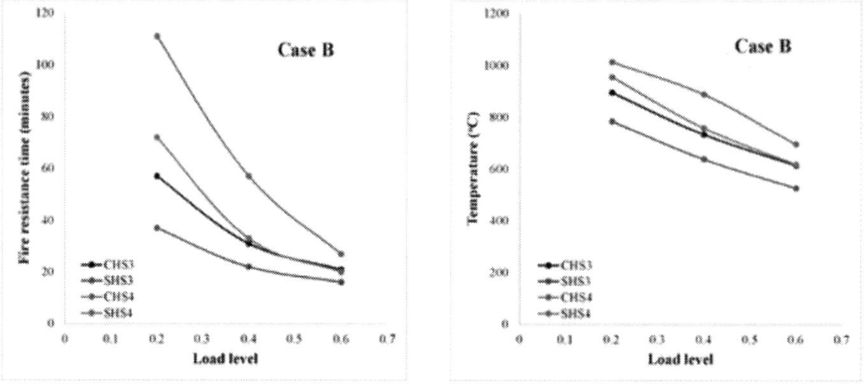

(a) Fire resistance (b) Critical temperature

Fig. 3. Relationship of load ratio to fire resistance and critical temperature for Case B.

3.3 Column with Equal Cross-Sectional Area of Concrete

Fig. 4 shows the relative comparison of the fire resistance and critical temperature for columns with equal cross-sectional area of concrete. Fig. 4(a) shows that the fire resistance in circular columns is slightly higher than the square columns for load level of 0.2. However, both columns failed around the same time for load level of 0.4. Then, it was found that the square column has longer fire resistance than the circular column at the load level of 0.6. Nevertheless, the results difference between the two shapes are almost negligible, thus it can be assumed that the influence of concrete on the fire resistance and critical temperature of the CFHSS columns are more prominent than the steel tube.

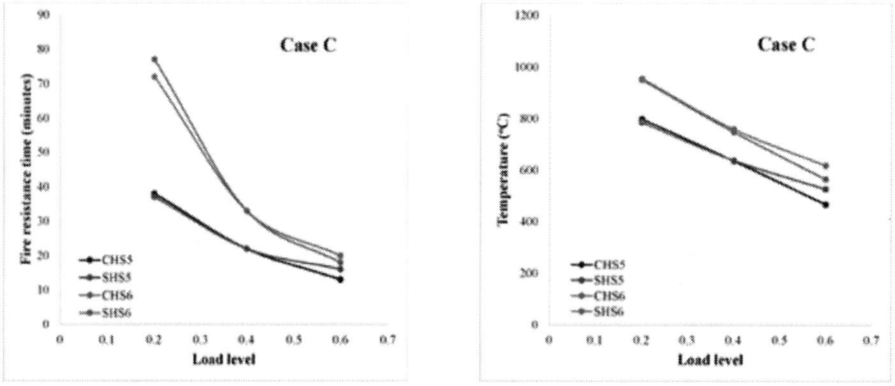

(a) Fire resistance (b) Critical temperature

Fig. 4. Relationship of load ratio to fire resistance and critical temperature for Case C.

3.4 Column with Different Types of Infill

Fig. 5 shows the effect of different types of infill on the fire resistance and critical temperature of between circular and square CFHSS columns. The results show a vast improvement on fire resistance for steel tube columns with concrete infill (CHS8 and SHS8) compared to void columns (CHS7 and SHS7). In addition, both void columns failed at the load level of 0.6 before they were subjected to fire. This indicates that presence of concrete towards the structural performance of hollow steel section column is prominent.

Moreover, the addition of reinforcing bars to concrete infill (CHS9 and SHS9) has also increased both fire resistance and critical temperature of the columns in which the reinforced effect increases with larger load levels.

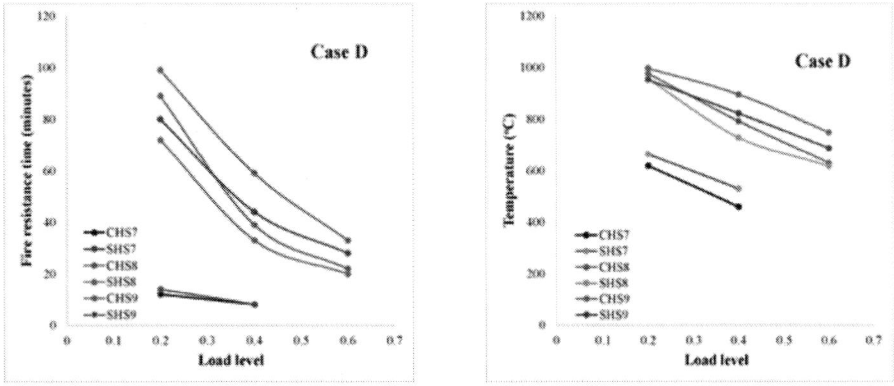

(a) Fire resistance (b) Critical temperature

Fig. 5. Relationship of load ratio to fire resistance and critical temperature for Case D.

3.5 Column with Different Thickness of Fire Protection

Fig. 6 shows the fire resistance and critical temperature of the CFHSS column with different thickness of fire protection. In this study, only load level of 0.6 was applied to the column. The results show that when the thickness of the fire protection is 10 mm (CHS11, and SHS11), it can delay the fire resistance of the unprotected CFHSS column (CHS10 and SHS10) by eight times. A similar trend was observed for the critical temperature for the columns in Case E, as illustrated in Fig. 6(b).

It should be noted that the fire resistance of the CFHSS columns with 25 mm protection thickness (CHS12 and SHS12) was actually greater than 200 minutes, which is the maximum fire exposure time fixed in this study. Thus, the fire resistance for CHS12 and SHS12 were set to 200 minutes for the recording purpose, as illustrated in Fig. 6(a).

This study indicates that the effect of fire protection on the fire resistance of the CFHSS column is evident. However, it is also important to consider the practicability of using fire protection especially when it involves the overall costs of a construction.

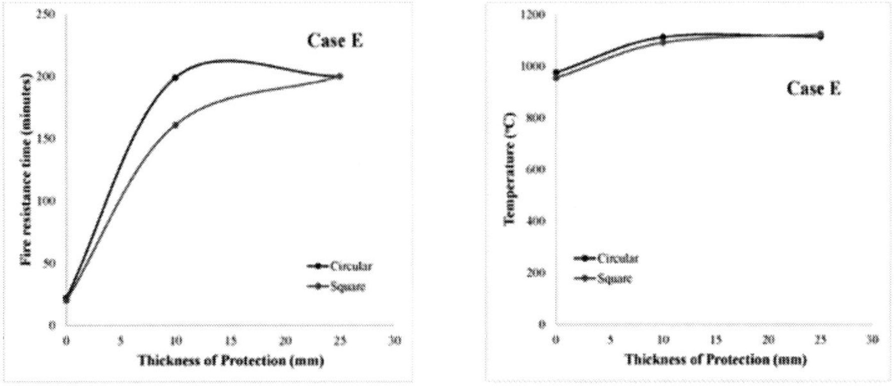

(a) Fire resistance (b) Critical temperature

Fig. 6. Relationship of load ratio to fire resistance and critical temperature for Case E.

4 Conclusions

From the analysis, the following conclusions are drawn: (a) Circular column generally has higher fire resistance the square column. (b) Concrete has significant influence on the fire performance of the CFHSS columns compared to the steel tube. (c) The use of reinforcing bars on the concrete fill has improved the fire resistance and critical temperature of the CFHSS columns especially at larger load levels. (d) The application of fire protection has slowed the temperature development of the steel section, hence higher fire resistance on the protected CFHSS columns.

References

1. A. Cote, *History of Fire Protection Engineering* (2003)
2. S. Hong, A.H. Varma, J. Constr. Steel Res. **65,** 1, 54 (2009)
3. A. Espinos, M.L. Romero, A. Hospitaler, J. Constr. Steel Res. **66,** 1030 (2010)
4. X.H. Dam, D. Lam. J. Constr. Steel Res. **73,** 117 (2012)
5. A.N. Rizalman, M. Md Tahir, S. Mohammad, A. Sulaiman, Appl. Mech. Mater. **752-753,** 507 (2015)
6. M. Md Tahir, A.N. Rizalman, P.N. Shek, M.A. Ab Kadir, J. Mirza, C.C. Lim, M.H.N. Shukor, M.F. Mat Din, J. Civil Environ Eng. **6,** 5 (2016)
7. *BS EN 1993-1-2. Eurocode 3: Design of Steel Structures* (British Standards Institution, London, 2005)
8. T.T. Lie, M. Chabot. J. Fire Prot. Eng. **2,** 4, 111 (1990)
9. *BS EN 1994-1-2. Eurocode 4: Design of Composite Steel and Concrete Structures* (British Standards Institution, London, 2005)
10. *ISO. Fire Resistance Tests, Elements of Building Construction* (ISO 834, Switzerland, 1980)
11. *BS EN 1991-1-2. Eurocode 1: Actions on Structures*, (British Standard Institution, London, 2002)

Compaction and Plasticity Comparative Behaviour of Soft Clay Treated with Coarse and Fine Sizes of Ceramic Tiles

Mohammed Ali Mohammed Al-Bared[1], *Aminaton* Marto[2, 3,*], *Indra Sati Hamonangan* Harahap[1], and *Fauziah* Kasim[3]

[1]Department of Civil and Environmental Engineering,Universiti Teknologi PETRONAS , 32610 Perak, Malaysia
[2]Disaster Preparedness and Prevention Centre, Malaysia-Japan International Institute of Technology, Universiti Teknologi Malaysia (UTM) Kuala Lumpur, 54100 Kulala Lumpur, Malaysia
[3]Faculty of Civil Engineering, Universiti Teknologi Malaysia, 81310 UTM Johor Bahru, Johor, Malaysia

Abstract. Recycled blended ceramic tiles (RBT) is a waste material produced from ceramic tile factories and construction activities. RBT is found to be cost effective, sustainable, environmental-friendly and has the potential to be used as an additive in soft soil stabilization. Recent reports show that massive amounts of RBT are dumped into legal or illegal landfills every year consuming very large spaces and creating major environmental problems. On the other hand, dredged marine clay obtained from Nusajaya, Johor, Malaysia has weak physical and engineering characteristics to be considered as unsuitable soft soil that is usually excavated, dumped into landfills and replaced by stiff soil. Hence, this study investigates the suitability of possible uses of RBT to treat marine clay. Laboratory tests included Standard proctor tests and Atterberg limits tests. The plasticity of marine clay was evaluated by adding 10%, 20%, 30% and 40% of 0.3 mm RBT. In addition, the compaction behaviour of treated marine clay was compared by adding two different sizes (0.3 mm and 1.18 mm diameter) of RBT. For both coarse and fine sizes of RBT, 10%, 20%, 30% and 40% of the dry weight of the soft clay were added. The mixture of each combination was examined in order to evaluate the Maximum Dry Density (MDD) and the optimum moisture content (OMC) for the treated soft clay. MDD and OMC for soft untreated samples were 1.59 Mg/m^3 and 22%, respectively. Treated samples with 10%, 20%, 30% and 40% of 0.30 mm size RBT resulted in a significant reduction of OMC ranged from 19 to 15% while MDD resulted in increment ranged from 1.69 to 1.77 Mg/m^3. In addition, samples treated with 10%, 20%, 30% and 40% of 1.18 mm size RBT resulted in major reduction of OMC ranged from 15 to 13.5% while MDD increased effectively from 1.75 to 1.82 Mg/m^3. For all mix designs of soft clay-RBT, MDD was gradually increasing and OMC was sharply reducing with further increments of both sizes of RBT.

* Corresponding author: aminaton@utm.my

1 Introduction

Continuous and fast development in construction industry reduces the natural available resources and induced pollutions to land, water and air. The necessity to provide alternatives to natural resources is the concern of recent research in civil engineering. Fast development in civil engineering projects resulted in a corresponding reduction in all-natural resources such as lands that force engineers and contractors to seek alternatives. One of the available options is the construction of buildings and other engineering structures on soft soils, which are weak and problematic in nature and they are not capable to sustain the imposed load of the sub and super structures of the buildings [1]. According to [2-4], soft soil is associated with high settlement and instability, low permeability and shear strength, high plasticity, and high-water content. Moreover, the major problem that makes soft soil unsuitable to be a foundation soil for construction is its swell-shrink behavior. Soft soil swells when it gains water content and shrinks when water is dried out [5]. Various kinds of cracks induced by excessive settlement may appear on railways, highways and other engineering structures founded on soft soils due to the swell-shrink phenomenon during wet and dry seasons [6, 7]. Besides, soft soils are very sticky when they are wet and hard to very hard when they are dry which make it difficult to be compacted [6]. The defects and the damages of roads and engineering structures caused by the soft soil associated problems require high budgets to repair. Therefore, seeking solutions to soft soil problems have been studied by many researchers.

Recently, wide ranges of chemical stabilization additives are implemented in ground stabilization. Traditional and non-traditional chemical additives resulted in significant improvements of soft soil. Lime, cement, calcium carbide and fly ash are examples of the chemical additives adopted successfully in soft soil improvements [8-12]. However, from environmental prospective, chemical additives are generally classified as non-environmental friendly, expensive and require advanced instrumentation during application at site. Besides, underground water contamination is an example of the several problems created by the implementation of chemical additives in soft soil improvement. On the other hand, ceramic tiles factories have high quantity of wastes during production ranged from 7% - 30% reported by various prominent researchers [13, 14]. This powered waste usually accumulate in lands near to the factory area and by the effect of atmosphere, it becomes dried and carried by the wind. Ceramic dust transportation by wind helps to spread the waste that affects the health of human beings as well as causing environmental problems. Furthermore, those cracked, smashed, or broken tiles at the construction sites are not accepted for installation that end up as a waste [15]. Utilizing recycled materials in soil stabilization is an alternative way that promotes a sound environment [16].

Soil stabilization by compaction is one of the widely used mechanical methods adopted by many researchers all over the globe. Soil compaction can be described as increasing the particle density of the soil and decreasing the air spaces within the particles at constant water content. This will increase the shear strength and decrease the associated settlement of the compacted soil as it becomes in an impermeable state [17]. Compaction methods differ between cohesive and cohesionless soils. Sheep foot compactor can be used to compact cohesive soils as it contains sufficient amounts of silt and clay that make it in an impermeable state when compacted properly. While cohesionless soils need a vibrating compacter for compaction as it contains sand and gravel [18].

Hence, this research is aimed to treat soft soil with an environmental, sustainable and economical additive that is still not widely investigated in this area of research. Recycled blended ceramic tiles (RBT) are introduced in this study as an environmental additive to be used in soil stabilization. The influence of different percentages and sizes of RBT on plasticity and compaction behavior of marine clay is assessed.

2 Materials

2.1 Marine clay

Marine clay is the soil being investigated in this study, due to its insufficient strength and poor engineering properties. This soil is usually excavated, transported and dumped into legal and illegal landfills as a waste material and then replaced at site by a stiff soil that has the minimum requirements to sustain the construction loads. Dredged marine clay was collected from Kota Iskander, Nusajaya, Johor, Malaysia to be studied and tested. It was collected from a depth of one meter and it was in wet condition with natural moisture content of 59%. The disturbed soil was transported to the laboratory and exposed to room temperature until it dried thoroughly. Fig. 1 shows the location of the sampling site of the marine clay.

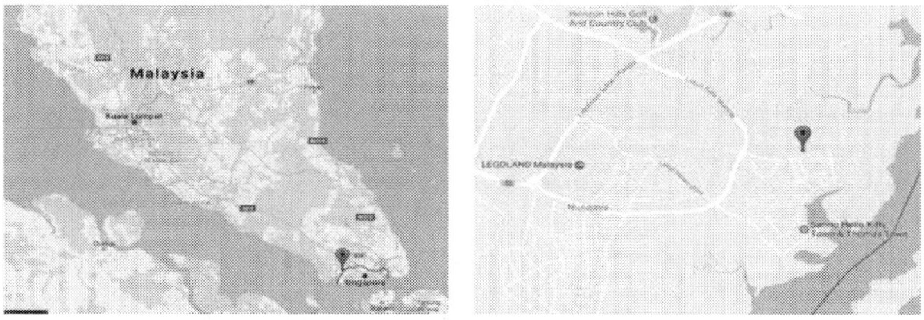

Fig. 1. Location of the site from which marine clay was collected.

When the marine clay reached to the laboratory, it was naturally dried, grinded and sieved for compaction tests. Soil was first oven dried at 105°C for 24 hours to make sure the soil has zero water content before testing. To obtain the physical properties of marine clay, sieved samples were used for the specific gravity test, the compaction test, the Atterberg limits tests, the organic test and the natural moisture content test. All basic tests were conducted in accordance with BSI 1377, Part 2: 1990 [19]. The average values of all the physical tests conducted on marine clay are demonstrated in Table 1. According to Wong et al. [17], soils that have an organic content of less than 10% are considered inorganic. Therefore, the tested marine clay is considered as inorganic soil. Moreover, based on the results of liquid limit and plasticity index and in accordance with [20], and by referring to the plasticity chart, the soil is classified as CLAY with intermediate plasticity. It is given the group symbol CI.

Table 1. Physical characteristics of untreated marine clay.

Soil properties	Standard	Average value
Natural moisture content (%)	BS 1377-1	59
Specific gravity (G_s)	BS 1377-2	2.52
Organic contents (%)		2.74
Atterberg limits		
Liquid limit (%)	BS 1377-1	41
Plastic limit (%)	BS 1377-1	22
Plasticity index (%)	BS 1377-1	19
Proctor test		
Optimum moisture content (%)	BS 1377-4	22
Maximum dry density (kg/m^3)	BS 1377-4	1590

2.2 Recycled ceramic tiles (RBT)

RBT in this research refers to the wastage of ceramic tile that is blended into fine and coarse powder to be used for the stabilization of the marine clay. Taman Ponderosa, Impian Heights Golf and Country Club and Taman Pelangi, Johor were the construction sites from which RBT was collected to be utilized in this research. Prior to RBT collection, an investigation was made to check on the availability of RBT wastage in those specific construction sites. Meanwhile, presence of cracks and smashes on the surface of new tiles delivered to construction sites and design problems / failures are the main reason for the high amount of tiles wastages at construction sites. This is due to the improper handling, casing, low quality and shaking during transportation. Tiles collected for this research are those that has been rejected at sites as well as those been hacked for renovation activities. Besides, the wastage produced by cutting tiles into exact sizes at site. The ceramic tiles used in this research were prepared in several steps started by grouping them into different groups based on colour and type. The major group of ceramic tile waste was cleaned to remove cement, dust and other substances stick on it. Then, the tiles were manually crashed into smaller pieces using hammer in order to be able to fit into the crushing machine. The crushing machine was used to obtain RBT of 5 mm size. Next, the 5 mm sized RBT were transferred into Los Angeles Abrasion machine in order to blend it into powder and this process took almost 24 hours. The period of time was chosen to be 24 hours only in order to obtain a powder that has a mixture of coarse and fine content. After that, tiles were sieved using 1.18 mm and 0.3 mm sieve meshes. The powder that passed 1.18 mm sieve and retained on 0.3 mm sieve was used as 1.18 mm sized RBT. While the powder that passed 0.3 mm sieve and retained on the pan was used as 0.3 mm sized RBT. For both sizes of RBT, a mix design of 10%, 20%, 30%, and 40% of the dry weight of marine clay was used in this study. This mix design was chosen based on previous study that used almost similar quantities of ceramic waste to replace Portland cement in concrete [21].

2.3 Samples preparation for standard proctor compaction tests

Sample preparation for all proctor tests followed the BSI 1377: Part 4: 1990 [22] for all the methods and procedures implemented. The compaction tests were performed in order to obtain the density-moisture relationships for various mix designs of RBT stabilized marine clay. These tests measure the maximum dry density and optimum moisture content of the untreated and the treated marine clay. In order to conduct the compaction tests, approximately 90 kg of untreated marine clay was oven dried at 105°-110°C for 24 hours and stored in air tight containers. Prior to preparing the admixture of marine clay and RBT, marine clay was sieved through 2 mm sieve mesh for the purpose of removing coarse particles, plants and roots. First series of tests were conducted for natural marine clay in order to determine its compaction parameters. Then, sieved marine clay was mixed with the predetermined amount of RBT before mixing with water. The admixture of marine clay and RBT was mixed thoroughly with water until uniform color was achieved. The mixture was kept in an air tight plastic bag for minimum 24 hours in order for the moisture to be well distributed. To perform the test, the admixture was transferred into the compaction mold (115.5 mm height and 105 mm diameter) that was attached to a removable base plate prior to weight measurement. Before the admixture was placed, a detachable steel extension collar was installed at the top of the compaction mold. The admixture was placed inside the compaction mold in three equal layers and compacted 27 blows for each layer using 2.5 kg steel rammer. After the three layers were compacted, the extension top part of the mold was dismantled and the admixture was cut to remove the excess marine clay-RBT admixture. A digital balance was used to weigh the admixture of marine clay-RBT together with the mold and the weight was recorded to the nearest three decimal places. In order to determine the moisture content of the mix, two samples were collected from the top and bottom of the compacted admixture. The same procedures were repeated five times in order to be able to draw a proper compaction curve.

2.3 Sample preparation for Atterberg limits tests

The physical properties of the combination of RBT and marine clay were assessed by conducting liquid limit and plastic limit tests. Sample preparation started by sieving the untreated marine clay through 0.425 mm sieve mesh. Due to the large size of 1.18 mm RBT, only 0.3 mm RBT could be used for Atterberg limit tests. The soil was air dried and mixed with various mix designs (10%, 20%, 30% and 40%) of 0.3 mm RBT. The mixture of marine clay and RBT was mixed with water and kept in an airtight plastic bag for a minimum of 24 hours. The liquid limit and plastic limit tests were then conducted using the cone penetrometer method in accordance with the BSI 1377: Part 2: 1990 [19].

3 Results and discussions

3.1 Atterberg limits results

The results of liquid limit, plastic limit and plasticity index for marine clay and 0.3 mm RBT treated marine clay are presented in Table 2. It can be observed that 0.3 mm RBT was able to decrease the plasticity index of the soil. Besides, increments of 0.3 mm RBT resulted in decreasing the plasticity index as it was expected [23-25]. The reduction in the value of plasticity index with the increments of RBT is due to the replacement of marine clay fine particles with RBT coarse particles. Comparison between the effect of RBT and palm oil fuel (POFA) as studied by [26] showed that both waste materials were able to decrease the plasticity of the stabilized soft soil. As shown in Fig. 2 (Left), the liquid limit of marine clay decreased sharply from 40.4% to 32.5% when only 10% of 0.3 mm RBT was added. Further increments in the percentage of 0.3 mm RBT (20-40%) decreased the liquid limit of marine clay. In addition, as shown in Fig. 2 (Right) the plastic limit of treated marine clay also reduced when 0.3 mm RBT was added. Further increments of 0.3 mm RBT resulted also in decreasing the plastic limit [27]. The reduction trend of both liquid limit and plastic limit resulted in decreasing the plasticity index of marine clay as shown in Fig. 3. The plasticity of marine clay reduced significantly when 0.3 mm RBT was added. Meanwhile, 0.3 mm RBT was able to reduce the plasticity of marine clay from intermediate plasticity to low plasticity when 40% 0.3 mm RBT was added. It is also found that the effect of 0.3 mm RBT depends on its percentage. When the percentage increases, further reduction in the plasticity index is achieved.

Table 2. Atterberg limit results for 0.3 mm RBT treated marine clay.

Type	0.3 mm RBT (%)	Liquid limit (%)	Plastic limit (%)	Plasticity index (%)
Marine clay	0	40.4	21.83	18.57
Marine clay - RBT	10	32.5	18.51	13.99
Marine clay - RBT	20	30.1	17.45	12.65
Marine clay - RBT	30	28.6	16.98	11.58
Marine clay - RBT	40	25.9	15.33	10.57

Moreover, the statistical observation of the results obtained by the addition of different mix designs of 0.3 mm RBT is expressed in Figs. 2 and 3. It is shown clearly that the value

 https://doi.org/10.1051/e3sconf/20183401012

of Coefficient of determination (R^2) ranged between 0.87 and 0.90 that means the inconsistency in the linearity of the achieved results is very insignificant. This confirms an appropriate interpretation of 0.3 mm RBT mix designs.

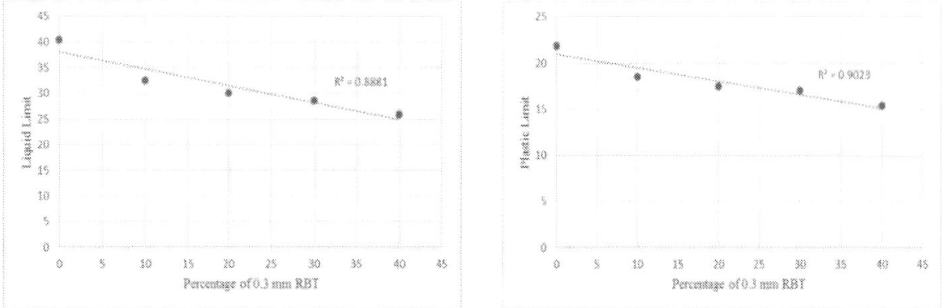

Fig. 2. Liquid limit (Left) and plastic limit (Right) of RBT treated marine clay at various mix designs.

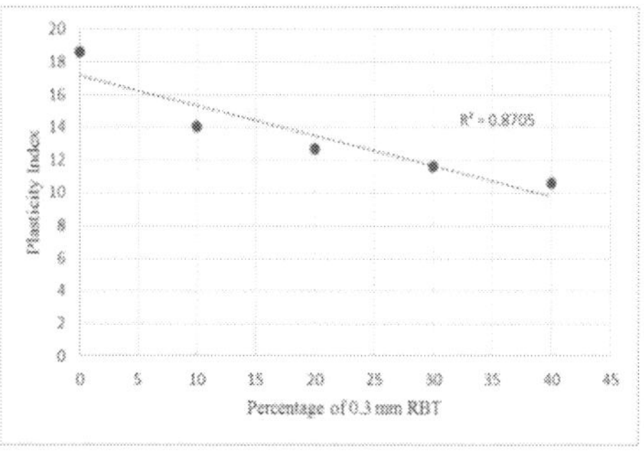

Fig. 3. Plasticity index of RBT treated marine clay at various mix designs.

3.2 Influence of RBT on compactability

The results of the maximum dry density (MDD) and the optimum moisture content (OMC) of untreated marine clay, 0.3 mm and 1.18 mm RBT treated marine clay are shown in Table 3. The results clearly showed continuous increases of MDD and continuous decreases in OMC as the percentage of RBT was increased from 10% to 40%. The reason of the increments of MDD can be explained due to the replacement of RBT particles of high specific gravity (2.57) with marine clay particles of low specific gravity (2.52). Whereas the reduction in OMC is due to the reduction of attraction for water molecules after the RBT particles are substituted with soil particles.

Table 3. Standard Proctor compaction test results of RBT treated marine clay.

Type	RBT size (mm)	RBT percentage (%)	Optimum moisture content (OMC) (%)	Maximum dry density (MDD) (Mg/m^3)
Marine clay	-	-	22	1.59
Marine clay - RBT	0.3	10	19	1.69
Marine clay - RBT	0.3	20	16.7	1.76
Marine clay - RBT	0.3	30	16.5	1.75
Marine clay - RBT	0.3	40	15	1.77
Marine clay - RBT	1.18	10	15	1.75
Marine clay - RBT	1.18	20	15	1.78
Marine clay - RBT	1.18	30	14	1.8
Marine clay - RBT	1.18	40	13.5	1.82

Fig. 4 shows the compaction curves of both 1.18 mm and 0.3 mm RBT treated marine clay. As can be seen in Fig. 4 (Left), addition of 1.18 mm RBT increased the MDD from 1.59 Mg/m^3 to 1.82 Mg/m^3, whereas OMC was reduced from 22% to 13.5% at 40% RBT. Besides, as can be seen in Fig. 4 (Right), the addition of 0.3 mm RBT increased the MDD from 1.59 Mg/m^3 to 1.77 Mg/m^3, while decreased the OMC from 22% to 15% at 40% RBT. The additions of RBT reduced the affinity of soil for water molecules and thus decreased the OMC. This was in agreement with [24, 25] who used ceramic tiles waste to treat soft soil. The influence of RBT on the compactability of soft soil can be compared to that achieved with the treatment using POFA. The addition of POFA to stabilize soft soil resulted in increasing MDD and decreasing OMC [26]. However, the addition of POFA without an additional binder such as cement or lime did not result in significant improvement of soft soil [26, 28].

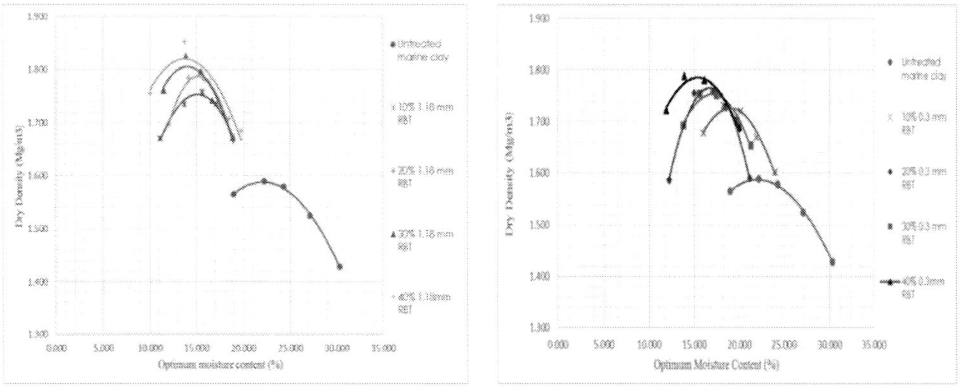

Fig. 4. Compaction behaviour of untreated and treated marine clay with different mix designs of 1.18 mm RBT (Left) and 0.3 mm RBT (Right).

3.2.1 Comparing the effect of RBT size

Fig. 5 (left) illustrates a comparative analysis of MDD for 0.3 mm and 1.18 mm RBT treated marine clay. It is clearly seen that 1.18 mm RBT resulted in higher MDD compared to 0.3 mm RBT that is due to bigger particle size. The 1.18 mm RBT was coated by the clay particles creating large size aggregate that occupied bigger area. Thus, the MDD was increased significantly and this is an indication of improvement of marine clay properties [26]. In addition, Fig. 5 (right) shows an analysis of OMC results of both sizes of RBT and it can be seen the lowest value of OMC was obtained at 40% 1.18 mm RBT. The larger the size of RBT, the lower the moisture content [29]. Furthermore, the statistical observation obtained by adding 10%, 20%, 30% and 40% of 0.3 mm and 1.18 mm RBT is expressed in Fig. 5. The coefficient of determination had a value of R^2 ranged between 0.91 and 0.67 that indicates insignificant discrepancy in the linearity of the achieved results. This confirms suitable interpretation of RBT mix designs.

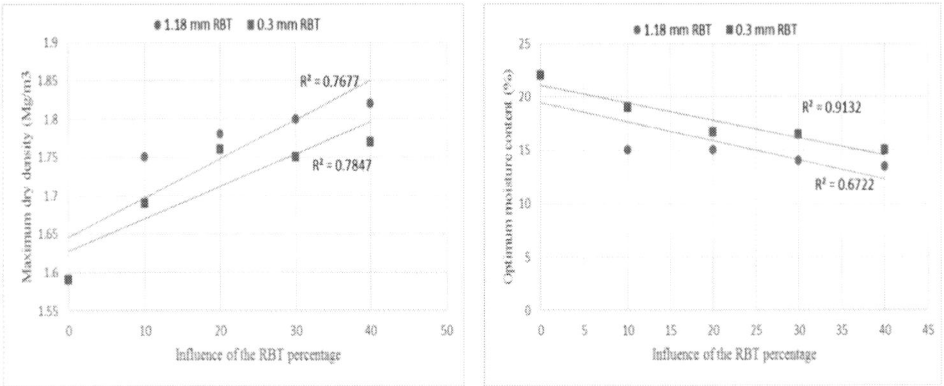

Fig. 5. Effect of various RBT mix designs on the MDD (Left) and OMC (Right) of marine clay.

4 Conclusions

The investigation of using ceramic tiles in soft soil stabilization was for the purpose of achieving the status of green technology, cleaner production and economical approaches. A series of laboratory tests were conducted to evaluate and check the suitability of recycled blended ceramic tiles (RBT) on treating marine clay. The tests included liquid limit, plastic limit, plasticity index, and compaction parameters. The following conclusions can be drawn ; 1. A decreasing trend of liquid limit, plastic limit and plasticity index was observed with increasing amount of 0.3 mm RBT. The plasticity of the marine clay was reduced from intermediate plasticity to low plasticity at 40% 0.3 mm RBT. 2. Additions of both sizes of RBT increased the MDD and decreased the OMC of the marine clay significantly. 3. In terms of size influence, 1.18 mm RBT was more effective than 0.3 mm RBT that act as a strengthening agent that bonded the marine clay particles together resulting in higher density and lower moisture content. The larger the size of RBT, the more effective the compaction parameters of the treated marine clay.

The authors would like to express their gratitude to UTP for the generous support given for the publication of this paper using the Graduate Assistantship Scheme. Besides, the authors express their gratitude to the Research University Grant Scheme Q.J130000.2522.16H04 from UTM.

References

1. Y. Cai, B. Shi, C.W.W. Ng, C. sheng Tang, Eng. Geol. **87** (2006)
2. M.A.M. Al-Bared, A. Marto, Proc. 2nd Int. Conf. Sep. Technol. 2017 Rev. (2017)
3. J.S. Yadav, S.K. Tiwari, J. Build. Eng. **9** (2017)
4. Y. Yi, L. Gu, S. Liu, Appl. Clay Sci. **103** (2015)
5. S. Vichan, R. Rachan, 53 (2013)
6. A. Mahamedi, M. Khemissa, Hous. Build. Natl. Res. Cent. J. **11** (2014)
7. L.C. Dang, B. Fatahi, H. Khabbaz, Procedia Eng. **143** (2016)
8. N. Yoobanpot, P. Jamsawang, S. Horpibulsuk, Appl. Clay Sci. **141** (2017)
9 V. Anggraini, A. Asadi, B.B.K. Huat, H. Nahazanan, Measurement **59** (2015)
10. A. Zukri, Procedia Eng. **53** (2013)
11. R.K. Etim, A.O. Eberemu, K.J. Osinubi, Transp. Geotech. **10** (2017)
12. T.-C. Ling, K.H. Mo, L. Qu, J. Yang, L. Guo, Constr. Build. Mater. **146** (2017)
13. H. Elçi, J. Clean. Prod. **112** (2016)
14. A.M.M. Al Bakri, M.N. Norazian, H. Kamarudin, C.M. Ruzaidi, Malaysian Univ.

Conf. Eng. Technol. (2008)
15. N. Anting, M.F. M. Din, M. Ponraj, K. Jungan, S. Mat Taib, K. Iwao, S. Chelliapan, N. Ubukata, Malaysian J. Civ. Eng. **26** (2014)
16. A. Ahmed, U.H. Issa, Soils Found. **54** (2014)
17. L.S. Wong, S. Mousavi, S. Sobhani, S.Y. Kong, A.H. Birima, N.I. Mohd Pauzi, Measurement. **94** (2016)
18. K. Xia, Comptes Rendus Mécanique **342** (2014)
19. *British Standard Institution, British Standard Methods of Test for Soils for Civil Engineering Purposes* (BS1377, London, 1990).
20. *The British Standards Institution, British Standard Code of practice for site investigations* (BS 5930, London, 1999).
21. M. A. Mas, J. Monzó, Paya, L. Reig, and M. V. Borrachero, Institution of Civil Engineers, (2015).
22 *British Standards Institution, British standard methods of test for soils for civil Engineering Purposes* (BS1377, Milton Keynes, London, U.K. 1990).
23 A.K. Sabat, Electron. J. Geotech. Eng. **17** (2012)
24. C. Neeladharan, V. Vinitha, B. Priya, S. Saranya, Int. J. Innov. Res. Sci. Eng. Technol. **6** (2017)
25. J.A. Chen, Idusuyi, F. O, Int. J. Eng. Mod. Technol. **1** (2015)
26. S. Pourakbar, A. Asadi, B.B.K. Huat, M.H. Fasihnikoutalab, Transp. Geotech. **3** (2015)
27. G. R. otoko, I. F. manuel, I. S. chinweike, P. harcourt, Journal of Multidisciplinary Engineering Science and Technology (JMEST), **3**(5) (2016).
28. Tg. Rani, C. Shivanarayana, D. Prasad, D. Prasada Raju, Int. J. Eng. Res. Dev. **10** (2014)
29. N. Ameta, A. Wayal, P. Hiranandani, Am. J. Eng. Res. **2** (2013)

Analysis of Factors Influencing Building Refurbishment Project Performance

Nurfadzillah Ishak[1,*], *Fazdliel Aswad* Ibrahim[2], and *Muhammad Azizi* Azizan[2]

[1]School of Environmental Engineering, Universiti Malaysia Perlis, Kompleks Pengajian Jejawi 3, 02600 Arau, Perlis
[2]Faculty of Engineering Technology (Civil), Universiti Malaysia Perlis, Kampus UniCITI Alam, Sungai Chuchuh, Padang Besar 02100 Perlis, Malaysia

Abstract. Presently, the refurbishment approach becomes favourable as it creates opportunities to incorporate sustainable value with other building improvement. In this regard, this approach needs to be implemented due to the issues on overwhelming ratio of existing building to new construction, which also can contribute to the environmental problem. Refurbishment principles imply to minimize the environmental impact and upgrading the performance of an existing building to meet new requirements. In theoretically, building project's performance has a direct bearing on related to its potential for project success. However, in refurbishment building projects, the criteria for measure are become wider because the projects are a complex and multi-dimensional which encompassing many factors which reflect to the nature of works. Therefore, this impetus could be achieve by examine the direct empirical relationship between critical success factors (CSFs) and complexity factors (CFs) during managing the project in relation to delivering success on project performance. The research findings will be expected as the basis of future research in establish appropriate framework that provides information on managing refurbishment building projects and enhancing the project management competency for a better-built environment.

1 Introduction

Building industry is one of the biggest contributors to pollution and waste through its life cycle [1]. Reference [2] has counselled the building industry to reduce the greenhouse emissions at least 50% in less than forty years and needs to have achieved at least 25% reduction in eleven years to avoid worst-case scenario of climate change. If nothing is done, greenhouse gas emissions from buildings will more than double in the next twenty years. In addition, towards the end of the 20th century, the major issue faced by building industry is the ratio of existing buildings to new construction is become overwhelming [3]. In short, it will affect the inequities huge stock of existing, which makes up the bulk of the

* Corresponding author: fadzishak@gmail.com

construction market not sustainably built. It will show that the physical development by building industry could not promise to pursuit the quality of life for future generation.

At this point, by considering the refurbishment principles from vary perspectives in building industry will create the stability environment for current and future generation in holistic long-term benefit. As stated by [4], upgrading an existing building for reuse can represent a 10% to 12% cost savings as opposed to a constructing new building. Thus, it can be concluded that long term benefit can only be achieved by integrate this principles and in order to survive, the refurbishment industry should increase its performance by defining correct strategies and measures regarding with the factors influencing project performance.

In conjunction with this, it might be conflict in concentrated particularly on defining the factors influencing refurbishment project performance because it well knows that refurbishment building projects is more complex compared to new construction [5, 6]. Therefore, addressing this gap by develop an appropriate framework has been the main motivation for this study.

2 The growth and performance of refurbishment building projects

Refocusing on refurbishment principles, there is no specific definition of it. [7] offers the definition of refurbishment is "to make neat, clean or complete as by renovating, re-equipping or restoring". However, [8] define refurbishment as synonym of rehabilitation has supported the terms conversion and modernization implying an approach that embraces the wider localized environment as well as individual structures. While, [9] reported that refurbishment presents a means of improving the performance of existing buildings without the economic and environmental coast associated with new build schemes.

Today, there is a widespread awareness of the needs to refurbish existing building and to modernize with emphasis on latest methods to accommodate modern building services and energy saving techniques as well as meets the sustainable agenda. Previous literature claimed that by demolishing buildings and preparing the site for the new build involves high embodied energy and costs [10-12]. Also, been supported by [13] which reported that at this rate of new built and demolition, about 70% to 80% of current existing stock will still be standing in year 2050.

Numerous studies have been promotes integration between these two fields; refurbishment and sustainability which will offer a huge potential to improve the performance of existing building stock in sustainably efficient [10, 14, 15]. According [16] discover that "sustainability cannot be achieved without addressing the existing building stock even every new building was a sustainable building, their impact on sustainability as a whole will be minimal for some time". Nevertheless, there is a missing link in refurbishment's research agenda. Yet, the majority refurbishment building projects have still been developed in an unsustainable manner and lack of implementing quality and performance management techniques [5, 10]. Despite, the vast number of research initiatives designed to better understand refurbishment industry aimed at improving its performance, therefore, the specific success factors has not been addressed properly.

In Malaysia scenario, refurbishment principles may regard as new phenomenon, which less approach in utilizes it. Particularly in Malaysia, the concept is still new in the field of building industry due on limited empirical research advocates this new area of management compared to sustainable new built [5]. Therefore, the studies on practices applied to the management of this complex construction environment are scarce. Also, the effectiveness of refurbishment management still in question and need to be improve because it would reflect the performance of project.

3 Factors influencing refurbishment building project performance

3.1 Critical success factors of refurbishment building projects

Critical success factors (CSFs) are necessary behind factors in order to assessing the success on the project performance. This is also affirmed [17] that understanding the CSFs could increase the chance of a project's success. Although, there have been several research studies on CSFs in building project success performance, rarely specifically looked at the issues relating to refurbishment projects. It has been supported by [18] suggest that there is need to develop project performance success indicator for refurbishment projects in order to add value to the existing building. Much of discussion in this subject is tended to focus on issues related to development performance of new build construction only [6]. Even though the refurbishment approach have identified as logical solution to reduce the environmental impact preferably [19-21], however little attention has been paid to this area.

There is an evident to address this gap by providing empirical evidence of CSFs influence in the context of refurbishment project performance. In extent literature, many studies have attempted to identify the success factors that could influence a successful implementation [22, 23]. The concept of CSFs is being widely discussed in management literature and has been central theme to the literature of project management. Many CSFs categorizations and frameworks are approached and integrated with the project success criteria in order to provide a comprehensive framework for evaluating the project based on CSFs. According to [24], several studies suggest that for a successful performance management framework, CSFs and performance management should be linked. It is because CSFs are defined as inputs that enhance and direct the project to be achieved successfully whereas, performance measurement are used to judge project success or failure. However, success is 'relative' when the meaning of success differs between different groups of users.

CSFs were first studied by [25]. Further, [26] come out with a list 13 factors affecting project success related to project area, human area and general management area. However, [27] project success factors remain the basis for project management success factor today. Consequently, [28] come out with four main areas of critical success factors related to: the development projects factors; the project manager and project team factors; the organization factors; the external environment factors. Reference [29] argued that the project management success factors are: project team promise; client's competencies; risk and liability assessments; client's competencies; end user's needs; end users imposed restrictions. Most of project success studies hypothesize that the absence of CSFs would lead to failure of project [30].

According to [31] claimed that CSFs relates to internal actions and the fact that project success is more about the holistic view of a project. References [32, 33] the CSFs are something that could be measured over the project lifecycle and relates to performance of the conventional criteria (time, cost, quality). In order to evaluate the success of refurbishment project performance it is important to identify it's CSFs of refurbishment building projects.

3.2 Complexity factors of refurbishment building projects

The significance of refurbishment is aimed at long term remodelling of a building, addressing constraints such as circulation and maximizing the potential offered by the site and building consent. One of the most often cited barrier and challenges for refurbishment as opposed to new build is the inherent complexity of the works themselves [34-36]. The complexity is measured against the process as a whole, but it must be noted that the challenges faced in this aspect also have singular significance, respectively to the project team that carrying out the works. As for refurbishment building projects, its nature and characteristics are again different due to the complexity of the extent of works as it usually risky and complicated, possibly involving existing and unsecured site conditions.

The complexity of refurbishment involved technical, technological, ecological, social, comfort and esthetical. Generally, the complexity are mainly derives during the planning stage in refurbishment project is reflected from difficulty to obtain the specific information to project organization. In 'Managing Refurbishment Project' book titled by reference [6] pointed out that the degree of complexity shows the gap between information needed and the availability of information, is greatest at the beginning of the project, such as during design stage.

While, lack coordination, delay the decision process and fragmentation is several issues that contributed the complexity during the implementation stage. It is because the occurrences of communication gap between the project organization, ineffective collaboration and limited understanding the nature of the complexity and in addition to the interdisciplinary nature of the planning process [37]. Despite this, the various parties are often to concentrate on their own area of specialization in refurbishment building projects [38]. This can caused the 'over the wall syndrome' that made up of separate parties from diverse professions that need to make a one decision in design and construction stage in refurbishment building project.

Reference [39] discovered that another refurbishment projects barrier which are the occurrence of large numbers of variation orders to the project. The reasons of the changes design are influence by variations from client requirement, designer, amendment of statutory regulations and problem on site environment. References [40, 41] discovered that the difficulty in accessibility also could contribute the level of complexity. Most of the refurbishment building projects site is located inside and the façade of the existing building, the designers need to be smartly planned and consider the available access elements such as door, window, stair and lift to be an access on the site which needs to acquire extra protection to maintain site safety, dust and noise problem.

4 Methodology

The tool used to achieve the relationship between critical success factors (CSFs) and complexity factors (CFs) during managing the project in relation to delivering success on project performance in this research by developing a conceptual framework. In the field of project management, the subject of project success is the heart of the field where it varies with project phases. In fact, though much literature has attempted to list CSFs, it is impossible to list a definite set of all the CSFs for refurbishment projects the refurbishment projects have different success criteria from other new construction projects since the targets and aims of the applications in refurbishment are different from those in other projects. A success criterion has long been considered the ability to fall within time, cost, quality constraint or known as 'iron triangle'. Therefore, for the purpose of this study, the success criteria covered the integrating the three pillars of sustainability in success criteria in order to achieve the goal of attaining sustainable structure. The impact of the energy efficient upgrade was not considered in this study. Same goes to the complexity factors which identified that related with the scope study. The building industry comprises of multidisciplinary professionals who are employed by different organization and specialization. Therefore, this study only covers the complexity factors that been faced by construction organization during managing the refurbishment projects.

5 Discussions

The conceptual framework (refer Fig. 1.) for this study integrates the CSFs and success dimension for refurbishment projects performance variables. The framework is extended and modify from the model that integrates different elements that were built as the basis for a project management theory. The framework is not typically used for explaining complicated process, but it does assist by simplifying the process and making it more understandable [42].

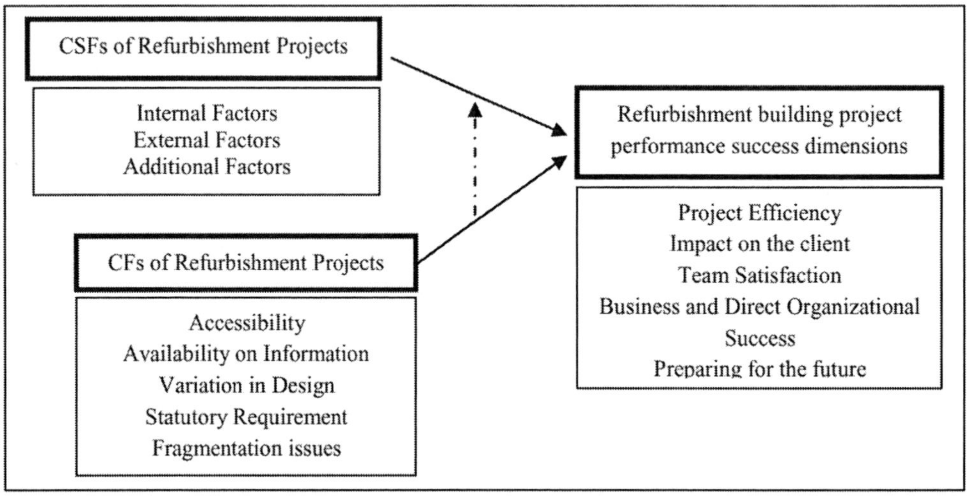

Fig. 1. The conceptual framework.

For the purpose of this study, the independent variables which consists three categories critical success factors (CSFs) are synthesized and re-classification created in order to simplify the conceptual framework of this study which will give influence the success dimension for refurbishment project performance. Therefore, in this study three CSFs have been classified namely internal factors (project management, top management, project team), external factors (environment, economic, social, policy and technological) and additional factor (project characteristic). These CSFs are actually a new classification inspired by the empirical studies of [43, 29, 28]. Meanwhile, for dependent variable, this study selected [44] multidimensional performance of projects model were adapted and presented as a reflective second order factor to capture the multifaceted nature of project success. These success dimensions are comprised of project efficiency, impact of the client, team satisfaction, business and direct organizational success and also preparing for the future. The main reason for adopting this construct as dependent variable was because of its coverage on the basic measures of project success, which have been developed and tested as a generalized success measure.

As mentioned before, that refurbishment project is more complex than new build scheme. Therefore, the moderating variables are taken into consideration in this study in order to recognize the complexity factors (CFs) that been occurred in refurbishment projects. A moderating variables is a third independent variable that causes the relationship between the independent and dependent variables to change where change could be characterized as the direction and strength of the [45, 46]. Five moderating variables are taken into consideration in this study in order to recognize the complexity factors that been occurred in refurbishment building projects and identified which having an effect on the refurbishment project performance success. Determination of the factors that are presents is the preliminary stage of an on-going research and needs further investigation for final model will be formed.

6 Conclusions

In order to survive and provide a sustainable strategic position in the building industry, a construction organization should increase its project success by seizing the opportunities in either domestic or international market. Moreover, considering the fact that refurbishment building projects have different objective and specialities compared to other new built, it is crucial to determine the factors that will bring success in refurbishment building projects.

For this reason, result from this study; it is aimed to determine valid performance measurement which leads refurbishment building projects to success in order to obtain integration with sustainable approach especially in current trend building industry in Malaysia. Thus, help to bring a rich insight by exploring the integrating ideas between sustainability concept and refurbishment works and its application to project management in Malaysia.

The authors would like to extend sincere gratitude to Universiti Malaysia Perlis for the support on carrying out this research.

References

1. A. Horvath, Annu. Rev. Env. Resour. **29,** 1, 181 (2004)
2. UNEP SBCI. *United Nations Environmental Programme,* (Sustainable Buildings and Climate Initiative, Paris, 2009)
3. A. Durmus-Pedini, B. Ashuri, Int. J. Facility Management **1**, 1 (2010)
4. R. Shipley, S. Utz, M. Parsons, Int. J. Herit. Stud. **12**, 6, 505 (2006)
5. A.S. Ali, S.N. Kamaruzzaman, A.S. Zulkiflee, M.R. Pitt, J. Asian Archit. Build. **9,** 2 (2010)
6. I. Rahmat, *Managing Refurbishment Projects*, (University Publication Centre (UPENA), 2008)
7. *Collins English Dictionary* (Wiliam Collins and Son, Glasgow, UK. 1989)
8. J.R. Mansfield, J. Property Management, **20**, 23 (2002)
9. Corus. *Refurbishment solutions for non-domestic buildings*, (2010).
10. S.I. Bhuiyan, K. Jones, N. Wanigarathna. *Procs 31st Annual ARCOM Conference*, 1093 (2015)
11. A.A. Sezer. *Contractor Monitoring of Productivity and Sustainability in Building Refurbishment* (Department of Technology Management and Economics Chalmers University of Technology, Göteborg, Sweden, 2014)
12. T. Häkkinen, Civ. Eng. Environ. Syst. **24**, 4, 247 (2007)
13. National Statistics. *English Housing Survey: Household Report 2010-11.* (Communities and Local Government, 2012).
14. A. Power. *Proceedings of the ICE - Urban Design and Planning*, **163**, 4, 205 (2010)
15. S.H. Hong, T. Oreszczyn, I. Ridley, Energ. Buildings **38,** 10, 1171 (2006)
16. B. Wood. Facilities. **24**, 1/2, 61 (2006)
17. A. Mobey, D. Parker. Work Study **51**, 4, 202 (2002)
18. A.A. Olanrewaju, M.F. Khamidi, A. Idrus. Engineering and Technology **56,** 726 (2011)
19. R. Jagarajan, M.N. Abdullah, J.Y. Lee, M.N. Jaafar. J. Technol. **5,** 85 (2015)
20. Y. Cronhjort, S. Le Roux. *World Congress on Municipal Engineering* (2012)
21. E. Miller, L, Buys. J. Property Investment and Finance **26**, 6, 552 (2008)

22. S. Dezdar, A. Sulaiman, Industrial Manag. and Data Systems **109**, 8, 1037 (2009)
23. S. Finney, M. Corbett, Bus. Proc. Manag. J. **13**, 3, 329 (2007)
24. E. Westerveld, Int. J. Proj. Manag. 411 (2003).
25. I. Rubin, W. Seeling. IEEE Trans. Eng. Management **14,** 3, 131 (1967)
26. D. Cleland, W. King, *Systems analysis and project management* (McGraw Hill, New York, 1983)
27. J.K. Pinto, D.P. Slevin, Technology Management **31** (1989)
28. W. Belassi, O. Tukel, Int. J. Proj. Manag. **14**, 3, 141 (1996)
29. A. Chan, D. Scott, A. Chan. J. Constr. Eng. Manag. **130**, 1, 153 (2004)
30. P. Poon, C. Wagner. Decision Support Systems **30**, 4, 393 (2001)
31. L.A. Ika. Proj. Manag. J. **40,** 4, 6 (2009)
32. T.J. Cooke-Davies. Int. J. Proj. Manag. **20**, 3, 185 (2002)
33. W.S. Han, A.M. Yusof, S. Ismail, N.C. Aun, Int. J. Business Manag. **7**, 90 (2012)
34. A. Azlan-Sha, Int. J. Phys. Sci. **5**, 6, 768 (2010)
35. Y.K. Juan, P. Gao, J. Wang. Energ. Buildings **42,** 3, 290 (2010)
36. C. Egbu, C.C.T. Lee. *21st Annual ARCOM Conference, Soas,* **2**, 865 (2006)
37. M.L. Chiu, Design Studies **23**, 187 (2002)
38. A.S. Ali, I. Rahmat. J. Building Appraisal Palgrave, 87 (2009).
39. G. Wilson, M. Kishk, R. Laing, Public Health Frontier **2,** 2, 114 (2013)
40. J.D. Gilleard, Y.C. Lee. *Proceeding of international symposium on management, maintenance and modernization of building facilities, Singapore,* (CIB, 213, 1998)
41. CIOB (Chartered Institute of Building) *Project management in building, 2nd Ed.* (Ascot, UK, 1988)
42. B.R. Hergenhahn, M.H. Olson. *An Introduction to theories of learning, 6th Ed.* (Prentice-Hall Inc, New Jersey, USA 2001)
43. P.L. Sui, G.W. Shang, T. Lin. Structural Survey **32**, 5, 413 (2014)
44. A.J. Shenhar, D. Dvir. *Reinventing Project Management: The Diamond Approach to Successful Growth and Innovation*, (Harvard Business School Press, 2007)
45. J.F. Hair, R.E. Anderson, B.J. Babin, W.C. Black. *Multivariate data analysis, 7th Ed.* (Prentice-Hall, NJ, 2010)
46. R.Y. Cavana, B.L. Delahaye, U. Sekaran. *Applied Business Research: Qualitative and Quantitative Methods,* (John Wiley and Sons Australia, Milton, Queensland, Australia, 2001)

Panel Board From Coconut Fibre And Pet Bottle

Norhayati Ngadiman[1,*], *Masiri* Kaamin[1], *Aslila* Abd. Kadir[1], *Suhaila* Sahat[1], *Aziza* Zaini[1], *Siti Raihana* Nor Zentan[1], *Nur Ain* Ahmad[1], and *Wan Haizatul Aisyhah* Wan Amran[1]

[1]Center for Diploma Studies, Universiti Tun Hussein Onn Malaysia (UTHM), Malaysia

Abstract. The rate of global deforestation and its impact on the environment has led particle board manufacture to search for alternative feedstock, especially in countries where wood is less available compared to other cellulosic natural product. Based on the properties of coconut fibre and PET bottle, these two materials can be recycle as raw material for manufacture of panel board. As for this study, the coconut fibre were used as the filler and PET bottle as outer lining of the panel board. Two types of coconut fibre were used which are grinding and un-grinding coconut fibre. At first, the coconut fibre are undergoes softening, grinding, drying and sieving process, while PET bottle was cleaning, shredding, sieving before compacted using hydraulic hot press machine. There are four types of testing that been carried out which are swelling, water absorption, Modulus of Elasticity (MOE) and Modulus of Rupture (MOR). The result show the conventional board has the highest value for MOE test, so it's indicate that the conventional board is less strength from the coconut fibre board. As for water absorption test, the average water absorption of coconut fibre based panel board is less than conventional board. Overall, the coconut fibre board is better than conventional panel board because coconut fibre board are less swelling, has low water absorption, high modulus of rupture and low modulus of elasticity. Based on the finding, this coconut fibre panel board has potential as a stronger and long-lasting panel board than the conventional board in the market. Other than that, the panel also have their own aesthetic value since the recycled plastic bottle used as outer lining is colourful and giving aesthetic value.

1 Introduction

The use of renewable resources such as agriculture residues is gaining increase interest in production of particle board. Coconut is a renewable resource and a natural fibre extracted from the husk of coconut and used in product such as floor. Coconut fibre, also known as coir, comes from the inner husk of coconuts and it most widely grown nut in the world and contribute significantly to the economy of many tropical areas. It's used in manufacturing map, rope and landscaping. However, most of coconut fibre waste are not be utilize and it end up as waste that can effect on the water surface, groundwater and soils because it contain salts and others chemical.

* Corresponding author: nryati@uthm.edu.my

1.1 Coconut Fibre

Coconut fibre also known as coconut coir is a fibrous material extracted from the husk of coconut. Commonly the coconut fibre is separated from the husk by mechanical method even though it's also can be separated using chemical and biological method [1]. Brown fibres fibre extracted from matured coconuts are thick, strong and have high abrasion resistance, while white fibres are smoother, finer and weaker [2]. Table 1 is showing the chemical composition of coconut fibre. Coconut fibre is the hardest natural fibre because high content lignin in coconut fibre up to more than 30 % that make the fibre tougher and stiffer by providing plant tissue and individual fibre with compressive strength and stiffen the cell wall of the fibre in order to protect the carbohydrates from chemical and physical damage [3]. According to Ali (2010), coconut fibres are capable of taking strain 4-6 times more than that of other fibres. Moreover, coconut fibre does not loose strength on storage and expose to sunlight [1]. Table 2 shows the physical properties of coconut fibre. The value density of coconut fibre is important to determine the total mass needed to produce the coconut fibre panel board.

Table 1. Chemical composition coconut fibre [3].

Chemical Composition	Percentage (%)
Halocellulose	56.3
α-cellulose	44.2
Lignin	32.8
Ash	2.2

Table 2. Physical properties of coconut fibre [4].

Physical Properties	Coconut Fibre
Density (g/cm^3)	1.2
Elongation at break (%)	30
Tensile strength (Mpa)	175
Young Modulus (Gpa)	4-6
Water Absorption (%)	130-180

1.2 Polyethylene terephthalate (PET)

PET bottle widely used drink bottle and the PET bottle waste took long times to decay. Polyethylene terephthalate or also known as polyethylene terephthalate, commonly abbreviated PET, PETE, or the obsolete PETP or PET-P, is the most common thermoplastic polymer resin of the polyester family and consists of polymerized units of the monomer ethylene terephthalate, with repeating ($C_{10}H_8O_4$) units [5]. Recycle PET is has the number "1" as and recycling symbol. PET is colourless in its natural state. Based on how it is processed, PET can be semi-rigid to rigid, and it is very lightweight. PET becomes white when exposed to chloroform and also certain other chemicals such as toluene [6]. About 60% crystallization is the upper limit for commercial products, with the exception of polyester fibres [7]. Due to its significant water and moisture resistance, it is widely used in making plastic bottles for soft drinks as shown in Fig. 1 [8]. One of the most important characteristics of PET is referred to as intrinsic viscosity (IV) [9] that, shows the thickness of the carbonated soft drink is bigger than plastic bottle. As the cover of the panel board, the PET plastic must be rough to prevent the filler from the termite while the carbonated soft drink also can protect the filler from the water because the PET plastic has lower water absorption.

The use of coconut fibre in the manufacture of furniture is increasingly widespread due to its properties. Previous research shows coconut fibre is stronger than wood fibre. Meanwhile, PET bottles are excellent barrier materials and are widely used for soft drinks. Therefore, properties coconut fibre of and PET bottle have been study carefully in order for the researcher to make sure these material as panel board main raw material. Thus, this project used coconut fibre to replace the wooden fibre as the filler of the panel and recycled PET bottle as a cover for the panel. The purpose of this panel board is to reduce waste of coconut fibre and plastic bottle (PET) and transform into useful product. In the other side, one of the problems for existing board now is easily expand and cannot withstand to water. This project is assign to; (1) design and produce a panel board using coconut fibre as filler and plastic as the outer lining and (2) test the Modulus of Elasticity, Modulus of Rupture, swelling and water absorption of the coconut fibre panel board.

Fig. 1. Recycled plastic Bottle (PET).

2 Methodology

2.1 Flowchart Process

According to the flowchart in Fig. 2, the first phase is to produce the panel board is preparing raw materials before undergoes the softening process. Then, the softening process took 24 hours and after that, it undergoes the cleaning process. Next, crushing and grinding the coconut fibre. Lastly, sieve the grinding coconut fibre to remove the dusk. While the PET plastic bottle need to be cleaned before shredding used the plastic granulator machine. The second phase is making the panel board. There is only one process which is compact the grinded coconut fibre and un-grinded coconut fibre. Then, the PET plastic will be compact as the outer layer of the panel board. In the third phase the project, the panel board sample are cut in smaller size sample before testing were carried out. After that, undergoes the finishing process the sample were tested according to the standard test applied on the panel which are modulus of elasticity, modulus of rupture, swelling and water absorption. The result will be compare to the conventional panel.

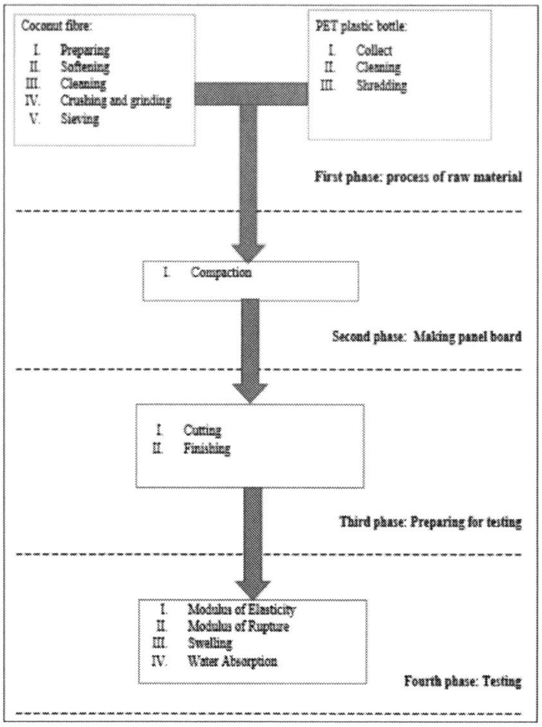

Fig. 2. Flowchart of panel board process.

2.2 Coconut fibre process

First step to produce the coconut fibre as the raw material is prepare the 5 kg of coconut fibre. Then, the coconut fibre need to be softening so it will be easier to use. The softening process will take 24 hours to be soak into the water and then remove the outer layer and cut the coconut husk into a small size. The coconut husk will be dried into the oven with 80 ° C for 20 minutes to ensure that all the water content is fully condense and dried coconut husk before put into the fibre process machine to separate the coconut peat and the coconut fibre. Coconut fibres are commercial available in three forms, namely bristle (long fibres), mattress (relatively short) and decorticated (mixed fibres) [2]. Therefore, the coconut fibre was divide into two equal weight because half of coconut fibre will undergo the grinding process while the other half will remain the original size as shown in Fig. 3 and Fig. 4. The physical appearance to make two different sample according to the physical appearance of coconut fibre. Then, sieve the grinding coconut fibre used 300 mm and 63 mm sieve size to remove the dusk. After that, the coconut fibre and glue is mixing each other using the ratio 70% of the coconut fibre while 30% of the glue from the mass of the panel board and need to be compact to get a thin layer of coconut fibre. Lastly, cut the layer into 50 mm x 50 mm measurement and 12 mm thickness.

Fig. 3. Coconut fibre before grinding.

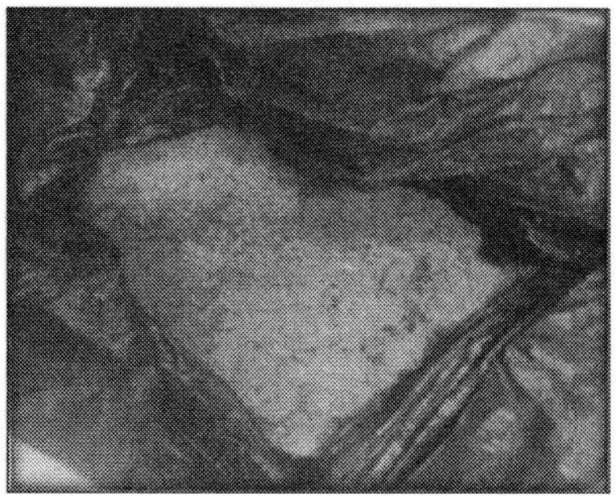

Fig. 4. Coconut fibre after grinding.

2.3 Polyethylene Terephthalate (PET) Bottle Process

For the first process of PET plastic bottle is collecting the bottle until it weight 1 kg and then, cleaning the bottle with wash used water and dried it naturally, it will then cut into smaller pieces. This is because the Plastic Granulator Machine cannot be operated with large chunk of plastic. As for the precaution the plastic is need to be clean and dry to prevent the damage to the machine. Fig. 5 shows the shredded the recycle PET bottle.

Fig. 5. Shredded recycle PET.

2.4 Panel board process

Process of making of panel board are including compaction, cutting into standard size and finishing process. Firstly, the compaction process will compact the grinded coconut fibre used hydraulic type hot or cold machine and set the machine with 170 ° C. For one panel, the mass of coconut fibre is 521.58 grams and mass of urea gam is 223.53 grams with 308 mm x 288 mm x 12 mm size of the mould. The same process same goes to the un-grinding coconut fibre. There are two panel for each sample which are one will covered with PET plastic while another one will not covered with PET plastic. Then, for the sample that covered with PET plastic, the PET plastic will be compact as the outer layer of the panel board with the same machine and same temperature to produce the aesthetic value. For the preparing panel for testing, the sample will be cut into area of 50 mm x 50 mm for water absorption and swelling test while cutting the sample into area of 240 mm x 50 mm for modulus of rupture and modulus of elasticity test. Then, the finishing process for the sample is shellac the outer layer for the 50 mm x 50 mm sample to undergo the water absorption and swelling test and want to compare with the conventional panel with the same condition.

3 Results and Discussions

3.1 Modulus of Elasticity (MOE)

This method of testing is used to determine the coconut fibre with grinding process, the coconut fibre with un-grinding process and standard panel board behaviour under an axial stretching load. The maximum load of this three sample being compared. The area of the panel board sample is 50mm x 240 while the volume is 50 mm x 240 mm x 12 mm. The Fig. 6 show the graph of the average of the modulus of elasticity of grinding sample is 1721.33 MPa, un-grinding sample is 1163.90 MPa and conventional sample is 2720.33 MPa. The un-grinding sample has the lowest modulus of elasticity followed by the grinding sample and conventional sample. The lower the modulus of elasticity, the higher the stiffness of the sample. The stiffness of un-grinding sample is the highest because it has long physical appearances that bond together to improve the strength of the panel board.

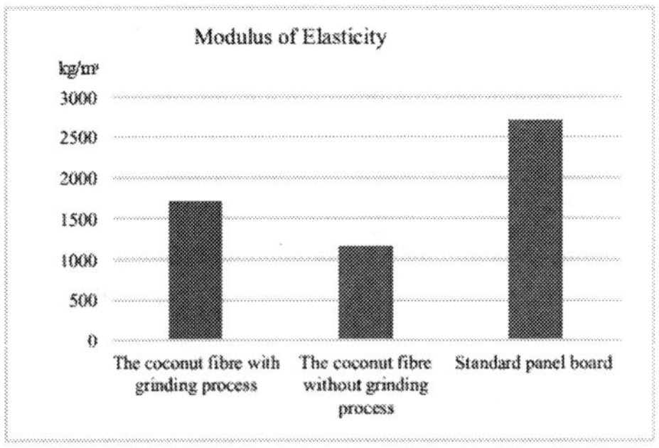

Fig. 6. Graph modulus of elasticity.

3.2 Modulus of Rupture (MOR)

The Fig. 7 shows the graph of average of the modulus of rupture of grinding sample is 1.914 x 10^{-5} kN, un-grinding sample is 2.205 x 10^{-5} kN and conventional sample is 1.881 x 10^{-5} kN. The un-grinding sample has the highest modulus of rupture followed by the

 https://doi.org/10.1051/e3sconf/20183401014

grinding sample and conventional sample. The un-grinding sample has the long physical appearance which will bond to each other and cause the higher strength compared to the other sample. The highest the modulus of rupture, the highest the strength of the sample. The un-grinding coconut fibre is stronger to load bearing capacity than the conventional board and grinding coconut fibre.

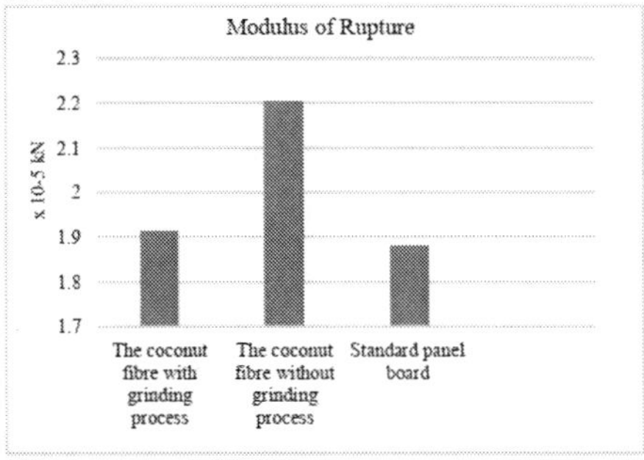

Fig. 7. Graph modulus of rupture.

3.3 Water Absorption Test

The Fig. 8 show the graph that the highest water absorption is conversional panel board with the average percentage of water is 67.08 %. The second highest is un- grinding sample of panel board is 59.84 % while the lower average water absorption is grinding sample of panel board is 56.21%. The grinding sample less absorption of water followed by the un-grinding sample panel board and the conventional panel board. In the other shell, grinding sample is the lower water absorption because the physical appearances of the fibre was grinding into the small size and the bond between coconut fibre bindings more strong to reduce the water absorb and fill in the pores. The grinding coconut fibre does not absorb more water because this panel is more compact.

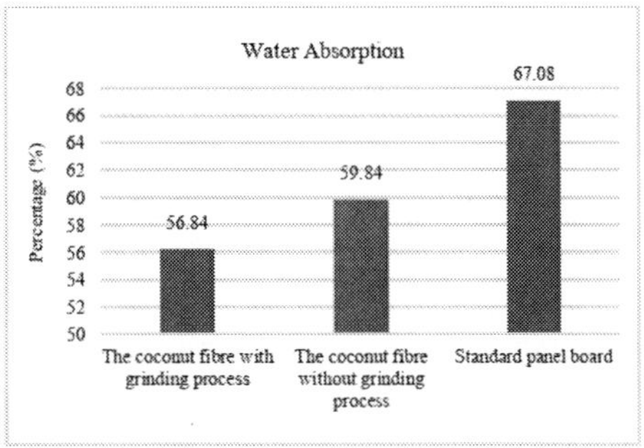

Fig. 8. Graph of water absorption.

3.4 Swelling Test

The Fig. 9 show the graph of the depth of the conventional panel board swelled is 1 mm compared to the grinding sample and un-grinding sample. The size of the grinding sample and un-grinding sample after immersed remain same as before immersed because lignin of the coconut fibre reduce the thickness swelling of the sample.

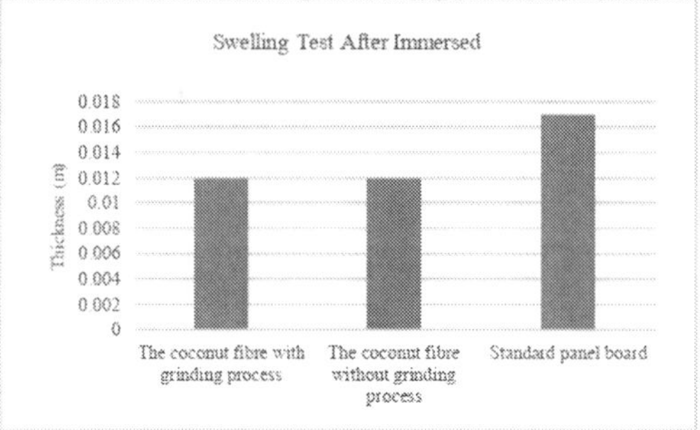

Fig. 9. Graph swelling test after immersed.

3.5 Discussions

For overall, according to the result from the four standard test that have been done, the un-grinding panel board is the stronger followed by the grinding coconut fibre and conventional panel board. The un-grinding panel board is the best because it have no swelling, less water absorption, high modulus of rupture and low modulus of elasticity. All of this test are used few constant condition such as, the size of the panel board, time taken to immersed the panel board and the room temperature of the place experiment. This project was achieved the expected outcome to become stronger than the normal panel in market. Table 3 show the advantage of panel board using coconut fibre and recycled plastic compared to wooden panel board. The coconut fibre panel board able to bear huge load in certain range which is larger than the conventional panel could do because high content of lignin in coconut fibre. Moreover, it is also long-lasting and has low absorption of water since plastic are not permeable to water. Lastly, this panel board has its own aesthetic value.

Since, the PET plastic were used as the outer lining so it has colourful design and give the aesthetic value as the cover most similar to mosaic. Other than that, this panel is also slightly rough but not harmful to the user. The temperature of the hydraulic type hot or cold machine should be set up to be higher than 170 °C to make sure the PET plastic will bonding together with the panel board as the cover layer. Other than that, the plastic also should be undergoes the layering process to produce the smooth of the outer layer and fully melt to gain more aesthetic value. To overcome the glue to be spot at one point on the panel board in the mixing process, spray the urea glue used the air compressor on the coconut fibre to make sure the urea glue spread evenly to all part of coconut fibre.

Table 3. Advantage of panel board using coconut fibre and recycled plastic compared to wooden panel board.

Wooden panel board	Panel board using coconut fibre and recycled plastic
Bear a huge load in certain range is small	Bear a huge load in certain range is large
Not long-lasting	Long-lasting
High absorption of water	Low absorption of water
No aesthetic value	Has aesthetic value
Smooth surface	Slightly rough surface

4 Conclusions

The objective of the study was achieved which is the reason for the coconut fibre as the filler for the panel board and PET plastic as the cover is because its own strength and weakness which can be improvised. Based on the criteria to get a good panel board, test on the strength and durability of the panel can be derived. Thus, the panel board which has an aesthetic value can replace the market of wooden panel board and solve the issue about water pollution in the country. The application of the project is can be replace the medium fibre board in the market according to the size and finding such as closet, cabinet, table and the relevant product to the medium board. The product of the study also can be commercial with the low cost and more advantages than the conventional panel board. There are possibility of this coconut fibre panel board has lower cost than the normal panel board since it is using the recycled waste product and do not need any chemical treatment for processing raw materials.

The authors would like to express our deepest gratitude to all parties who have contributed to this research, especially to Short Term Grant (STG) Vote U387, Office for Research, Innovation, Commercialization and Consultancy Management, (ORICC), Center of Diploma Studies (CeDS) UTHM, Zi Yu Recycle Resources Industrial Sdn. Bhd., laboratory staff of Fakulti Awam dan Alam Sekitar (FKAAS) and Fakulti Mekanikal and Pembuatan (FKMP).

References

1. S. Sengupta Basu, G. Chakraborty, R.C.J. Thampi, Indian J. Fibre Text. **39**,14 (2014)
2. M. Ali, *Second International Conference on Sustainable Construction Materials and Technologies*, **3**, 3, 1441 (2010)
3. H.P.S. Abdul Khalil, M. Siti Alwani, K. Mohd Omar, BioResources **1**, 2, 220 (2006)
4. I.Z. Bujang, M.K. Awang, A.E. Ismail, *Regional Conference on Engineering Mathematics, Mechanics, Manufacturing & Architecture (EM*ARC)* (2007)
5. Wikipedia. Polyethylene terephthalate (2017)
6. *Disposable Products Manufacturing Handbook*, (NIIR Project Consultancy Services, Delhi, 2014)
7. L.N. Ji, Appl. Mech. Mater. **312,** 406 (2013)
8. F. Shady, R.K. Konda, B. Arijit, Abraham, Nature **180**, 4577, 141 (1957)
9. U.K. Thiele, *Polyester Bottle Resins, Production, Processing, Properties and Recycling*, (Heidelberg, Germany, 2007)

Preliminary Investigation of Acoustical Properties of Concrete Containing Oil Palm Shell as an Aggregate Replacement

J. Zanariah[1,*] *H.* Zaiton[1], *Y.* Musli Nizam[2], *Y.* Khairulzan[1], *M.* Dianah[1], *D.* Nadirah[1], and *O.* Mohd Hanifi[3]

[1]Faculty of Civil Engineering, Universiti Teknologi Malaysia, 81310 Johor Bahru, Johor, Malaysia
[2]Faculty of Mechanical and Manufacturing Engineering, Universiti Tun Hussien Onn, 86400 Parit Raja, Batu Pahat, Johor, Malaysia
[3]Faculty of Civil and Environmental Engineering, Universiti Tun Hussien Onn, 86400 Parit Raja, Batu Pahat, Johor, Malaysia

Abstract. Research has been so far focused extensively on mechanical properties of oil palm shell (OPS) concrete but less on sound properties. Thus, the objective of this study is to investigate whether concrete containing OPS can be applied in the field of road noise barrier. The acoustic properties of the samples were determined by using an impedance tube connected to a sound source. The noise reduction coefficient (NRC) and weighted sound absorption coefficient (α_w) which is more commonly use in the road traffic noise barrier field were calculated according to BS EN ISO 11654:1997. Compressive strengths of samples were also determined by using compressive test. The results presented that the compressive strength of the OPS composites decreased as increased in w/c wit minimum of 20.44 N/mm^2 at 28 days for w/c = 0.6 but still satisfactory for structural use. The sound absorption coefficient demonstrated that they were decreased as the w/c are higher with typical curve of two peaks at 315Hz and 1000Hz. All samples were then can be classified as class E as 0.5< α_w < 0.25 and should be classified as L due to favourable deviation higher than 0.25 for 250 Hz.

1 Introduction

Palm oil has made impressive and sustained growth in the global market, the average annual production of palm oil in Malaysia will reach 15.4 million tonnes. Thus large amount of wastes from palm mill such as oil palm shell has been generated. Based on the fundamental principles of waste management, the waste generated from palm oil industry should be minimise, recycle the waste or reuse as alternative. For this reason, the need for seeking new application in other industries where the oil palms shells can be recycled.

Another environmental pollution, which is becoming an important problem to human, is noise pollution. Exposure to high intensity noise would lead to hearing loss [1]. Exposure

* Corresponding author: riah_yaya@yahoo.com

to lower intensity but long term such as transportation noise is generally bring the implication of various health problems such as stroke, heart disease, stress and sleeping disorder. There are many alternative mitigate noise generate from road such as by using noise barrier. Generally, noise barrier is functioning by obstructing the propagation path between the noise source and receiver. Concrete barrier is the most frequently used across Europe. Following this footstep, Malaysia has also increased the usage of concrete noise barrier in recent years.

The higher demand of concrete noise barrier will drastically reduce natural stone deposits. To tackle this problem, an alternative method or material such as recycled or waste materials, needs to be identified to minimize the decrease of natural aggregates. In the field of concrete research, the replacement of aggregate in concrete with waste is a popular topic. As a result, the oil palm shell (OPS), which are light and naturally sized was ideal for substituting aggregates in concrete construction. For example replacement of aggregate with oil palm shell (OPS) in concrete, which is suitable for lightweight concrete [2-8]. Studies by [8] found that oil palm shell was the organic aggregate which is better impact resistance compared to normal weight aggregate. Besides, OPS contain many pores and the water absorption is high compared to normal weight aggregate. Increasing voids in concrete would probably increase the porosity of the concrete [2]. Referring to AS 3600 minimum compressive strength for solid loadbearing block is 10 MPa and for solid non-loadbearing block is 3.0 Mpa.

Thus, the aims of this research are to investigate the acoustical properties oil palm shell, which can reduce traffic noise pollution levels and to recycle an industrial waste. The objective of this study is to investigate whether concrete with OPS is possible to use for acoustic purpose such as noise barrier.

2 Materials and methods

2.1 Collection and preparation of oil palm shell

The oil palm shell was obtained from oil palm mill located at Kulai Johor, Malaysia. They were freshly discarded and were irregularly shapes. On this study, most of the shells were within the thickness 0.15-0.5 mm and the surface texture of shell fairly smooth for concave faces. Pre-treatment is necessary to remove dust and other impurities before used in concrete. In this research, the shells were soaked in hot salty water for about 10 min to remove all the impurities and to avoid microbial deterioration. Then the OPS were sun-dried until there is no water in shell. After that, these dried shells were sieved to remove the fibres and all the shell sizes range 5mm to 10mm were collected and kept in dry room condition. Fig. 1 shows the OPS after all impurities were removed.

Due to the high water absorption of OPS, pre-soaking of OPS is necessary. The shell was soaked in water for 24 h, and then air dried in the laboratory for 48 h before use to ensure the shell in the saturated surface dry condition.

2.2 Preparation of test specimen

The proportion of OPS concrete and control concrete mixes were obtained from the trial mix by using the design of experiment (DOE) method to design the concrete with grade 25. The oil palm shell (OPS) concrete samples using different water-cement ratio of 0.4, 0.5 and 0.6 as a matrix and were replace 50% of coarse aggregate with OPS. The specimens were combination of cement, water, fine and coarse aggregate and oil palm shell. Cement, aggregate and OPS was placed in the mixer and were mix until a homogenous mixture was achieved.

Fig. 1. OPS sizes 5 – 10 mm.

Then, water was added to the mixture and re-mixed until a homogenous paste was obtained. When the mixing was completed, the slump test was carried out to measure the workability of concrete. The workability of concrete is important to describe the ease of flow in a simple index, which can reduce the void from destroying the stability of concrete [9]. All mixes of concrete should achieve a good workability, so that no segregation and bleeding occurs and improves the strength of concrete.

Then, the paste was placed in moulds of cubes (100x100x100mm), cylinders (98mm diameter) and cubes (50x50x50mm). The paste was compacted using vibrating table when the paste half-full in moulds and full in moulds. After 24 h, the concrete paste was taken out from the moulds and was cured in water.

2.3 Test methods

2.3.1 Compressive test

Concrete cubes (100 x 100 x 100 mm) were tested in compression at the end of 7, 14 and 28 days by using compression machine with maximum capacity 3000kN with control rate 3.0 kN/s. The averages of three samples were calculated to obtain the compressive strength value. Compressive strength can be calculated by using the formula:

$$C(s) = P/A \qquad (1)$$

where Cs = compressive strength; P = maximum applied load (kN); A = cross sectional area (mm).

Before that, the dry density of the sample needs to measures. The decrease of the density will affect the compressive strength of the OPS concrete. According [10] as the dry density of concrete increased, the percentage of porosity in concrete is exponentially dropped [10]. This follows the fundamental relationship between strength of solid and porosity [9] In this study, porosity of the OPS concrete was determined by using vacuum saturated method. The sample uses for this porosity test are sized 50x50x50 mm.

2.3.2 Impedance tube test

To determine the acoustic properties of the samples, the sound absorption coefficient at normal incidence was determined by the transfer-function method described in EN ISO 11654-2 [11]. Transfer-function method often called the two-microphone method and the advantages of using this approach is that it obtains the absorption coefficient and impedance of the surface for all frequencies [12]. Thus, impedance tubes were used to measure the sound absorption of materials. Fig. 2 shows the impedance tubes for measuring the sound transmission loss consist of three parts: an upstream tube with a loud speaker, a removable test sample holder and a downstream tube with a semi-anechoic termination. The diameter of the tube used in this study is 100mm and reliable frequency range is 350Hz to 1500Hz for low frequency. The frequency range corresponds to diameter of the tube. For high frequency, the tube with smaller diameter is used.

Basically, when the sound wave strikes a material that inserted in tube, a portion of sound wave is absorbed and other portion is reflected back. As a result, the ratio of absorbed energy to total incident energy can be determined and it called as sound absorption coefficient (α). In this study, three samples for each composition have been measured to ensure the value obtained more accurate.

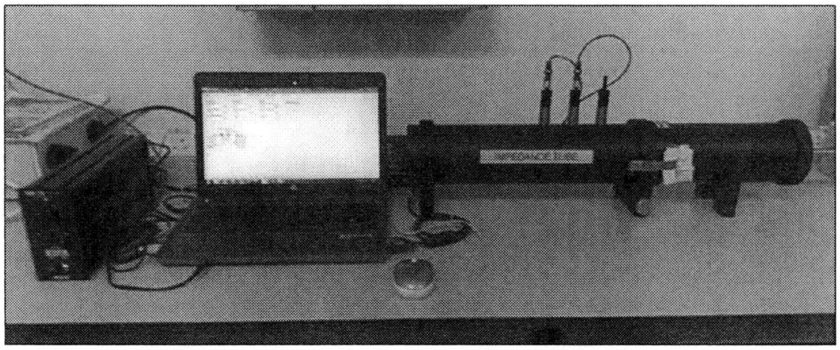

Fig. 2. Impedance tube equipment.

2.3.3 Data analysis

The noise reduction coefficient (NRC) was determined by calculating the arithmetic mean of the absorption coefficient at 250, 500, 1000 and 2000 Hz (Equation 2), in order to have single numeric data of the sound absorption coefficient to make comparisons.

$$NRC = \frac{\alpha_{250} + \alpha_{500} + \alpha_{1000} + \alpha_{2000}}{4} \quad (2)$$

However, for road traffic noise barrier field, the weighted sound absorption coefficient (α_w) is commonly used [12] and was calculated by referring [11]. A weighted sound coefficient (α_w) is a result from comparison between the sound absorption coefficient values at standard frequencies and reference curve in accordance with ISO 11654. The α_w also used to classify the sound absorption materials based on the measured absorption curves to categories from A to E.

3 Results and discussions

3.1 Workability

In this study slump test was used to determine the workability of concrete. The results obtained from slump test are presented in Table 1. Data presented in table reveals that workability gradually increased with increase water cement ratio. Water cement ratio has much effect in workability. Workability is directly proportion to water cement ratio.

 https://doi.org/10.1051/e3sconf/20183401015

Table 1. Slump test of OPS concrete.

Water/cement ratio	0.4	0.5	0.6
50 % OPS	20mm	25mm	35mm

The measured slump values for concrete with water cement ratio 0.5 and 0.6 were 25mm and 35mm, respectively as shown in Table. This is due to the increase the specific surface as a results increase in the quantity of OPS, requiring more water to make the mix workable. Besides, according to [4], the workability of fresh concrete influenced significantly by physical properties such as roughness, shape and texture of aggregate. The smooth surface OPS may lead to a better workability. In this study, the OPS have varies irregular shape and porous texture. The shape and surface texture of OPS control the void content and affect the water requirement of concrete. However, all the measured slumps were true slumps. Higher slump indicate higher workability of fresh concrete while lower slump indicate low workability.

3.2 Compressive strength

The trends of compressive strength for each sample of the same percentage of OPS with different water cement ratio increased as the curing period increased as shown in Fig. 3. The OPS concrete with water/cement ratio 0.4 attained the highest compressive strength with 38.71 MPa within 28 days. The mix was dry and more efforts of compaction were needed during the concrete casting.

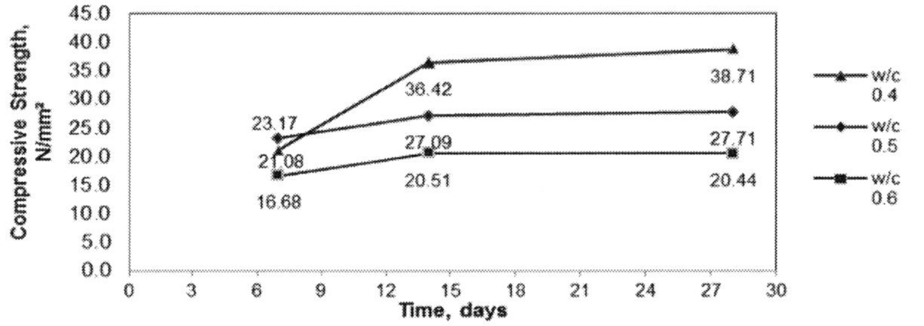

Fig. 3. Compressive strength of OPS concrete.

In contrast, the oil palm shell with highest water cement ratio has the lowest compressive strength. The excessive amount of water in the concrete mix will evaporate and thus causing void development in concrete. Based on Fig. 4 the dry density of OPS concrete decreased as the water cement ratio increased. Besides that, Fig.5 shows the porosity of the OPS concrete increased as the water cement ratio increased. That means, the void ratio in oil palm shell concrete samples reduce the strength of concrete and makes the concrete porous. Hence, the increasing w/c ratio in concrete, the samples become more porous and lighter. The compressive strength for water cement ratio 0.6 was found to be lower than the design concrete. Therefore, the most porous materials had the weakest compressive strength.

 https://doi.org/10.1051/e3sconf/20183401015

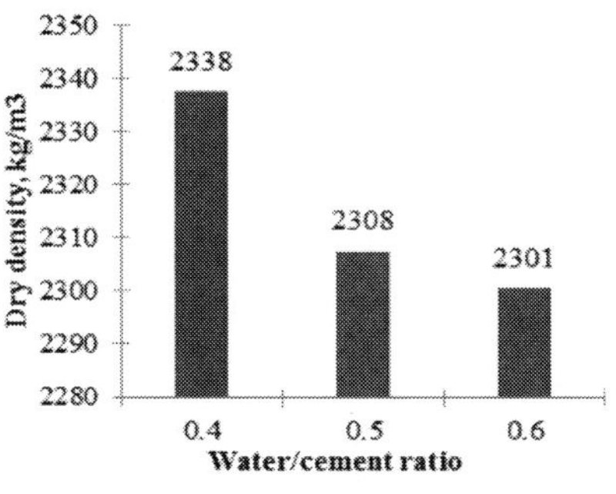

Fig. 4. Dry density versus water cement ratio.

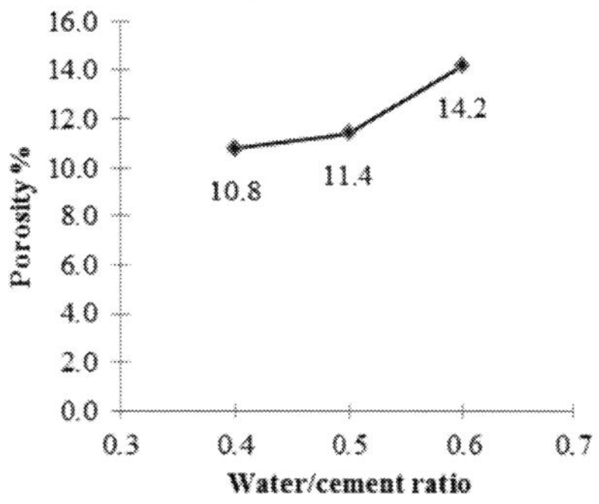

Fig. 5. Porosity of OPS concrete.

3.3 Sound absorption coefficient

The average the acoustic experiment results from the three samples for each w/c were plotted. Fig. 6 shows the sound absorption curves for 50% OPS replacement in concrete with different water cement ratio of 0.4, 0.5 and 0.6, represented in one-third octave bands. The sound absorption coefficient curves for all three different water cement ratio reach peak absorption at different frequency. The samples with water cement ratio 0.4 and 50% of OPS have the maximum sound absorption coefficient of 0.49 at 315 Hz (at 1st peak) and 0.26 at 1000 Hz (2nd peak). While the samples of w/c ratio 0.6 and 50% OPS has the lowest sound absorption coefficient at frequency 315 and 1000 Hz, respectively for 1st and 2nd peak. Due to the same amount of OPS content, the density of the sample depends on the water cement ratio. Thus, water cement ratio shows the great influence on the sound absorption coefficient

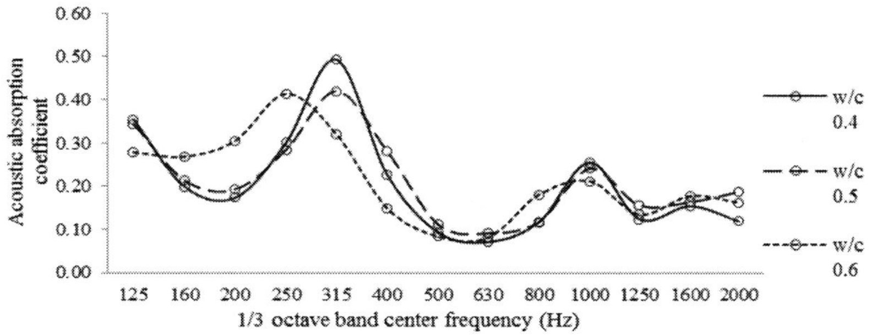

Fig. 6. Relationship between water cement ratio and sound absorption (50% OPS).

3.4 Noise reduction coefficient

Noise reduction coefficient (NRC) represents a single number, which is the average value of the absorption coefficients of the samples at frequencies 250, 500, 1000 and 2000 Hz. The NRC is the percentage of sound that a surface absorbs. Fig. 7 shows the calculated NRC for the three different water cement ratio with the same amount of OPS. It can be observed from this figure that NRC increased with increasing values of water cement ratio. The highest sound absorption was obtained at water cement ratio 0.6 with the value 0.22 but all samples have NRC of 0.2. This result occurs because the higher porosity (14.2%) compared to the other w/c as shown in Fig. 5.

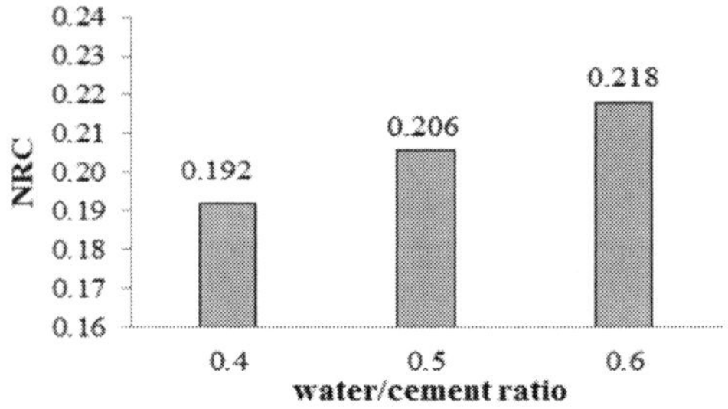

Fig. 7. Variation of NRC with water/cement ratio.

3.5 Weighted sound absorption coefficient

Table 2 shows the calculated value of practical sound absorption coefficient, α_p for all samples and the values of α_w. The α_p is plotted against the frequencies and compared to a reference curve. The weighted sound absorption coefficient α_w is the defined as the value of the reference curve at 500 Hz.

Table 2. α_w for different water cement ratio.

Absorption Frequency [Hz]	w/c = 0.4			w/c = 0.5			w/c = 0.6		
	α_s 3rd	α_p Octave	α_w	α_s 3rd	α_p Octave	α_w	α_s 3rd	α_p Octave	α_w
200	0.17			0.19			0.31		
250	0.30	0.30		0.28	0.30		0.41	0.40	
315	0.49			0.42			0.32		
400	0.23			0.28			0.15		
500	0.09	0.10	0.2 (Class E)	0.11	0.20	0.2 (Class E)	0.08	0.10	0.2 (Class E)
630	0.07			0.09			0.08		
800	0.12			0.12			0.18		
1000	0.26	0.20		0.24	0.20		0.21	0.20	
1250	0.12			0.16			0.14		

All samples were then can be classified as class E as 0.15< α_w<0.25 and should be classified as L due to favourable deviation higher than 0.25 for 259 Hz.

3.6 Relationship between compressive strength and NRC

The effect of compressive strength and NRC also was investigated as a method to obtain the optimum mix design of OPS concrete. Fig. 8 shows the compressive strength of OPS concrete was decreased with increasing w/c ratio, while NRC was increased with increasing w/c ratio with the same replacement of OPS. It is found that the OPS concrete have better sound absorption due to more pores of concrete. However, the more pores in OPS concrete lead to the reduction of strength of concrete strength. Okpala (1990) [2] also found that the OPS concrete has better sound insulation with increasing w/c ratio for the same mix design. Based on Fig. 8, the optimum design was at w/c ratio 0.49.

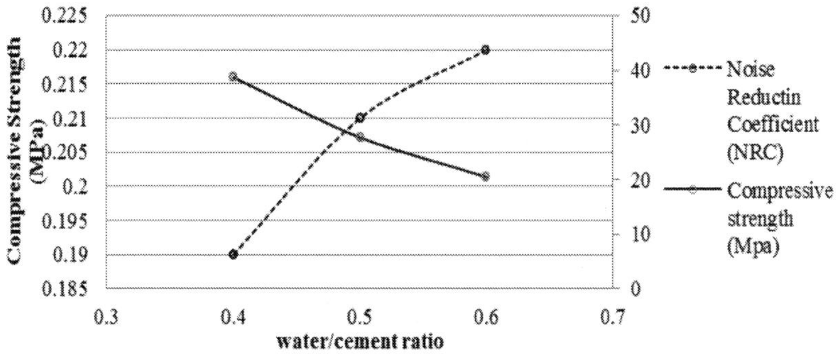

Fig. 8. Relationship w/c ratio with compressive strength and NRC values.

4 Conclusions

The 50% oil palm shell replacement as aggregate in concrete mixture with w/c 0.4 to 0.6 decreased the workability and strength of the concrete. Despite this, the compressive strength obtained was greater than minimum strength required by concrete block for noise barrier. Moreover, all samples can be classified E class according to BS ISO 11654 with w/c 0.6 demonstrated better sound absorption due to the higher porosity of the concrete. Therefore, the OPS concrete may be considered as new material for barrier in the future in view of reducing the problem related to the noise pollution and recycle waste.

References

1. S. Wachasunder, Int. J. Environ. Stud. **61(4)**, 459-470, (2004)
2. D. C Okpala, Build. Environ. **125**, 29—296, (1990)
3. H. Basri, M. Mannan, M. F. Zain, Cem. Concr. Res. **29**(4**)**, 619–622, (1999)
4. M. Mannan, C. Ganapathy, Constr. Build. Mater. **16(**1), 29–34, (2002)
5. D. C. L Teo, M. Mannan, V. J. Kurian, C. Ganapathy, Build. Environ. **42**(7), 2614–2621, (2007)
6. P. Shafigh, M.Z. Jumaat, H. Mahmud, U. J. Alengaram, Mater. Des. **32**(10), 4839–4843, (2011)
7. P. Shafigh, M.Z. Jumaat, H. Mahmud, N. Anjang, A. Hamid, Constr. Build. Mater. **27**(1), 252–258, (2012)
8. U.J. Alengaram, B.A.Al Muhit, M.Z. Jumaat, Constr. Build. Mater. **38**, 161–172, (2013)
9. P.M. Metha and P.J.M. Monteiro, Concrete Structure **3**, (2006)
10. E.P. Kearsley, Cem. Concr. Res. **32**, 233 – 239, (2002)
11. BS EN 1SO 11654:1997 Acoustics-Sound absorbers for use in buildings-Rating of sound absorption
12. J. Trevor Cox, P. D'Antonio, *Acoustic Absorbers and Diffusers – Theory, Design and Application* **2** (2004)
13. C.Arenas, L.F. Vilches, C. Leiva, B. Alonso-Fariñas, M. Rodríguez-Galán, Material De Construction **66** (2016)

E3S Web of Conferences **34**, 01016 (2018) https://doi.org/10.1051/e3sconf/20183401016
CENVIRON 2017

Strength Performance of Blended Ash Based Geopolymer Mortar

Zaidahtulakmal M. Zahib[1,*], *Kartini* Kamaruddin[1], and *Hamidah* M. Saman[1]

[1]Faculty of Civil Engineering, Universiti Teknologi Mara, 40450 Shah Alam Selangor, Malaysia

Abstract. Geopolymer is a based on inorganic alumino-silicate binder system. Geopolymeric materials are formed using materials that containing silica and aluminium such as fly ash and rice husk ash, which activated by alkaline solution. This paper presents the study on the effect of replacement of SSA in RHA based geopolymer, types of curing and different molarity of NaOH used on the strength of Sewage Sludge Ash (SSA) and Rice Husk Ash (RHA) based geopolymer mortar incorporating with three (3) different mix proportions. Based geopolymer mortar was synthesized from treated sewage sludge and rice husk undergoing incineration process in producing ashes, activated with sodium silicate and sodium hydroxide solution by ratio of 2.5:1 and solution to ash ratio of 1:1. Molarity of 8M and 10M NaOH were used. The percentages of SSA replacement were 0%, 10% and 20% by weight. Compressive strength was conducted at age 7, 14 and 28 days to see the development of strength with two curing regimes, which are air curing and oven curing (60°C for 24 hours). From the research conducted, the ultimate compressive strength (6.28MPa) was obtained at zero replacement of SSA taken at 28 days of oven curing with 10M of NaOH. This shows that RHA, which is rich in silica content is enough to enhance the strength of geopolymer mortar especially with high molarity of NaOH.

1 Introduction

Portland cement is a major construction material used worldwide. The crucial issue in manufacturing cement was the calcination of limestones. It is responsible to contribute on the amount of CO_2 emission to the environment. Manufacturing of 1 ton of Portland cement generates 1 ton of CO_2 emission to atmosphere [1]. This issue become critical by years because of the rapid construction development to fulfil the needs. Moreover, waste generated is also one of the crucial issues that have being concern nowadays. Rapid economic development and population growth, inadequate infrastructure and land scarcity was Malaysia's most critical issues in managing the municipal waste. As stated by [2], the per capita generation rate of municipal waste is about 0.5 to 0.8 kg/person/day, which domestic waste is the primary sources. Agriculture waste, industrial waste, domestic waste and construction waste were wastes that are trying to be developed by researchers and commercialize as construction materials in replacing cement.

* Corresponding author: zaidahtulakmal@gmail.com

Aluminosilicate polymer also known as geopolymer is aluminosilicate network structure that form between the reaction of solid amourphous aluminosilicate (Si-Al) with highly concentrated alkali hydroxide or silicate solution to activate reactive Si and Al [1, 3, 4]. This geopolymer consists of repeating tetrahedral unit (Si-O-Al-O). The sources of materials used in geopolymer production are any pozzolonic materials that containing Si and Al to react with alkaline activators. Mohd Mustafa *et al.* indicated that curing process and activator used were the most important factors that affect the strength performance of geopolymer [5]. Their statement is in line with the study conducted by [6] stated that molarity and curing temperature affected the strength of geopolymer. Commonly, the curing temperatures used in geopolymer were at range 60°C to 120°C for 24 to 90 hours [7-11].

Geopolymer technology is seen to have a potential in replacing conventional concrete and has good opportunity to convert waste into beneficial product in construction industry. Therefore, RHA and SSA are seen as potential waste to use as precursor geopolymer mortar since they have reactive silica and aluminium. With that in mind, the objective of this study is to investigate the effect of SSA replacement in RHA based geopolymer mortar with different molarity of NaOH under two curing regimes.

2 Experimental Programmes

2.1 Materials

Rice husk ash (RHA) and sewage sludge ash (SSA) were used as raw materials in geopolymer mortar. Table 1 presented the total chemical compositions of RHA and SSA used in this study respectively. From that table, it can be seen that these two materials can be combined in order to obtain more reactive silica and aluminium and thus enhanced the geopolymerisation process. Based on the chemical compositions in Table 1, contaminations of silica in RHA is higher than SSA with 74.13% and 39.21% respectively, while SSA gave aluminium content with 19.71% compared to RHA with 0.13%.

The rice husk was obtained from rice mill and was burnt in the ferrocement furnace and lefts for 24 hours under uncontrolled burning. The next 24 hours were used to make sure the ash cool down before collecting process take place and ashes were taken for grinding process using Los Angeles Machine. RHA was found to be amourphous based on the grey colour obtained, while sewage sludge cake was taken from Wastewater Treatment Plant, KLIA Sepang. Sewage sludge cake were dried before burning process takes place at 900°C for 5 hours by using furnace gas. Then, the burnt sewage sludge cake was taken out from the furnace and undergoing grinding process to obtain the fineness.

Sodium hydroxide and sodium silicate were used as alkaline activator in this research study. The ratio of sodium silicate to sodium hydroxide was 2.5:1 by weight. Local sand was used as fine aggregates having a fineness modulus of 3.

2.2 Experimental methods

In the production of the geopolymer mortar, alkaline activators were prepared 24 hours before the casting process takes place so that the solutions sufficiently combined. The activator solutions consist of sodium silicate solution (Na_2O = 10.6%, SiO_2 = 26.5% and H_2O = 62.9% by mass) and either 8M or 10M of sodium hydroxide. The mixes were prepared by using a Hobart mixer and cast in 50mm cubes mould and mechanically compacted using vibrating table. The ratio of RHA to sand was 1:2 and RHA was replaced by 0%, 10% and 20% of SSA by weight. Additional water was added during mixing to make sure the homogeneity of the mortar. The samples were sealed using wrapping plastic before subjected to oven curing at 60°C for 24 hours, while the other batch of samples was exposed to air curing. Compressive strength test was conducted at the age of 7, 14 and 28 days.

Table 1. Total chemical compositions of the used RHA and SSA.

Chemical Composition	RHA	SSA
SiO_2	74.13	39.21
Al_2O_3	0.13	19.71
Fe_2O_3	0.32	6.59
CaO	0.65	2.30
MgO	0.66	1.25
SO_3	0.53	2.63
K_2O	2.87	2.27
Na_2O	0.03	0.24

3 Experimental Results and Discussion

3.1 Effect of SSA replacement in strength of RHA based geopolymer mortar

The strengths of different replacement of RHA and SSA geopolymer mortar cured at 60°C for 24 hours with 10M of NaOH were illustrated in Fig. 1. For 100% of RHA based geopolymer mortar at the age 7, 14 and 28 days, the strengths were 2.70, 5.70 and 6.28 MPa respectively. However, with 10% replacement of SSA, it displays strength of 1.93, 3.43 and 3.24 MPa, while the strength for 20% SSA replacement was 2.10, 3.54 and 3.98 MPa at ages 7, 14 and 28 days respectively. It is clearly showed that there are significant increases in strength for 100% RHA, while with replacement 10% and 20% of SSA exhibited decreasing in strength and the strength increased linearly due to age of curing. RHA is rich in silica contains as compared to SSA up to 74.13% as in Table 1. In geopolymerisation, silica content is one of the crucial factors that affect the creation of Si-Al network. Meanwhile, SSA with high in alumina contains compare to RHA is one of the potential waste in combination with RHA (high in silica) as based in geopolymer since it has a quality in strength.

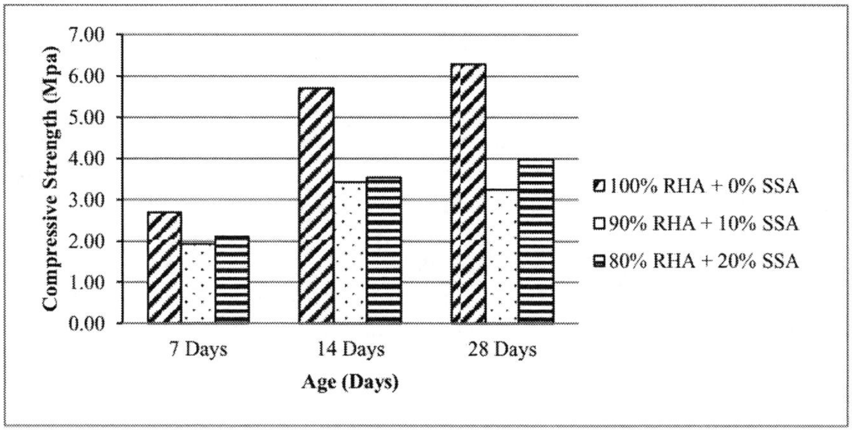

Fig. 1. Compressive strength of RHA and SSA based geopolymer mortar with 10M NaOH and oven cured.

3.2 Effect molarity of NaOH on the compressive strength of geopolymer mortar

The effect of molarity on the RHA and SSA based geopolymer mortar are discussed. Fig. 2 illustrated the compressive strength of different SSA replacement with different molarity of NaOH at age 28 days of air curing conditions. Each SSA replacement showed that 10M of NaOH gave high compressive strength as compared to 8M with 5.8MPa (100%RHA 0%SSA), 4.5MPa (90%RHA 10%SSA) and 3.5MPa (80%RHA 20%SSA), while 4.8MPa (100%RHA 0%SSA), 3.42MPa (90%RHA 10%SSA) and 3.20MPa (80%RHA 20%SSA) respectively. The results of this study were in line with [13], stated that higher concentration of NaOH solution resulted in higher compressive strength. It is suggested that by increasing the molarity of NaOH, the strength of geopolymer mortar increase with increasing age of curing. Nevertheless, too high molarity of NaOH solution and too high temperature induce in curing process could lower the strength because of the break of Si-Al bonding. It is suggested that the strength decreasing with higher SSA replacement in RHA based geopolymer mortar however there are an increment due to age of curing for each replacement.

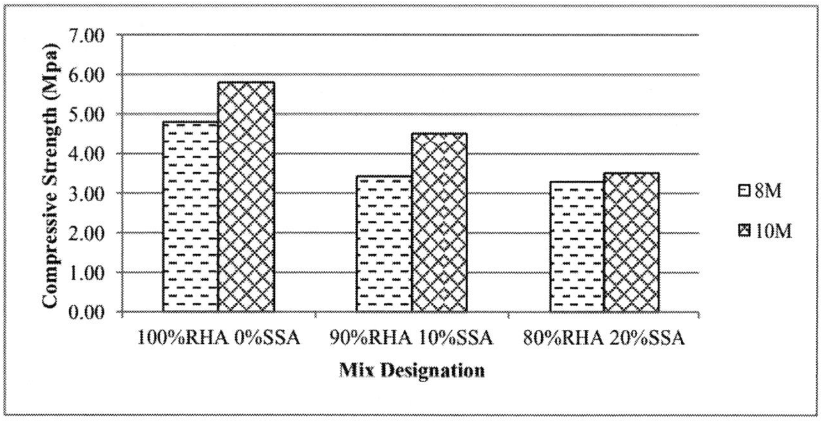

Fig. 2. Variation of compressive strength at age 28 days of air curing.

3.3 Effect of curing condition on the compressive strength of geopolymer mortar

The strength of RHA geopolymer mortar cured at 60°C for 24 hours and air curing with molarity of 8M and 10M of NaOH were illustrated in Fig. 3. For 8M NaOH at the age of 7, 14 and 28 days, the strengths were 2.00MPa, 4.53MPa and 4.80MPa (air cured) and 2.54MPa, 5.12MPa and 5.72MPa (oven cured) respectively. For molarity of 10M NaOH, the strengths were 2.19MPa, 4.81MPa and 5.80MPa (air cured) and 2.70MPa, 5.70MPa and 6.28MPa (oven cured). It is clearly showed that there are linear increments on the strength of RHA based geopolymer mortar on age and method of curing. On the other hand, with higher molarity, the compressive strength increased.

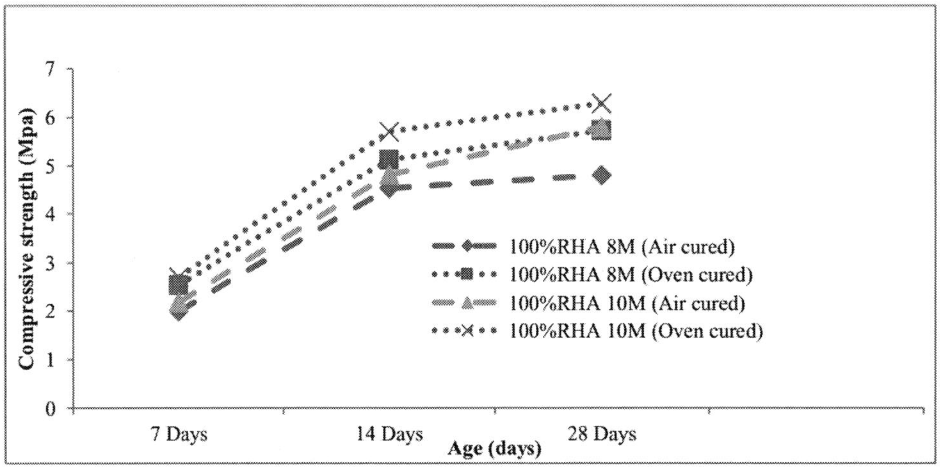

Fig. 3. Compressive strength of 100%RHA 0%SSA based geopolymer mortar.

From the Fig. 3, it can be clearly seen that the specimens cured at 60°C for 24 hours exhibited higher compressive strength as compared to air cured specimens with a relative difference of 16%. It can be seen, by inducing heat to the specimens during curing process, polymerisation effectively could gain strength at early age as low as at 7 days [12]. For the specimens cured at oven cured with 10M NaOH shows significant increase in strength than 8M NaOH along the age of curing process with 6.28MPa and 5.72MPa taken at 28 days respectively. While the specimens under air cured was 5.80MPa (10M) and 4.80MPa (8M) respectively. This suggested that increasing in molarity of NaOH with inducing heat could increase the polymerisation rate reaction in RHA based geopolymer mortar. These two parameters gave significant effect on the strength of geoplymer mortar which is giving high rate of polymerisation reaction and it is in line with finding by [14], in which their suggested that increasing in molarity and heat increase the strength of geopolymer mortar.

Fig. 4 illustrated the compressive strength of 10% replacement of SSA by weight in RHA based geopolymer mortar with respect to ages and two curing regimes. By replacing with 10% of SSA, the specimens' exhibited lower strength as compared to 0% replacement of SSA, however there was a linear increased in strength up to age 28 days. Based on the Fig. 4, specimens cured in air curing with 10M of NaOH gave higher strength than specimens cured at oven with 4.5MPa and 3.4MPa respectively at 28 days. By replacing RHA with SSA it is can be seen that air curing gave high strength than oven curing.

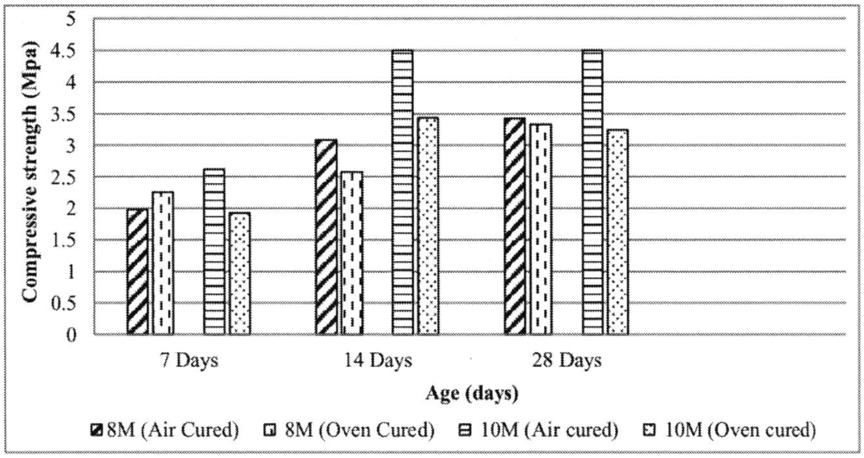

Fig. 4. Compressive strength of 10% SSA with RHA based geopolymer mortar.

4 Conclusions

From the study conducted, the following conclusions were made:

a. The influences of SSA replacement in RHA based geopolymer showed decreasing in value of compressive strength. . However, it is clearly showed that there still an incremental value of strength with prolong periods of curing.
b. The compressive strength of RHA based geopolymer mortar gave higher strength than SSA and RHA based geopolymer.
c. The compressive strength of the geopolymer increased with increasing in the molarity of NaOH solutions up to 10M.
d. Inducing heat through oven curing on the geopolymer mortar plays a major role for the strength development of geopolymer mortar.

References

1. J. Davidovits, Journal of Thermal Analysis **37**, 1633-1656 (1999)
2. L.A. Manaf, M.A. Samah, N.I. Zukki. Waste Management **29**, 2902-2906 (2009)
3. P. Duxson, J.L. Provis, G.C. Lukey, J.S.J. van Deventer, Cement and Concrete Research **37(12)**, 1590-1597 (2007)
4. D. Bondar, C.J. Lynsdale, N.B. Milestone, N. Hassani, A.A. Ramezanianpour, Engineering Properties of Alkali Activated Natural Pozzolan Concrete (2010)
5. A. Mohd Mustafa, H. Mohammed, H. Kamarudin, I. Khairul Niza, Y. Zarina. Journal of Engineering and Technology Research **3** (1), 1-4 (2011)
6. P. Kamhangrittirong. *33rd Conference on Our World in Concrete & Structures* (2008)
7. M.B. Satpute, M.R. Wakchaure, S.V. Patankar. International Journal of Engineering and Innovative Technology (IJEIT) **1** (5), 152-155 (2012)
8. A.A. Patil, H.S. Chore, P.A. Dode, Advances in Concrete Construction **2** (1), 29-37 (2014)
9. A.M. Fareed, N. Muhd Fadhil, D. Samuel, S. Nasir. World Academy of Science, Engineering and Technology **5**, 571-574 (2011)
10. N. Ranjbar, M. Mehrali, U.J. Alengaram, H.S.C. Metselaar, J. Muhd Zamin Construction and Building Materials **65**, 114–121 (2014)
11. X. Guo, H. Shi, W.A. Dick. Cement Concrete Comp **32** (2), 142-147 (2010)

12. M.A. Salih, N. Farzadnia, A.A. Abang Ali, R. Demiboga. Constr Build Mater **94**, 116-125 (2015)
13. C.K. Madheswaran, G. Gnanasundar, N. Gopalakrshnan. Civil and Structural Engineering **4** (2), (2013)
14. A. Palomo, M.W. Grutzeck, M.T. Blanco. Cement Concrete Res **29,** 1323-1329 (1999)

Microstructural aspects in steel fiber reinforced acrylic emulsion polymer modified concrete

Dayang *Hazimmah*[1*], Afizah *Ayob*[2], Lau *Sie Yee*[1], and Wong *Chee Cung*[1]

[1]School of Engineering and Technology, University College of Technology Sarawak, 868 Persiaran Brooke, 96000 Sibu, Sarawak
[2] Programme of Civil Engineering, School of Environmental Engineering, Universiti Malaysia Perlis, 02600 Arau, Perlis, Malaysia

Abstract. Scanning electron microscope observations of polymer-free and polymer-modified cements have shown that the polymer particles are partitioned between the inside of hydrates and the surface of anhydrous cement grains. For optimum dosage of acrylic emulsion polymer with 2.5%, the C-S-H gel in this structure is finer and more acicular. Some polymer adheres or deposit on the surface of the C-S-H gel. The presence of acrylic emulsion polymer confines the ionic diffusion so that the $Ca(OH)_2$ crystallized locally to form fine crystals. The void in the structures seems to be smaller but no polymer films appears to be bridging the walls of pores although many polymer bonds or C-S-H spread into the pore spaces. In addition to porosity reduction, acrylic emulsion polymer modified the hydration products in the steel fiber –matrix ITZ. The hydration product C-S-H appeared as a needle like shape. The needle-shaped C-S-H increases and gradually formed the gel, with needles growing into the pore space. The phenomenon is more obvious as curing age increased.

1 Introduction

Polymer modified cementitious mixtures (PMC) have been called by various names such as polymer Portland cement concrete (PPCC) and latex modified concrete (LMC) [1]. Polymer modified cementitious mixtures includes polymer modified cementitious slurry, mortar and concrete. The improvement from adding polymer modifier to concrete includes increased bond strength, freezing and thawing resistance, abrasion resistance, flexural and tensile strength and reduced permeability and elastic modulus [1]. A reduced elastic modulus will result in reducing the stresses developed due to differential shrinkage and thermal strain that would reduce the tendency of the material to crack. PMC can also have increased resistance to penetration by water and dissolve salts and reduced need for

* Corresponding author: siti.hazimmah@ucts.edu.my

sustained moist curing. The improvements are measurably reduced when PMC is tested in the wet state [2].

Fiber reinforced concrete (FRC) is an alternative building material as it can strengthen the crack resistance and the toughness of the concrete effectively, and it is capable of reinforcing the impermeability and effectively eliminating the stress concentration of concrete as well [3]. Whereas, with the progression of the research and application, the existing issues of FRC are equally striking. Inside FRC, especially in the interface transition zone (ITZ) between fiber and cement substrate, there tend to be a large quantity if harmful pores, giving rise to the weak bonding among cement hydrates, fibers and aggregates [14]. Herein, the role of the fiber is deteriorated in enhancement [4, 5].

Fortunately, with the polymeric materials science and in-depth insights of the correlations and causations between material structures and properties progressing, large quantities of excellent properties possessed by polymer, such as waterproof, filling, flocculation and thickening effects were discovered. It is to be noted that polymer has been already implemented or gradually applied to concrete field currently [6, 7]. With the purpose of making concrete capable of meeting the requirements of structural applications, the inclusion of polymer in concrete has become one of the focused research topics in engineering circles [8].

Different amounts of polymer addition are required to gain various improved properties of mortar or concrete. As far as mechanical properties or impermeability is concerned, a polymer addition of 3% – 20% is usually needed. With less than 3% of polymer addition, the mechanical properties of impermeability hardly change. Impermeability can only be improved when the polymer forms a semi-continuous film in the matrix of hardened cement hydrates. Hence, a minimum polymer addition is required to ensure film formation, which is usually larger than 3% of p/c ratio [9]. After the film formation of the polymer and the hardening of the mortar, a gradient polymer distribution is generated from the surface to the interior region of the mortar. A significantly condensed surface is obtained *in situ*, which provides a radically effective protective layer. These research objectives are to study the effect of acrylic emulsion polymer and steel fiber addition on mechanical properties and thermal properties of steel fiber reinforced acrylic emulsion polymer modified concrete (SFRPMC).

2 Experimental

The process of mixing and casting of the composites were carried out following [10]. The amount of acrylic emulsion polymer was kept ready in suitable container. The quantity of water was then measured with a measuring cylinder and kept in suitable container. All polymers were mixed with half of water that measured into another container and put into the pan mixer. The mixing time was let to be 1 minute to 2 minutes.

Then the amount of cement, silica fume and aggregates was kept ready in suitable container. Sand was fed into the pan mixer, a part of the coarse aggregates, cement, silica fume and the materials were mixed for 2 minutes. The other half of water with some of the remainder of the coarse aggregates was fed so as to break up any modules of mortar. After all ingredients were added, the mixing time was set to be 3 to 5 minutes till homogeneous obtained. The calculated steel fibers were measured and kept in suitable container. The steel fibers with 60 mm length sized hooked end were added in percentage by weight (of total wet solid). The fibers were added in small increments by sprinkling them onto the surface of the mix until all the fibers absorbed into the matrix. This technique was performed to prevent 'balling' or 'interlocking' of the fibers.

From each experiments of steel fiber reinforced polymer modified cement concrete, there were consisted of three cylinders (150x300 mm), three cubes (100 mm) and three beams (100x100x500 mm). After conducting slump test, the concrete mix is placed inside the prepared molds. Then, all specimens were compacted by means of a vibrating table. After that, the specimens were demolded after 24 hours and subjected to water curing until the testing date.

With the same percentages of mix proportions in different curing aging of 1 hour, 24 hours and 168 hours, a few small specimens of steel fiber reinforced acrylic emulsion polymer modified concrete with 1 mix of basic concrete were also tested for microstructural performance by using Scanning Electron Microscopy (SEM) with Energy Dispersive X-ray Analysis (EDX). In this experiment, for each formulation 6 specimens were fabricated for each test. The micrographs of the specimens of were recorded using SEM (Hitachi with S-3400N) and ZEISS SUPRA 55VP connected to an analysis system with EDX.

3 Results and discussions

Morphology examination by using SEM has been conducted on composite system that had been produced. The microstructures of tested specimens are shown in Fig. 1 to 3. It can be seen that due to the presence of polymer film in all the specimens, the bond among steel fibers, aggregates and cement is greatly improved. Obviously, the porosity and pore size distribution are influenced by the corporation of polymer. The overall porosity increases with the increasing dosage of polymer [11].

The SEM image of polymer modification cement shows a quite different morphology. It can be seen the fracture surface has more fibrillar outgrowths, or more recticular networks. Fig. 1 shows well defined hydrates with some C-S-H gel and the $Ca(OH)_2$ crystals in the layer form, some C-S-H gel in the fiber form. It seems that they spread into the pore space, but do not completely fill the space and leave large voids.

Fig. 1. Microstructure on mix 26 (1.0% acrylic emulsion polymer, 8.0% silica fume, 1.5% steel fiber and 0.50 w/c ratio).

Fig. 2 shows the gross structure of polymer-cement paste that can be seen and the C-S-H gel in this structure is finer and more acicular. It seems that some polymer adheres or deposit on the surfaces of the C-S-H gel. Needles of C-S-H appear to spread out from a cluster of hydrates and interweave together forming a crossing network. An additional observation of these figures is the morphology and dimension of $Ca(OH)_2$ crystals is shown in Fig. 2. A group of fine $Ca(OH)_2$ crystal are observed in the limited areas. The presence of polymer confines the ionic diffusion so that the $Ca(OH)_2$ crystallized locally to form fine crystals. The void in the structures seems to be smaller, but no polymer films appears to be bridging the walls of the pores although many polymer bonds or C-S-H gel spread into the pore space. Due to the presence of polymer films in specimens, the bond among steel fibers, aggregates and cement is greatly improved [11] and leads to the highest strength for mix 43.

Fig. 2. Microstructure on mix 43 specimen (2.5% acrylic emulsion polymer, 8.0 silica fume, 1.5% steel fiber and 0.50 w/c ratio).

As shown in Fig. 3, the concrete constituents are not compactly joined leads to lower flexural strengths. The acrylic emulsion polymer cement paste (Fig. 3) displays a fishing-net or honeycomb structure. There are ettringite or big pores are present. Furthermore, the polymer films in Fig. 3 are thicker and more coherent than Figure 2. Thus steel fibers can be pulled out more easily in steel fiber reinforced acrylic emulsion polymer modified concrete-21 (mix 63), and flexural strengths then become lower.

Fig. 3. Microstructure on mix 63 (4.0% acrylic emulsion polymer polymer, 8% silica fume, 1.5% steel fiber and 0.50 w/c ratio).

Fig. 4 and Fig. 5 show typical SEM microstructures through the combination of several photos, and some inkling of transition zone (ITZ) can be discovered. The results for various ages from SEM are explained as follows [12].

https://doi.org/10.1051/e3sconf/20183401017

Fig. 4. Spot 1 for mix 43 (optimum) at 1 day curing with three different spots.

 https://doi.org/10.1051/e3sconf/20183401017

Fig. 5. Spot 1 for mix 43 (optimum) at 7 days curing with three different spots.

Morphology examination by using SEM has been conducted on composite system that had been produced. At the edge of the aggregate, there are much unhydrated cement particles and products of initial hydration. From SEM inspection, a few C-S-H gels can be seen, but the hydration products of C-S-H increases quickly with the increase in curing age.

At age of 7 days, the unhydrated cement particles and products at the age of 1 day are transformed into C-S-H at the age of 7 days. The needle-shaped C-S-H increased and gradually formed the gel, with needles growing into the pore space just like mesh. At this age, silica fume particles still does not react, but among the hydration products, pore existed.

At the edge of the aggregate, there will still much hydration products caused by the pozzolanic reaction. Due to the mutual stack of these hydration products, it contained plenty of pores. Beyond this range, it can be found that the previous pores are now filled with the hydration products of pozzolanic reaction, making the previous mesh structure turn into plain structure. This phenomenon is more obvious as curing age increased. Moreover, from SEM microstructures, the porosity decreased as the distance from the aggregate edge increased. From the observation silica fume particles kept on breaking and dropping off, and needle-shaped C-S-H is growing in it, in addition of acrylic emulsion polymer into the concrete.

The addition of emulsion polymer causes a plasticizing effect on the cement paste to some extent [13]. Along with cement hydration, the agglomeration of emulsion polymer particles and film formation absorbed on cement occur, which effectively changes the microstructure by compacting the interfacial zones among phases. Finally, a structure of dense interpenetrating network is formed, involving cement hydrates, fine aggregates, and emulsion polymer membranes. From the microstructural analysis of composite, it is observed that using acrylic emulsion polymer modification, through application of high performance 8% of silica fume, caused a reduction in the porosity of aggregate-matrix ITZ region, compared to unmodified concrete.

4 Conclusions

In addition to porosity reduction, it is observed that polymer modified the hydration products in the steel fiber-matrix ITZ. The modification is simpler in concrete with richer cement matrix that is concrete elaborated with silica fume and steel fiber reinforcement improve the bonding strength and this influences directly on mechanical properties of SFRPMC specimens. In addition, the acrylic emulsion polymer can form continuous polymer films inside the concrete, which boosted toughness and compact degree of interface transition zone. The acrylic emulsion polymer could trigger steel fiber and the cement paste to a tight mutual connection. It as well alleviated the macro and micro defect of interface transition zone between fiber and cement paste, strengthening the role of fiber regarding toughness and crack resistance properties. For the SFRC with acrylic emulsion polymer, the impermeability is modified due to the combined effects of plasticizing and filling. At very low p/c (<2.5%), adding emulsion polymer has little effect on the impermeability. Further addition of emulsion polymer leads to a significant increase in the impermeability. It is believed that the filling effect plays a dominant role in increasing the impermeability when their dosages are high.

This work has been gratefully funded by the University College of Technology Sarawak (UCTS) Grant.

References

1. ACI Committee 548, *Guide for use of polymers in concrete*, (ACI International, 2003)
2. S. Popovics, *Polymer-modified concrete Detroit*, (Concret Institute (ACI), Michigan, 1987)

3. N. Banthia, *Proceedings of FIBCON2012,* (ICI, Nagpur, India, 2012)
4. M. Victor, A.E. Flavio, Cement Concrete Res. **42**, 1417-1427 (2012)
5. R. Wang, C. Meyer, Cement Construction Composite **34,** 975-981 (2012)
6. S. Zhong, H. Yuan, Chemistry Industry Press (2003)
7. Y. Ohama, *Proceedings of the Third Asia Symposium on Polymer in Concrete*, (Shanghai, 2000)
8. V. Dionys, C. Lech, Cement Construction Composite. **27**, 926-33 (2005)
9. X.M. Kong, C.C. Wu, Y.R. Zhang, J. Li, Constr. Build. Mater. **38**, 195-203 (2014)
10. M. Aziz, P. Paramasivam, S. Lee, International Journal Cement Composite Lighweight Concrete **3,** 123-32 (1981)
11. G. Li, X. Zhao, C. Rong and Z. Wang, Constr. Build. Mater **24** (7), 1201-1206 (2010)
12. Y.L. Kuo, K.C. Ping, N.P. Yaw, C.Y. Chih, Cement and Concrete Research. **34,** 977-989 (2004)
13. J. Yang, P. Yan, X. Kong, X. Li, Sci. China Ser. E. **53** (5), 1406-1412 (2010)
14. A. Ayob, M.E. Razali, S. Alias, A.G. Ahmad, D.S.H. Ali, *MATEC Web Conf. The 9th International Unimas Stem Engineering Conference (ENCON 2016)* **87**, 01002, (2017)

 https://doi.org/10.1051/e3sconf/20183401018

Characterization of Fly and Bottom Ashes Mixtures Treated using Sodium Lauryl Sulphate and Polyvinyl Alcohol

C.G. Robert[1], *A.* Ayob[1*], *M.F.* Muhammad Zaki[1], *M.E.* Razali[1], *E.V.* Lew [1], and *P.Y. Hong* [1]

[1]School of Environmental Engineering, Universiti Malaysia Perlis, 02600 Arau, Perlis, Malaysia

Abstract. Malaysia promotes coal as an option for solid fuel in electric power generation. Demanding of electricity needs, therefore, has led to increase the coal consumption and thus producing more coal waste products. The disposal of coal waste ashes has been a main concern to power generation station due to the need of disposal sites and operational costs. This study investigates the composition of fly ash (FA) and bottom ash (BA) mixtures with difference component percentage treated with sodium lauryl sulphate (SLS) and polyvinyl alcohol (PVA) at 1.5 and 2.5 wt% solutions and examined in terms of specific gravity, pH, maximum dry density properties, and its surface morphology. Although the chemical composition of the SLS and PVA treated fly and bottom ashes studied in this current work is not altered extensively, significant changes could be observed in its physicochemical properties. Chemically treated fly and bottom ashes mixtures with SLS and PVA at 1.5 wt% solution exhibited specific gravity of 1.97 to 2.92 and high pH values within range of 9.28 to 10.52. The mixture of BA:FA=0:1 ratio depicting high maximum dry density of 1.35 to 1.56 g/cm^3 in both SLS and PVA solutions at 1.5 and 2.5 wt%. Scanning electron microscopy image shows distinct surface morphologies of SLS-treated fly and bottom ashes mixture that the particles are packed closely, strongly bonded similar to popcorn shape due to the effect of active silanol groups acted on coal ashes surface with the presence of Al-O/Si-O/other oxides. These findings suggest that higher level of chemical interaction between the fly and bottom ashes particles, significantly enhances pozzolanic reactions such as shear strength, plasticity, cementing properties, and thus other engineering properties.

1 Introduction

Coal power plant which generates electricity consumed coal and producing residuals which are fly ash (FA) and bottom ash (BA). Approximately the by-products ratio of coal combustion is 80% fly ash and 20% bottom ash [1]. It is estimated both chemically toxic ashes produced as much as 100 million tons in US in the year 1999 [2]. These most complex and abundant of anthropogenic materials, if not properly disposed of, can lead water and soil pollution, disrupt ecological cycles and pose environmental hazards [3].

* Corresponding author: afizah@unimap.edu.my

Fly ash is mostly light, fine particle, spherical in shape and grey while bottom ash is heavy, coarse particles, angular and blackish. A dark grey to black colour is typically indicated to elevate unburned carbon content [4]. Fly ash is most often used term to define fine fly ash, particles of which are captured via flue gas and collected by electrostatic of mechanical precipitation. Bottom ash refers to the ash that falls down through the airflow to the bottom of the boiler and is mechanically separated [3]. However, the chemical constituents of both two materials are very similar, with the major difference in their particle shape and size [5].

Engineers been have trying to reuse these by-products in construction includes as road base construction, soil amendment, and apply as a filler in polymer; solving the waste management problem and reducing construction cost by using recycle material. In 2001, researchers suggested that fly ash could increase the strength and durability of concrete [6]. Both fly and bottom ashes have been used in wastewater treatment system as reducer of heavy metals from water by absorption mechanism [7]. In many cases, fly ash alone cannot be used effectively which therefore binding materials or admixtures can be added such as organic polymer, or lime [5]. The use of polymeric materials in settling soil improvement is growing daily, for example application of aqueous polymer modification on soil-polymer interaction by Naeini et al. (2013) [8] and Ates (2013) [9].

However, among the various uses of fly ash, its bulk utilization is feasible only in geotechnical engineering application. The best usage could be as an embankment and road construction. One of the important perimeter in geotechnical is density. Density affect the strength, stability and permeability of a substance to bear load. Previous studies have been conducted in which the characteristics of soil with fly ash mixes were determined in the laboratory. Sahu (2001) [10], stated that maximum dry density decreased with the increase in fly ash proportion and there was hardly any effect on optimum moisture content, as the specific gravity of fly ash was relatively lower than that of soils as well as it counterpart, bottom ash. Additionally, increasing lime content in stabilization of fly ash-soil mixtures, will significantly decrease the maximum dry density of soil-lime mixes, and lowered the optimum moisture content [11]. Santos et al. (2011) [12] suggested that fly ash-soil mixtures exhibits relatively well defined moisture density relationship, varying with mixture ratios. The dry unit weight of fly ash-soil mixtures was lower than those of typically compacted soils. This research study the characterization of fly and bottom ashes mixture treated using sodium lauryl sulphate (SLS) and polyvinyl alcohol (PVA) in regards to its physico-chemical properties, maximum dry density and its surface morphology.

2 Material and methods

Fly ash and bottom ash used for this research has been acquired from Jana Manjung coal power plant in Perak, Malaysia. SLS and PVA of molecular weight ~ 20,000 and ~288, respectively, that fully hydrolyzed were purchased from Sigma Adrich and used as recieved. The mix proportions in percentage of weight were denoted as 1:0, 3:2, 2:3 and 0:1 for BA:FA ratios, were examined for its grain size particle and specific gravity. The SLS and PVA were dissolved in distilled water at 80°C to prepare about 1.5 and 2.5 wt% solution. pH values were determined after stirring the samples on the mixed solutions for 15 mins to allow the pH of the slurry to stabilize.

These coal ashes treated with the SLS or PVA solutions were then placed in a stirrer with continuous shaking at 130 revolutions per minute. Thereafter, the 20 individual mixtures tested at various water content of 8%, 12%, 16%, 20%, and 24% were filtered and washed numerous times with distilled water, poured into square molds of 100 mm × 100 mm × 100 mm, and dried for 2 days at ambient curing, before undergo Proctor compaction test adhering to ASTM D689 standard [13]. The morphological characterization of 28 days curing specimen was studied through scanning electron microscope (JEOL JSM 6460 LA).

3 Results and discussions

3.1 Grain size distribution

Fig. 1 shows the grain size distributions of the fly and bottom ashes from Jana Manjung, and their varying BA:FA ratio of mixtures. Generally, the fly ash was well graded, ranging from mostly silt to fine sand size, in a range between 0.075 mm to 0.2 mm. The small size of spherical fly ash particles gives to a better packing of aggregate materials, which decrease porosity, and therefore significantly increase the compressibility and workability of fly-bottom ashes mixtures for further engineering application. While for the bottom ash, its sizes occurred in a range between 2.0 mm to 4.75 mm that mostly coarse sand to gravel size.

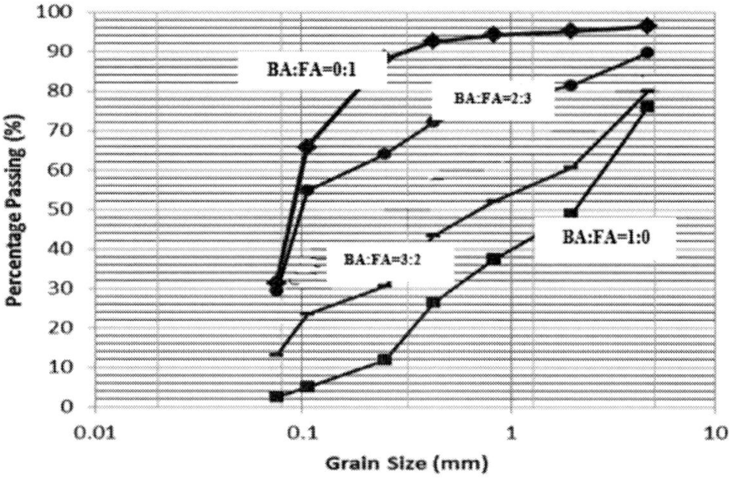

Fig. 1. Grain size distribution of untreated Jana Manjung coal ashes.

3.2 Specific gravity and pH values

Table 1 shows the range of specific gravity values for fly and bottom ashes mixtures recorded between 1.97 to 2.86 and 2.15 to 2.5 treated with 1.5 wt% solution of SLS and PVA, respectively.

Table 1. Specific gravity and pH vale of BA:FA mixtures with 1.5 wt% solution of SLS and PVA.

Parameter	Sample (BA:FA)							
	Treated with 1.5% SLS				Treated with 1.5% PVA			
	1:0	3:2	2:3	0:1	1:0	3:2	2:3	0:1
Specific gravity	2.92	2.00	2.86	1.97	2.50	2.48	2.38	2.15
pH value	9.37	9.28	9.41	9.56	10.58	10.35	10.52	9.50

The difference in the range of these values could be due to two factors; (i) the chemical composition and (ii) presence of hollow fly ash particles or either bottom ash particles with their porous or vesicular structure. In both treatment methods, bottom ash mixtures

(BA:FA=1:0), depicted higher specific gravity values of 2.92 and 2.50, than that of fly ash alone (BA:FA= 0:1; 1.97 and 2.15), this could be due to the higher contents of iron oxide and the larger and irregular bottom ash particles in the first mixture.

In the coal ash constituent, fly ashes contain of amorphous ferro-alumino silicate, in which these aluminum, iron, and manganese oxides will absorb the trace element [5]. Therefore, during the pH testing, the solubility of these oxides determined the release of elements to the solution. Both treated mixtures with SLS and PVA at 1.5 wt% are alkaline, with the ranging pH value from 9.28 to 10.52 (Table 1). The pH value of ash-water system determine mainly on the calcium/sulphur molar ratio in ash, yet other minor alkalis or alkaline earth cations do give to the balance.

3.3 Effect of SLS and PVA treatment to maximum dry density value of fly and bottom ash mixtures

All compaction data have been plotted into dry density against moisture content as shown in Fig. 2.

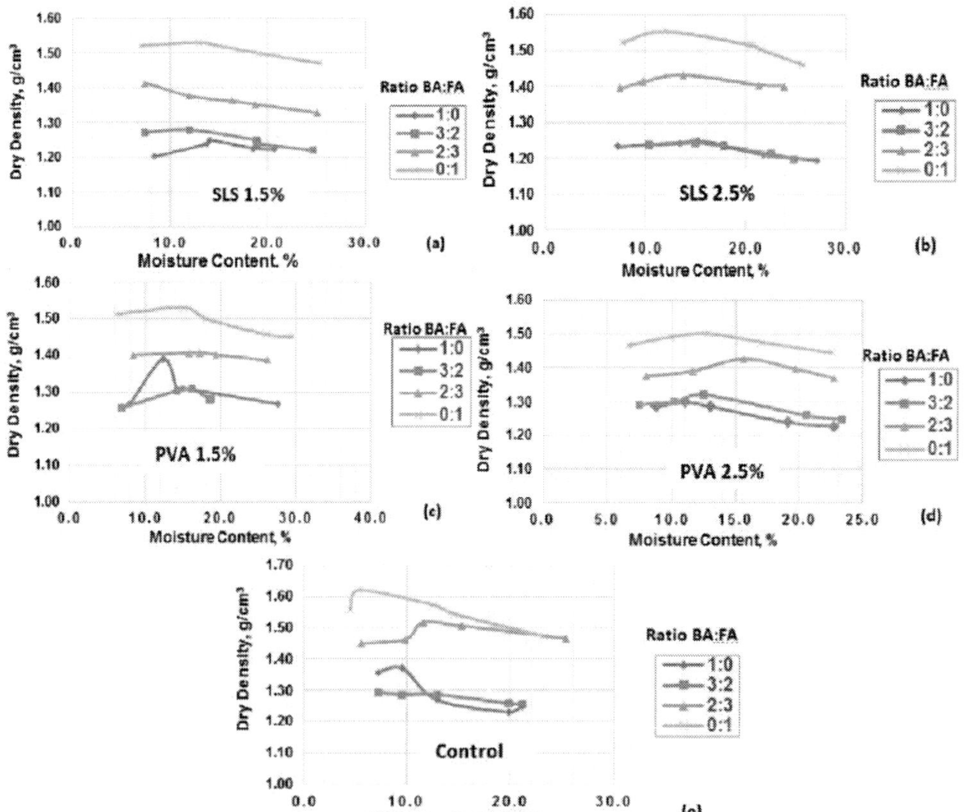

Fig. 2. Maximum dry density value for specimen treated with SLS solution at (a) 1.5% (b) 2.5%, and PVA solution at (c) 1.5%, (d) 2.5%, and untreated (control) specimen for the mixing BA:FA ratio of 1:0, 3:2, 2:3, and 0:1.

All test samples were showing curve similar to the non-cohesion soil. Fly ash alone (BA:FA= 0:1) showing high maximum dry density of 1.35 to 1.56 g/cm³ in both SLS and

PVA solutions of 1.5 and 2.5 wt%, while bottom ash (BA:FA=1:0) depicting the lowest values (1.19 to 1.30 g/cm^3). These trends continue as less bottom ash (BA:FA = 2:3) were added into the mixture, the maximum dry density decreases. The gap between highest and lowest dry density for each mixes is rather equivalent for both treatment methods of SLS and PVA. As the fly ash content decreases (BA:FA=3:2) the maximum dry density is decreasing and the moisture content values recorded 18.5% to 24%.

The inclusion of low weight fly ash and irregular coarser bottom ash in the SLS or PVA chemical solution can make the mixed samples comparatively coarser and decrease the overall weight. These decreases resulted from the flocculation and agglomeration of ash particles in present of sufficient chemical solution leading to increase in voids and thus, decrease in dry densities (BA:FA= 3:2; Fig. 2b, 2c). At the same time, the reason for the increasing moisture content could be due to the affinity of the soil-fly ash mixed samples for more water to complete the cation exchange reaction and with the applied of compaction effort the voids are occupied by more water (BA:FA=0:1; moisture content =29.5%; Fig. 2c). In addition, for the control specimen (Fig. 2e), some weak and large cenosphere fly ash particles (BA:FA=0.1) are crushed into finer particles during the compaction, thus lowering air content, because the fines fill up the voids between the matrix, and significantly contributing to higher maximum dry density (1.48 to 1.63 g/cm^3) and decrease in moisture content values (5% to 22%). This finding is consistent to Pal and Gosh (2010) [14], in their study of influence physical properties of class F fly ash.

In comparison between the effect of SLS and PVA chemical solutions to the maximum dry density of fly and bottom ashes mixture, qualitative analysis has been conducted (Table 2). Calculated standard deviation of 5.91% to 9.43% and 2.03% to 4.45%, respectively for varying maxing ratios of BA:FA and manipulating the percentage of solution at 1.5 and 2.5 wt%. The standard deviation values indicate low significance effect of two difference methods of chemical treatment (SLS vs. PVA; 1.5 and 2.5 wt%) for the maximum dry density values.

Table 2. Standard deviation calculated from maximum dry density values for each mixes of diference BA:FA ratios treated by using SLS and PVA solution.

Samples	Calculated standard deviation			
	BA:FA (1:0)	**BA:FA (3:2)**	**BA:FA (2:3)**	**BA:FA (0:1)**
Untreated	1.37	1.30	1.52	1.62
Treated with 1.5% SLS	1.25	1.28	1.41	1.53
Treated with 2.5% SLS	1.25	1.25	1.43	1.55
Treated with 1.5% PVA	1.39	1.25	1.41	1.53
Treated with 2.5% PVA	1.30	1.32	1.43	1.50

3.4 Morphological characterization of fly and bottom ashes mixtures treated with SLS and PVA solution

SEM images show the surface morphologies of untreated and treated fly bottom ashes mixture in 1.5 wt% solution of SLS and PVA after 28 days of curing. SEM images shows that the untreated botttom ash of BA:FA=0:1 has rough, irregular and complex in shape particles (Fig. 3a), while the mixture with higher fly as (BA:FA = 2:3, Fig. 3b), exhibited smooth surface and good sphericity, eventhough some particles were observed agglomerated into a large particle tightly attached to each other.

For the treated samples, a distinct morphological difference were observed in both mixtures, (Fig. 2c, 2d), where the fly and bottom ashes particles are packed more closely and strongly bonded similar to popcorn shape due to the effect of SLS by the presence of active silanol groups on the coal ashes surfaces (Al-O/Si –O/other oxides), that cause strong filler-filler interactions leading to more adhesive specimen (BA:FA = 3:2, Fig. 2c). Therefore this observation suggests higher level of chemical interaction between the coal ash particles, which significantly enhances pozzonalic reaction, increase its cementing properties, and thus its engineering properties.

Fig. 3. SEM surface morphology of untreated (a) BA:FA =1:0, (b) BA:FA = 2:3, treated speciemen with 1.5% wt. solution of (c) SLS for BA:FA = 3:2, and (d) BA:FA = 2:3 for 28 curing days.

4 Conclusions

Grain size particles of fly ash is fine and nearly spherical, mostly within the range of 0.01 to 2.0 mm, the bottom ash (0.08 to 5.0 mm) is consists of coarse-grained solid particles. The specific gravity and pH recorded for fly ash and bottom ash treated mixtures with 1.5% wt. solution of SLS and PVA are within the range of 1.97 to 2.92, and 9.28 to 10.58, respectively. No specific trend of relationship between grain-size distribution and specific gravity. Treated fly ash alone (BA:FA= 0:1) in SLS and PVA of 1.5% to 2.5% wt. solution contribute to higher maximum dry densities of 1.35 to 1.56 g/cm^3. However, as the mixture ratio become BA:FA=3:2, total comparative weight decrease, resulted from particles flocculation and agglomeration in the SLS and PVA solution, that increase voids, and consequently lower the maximum dry density. In comparison of two difference chemical solutions of SLS and PVA, no significance difference has been observed in regards to maximum dry density for treated fly ash and bottom ash mixtures. The chemically treated fly ash and bottom ash sample with SLS shows a distinct morphological surface difference as the particles are agglomerated, packed more closely, and strongly bonded due to the effect of active silanol groups, which enhances the pozzolanic chemical reactions and the engineering properties such as the plasicity and shear strength thus leading possibility for futher geotechnical applications.

This research is fully supported by Research Acculturation Grant Scheme (RAGS No: 9018-00078). The authors fully acknowledged Ministry of Higher Education (MOHE) and Universiti Malaysia

Perlis (UniMAP) for the approved fund which makes this important research viable and effective. Thank you to the Jana Manjung Power Plant, Perak for their approval and assistance of getting the sample from the site.

References

1. A. Ayob, M.Z.A.M. Zahid, M.F.M. Zaki, S.H.A. Hamid, M.A.M. Yussuf, A.N.M. Yunus, Adv. Environ. Biol. **8**(22), 25-29, (2014)
2. E.A. Okunade, Mod. App. Sci. **4**(12), (2010)
3. Z.T. Yao, X.S. Ji, P.K. Sarker, J.H. Tang, L.Q. Ge, M.S. Xia, Y.Q. Xi, Earth Sci. Rev. **141**, 105-121, (2015)
4. D. Shaunik, S.S. Gupta, Int. J. Basis App. Agr. Res. **12**(1), (2104)
5. C.G. Robert, A. Ayob, M.F.M. Zaki, M.Z.A.M. Zahid, L. Ee Von, J. Adv. Res. Mat. Sci. **10**(1), 12-25, (2015)
6. I. Siddique, M.I. Khan, Supplementry Cementing Materials, Springer Science & Business Media, London, (2001)
7. A.R. Awang, A. Marto, A.M. Makhtar, EJGE **16**, 1515-1531, (2011)
8. S.A. Naeini, B. Naderinia, E. Izadi, KSCE J. Civil Eng. **6** (16), 943-949, (2012)
9. A. Ates, Int. Journal Polymer Sci. 356214, (2013)
10. B.K. Sahu. In Symposium on Int. Ash Utilization. Kentucky, USA, Centre of App. Energy Res. (2001)
11. A. Kumar, B.S. Walia, A. Bajaj, J. Mat. Civil Eng. **19**(3), 242-248, (2007)
12. F. Santos, L. Li, Y. Li, F. Amini, *In Proceedings og World of Coal Ash Conference,* Denver, USA, May 9-12 (2011)
13. ASTM Standard C698, Standard Test Methods for Laboratory Compaction Characteristics of Soil Using Standard Effort, ASTM International, West Conshohocken, PA, 2015. www.astm.org. (2012)
14. S.K. Pal, A. Gosh, *Indian Geo. Conference*, Dec. (2010)

Application of GNSS Methods for Monitoring Offshore Platform Deformation

Khin Cho Myint[1,*], *Abd Nasir* Matori[1], and *Adel* Gohari[1]

[1]Department of Civil and Environmental Engineering, Universiti Teknologi PETRONAS, 32610 Tronoh, Malaysia

Abstract. Global Navigation Satellite System (GNSS) has become a powerful tool for high-precision deformation monitoring application. Monitoring of deformation and subsidence of offshore platform due to factors such as shallow gas phenomena. GNSS is the technical interoperability and compatibility between various satellite navigation systems such as modernized GPS, Galileo, reconstructed GLONASS to be used by civilian users. It has been known that excessive deformation affects platform structurally, causing loss of production and affects the efficiency of the machinery on board the platform. GNSS have been proven to be one of the most precise positioning methods where by users can get accuracy to the nearest centimeter of a given position from carrier phase measurement processing of GPS signals. This research is aimed at using GNSS technique, which is one of the most standard methods to monitor the deformation of offshore platforms. Therefore, station modeling, which accounts for the spatial correlated errors, and hence speeds up the ambiguity resolution process is employed. It was found that GNSS combines the high accuracy of the results monitoring the offshore platforms deformation with the possibility of survey.

1 Introduction

Offshore structures are usually deployed several kilometres into the sea and at a given depth [1]. In the last several decades, deformation of offshore platforms was observed to be due to the presence of shallow gas reservoir compaction phenomena. It can cause the structural subsidence and deformation, which will affect the integrity of these structures. In addition, such excessive subsidence may result in the loss of human lives as well as infrastructure [2].

Offshore structural subsidence is monitored regularly in order to anticipate and prevent the consequences there in. However, due to their physical locations, accurate measurement of such subsidence becomes a challenge [3]. Conventional GNSS measurement is a method that can be used to obtained centimetre-level accuracy measurement by effectively taking away biases and errors caused by atmospheric interference [4]. The use of reference station networks has become the global solution for high precision satellite positioning applications. Therefore, this research aims to determine the deformation magnitude and

* Corresponding author: khinchomyint07@gmail.com

direction of the monitored offshore platforms using high precision GNSS coordinates for the three (3) stations on the platform.

1.1 Global Navigation Satellite Systems (GNSS)

Within the last three decades, GNSS techniques have been used in land survey for photo control and geodetic control networks. The use of GNSS usually relies on some precise carrier-phase measurements where ambiguities were successfully fixed [5]. The advancements in modern technology have further enhanced the use of such technology in which 24-hour usage with a full complement of satellites is possible. Currently, GNSS technology are developed for mapping and surveying tasks such as real-time deformation monitoring, setting out survey as well as camera positioning aerial photography. Therefore, ambiguity resolution (AR), which comprises of ambiguity estimation and evaluation has become the key for precision in GNSS applications [5].

In GNSS applications, the successful resolutions and integer uncertainties are both important and challenging undertakings most especially in terms of GPS positioning [6][7]. At present, the major topic of concern for high precision GNSS application is the carrier phase ambiguity resolution [8]. Due to memory and speed constraints, the efficiency of the calculation is of major importance as time taken for the ambiguities to be resolved in to consideration [9] [10]. Under different navigational environments, accurate and dependable positions solutions involve a number of procedures [11] [12] namely; message decoding, inter ambiguity resolution, GNSS positioning, navigation and augmentation data availability monitoring, data correction and position calculation [13].

1.2 Offshore deformation

Offshore structural deformation such as subsidence poses a serious and potentially dangerous consequences. It is reported in [14] that the process of hydrocarbon extraction may cause reservoir compaction which in turn, could result in severe structural subsidence. For example, reservoir compaction was reported to have caused a subsidence of up to 50cm by Bedrige oil field in which a revenue loss of up to USD200 million was recorded [15]. Offshore structures are normally located several kilometers offshore, one of the famous techniques for their monitoring is the use of GPS [16, 17]. The method proves to be a means for monitoring of system deformations where both human safety as well as cost saving is achieved.The object of this study is to observe the deformation of offshore platforms using Global Navigational Satellite System (GNSS).

2 Methodology

2.1 Study Area

The study area selected as West Lutong offshore platform, which is approximately 20 km from Miri coastline in Sarawak. Offshore platforms are subject to deform in horizontal and vertical directions over time, and these are due to instability of the platform and other factors such as plate motion or post-seismic motion of an earthquake. Platform deformation monitoring is one of the routine which implementing by Oil and Gas Company to detect and estimate the motion rate and the displacement magnitude between epochs.

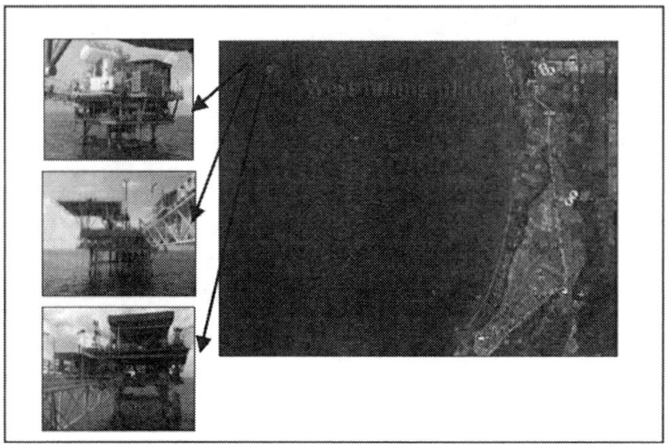

Fig.1. West Lutong platform is located 20 km from Miri coastline in Sarawak.

2.2 Materials and Method

The material includes GNSS receivers, and GNSS data processing using Bernese and Matlab programming software. The method has been divided into two parts in this study. Firstly, it includes experimental work; conducting field measurement trials to study as well as collect GNSS data- considering different navigation settings and measurement scenarios. Secondly, methodology which towards the developing long baseline based on GNSS result on offshore platforms.

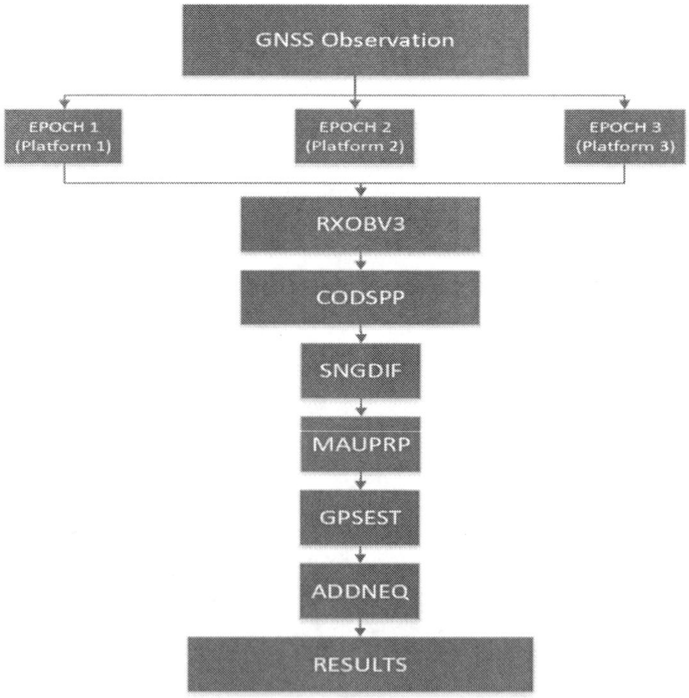

Fig. 2. The processing of the BERNESE Software.

3 Results and discussions

GNSS data was observed from 2013 and 2014 in West Lutong platform. The survey of observation was conducted 24 hours and 15 intervals second. During the time of observation, the data were received the position and type of satellite in GNSS receiver. Data processing and adjustment were carried out using appropriate scientific Bernese processing software. The Secondary Layer of the processing especially on stations repeatability analyses as well as for the final coordinate adjustment. The result of Platform deformation monitoring is to detect and estimate the motion rate and the displacement magnitude between epochs. It investigates the existing processing strategies to identify the preferred processing approach that can support precise positioning for offshore platform deformation monitoring. The results of Ellipsoid Height for each platform are shown in Table 1.

Table 1. Ellipsoid Height (M) of West Lutong Platforms.

Station	Parameter	2013	2014
WL01	Ellipsoid Height (M)	62.520	62.523
WL02	Ellipsoid Height (M)	64.302	64.320
WL03	Ellipsoid Height (M)	49.635	49.663

Table 2. Data of GNSS result for West Lutong Platforms.

Station	2013	2014
WL01	0.0	-0.003
WL02	0.0	-0.018
WL03	0.0	-0.028

The offshore platforms are located in marine environment; the GPS signals will be subjected to more influence of multipath due to sea surface reflection of the signals. Based on the GNSS observation data Figure 3 shows the plot of the relative deformation of all the three platforms between 2013 and 2014 as obtained. It can be seen from the result that the deformation up to 0.028m was recorded for platform 3 while deformation of platform 1 was deform less than 0.005m. Platform 2 movement was shown between 0.005m and 0.02m during the observation. Although vertical movements can be occurred for the platforms due to uplift force from wave and win action, it may also be due to GNSS measurement errors such as satellite multipath errors, orbit errors, ionospheric error and tropospheric errors.

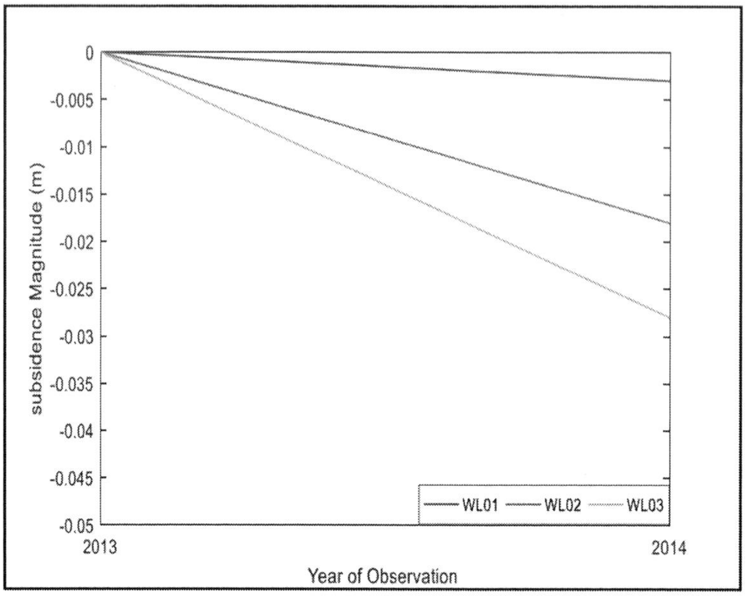

Fig. 3. Deformation observed for West Lutong Platform 2013 and 2014.

4 Conclusions

GNSS observations are widely used for monitoring deformation applications, geodesy, agriculture, Transportation and others surveying. Continuously Operating References Stations (CORS) are being installed in federal, state and private infrastructue in order to obtain the precise position. However, consequent upon the fact that most of the offshore platform in Malaysia waters are loactaed up to 200km offshore. The application of GPS in offshore monitoring is become very limtted. Therefore, there is strong need for the development of GNSS method for monitoring of offshore platform in deeper waters. In addition, elimination of systematic error such as multipath and atmospheric effectt as wel as satellite orbic error become difficult in the processing of the algorithms of baseline GNSS data. Consequently, these errors becomes problem in high precision applications as the results of their effect in ambiguity resolution process. Hence it becomes important to propose a better algorithm for data processing in which a good precision can be obtained in offshore deformation studies.

References

1. J., Peuchen, Geotechnical and geophysical site characterization **4**, 83-112 (2012).
2. A., Badellas, P., Savvaidis, Papageorgiou Publ. Co, 257 (1990).
3. H., Euler, C., Keenan, B., Zebhauser, G., Wübbena, ION GPS, 11-14 (2001)
4. J., Aponte, X., Meng, C., Hill, T., Moore, M., Burbidge, A., Dodson, Journal of Applied Geodesy **3**, 25-34 (2009).
5. N. Widjajanti, *Universiti Teknologi Petronas* (2010).
6. J. Haasdyk, C. Roberts, and V. Janssen, *FIG Congress*, Sydney, Australia, 11-16 (2010)
7. C. Rizos, S. Han, Wuhan University Journal of Natural Sciences **8**, 566-574 (2003).

8. P. Loomis, L. Sheynblatt, and T. Mueller, *Proceedings of the 4th International Technical Meeting of the Satellite Division of the Institute of Navigation (ION GPS 1991)*, 511-520 (1991)
9. R. B. Langley, GPS world **8**, 40-45 (1997).
10. T. Yan, Positioning **1**, (2005).
11. L. Wanninger, *Proceedings of the 8th International Technical Meeting of the Satellite Division of the Institute of Navigation*, ION GPS-95, 55-62 (1995)
12. C. J. Deckert, P. V. Bolstad, Southern Journal of Applied Forestry **20**, 81-84 (1996).
13. D. N. Dong, Y. Bock, Journal of Geophysical Research: Solid Earth **94**, 3949-3966 (1989).
14. E. J. Fielding, R. G. Blom, R. M. Goldstein, Geophysical Research Letters **25**, 3215-3218 (1998).
15. A. Sroka, R. Hejmanowski, *Proceedinds 12th FIG Symposium*, Baden (2006).
16. H. Setan, R. Othman, Positioning **1** (2006).
17. A.N. Matori, H. Setan, J. Platform **1**(2000).

Infrastructure Joint Venture Projects in Malaysia: A Preliminary Study

Norsyakilah Romeli[1*], *Faridah Muhamad* Halil[1], *Faridah* Ismail[1], and *Muhammad Sufian* Hasim[1]

[1] Faculty of Architecture, Planning and Surveying, Universiti Teknologi Mara, Shah Alam, Malaysia

Abstract. As many developed country practise, the function of the infrastructure is to connect the each region of Malaysia holistically and infrastructure is an investment network projects such as transportation water and sewerage, power, communication and irrigations system. Hence, a billions allocations of government income reserved for the sake of the infrastructure development. Towards a successful infrastructure development, a joint venture approach has been promotes by 2016 in one of the government thrust in Construction Industry Transformation Plan which encourage the internationalisation among contractors. However, there is depletion in information on the actual practise of the infrastructure joint venture projects in Malaysia. Therefore, this study attempt to explore the real application of the joint venture in Malaysian infrastructure projects. Using the questionnaire survey, a set of survey question distributed to the targeted respondents. The survey contained three section which the sections are respondent details, organizations background and project capital in infrastructure joint venture project. The results recorded and analyse using SPSS software. The contractors stated that they have implemented the joint venture practice with mostly the client with the usual construction period of the infrastructure project are more than 5 years. Other than that, the study indicates that there are problems in the joint venture project in the perspective of the project capital and the railway infrastructure should be given a highlights in future study due to its high significant in term of cost and technical issues.

1 Introduction

The circumstance of the construction joint venture has been a standout amongst the most basic issues in the industry development. Many finished ventures were accounted for costs considerably higher than the genuine contract whole because of low quality. In Malaysia, the development part contributes altogether to the financial development for enhancing the personal satisfaction [1, 2, 4]. The development area has been assuming a critical part in total economy of the nation in term of its commitment to income era which energizes the total national output and the financial improvement of Malaysia. The construction industry

* Corresponding author: kilaromeli@gmail.com

in Malaysia unanimously dominated by four pioneer fields which are construction of industrial, residential, commercial and infrastructure. The research focused on the infrastructure development as the construction cost for the infrastructure development is quite large it lead to the economic significant in Malaysian economic transformation history. Fig. 1 shows the infrastructure division in construction industry. The railway, highway, seaports, bridges, airports and dam are the example of the types of the buildable infrastructure elements. The infrastructure implementation is usually via joint venture, partnership, privatisation, and outsourcing [16, 18]. Through the JV, a imperative measures usually discussed in IJVP is a collaboration pattern, time frame, infrastructure types, procurement selection, benefits from the collaboration, sharing elements, and financial determination [3, 4, 6, 20, 21].

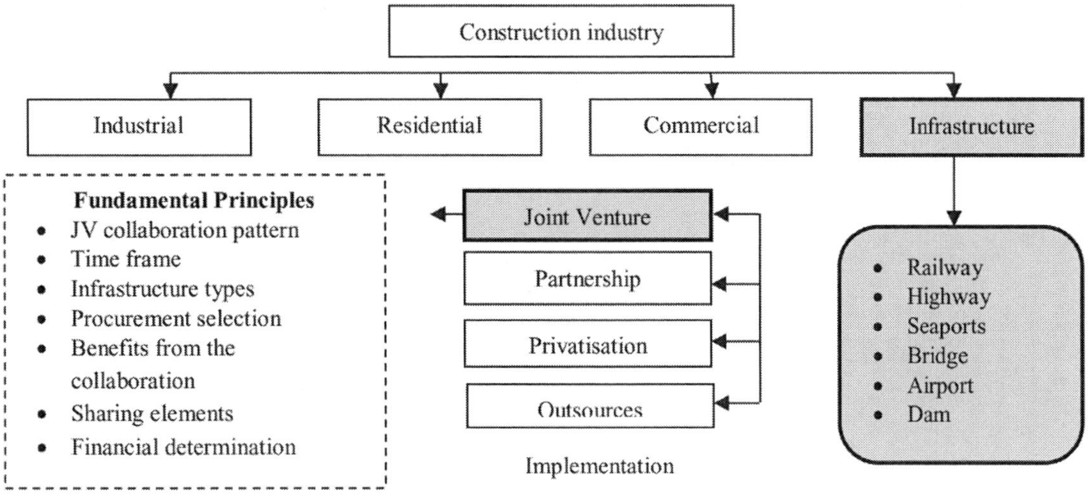

Fig. 1. The Infrastructure Division in Construction Industry.

2 Landscape of Malaysian Infrastructure Joint Venture Projects

The IJVP typically adhere with the involvement of private concessionaire in the delivery of public infrastructure worldwide is apparent in reducing the budgetary burden on the government's part specifically due to the major downturn in the global economy [16,19]. Their involvement varies from concessionaire, privatization to partnerships. [13,14,18]. Looking at the most current type of project delivery approach procured around the globe, which is known as Public Private Partnerships (PPP), although the opportunities are widely opened for the private concessionaires to partake in the delivery of public infrastructure projects with numerous incentives by the government, their responses are still minimal [9-11]. The clients appoint the Project Delivery Partners or Special Purpose Vehicle to manifest the project from the monetary acquisition, procurement path, tendering process and claim from the awarded contractors. A huge responsibility portray by the PDP to ensure the successfulness of the project [17,19,20]. Meanwhile, in certain project the IJVP occurred with the establishment of the new company for example ABJV to remarks the new legal organisation that will manifest the IJVP projects. The demand of the IJVP projects due to transferring the foreign technology to Malaysia for example the tunnelling process which requires expertise as in the application of New Austrian Tunnelling Method (NATM) and many others outstanding technologies in various infrastructure project [5-7]. Joint venture implementation is not new in Malaysia according to the current foreign direct investment to Malaysia which currently above the average levels at 67 per cent by year 2015 which indicated the sustainable implementation [12]; however there is relatively small exposure from the academic writing to uncover these matters [1,15]. Most of the knowledge gained from the Western reference via various research engines, however the research currently made at the Western countries and the legal and geographical factors may be different from the IJVP implementation in Malaysia.

Table 1. Research Trends on IJVP for past decades.

Elements	List of Authors
Involvement of the Public Sectors / Privatisation/ Procurement Path	Saleh & Siong (2008), Oyegoke et al (2009), Molavi & Barral (2016),
Risk Sharing / Risk Discourse / Risk Implementation / Risk Mitigation	Annemiek et al (2015), Adnan et al (2008), Ravindhar et al (2015), Zhang (2000), Thuyet et al (2004), Larimo et al (2016)
Contractual Issues / Legal/ Dispute Resolution / Conflicts	Ismail et al (2012), Marani et al (2014), Carmichael et al (2015), Welch et al (2000),
Key Performance Indicator / Critical Success Factor / Technology Transfer / Knowledge Management	AS Bakri et al (2009), SS Khaderi et al (2009), Famakin et al (2012), Masrom et al (2015), Ozorhon (2014), Takim & Akintoye (2008), AbdulAziz et al (2009), Ondo (2012), Ling et al (2012)
Urban Growth / Development / Implementation	Ma & Voo (2011), Famakin et al (2012), Charlebois (2008)
Behaviour / Organisational Management/ Human resources formation	Savrina & Rimsa (2013), Cheung (2009), Shaoul et al (2013), Henry (2010)
Project Capital/ Financing / Investment/ Economical Challenges	Flyvbjerg et al (2004), Khang & Thao (2015), Howes & Robinson (2005), Merna & Njiru (2002), Romeli et al (2016), Dziadosz (2015), Bakar et al (2012), Wonglimpiyarat (2016), Hwang et al (2003)

Other than that, there is depletion in information on the actual practise of the infrastructure joint venture projects in Malaysia. The depletion can be seen through the research trend analysis for past decades at Table 1. The issues on the project capital and financing been explored by majority of the foreign researcher and only few from the local. Less highlights given to the IVP in Malaysia on financial matters compares to risk, key performance indicator, critical success factors and knowledge management. Therefore, the circumstances suited the purpose of this study which attempt to explore the real application of the joint venture in Malaysian infrastructure projects.

3 Methodology

The methodology used in this research is quantitative methods. Therefore an instruments used is questionnaire survey. The respondent sample is selected based on the purposive sample. The list of contractors derived from the Construction Industry Development Board (CIDB), later on 35 numbers of infrastructure contractors selected according to the higher class in the contractor's registration with years of the establishment of the organisation more than 30 years. The purpose of selection is due to the higher years of the company establishment represent the validity of the response based on their experience in the infrastructure development. The questionnaire survey are divided into three (3) sections as Section A is for The Respondent Details, Section B is for Overview of the IJVP and Section C is for The Project Capital in Infrastructure Joint Venture Project. In within 6 months, the response recorded and keyed in the Statistical Package for Social Science

(SPSS) software and then analysed via the frequency analysis. The findings of the research documented as chapters below.

4 Results and discussions

4.1 Section A: The Respondent Details

The outcome of this section can be view at the Fig. 1 to 6 below. For the professional background, there are 40 per cent of the respondents are quantity surveyors, followed by 27 per cent engineer, 23 per cent project manager and 10 per cent is architect. As for the working experience there are 34 per cent of them with 10 to 15 years of working experience, 33 per cent at 5 to 10 years and 23 per cent less than 5 years. The results show that the respondents are among the professional practitioner that valid to answer the questions. Therefore the results shows that findings of this research from the reliable sources. The questions also asked about the preferable sharing ration in their company. The results indicated that 46 per cent of them agreed the JV sharing ration is 50%:50%, followed by 27 per cent respectively for ratio of 30%: 70% and 60%:40%. The respondents also had more experience implemented JV with local partner at 66.67 per cent rather than foreign partner at 63.33 per cent. Meanwhile, the respondent collaboration partner's countries are highest with the Asian partners at 96.67 per cent, followed by Middle East at 33.33 per cent, European at 25.42 per cent, and Oceania at 8.47 per cent, and America at 0.00 per cent and Africa at 0.00 per cent. This shows that the JV can be a continent approach for the Asian countries as the collaboration term and geographical factor slightly acceptable rather than other continents in the world. For types of JV partners the respondents agreed that the preferable JV approach implemented is among the client or owner at 76.67 per cent, followed by supplier at 70.00 per cent, manufacturer at 70.00 per cent, sub contractors at 63.33 per cent, specialist at 56.67 per cent and consultant at 16.67 per cent. The results indicated that the flows of the joint venture may smooth and practical with the collaboration partners of owner and clients compared to consultants. It is understood that from the Section A that the characteristics of the IJVP practitioners are identified and can be the guideline to the future researcher to acknowledge the respondents in this study area.

4.2 Section B: The Overview of the Infrastructure Joint Venture Projects

The questionnaire gets into the deeper essence of the knowledge where the enquiries on overview of the infrastructure joint venture project asked in the survey sessions. The questions are comprises of the types of JV implemented, the typical IJVP construction period, the IJVP construction cost, the classification of the IJVP, the IJVP procurement selection, benefits from the collaborations and the sharing IJVP elements. The outcome of the research is presented at the Figure 7 to 13 below. The results indicated the typical types of JV implemented are incorporated joint venture at 83.33 per cent compared to unincorporated joint venture at 63.33 per cent. The number shows that the respondents prefer to implement the JV via establishing a new company rather than appointing the project delivery partner as parts of the JV structure.

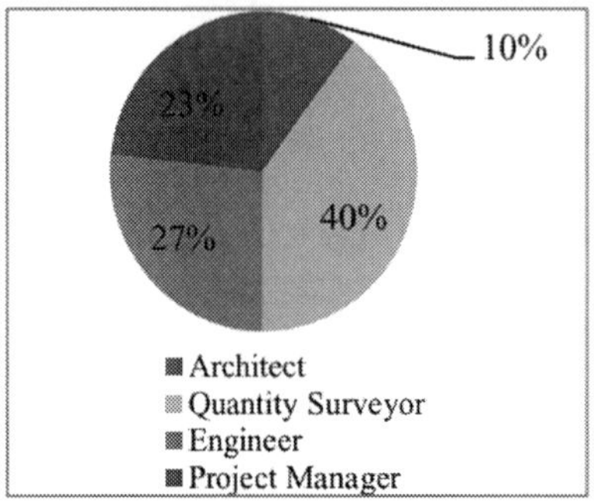

Fig.2.Professional Background.

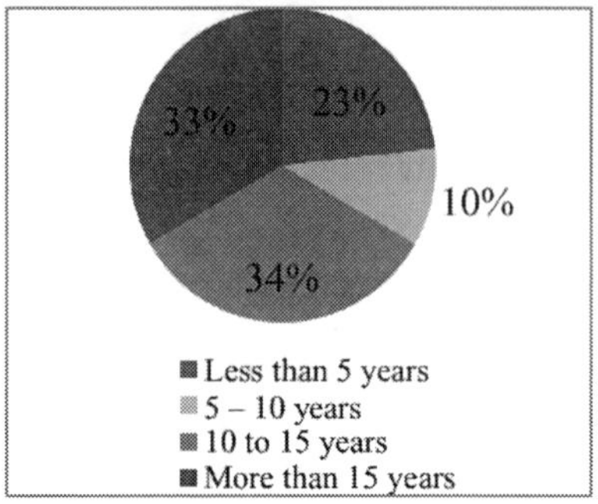

Fig.3.Working Experience.

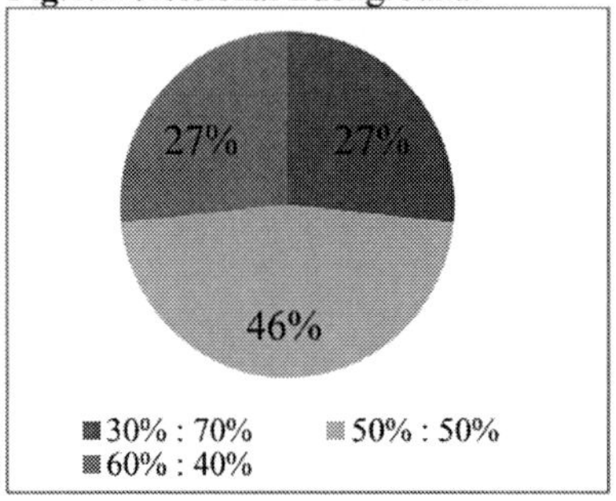

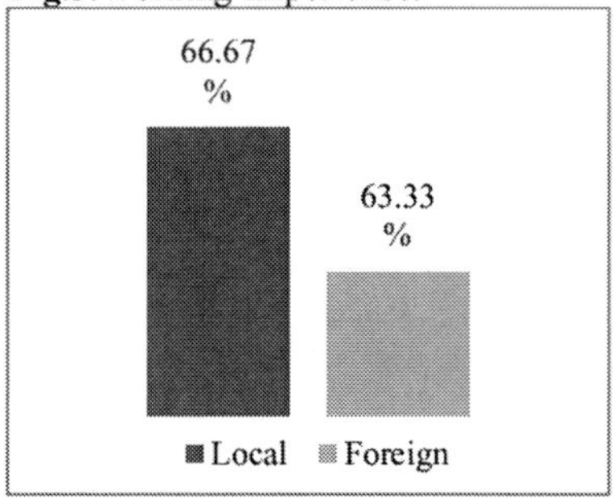

Fig.4.JV Sharing Ratio.

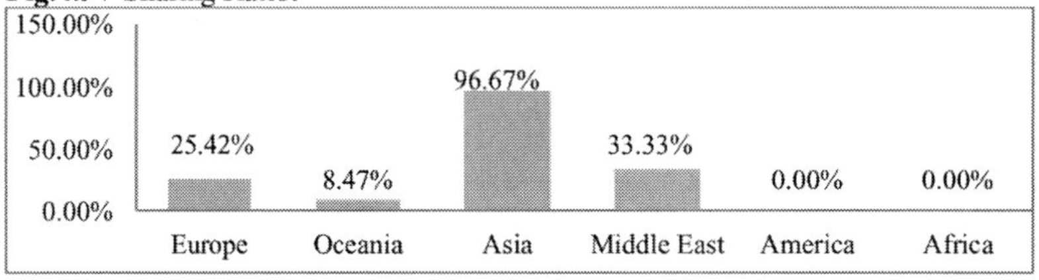

Fig.5. Collaboration Partner Range.

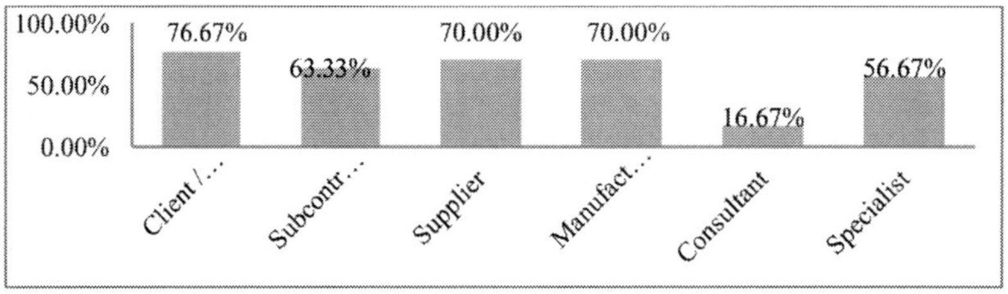

Fig.6. Types of JV Partners.

For the IJVP period of construction the 63.33 per cent answer recorded on more than 5 years, followed by 33.33 per cent for 3 to 5 years, 26.67 per cent for one to 3 years, and 0.00 per cent for less than a year. The results indicated that the IJVP construction period took longer than normal construction period due to the complexities of the projects. For the IJVP construction cost, the highest poll is on the MYR 50 Billion to MYR 100 Billion at 66.67 per cent, followed by 33.33 per cent for MYR 10 Billion to MYR 50 Billion, 26.67 per cent for below MYR 10 Billion and 13.33 per cent for more than MYR 100 Billion. The results shows that the IJVP required a big financial fund due to the big infrastructure development, material, manpower as well as the new technology adopted making the project required large financial resources. For the classification of the IJVP, the respondent agreed that the IJVP usually held for the highway project at 66.67 per cent, followed by bridges at 33.33 per cent, dam, airport and seaports at 16.67 per cent and railway at 13.33 per cent. The results show that the IJVP commonly focus on the highway infrastructure rather than railway infrastructure. Therefore, the concentration towards railroad infrastructure been given fewer highlights in Malaysian joint venture framework. For the procurement path, the respondent agreed Design and Build has been the typical procurement path available for the IJVP at 83.33 per cent, followed by Engineering, Procurement and Management Contract at 33.33 per cent, Traditional Method at 16.67 per cent, Turnkey at 10.00 per cent, and Emerging Cost Contract at 6.67 per cent. The results shows that the IJVP practitioner perceived infrastructure construction via the Design and Build approach are likely more doable rather than Emerging Cost Contract.

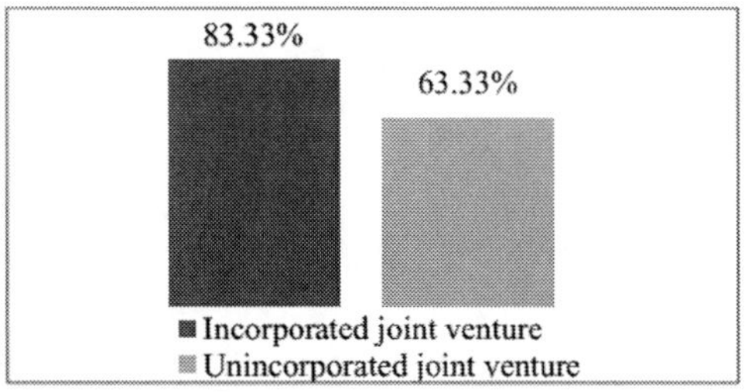

Fig.7.Types of JV.

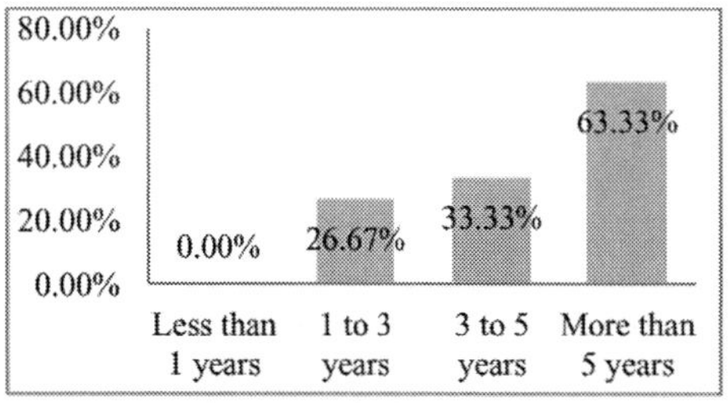

Fig.8. Typical IJVP Period.

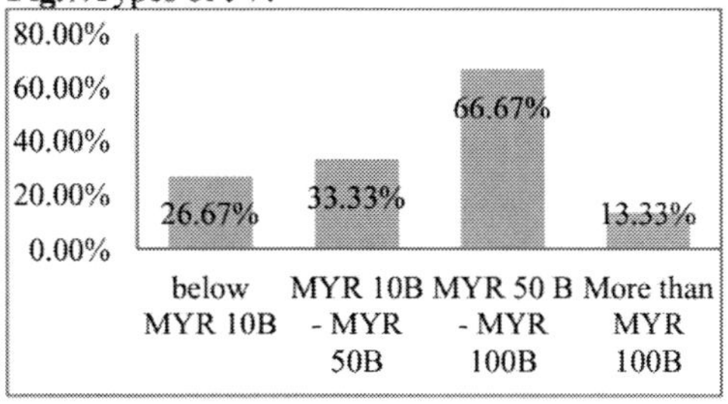

Fig.9.IJVP cost.

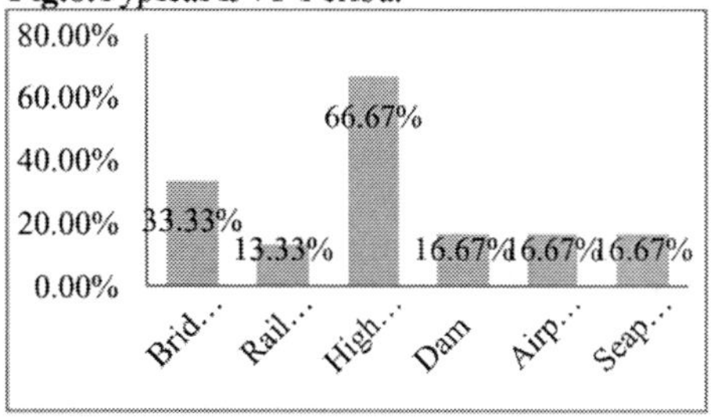

Fig.10.Classification of IJVP.

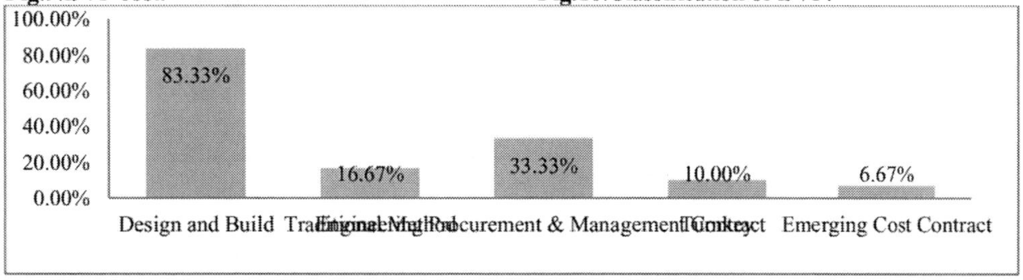

Fig.11. IJVP Procurement Path.

For the perspective of the benefits of the collaboration the respondent agreed that the implementation of the IJVP are based on the legal reason which the highest score is at 83.33 per cent, followed by sharing benefits at 70.00 per cent, minimizing cost at 63.33 per cent, employment at 56.67 per cent, strategic goals and purpose of survive at 46.67 per cent respectively. The results show that most of the establishment of IJVP in Malaysia are based on the legal procedure in the country itself. This has supported by the implementation of the Construction Industry Transformation Program which allow the joint venture implementation in Malaysia establish from 35 percent of the non – Malaysian in any joint venture infrastructure development in Malaysia. Apart from the benefits of the collaboration, the outcome on the preferable sharing elements is derived from this research. There are 93.33 per cent response recorded on risk, followed by market at 83.33 per cent, tax at 80.00 per cent, property at 76.67 per cent, manpower at 73.33 per cent, services for 66.67 per cent, technology and monetary at 40.00 per cent respectively, managerial style at 33.33 per cent, and culture at 30.00 per cent. The results indicated that the preferable elements to be share among the IJVP practitioner are risk as they perceive the risk is imperative rather than the culture. This is aligning with the research by local author Adnan (2008) on her research in construction risk implementation management in IJVP Malaysia. The research focused on the perception of the contractors in Malaysia that concern the contractors the most on the risk implementation process. The researches however generally discuss risk, and in depth study are required to explore the risk stated by the author especially on the technical and financial risks. Hence, the issues on the IJVP financial then explored at the Section C.

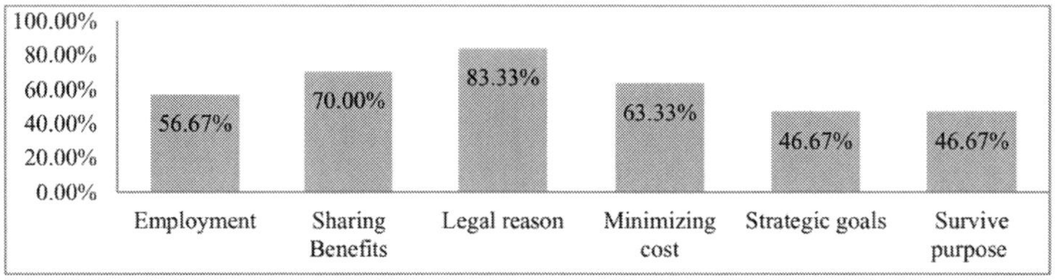

Fig.12.Benefits from the collaboration.

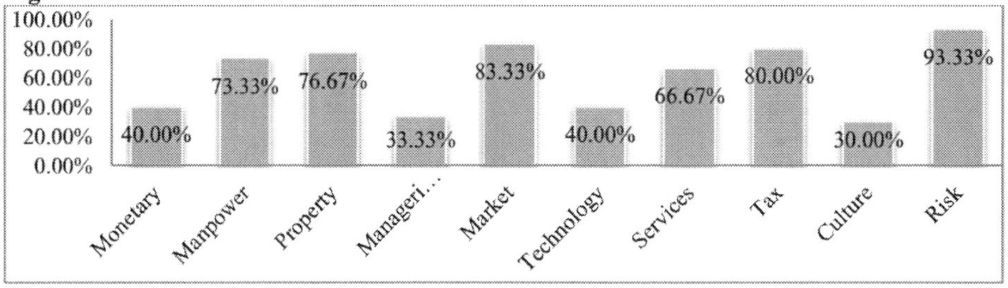

Fig.13.The JV sharing elements.

4.3 Section C: The Project Capital in Infrastructure Joint Venture Project

The outcome from this section explored on the financial side of the IJVP. The financial issues that mostly top of the practitioner's worries are the project capital as it is the start up instrument to commence the project. Hence, the question on the barriers in project capital acquisition, benefits of comprehensible project capital acquisition and attributes to project capital improvement been asked. The results from the questionnaire survey are as follows.

4.3.1 Barriers in Project Capital Acquisition

Fig 14 shows that the respondent response on the barriers in project capital acquisition in the IJVP. The results shows that the highest response are on the lack of financial aid at 83.30 per cent, followed by hard to control the project capital at 70.00 per cent, less understanding on the project capital contractual term and condition at 33.33 per cent, and the project capital expenditure accuracy at 30.00 per cent. The results show that the resources of the project capital are lacking in the IJVP. Although the government provides huge funding aid to support the projects, it is still insufficient as the normal infrastructure project would cost around MYR 50 Billion to MYR 100 Billion. Other than that, the difficulties to control the project capital can lead to the damage of the project financial flows. There are the needs for the better control tools or instrument to manifest the flow of the project capital at the initial, recurring and completion of the project. It is advisable for the project practitioner to control their tools via Code of Accounts Structure, Definitive Cost Estimate, Engineering Labour Report, Field Labour Report, and Project Cost Summary. These tools are imperative to estimate the outflow and inflow of the project capital in IJVP. The understanding of the project capital term and condition must be interpret by the legal practitioner to avoid the breach of the contract, upon the project capital utilisation phase as if the project capital is in term of the bank loan, the practitioner are required to pay the loan in within payback period with certain interest rates. The failure to comply with the contract leads the IJVP practitioner into the financial fatality stage. The accuracy of the project capital expenditure are among the problems do jeopardise the performance of the project capital. Hence, the better project cost planning and framework are required to sustain the project capital expenditure in the future.

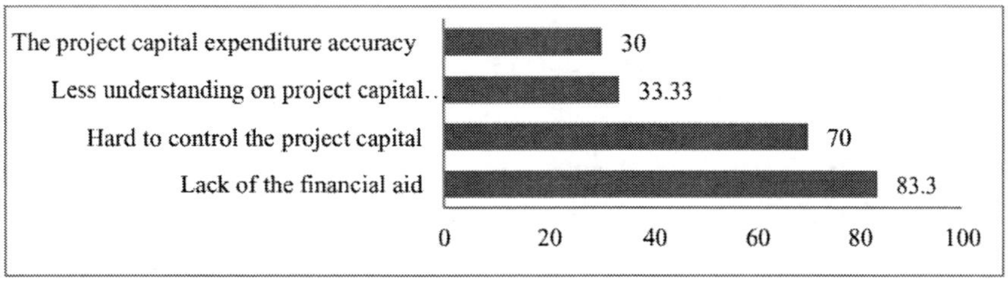

Fig.14. Barriers in Project Capital Acquisition.

4.3.2 Benefits of Comprehensible Project Capital Acquisition

Fig. 15 shows the results on the benefits of comprehensible project capital acquisition in IJVP. The results indicated that the benefits of the comprehensible project capital acquisition are as the structure for partner fit at 73.33 per cent, followed by financial stability at 66.67 per cent, to avoid dispute at 40.00 per cent and shows the mutual interest at 36.67 per cent. The respondents agreed that the structure fit are among the most benefits of having a comprehensible project capital acquisition as the foundation of having the IJVP is to share the monetary matters and distribute the risk accordingly and the measurement towards the successful joint venture project are having the project successfully fits. Meanwhile the project capital strength is act as the financial stability in ones project which the contingencies cost can be overcome if the organisation obtains the financial stability. Besides that, the dispute can be handling wisely with the existence of the stable project capital as the common dispute occurred due to the financial crisis in the organisation. The mutual interest in this research score the least as the respondent perceived that there is other matters in the IJVP can represent the mutual interest among the collaboration partners.

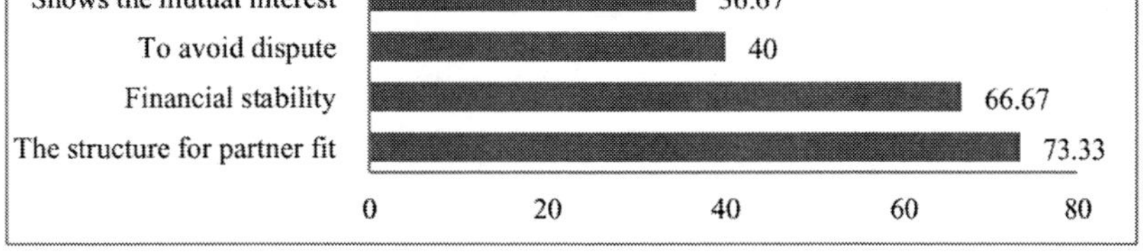

Fig.15. Benefits of Comprehensible Project Capital Acquisition.

4.3.3 Attributes to Project Capital Improvement

Fig. 16 shows that the results on the attributes to project capital improvement in IJVP. The results indicated that the respondents agreed that the attributes are to have better tools to supervise the flow of the project capital at 74.00 per cent, followed by concentration on project capital contractual form at 67.00 per cent, adequate funding base at 40.00 per cent, and a good networking skills at 37.00 per cent, and better construction cost planning at 21.00 per cent. The results indicated that the tools to supervise the flow project capital are required in this industry of IJVP. Hence, the improvement of the project capital shall focus on the control tools. Besides that, the understandable contractual form is needed to improve the project capital performance and stability. Hence, the capital project can be improved with the deeper understanding of the practitioner while handling the IJVP projects. The characteristic of adequate funding base a likely preferably by the respondents too, as the availability of the project capital are lack of the source so the IJVP practitioner shall do networking, do target the investor and investigate the investors background to improve the project capital. Apart from that, the IJVP practitioners may need to do the background check on the investors due to the term and condition from the lenders may distinguish from another. Some of the IJVP may also do come up with the better construction cost planning and development proposal to attract multi – types of investor towards the successful funding of the IJVP. From the project capital acquisition process, the research found that there is so much the strategies are needed to improve the project capital as the stability of the project capital are uncertain, therefore there is an art to lever it.

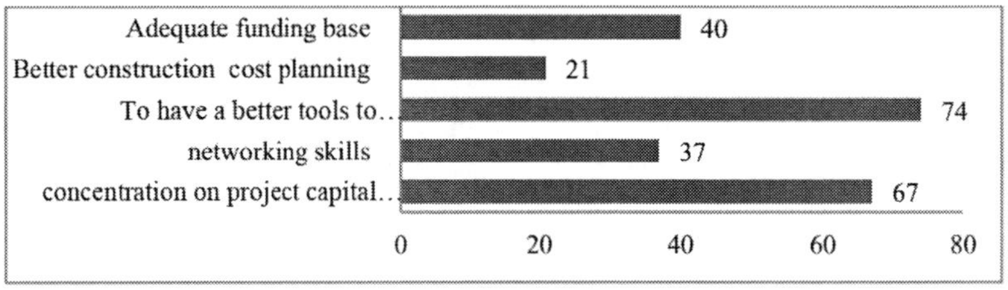

Fig.16. Attributes to Project Capital Improvement.

5 Conclusions

As the conclusion the IJVP implementation are mostly in between contractors and client with the usual construction period of the infrastructure project are more than 5 years with the project cost are high and the procurement path also complex due to the project needs. The collaboration made among the local practitioners and for the foreign IJVP the construction mostly among the Asian countries and the typical infrastructure development are highways instead of railways. The collaboration made based on the legal reason with the high level of risk sharing and mitigation awareness as the concern of the collaboration is on the monetary perspectives. Other than that, the study indicates that there are problems in the joint venture project in the perspective of the project capital and the railway infrastructure should be given a highlights in future study due to its high significant in term of cost and technical issues. Hence, the research provides an ample justification on the current practise of the infrastructure joint venture project in Malaysia. This research would deliver a general picture on the joint venture in infrastructure project in Malaysia to the body of knowledge and as well as the construction practitioner.

The authors would like to thank all of the infrastructure contractors for the good feedback received as well as the best commitment given by them. Special thanks to the personnel from the UEM Berhad and MMC – Gamuda for the warmest treatment during the poll session.

References

1. Z.Abu Bakar. UTM. (2008)
2. Abdul-Aziz, A.-R., Lee, K. Z. Ltd Intl J of Enrgy Sect Mgmt **1** (2009)
3. Adnan, H. UiTM, 99–106. (2008)
4. Bakri, A. S., Ingirige, Amaratunga D. Salford Univ, UK. Schl of Blt Envrnmt. (2008)
5. Famakin, I. O., Aje, I. O., & Ogunsemi, D. *J* of Fncl Mgmt of Prprty & Cons **17** (2012)
6. Howes, R., & Robinson, H. Elsevier B.V. (2005)
7. Ismail, S., Md Yusof, A., Soon Han, W. IBIMA Bis Rev, **11** (2012)
8. M.A.Ravindhar Raja, M.Gayathri, Knight, G. M. S. Intl J of Civ Eng &Tech **6** (2015)
9. Ma, T., & Voo, M.. A 2014 EPPM Conf at Por Elizbth, ZA, (2014)
10. Marani, D. A. A., Danuri, M. S. M., & Mohamed, O. Intl J of Prprty Sc **4** (2014).
11. Masrom, M. A. N., Rahim, M. H. I. A., Mohamed, S., Chen, G. K., & Yunus, R. Procedia Eng **125** (2015)
12. MIDA, e-newsletter, 2 (2016)
13. Molavi, J., & Barral, D. L. A. Procedia Eng **145** (2016).
14. Oyegoke, A., Dickinson, M., Khalfan, M., McDermott, P., & Rowlinson, S. Intl J of Mg Prjct in Biz **2** (2009)
15. Romeli, N., Halil, F. M., Ismail, F., Abd Shukor, A. S. Procedia - Soc Behav Sci **234** (2016)
16. Salleh, D., & Siong, H. C. J Alam Bina **11** (2008)
17. State, O. J Cons Prjct Mgmgt & Innov **2** (2012)
18. Takim, R., Adnan, H. Asian Soc Sci **4** (2008)
19. Thuyet, N. Van, Ogunlana, S. O., Dey, P. K., Pongsiri, N. Intl J Public Sect Mgmt J of Biz **17** (2004)
20. Van Os, A., Van Berkel, F., De Gilder, D., Van Dyck, C., Groenewegen, P. Intl J of Prjct Mgmt **33** (2015)
21. Zhang, S. Kyoto Univ, (2000)

E3S Web of Conferences **34**, 01021 (2018) https://doi.org/10.1051/e3sconf/20183401021
CENVIRON 2017

Laboratory Study on the Fatigue Resistance of Asphaltic Concrete Containing Titanium Dioxide

Rosnawati Buhari[1,*], *Mohd Ezree* Abdullah[1] , *Mohd Khairul* Ahmad[2] , *Saiful Azhar* Tajudin[1] and *Siti Khatijah* Abu Bakar[1]

[1]Faculty of Civil and Environmental Engineering, Universiti Tun Hussein Onn Malaysia, 83000, Batu Pahat, Johor Malaysia
[2]MINT-SRC, Universiti Tun Hussein Onn Malaysia, 86400, Parit Raja, BatuPahat Johor, Malaysia

Abstract. This study aims to evaluate the fatigue performance of modified asphalt mixture using Indirect Tensile Fatigue Test. Titanium Dioxide (TiO_2) powder in a form of rutile was used for producing asphalt concrete with lower mixing and compaction temperature compared to conventional hot mix asphalt without reducing its physical and mechanical also resistance to fatigue. The characteristic of the asphalt and modified asphalt was evaluated using penetration test, softening test and rotational viscosity test. Titanium dioxide of 2%, 4%, 6%, 8% and 10% by weight of asphalt has been incorporated into unaged 80/100 asphalt mix in order to improvise its performance and to fulfill the objectives of this experimental study. As a result, TiO_2 as an additive is potential to decrease the penetration and increasing the softening point of the asphalt. In terms of fatigue performance testing, addition TiO_2 additive does help in improving the fatigue properties as it shows greater result than the control asphalt. In conclusion, TiO_2 is great in improving fatigue properties.

1 Introduction

Warm Mix Asphalt (WMA) is one of the technologies that has been gaining attention in recent years, is used to reduce the heating, mixing and compaction temperature of the asphalt mixture, improving the working condition at site, reduced time of compaction in site and minimizes the negative effects on the environment. These technologies can reduce the production temperature of the Hot Mix Asphalt (HMA) by 16°C to over 55°C [1]. It was first applied in Europe in the year 1997, followed by United States in 2002 [2]. Exposure of WMA mixture is relatively new in Malaysia, however, it is becoming a mainstay in construction and pavement industry. To date, long term performances of WMA mixture are still being reviewed and there is no pavement failures or concerns have been observed. WMA is an alternative for hot mix asphalt (HMA) and is produced at a temperature ranging from 120°C to 140°C, which is 10°C to 40°C lower than that of HMA [3]. This environmental friendly technology has claimed to contribute to

* Corresponding author: rosna@uthm.edu.my

lower emission of fumes, less energy usage, better working environment and longer hauling time without compromising the performance of the pavement. In order to attain lower asphalt production and compaction temperature whilst maintaining its feasible viscosity, additives are needed to be incorporated in the mixture. According to Kilas et.al, three types of commonly found WMA technology in the worldwide are foam asphalt, organic additives and chemical additives [3]. These additives will help in lowering the viscosity of asphalt during the production process of WMA, which is a crucial factor in obtaining high workability of the asphalt. For example, Abdullah et. al was conducted a study on performance of WMA mixture using nanoclay modified asphalt binder called Nano Clay Modified Binder (NCMB) in terms of moisture susceptibility, rutting and fatigue potential [4]. The prepared the WMA samples using mixing temperature 135°C, 125°C and 115°C and less 10°C for compaction temperature. As a result, the fatigue life was decreased as it is decreasing in mixing and compaction temperature. Also, Moatasim et al. conduct on study WMA with modified asphalt using HDPE as an additive, the objective of the study is to evaluate the performances of the asphalt concrete mixtures that using modified asphalt 80/100 with HDPE [5]. From the result, the performance of the concrete mixtures improve when used 5% of HDPE by weight in asphalt blend. Then, Yu Kuang et al., did a researched with two main objective, first is to evaluate on behaviour of Evotherm-J1 and M1 as additive in compaction technology and second objective is to study the effect of Evotherm-J1 and M1as additive in moisture anti strip [6]. The study carried by Diefenderfer et al. (2008), using sasobit as an additive in asphalt with two different mixtures in the study.

In this study, chemical additive known as titanium dioxide (TiO_2) was used to evaluate the properties and performance of the asphalt pavement. The objectives of this study are to; (i) evaluate the physical properties of asphalt modified with Titanium dioxide based on Malaysia Standard Specification and Superpave binder specification and (ii) evaluate and compare the fatigue characteristics of asphalt modified with TiO_2 at warm mix asphalt compacted temperature with conventional hot mix asphalt.

2 Methodology

2.1 Materials

There are several sizes of sieve typically used for Superpave mix design which includes sieve size 37.5mm, 19mm, 12.5mm, 9.5mm, 4.75mm, 2.36mm, 1.18mm, 0.600mm, 0.300mm, 0.150mm and 0.075mm.The gradation size was determinate follow the Superpave gradation limit. The TiO_2 used are in form of rutile particles size ranges from 40 nm to70 nm as shown in Fig. 1 and the binder used are asphalt penetration grade 80-100. The selection of rutile form are based on the purity of titanium dioxide substance in its composition and higher surface area as the smaller sizes of particle, the surface area are higher.

2.2 Sample preparation

The modified asphalt was prepared using asphalt of penetration grade 80/100 as base asphalt and several percentage of TiO_2 in powder form. It is started by placing the 500 g of base asphalt in an aluminum can and then placed in the oven at 155°C for approximately half an hour. It was then put on top of a hot plate and continuously observes to ensure that the asphalt stayed constant at 155°C. Next, a specified percentage of TiO_2 by weight of asphalt are gradually added into the asphalt as the high shear asphalt mixer is rotating at 5200 rpm for duration of 30 minutes. The blending parameter used in this task are follows the procedure that came out with the homogeneous mixture in previous study [7].

To evaluate the fatigue characteristic of the mixture, the Optimum asphalt Content (OBC) was determined in the first place. The aggregate gradation with nominal size 9.5 mm was used. The maximum content of the aggregate is 4.75 mm sieve size and the filler content is 2% from total weight of 1200 g. A HMA mixture without TiO_2 content was prepared using mixing and compaction temperature that are 165°C and 155°C respectively. It was denoted as control mixture and WMA mixture that containing TiO_2 were prepared by using mixing and compaction temperature of 135°C and 125°C.

Fig. 1. Morphology of the TiO_2.

2.3 Laboratory testing

A laboratory testing was conducted based on American Society for Testing and Material (ASTM), American Association of State Highway and Transportation (AASHTO), Asphalt Institute Superpave Asphalt Binder Test Methods and Asphalt Institute Superpave Mix Design (SP-2). A penetration test was conducted followed ASTM D5 while softening point test following ASTM D36 [8-9]. Brookfield Viscometer was used in rotational viscosity (RV) test to measure the flowing resistance of asphalt materials [10]. The Superpave™ specification requires that the maximum viscosity of asphalt binder is no greater than 3 Pa.s at 135 °C for the convenience of storage and pumping in construction period [11]. Sieve analysis was conducted to find the aggregates gradation that followed the Superpave aggregates gradation limit [11]. The determination of the OBC was following the trial asphalt content based on 9.5 mm aggregates size with 5%, 5.5%, 6.0% and 6.5% trial asphalt content. After OBC was determined, mixtures sample for Indirect Tensile fatigue test was conducted. Fatigue cracking is one of prominent pavement distress in tropical countries after several years of construction due to weathering and repeated numbers of traffic loads. In laboratory, for asphalt mixture the Indirect Tensile Fatigue test (ITFT) was conducted using Universal Testing Machine (UTM) to evaluate the fatigue performance of modified asphalt mixture with TiO_2 powders. Before testing is conducted, sample and related testing accessories were conditioned at 20°C for a minimum 4 hours in the provided environmental chamber to reach a uniform temperature throughout the specimen. After that, testing is run at the same temperature and the specimen was exposed to a repeated compressive load 45 equal to 600 kPa with a haversine load signal through the vertical diametral plane followed by a loading time of 0.1 s and rest period of 0.4 s. The pattern of loading applied to the specimen is relatively uniform tensile stress perpendicular to the direction of the applied load. After certain period, the testing is terminated and the total number of cycles was recorded.

3 Results and discussions

3.1 Physical and flow behaviour

Softening point test and Penetration test were performed to determine the physical properties of unaged virgin and modified binder. Table 1 shows that the penetration value is gradually getting lower as more additive is added into the binder. This shows that addition of additive does help in improving the hardness of the binder, which is a desired factor in modified binder property. Softening Point test measured the temperature where asphalt material starts to change to liquid state [12]. Higher softening point indicates stiffer asphalt and lower temperature susceptibility at high temperature, which is more preferred in hot climates. As seen in Table 3, the softening point value started to increase for modified binder added with 2% TiO_2 until 8% TiO_2. It can be concluded that the penetration value decreases as higher percentage of additive was added to the binder. This indicates that proper amount of TiO_2 added can help in improving the stiffness of the asphalt.

Table 1. Results of Softening Point and Penetration Test.

Sample (% TiO_2)	Softening point (°C)	Penetration test (mm)
0	46.4	99.8
2	47.7	96.1
4	49.3	94.2
6	50.4	89.4
8	51.1	83.2
10	51.9	79.6

Fig. 2 shows viscosity test results. Conventional asphalt is known to have a higher mixing and compaction temperature compared to modified asphalt. As shown in the table, enhancement the reading by increasing the percentage of additive while the result become decreasing when the temperature raised until 180°C. As a conclusion, this additive can increasing the viscosity of asphalt binder.

The analysis continue with plotting graph in Fig. 3 for determination of recommendation of compaction and mixture temperature value. According to Superpave Asphalt Binder Test Method (SP-1), references line for compaction range and mixing range which are 0.25 to 0.31 for compaction and 0.15 to 0.19 for mixing. It was found that the minimum compaction temperature value for warm mix asphalt has larges value than control, while the minimum mixing temperature value is better than control value, this can be expressed the modified binder is using less energy and fuel than control binder for preparation of binder.

 https://doi.org/10.1051/e3sconf/20183401021

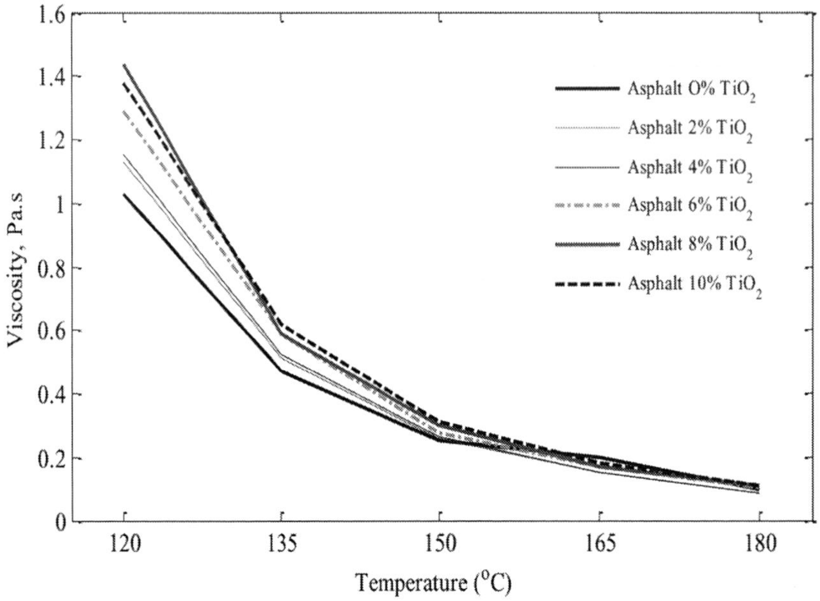

Fig. 2 Rotational viscosity of original asphalt 80/100 and Modified asphalt with TiO_2.

3.2 Resistance to fatigue

Fatigue is a major distresses normally occurred in asphalt pavement. Fatigue occurs at low to intermediate temperature while rutting occurs at higher temperature. The initiation and propagation of cracking generally relates to the magnitude of dissipated energy produced by external loading. In this study, results of ITFT according to the mixing and compaction temperature for the mixture are shown in Table 2 and the trend of fatigue resistance of the mixture is illustrate in Fig. 3. The trend for mixtures compacted at 125°C is showing an increasing pattern as the percentage of TiO_2 increased. This could indicate that proper amount of TiO_2 addition will help to improve the reaction between the binder and the aggregates, which eventually increases the strength of the mixture as a whole and completly increased the performance of the mixture againt fatigue failure.

Table 2. Fatigue life cycle

Sample (% TiO_2)	Mixing Temperature, (°C)	Compaction Temperature, (°C)	Fatigue life cycles
0	165	155	11416
2	135	125	12051
4	135	125	12578
6	135	125	13145
8	135	125	13463
10	135	125	13866

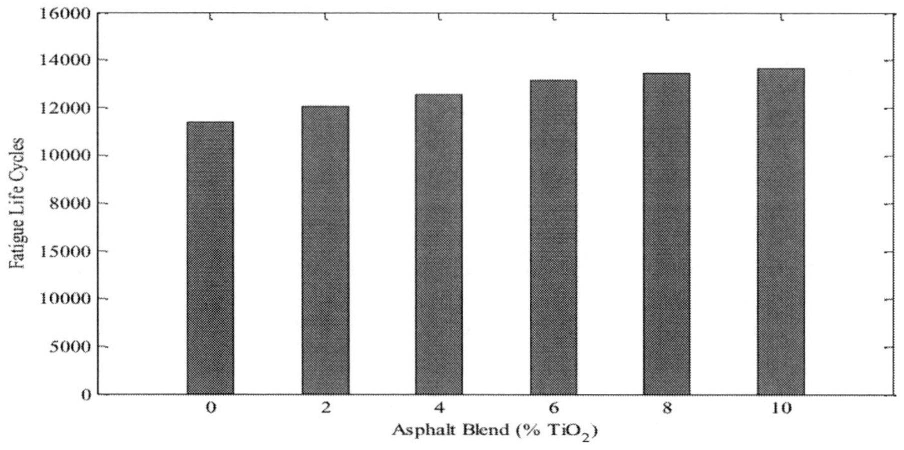

Fig. 3. Number of cycles of modified asphalt with TiO_2.

4 Conclusions

In this study, a TiO_2 in a powder form of rutile particles of 2%, 4%, 6%, 8% and 10% were added to the base binder to investigate for the effectiveness of the additives in improving the properties of the binder. Based on the results presented, the following findings can be deduced with respect to the applications:

- The addition of TiO_2 has significantly improved the penetration and softening point of the asphalt. This demonstrated the improvement of stiffness of the binder and also better temperature susceptibility.
- Addition of TiO_2 has the potential in reducing the mixing temperature of the mixture.
- In terms of fatigue cracking performance of mixture, addition of modified TiO_2 does aid in controlling the effectiveness of mixture towards fatigue resistance.

In conclusion, higher percentage of TiO_2 in form of rutile particles is a good substance to be used as additive in Warm Mix Asphalt as it improved the physical and fatigue performance characteristics of the binder. It is also reducing the temperature of the mixing.

The authors wish to thank the Universiti Tun Hussein Onn Malaysia. This study was funding by Ministry of Higher Learning Malaysia under FRGS Grant (1458).

References

1. D'Angelo, J., Harm, E., Bartoszek, J., Baumgardner, G., Corrigan, M., Cowsert, J., Harman, T., Jamshidi, M., Jones, W., Newcomb, D., Prowell, B., Sines, R., Yeaton, B. *Federal Highway Administration (FHWA)*, Alexandria, FHWA Report 08-007 (2008)
2. Kumar, M., Vaitkus, A., & Paliukait, M.. *10th International Conference of Modern Building Materials, Structures and Techniques*, Lithuana, (2010)
3. Kilas, M., Vaitkus, A., & Paliukait, M, *Warm Mix Asphalts Research, Analysis and Evaluation. 10th International Conference*, (2010)
4. Abdullah, M.E., Zamhari, K.A., Buhari, R., Bakar, S. K. A., Kamaruddin, N. H. M.,Nayan, N., Hainin, M. R., Hassan, N. A., Hassan, S. A., Yusoff, N. I. M., Journal Technology (Sciences & Engineering) **71** (2014)
5. Moatasim, A., Cheng, P.F., Al-Hadidy, Constr Build Mater **25** (2011)

6. Al-Mansob, R. A., Ismail, A., Alduri, A. N., Azhari, C. H., Karim, M. R., & Yusoff, N. I. Constr Build Mater **63** (2014)
7. Buhari. R, Chong, A.L., Abdullah M. E., Abu Bakar S. K., Mohd Kamarudin N. H., Shamsudin M.K., Ahmad M.K., Puteh S., Jurnal Teknologi **5** (2016)
8. ASTM D36, *Annual Book Of American Society For Testing And Materials (ASTM) Standards*, West Conshohocken, United States (2005)
9. ASTM D5, *Annual Book of American Society For Testing And Materials (ASTM) Standards*, West Conshohocken, United States (2005b)
10. AASHTO T315, *Determining The Rheological Properties Of Asphalt Binder Using A Dynamic Shear Rheometer* (DSR), (2010)
11. Asphalt Institute, *Superpave Mix Design*, Superpave Series No.2 (SP-2)(2001)
12. Asphalt Institute, *Superpave Performance Graded Asphalt Binder Specifications and Testing*, Series No. 1 (SP-1) (2003)

E3S Web of Conferences **34**, 01022 (2018) https://doi.org/10.1051/e3sconf/20183401022
CENVIRON 2017

Effects of the Fuel Price Increase on the Operating Cost of Freight Transport Vehicles

Adel Gohari[1,*], *Nasir* Matori[1], *Khamaruzaman* Wan Yusof[1] , *Iraj* Toloue[1] , and *Kin* Cho Myint[1]

[1]Department of Civil and Environmental Engineering, Universiti Teknologi PETRONAS, 32610 Perak, Malaysia

Abstract. One of the most important criteria in freight modal choices is the transport operating cost in which fuel price changes has a significant effect on it. This paper presents the impact of fuel price increases on the operating cost of the different transport modes for the containerized freight transportation. In this study, an operating cost equation was applied to compare the operating cost of different freight transport vehicles as well as evaluation of the operating cost changes across a range of fuel prices between the current price and one-hundred percent increase. The equation consists of influential parameters such as fuel cost, driver wage and maintenance cost of a vehicle. It has been concluded that the effect of the fuel price increase on the operating cost of different freight transportation modes is not in the same rate. According to equation and effective parameters considered, comparing the results showed that truck has the highest cost, train has the largest increase in price. Finally, the ship is the most influenced vehicle in terms of operating cost percentage increase when the rate of fuel price increase, followed by train and truck.

1 Introduction

Freight transportation is one of the backbones of any industrial system and it is a major component of the supply chain to ensure the efficient movement and timely availability of raw materials and finished products in local and international markets [1]. Freight transport is considered more censorious and critical than the past [2,3,4,5,6]. It involves the movement of commodities in a transport network from the origin to the destination. This movement needs infrastructures like roads, railways, waterways and airways and vehicles such as truck, train, ship and aircraft.

Freight transportation includes internal and external costs. Internal costs are the direct expenses incurred by providers of freight transportation [7]. It includes operating costs and investment in capital facilities. Energy, wages, maintenance, user charges, depreciation, and insurance are parameters that can be considered as the total operating cost. More than fifty percent of total operating cost for the transportation industry comprised of fuel price [8]. Hence, the fuel price has taken a large share in freight transportation operation cost

* Corresponding author: adel.gohari@gmail.com

across all modes. The fuel price depends on variety of parameters such as international brent price, oil refining costs or import taxes, fuel marketers margins, distribution costs to deliver the refined products to the final customers, and the national petroleum taxes [9]. External costs are defined as social costs imposed on others, but not paid for, by the infrastructure user [10]. The most important external costs of transport are accidents, noise, air pollution, climate change, and congestion [11].

The contribution of this work is to determine the percentage increase of the operating costs of each transport mode based on the fuel price increase. In this study, Truck, train and ship were considered as freight transport vehicles. The operating cost equation was applied to identify the operating cost changes. This consists of three influential factors such as fuel cost, driver wage and maintenance cost of the vehicle. The average speed is assumed to be constant between origin and destination. Equal travel distance is considered for operating cost computation of all transport vehicles.

2 Literature review

The supply chain is a system in which different organizations, people, activities, resources and information are connected to enable movement of a specific product or service from supplier to the customer. The classic graphical overview of the supply chain is depicted in Figure 1 [12]. In this system, the integrity of the chain depends on the information flow between supplier and customer. The purpose of this process is to deliver the product in high quality and the least possible cost. A transportation chain consists of pre-haul, line-haul and end-haul parts. Pre-haul and end-haul segments mostly are carried out by road mode, while any of the road, rail, sea, air or their combinations could be considered as the transport mode for the line-haul part. The line haul transportation is represented as the central link in Figure 1.

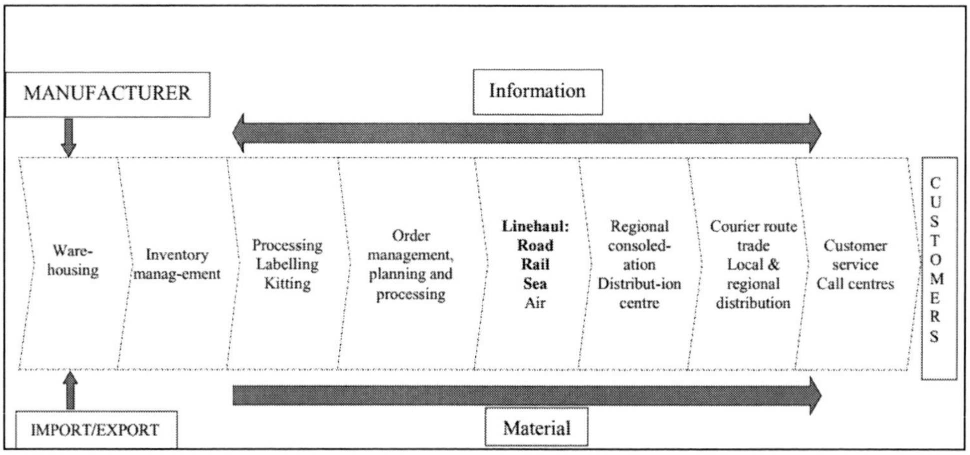

Fig. 1. A view of the supply chain.

There have been some studies related to the effect of the fuel price changes on freight transportation. A model was developed for calculating comparable combined internal and external costs of intermodal and road freight transport networks, and also conducted the sensitivity-based analysis to show the dependence of the average external, internal and full costs of the intermodal and road transport network on the door-to-door distance [13]. The impact of oil prices on U.S. domestic freight transportation was evaluated. It forecast the potential range of oil prices in the short and long terms, and then assesses how such prices would impact transportation logistics chains and evaluates likely changes [8]. The truck and rail demand split prediction under different West Texas Intermediate (WTI) crude oil prices estimated. When crude oil price was 50 $/barrel, the prediction for truck and train share were 68% and 32% respectively. While crude oil price changed to $200/barrel, the prediction for truck and train share became 12% and 88%, respectively [14]. The effects of changes in fuel prices on sea freight in Khorramshahr port examined in order to determine the effectiveness level of each factor in changes of transportation costs [15]. In spite of extensive literatures on freight transportation studies, yet a detailed and specific study about the effect of fuel price increases on operating transport cost of containers has not been investigated. The research question of this study is how many percent operating cost of each transport mode increase when fuel price increase by 100%.

3 Methodology

In this study, five-hundred containers were considered to be moved from the origin to the destination. The distance between origin and destination was considered as one-thousand kilometres for all freight transport modes. The current diesel price in Malaysia for the truck, train and ship were considered as MYR 1.70, 1.75 and 1.55 per litre, respectively. Fuel consumption for truck, train and ship were considered as 0.193, 4.25 and 51.476 litres per kilometre respectively [16]. Driver wage is on an hourly basis and it varies for different modes of transport. Maintenance cost of transport vehicles was considered per kilometre of travel. Assumed capacity for truck, train and ship were 2, 40 and 500 TEUs per vehicle, respectively. TEU is the twenty-foot equivalent unit and one TEU indicates one 20-foot ISO-standard intermodal container [17]. The average speed for truck, train and ship were assumed as 90, 60 and 30 kilometres per hour respectively. The Capacity of vehicles was considered on the basis of the number of TEUs that each one of them can carry. Total operating cost (TC) was calculated using the Equation 1.

$$TC = [(Fp \times Fc \times Td) + (Wa \times Tt) + (Ma \times Td)] \times [Nc/Cv] \quad (1)$$

where TC, Fp, Fc, and Td are transport cost (MYR), fuel price (RM/liter), and fuel consumption (liter/km), travel distance (Km), respectively. Wa, Tt, Ma, Nc, and Cv represent wage (RM/h), travel time (h), maintenance (RM/Km), the number of containers, and capacity of the vehicle (TEU) consequently.

4 Results and discussions

The result of operating cost calculation for each transportation mode are depicted in Tables 1-3, and their related graphs are shown in Fig. 2. The operating costs were calculated based on current diesel price, 20%, 40%, 60%, 80% and 100% price increase for all freight transport vehicles. In the aforementioned tables, the cost of one vehicle, total cost and cost per TEU were calculated. The cost of one vehicle is related to the generated transport cost for the movement of a number of containers according to the capacity of each type of vehicle from the origin to the destination. The total cost is referred to the cost for the shipment of all containers according to a number of necessary certain vehicles. The cost per TEU indicates the cost for the movement of one container from the source to the destination. Comparing Tables 1-3 indicated that freight transport by truck was more expensive than the train and ship. According to current fuel price, operating cost of one container movement by truck was RM 702.93, while cost for train and ship were MYR 292.18 and MYR 181.57 respectively. Also, when fuel price changed to two times more expensive than the current price, operating cost per container by train was increased to RM 185.93, and truck and ship cost increased to RM 164.05 and 159.57 respectively. Fig. 2 show the relationship between fuel price as the independent variable and operating cost per TEU as the dependent variable for the truck, train and ship, respectively. These relationships for all transport vehicles were linear. The changes of operating cost of transport in percentage for different modes of transport against percentage of fuel price increase were compared and showed in Fig. 5.

Table 1. Operating cost of truck across fuel price increase.

Fuel Price (RM/L)	Wage (RM/H)	Maintenance (RM/Km)	Cost of one Vehicle (RM)	Total Cost (RM)	Cost/TEU (RM)
1.70	7	1	1405.87	351469.44	702.93
2.04	7	1	1471.49	367874.44	735.74
2.38	7	1	1537.11	384279.44	768.55
2.72	7	1	1602.73	400684.44	801.36
3.06	7	1	1668.35	417089.44	834.17
3.40	7	1	1733.97	433494.44	866.98

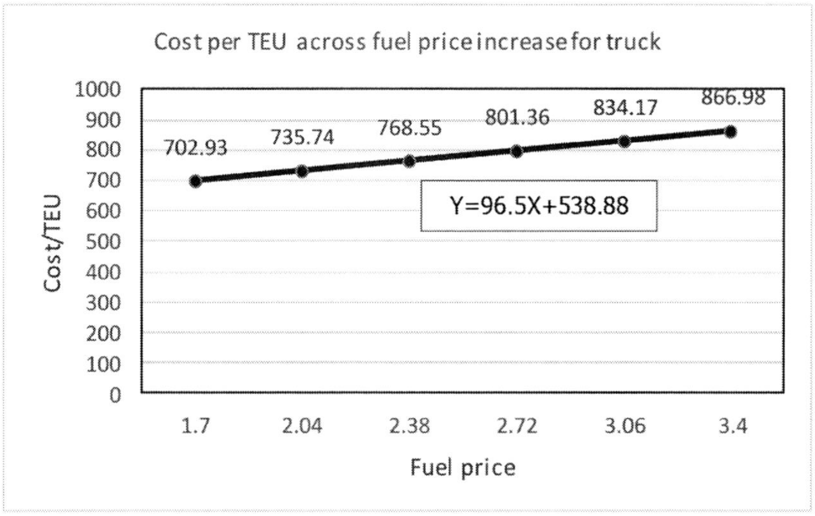

Fig. 2. Cost per TEU across fuel price increase for truck.

Table 2. Operating cost of train across fuel price increase

 https://doi.org/10.1051/e3sconf/20183401022

Fuel Price (RM/L)	Wage (RM/H)	Maintenance (RM/Km)	Cost of one Vehicle (RM)	Total Cost (RM)	Cost/TEU (RM)
1.75	15	4	11687.5	146093.75	292.18
2.10	15	4	13175.0	164687.50	329.37
2.45	15	4	14662.5	183281.25	366.56
2.80	15	4	16150.0	201875.00	403.75
3.15	15	4	17637.5	220468.75	440.93
3.50	15	4	19125.0	239062.50	478.12

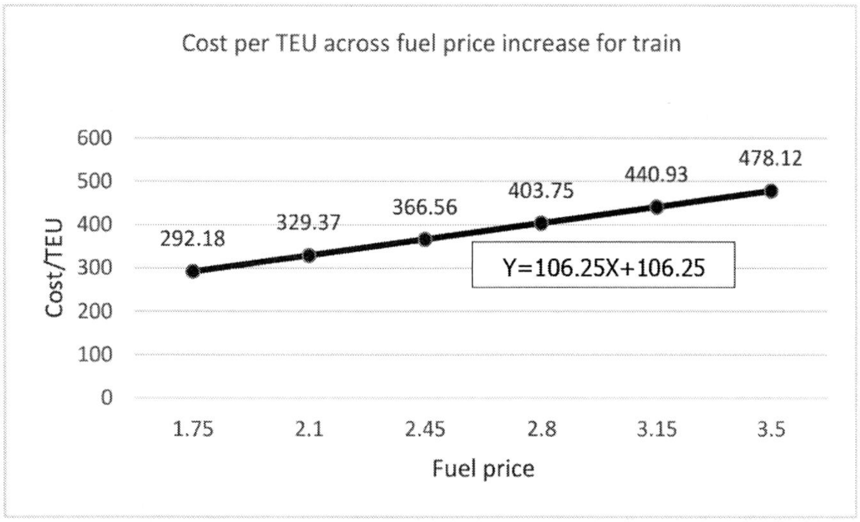

Fig. 3. Cost per TEU across fuel price increase for train.

Table 3. Operating cost of ship across fuel price increase

Fuel Price (RM/L)	Wage (RM/H)	Maintenance (RM/Km)	Cost of one Vehicle (RM)	Total Cost (RM)	Cost/TEU (RM)
1.55	30	10	90787.80	90787.80	181.57
1.86	30	10	106745.36	106745.36	213.49
2.17	30	10	122702.92	122702.92	245.40
2.48	30	10	138660.48	138660.48	277.32
2.79	30	10	154618.04	154618.04	309.23
3.10	30	10	170575.60	170575.60	341.15

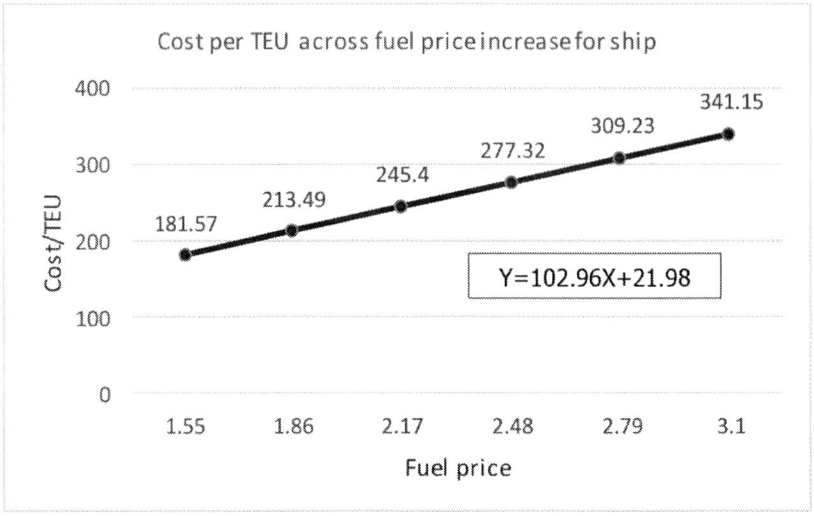

Fig. 4. Cost per TEU across fuel price increase for ship.

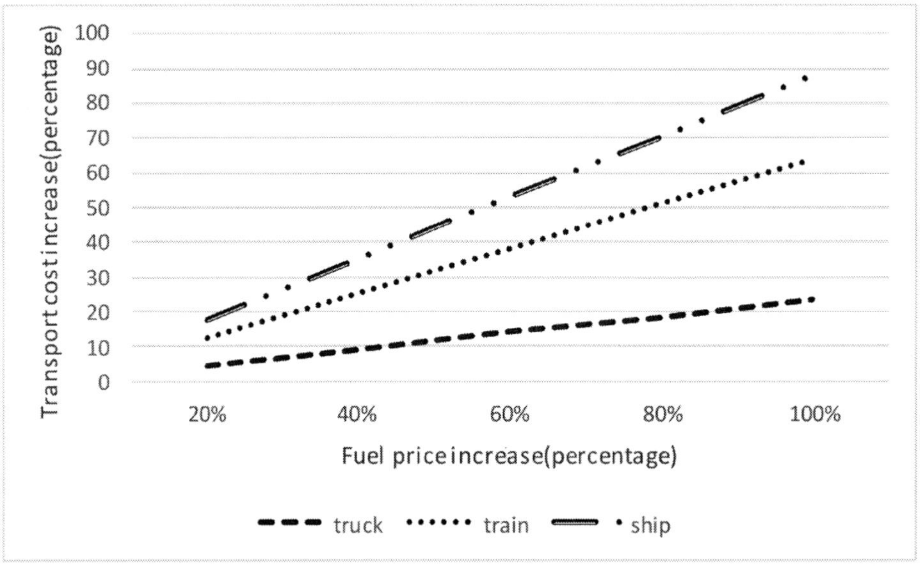

Fig. 5. Comparison of cost of transport modes against fuel price increase.

5 Conclusions

In this paper, the operating cost equation of freight transport was used to evaluate transport cost of freight container across fuel price percentage increase for truck, train and ship. Based on the operating cost equation and assumptions considered in this study, the following can be concluded. The truck was the most expensive vehicle and its cost per TEU was about 2.5 times more than the train and almost more than four times compared to the ship. Moreover, there was linear relationship between fuel price and the operating cost for different freight transport vehicles. The highest operating cost increase was related to train and followed by truck and ship. The rate of fuel price increase has the most significant effect on the operating cost of the ship. The operating cost of the ship increased by 87.8%

when fuel price increases by 100%. Whereas, for train and truck, 100% fuel price increases led to an increment in operating cost by about 63.6% and 23.3% respectively.

References

1. T.G, Crainic, *Handbook of Transportation Science*, Springer US, (2003).
2. E. Kreutzberger, C., Macharis, L., Vereecken, J., Woxenius, *Nectar Conference* **7** (2003).
3. T.R. Lakshmanan, X., Han, Transportation Research Part D: Transport and Environment **2**, 1-5 (1997).
4. P.J., Marcotullio, E., Williams, J.D., Marshall, The Journal of Environment & Development **14**, 125-48 (2005).
5. A.M., Hricko, Environmental Health Perspectives, **114**, 204-205 (2006).
6. J., Léonardi, M., Baumgartner, Transportation Research Part D: Transport and Environment **9**, 451-64 (2004).
7. A., Ricci, I., Black, Research In Transportation Economics **14**, 245-85 (2005).
8. Transportation Economics and Management Systems (TEMS), *Maritime Administration, Department of Transportation* (2008).
9. C. Macharis, E., Van Hoeck, E., Pekin, T., Van Lier, Transportation Research Part A: Policy And Practice **44**, 550-61 (2010).
10. Danish Ministry of Transport, *Review of European Studies* (2004).
11. C. Schreyer, C., Schneider, M., Maibach, W., Rottengatter, C., Doll, D., Schmedding, *Final report,* INFRAS/IWW, Zurich (2004).
12. Economic and Social Commission for Asia and the Pacific (ESCAP), *Review of Developments in Transport and Communications in the ESCAP Region 1996-2001*, New York, (2001).
13. M., Janic, Transportation Research Part D: Transport and Environment **12**, 33-44 (2007).
14. T., Hwang, Y., Ouyang, Journal Of The Air & Waste Management Association **64**, 2-12 (2014).
15. R.D., Bahmanshir, M.A., Falahi, Advances in Management and Applied Economics **5**, 87 (2015).
16. P.D., Cenek, R.J., Kean, I.A., Kvatch, N.J., Jamieson, *NZ Transport Agency Research Report 497*, Wellington, Oct. (2012).
17. P., Brodie, *Dictionary of Shipping Terms*, ed. 2, Lloyd's of London Press LTD (1994).

E3S Web of Conferences 34, 01023 (2018) https://doi.org/10.1051/e3sconf/20183401023
CENVIRON 2017

A Case Study on the Walking Speed of Pedestrian at the Bus Terminal Area

Mohd Firdaus Mohamad Ali[1], *Muhamad Salleh* Abustan[1,*], *Siti Hidayah* Abu Talib[1], *Ismail* Abustan[2], *Noorhazlinda* Abd Rahman[2], and *Hitoshi* Gotoh[3]

[1]Faculty of Civil and Environmental Engineering, Universiti Tun Hussein Onn Malaysia, 86400 Parit Raja, Batu Pahat, Johor, Malaysia
[2]School of Civil Engineering, Universiti Sains Malaysia, 14300 Nibong Tebal, Penang, Malaysia
[3]Department of Civil and Earth Engineering, Kyoto University, Japan

Abstract. Walking speed is one of the factors in understanding the pedestrian walking behaviours. Every pedestrian has different level of walking speed that are regulated by some factors such as gender and age. This study was conducted at a bus terminal area with two objectives in which the first one was to determine the average walking speed of pedestrian by considering the factors of age, gender, people with and without carrying baggage; and the second one was to make a comparison of the average walking speed that considered age as the factor of comparison between pedestrian at the bus terminal area and crosswalk. Demographic factor of pedestrian walking speed in this study are gender and age consist of male, female, and 7 groups of age categories that are children, adult men and women, senior adult men and women, over 70 and disabled person. Data of experiment was obtained by making a video recording of the movement of people that were walking and roaming around at the main lobby for 45 minutes by using a camcorder. Hence, data analysis was done by using software named Human Behaviour Simulator (HBS) for analysing the data extracted from the video. The result of this study was male pedestrian walked faster than female with the average of walking speed 1.13m/s and 1.07m/s respectively. Averagely, pedestrian that walked without carrying baggage had higher walking speed compared to pedestrian that were carrying baggage with the speed of 1.02m/s and 0.70m/s respectively. Male pedestrian walks faster than female because they have higher level of stamina and they are mostly taller than female pedestrian. Furthermore, pedestrian with baggage walks slower because baggage will cause distractions such as pedestrian will have more weight to carry and people tend to walk slower.

1 Introduction

Every person has their own preferred walking behaviour and the average walking speed of each person is different based on several factors such as age, gender, residential or working

* Corresponding author: sallehb@uthm.edu.my

area, health etc. [1-4, 11]. These factors can give impacts on the evacuation process during natural disaster or emergency situations in either confined or unconfined area. Other than that, pedestrian crossing time will also be influenced by the aforementioned factors that the timing for signalized crossing does not enough for certain pedestrian. Different people have different rate of walking speed and each person tend to walk at a speed that they find most comfortable for themselves and according to [5] the average walking speed of Malaysian was 1.16m/s at crosswalk. A study on the average walking speed can also be implemented at the bus terminal area to obtain the more accurate value of the walking speed of pedestrian. The difference that will be obtained in comparison with the previous study that was done by [5] can be useful for the improvement in traffic system, road design for pedestrian at signalized crossing and for the safety of building design for evacuation process during emergency situations.

In Malaysia, accidents are the third highest cause of death after heart and cancer diseases [6]. According to [7], it was reported that Malaysian pedestrian died due to traffic accidents were at the third rank in which the total of Malaysian pedestrian that were killed in 10 years from 2007 to 2016 was 5,472 and in 2016 alone was 511 as reported by [7]. The fatalities of pedestrian are usually caused by hit-and-run accidents, jaywalking, crossing busy roads and dangerous way of driving by road users and rank of pedestrian fatalities are at third rank after motorist and motorcyclist. Numbers of pedestrian accidents are rising due to the increasing number of vehicles and drivers yearly.

Pedestrian have different value of walking speed while walking at different locations as they walk based on the purpose and the environment of the walkway. Walking behaviour can be measured by using several parameters such as walking direction, walking speed, walking experience and other parameters that are considerable to be used [4]. In addition, there are several factors that influence the pedestrian walking behaviour and can be put into consideration in this field of research studies and they are pedestrian speed-density relationship, speed-flow relationship, pedestrian space requirement, pedestrian walking speed and pedestrian start-up time and capacity. Some people preferred to walk faster than others in situations like catching a public transport, rushing to work and those who do not like wasting time by walking slowly. Hence, it is beneficial to study about the walking behaviour to measure the walkability level in bus terminal area.

This research focuses on pedestrian at bus terminal that considers the pedestrian or the traveller that are using the terminal for transits or pedestrian who uses public transport. The data were analysed based on factors of gender, age, with and without carrying baggage. Therefore, this study aims to obtain the average walking speed of pedestrian with baggage, without baggage and the difference between average walking speed at terminal and at crosswalk. Density of vehicles on highway are categorized by the measurement of traffic flow and the traffic level based on speed of vehicles, density, type of vehicles, number of lanes, traffic phase etc. was used by traffic quality services are measured using qualitative measure using Level of Service (LOS). LOS for pedestrian facilities, the effectiveness is measured by pedestrian per unit area and the methods used to analyze is using Highway Capacity Manual (HCM). Based on T. V. Mathew (2014) [8], LOS of pedestrian consist of LOS A (Pedestrian Space > $5.57m^2/p$), LOS B (Pedestrian Space > $3.716\text{-}5.574m^2/p$), LOS C (Pedestrian Space > $2.230\text{-}3.716m^2/p$), LOS D (Pedestrian Space > $1.394\text{-}2.230m^2/p$), LOS E (Pedestrian Space > $0.734\text{-}1.394m^2/p$) and LOS F (Pedestrian Space = $0.734m^2/p$).

2 Methodology

During the experiment, a camcorder was used as the equipment for collecting the data of pedestrian and for the analysis part; it was done by using Human Behaviour Simulator (HBS) software that was plugged into Autodesk Maya software. The observation was conducted at the selected walkway that has high number of pedestrian. The age categories were categorized into 7 groups that are children (<9 years), adult men (10~39 years), adult women (10~39 years), senior adult men (40~69 years), senior adult women (40~69 years), over 70 years and disabled person. The pedestrian movement were recorded while walking at the pedestrian walkway or pedestrian crossing area and the video had been analysed by using Autodesk Maya and Human Behaviour Simulator (HBS) software to get the walking speed of each randomly selected person. The age groups were defined based on pedestrian attire and physical size that were done by visual judgment of author in analysing the data collected from video tape camera. Figure 1 shows image of female adult pedestrian tracking and the data of pedestrian coordinates was obtained from HBS tools.

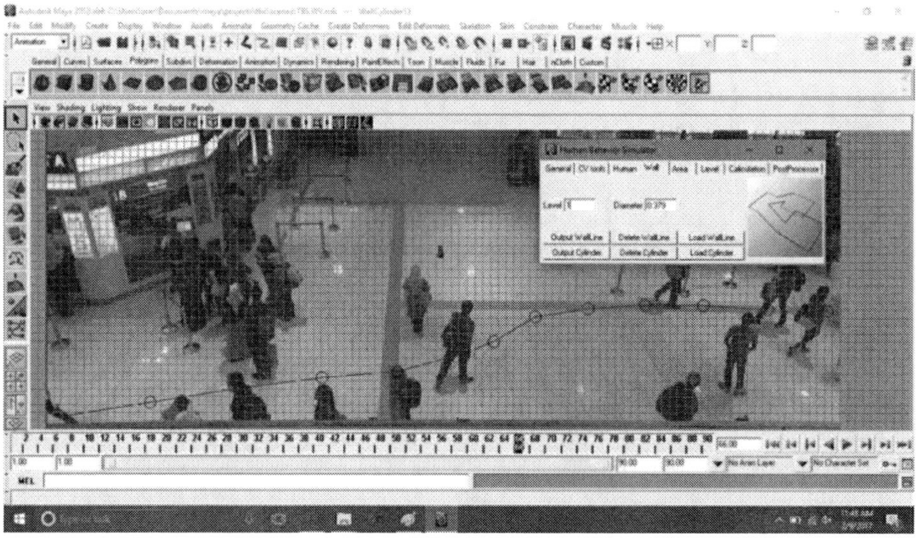

Fig. 1. Example of pointing out the coordinates of pedestrian movements.

There are 7 groups of pedestrians which are child, male adult, female adult, male senior adult, female senior adult, over 70 and disabled person and the descriptions for each category are shown in Table 1.

Table 1. Age categories' descriptions.

Category	Description of age classification
Children	Smaller body size (height), childish attire, normally walking with family members.
Adult (Men and women)	Can be seen from the physical appearance, attire and body movement that normally walk faster than "Over 70" category.
Senior adult (Men and women)	
Over 70 (Men and women)	Physical appearance, mostly walk slower than "Senior adult" category and some walk by using sticks.
Disabled person	Can be seen from their unusual gait and their appearance.

Data of experiments were analysed by using Human Behaviour Simulation (HBS) that was plugged into Autodesk Maya software. The video was converted from (avi.) format to (jpeg.) format by using Adobe after Effects CS4 software. The image sequences were imported to Autodesk Maya software and were analysed by every 1 second of video duration and the calculation of the data obtained from Autodesk Maya of walking speed among the selected pedestrian was done by using Microsoft Excel. Hence, the data had been classified in groups by categories of gender, age, person carrying baggage and person without carrying baggage.

3 Results and discussions

The results of this study are presented in Table 2 until Table 7.

Table 2. Pedestrian: average walking speed without carrying baggage.

Category	Number of data	Average (m/s)	Standard deviation (m/s)	Range	
				High	Low
Child (below 9)	2	0.63	0.17	0.75	0.51
Male adult (10-39)	186	1.20	0.07	1.94	1.10
Female Adult (10-39)	146	1.16	0.10	1.90	0.91
Male senior adult (40-69)	48	1.16	0.11	1.80	0.93
Female senior adult (40-69)	19	1.07	0.20	1.17	0.82
Over 70	0	0	-	-	-
Disabled person	1	0.87	-	-	-
Average	-	1.02	0.13	-	-
Total	402	-	-	-	-

The average walking speed of the observed pedestrian without carrying baggage was 1.02 m/s. Table 2 shows that male adult pedestrian group walked faster than others. The speed of this group was also the highest as this group covers teenage and matured pedestrian that age from 10 to 39 years old with walking speed of 1.20 m/s. Secondly, it is logical as walking speed of pedestrian tends to decline as the pedestrian becomes older. The average of the lowest range of speed was 0.63 m/s as obtained from child category.

Table 3. Pedestrian average walking speed with baggage.

Category	Number of data	Average (m/s)	Standard deviation (m/s)	Range	
				High	Low
Child (below 9)	0	-	-	-	-
Male adult (10-39)	21	1.15	0.04	1.20	1.06
Female Adult (10-39)	43	1.05	0.07	1.12	0.86
Male senior adult (40-69)	13	1.00	0.10	1.09	0.83
Female senior adult (40-69)	8	0.98	0.10	1.09	0.83
Over 70	0	0	-	-	-
Disabled person	0	0	-	-	-
Average	-	0.70	0.31	-	-
Total	85	-	-	-	-

Average walking speed of pedestrian carrying baggage from the observations at bus terminal area was 0.70 m/s. Table 3 shows female adult pedestrian walked slower than male adult pedestrians which are 1.15 m/s and 1.05 m/s respectively and same goes to the comparison between male and female senior adult in which the walking speed rate of male senior adult is slightly higher than senior female adult pedestrians with the speed of 1.00 m/s and 0.98 m/s respectively. That shows both male pedestrian groups walked faster than female pedestrian groups. Therefore, the highest walking speed of this group was male adult pedestrian with 1.15 m/s.

Table 4. Average walking speed of pedestrian with and without carrying baggage correspond to age.

Category	Percentage (%)	Average walking speed (m/s)	
		With baggage	Without baggage
Child (below 9)	-	0	0.63
Male adult (10-39)	4	1.15	1.20
Female Adult (10-39)	9	1.05	1.16
Male senior adult (40-69)	13	1.00	1.16
Female senior adult (40-69)	8	0.98	1.07
Over 70	-	0	0
Disabled person	-	0	0.87
Total	34	-	-

Table 4 shows the reduction of average walking speed when pedestrian carrying baggage while walking at bus terminal. The result shows that carrying baggage was one of the reduction factors on walking speed of pedestrian because of the size and weight of baggage. Male senior adult category had the highest reduction walking speed with 13% off than walking speed of pedestrian without baggage. It's followed by female adult pedestrian and female senior adult with 9% and 8% drop respectively. Male adult pedestrian category had the lowest reduction on walking speed, with only 4% drop and male adult still walked faster compared to other age groups. Therefore, pedestrian that walk while carrying baggage have slower walking speed as they need to carry more weight that will distract them to walk faster. This can be supported by [12] that have observed pedestrian who walk with baggage tend to walk slower than non-carrying baggage that corresponds to the size and weight of baggage. Consequently, the total reduction of the average walking speed between pedestrian carrying baggage and without carrying baggage was 34%.

Table 5. Pedestrian walking speed correspond to gender.

Gender	Number of data	Average (m/s)	Standard deviation (m/s)	Range	
				High	Low
Child	2	0.63	0.17	0.75	0.51
Male	234	1.13	0.08	1.00	0.08
Female	165	1.07	0.12	0.64	0.14
Over 70	0	0	-	-	-
Disabled person	1	0.87	-	-	-
Average	-	0.93	0.12	-	-
Total	402	-	-	-	-

Table 5 shows that male pedestrian was found to walk faster than female pedestrian with the speed of 1.13 m/s and 1.03 m/s respectively. These findings support the results of previous studies [2, 9]. According to [10], female pedestrian walked slower with the speed of 1.07 m/s compared to male that walked with the speed of 1.24 m/s. Female pedestrian specifically in Malaysia tend to walk slower compared to male pedestrian because of their attire in which some of them are wearing *Baju Kurung* which is one of the Malaysian female traditional attires, skirts and high heels also can cause them to walk slower. These findings also supported by the results from previous study of [10] that found female pedestrians walked slower than male pedestrian by nearly 0.17 m/s of difference.

Table 6. Average walking speed of Malaysian pedestrian at crosswalk [5].

Category	Number of data	Average walking speed (m/s)	Standard deviation (m/s)
Child (below 9)	90	1.06	0.15
Male adult (10-39)	337	1.35	0.16
Female Adult (10-39)	351	1.20	0.15
Male senior adult (40-69)	175	1.14	0.15
Female senior adult (40-69)	139	1.04	0.13
Average	-	1.16	0.15
Total	1,092	-	-

Table 7. Average walking speed of Malaysian pedestrian at bus terminal area.

Category	Number of data	Average walking speed (m/s)	Standard deviation (m/s)
Child (below 9)	2	0.63	0.17
Male adult (10-39)	186	1.20	0.07
Female Adult (10-39)	146	1.16	0.10
Male senior adult (40-69)	48	1.16	0.11
Female senior adult (40-69)	19	1.07	0.20
Average	-	1.04	0.13
Total	401	-	-

The average walking speeds of pedestrian are varying at different locations because of the surroundings, environment, walking space etc. For this study, the comparison was made between the average walking speed at bus terminal and pedestrian crosswalk. The previous study that had been conducted by [5] was to determine the average walking speed of pedestrian at crosswalk area and the result is shown in Table 6. Table 7 shows the results of the average walking speed of pedestrian at bus terminal area as conducted for this research. The results of the average walking speed of both locations have differences in which some pedestrian at crosswalk walked faster if compared to those at bus terminal. The highest difference of percentage on average walking speed at crosswalk and terminal area was child with 40% reduction percentage. The difference of percentage is higher because the number of sample of child at bus terminal as observed was lower than crosswalk area. Besides, both male and female adult pedestrian at crosswalk had higher walking speed than male and female adult pedestrian at bus terminal with 11% and 3% differences respectively. On the contrary, male senior adult and female senior adult at bus terminal had higher walking speed compared to male senior adult and female senior adult pedestrian at crosswalk with 2% and 3% differences respectively.

4 Conclusions

From the data analysis, there were slight differences in walking speed of the aforementioned factors. Pedestrian without baggage had the walking speed of 1.02 m/s while the walking speed of pedestrian with baggage was 0.70 m/s. Moreover, the average walking speed of pedestrian by gender both male and female were 1.13 m/s and 1.07 m/s respectively. The obtained results have proven that pedestrian with baggage tend to walk slower than pedestrian without baggage and male pedestrian walked faster than female pedestrian. The data from "Child", "Over 70", and "Disabled Persons" were not being able to be obtained precisely because most of the pedestrian at the location were not among those three categories. In comparison, pedestrian at crosswalk walked faster than pedestrian at the bus terminal area due to some factors that give influence to the decision of every pedestrian to either walk faster or slower such as, this situation will happen when the LOS of pedestrian increases in which the higher density of pedestrian at the bus terminal will cause the pedestrian to walk slower as the walkway or their personal space of walking are smaller. Other than that, time constrain can be considered as a factor that affect the walking speed of pedestrian at crosswalk to be higher than the walking speed at bus terminal. For example, some pedestrian at crosswalk tend to walk faster when catching the regular public buses at one stop to another and it is different than the situation of the buses at bus terminal area that are time-based centralized that helps people in estimating their time for the departure. Additionally, people tend to walk faster at crosswalk because of the hot or rainy weather and due to time limitation given for crossing the road. Therefore, this study can be useful as a reference for understanding the walking speed of Malaysian specifically at the bus terminal area and this field of research can be expanded in understanding the Malaysian walking speed in more locations of the pedestrian walking facilities for future improvements.

This research is funded by the Ministry of Higher Education (MOHE) of Malaysia under Fundamental Research Grant Scheme (FRGS) (Vot 1526).

References

1. T.J. Gates, D.A Noyce, A.R. Bill, N. Van Ee, T.J. Gates, *Proceeding of TRB 2006 Annual Meeting* (2006)
2. K.K. Finnis, D. Walton, *2nd International conference on sustainability engineering and science*
3. R. Rastogi, S. Chandra, *European Transport* (Trasporti Europei (EUT Edizioni Università di Trieste di Trieste, 2013)
"*Pedestrian Flow Characteristics for Different Facilities and Situations,*" (2013)
4. D. I. Azmi, H. A. Karim, M. Z. M. Amin, Journal of Asian Behavioral Studies, **11**, 97-110 (2013)
5. M.S. Abustan, E. Harada, H. Gotoh, *Proceeding of Coastal Engineering, JSCE*, **3**, 56-60 (2012)
6. B. H. Goh, K. Subramaniam, Y. T. Wai, A. A. Mohamed, A. Ali, International Journal for Traffic and Transport Engineering **2,** 323-332 (2012)
7. JKJR. (2016). *Jabatan Keselamatan Jalan Raya Malaysia.* Retrieved from http://www.jkjr.gov.my/ms/maklumat_keselamatan/statistik/func-startdown/125/
8. T. V. Mathew, *Indian Institute of Technology Bombay* (2014)
9. B. P Devkota, *Proceeding of IOE Graduate Conference*, 185 (2014)
10. K. Rahman, N. A. Ghani, A. A. Kamil, A. Mustafa, Research journal of applied sciences, engineering and technology **4**, 4299-4304 (2012)
11. F. Ghani, J.N., Rachele, S., Washington, G., Turrell, Prev Med Rep. **4**, 75–80 (2016)
12. K.K, Laxman, R. Rastogi, S., Chandra, Journal of Urban Planning and Development, **136,** 23–33 (2010)

Structural Effects of Reinforced Concrete Beam Due to Corrosion

Hamidun Mohd Noh[1,*], *Nur'ain* Idris[2], *Nurazuwa* Md Noor[2], *Norliana* Sarpin[1], *Rozlin* Zainal[1], and *Narimah* Kasim[1]

[1]Department of Construction Management, Faculty of Technology Management and Business, Universiti Tun Hussein Onn Malaysia, 86400, Batu Pahat, Johor
[2]Department of Structure and Material, Faculty of Civil Engineering and Environment, Universiti Tun Hussein Onn Malaysia, 86400, Batu Pahat, Johor

Abstract. Corrosion of steel in reinforced concrete is one of the main issues among construction stakeholders. The main consequences of steel corrosion include loss of cross section of steel area, generation of expansive pressure which caused cracking of concrete, spalling and delaminating of the concrete cover. Thus, it reduces the bond strength between the steel reinforcing bar and concrete, and deteriorating the strength of the structure. The objective of this study is to investigate the structural effects of corrosion damage on the performance of reinforced concrete beam. A series of corroded reinforced concrete beam with a corrosion rate of 0%, 20% and 40% of rebar corrosion is used in parametric study to assess the influence of different level of corrosion rate to the structural performance. As a result, the used of interface element in the finite element modelling predicted the worst case of corrosion analysis since cracks is induced and generate at this surface. On the other hand, a positive linear relationship was sketched between the increase of expansive pressure and the corrosion rate. Meanwhile, the gradient of the graph is decreased with the increase of steel bar diameter. Furthermore, the analysis shows that there is a significant effect on the load bearing capacity of the structure where the higher corrosion rate generates a higher stress concentration at the mid span of the beam. This study could predict the residual strength of reinforced concrete beam under the corrosion using the finite element analysis. The experimental validation is needed on the next stage to investigate the quantitative relation between the corrosion rate and its influence on the mechanical properties.

1 Introduction

Corrosion reinforcement is one of the main causes of structural deterioration and the most predominant degradation mechanisms in the reinforced concrete structures. It is usually associated with carbonation phenomenon or chlorides penetration, which generally induces uniform and localized attacks respectively. Corrosion of steel in concrete causes internal

* Corresponding author: hamidun@uthm.edu.my

damage to reinforced concrete elements, owing to the loss of steel area and the formation of associated expansive corrosion products.

The properties of oxide layers as the corrosion products have a great influence on corrosion crack in the concrete. The most influential factor is the expansion ratio of the oxide which depends on the specific type of oxide formed [1-2]. Depending on the level of oxidation, the volume increase due to rebar corrosion is commonly around 2.0 [3-4] and would be up to 6.5 times than the original iron volume which get consumed by the corrosion process because of the formation of various corrosion products [5]. Nevertheless Molina [4] claimed that the effective expansion ratio maybe less than that corresponding to a given type if the oxide diffuse in the porous structure of the concrete. Table 1 gives the typical characteristic relative volume ratios for different corrosion products [2].

When corrosion of reinforcement develops significantly, the corrosive products expand continuously and generate internal pressure to concrete around the steel bar. The continuous process of reinforcement corrosion does not only affects in structural serviceability by cracking, or even spalling the concrete cover, but also give serious impact on the structural safety by decreasing the load-bearing capacity. Besides, the physical effects of corrosion include loss of steel area, loss of bond strength between steel reinforcing bars and concrete, and reduce of concrete strength due to cracking.

Table 1. Correlation between 'α' and 'α_1' for various corrosion products.

	Name of corrosion products					
	FeO	Fe_3O_4	Fe_2O_3	$Fe(OH)_2$	$Fe(OH)_3$	$Fe(OH)_3.3H_2O$
α'	0.777	0.724	0.699	0.622	0.523	0.347
α_1'	1.80	2.00	2.20	3.75	4.20	6.40

where: α, ratio of molecular weight of iron to the molecular weight of corrosion products; α1, ratio of volume of corrosion products to the volume of iron consumed in the corrosion process.

In the present study, non-linear finite element analysis using MARC software is used to investigate the potential effects of corrosion damage on the performance of reinforced concrete beam. The main effects of corrosion are recognized and considered in the analysis. A series of corroded reinforced concrete beam are analyzed. Then, the numerical models are used in the parametric study to assess both the influence of different level of corrosion and their associated assumptions on the predicted response.

A reinforced concrete beam with size of 150mm x 200mm x 2300mm was modelled consisting of 8mm diameter steel bars as the upper rebar and the 10mm steel bars as the lower rebar. Meanwhile, 6mm diameter of stirrup was used to be the shear reinforcement and arranged at every 170mm intervals. The detail dimension and the rebar's location of reinforced concrete beam are shown in Fig. 1.

2 Numerical Simulation

2.1 Finite element modelling of beam specimen

A three-dimensional finite element model is adopted in this study (Fig. 2) using the non-linear finite element code of MARC 2010. The reinforced concrete beam was modelled in ½ from the actual dimension by taking into account the symmetry condition along the y-axis. By using the 8 nodes of hexahedral elements, concrete and the main reinforcements was classified as solid while the shear reinforcement was designed using one-dimensional (axial) truss element. By considering the changes in the adhesion strength between the reinforcement bar and concrete, the interface element was applied by using 8 nodes hexahedral elements.

2.2 Corrosion damage model

Fig. 3 shows the material properties of concrete, rebar and stirrup which summarized in Table 2. The rebar was assumed as complete elasto-plastic body and constitutive law based on the von Mises yield condition is adopted. As for the concrete, elastic-plastic constitutive law based on the Drucker-Prager yield condition is applied. The degradation ratio due to the corrosion is defined according to the degree of damage in three principal axis direction as $\emptyset_x, \emptyset_y$ and $\emptyset_z$ by assuming the tension softening. To adequately assess the progress of cracks in the concrete, it is necessary to make the tension softening stress released by subsequent tensile strength. Then, only the extracted positive principal strain component, the degree of damage in principal strain direction D_x^L, D_y^L and D_z^L are converted to the whole coordinate axes of D_x^G, D_y^G and D_z^G. The degradation ratio of each coordinate axis direction was calculated by the following Equation (1).

$$\begin{aligned}
\emptyset_x &= \sqrt{1-D_x^G}, \quad \left(D_x^L = \frac{\varepsilon_x^L}{\varepsilon_u}\right) \\
\emptyset_y &= \sqrt{1-D_y^G}, \quad \left(D_y^L = \frac{\varepsilon_y^L}{\varepsilon_u}\right) \\
\emptyset_z &= \sqrt{1-D_z^G}, \quad \left(D_z^L = \frac{\varepsilon_z^L}{\varepsilon_u}\right)
\end{aligned} \tag{1}$$

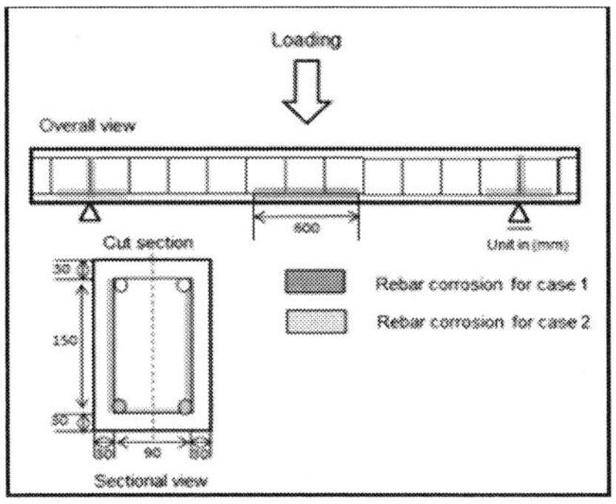

Fig. 1. Detailed dimension of Beam.

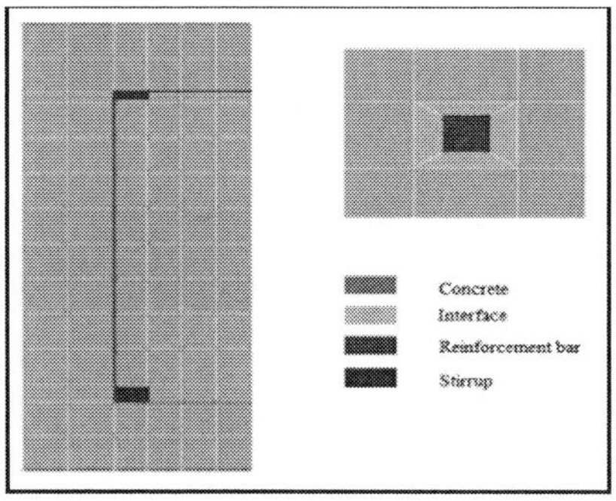

Fig. 2. Beam cross-section.

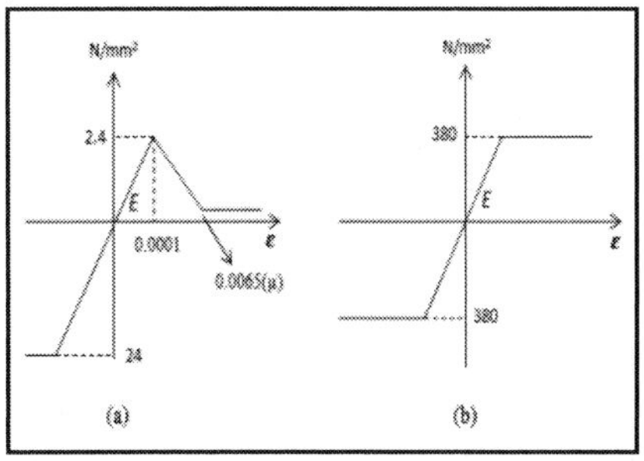

Fig. 3. Material properties of (a) concrete (b) reinforcement steel.

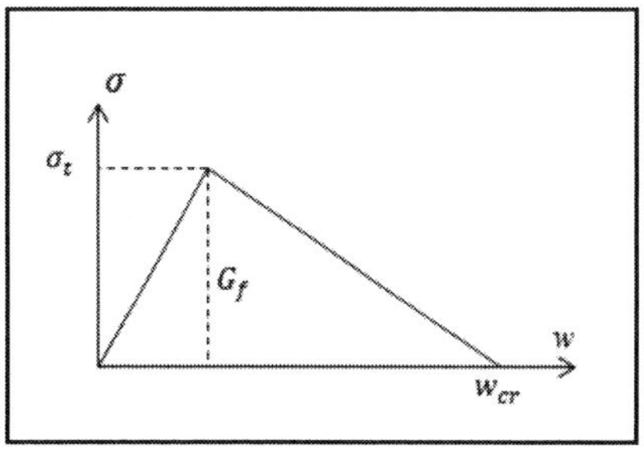

Fig. 4. Bond-slip between bond strength and the relative displacement.

Table 2. Material constants.

	Compression (N/mm^2)	Tension (N/mm^2)	Young modulus (N/mm^2)	Poison ratio
Concrete	24	2.4	25000	0.2
Reinforcement bar	380	380	200000	0.3
Stirrup	350	350	157000	0.3

Equation (1) shows the relationship between $\emptyset_x, \emptyset_y$, $\emptyset_z$ and D_x^G, D_y^G and D_z^G with the definition of D_x^L, D_y^L, D_z^L. Here, ε_u is defined as tensile strain limit. In addition, the relationship between the stress rate and the degradation ratio associated with the tension softening is defined in Equation (2).

$$\alpha_{xx} = \phi^2{}_x, \quad \alpha_{yy} = \phi^2{}_y, \quad \alpha_{zz} = \phi^2{}_z \tag{2}$$

Furthermore, in this analysis the degradation ratio was multiplied by elastic stiffness matrix and anisotropic constitutive law with damage influence is obtained as shown in Equation (3).

$$\{\sigma_{ij}\} = \{[D^e] - [D^p]\}\{\varepsilon_{kl}\}$$

$$[D^e] = \begin{bmatrix} \phi_x^2(\lambda+2\mu) & \phi_x\phi_y\lambda & \phi_x\phi_z\lambda & 0 & 0 & 0 \\ \phi_y\phi_x\lambda & \phi_y^2(\lambda+2\mu) & \phi_y\phi_z\lambda & 0 & 0 & 0 \\ \phi_z\phi_x\lambda & \phi_z\phi_y\lambda & \phi_z^2(\lambda+2\mu) & 0 & 0 & 0 \\ 0 & 0 & 0 & \phi_x\phi_y 2\mu & 0 & 0 \\ 0 & 0 & 0 & 0 & \phi_y\phi_z 2\mu & 0 \\ 0 & 0 & 0 & 0 & 0 & \phi_z\phi_x 2\mu \end{bmatrix} \tag{3}$$

In order to avoid unrealistic local failure in the analysis, adequate element length is determined by using the fracture energy of concrete G_f. In this study, the maximum dimension of the material is 20mm, compressive strength of concrete is assumed to be 24 N/mm^2 and thus G_f is calculated as 78.3 (N/m) by Equation (4).

$$G_f = 10 \text{ x } (d_{max})^{1/3} \text{ x } (f`_{ck})^{1/3} \tag{4}$$

$$G_f = \frac{1}{2}\sigma_t\, w_{cr} \tag{5}$$

Using this G_f, corresponding w_{cr} is obtained as 0.06425mm and we finally determined element length is 10mm and corresponding tensile strain limit is 6500 (μ) by equation (6).

$$\varepsilon_u = \frac{w_{cr}}{d} \tag{6}$$

3 Mechanism of deformation process due to steel corrosion

3.1 Corrosion and expansion pressure

The corrosion in concrete causes volume expansion of steel bar. Generally, the expansive pressure due to the volumetric change of the steel bar in concrete induces the tensile strains in the surrounding concrete and they are proportional to the degree of corrosion. As the expansion process begins continuously, the tensile strains developed and induces cracks in the surrounding concrete and also at the surface of concrete cover. In order to analyze the cracking of concrete cover due to steel corrosion, it is necessarily to know the relation between the amount of corrosion of steel bar and the internal pressure arises from the corrosion. Therefore, a realistic relation between the amount of corrosion and the internal expansion pressure can be investigate in this study.

3.2 Deformation of corrosion layer

Kim [6] proposed a deformation model of steel bar corrosion. It was detailed out in Fig. 5 which consist r_0 as the initial radius of steel bar and x_p is the loss radius of steel bar due to corrosion. Meanwhile, the radius increment of steel bar due to corrosive expansion under restraint and unrestraint condition (mm) represent by Ur and Δr_b, respectively while the degree of corrosion of the steel bar which based on the ratio of weight loss due to corrosion to the initial weight of steel bar is known as w_{corr}.

The corrosion product layer is compressed by the strain ε_{rust} due to expansive pressure, P. Lundgren [8] derived the relationship between normal stress versus strain of rust evaluated from a combination of experimental results and analysis together with the chosen parameter K_{rust} and p. According to their research, analysis results have a good agreement with the experimental results using the value of K_{rust} = 7.0 GPa and p = 7.0. Thus the strain ε_{rust} and the expansive pressure P is defined as follows.

$$\varepsilon_{rust} = \frac{u_r - \Delta r_b}{x_p + \Delta r_b} \tag{7}$$

$$P = K_{rust}\varepsilon_{rust}{}^{p} \tag{8}$$

3.3 Mechanical behavior of the rust

The mechanical behavior of the corrosion products needs to be known in modelling the corrosion layer. Molina [4] assume that the rust is elastic but the mechanical properties of the rust should be replaced by initial property of steel. Meanwhile, a scratching test was used by Petre-Lazar & Gerard [10] to explore the mechanical properties of the corrosion products. They conclude that rust is cohesion-less assemblage of incompressible crystals. Al-Sulaimani [11] and Cabrera & Ghoudoussi [12] have carried out pullout tests on corroded reinforcement bars concentrically placed in concrete blocks. In their study, the reinforcement corroded until the specimen was cracked and corrosion penetration was measured by the weight loss method. In addition, they performed axisymmetric finite element analyses of the test specimens and only the concrete was modelled with expansive normal stress. In this study, we also considered only the concrete domain with a constitutive model based on elastic-plastic theory.

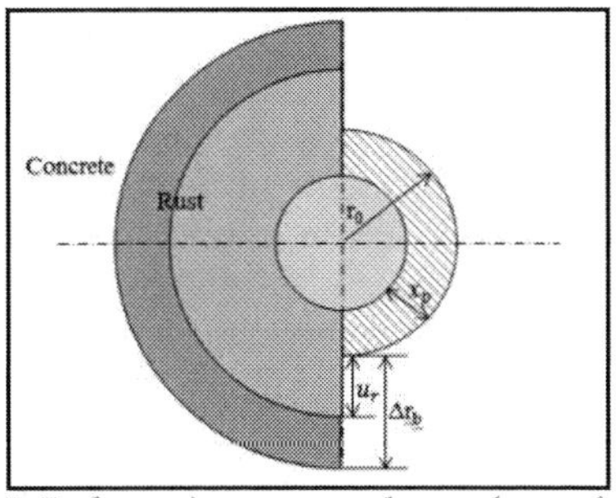

Fig. 5. Deformation process due to the evaluated steel expansion.

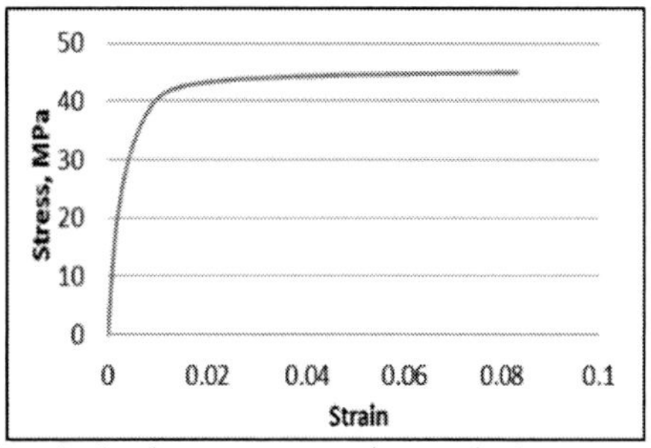

Fig. 6. Strain versus normal stress in the rust.

Since the Young's modulus of steel is much larger than concrete, Lundgren [8] assumed that the deformation of the reinforcement bar is negligible. Furthermore the stiffness of the bond layer is large enough and so the deformation of the bond layer is also negligible. Thereby, Lundgren [8] concluded that the deformation at the hole approximately equals the deformation in the corrosion layer, U_r. Therefore, the value of U_r can be calculated as the value of strain in the rust using Equation (7). This strain in the rust corresponds with the applied normal stress when cracking occurred in the analysis. The results are summarized in Fig. 6, where the normal stress versus strain in the rust is plotted. This figure indicate that the rust does not have a linear elastic behavior. In fact, the stress and strain increases monotonically until reach a point where the stress becomes constant in increasing of strain. This might be cause by the stiffness of the rust change and deteriorated. This mechanical behavior of the rust could be described by Equation (8).

4 Results and analysis

4.1 Influence of adhesion strength reduction due to rebar corrosion

In general, cracks and rebar corrosion of the reinforced concrete structures are caused by various environmental factors and reduction of adhesion strength is also observed. In this study, the adhesion strength of rebar was also considered with reference to the research by Lee [13]. The relationship of the corrosion rate and the adhesion strength of the corroded reinforcing bars were stated in Equation (9) and (10).

$$\tau_{max} = 0.34\sigma_c - 1.93 \quad (9)$$

$$\tau_{max} = 5.21e^{-0.056\Delta w} \quad (10)$$

Here, σ_c is the compressive strength of the concrete and Δw represents the corrosion rate of reinforcing steel (%). Corrosion of the rebar was analyzed with respect to the 3 levels of degradation state which are 0%, 20% and 40%. The bond strength for each corrosion rate was summarized in Table 3. In general, a reinforced concrete beam is considered to have two types of damage state with different corrosion position. For case 1, the corrosion is modelled in the main rebar with 600mm length at the mid span. Case 2 takes into account the corrosion of shear reinforcement at both ends (Fig. 1).

4.2 Analysis results

At first, the influence of adhesion condition between concrete and main reinforcement is investigated by proposed analysis. Fig. 7 shows the load-displacement relationships using adhesion model (with interface) and in complete adhesion condition (without interface). From this Fig., the strength without interface has a higher strength compared to the strength with interface elements, however the presence or absence of the interface does not give a big effect to the bond strength under the pure bending condition. In general, the interface elements describe a relation between traction and the relative displacement which included both the frictional bond model and the corrosion model[9] and the mechanical behavior and the volume of the corrosion products were given as input for the corrosion layer. So that, structure with interface elements is weaker since the cracks is induced and generate at this surface. Then, the bond-slip may occur from the interaction of internal pressure by the reinforcement steel surface, which cause cracking to the concrete layer.

Table 3. Relationship between corrosion degree and bond strength.

Corrosion Degree (%)	Bond Strength (Mpa)
0	6.23
20	1.70
40	0.55

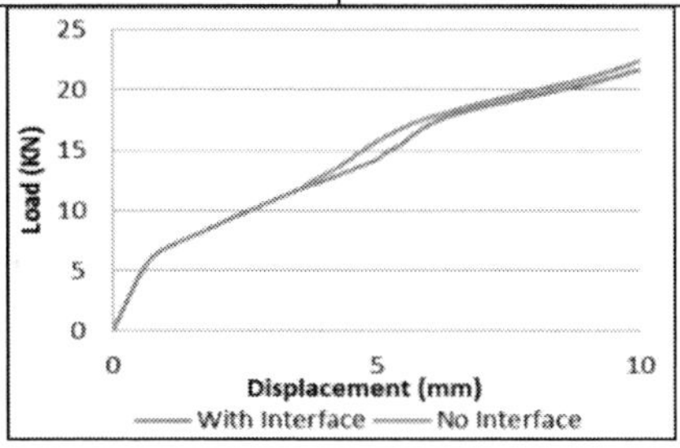

Fig. 7. The influence of the presence and the absence of interface.

4.3 Volume expansion of rust in corroded concrete

When a rebar starts to corrode, a gradual decrease of its diameter (axisymmetric corrosion is assumed) is produced, together with the generation of rust. The increase in volume induce the formation of cracks when the concrete's tensile strength is surpassed. The cracks are generated first at the bar-concrete interface and then propagate radially. As can be seen from Fig. 8, it is presented the rust expansion of 8mm diameter steel bar at 10% of corrosion degree. Under the unrestraint condition, Δr_b at α value of 0.523 and 0.622 correspond to the free rust expansion of 0.7135mm and 0.5445mm respectively. Meanwhile, under restraint concrete condition, the value of Ur at α value of 0.523 and 0.622 are 0.711mm and 0.5425mm respectively.

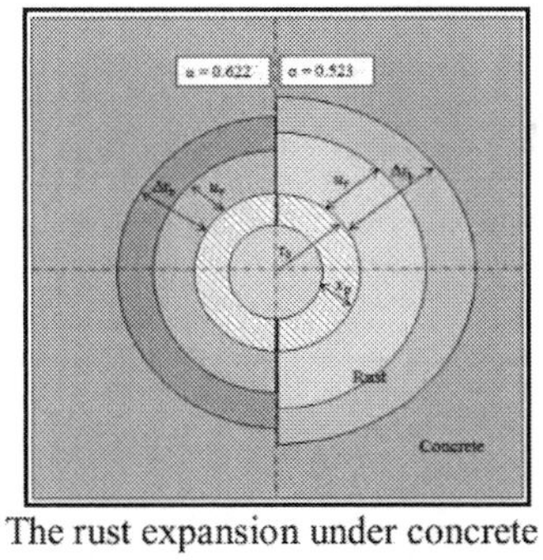

Fig. 8. The rust expansion under concrete constraint.

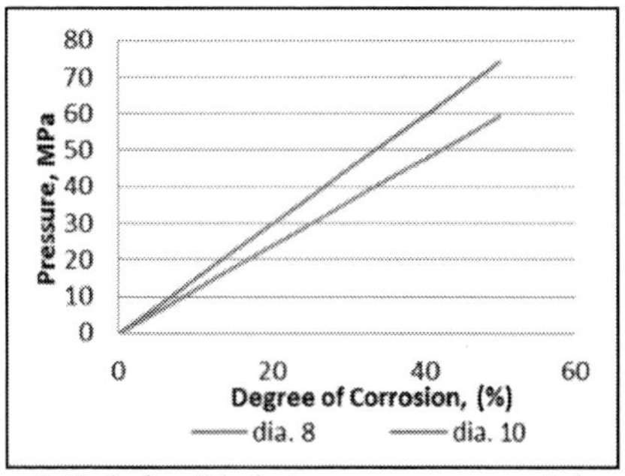

Fig. 9. Relationship between pressure and degree of corrosion.

The relation between pressure and the degree of corrosion could be expressed as a linear line shown in Fig. 9. It is also found that the gradient of this graph is decreased with the increase of steel bar diameter. This is because, the pressure created from 8mm steel bar diameter is higher than the pressure of 10mm diameter at the same degree of corrosion. The relationship between the pressure in function of degree of corrosion is stated in Equation (11) where ν_r is identified to be the relative ratio of steel density to corrosion product density.

These results coincide with the results of previous research [14, 15] which reported a decrease in maximum pressure with an increase of internal diameter. On the other hand, if the specimens have the same cover thickness with different bar sizes, it is clear that the cover of the smaller bar specimen is stiffer and if the same load is applied to both specimens, the deflection in the cover of the larger specimen will be greater than that in the smaller bar specimen.

$$P = K_{rust}\left[\frac{u_r - r_b(\sqrt{1+(v_r-1)w_{corr}}-1)}{r_b[(1-\sqrt{1-w_{corr}})+(\sqrt{1+(v_r-1)w_{corr}}-1)]}\right] \quad (11)$$

4.4 Influence of structural strength reduction due to rebar corrosion

Further analysis was conducted to investigate the influence of corrosion to the structural strength in two cases. For case 1, the corrosion is considered in the middle of main bar with 600mm length while case 2 considered the corrosion of shear reinforcements at both ends.

4.3.1 Case 1

The analysis shows that there is a significant effect on the load bearing capacity of the structure. For case 1, the bearing force of the beam is decreasing in increasing of corrosion degree. It is found that the curve of 0% corrosion have a bearing force with 28.26KN. Meanwhile, the bearing force drop until 20.72 KN and 19.78 KN for the 20% and 40% curve, respectively. This results indicates that the corrosion in mid span will cause severe condition since the stiffness decrease due to the corrosion and caused noticeable strength reduction as shown in Fig. 10. It is also found that the 40% degree of corrosion case has almost the same bearing force with the 20% degree of corrosion case. This is because, higher corrosion rate generate higher stress concentration at mid span as shown in Fig. 11 and this beam has enough bending capacity, thus it get off the serious failure under this displacement range.

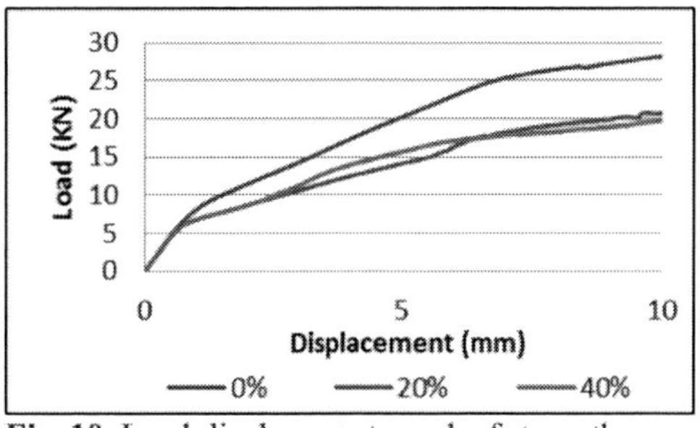

Fig. 10. Load-displacement graph of strength reduction for Case 1.

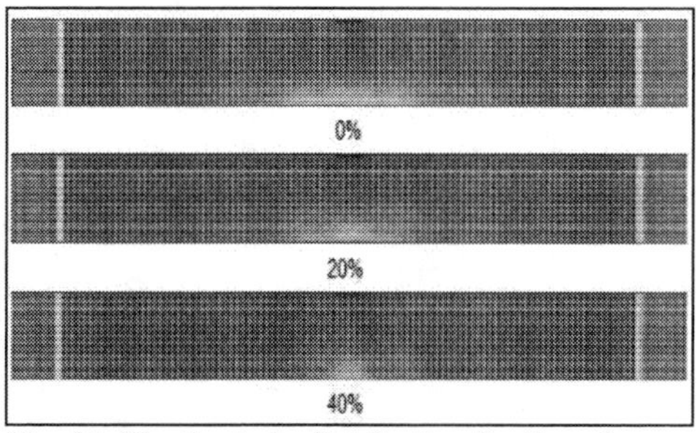

Fig. 11. Plastic strain distribution for Case 1.

4.3.2 Case 2

For the case 2 which considered corrosion state at the shear reinforcement of both ends, the bearing forces with 20% and 40% corrosion rate show around 20 percent decrease compared to the 0% corrosion case (Fig. 12). The distribution of stress at the cross section of the beam is presented in Fig. 13. Since the corroded areas consist of shear and lower reinforcement only, the stress distribution is concentrated at the lower reinforcement as the stress level increasing when the corrosion degrees increase. Overall, for this case, the corrosion at shear reinforcement have a better condition compared to the corrosion at the mid span. In general, the stress condition of the beam is influenced by the span length and cross-section. For longer span that is expected bending failure, case 1 is in danger while for shorter span that is expected shear failure, case 2 might be potential to have a failure earlier since the shear stress of the reinforcement is dominant.

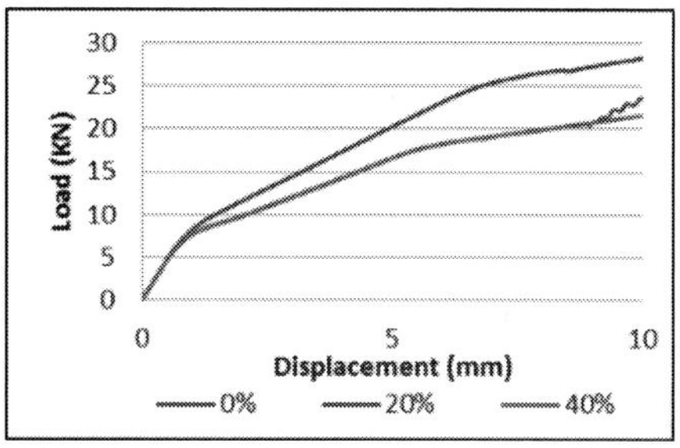

Fig. 12. Load-displacement graph of strength reduction for Case 2.

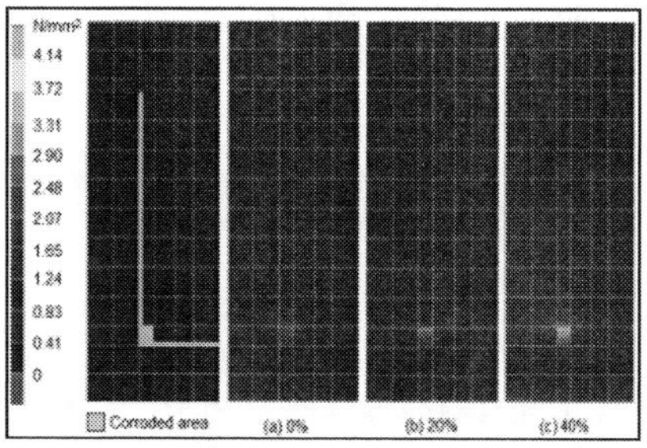

Fig. 13. Stress distribution for Case 2.

5 Conclusions

In this study, we could predict the residual strength of RC beam under the corrosion condition using the finite element analysis. However, the quantitative relation between the degree of corrosion and its influence on the mechanical properties are still needed to investigate. Thus, we are going to do the experimental validation in the next stage.

References

1. K. Tutti, *Corrosion of steel in concrete*. (Swedish Cement and Concrete Research Institute, Stockholm, Sweden, 1982)
2. P.K. Mehta, P.J.M. Monteiro, *Concrete Microstructure, Properties and Materials*, (Indian Concrete Institute, Chennai, India, 1997)
3. C. Andrade, C. Alonso, F.J. Molina, Mater. Struct. **26**, 453-464 (1993)
4. F.J. Molina, C. Alonso, C. Andrade, Mater. Struct. **26**, 532-548 (1993)
5. K. Bhargava, A.K. Ghosh, Y. Mori, S. Ramanujam, Eng. Struct. **28**,1093-1109, (2006)
6. K.H. Kim, S.Y. Jang, B.S. Jang, B.H. Oh, ACI Material Journal **107**, 14, 106-113, (2010).
7. C. Alonso, C. Andrade, J. Rodriguez, J.M. Diez, Mater. Struct. **31**, 435 – 441 (1998)
8. K. Lundgren, *Bond between corroded reinforcement and concrete*. (Chalmers University of Technology, Göteborg, Sweden, 2001)
9. D. Coronelli, P.G. Gambarova. *Proceedings EM2000*, (Fourteenth Engineering Mechanics Conference, Austin, Texas, USA, May 2000)
10. I. Pètre Lazar, B. Gerard, *Proceedings EM2000*, (14th engineering mechanics conference ASCE, Austin, TX, USA, 2000)
11. G.J. Al-Sulaimani, M. Kaleemullah, I.A. Basunbul, Rasheeduzzafar, ACI Struct. J. **87** (2), 220–231 (1990)
12. J.G. Cabrera, P. Ghoddoussi, *Bond in Concrete*, (Proceedings of an International Conference, CEB, Riga, 1992)
13. H.S. Lee, T. Noguchi, F. Tomosawa, Cement and Concrete Research **32**, 1313-8 (2002)
14. S.J Williamson, L.A. Clark, Magazine of Concrete Research, **52**, 6, Dec, 455-467 (2000)
15. Morinaga S., *Prediction of service lives of reinforced concrete buildings based on the rate of corrosion*, (Institute of technology, Shimuzu Corporation, special report No. 23, 1988)

E3S Web of Conferences **34**, 01025 (2018) https://doi.org/10.1051/e3sconf/20183401025
CENVIRON 2017

Investigation on factors that contribute to the abandonment of building in construction industry in Malaysia

Nur Farhayu Ariffin[1*], *Mohd Faizal* Md Jaafar[1], *Mohamad Idris* Ali[1], *Noram Irwan* Ramli[1], *Khairunisa* Muthusamy[1], and *Nor Hasanah* Abdul Shukor Lim[2]

[1]Faculty of Civil Engineering and Earth Resources, Universiti Malaysia Pahang, Pahang, Malaysia
[2]Faculty of Civil Engineering, Universiti Teknologi Malaysia, Johor, Malaysia

Abstract. In Malaysia, the construction sector is one of the important sectors that contribute to economic growth and employments. However, a major concern facing the construction industry is the growing rate of delays in project delivery. In the worse cases, the projects were abandoned due to some reasons when the contract period ended. Abandoned building defines as construction work that has been continuously stalled for 6 months or more, during the project completion period or beyond the scheduled date of completion. When the projects become abandoned, it gives an adverse effect on many parties such as the developer, contractor, consultant and also client. According to previous researchers, the abandonment of building causes a serious problem and need some mitigation plan to avoid this problem from occurring. This study will investigate the fundamental factors that contribute to the abandonment of building and projects in Malaysia based on the current data of the abandoned building in most states in Malaysia. The data was collected from the respondents who is in the construction industry and had experience working with abandoned housing project. Form the respondents perspective, it shows that the main factor contribute to the building abandonment is due to the financial problem facing by the developer company.

1 Introduction

The construction industry plays a very dominant role in the economy of any nation. A healthy economy usually experiences an increase in construction activities, but in a depressed economy, the incidence of project abandonment tends to be more prevalent. Nowadays, the issue related to the building abandonment had been arising frequently due to its negative effect on the country. The building abandonments had occurred not only in the commercial building but also housing project by the government and the private company. The Ministry of Housing and Local Government's with regards to the abandoned housing projects as in June 2015, Malaysia records 10,403 house buyers being affected from 15,206

* Corresponding author: farhayu@ump.edu.my

abandoned housing units on 53 housing projects [1]. According to The Star, [2] if the project has passed its promised delivery date by 10 %, it is considered late; if the delay stretches beyond 10 % to 30 %, then it is considered 'sick'; and finally, if no work has been carried out or no workers are on the project site for up to six months, it is deemed abandoned [2]. Another definition as found in an Enactment (1978) is; "Abandoned housing development means where a licensed housing developer had refused to carry out, delayed, suspended, stopped or ceased work continuously for a period of more than six months from the expected date of completion/delivery as agreed upon the contract and stated in the sales and purchase agreement" (Housing Development (Control and Licensing) Enactment, 1978). The abandonment of development projects is the act of discontinuing any activities or maintenance works on such development project within a time frame of the contract agreement and with no intention of returning back to the development [3].

The abandonment of projects gives adverse effects on parties such as the developer, contractor, consultant and client in term of relationship, reliability and reputation. There are many factors contributes to the building abandonment such as mismanagement by the company, financial problem, government policy, unskilled personnel, inadequate planning, and corruption. Other than that, problems might also come from the inaction or action on the part of owner, contractor, subcontractors, consultants or the government. When the project of the building has become abandoned, this can give adverse effects on parties such as the developer, contractor, consultant as well as client.

As we know, when the building becomes abandoned it can give negative effect on both parties involved and also to the surrounding environment. For the parties involved, they may loss the profit and also loss the belief from the citizen especially for the project that was funded by the government. As a developer company, they should know the objectives of the project so that the completion of the projects can be achieved. Besides that, some of the building was abandoned due to the problem arise from the design and structural aspect. The longer the building become abandoned, the cost of the project will increases. From the perspective view of the society, the abandonment of building can also become the place for the crimes to occur.

Therefore, regarding this matter, this study was conducted to study the factor that contributes to the abandoned building project in Malaysia and the mitigation process to prevent this problem to occur in the future.

2 Literature Review

The previous Prime Minister, Abdullah Ahmad Badawi was on 22.11.2005, The Star [2] quoted as saying, "If the projects have been monitored on a regular basis from the start, any sign of them being abandoned could have been detected and the salvaged". Usually, the most building that had been abandoned is housing building. This is due to the price and land properties itself. If the house was built in the flooding area, the buyer's rate will be decreased. Moreover, the house area near the mangrove or river should be avoided due to the safety factor and weathering effect. Despite the easily available banking home financing facilities and the strong willingness of the Malaysian government to develop the housing sector, there are still spill overs which are causing a recurring problem for the nation. There are cases of house buyers took financing from banks and bought houses under construction, but these houses were later declared "Abandoned". The developer failed to complete the housing project and declared for bankruptcy. There are also more extreme cases where the contractors abscond and are neither traceable nor contactable. The poor house buyer is now left with an unfinished house and a huge debt to repay. Figure 1 shows a statistic of an abandoned housing project in Malaysia in year 2009.

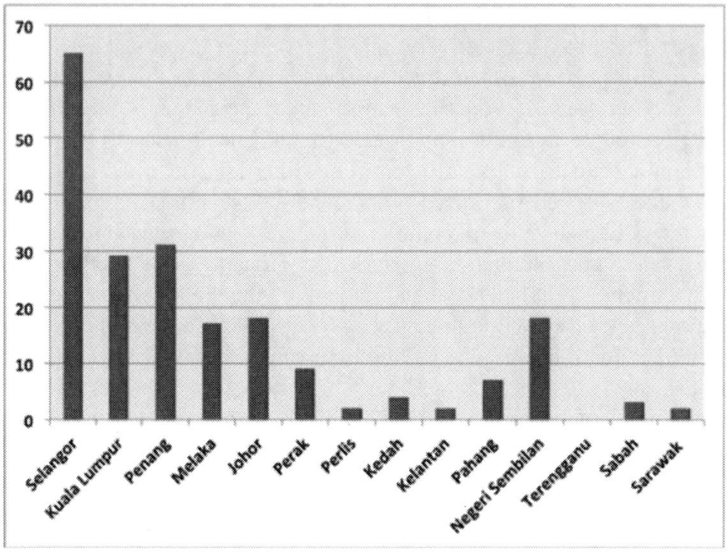

Fig. 1. Abandoned Building Projects in Malaysia in the span of 20 years, showing the comparison of a number of projects in each state [5].

3 Methodology

General causes and effects of building abandonment were analyzed by conducting a questionnaire with the respondents from quantity surveyors, civil engineers, architects, builders, contractors and others public workers. The questionnaires were distributed evenly to the expertise and were analyzed by using the Relative Importance Index (R.I.I.) technique based on the work of [4]. In this study, the questionnaire was divided into four Sections, where section A is meant for the respondent background identification while in Section B, the question asked about the knowledge of the respondent regarding abandoned building project. Meanwhile, in Section C, the question discussed factor and cause of the building abandonment. The last section which is Section D is a suggested solution from the respondent to solve the abandoned building project issue.

The data obtain from the questionnaire was analyzed using the average index method and the multiple choice question is made based on Likert's Scale for five ordinal measure of the agreement. The Relative Importance Index, (RII) is as Equation 1;

$$\text{R.I.I.} = [(5n_5 + 4n_4 + 3n_3 + 2n_2 + n_1) / 5N] \quad (1)$$

where; n_5 = Strongly Agree (SA); n_4 = Agree (A); n_3 = Unsure (US); n_2 = Disagree (DA); n_1 = Strongly Disagree (SDA); and N = number of respondents.

4 Results and discussions

4.1 Section A: Respondents background

There were 56 respondents had been answered the questionnaire. From the respondents' background information, there were 45% males and 55% females which the age is between

25 to 50 years old. Based on the respondents' background, the majority of respondents come from different scope of work in of civil engineering such site engineer, consultant, developer and contractor. From that, it was divided into two sectors which are government sector and private sector. It was analyzed that 55% of the respondents come from government sector while the rest is in the private sector as shown in Figure 2. It is important to have a respondent that understand the topic and in the same scope of work because, in order to have a reliable data, the respondent background should be in the same area of work.

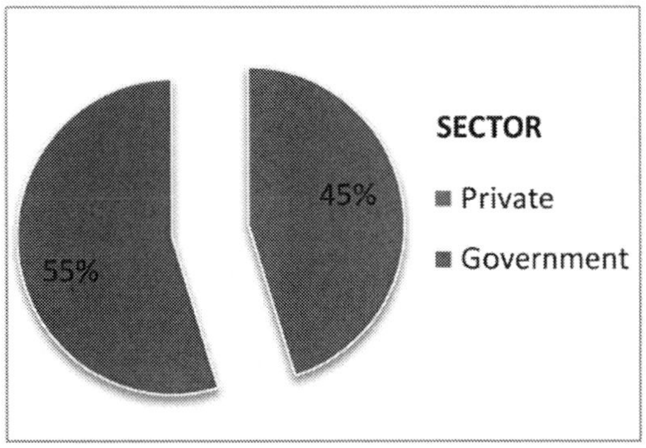

Fig. 2. Sector of the respondent represents.

4.2 Cause of Building Abandonment

A construction project usually involves many parties such as developer, supplier, engineer and buyers. Once the project is abandoned, all parties will be affected by it. These include suppliers of construction materials, transportation companies, manufacturers of plant and machinery and others. From the conducted survey as shown in Figure 3, it shows that the main cause of abandoned building is by delay of payment to the subcontractor which relative important index is 0.75. This also can be noted as a financial problem facing by the developer company. When the company has a financial problem, it will affect the progress of the development. The developer should have a stable financial in order to have a smooth and non-delay project. From the figure also, another cause that should be highlighted is mismanagement of the lead company. The company should have a proper scheduling plan so that the progress of the construction can be carefully monitored and it can prevent a delay. Once the delay occurred, the possibility of the project to be abandoned is higher.

The same results had shown from studies conducted by [5,6]. From the investigation, it revealed that the lack of adequate financing in the process of construction management leads to severe crisis. As a result, developers failed to meet the construction cost and to repay loan, which ultimately leads to abandonment of the project. The report also revealed that mismanagement, conflicts and insufficient coordination between developers and authority are also the major causes of abandoned project.

On the other hand, [7] revealed that financial problem, abuse of government incentive for small contractors, mismanagement and inefficiency of management team and the role of government has major implication towards abandoned building project [7].

https://doi.org/10.1051/e3sconf/20183401025

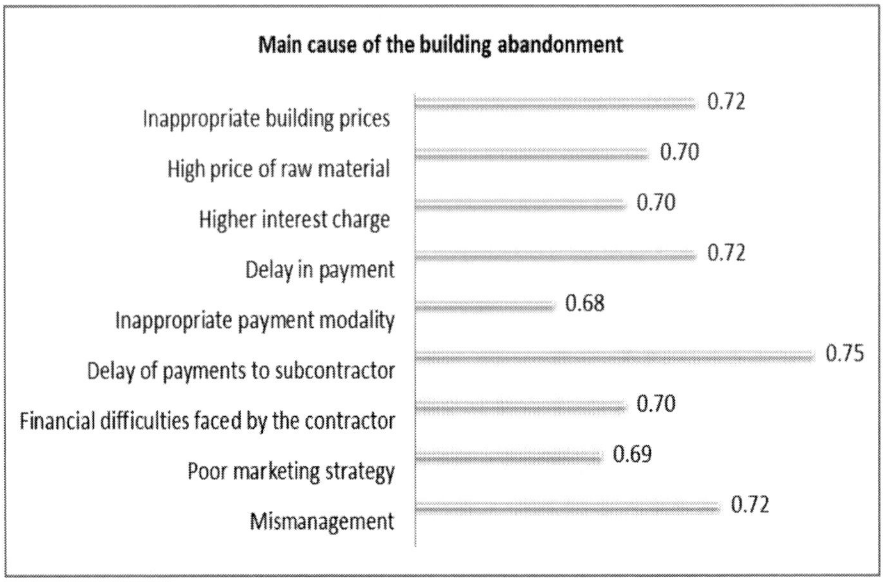

Fig. 3. The main cause of the building abandonment.

4.3 Mitigation Plan

The survey was concluding with the certain suggestion and recommendations that have been given to overcome or prevent the cases of abandoned projects from happening. This is to secure the image of the construction industry and most importantly to save and prevent loses of money and valuable resources from happening. Figure 4 shows a relative important index of the suggestion solution for abandoned building project. From the figure, the respondents' recommend to rehabilitating the abandoned building by doing a partnership project between public-private company or vice-verse. With this partnership, it will help the problematic company to continue the project until finish. By doing this, the numbers of abandoned building project can be reduced as well as the negative impact of this problem can be also be reduced. Besides, the respondents' also suggested to amends the Housing Development Act that had been implementing to this country since years ago. By amends this Act, the complication in term of built and sell can be lessen.

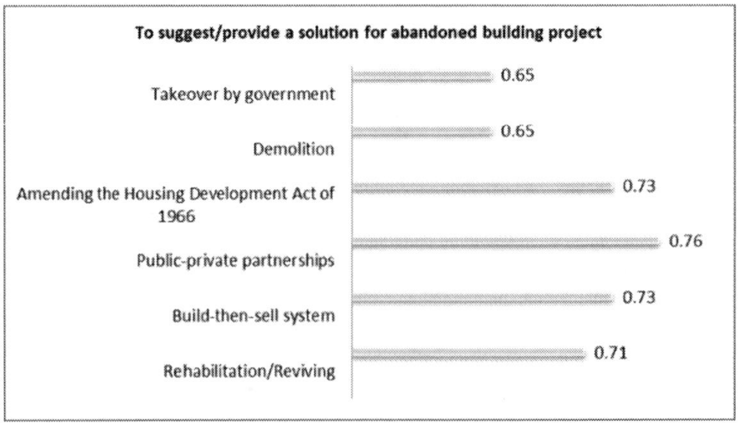

Fig. 4. The suggested solution for abandoned building project.

The proper solution that could be applied in preventing the occurrence of abandoned building project will help not only the developer, but to protect the welfare of the buyers as well. If any sort of prevention methods is successful, it will save money in millions, which ultimately contributes towards the growth of the country's economy [8].

5 Conclusions

Several conclusions can be made from the obtained results. The conclusions are:

1. The main cause of the abandoned building in Malaysia is cause by the financial problem of the developer company.
2. The suggestion of solution that has been chosen by the respondent is by doing a public-private partnership system.

References

1. R. Haron, D. A. Razak, J. H. Babatunde, Journal of Global Business and Social Entrepreneurship **1,** 104 (2016)
2. A. Ng. The Star Newspaper **18 July** (2009)
3. W. Spelman, Journal of Criminal Justice **21** (1998)
4. E.C. Lim, J. Alum, International Journal of Project Management **13,** 51 (1995)
5. M. Dahlan, N. Hilal, The Malayan law journal **6,** 1 (2006).
6. N. H. M. Dahlan, Malayan Law Journal **5,** 12 (2009)
7. Y. E. Hoe, *Causes of abandoned construction projects in Malaysia, Doctoral dissertation*. (2013)
8. S. V. Doraisamy, Applied Mechanics and Materials **773**, 979 (2015)

Performance Evaluation of Hot Mix Asphalt with Different Proportions of RAP Content

Ahmad Kamil Arshad[1,*], *Haryati* Awang[1], *Ekarizan* Shaffie[2] , *Wardati* Hashim[2], and *Zanariah* Abd Rahman[2]

[1]Institute for Infrastructure Engineering and Sustainability Management (IIESM)/Faculty of Civil Engineering, University Teknologi MARA, Shah Alam, Selangor, Malaysia
[2]Faculty of Civil Engineering, Universiti Teknologi MARA, Shah Alam, Selangor, Malaysia

Abstract. Reclaimed Asphalt Pavement (RAP) is old asphalt pavement that has been removed from a road by milling or full depth removal. The use of RAP in hot mix asphalt (HMA) eliminates the need to dispose old asphalt pavements and conserves asphalt binders and aggregates, resulting in significant cost savings and benefits to society. This paper presents a study on HMA with different RAP proportions carried out to evaluate the volumetric properties and performance of asphalt mixes containing different proportions of RAP. Marshall Mix Design Method was used to produce control mix (0% RAP) and asphalt mixes containing 15% RAP, 25% RAP and 35% RAP in accordance with Specifications for Road Works of Public Works Department, Malaysia for AC14 dense graded asphalt gradation. Volumetric analysis was performed to ensure that the result is compliance with specification requirements. The resilient modulus test was performed to measure the stiffness of the mixes while the Modified Lottman test was conducted to evaluate the moisture susceptibility of these mixes. The Hamburg wheel tracking test was used to evaluate the rutting performance of these mixes. The results obtained showed that there were no substantial difference in Marshall Properties, moisture susceptibility, resilient modulus and rutting resistance between asphalt mixes with RAP and the control mix. The test results indicated that recycled mixes performed as good as the performance of conventional HMA in terms of moisture susceptibility and resilient modulus. It is recommended that further research be carried out for asphalt mixes containing more than 35% RAP material.

1 Introduction

A pavement which deteriorates gives poor riding quality, reduce vehicle traction due to lack of surface friction and increase vehicle users' maintenance costs. Timely maintenance is required to maintain the pavement at a certain desirable condition. Pavement recycling can be combined with other preventive pavement maintenance treatments to provide a long lasting pavement that performs at a high level of serviceability [1]. Pavement recycling

* Corresponding author: drahmadkamil@salam.uitm.edu.my

involves the reuse of reclaimed asphalt material (RAP) obtained from deteriorated roads by milling or ripping/crushing operation. There are several advantages in reusing RAP as pavement materials in terms of environmental, economic and engineering perspectives. The primary environmental advantages include conservation of resources and saving landfill space. It is also economical as the material cost is reduced when RAP is used in HMA [2][3].

In the United States, approximately about 80% of asphalt pavement removed during highway rehabilitation or reconstruction is recycled back into new hot mix asphalt (HMA), disposing of only 20% through normal waste stream channels [4]. These materials still possess desirable properties to be used for the surfacing layers, subject to the limitations set in the specification used. In the hot mix recycling process, RAP material is combined with new aggregate and asphalt binder or recycling agent in a hot mix plant. With proper materials evaluation and mix design, similar or better performance can be achieved for hot mix asphalt combined with RAP compared to mixtures made with virgin materials. However, although recycling is beneficial in reducing the consumption of virgin materials, the performance of the highway pavement should not be compromised for cost reduction [5]. It has been accepted that RAP can be a feasible constituent in HMA pavements and if properly designed and constructed, HMA mixtures with RAP can perform as well as conventional mixtures [6].

Studies in Europe and the United States have concluded that over 80% of the recycled material is reused in the construction of roads, but regulations are still strict in allowing the inclusion of RAP in proportions ranging between 5 and 50% for the production of new HMA mixtures [7]. Some studies indicate that incorporating low percentages of RAP (up to 15% RAP) have no significant changes in the performances of mixes [8][9]. Previous researchers also stated that RAP replacement at proportions below 50% is feasible to produce new HMA mixtures, obtaining satisfactory results in the mechanical properties. In addition, RAP replacement at proportions above 50% is feasible to produce new HMA mixtures, obtaining satisfactory results in the mechanical properties [10][11]. Partial blending of RAP binder occurs in the hot mix asphalt when RAP is mixed with virgin mineral materials and virgin binder. The amount of RAP has been limited to certain portions because the degree of blending between the RAP and the virgin materials is not known. If the weight by percent of all materials and their recycling technology are properly selected, the composition and properties of RAP HMA are improved.

Little research has been carried out on mixes containing RAP exceeding 25% as higher amount of RAP produced variability in Marshall Properties and performance of HMA. The objective of this study is to evaluate the performance of HMA mixtures with variable RAP content (15%, 25% and 35% RAP), in particular the Marshall properties and performance in terms of resilient modulus, moisture susceptibility and rutting resistance.

2 Methodology

This study is based on laboratory experimental work. The samples were prepared using the Marshall Test method in accordance with ASTM D1559. The mix design used AC14 aggregate gradation in accordance with PWD Malaysia's Specification for Road Works [12], with the range of binder content from 4.0% to 6.0%. A total of 15 samples were prepared for the control mix (100% virgin aggregates) and the asphaltic concrete mixes with RAP contents of 15%, 25% and 35%. All the samples were compacted with 75 blows on both sides of the samples using Marshall Compactor. Volumetric properties and analysis were carried out to determine the optimum binder content (OBC) for control mix and recycled mix. The optimum bitumen content value was used for the performance tests, including resilient modulus, moisture susceptibility and rutting. The Modified Lottman test was performed to evaluate the moisture susceptibility of the mixes in accordance with the procedure described in AASHTO T 283. Samples for the test were prepared at 7 + 0.5 percent air voids. The resilient modulus test was performed in accordance with ASTM D 4132. Testing temperatures of 25 and 35°C were used and loading of 1300N with a load pulse of 0.1 seconds and rest period of 0.9 seconds were set. The Hamburg Wheel Tracking test was carried out in accordance with AASHTO T324 to evaluate the rutting susceptibility of mixtures under high temperature. The equipment used for this test was the Asphalt Pavement Analyzer (APA) that is capable of running Hamburg Wheel Tracking tests. The test was run at a rate of 50 passes of a steel wheel per minute with a load of 705 + 22N. This test was conducted until the maximum number of 8,000 passes has occurred or until the maximum rut depth value of 20mm has been reached.

3 Results and discussions

The initial laboratory tests have been conducted to determine physical properties of the materials used. The Marshall mix design was conducted to determine the marshall properties and the optimum binder content (OBC) for control mix and rap-asphalt mixes. The results of the moisture susceptibility test, resilient modulus and rutting test as performance tests were then compared between the control mix and RAP-Asphalt mixes.

3.1 Comparison of RAP aggregate and virgin aggregate properties.

RAP aggregate properties assist in evaluating the suitability of the aggregates to be used in road construction. The basic properties tests that were conducted for both virgin and RAP aggregates consist of Los Angeles abrasion test, aggregate impact test and specific gravity. The detailed test results for both types of aggregate are shown in Table 1.

In this study, based on the results as shown in Table 1, the LAAV, AIV, Specific Gravity for RAP aggregate is higher than virgin aggregate. This indicates that the strength of the recycled aggregate is lower than virgin aggregate. However, the value complied with PWD Malaysia's specification requirements which must not exceed 25% for LAAV and AIV. This is due to the fact that the recycled aggregate was exposed to traffic loading for a long period that reduced the strength of aggregate, unlike the virgin aggregate which has not been exposed to traffic. Since both types of aggregate complied with the specification requirements, the aggregate is suitable to be used in the asphalt mixture preparation.

Table 1. Basic properties of Virgin Aggregate and RAP Aggregate.

Parameter	Virgin Aggregate	Reclaimed Asphalt Aggregate (RAP)	PWD Malaysia's Specification requirement
Los Angeles Abrasion Value (%)	24.00	24.91	< 25
Aggregate Impact Value (%)	14.99	19.98	< 25
Bulk Specific Gravity, Gsb (g/cm^3)	2.648	2.625	-
Apparent Specific Gravity (g/cm^3)	2.705	2.711	-
Water Absorption (%)	0.626	1.063	< 2.0

3.2 Marshall properties

Marshall Test was conducted to determine the volumetric properties of the mixture. The volumetric properties determined included bulk specific gravity, air void, voids in the mix (VMA), stability and flow. The Optimum Binder Content (OBC) for the control and RAP-Asphalt mixes were determined from the individual plots of bulk density, voids in the total mix, voids in the mix, flow and stability versus percent asphalt content.

Mix properties of virgin materials and varying different percentage of RAP showed that there is a significant effect on the Marshall Properties with the inclusion of RAP in the asphalt mixes. Marshall Test results showed that the Marshall flow, VFB and Bulk SG increased as the RAP content increased. However, Marshall Stability value and voids in mix showed a decreasing trend with increasing RAP content. The trends are shown in Fig. 1 to Fig. 5 below. This result could be attributed to the fact that the virgin bitumen failed to rejuvenate the RAP binder as its percentage increases. However, the results and analysis showed that there is not much variation between conventional mixes with RAP-Asphalt mixes and all the mixes satisfy the requirements of the Specifications for Marshall Test.

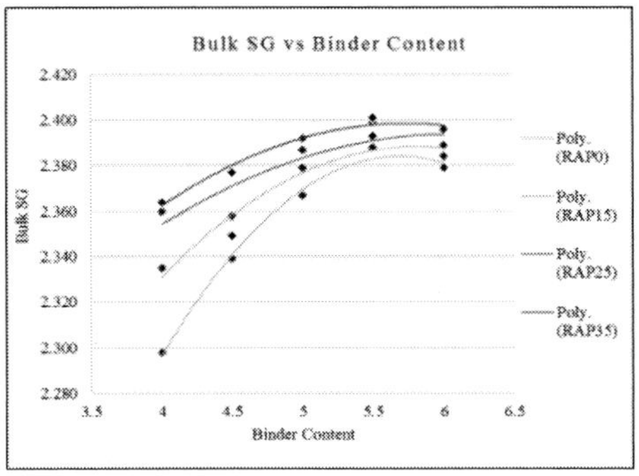

Fig. 1. Bulk Specific Gravity versus Binder Content for Different RAP Contents.

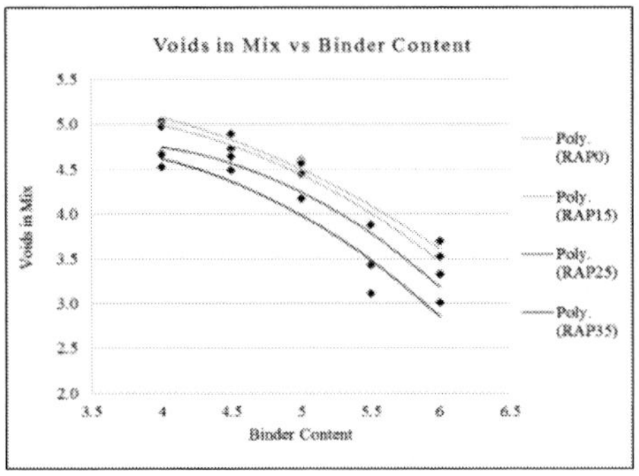

Fig. 2. Voids in Mix versus Binder Content for Different RAP Contents.

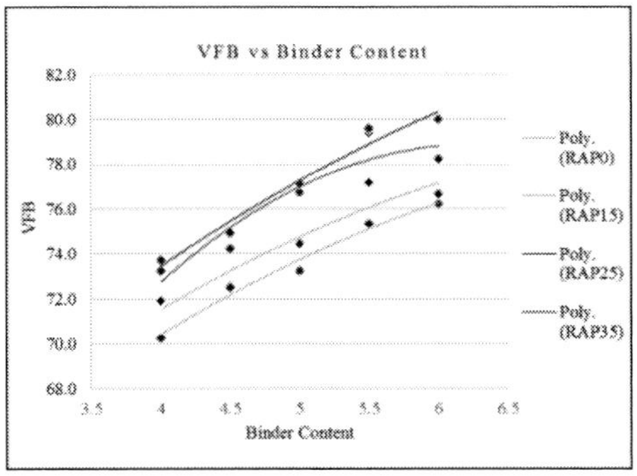

Fig. 3. VFB versus Binder Content for Different RAP Contents.

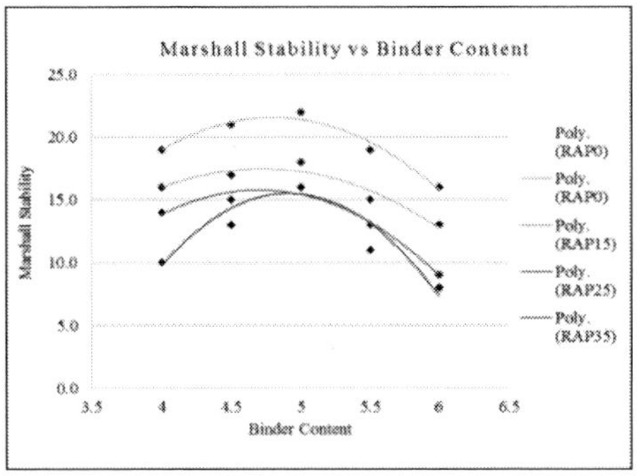

Fig. 4. Marshall Stability versus Binder Content for Different RAP Contents.

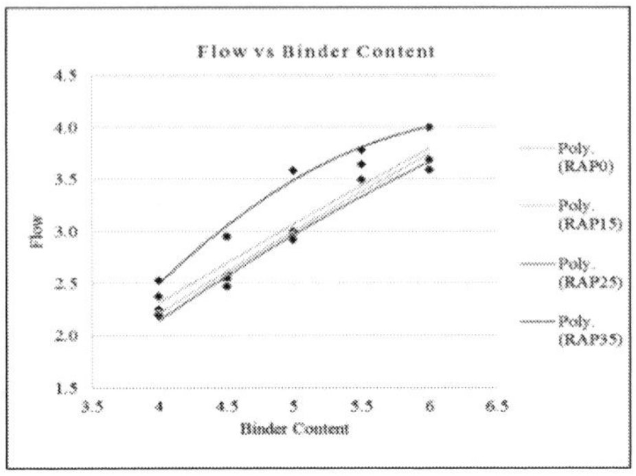

Fig. 5. Flow versus Binder Content for Different RAP Contents.

3.3 Marshall properties and determination of optimum binder content

The OBC for control mix and RAP-Asphalt mixes were taken as the arithmetic mean of binder from the five smooth curves plotted and the value was checked with the limits set by the Public Works Department of Malaysia specification. Table 2 showed the Optimum Binder Content (OBC) for the control and RAP-Asphalt mixes. The OBC decreased as the RAP content increased. The OBC for control mix, RAP15 and RAP25 and RAP35 are 5.34 %, 5.22 %, 5.10 % and 4.88 % respectively. The result obtained showed that there were no substantial differences in Marshall Properties. However, the lower values of OBC with increasing RAP content indicates the possibility that the existing binder in the RAP interacts with the virgin binder and therefore lowered the OBC value.

Table 2. Mix Design selection Criteria.

Properties	Control Mix	RAP15	RAP25	RAP35	PWD Specification
OBC (%)	5.34	5.22	5.10	4.88	-
Bulk S.G	2.388	2.393	2.399	2.401	-
VIM (%)	4	4	4	4	3-5
VFB (%)	75.33	75.20	75.20	75.10	70-80
Stability (kN)	22	18	16	18	>8
Flow (mm)	3	2.95	2.92	2.95	2-4

3.4 Moisture Susceptibility

The results of indirect tensile strength (ITS) for control mix and RAP-Asphalt mixes are shown in Fig. 6. It could be seen that the increase of RAP in the asphalt mix reduces the tensile strength values compared with the control mix. The results show that the ITS values of wet condition and dry condition decreased with increasing RAP content. The values of Control Sample, RAP15, RAP25 and RAP35 reduced by 4.29 %, 12.16 %, 16.13 %, and 20.0 % respectively, after the wet conditioning process. The highest value of ITS in dry condition is 0.74 MPa for RAP15 and 0.67 MPa in wet condition for control sample.

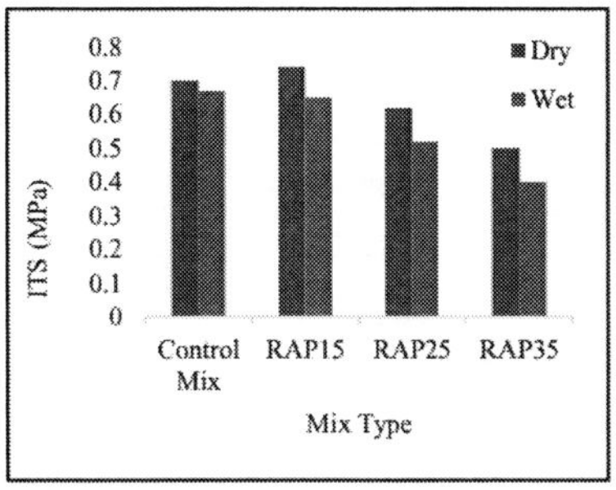

Fig. 6. Indirect Tensile Strength (ITS) for Variable RAP Content.

The Tensile Strength Ratio (TSR) values for control mix and RAP-Asphalt mixes are shown in Fig. 7. From the results, it shows that the TSR values decreased with increasing of RAP content. The highest TSR value is 95.96 % for Control mix while the lowest TSR value is 80.14 % for RAP35. The results indicate that Control Mix to be less susceptible to moisture damage compared to RAP-Asphalt mixes. However, all the mixes have the potential to resist moisture damage since the TSR values obtained are more than 80%.

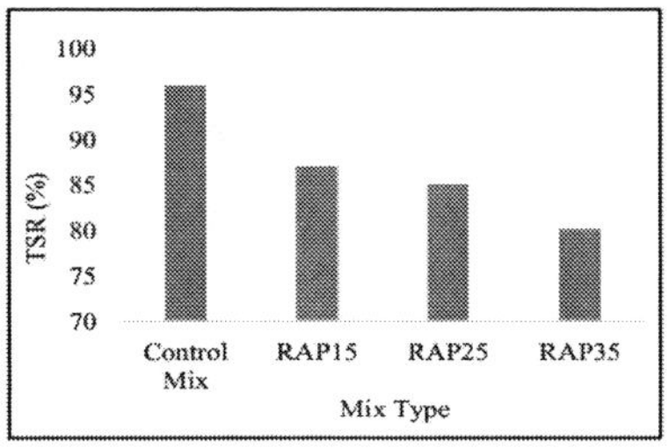

Fig. 7. Tensile Strength Ratio (TSR) for variable RAP Content.

3.5 Resilient Modulus of RAP Mixtures

The results of the resilient modulus test for control mix and RAP-Asphalt mixes at temperatures of 25°C and 35°C are shown in Fig. 8. The resilient modulus for all mixes at temperature 25°C is higher for all mixes than at temperature of 35°C. The control sample has the highest resilient modulus under both testing temperatures compared to RAP-Asphalt mixes. The resilient modulus of control sample at temperatures of 25 °C and 35 °C are 3953 MPa and 2842 MPa respectively. This is followed by RAP15 with values of 3784 MPa and 1991 MPa at temperatures of 25°C and 35°C respectively. RAP25 and RAP35 have lower resilient modulus values compared to RAP15. The results showed that the resilient modulus of the mixes decreases with increasing RAP content. This decrease in resilient modulus value may possibly be due to the effect of the existing binder in the RAP which decreases the stiffness of the mixes.

3.6 Rutting Resistance of RAP Mixtures

The result of the rutting test for control mix and RAP-Asphalt mixes is shown in Figure 9. The rut depth increased with the increase in the RAP content of the asphalt mix. The control sample exhibit the highest resistance to rutting compared to RAP-Asphalt mixes with a depth of 3.4 mm at 8000 wheel passes. RAP15 reached a rut depth of 3.7 mm at 8000 wheel passes, followed by RAP25 and RAP35 with rut depths of 4.2 mm and 4.4 mm respectively. The higher rut depth indicates that the mixes with higher RAP content have lower resistance to rutting. This increase in rut depth may possibly be due to the effect of the existing binder in the RAP which increases the deformation of the mix and increases the rut depth.

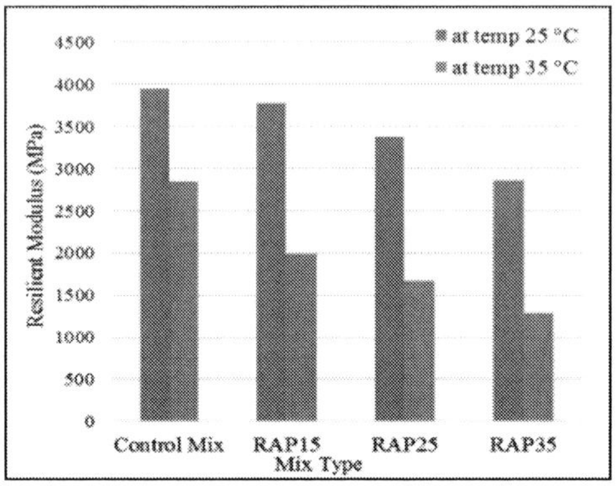

Fig. 8. Resilient Modulus for Variable RAP Content at Temperature of 25°C and 35°C.

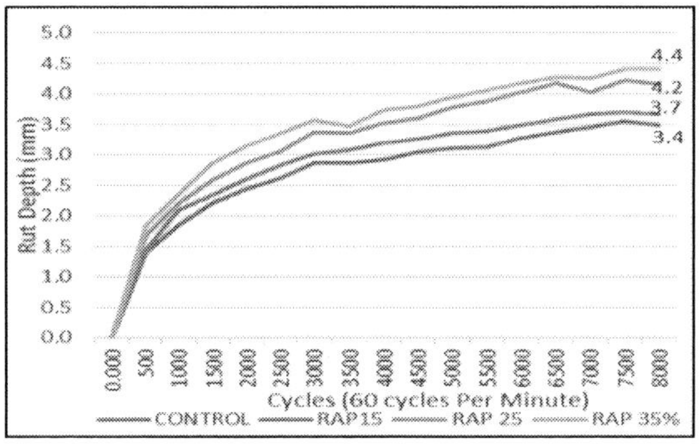

Fig. 9. Rut depth for Hamburg wheel tracking test for different RAP content.

4 Conclusions

From this study, the following can be concluded:

i. The laboratory results on the physical properties of virgin aggregate and RAP shows that the aggregates used in this study complied with the requirements in the Public Works Department specifications. The recovered aggregate from RAP has lower strength compared to virgin aggregate. This is due to the RAP material which is exposed to the environment and traffic for a certain period of time.

ii. The OBC of control mix, RAP15, RAP25 and RAP35 showed a decreasing trend with increasing RAP content. The OBC of control mix, RAP15, RAP25 and RAP35 are 5.34 %, 5.22 %, 5.10 % and 4.8 % respectively. This indicates that there was partial blending of aged binder from RAP, but the effect is not as significant for low percentages of RAP in the mixes.

iii. Dry and wet ITS tests were performed to evaluate the moisture susceptibility performance of recycled mixes. The ITS and TSR value of the mixture decreases as the RAP content increases under both conditions. RAP-Asphalt mixtures satisfied the minimum required value for the moisture susceptibility test.

iv. The resilient modulus of the RAP-Asphalt mixtures decreased with increasing RAP content and temperature.

v. The Hamburg wheel tracking test showed that the rutting resistance also decreased when the RAP content increased. However, the difference in rut depth between the control sample and RAP mixes is minimal, indicating that the mixes with RAP have comparable rut resistance compared to control mix.

Based on the findings of the study, it can be concluded that it is possible to design acceptable quality asphaltic concrete mixes with RAP that meets the required volumetric and desired performance criteria. As the performance of RAP mixes satisfied the minimum requirements of PWD Malaysia, the use of recycled mixes should be encouraged as the use of RAP likely to be cheaper than normal mix. The results of this study indicate that the addition of RAP to HMA is feasible as the performance is comparable to conventional asphaltic concrete mixes. The use of higher percentages of RAP content requires a detailed study and investigation to be carried out to ensure the RAP suitability and compliance with the specification requirements.

The authors would like to express gratitude to the Institute for Engineering and Sustainability Management (IIESM) and The Ministry of Higher Education, Malaysia for the financial support through the FRGS Research Grant: FRGS/1/2015/TK08/UITM/02/3.

References

1. Asphalt Recycling and Reclaiming Association, *Basic Asphalt Recycling Manual*, ARRA, Annapolis, (2001)
2. K. Kuehl, J. Korzilius, M. Marti, *Synthesis of Recycled Asphalt Pavement (RAP) Material*, Minnesota Department of Transportation, St. Paul, (2016)
3. A.M. Abu Abdo, Int. J. Thermal & Env. Eng. **12**, 61-66 (2016)
4. A. Carter, *Development of A Non-Solvent Based Test Method for Evaluating Reclaimed Asphalt Pavement Mixes*, PhD Thesis, Auburn University, Alabama, USA, (2004)
5. L. Mohammad, I. Negulescu, Z. Wu, C. Daranga, W. Daly, J. Assoc.Asphalt Pav. Tech. **72**, 551-594 (2003)
6. B. Huang, X. Shu, D. Vukosavljevic, ASCE J. of Mat. in Civil Eng. **23**, 1535-1543 (2011)
7. W. Mengqi, W. Haifang, W. Muhunthan, N. Kalehiwot, N., *Proc. of the 91st Trans. Res. Board Mtg. Transportation Research Board*, Washington D.C., (2012)
8. H.R. Paul, *Proc. of The Assoc. of Asphalt Paving Tech.* **65**, 231-254 (1996)
9. Austroads, *Maximising the Re-use of Reclaimed Asphalt Pavement (Field Validation)*, Austroads Ltd, Sydney, (2016)
10. R. West, J.R. Willis, M. Marasteanu, *Improved Mix Design, Evaluation, and Materials Management Practices for Hot Mix Asphalt with High Reclaimed Asphalt Pavement Content*, Transportation Research Board, Washington D.C. (2013)
11. K. Vislavicius, H. Sivilevicius, J. of Civil and Mech. Eng. (2013)
12. Public Works Department of Malaysia, *Standard Specification For Road Works - Section 4: Flexible Pavement (JKR/SPJ/2008-S4)*, PWD Malaysia, Kuala Lumpur, (2008)

A novel approach to enhance the accuracy of vibration control of Frames

Iraj Toloue[1*], *Mohd Shahir* Liew[1], I.S.H Harahap[1], and H.E. Lee[1]

[1]Civil and Environmental Engineering Department, Universiti Teknologi PETRONAS, Bandar Seri Iskandar, 32610 Seri Iskandar Perak, Malaysia

Abstract. All structures built within known seismically active regions are typically designed to endure earthquake forces. Despite advances in earthquake resistant structures, it can be inferred from hindsight that no structure is entirely immune to damage from earthquakes. Active vibration control systems, unlike the traditional methods which enlarge beams and columns, are highly effective countermeasures to reduce the effects of earthquake loading on a structure. It requires fast computation of nonlinear structural analysis in near time and has historically demanded advanced programming hosted on powerful computers. This research aims to develop a new approach for active vibration control of frames, which is applicable over both elastic and plastic material behavior. In this study, the Force Analogy Method (FAM), which is based on Hook's Law is further extended using the Timoshenko element which considers shear deformations to increase the reliability and accuracy of the controller. The proposed algorithm is applied to a 2D portal frame equipped with linear actuator, which is designed based on full state Linear Quadratic Regulator (LQR). For comparison purposes, the portal frame is analysed by both the Euler Bernoulli and Timoshenko element respectively. The results clearly demonstrate the superiority of the Timoshenko element over Euler Bernoulli for application in nonlinear analysis.

1 Introduction

Smart structures are those structures equipped with sensors and computers, which enable them to sense the environmental dynamic loadings such as earthquakes. Additional devices such as actuators along with installed sensors help structures to modify their behavior in real time and bear more loads [1]. This method is typically in relative contrast to the traditional body of knowledge of structural engineering [2]. It is worth nothing that in the past thirty years, the interest on integrating actuators with structures to provide controlled forces to compensate applied external loads and minimize structural responses have increased [3]. All of these structures required a suitable algorithm to determine the magnitude of the required forces to reduce the structural response in real time.

* Corresponding author: toloue.iraj@gmail.com

Active tendon systems (ATS), are composed of pre-stressed cables, which connect the floors of the structures in a crosswise manner with an actuator. The Actuator is connected to the pre-stressed cables, and its function is primarily to vary the level of the tension in the cables (see Figure.3). As a result the response of the structure is minimized in this system [4].

Rodllar et.al, by employing an ATS on a cable steady bridge attempted to show the system's efficiency. The linear controller was derived based on the algebraic Lyapunov equation. A finite element numerical example of a 142.5m long pedestrian cable steady bridge, subjected to the 1952 Taft earthquake, showcased a significant reduction in the bridge response [5].

Alavinasab et,al. proposed a new approach to optimize the linear quadratic regulator (LQR) controller employed in an active tendon control system to achieve minimum top node displacement. Their method was derived similar to the LQR method, but based on the energy technique. Numerical simulations comprising, a two dimensional and a three-story frame with two ATS, subjected to variety of earthquakes were performed. The study used lumped mass system to derive the gain matrices and retrieved with notable displacement attenuation [6].

Atig et,al. showed the possibility of controlling the instability of tensegrity systems even during nonlinear analysis by ATS. To obtain the nonlinear analysis the iterative Newton Raphson method is used [7]. Therefore, it can be deduced that nonlinear analysis is one of the key methods in vibration analysis, which involves complex algorithms from being computationally expensive. Consequently, advanced and powerful computers are required for such analysis. To circumvent this setback there are notably new alternative methods, which can replace the traditional methods to run faster analyses with sufficient accuracy. This paper highlight the use of Force Analogy Method (FAM) introduced by Lin in 1968 for inelastic behavior in continuum mechanics[8]. FAM was later further developed by Wong and Yang for structural analysis purpose which instead of tracking the changes in stiffness, used the displacement variations to represent the inelastic behavior of the system [9].

To demonstrate the accuracy of the FAM, Wong and Yang, applied the method on a two- dimensional six story hospital building subjected to 1994 Northridge earthquake. Uniform damping matrix, lumped mass system, and the general assumption of the FAM applied in the simulation. The study's results were proven to be sufficient to compute the true plastic hinges [10]. Wang and Wong, introduced the FAM for stochastic dynamic analysis of structures at the inelastic zone. They used the structural responses to generate the variance covariance function to enable the method to be in good agreement with earthquake analysis [11]. Qu and Li, applied the FAM to incorporate weighting coefficients in developing a new location index model. This approach intends to reveals optimized location of dampers. The performance index was based on the plastic energy of each member as computed from the derived FAM. The accuracy of the FAM was shown in retrieving hysteric energy for a structural system equipped with passive control devices [12].

Wong and Li also proposed a few methods to demonstrate the implementation of FAM for frames equipped with passive, semi active, and active controllers. All of the models were verified for their representation accuracies of the systems [13]. The structural responses such as P-Δ effect, and calculated displacement by FAM, are typically computed based on the Euler Bernoulli beam element [14]–[16]. The

Euler Bernoulli element is derived based on flexural considerations only while rotary inertia and shear effects are both neglected [17]. The Timoshenko element, on the other hand includes the aforementioned neglected assumptions [18]. Many researches including Clough believed that the Euler Bernoulli element, is sufficient for accuracy requirements of general civil engineering analysis [19]. Dixit, who performed structural damage detection, which required accurate analysis compared results based on Timoshenko element of a damaged single beam with that of Euler Bernoulli. The results indicate differences in computed natural frequencies of the elastic structure [20]. Mei performed vibration analysis on a multi-story frame based on the Timoshenko element. The natural frequency responses of the analyzed frame was studied for "L", and "T" shape joints. The results are in good agreement with classical elements [21].

In this paper, the computed difference from using the Euler Bernoulli, and Timoshenko element on nonlinear active vibration controlled frames analyzed by FAM is given full treatment. An LQR controller and state space method coded in MATLAB is developed to perform the nonlinear analyses. An iterative loop is built into the algorithm to uncover the phase change effect, when the wrong element selected.

2 Methodology

The equation of motion for a structure equipped with ATC, is shown in Equation 1.

$$M\ddot{x} + C\dot{x} + Kx = F_e + F_c \qquad (1)$$

Where M, C, K, F_e, and F_c are mass, damping, stiffness, external force, and control force matrices. The state space form of Equation 1 is written as Equation 2.

$$Z_{t+\Delta t}(t) = F_d Z_t(t) + H_d a(t) + G_d X^{''} + G_c f_{ck} \qquad (2)$$

Where $Z_{t+\Delta t}(t)$ is the state matrix at time "t", a(t) is the base acceleration, "D" is the base acceleration and the ATC distribution control matrix, $X^{''}$ is the inelastic displacement, and f_{ck} is the ATC force matrix which are numerically denoted in the following: $F_d = e^{A\Delta t}, A = \begin{bmatrix} 0 & I \\ -K/M & -C/M \end{bmatrix}, G = \begin{bmatrix} 0 \\ K/M \end{bmatrix}, G_c = F_d D\Delta t\,, G_d = F_d G\Delta t, H_d = F_d H\Delta t.$

The fundamental concept of the FAM, is shown in Fig. 1. The initial stiffness, post yielding stiffness, yielding force, yielding displacement, total force, and total displacement denoted by K_i, K_{py}, F_y, X_y, Ft, X, respectively.

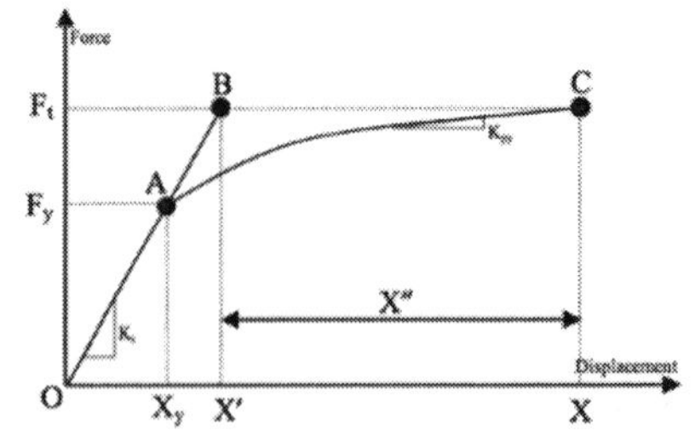

Fig. 1. Force –Displacement relationship.

By extending the OA line from yielding force to the total force (point B), the elastic displacement is determined. The difference between the total displacement and the elastic displacement as shown in Equation 3 , defines the inelastic displacement.

$$X'' = X - X' \tag{3}$$

The same concept for Equation 3, is applicable for moment which leads to the Equation 4.

$$m'' = m - m' \tag{4}$$

Equations 3 and 4 show that it is possible to ascertain the nonlinear behaviour of a system by solely relying on the initial stiffness. This is extremely useful as it enables avoidance of additional computational effort to repetitively calculate the post yielding stiffness matrix.

The plastic rotation at individual members of a structure can be replaced by a set of imaginary moments and forces, which are shown in Equations 5 and **Error! Reference source not found.**, respectively.

$$F_{RForce} = -K'\theta'' \tag{5}$$

$$M_{RForce} = -K''\theta'' \tag{6}$$

The combination of the Equations 3, 5 and the stiffness method results in a fully linear method to ascertain the inelastic displacement, which is shown in Equation 7.

$$X'' = K^{-1}K'\theta'' \tag{7}$$

In a same way the inelastic moment can be determined as shown in Equation 8.

$$m'' = -\left(K'' - K'^{T}K^{-1}K'\right)\theta'' \tag{8}$$

Finally by combining the Equations 4, **Error! Reference source not found.**, 7, and 8 a relation to compute the moment at the predefined plastic hinge locations (PHL), is shown in Equation 9.

$$m = K'^{T}x - K''\theta'' \tag{9}$$

The plastic rotation vector θ'' is zero at the initially and if the moment computed via in Equation 9, exceeds the material yielding limit, the rotation vector will be calculated based on Equation 10.

$$\theta'' = K''^{-1}\left(K'^{T}x - m\right) \tag{10}$$

Where based on the Timoshenko element the stiffness matrices are illustrated in Equations 11 and 12.

$$K = \begin{bmatrix} \frac{12EI}{L^3(1+\varphi)} & \frac{6EI}{L^2(1+\varphi)} & \frac{-12EI}{L^3(1+\varphi)} & \frac{6EI}{L^2(1+\varphi)} \\ \frac{6EI}{L^2(1+\varphi)} & \frac{(4+\varphi)EI}{L(1+\varphi)} & \frac{6EI}{L^2(1+\varphi)} & \frac{(2+\varphi)EI}{L(1+\varphi)} \\ \frac{-12EI}{L^3(1+\varphi)} & \frac{6EI}{L^2(1+\varphi)} & \frac{12EI}{L^3(1+\varphi)} & \frac{6EI}{L^2(1+\varphi)} \\ \frac{6EI}{L^2(1+\varphi)} & \frac{(2+\varphi)EI}{L(1+\varphi)} & \frac{6EI}{L^2(1+\varphi)} & \frac{(4+\varphi)EI}{L(1+\varphi)} \end{bmatrix} \tag{11}$$

$$K' = \begin{bmatrix} \frac{6EI}{L^2(1+\varphi)} & \frac{6EI}{L^2(1+\varphi)} \\ \frac{(4+\varphi)EI}{L(1+\varphi)} & \frac{(2+\varphi)EI}{L(1+\varphi)} \\ \frac{-6EI}{L^2(1+\varphi)} & \frac{-6EI}{L^2(1+\varphi)} \\ \frac{(2+\varphi)EI}{L(1+\varphi)} & \frac{(4+\varphi)EI}{L(1+\varphi)} \end{bmatrix}, K'' = \begin{bmatrix} \frac{(4+\varphi)EI}{L(1+\varphi)} & \frac{(2+\varphi)EI}{L(1+\varphi)} \\ \frac{(2+\varphi)EI}{L(1+\varphi)} & \frac{(4+\varphi)EI}{L(1+\varphi)} \end{bmatrix}, \varphi = \frac{12EI}{G\frac{A}{\kappa}L^2} \tag{12}$$

In the stiffness matrix, modulus of elasticity and moment of inertia are shown by "E", and "I". Length, cross section area, shear modulus, and Timoshenko shear coefficient are respectively represented by "L", "A", "G", and "κ".

The Variable f_{ck} in Equation 2 is the required actuator force to control the system's vibration based on the LQR system. The LQR is modelled by defining a cost function to minimize the system's response, as shown in Equation 13.

$$J = \frac{1}{2}\int_0^\infty Z^T QZ + f_c{}^T R f_c \tag{13}$$

Control and state variables are represented by "f_c", and "Z". The weighting matrices "Q", and "R" are usually obtained through an iterative trial and error process. In this study the weighting matrices are chosen based on the previous literatures. The minimization of cost function, and discrete algebraic RICCATI equation (DARE), the "f_{ck}" is obtained by Equation 14.

$$f_{ck} = \frac{G_c{}^T \times P_B \times f_d \times Z_0}{G_c{}^T \times P_B \times G_c + R} \tag{14}$$

"P_B" is the result of the DARE. The process algorithm of this study is shown in Fig. 2.

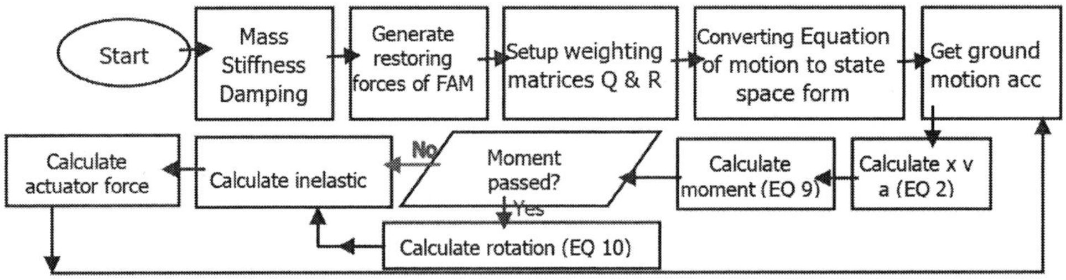

Fig. 2. Flowchart for proposed algorithm.

3 Applications

The portal frame and the node numbering are shown in Fig. 3. This frame model is similar to previous studies for convenience of benchmarking [22-23]. As illustrated, the actuator is connected to the Nodes number 2 and 4. There are two PHL defined on both end of each member. The actuator movement to the right side considered as positive. The section details for both beam and columns are a cubic hallow section and it is shown in Fig. 4.

 https://doi.org/10.1051/e3sconf/20183401027

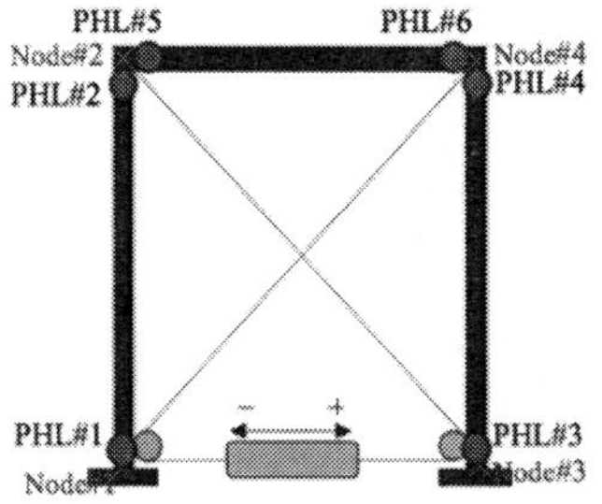

Fig. 3. Proposed model.

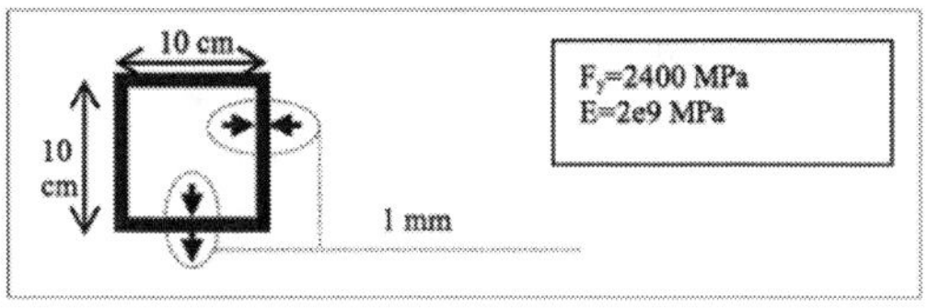

Fig. 4. Section details.

Finally the frame model in Fig. 3 is subjected to ground motion from the 1940 El-Centro earthquake which is shown in Fig. 5.

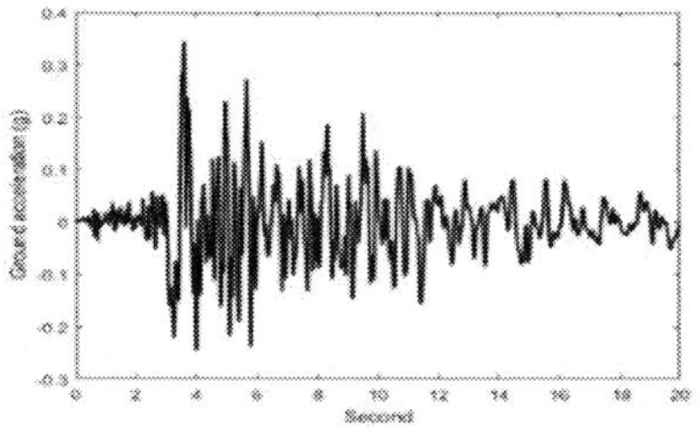

Fig. 5. Ground acceleration.

4 Results and discussions

The proposed frame is analyzed and controlled by both Timoshenko, and Euler Bernoulli elements. The non-linear rotations of the frame at all of the defined PHLs are shown in Figure 6. Each of the graphs, separately show the two end plastic hinges results for one element (two columns and one beam).

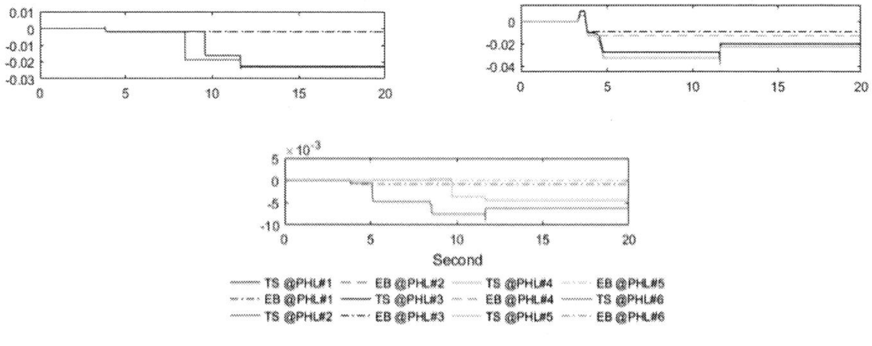

Fig. 6. PHL at all 6 nodes with different element type.

It is shown in the previous study that the results of the analysis based on the Timoshenko (TS), and the Euler Bernoulli (EB) elements are not similar. Moreover, the differences between these two elements are not significant in terms of magnitude. The maximum differences between EB and TS element are shown in Table 1. The differences between TS, and EB, in respect to EB element, are shown in last column of Table 1, and Figure 7.

Table 1. Nonlinear rotations in respect to EB element.

PHL	EB (rad)	TS (rad)	Difference (%)
#1	-0.0227	-0.0019	91.746
#2	-0.0353	-0.0090	74.5410
#3	-0.0242	-0.0020	91.5803
#4	-0.0391	-0.0126	67.8186
#5	-0.0045	0	100
#6	-0.0091	-0.008	90.6776

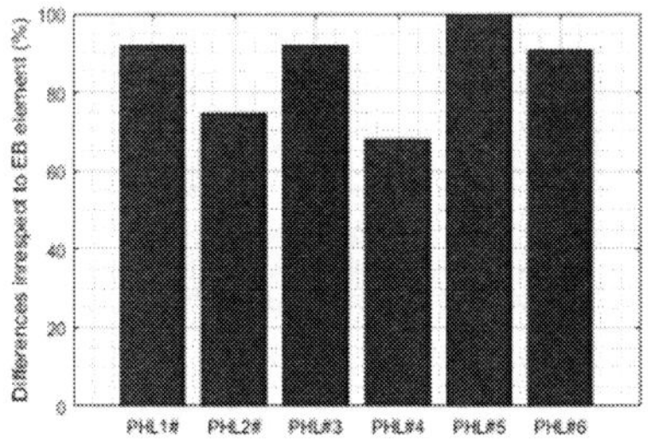

Fig. 7. Differences between EB and TS element.

It was expected due to the symmetrical frame shape, the results on PHL#1, 2, and 5, be almost the same as the PHL#3, 4, and 6, respectively. However, Fig. 7, does not substantiate this hypothesis, and proving that all of the individual PHLs are required to be calculated separately. On the other hand, scrutinizing to the applied control force, as it is shown in Fig. 8, the changed phase between the Euler Bernoulli and Timoshenko element can be clearly seen. A close look at the control force between 7.9 to 8.1 seconds show that the phases of the control force based on these two elements are in the opposite direction of each other. In other words, the actuator based on the EB element moved towards the left with -1kN, while the actuator based on TS moved to the right with +1kN. For the time frame between 9.75 to 11.0 seconds, the EB case actuator saturated at the left side with -1kN, while TS case actuator force is fluctuates between the maximum levels of both left and right directions.

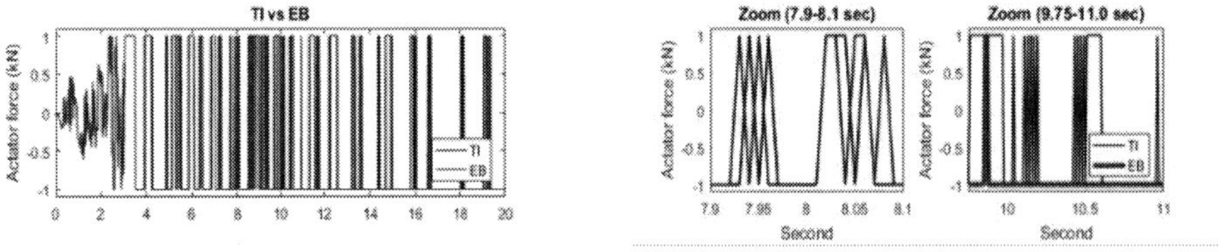

Fig. 8. Control force with different element type.

The displacement results of the system in X and Z directions are reported in Fig. 9 and Fig. 10, respectively. Although the analysis results represent a shift in the displacement, but there are only minor differences between the amplitude of vibration of these two elements. According to Clough this small amount of difference is negligible in general practice of structural engineering [24].

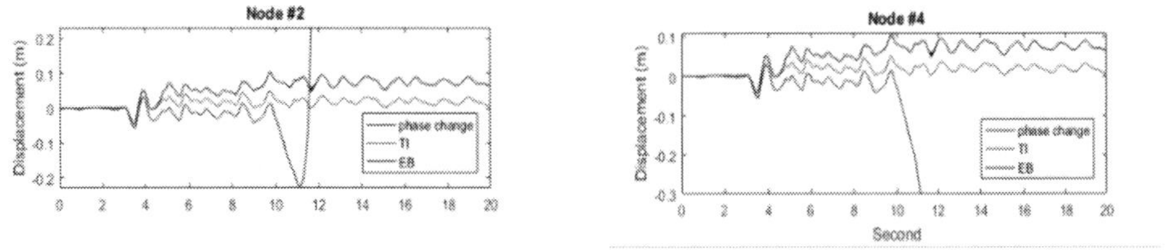

Fig. 9. Displacement in X-direction with different element type.

The frame response is computed again with TS element in lieu of EB for more accuracy. The actuator, as a common practice for FAM, is controlled by the EB element rather than TS. The results indicated structural collapse. The failure occurred when the phase change of the actuator is taken into account. Figure 8 showcases the start of collapse which can be deduced to be at the time 7.9 seconds.

It is much clearer to see the differences between these two elements in Fig. 10. The highlighted periods at Fig. 8, at the time 7.9sec, and 9.75sec, are obvious in Figure 10. Lastly, it can be additionally remarked for this study that the recorded response in the Y direction is too small to pose any threat to the structure's health.

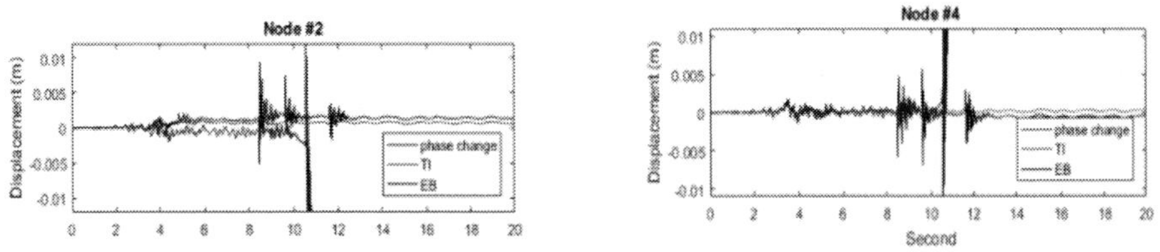

Fig. 10. Displacement in Z-direction with different element type.

As the Timoshenko element, unlike the Euler Bernoulli, consider the shear deformations, the accuracy of the FAM on calculation of rotations at PHL increased. Moreover, the small changes in displacement amplitude return a huge difference in the phase of the actuator force. This phase change continued to some extend that the state space results in prediction of the structural response in time "k+1" to return error and trigger structural collapse.

5 Conclusions

In this study, a numerical 2D portal frame is equipped with an actuator is subjected to the 1940 El-Centro earthquake. The nonlinear analysis on this frame was performed by FAM, which is based on the EB element and TS elements respectively. Minor differences in structural response between the use of EB and TS elements were detected when the actuator phase changes are neglected. As the TS consider the neglected shear deformations of the EB element, such small changes were captured by the model which resulted in the phase of the actuator for EB and TS element to be acting in the opposite direction of each other. When the phase change of the actuator was taken into consideration the structure was noted to have collapsed.

References

1. H. Adeli, X. Jiang, *Intelligent infrastructure: neural networks, wavelets, and chaos theory for intelligent transportation systems and smart structures.* (CRC Press, 2009)
2. L.-Y. Lu, G.-L. Lin, C.-H. Lin, Comput. Civ. Infrastruct. Eng. **24**(1), 41–61, (2009)
3. H. Adeli, A. Saleh, *Control, Optimization, and Smart Structures—High-Performance Bridges and Buildings of the Future*, (John Wiley and Sons, 1999)
4. T. E. Saaed, G. Nikolakopoulos, J.-E. Jonasson, H. Hedlund, J. Vib. Control **21**(5), 919–937, (2013)
5. J. Rodellar, V. Manosa, C. Monroy, J. Struct. Control, (2002)
6. A. Alavinasab, H. Moharrami, A. Khajepour, Comput. Civ. and Infrast. Eng. **21**(8), 605–611 (2006)
7. M. Atig, M. H. El Ouni, N. Ben Kahla, Eur. J. Environ. Civ. Eng. **8189** (March), 1–18, (2017)
8. T. H. Lin, *Theory of Inelastic Structures.* (New York, NY, USA: John Wiley & Sons, 1968)
9. K. K. F. Wong, R. Yang, J. Eng. Mech., **125**(10), 1190–1199 (1999)
10. K. K. F. Wong, R. Yang, J. Eng. Mech., **128**, 308–317 (2002)
11. Z. Wang, K. K. F. Wong, J. Eng. Mech., **133**(4), 414–421, (2007)
12. J. Qu, H. Li, Math. Probl. Eng., **2013**, 10, (2013)
13. Y. Chuan-guo, L. Chun-guang, X. Chun-xu, J. Water Resour. Archit. Eng., **3**(3), 72–75, (2014)
14. Y. Zhang, H.N. Li, G. Li, Int. J. Struct. Stab. Dyn., **16**(5), 1550012, (2016)
15. G. Li, H.N. Li, Y. Zhang, Struct. Des. Tall Spec. Build., **24**, 59–72 (2015)
16. G. Li, L. A. Fahnestock, H.N. Li, J. Struct. Eng., **139**(4), 500, (2012)
17. T. M. Wang, T. A. Kinsman, J. Sound Vib., **14**(2), 215–227, (1971)
18. R. J. Astley, *Finite elements in solids and structures. An introduction.* (Chapman & Hall (Springer), 1992)
19. R. W. Clough, Bull. Seismol. Soc. Am., **45** (4), 289-301, (1955)
20. A. Dixit, J. Sound Vib., **333(**18), 4341–4353, (2014)
21. C. Mei, J. Dyn. Syst. Meas. Control, **136(**5), 51004, (2014)
22. K. K. F. Wong, G. Li, *Theory of Nonlinear Structural Analysis: The Force Analogy Method for Earthquake Engineering*, (John Wiley and Sons, 2014).
23. G. C. Hart, K. Wong, *Structural Dynamics for Structural Engineers.* (New York, John Wiley and Sons, 2000)
24. R. W. Clough, J. Penzien, *Dynamics of structures*, Third. (California, 2003)

Submerged Glow-Discharge Plasma: An Economical Approach to Convert Construction Scrap Metal into Nanomaterials

Peter Nai Yuh Yek[1*], *Muhammad Rafiq Mirza* Julaihi[2], *Mohammad Shahril* Osman[1], *Tung Chuan* Tiong[1], *Wak Ha* Lee[3], and *Chern Leing* Lee[4]

[1] Department of Engineering, University College of Technology Sarawak, 96000, Sibu, Sarawak, Malaysia
[2] Faculty of Engineering, Swinburne University of Technology (Sarawak Campus), Jalan Simpang Tiga 93350 Kuching, Sarawak, Malaysia
[3] Faculty of Resource Science and Technology, Universiti Malaysia Sarawak (UNIMAS), 94300 Kota Samarahan, Sarawak, Malaysia
[4] Chemical Engineering Discipline, Monash University Malaysia, Jalan Lagoon Selatan, 47500, Bandar Sunway, Selangor, Malaysia

Abstract. Submerged glow-discharge plasma (SGDP) is relatively new among the various methods available for nanomaterials synthesis (NMs) techniques. This method allows great control over the production cost of nanomaterials synthesis. A lab-scale batch type SDGP technology has been constructed to produce nanomaterials and investigate the inter-relationship between plasma excitation voltages, electrodes submerged areas and electrolyte concentration. Metal oxide nanospheres has been synthesised from different electrolyte concentrations (1M-0.001M) and characterized by Scanning Electron Microscopy (SEM) and Energy Dispersive Spectroscopy (EDS). As the major results showed that the nanospheres are uniformly spherical with diameter size distribution are between 100 nm - 2μm. EDS analysis shown the nano-Iron Oxide have been formed. Scrap metal initially showed around 6.45% and 93.55% of Carbon and Iron composition respectively. After SGDP process to the scrap metal, Carbon content has increased to 34-35% and Iron content has reduced to around 15-40%. EDS results also shown the higher percentage of Iron amount has remained with lower electrolyte concentration and Current is proportionally related to submersion area of cathode.

1 Introduction

Construction is one of the most important contributors to overall socio-economic

* Corresponding author: peter.yek@ucts.edu.my

development. However, without proper construction waste management, it is also a significant polluter to the environment. Over the last two decades Malaysia construction industry contributed between 3 to 5 percent of the aggregate economy of gross domestic product (GDP) [1]. The construction industry also cause abundant of construction waste and facing challenges in effective management to minimize the environmental pollution [2]. Currently, common practice in construction by developers are 3R (Reduce, Reuse, Recycle) [2, 3]. Waste left over in construction site often contains bulky materials including concrete, wood, metals, bricks, glass, plastics and so on [4]. Metals such as aluminium, copper, and steel are widely being utilized in the construction industries. Among all type of construction waste generated, 1% is consist of scrap metal products and 0.27% of scrap metal can be reused on site [5] and other sold to scrap dealers. Reusing scrap metal on site and re-melting are commonly used as profitable ways and to maintain the life cycle of metallic objects. This research explores the alternative way of converting scrap metal into higher added value product, namely nanomaterials via submerged glow-discharge plasma (SGDP).

Plasma is generated from the ionization of neutral gases by applying extra heat or energy to free the electron from the covalent bond. The plasma contains an equal number of positive ions and free negative electrons to create a strong electric field. Furthermore, glow discharge plasma in liquid also generates active radicals, an effective reducing and oxidation agents to treat biological and chemical matter [6]. Currently, glow-discharge plasma in liquid has provide new scientific challenges and emerging plasma technology widely applied in drinking and waste water treatment [7-9], hydrogen peroxide production [10, 11], plasma medicine [12], surface engineering [13], nanomaterial synthesis [14-17] etc.

The interest of nanomaterials synthesis via SGDP is growing due to its applicability in the industry for large scale [15]. Submerged glow-discharge plasma has been researched as an innovative and economical method for nanoparticles synthesis [14, 17-19] and offers several advantages: (1) relatively simple in experiment setup and mass production , (2) no external gas supply is needed, and (3) easily available precursor of nanoparticles [20]. The field of nanomaterials synthesis via submerged glow-discharge plasma [14, 18] has grown significantly due to the growing demand for various types of nanotechnology developments. This paper discusses the generation mechanism of glow discharge plasma for the preparing nanoparticles from scrap metal submerged in liquid and physical characteristics of the produced nanoparticles.

2. Materials and methods

2.1 Precursor preparation

Scrap metal of iron nail in 3mm diameter was collected from a construction site then washed by distilled water to remove all the impurities attached on the scrap metal surface.

2.2 Submerged glow-discharge plasma setup

Fig. 1 shows the schematic diagram of SGDP experimental prototype with batch type reaction shown in using 500ml open top cylindrical funnel as SGDP reactor. During the plasma reaction, electrolyte has to be filled continuously to maintain the submerged area of cathode in SGDP reactor. Hence, nanospheres as product of plasma discharge will precipitate at the bottom and discharged for further purification process. SGDP reactor was supplied with electricity by using DC power supply (Kikusui, Tokyo, Japan) used to perform the plasma generation. A 300mm platinum wire with 0.3 mm diameter and 99.98 mass% purity (Nilaco, Tokyo, Japan) used as the anode; this wire was bent into a spiral along the length of a glass rod. Meanwhile, a scrap metal nail of 3mm diameter used as cathode. The distance between the electrodes was kept at 30 mm. The potassium carbonate (K_2CO_3) used as electrolytes with the concentrations ranging from 1 to 0.01 mol. The voltage was applied at a rate of 5 V s^{-1} until the formation of plasma and voltage will fixed at a constant value for 60 min. After the plasma discharge experiments, molten droplets of nanomaterial condensed and washed with distilled water. Finally, the nanospheres were dried in oven with 105°C for 24h for further analysis or stored in distilled water.

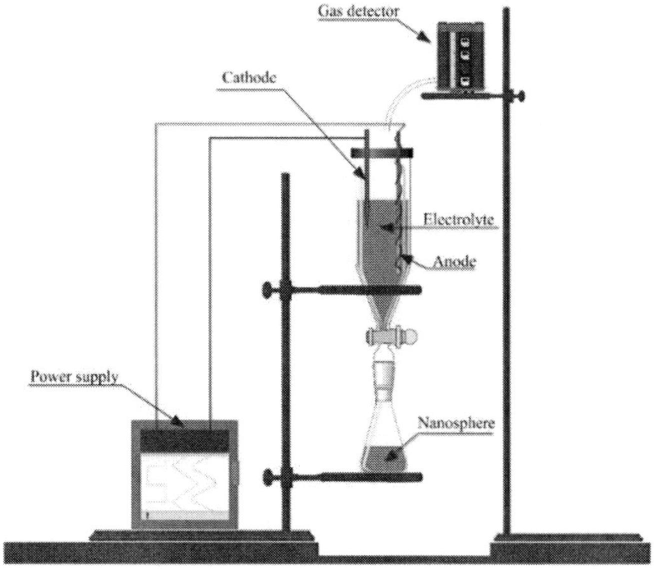

Fig. 1. Schematic diagram of submerged glow-discharge plasma.

2.3 Characterization of iron wire and nanosphere

The collected nanospheres from plasma process were analyzed via scanning electron microscope (SEM) technique to study the overall appearance of the produced nanostructures and to estimate the nanospheres sizes. SEM was performed using a JOEL JSM6010 PLUS/LV microscope operating at an accelerating voltage of 20kV. The microscope also fitted with an Oxford Instrument INCA energy dispersive X-ray spectroscopy (EDS) attachment. EDS technique was used to detect the chemical elements of nanospheres and identify the metallic and oxidized phases at the nanospheres.

3 Results and Discussion

3.1 Current – voltage (I-V) curves of submerged glow-discharge plasma

Fig. 2 shows the typical I-V curves of different electrolyte concentration and Fig. 3 shows the I-V curves of vary cathode submerged length. Experiments carried out to measure the I-V curves in Fig. 1 of scrap metal nail as cathode (⌀3mm) with constant submerged length (0.5cm) in various electrolyte concentration (1-0.01M) of potassium carbonate. However, I-V curves in Fig. 3 show the results from the experiment with various cathode submerged length (0.5 – 1.5cm) in constant 0.01M of electrolyte. Temperature of electrolyte is crucial

https://doi.org/10.1051/e3sconf/20183401028

to initial the generating plasma and make it more effective at lower power density [21]. CO gas was detected during plasma initiation. The reactions occurred are described in 4 stages; (1) electrolysis stage, (2) vaporization stage, (3) transition stage, and (4) full plasma stage.

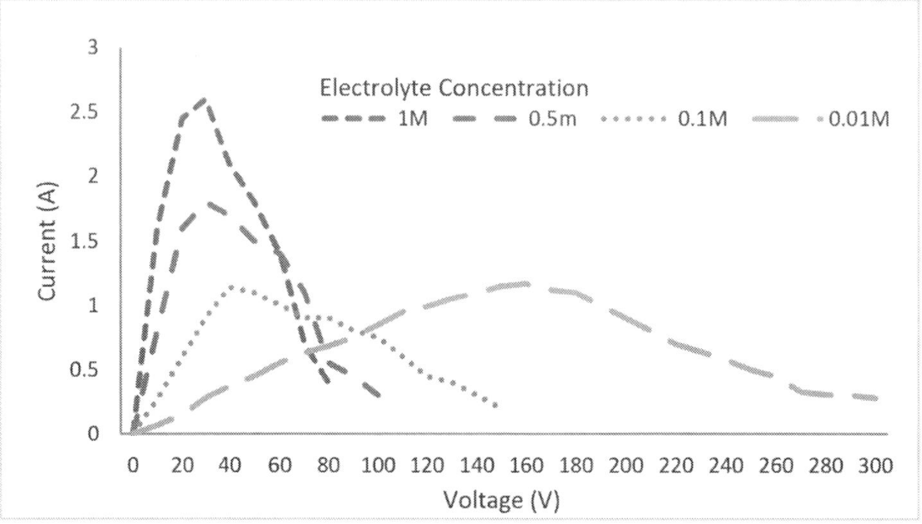

Fig. 2. Typical I-V curve for different electrolyte concentration.

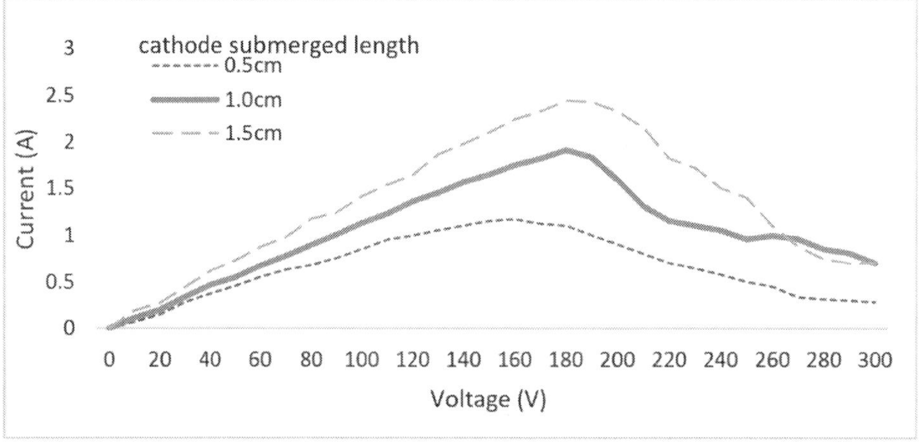

Fig. 3. Typical I-V curve for different cathode submerged length.

At low voltage of electrolysis stage (1), small bubbling was observed on both electrodes. During this stage, current increased along with voltage increase in accordance with Ohm's law. However, thermal loss was concentrated at the cathode side to heat up the electrolyte to boiling point during vaporization stage (2) with large bubbling sound. Hence, transition stage (3) with a thin layer of gas was surrounding the cathode and caused the conductivity to reduce, then the current reading start to reduce while fluctuating slightly. The collisions between electron and neutral atom produce excitation, ionization, dissociation with the emission of light. This phenomenon is known as "glow" discharge. The ions produced are accelerated toward the cathode, bombarding at the cathode surface, causing the release of secondary electrons. These secondary electrons causes more ionization collisions to ensure the glow-discharge as a self-sustained plasma [9]. Consequently, the discharge plasma can be generated at low current and make SGDP produce nanospheres with more economically.

During full plasma stage (4), the surface of the cathode is partially melted to produce nanoparticles owing to the concentration of current causes by the electro-thermal instability. [14]. Eventually, the cathode melts or ionizes at the spot to form nanoparticles by current concentration spot during discharge plasma, and rapidly cooled in the solution to form nanospheres due to surface tension and quenching effect of the electrolyte [16]. However, the amount of current measured was very much related to cathode submerged surface area and concentration of used electrolyte. Larger submerged area of cathode required more current to generate and maintain the plasma. Subsequently, this had an effect of increasing the nanosphere productivity.

3.2 Micrograph of produced nanosphere

Fig. 4 (a) is SEM of scrap metal at 850 times magnification and Fig. 5 (b), (c), (d) and (e) are iron oxide at 7000 times magnification produced at 0.01M, 0.1M, 0.5M and 1M of potassium carbonate concentration respectively after glow-discharge plasma synthesis and centrifuged to remove excessive electrolyte. SEM has been performed in order to confirm that the produced nanospheres have uniformly spherical shape due to the effect of surface tension and quenching of the molten metal by the surrounding electrolyte [22]. The particles were spherical in shape, with sizes ranging from 100nm to 2000nm. Higher electrolyte concentrations were able to reduce the voltage applied to generate plasma in liquid.

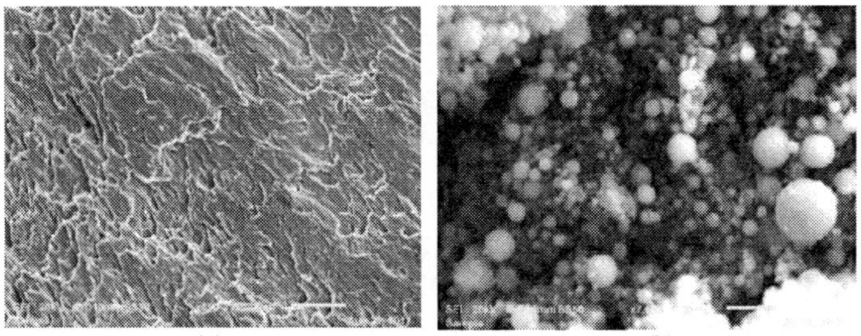

(a) (b)

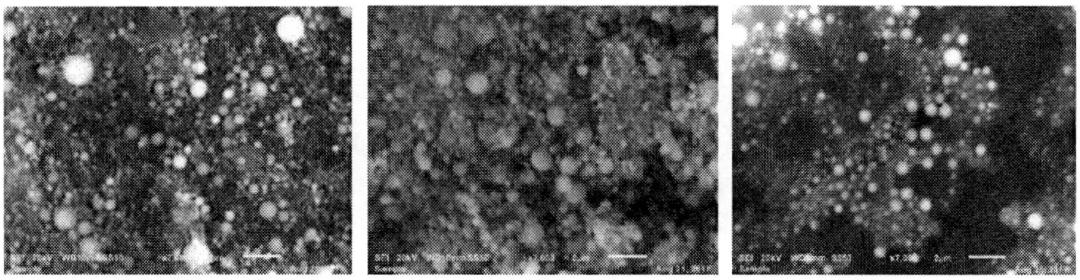

(c) (d) (e)

Fig. 6. SEM micrograph of (a) scrap metal, nanospheres produced from (b) 0.01M, (c) 0.1M, (d) 0.5M and (e) 1M of the electrolyte concentration.

3.3 Elemental composition of produced nanosphere

Table 1 shows the elemental composition of scrap metal nail and nanospheres produced via SGDP. Before the application of plasma, iron nail commonly consisted of Carbon (C) and iron (Fe) atoms. Scrap metal initially showed around 6.45% and 93.55% of Carbon and

Iron composition respectively. However, SGDP process has increased the Carbon content to 34-35% and Iron content has reduced to around 15-40%. The nanospheres were found to contain Oxygen (O) atom in addition to C and Fe atoms of its parent material. These results indicate that the surfaces of iron nanospheres were in their oxidised form after synthesis, and metal oxide produced from SGDP is commonly reported [16, 20]. Carbon content of the nanospheres produced from plasma process is much higher than raw scrap metal. Oxygen content is tended to increase with less concentration of electrolyte. EDS results also shown the higher percentage of Iron amount has remained with less concentration of potassium carbonate as electrolyte. It was found that at higher electrolyte concentration the plasma excitation voltage was reduced and do not have significant effect to the current flow. Current is proportionally related to submersion area of cathode.

Table 1. Elemental composition of the iron wire and nanosphere from plasma.

Electrolyte Concentration	Voltage plasma maintained	Elementary analysis (weight %)					Morphology
		C	Fe	O	Other	Total	
Scrap metal		6.45	93.55	-		100	
1.0M	80V	34.73	15.87	41.30	K 8.10	100	Sphere
0.5M	100V	34.48	19.58	39.52	K 6.42	100	Sphere
0.1M	150V	34.52	23.92	40.26	K 1.30	100	Sphere
0.01M	280V	35.26	40.54	24.21	-	100	Sphere

3.4 The economics of nanospheres production using SGDP

The price of a commercially available iron (II, III) oxide powder with 95% purity by Sigma-Aldrich is RM201.50/25g [23]. In order to perform a comparison, the calculation for the price of iron oxide nanospheres made using SGDP is shown in Equation 1.

SGDP Cost per 10g = Electricity cost per unit [24] + Electrolyte cost + Raw material cleaning cost

$$= RM(100V*1A*1hour*0.32) + RM2 + RM0.5$$

$$= RM34.5 \quad (1)$$

The equation 1 calculation corresponds to RM86.25 per 25g of nanospheres. It is clear that the iron oxide nanospheres made using SGDP is 57% cheaper compared to the cost of commercially available iron oxide powder. Therefore it shows that SGDP is an economical approach to nanomaterials production.

4 Conclusions

In this study, the effect of cathode submerged area and electrolyte concentration on plasma formation was studied. Nanospheres of Iron Oxide ranging from 100 to 2000 nm in size has been successfully produced. Higher electrolyte concentration was significantly reduce the initial plasma excitation voltage and further low electric power density (80V and 0.3A) is required to generate nanospheres. As a conclusion, submerged glow-discharge plasma has the potential as a low-cost nanosphere synthesis method.

The authors gratefully acknowledge the financial support provided by University College of Technology Sarawak under Grant scheme (Project No. UCTS/RESEARCH/<1/2016/03>(01)).

References

1. R.A. Khan, M.S. Liew, Z.B. Ghazali, Procedia - Procd Soc Behv. **109**, 507-513, (2014).
2. S.K. Lachimpadi, J.J. Pereira, M.R. Taha, M. Mokhtar, Resour., Conserv. Recycl. **68**, 96-103, (2012).
3. Z. Chen, J. Xu, Y. Chen, E.M. Lui, Constr. Build. Mater. **126**, 641-660, (2016).
4. S.H. Hassan, N. Ahzahar, M.A. Fauzi, J. Eman, Procd Soc Behv. **42**, 175-181, (2012).
5. R.A. Begum, C. Siwar, J.J. Pereira, A.H. Jaafar, Resour., Conserv. Recycl. **48**, 86-98, (2006).
6. B. Peter, L. Christophe, Phys D Appl Phys. **42**, 053001, (2009).
7. M.A. Malik, Plasma Chem. Plasma Process. **30**, 21-31, (2010).
8. B. Jiang, J. Zheng, S. Qiu, M. Wu, Q. Zhang, Z. Yan, et al., Chem. Eng. J. **236**, 348-368, (2014).
9. T. Selma Mededovic, R.S. Gunnar, D. Fei, L.B. Christopher, M.H. Thomas, G.B. Douglas, et al., **J** Phys D Appl Phys. **50**, 014003, (2017).
10. L. Wang, X. Jiang, Y. Liu, J Hazard Mater. **154**, 1106-1114, (2008).
11. P.J. Hansen, M.P. Hoggard, A.C. Rathwell, J. Reprod. Immunol. **10**, 157-166, (1987).
12. H.J. Forman, A. Bernardo, K.J.A. Davies, Arch. Biochem. Biophys. **603**, 48-53, (2016).
13. N. Pourali, G. Foroutan, Phys. Lett. A. **380**, 2422-2429, (2016).
14. M.R.M.B. Julaihi, S. Yatsu, M. Jeem, S. Watanabe, J. Exp. Nanosci. **10**, 965-982, (2015).
15. T.A. Kareem, A.A. Kaliani, Ionics. **18**, 315-327, (2012).
16. S. Yatsu, H. Takahashi, H. Sasaki, N. Sakaguchi, K. Ohkubo, T. Muramoto, et al. Arch. Metall. Mater. **58**, 425, (2013)
17. G. Saito, Y. Nakasugi, T. Yamashita, T. Akiyama, Nanotechnology. **25**, 135603, (2014).
18. G. Saito, Y. Nakasugi, T. Akiyama, J. Appl. Phys. **116**, 083301, (2014).
19. G. Saito, Y. Nakasugi, T. Yamashita, T. Akiyama, Appl. Surf. Sci. **290**, 419-424, (2014).
20. G. Saito, S. Hosokai, T. Akiyama, Mater. Chem. Phys. **130**, 79-83, (2011).
21. G. Saito, Y. Nakasugi, T. Akiyama, J. Appl. Phys. **118**, 023303, (2015).
22. Y. Nakasugi, G. Saito, T. Yamashita, T. Akiyama, J J. Appl. Phys.. **115**, 123303, (2014).
23. https://www.sigmaaldrich.com/catalog/product/aldrich/310069?lang=en&region=MY
24. https://www.sarawakenergy.com.my/index.php/customers/commercial-pricing-tariff

Mechanical Properties of Lightweight Concrete Using Recycled Cement-Sand Brick as Coarse Aggregates Replacement

Ilya Joohari[1,*], *Nor Farhani* Ishak[1], and *Norliyati* Mohd Amin[2]

[1]School of Environmental Engineering, Universiti Malaysia Perlis, 02600 Arau, Perlis, Malaysia
[2]Faculty of Civil Engineering, Universiti Teknologi MARA, 40450 Shah Alam Selangor, Malaysia

Abstract. This paper presents the result of replacing natural course aggregate with recycled cement-sand brick (CSB) towards the mechanical properties of concrete. Natural aggregates were used in this study as a control sample to compare with recycled coarse aggregates. This study was also carried to determine the optimum proportion of coarse aggregates replacement to produce lightweight concrete. Besides, this study was conducted to observe the crack and its behaviour development during the mechanical testing. Through this study, four types of concrete mixed were prepared, which were the control sample, 25%, 50% and 75% replacement of CSB. The test conducted to determine the effectiveness of recycled CSB as coarse aggregates replacement in this study were slump test, density measurement, compression test, and flexural test and. The strength of concrete was tested at 7 days and 28 days of curing. From the results obtained, the optimum proportion which produced the highest strength is 25% replacement of recycled CSB. The compressive and flexural strength has decreased by 10%-12% and 4%-34% respectively compared to the control sample. The presence of recycled coarse aggregates in sample has decreased the density of concrete by 0.8%-3% compared to the control sample.

1 Introduction

One of the significant problems nowadays is the accumulation and unsatisfactory management of construction and demolition waste. Therefore, it is increasing along with continuous spreading of industrialization and urbanization. However, it was stated that the construction and demolition waste can be recycled and reused as a raw material for new applications [1]. Usually, most of the waste materials are left behind at landfill or illegally dumped. In addition, the environmental impact can be reduced by making more sustainable use of this waste.

Waste management is one of the priorities of every community [2]. As a result, it has become an evident that good waste management can enhance better quality of life. The main principle of a quality waste management is by reducing the mass production of new

* Corresponding author: ilya@unimap.edu.my

waste, then find the way to recycled and reuse an existing waste, and safe and ecologically acceptable in depositing of unused waste.

Nowadays, research on lightweight aggregates has been conducted worldwide in a large range. The structural lightweight concrete has their own advantages such as superior heat, sound insulation characteristics due to air voids and lower coefficient of thermal expansion [3]. Moreover, the reduction in dead weight in concrete will reduce the cross section of structural members and steel reinforcement.

1.1 Problem Statement

Natural resources of this country are decreasing year by year. In this research, the natural coarse aggregates that will be used are gravel. Gravel is found naturally, mostly in lake, river, and ocean beds, where the constant movement of the water and waves keeps the gravel from settling and fusing with other rocks, sand is a form of micro-gravel, being tiny stones worn down by constant motion. Gravel deposits are mined and sifted to create commercial gravel products, which are used in many industrial applications as well as construction. To prevent the natural resources from decreasing, one of the ways is to recycle and reuse the waste materials such as the cement-sand brick that produced in construction site. Therefore, in order to prevent the natural coarse aggregates from depleting, the recycled brick will be used as a replacement.

The environment impact can be reduced also by making more sustainable use of this waste material. This research is conducted to investigate the effect of replacing the coarse aggregates with recycled cement-sand brick. By determining the optimum proportion of coarse aggregates replacement, this will not only determine the lightweight concrete, it also can determine which proportion will produce both lightweight and have the strength of a normal concrete.

2 Materials and Methodology

2.1 Materials

The materials used in this research are recycled CSB as a replacement of coarse aggregates. There are two types of mould used in this research such as cube and prism mould. The size of cube mould is 100 mm x 100 mm x 100 mm while the prism mould is 100 mm x 100 mm x 500 mm. Besides, the replacement of recycled CSB at 25%, 50% and 75% and the result is compared with the control sample without recycled CSB replacement. Table 1 below shows the concrete design mix for the respective concreting work.

Table 1. Concrete Mix Design.

Type of Mix	kg/m^3
Free water content	190.00
Modified free/water cement ratio	351.85
Fine aggregates content	534.28
Coarse aggregates content	1373.87

2.2 Methodology

Recycled Cement-Sand Brick (CSB) is the main material in this research. Firstly, the wastes were collected at Jejawi, Kangar. After collecting the waste material, the preparation of waste material is conducted. The preparations of waste material consist of crushing process, which consists of manual crushing process with the use of a hammer in order to obtain the desired size.

Before the mixing process, aggregates have undergone a sieving process. This method is called sieve analysis. It is a test that classified the aggregates to different sizes. For this research, the sizes of aggregates were between 10 mm to 14 mm. Through this test, course aggregate and fine aggregates was collected. By using recycled CSB as coarse aggregates replacement, it is expected to reduce the density of concrete. In this research, the grade of concrete designed was Grade 25. The maximum free-water/cement ratio for this design is 0.54.

After all the preparation for materials has completed, concreting the cubes and prism were conducted according to the design mix. Slump test was conducted at fresh state in order to determine the workability of the concrete mix. After 24 hours, the specimens were cured in water for a period of 7 and 28 days prior to mechanical testing. The testing involves compressive strength and flexural strength test to determine the mechanical behaviour of the concrete.

3 Results and Discussions

3.1 Slump Test

Slump test was the first test conducted on this research, with the design for slump to be in between 10 – 30 mm. It was carried out to study the effect of replacement of CSB on the workability. There are no slight different in slump for all proportion including control sample except for the 25% coarse aggregate replacement. The result shows that the consistency in grade of concrete is stiff. Density test was also conducted for this research. It was carried out to determine whether the concrete was in lightweight category or not. The optimum proportion to produce lightweight concrete is 75% of replacement of recycled CSB. The result shows that the density of concrete kept decreasing between the ranges of 0.8% to 3% from control sample. This is because the density of recycled CSB is 2070 kg/m^3 while the density of gravel is 2700 kg/m^3, which is why the density keeps decreasing when the substitution of recycled CSB is increasing.

Based on a previous research conducted, the density of recycled brick is lower about 8% to 17% than the ordinary sample [4]. The increasing of substitution of natural aggregates with crushed brick will cause the percentage of entrained air increases. It is because of a lower density of brick aggregates, concrete made with it should have lower density than the one with natural aggregates in concrete. The results for slump test are shown in Table 2.

Table 2. Result for Slump.

Samples	Slump (mm)
Control	22
25% Coarse Aggregate Replacement	18
50% Coarse Aggregate Replacement	20
75% Coarse Aggregate Replacement	20

3.2 Compressive and Flexural Test

These tests were carried out to study the effect of replacement of CSB on mechanical and physical properties of the concrete in term of strength development. The concrete was designed to be Grade 25. The result of compression test shows that the strength of concrete is decreasing by 10% to 14% from the control sample while the result of flexural test shows that the strength of concrete is decreasing by 5% to 35% from the control sample. The optimum proportion that has produced the highest compressive and flexural strength is 25% of replacement of CSB. The strength keeps decreasing because the percentage of substitution of recycled CSB is increasing, which is related to the lower density of CSB which is 2070 kg/m^3. The proportion of 25% replacement gives the highest strength among the three proportions because the percentage of gravel inside the concrete is 75%. The strength of concrete with 100% of gravel (control sample) is higher compare to the strength with 75% of gravel. As the percentage of proportion of replacement of recycled CSB increases, the strength for compression and flexural decreasing. The results of compressive and flexural strength are shown in Table 3.

Table 3. Result for Compressive and Flexural Strength Test.

Samples	Compressive Strength, MPa		Flexural Strength, MPa
	7 Days	28 Days	28 Days
Control	17.903	25.593	1.436
25% Coarse Aggregate Replacement	18.648	21.895	1.336
50% Coarse Aggregate Replacement	13.501	19.671	0.873
75% Coarse Aggregate Replacement	13.257	17.448	0.835

These both compression and flexural test result have been approved by previous researcher. It was said that after 28 days of curing process, the compressive strength is decreasing in range 10% to 35% for recycled coarse aggregates [4] in comparison with an ordinary concrete while according to another study [5], it was stated that the flexural strength of the concrete with crushed brick is about 8% to 15% lower than the one of the ordinary concrete. There have a correlation between flexural and compressive strength of concrete with recycled brick aggregates. The decreasing in flexural strength has a similar pattern with compressive strength [4].

3.3 Crack Behaviour

After compression and flexural test are conducted, the crack behaviour on the concrete is observed. This test is conducted to observe the crack and its behaviour. In this research, all the samples were tested after curing process for 7 and 28 days. The Fig. 1 shows the crack behaviour on the cube concrete after compression test. The concrete with 25% proportion of replacement of recycled CSB shows a cracking starts at the left and right of the concrete at the same time. The crack produced a double line from top to bottom of the concrete. The cracking pattern for 25% replacement is quite the same with concrete of 50% replacement. The cracking for 50% replacement starts at the left of the concrete that produced a single line from bottom to top. Lastly, the concrete with 75% proportion of replacement shows a cracking start at the left of the concrete. The crack produced a single line same as the concrete with 50% replacement but small cracking produced at the top of the concrete.

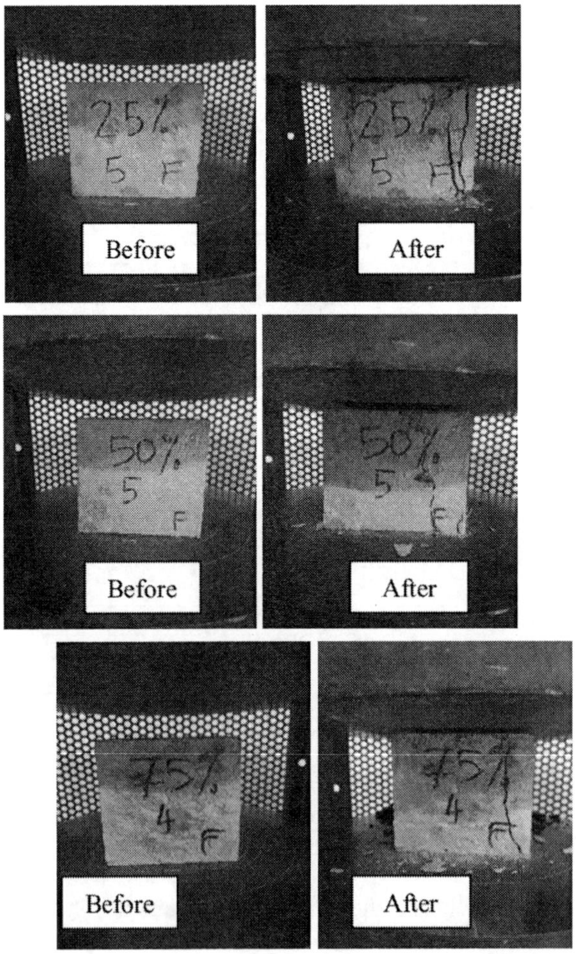

Fig. 1. The crack behaviour before and after compression test for all proportion.

Fig. 2 shows the crack behaviour on the prism concrete after flexural test. All proportion of replacement of recycled CSB shows a crack with single line at the middle of the

 https://doi.org/10.1051/e3sconf/20183401029

concrete. The crack starts from bottom to top. For the proportion of 50% and 75% of replacement, the crack split producing micro cracks at the bottom of the prism. There are different lengths of crack line for every percentage of proportion. The length of crack line for 25%, 50% and 75% are 90 mm, 95 mm and 50 mm respectively.

All crack pattern and behaviour are related with density, compression and flexural test. The least amount of cracks was produced on concrete is 25% replacement of recycled CSB. It is because the density of the concrete for this proportion is greater than the other two proportions by 0.8% to 2.5%. Besides, the percentage of gravel inside the concrete of this proportion is 75% higher compare to another two proportion that will make the concrete is stronger to cater the load. In the same way, the compressive and flexural strength for this proportion is the strongest among the other three proportions by 10% to 12% and 4% to 8% respectively. The other two proportions indicate the sample failed with a deep crack and then separated into two halves. When the density, compressive and flexural strength increases, the crack produce on the sample is less.

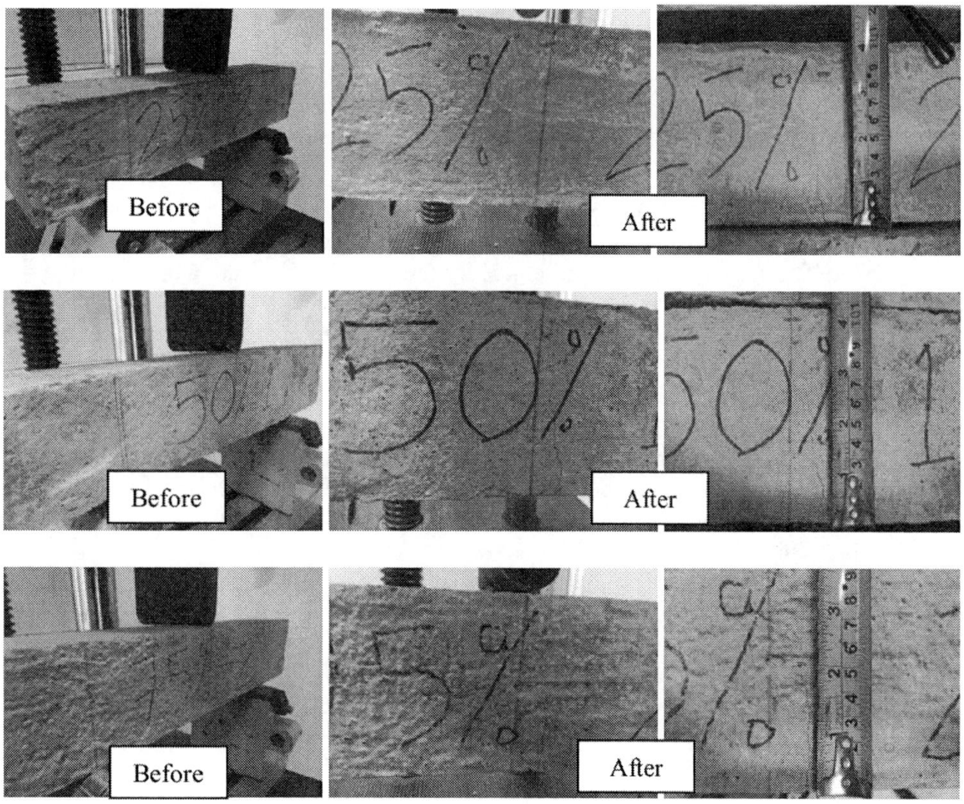

Fig. 2. The crack behaviour before and after flexural test for all proportion.

4 Conclusions

1. The 25% of replacement of recycled CSB gives an optimum effect on the physical and mechanical properties of concrete. The compressive strength of the concrete is decreasing by 10% to 12% when compare with control sample. The flexural strength is decreasing by 4% to 34% as the number of proportion increasing. The higher the

percentage of replacement of recycled CSB, the lower the compressive and flexural strength of the concrete.

2. The optimum proportion of coarse aggregates replacement in terms of producing lightweight concrete is 75% of replacement of CSB. It is because the density of CSB is 2070 kg/m^3 while the density of gravel is 2700 kg/m^3. It can be concluded that, the density of the concrete is decreasing by 0.8% to 3% from control sample. Even though the density of concrete is decreasing, but the density achieved is higher than required for lightweight concrete.

3. The fewest cracking produce on concrete for both prism and cube concrete is 25% of replacement of CSB. With the small cracking produce, it shows that the concrete is in high density, compressive and flexural strength too. It is because the percentage of gravel in this proportion is higher which was 75% replacement.

This study was conducted with the technical support from the Structural Lab at the School of Environmental Engineering, and with the financial support from Universiti Malaysia Perlis.

References

1. Batayneh, M.; Marie, I.; Asi, I., Use of Selected Waste Materials In Concrete Mixes, Waste Management **2**, 1870-1876 (2007)
2. Bjegovic, D., *Sustainability as a Condition for Development in Croatia*, Proceedings of International Conference on Sustainability in the Cement and Concrete Industry. Lillehammerpp. 2-16, (2007)
3. K. M. A Hossain, *Blended Cement and Lightweight Concrete Using Scoria: Mix Design, Strength, Durability and Heat Insulation Characteristics*, Ryerson University Toronto, Ontario, Canada, (2006).
4. Debied, F., Kenai, S., The Use Of Concrete And Crushed Bricks As Aggregate In Concrete, Construction and Building Materials **22**, 886-893 (2008)
5. Khalaf, F. M.; DeVenny, A. S., Recycling of Demolished; Masonry Rubble as Coarse Aggregate in Concrete: Review, Journal of Materials in Civil Engineering **16**, 331-340 (2004)

The effects of alkanolamide addition on cure characteristics, swelling behaviour and tensile properties of silica-filled natural rubber (NR) / chloroprene rubber (CR) blends

Indra Surya[1*], *Syahrul Fauzi* Siregar[1], and *Hanafi* Ismail[2]

[1]Department of Chemical Engineering, Faculty of Engineering, Universitas Sumatera Utara, Medan, Sumatera Utara, Indonesia
[2]School of Materials and Mineral Resources Engineering, Universiti Sains Malaysia, Penang, Malaysia

Abstract. Effects of alkanolamide (ALK) addition on cure characteristics, swelling behaviour and tensile properties of silica-filled natural rubber (NR)/chloroprene rubber (CR) blends were investigated. The ALK was synthesized from Refined Bleached Deodorized Palm Stearin (RBDPS) and diethanolamine, and incorporated into the silica-filled NR/CR blends as a non-toxic rubber additive. The ALK loadings were 0.0, 1.0, 3.0, 5.0 and 7.0 phr. It was found that the ALK exhibited shorter scorch and cure times and higher elongation at break of the silica-filled NR/CR blends. The ALK also exhibited higher torque differences, tensile modulus and tensile strength at a 1.0 phr of ALK loading and then decreased with further increases in the ALK loading. The swelling measurement proved that the 1.0 phr loading of ALK caused the highest degree in crosslink density of the silica-filled NR/CR blends.

1 Introduction

Magnesium and zinc oxides (MgO and ZnO) and ethylene thiourea (ETU) are widely utilised as the curative additives for chloroprene rubber (CR) [1-2]. Generally, MgO and ZnO are the vulcanising agents and ETU is the accelerator. ETU is a toxic material and is suspected to be carcinogenic [3-4]. Therefore, the appearance of an alternative vulcanisation accelerator for CR has been in demand.

Alkanolamide (ALK) is a relatively new rubber additive [5-11]. It can be prepared from refined bleached deodorised palm stearin (RBDPS), a by-product (waste) of palm cooking oil production, and ethanolamine. Since it comes from palm oil, ALK is a palm oil based rubber additive and it can be considered as a sustainable, (semi) renewable and non-toxic ingredient. ALK is a solid waxy material and it has some unique natures of the molecule which is structured by the combination of hydrophobic long chain hydrocarbon and hydrophilic amine. Due to its unique molecular structure, ALK has been utilised as a

* Corresponding author: indradanas@yahoo.com

curative additive in rubber compounding. The hydrophobic part of the molecule has the tendency to interact with rubbers, especially non-polar rubbers such as natural rubber (NR), styrene butadiene rubber, etc. [5-7]. The hydrophilic amine interacted not only with the polar filler such as silica [5-6] but also with polar rubber such as epoxidised natural rubber [7]. The ALK also cross-linked the CR chains of carbon black-filled CR compounds [9]. Such interactions affected the cure characteristics and tensile/mechanical properties of the rubbers compounds.

Based on the above explanations, ALK is likely to yield some good results in replacing the carcinogenic ETU as a vulcanising agent. Therefore, this study reports the utilisation of ALK as a vulcanising agent for silica-filled NR/CR blends. The effects of ALK on cure characteristics, swelling behaviour and tensile properties of silica-filled NR/CR blends were reported.

2 Materials and methods

2.1 Materials

NR grade SMR L was used and obtained from Guthrie (M) Sdn. Bhd., Seremban, Malaysia. Polychloroprene rubber (CR) [Skyprene B-30] was purchased from TOSOH Co. (Japan). Other compounding ingredients such as sulphur, zinc oxide, magnesium oxide, stearic acid, N-isopropyl-N'-phenyl-p-phenylenediamine (IPPD), N-cyclohexyl-benzothiazyl-sulfenamide (CBS), Tetramethylthiuram disulfide (TMTD) and precipitated silica (grade Vulcasil – S) were supplied by Bayer (M) Ltd., Sdn. Bhd., Petaling Jaya, Selangor, Malaysia. All materials were used as supplied. The ALK was synthesized in the laboratory using Refined Bleached Deodorized Palm Stearin (RBDPS) and diethanolamine. Its molecular structure is shown in Fig. 1.

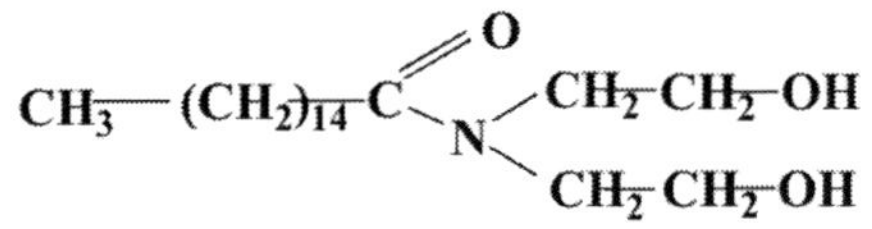

Fig. 1. The molecular structure of alkanolamide.

2.2 Compounding

A semi-efficient vulcanization system was used for the rubber blending. The recipe for the preparation of the NR/CR blend is given in Table 1. The blending procedure was done in accordance with the American Society for Testing and Material (ASTM) – Designation D 3184 – 80. The blending was done on a two-roll mill. Table 2 shows the designation and composition of the NR/CR blends used in this study.

Table 1. The composition of silica-filled NR/CR blends without/with ALK.

Ingredients	Content (phr)*
NR/CR	50/50
ZnO	5.0
MgO	4.0
Stearic acid	1.0
IPPD	1.0
CBS	1.0
TMTD	0.5
ETU	0.5
Sulphur	2.0
Silica	30.0
ALK	0.0; 1.0; 3.0; 5.0; 7.0

*parts per hundred parts of rubber blend

Table 2. The designation and composition of the silica-filled NR/CR blends without and with ALK.

Designation	Composition
NR/CR-0.0 (Control blend)	Silica-filled NR/CR blend without ALK
NR/CR-1.0	Silica-filled NR/CR blend with ALK (1.0 phr)
NR/CR-3.0	Silica-filled NR/CR blend with ALK (3.0 phr)
NR/CR-5.0	Silica-filled NR/CR blend with ALK (5.0 phr)
NR/CR-7.0	Silica-filled NR/CR blend with ALK (7.0 phr)

2.3 Cure characteristics

The cure characteristics of the rubber blends were obtained using a Monsanto Moving Die Rheometer (MDR 2000) which was employed to determine the scorch time (ts_2), cure time (t_{90}) and torque difference (M_H-M_L) value according to ISO 3417. Samples of the respective rubber blends were tested at 150 °C. The rubber blends were subsequently compression-moulded using a stainless steel mould at 150 °C, with a pressure of 10 MPa and applying a laboratory hot-press based on respective curing times.

2.4 Measurement of swelling behaviour

Swelling behaviours were performed in toluene in accordance with ISO 1817. Cured test pieces (30 mmx5 mmx2 mm) were weighed using an electrical balance and swollen in toluene until equilibrium, which took 72 h at room temperature. The samples were removed from the liquid, the toluene was removed from the sample surfaces, and the weight was determined. The change in mass is as follows:

$$\text{Swelling (\%)} = 100\% \, (W_2 - W_1) / W_1 \qquad (1)$$

where W_1 is the initial mass (g) and W_2 is the mass (g) after immersion in toluene.

2.5 Tensile properties

Dumbbell-shaped samples were cut from the moulded sheets according to ISO 37. Tensile tests were performed at a crosshead speed of 500 mm/min. Tensile tests were carried out with a universal tensile machine Instron 3366 to determine the tensile properties in terms of tensile strength (TS), stresses at 100% and 300% elongations (M100, M300) and elongation at break (EB).

3 Results and discussions

3.1 Effect of ALK on cure characteristics of the silica-filled NR/CR blends

Effect of ALK on the cure characteristics of the silica-filled NR/CR blends is showed in Table 3. The scorch and cure times decreased with increasing the ALK loadings. It was a cure enhancement which indicated ALK acted as a co-curing agent (secondary accelerator) in the curing. The amine content of the ALK caused the rubber blends more basic. Alkaline substances tend to improve the cure rate [12].

Table 3. The cure characteristics and tensile properties of silica-filled NR/CR blends without/with ALK.

Rubber blends	Cure characteristics			Tensile properties			
	ts_2, min.	t_{90}, min.	M_H–M_L (dN.m)	M100 (MPa)	M300 (MPa)	TS (MPa)	EB (%)
NR/CR-0.0	1.10	7.25	23.25	2.315	6.080	18.4	610.0
NR/CR-1.0	1.09	6.03	25.02	2.391	6.200	21.2	636.7
NR/CR-3.0	1.06	4.63	15.58	2.170	5.690	18.8	656.7
NR/CR-5.0	1.05	4.33	14.05	2.069	5.370	16.8	701.7
NR/CR-7.0	0.93	3.96	13.66	1.940	5.070	15.4	768.3

Note: M_L = Minimum torque; M_H = Maximum torque; M_H–M_L = Torque difference

The addition of 1.0 phr of ALK increased the torque difference (M_H–M_L). Further increases in the ALK loading decreased the torque difference. The torque difference represents shear dynamic modulus and it is believed that its value relates to the crosslink density of a rubber vulcanisate [8, 10, 13-14]. The addition of ALK affected the torque difference or the crosslink density of the silica-filled NR/CR blends. It was due to the action of the ALK as a secondary accelerator which enhanced the state of the curing process.

The reduction of crosslink density beyond 1.0 phr of ALK loading was due to the softening or lubricating effect of the excessive ALK which dissolved a part of the elemental sulphur [7-8] and hence reduced crosslink density.

3.2 Effect of ALK on swelling behaviour of the silica-filled NR/CR blends

Effect of ALK loading on the swelling percentage of the silica-filled NR/CR blends with the error bar is shown in Fig. 2. The five-cured test pieces (30 mmx5 mmx2 mm) of every rubber blend were weighed using an electrical balance and swollen in toluene until equilibrium, which took 72 hours at room temperature. The swelling percentage was measured by toluene uptake until equilibrium swelling was reached. It is widely used that the swelling is related to the crosslink density of a network chain [15], with less toluene uptake or penetration into the blends indicating a higher degree of crosslink density. Therefore, the swelling percentage of silica-filled NR/CR blend decreased with the addition of 1.0 phr of ALK and increased with further increases in the ALK loading. The increases in swelling percentage might be due to the excessive loading of ALK which reduced the degree of crosslink density [7-8].

The results of swelling behaviour also showed that the swelling percentage have not much different for ALK loadings 3.0, 5.0 and 7.0 phr. It was due to, beyond the 3.0 phr, the function of ALK as a curative agent was less pronounced and hence, the swelling percentage or crosslink density was not much different.

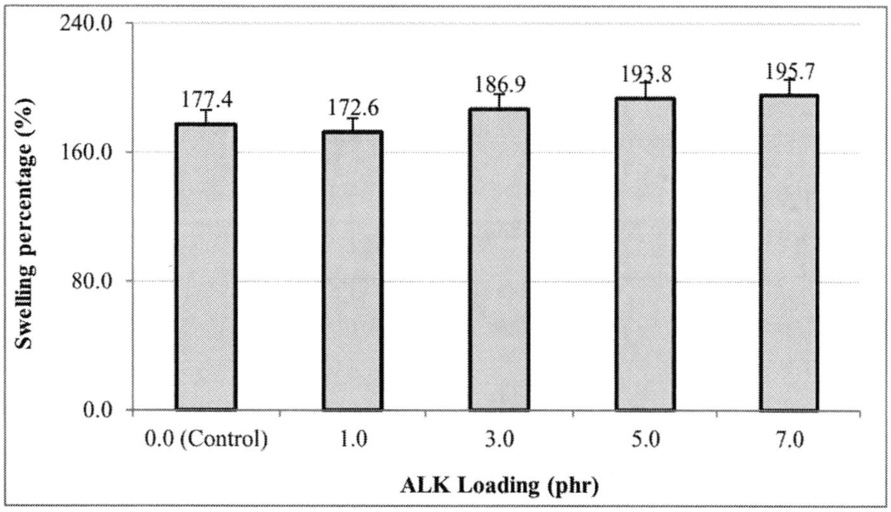

Fig. 2. Effect of alkanolamide loading on swelling percentage of silica-filled NR/CR blends.

3.3 Effect of ALK on tensile properties of the silica-filled NR/CR blends

Effect of ALK loading on the tensile properties of the silica-filled NR/CR blends is also shown in Table 3. The addition of 1.0 phr ALK produced an NR/CR-1.0 rubber blend with a higher EB. Increasing the ALK loading caused further increases in the percentage of EB or the extensibility of the rubber blends. This was due to the function of ALK as an internal plasticiser which modified flexibility or distensibility of rubber blends [5-6].

The addition of ALK increased the M100 and M300 up to the maximum level at a 1.0 phr and decreased with further increases in the ALK loading. The results of tensile strength also exhibit the similar trend. Tensile modulus and tensile strength are dependent only on the degree of crosslink [5, 16-17]. The improvement of those properties was attributed to a higher degree of crosslink (as indicated by a higher value of torque difference). The deterioration of those properties was due to the excessive amount of ALK which dissolved other curative and hence reduced the degree of crosslink density (as indicated by a lower value of torque difference). As discussed earlier, the crosslink density of a rubber compound is indicated by torque difference value. The greater the torque difference, the higher is the degree or total crosslink density [8].

4 Conclusions

From this study, the following conclusions were drawn:
1. Alkanolamide was a curative additive in silica-filled natural rubber/chloroprene rubber blends.
2. Alkanolamide enhanced the cure rate, reduced the scorch and cure times and improved the flexibility by increasing the elongation at break.
3. Alkanolamide enhanced the tensile properties and crosslink density of the silica-filled natural rubber/chloroprene rubber blend especially at a 1.0 phr of alkanolamide loading.

The authors would like to thank Universiti Sains Malaysia, for providing the research facilities for carrying out the experiment and for making this research work possible. Indra Surya and Syahrul Fauzi Siregar are grateful to Universitas Sumatera Utara for the Talenta research grant under contract No. 5338/UN5.1.R/PPM/ 2017 (22 May 2017).

References

1. M. Akiba, A. Hashim, Prog. Polym. Sci **22,** 475 (1997)
2. B. Rodgers, *Chemistry and Applications* (CRC, 2004)
3. R. Chhabra, S. Eustis, J. Haseman, P. Kurtz, B. Carlton, Fundam. Appl. Toxicol **18,** 405 (1992)
4. D.M. Smith, Br J Ind Med **41**, 362 (1984)
5. I. Surya, H. Ismail, A. Azura, Polym Test **32**, 1313 (2013)
6. I. Surya, H. Ismail, A. Azura, Polym Test **40**, 24 (2014)
7. I. Surya, H. Ismail, A. Azura, Polym Test **42**, 208 (2015)
8. I. Surya, H. Ismail, Polym Test **50**, 276 (2016)
9. I. Surya, H. Ismail, Plast Rubber and Compos **45,** 287 (2016)
10. I. Surya, M. Ginting, H. Ismail, IOP Conf. Ser., Mater. Sci. Eng **223,** 012006 (2017)
11. N. Hayeemasae, I. Surya, H. Ismail, Int J Polym Anal Charact **21,** 396 (2016)
12. H. Long, *Basic Compounding and Processing of Rubber* (USA, 1985)
13. P. Teh, Z. Mohd Ishak, A. Hashim, J. Karger-Kocsis, U. Ishiaku, Eur. Polym. J **40,** 2513 (2004)
14. H. Ismail, C. Ng, J Elastomers Plast **30,** 308 (1998)
15. H. Nabil, H. Ismail, A. Azura, Polym Test **32**, 385 (2013)
16. H. Ismail, H. Chia, Polym Test **17,** 199 (1998)
17. I. Surya, SF Siregar, H. Ismail, IOP Conf. Ser., Mater. Sci. Eng **223**, 012012 (2017)

The Ultimate Pile Bearing Capacity from Conventional and Spectral Analysis of Surface Wave (SASW) Measurements

Nor Faizah Bawadi[1]*, *Shamilah* Anuar[1], *Mustaqqim* A.Rahim[1], and *A.Faizal* Mansor[2]

[1]School of Environmental Engineering, Universiti Malaysia Perlis, Kompleks Pengajian Jejawi 3, 02600 Arau, Perlis, Malaysia
[2]Faculty of Engineering Technology (Civil), Universiti Malaysia Perlis, Aras 1, Blok S2, Kampus UniCITI Alam, Sungai Chuchuh, Padang Besar, Perlis, Malaysia

Abstract. A conventional and seismic method for determining the ultimate pile bearing capacity was proposed and compared. The Spectral Analysis of Surface Wave (SASW) method is one of the non-destructive seismic techniques that do not require drilling and sampling of soils, was used in the determination of shear wave velocity (V_s) and damping (D) profile of soil. The soil strength was found to be directly proportional to the Vs and its value has been successfully applied to obtain shallow bearing capacity empirically. A method is proposed in this study to determine the pile bearing capacity using V_s and D measurements for the design of pile and also as an alternative method to verify the bearing capacity from the other conventional methods of evaluation. The objectives of this study are to determine V_s and D profile through frequency response data from SASW measurements and to compare pile bearing capacities obtained from the method carried out and conventional methods. All SASW test arrays were conducted near the borehole and location of conventional pile load tests. In obtaining skin and end bearing pile resistance, the Hardin and Drnevich equation has been used with reference strains obtained from the method proposed by Abbiss. Back analysis results of pile bearing capacities from SASW were found to be 18981 kN and 4947 kN compared to 18014 kN and 4633 kN of IPLT with differences of 5% and 6% for Damansara and Kuala Lumpur test sites, respectively. The results of this study indicate that the seismic method proposed in this study has the potential to be used in estimating the pile bearing capacity.

1 Introduction

The valuable consideration on basic soil parameters should be implemented in evaluation of pile bearing capacity using various conventional methods laboratory or from in-situ testing. Generally, the in-situ seismic method commonly used for evaluating pile bearing

* Corresponding author: norfaizah@unimap.edu.my

capacity because of difficulty and cost of collecting a good sample for laboratory testing. The effectiveness of this seismic method obtained more stable and reliable when compared with those of the conventional methods [1]. This study was conducted to evaluate the ultimate bearing capacity of pile at sites located on residual soils in Damansara and Kuala Lumpur, Malaysia. The Spectral Analysis of Surface Wave (SASW) test has been conducted to measure shear wave velocity (V_s) and damping (D) values based on SASW data interpretation. Data from Instrumented Pile Load Test (IPLT) are back calculated with shear wave velocity and damping value, and can be used in evaluation of pile bearing capacity in residual soils. Indirectly, the results of the pile bearing capacity obtained using shear wave velocity and damping measurements was compared with IPLT's results.

2 Basic parameters of SASW measurement

It is a common practice to determine the bearing capacity of foundation based on the shear wave velocity and damping measurements carried out as part of the field testing. Basically, shear wave velocity only travel in the matrix of soil particles and does not travel in its liquid part. This characteristic of shear wave velocity enables it to measure the effective soil strength in which it travels. The good agreement has been shown with the travelling shear wave causes minute strain and its measurement is known to be elastic and hence the dynamic shear modulus [2]. The consideration of overburden pressure (σ'_v) in evaluation of penetration resistance [3], the measured V_s obtained from SASW testing should be normalized (V_{s1}) based on relation of atmospheric pressure (P_a) using Equation (1).

$$V_{s1}=V_s\,(P_a/\sigma'_v) \quad (1)$$

Initial shear modulus can be defined as shear modulus at infinitely small-strain amplitude. It is therefore necessary to obtain shear strain amplitude in soil deposits when the SASW test is to be conducted. The dynamic shear strain (γ) is defined as the ratio of peak particle velocity (υ) to shear wave velocity (V_s) as presented in Equation (2) [4-5].

$$\gamma=\mu/V_s \quad (2)$$

At small-strain conditions, particle motion resulting from propagation of shear waves is non-destructive. As shear strain increases, the initial shear modulus will decrease from the maximum small-strain value. In-situ tests have commonly been assumed to be small-strain and the measurement of shear wave velocity will be directly related to the maximum shear modulus [6]. The appropriate shear strain must be known, as well as the appropriate modulus reduction with increasing shear strain. Researcher state that damping refers to the energy dissipation within a soil mass during dynamic loading [7]. For small value of material damping, the quality factor and damping ratio can be defined as the resonance frequency divided by the bandwidth [3,8]. Originally, the half-power bandwidth method has been developed to determine the damping ratio of a structure from the width of the peaks in the structure's frequency response function in the field of structural dynamics.

The strain amplitudes affecting damping ratio of soil [9]. The determination of the damping ratio established [3,10,11,12]. The relationship of damping and shear strains [13], the characteristic of normal or shear strain can be determined using rearrangement of the Equation (3) as shown below:

$$D=D_{max}/\,[1+\ln\,(1+(e-1)\,(\gamma_r/2\gamma))] \quad (3)$$

where D_{max} is maximum damping at 33%, D is damping ratio; γ is shear strain at damping measured; e is constant and γ_r is characteristics shear strain.

3 Pile bearing capacity using SASW measurement

Soil dynamics tests under cyclic loading, both in normal triaxial or shear test have shown common hysteretic behavior. The skeleton of the backbone curve of the hysteresis to express form of hyperbolic stress-strain soil model [9]. Based on this approaches, the normal (σ) and shear stress (τ) of soil can be defined using the following equation:

$$\sigma = (E_0\varepsilon)/((1+(\varepsilon/\varepsilon_r))) \tag{4}$$

$$\tau = (G_0\gamma)/((1+(\gamma/\gamma_r))) \tag{5}$$

where E_0 and G_0 are elastic and shear modulus at small-strain (less than 10^{-4}%); ε and γ are normal and shear strain under pile load respectively. The ultimate pile bearing capacity (Q_{ult}) can be obtained by estimating the stress generated around the pile that contacted to the soil. Consideration of stress-strain behavior of soil has been used in evaluation of total skin resistance (Q_s) and end bearing (Q_b) of pile.

4 Methodology

This study is mainly dividing into two stages. The first stage deals with field configuration and data acquisition in order to determine the data of shear wave velocity and damping. The basic configuration of the source and receivers used in the field testing using the Common Array Profiling (CAP) [14]. There are two numbers of the geophone sensor used which having a natural frequency of 1 Hz and calibration factor is 400/V/ (m/s). Several assorted hammers comprise a small geology, rubber, steel and steel sledge hammer were employed to generate energy over the required frequency ranges. The output of SASW data analysis to obtain shear wave velocity profile, dynamic shear strain and damping ratio of soil.

Meanwhile, in the second stage, the back analysis on estimation of pile bearing capacity of a pile using shear wave velocity and damping measurements. The skin resistance and end bearing of the pile were then calculated [9] and the SASW data covered relevant shear wave velocity, damping and dynamic shear strain amplitude at measured damping. In addition, prediction of soil density and shear modulus has been correlated to the shear wave velocity [15]. By back calculated strain amplitudes data, the ultimate pile bearing capacity evaluated and the results of pile bearing capacity in second stage were then compared to the results of IPLT.

5 Results and discussions

The SASW data were carried out using the National Instrument USB6289 data acquisition system with the WinSASW 3.2.12 and analysis software. The measured shear wave velocity, damping and N-value for tested sites are shown in Table 1. The distribution pattern of V_s and N_{value} at the both tested sites indicate that the V_s profile similar to the SPT profile which the parameters increased proportionally with depth.

Table 1. Results of shear wave velocity, damping and N_{value} for site of Kuala Lumpur and Damansara.

Sites	Soil Parameters			
	Depth, z (m)	N_{value}	Shear Wave Velocity, V_s (m/s)	Damping, D (%)
Kuala Lumpur	0-3	8-20	100-250	0.065-0.069
	3-11	10-20	250-500	0.058-0.069
	11-25	18-50	500-800	0.058
Damansara	0-5	6-18	80-200	0.000002-0.108
	5-10	18-30	200-600	0.099
	10-16	30-50	600-800	0.099

In this study, the frequency-response curve obtained from WinSASW programmed has been selected to interpret the damping of soil from the signal spectrum recorded from SASW test using the half- power bandwidth method. Generally, SASW method predicted similar characteristics of strain and damping, which decreased with depth. Analysis of the performed SASW field data in this study shows that the strain amplitudes caused by several assorted hammer as seismic source are generally less than 10^{-6} %. From this findings, clearly shows that SASW method tends to have a more linear behavior of soil and dynamic properties at small strains (10^{-4} % or less), conditions as shown in Fig. 1.

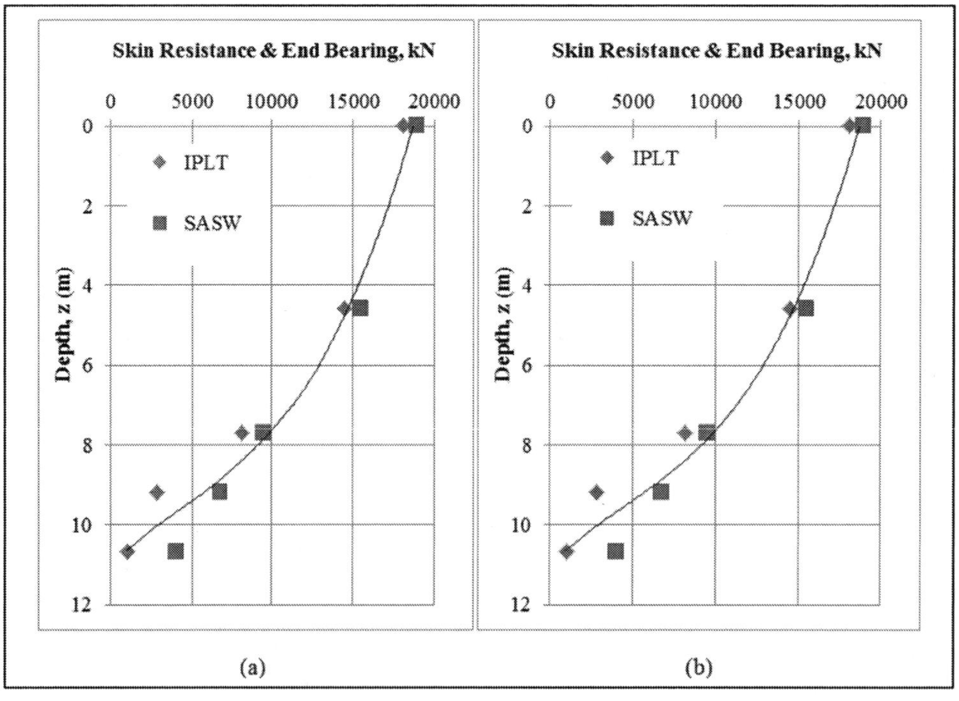

Fig. 1. Skin and end bearing resistances of pile at site of a) Kuala Lumpur and b) Damansara.

For foundation design based on hyperbolic model, the two most important parameters are the modulus of soil and strain amplitude. Back analysis shows that the ultimate pile bearing capacity is equal to 4947 kN and 18891 kN at small strain levels of about 10^{-5} to 10^{-6} % for sites of Kuala Lumpur and Damansara, respectively. The result of normal and shear stress of soil [9] was used to determine skin and end bearing resistance of pile as shown in Figure 3. This ultimate pile bearing capacity is in close agreement with the results of instrumented pile load test at site as tabulated in Table 2. The small difference shows that the estimated ultimate pile bearing capacity are in really good agreement and the difference might be due to stress oscillation and the average values used in this study.

Table 2. Comparisons of ultimate pile bearing capacity between IPLT and SASW methods.

Site	**Ultimate Pile Bearing Capacity (Q_{ult}) kN**		**Difference (%)**
	SASW	IPLT	
Kuala Lumpur	4947	4633	6
Damansara	18981	18014	5

5 Conclusions

Based on the results performed in this study, the shear wave velocity and damping measurements obtained using SASW test able to evaluate the ultimate pile bearing capacity. An empirical equation [9] is back calculated to predict the ultimate pile bearing capacity and the outcomes were compared with the conventional results. The 6% and 5% difference of the amount ultimate pile bearing capacity obtained in this study justified that ultimate pile bearing capacity can be measured using shear wave velocity and damping measurements for Kuala Lumpur and Damansara, respectively. Furthermore, this percentage is less than 10% of the standard acceptance in the design of conventional civil engineering structures.

References

1. S. Nazarian, K.H. Stokoe II, Austin, Texas, (1985)
2. C. P. Abbiss, Géotechnique **31**(1), 91–104, (1981)
3. G.J. Rix, C.G. Lai, M.C. Orozco, G.L. Hebeler, V. Roma, *Proceedings of the fifteenth International Conference on Soil Mechanics* **1-3**(3), 499–502, (2001)
4. A. Hara, T. Ohta, M. Niwa, S. Tanaka, T. Banno, Japaneses Society of Soil Mechanics and Foundation Engineering **14**(3), (1974)
5. T. Inazaki, *The Proceedings of IWAM04, Mizunami, Japan*, hlm. **S4**, (2004)
6. S.S. Tezcan, A. Keceli, Z. Ozdemir, Geotechnical and Geological Engineering **24**(1), 203–218, (2006)
7. R.V. Whitman, S. Liao, Vicksburg, Mississipi, (1985)
8. C.P. Abbiss, Géotechnique **36**(4), 565–580, (1986)
9. B.O. Hardin, V.P. Drnevich, Journal of the Soil Mechanics and Foundations Division, SM **7**(9006) 667 – 692, (1972)

10. S.A. Badsar, M. Schevenels, G. Degrande, ECCOMAS Thematic Conference on Computational Methods in Structural Dynamics and Earthquake Engineering, (June), (2009)
11. S. Foti, Géotechnique **53**(5), 455–461, (2003)
12. C.G. Lai, G.J. Rix, S. Foti, V. Roma, Soil Dynamics and Earthquake Engineering **22**(9-12), 923–930, (2002)
13. C.P. Abbiss, Géotechnique **33**(4), 397–405, (1983)
14. S.H. Joh, T.H. Kang, D.W. Jang, I.-W Lee, Journal of the Korean Geotechnical Society. Korean Geotechical Society **21**(4), 71–81' (2005)
15. A. Keçeli, Jeofizik **16**, 17–29, (2012)

Characteristics Buton Natural Asphalt-Rubber (BNA-R) on the Performance Improvement of Warm Mix Asphalt Using Natural Zeolite

Nurul Wahjuningsih[1,2,*], *Sigit Pranowo* Hadiwardoyo[1] , and *R Jachrizal* Sumabrata[1]

[1]Civil Engineering Department, Universitas Indonesia, Depok, Indonesia
[2]Civil Engineering Department, Institut Teknologi Bandung, Bandung, Indonesia

Abstract. The decrease in the ability of service of pavement can be caused by the durability factor in the pavement layer in receiving heavy traffic load and the temperature of the pavement. Permanent deformation is one of the criteria of failure of asphalt mixture. Performance assessment of the asphalt mixture can be observed from the rheological properties of asphalt binder. The use of BNA-R in this study is intended to modify the characteristics of bitumen penetration grade 60 / 70 used in warm mix asphalt. Warm mix asphalt with lower temperatures of mixing and compaction than conventional asphalt mixtures was chosen because it is more environmentally friendly. To reduce the temperature in this warm asphalt technology is achieved by using natural zeolite. Both of these materials are local materials that are widely available in Indonesia. The rheology of asphalt 60/70 modified with BNA-R indicates that the addition of BNA-R in the base asphalt increase the complex modulus value and decrease the phase angle value. These values were related to the performance of mixture in the permanent deformation criteria. Reducing the temperature of mixing and compaction should be balanced with modifying the asphalt binder used. Rutting due to permanent deformation can resulted in inconvenience to the passengers and can lead to high costs of road maintenance. To determine the permanent deformation of asphalt mix with material combinations was performed through the wheel tracking test machine with 3,780 cycles for 3 hours. The results shows that after test track over 7 thousand passes have seen permanent deformation characteristics of asphalt concrete mixture with a variation of the characteristics of bitumen.

1 Introduction

Permanent deformation is one of failure criteria of asphalt mixture [1]. On the pavement layer this condition will cause inconvenience of user roads and high cost maintenance. With the increase in traffic load and tire pressure, most of the permanent deformation occurs in the upper layers [1-2]. The improvement of permanent deformation of asphalt pavement layer will be important things to enhance the convenience of user road.

* Corresponding author: nwahjuningsih@gmail.com

Warm Mix Asphalt (WMA) is one of innovation in the manufacturing of asphalt mixture that reduce the mixing and compaction temperature about 20-40°C with involve the organic additives, chemical additives and water-based or water-containing processes [3], [4]. Many advantages of WMA as reported the researcher like reduced emission, lower energy consumption in production of asphalt mixture, better working condition because absence of harmful gasses, longer hauling distances, and extended paving window. Besides, the drawbacks of this mixture still found such are higher costs because there is still a certain reticence to using them, lack of data concerning their long-term performance, greater moisture susceptibility due to lower temperatures and coating and bonding problems [3], [5].

In the laboratory, the simulation of the deformation caused by wheel load were performed using wheell tracking test because this test could model the stress of pavement at the field [6]. Generally, the testing was done in the 1,260 trajectories or in 1 hour. The studies have been conducted showed that this trajectories number has limitation for advanced analysis. The result of the one hour cycle has not reached the secondary zone which the accumulation of permanent deformation has not occurred, so that increasing the number of cycle more than 1 hour is required [7].

This study aims to characterize the properties of bitumen samples involving BNA-R with the different content and knowing the influence to the permanent deformation of warm mix asphalt. For this purpose, conventional test methods as well as dynamic shear rheometer test at different temperatures have been performed to the base bitumen and modified bitumen.

2 Material and methods

Bitumen base used in this research was asphalt cement with penetration grade 60/70 which fulfilled Indonesian specification. The characteristics of base bitumen have shown in the Table 1.

Table 1. Characteristics of base bitumen.

No	Laboratory Test	Method	Unit	Indonesian Specification	Bitumen Penetration Grade 60/70
1	Penetration at 25°C	ASTM-D5	0.1 mm	60 - 70	64.2
2	Softening Point	ASTM-D36	°C	≥48	51.5
3	Flash Point (Cleveland)	ASTM-D92	°C	≥232	338
4	Ductility at 25°C	ASTM-D113	cm	≥100	>100
5	Specific Gravity at 25°C	ASTM-D70	kg/m3	≥1.0	1.037
6	Penetration after TFOT	ASTM-D5	0.1 mm	≥54	61.3
7	Kinematic Viscosity at 135 °C	ASTM-D2170-67	cSt	≥300	459.6

The aggregate consisted of coarse, medium and fine aggregate with the largest size of 19 mm and the composition of aggregate gradation shown in Table 2. The properties of each aggregates are shown in Table 3. And the sample of asphalt mixture was prepared for the AC-WC (Asphalt Concrete Wearing Course) type based on Indonesian specification 3rd revision 2010.

Table 2. Gradation of AC-WC.

Sieve Size		% passing
¾ in	(19.1 mm)	100
½ in	(12.7 mm)	95
3/8 in	(9.52 mm)	83.5
No. 4	(4.76 mm)	61
No.8	(2.38 mm)	43
No. 16	(1.18 mm)	30.5
No. 30	(0.59 mm)	22
No. 50	(0.279 mm)	15.5
No. 100	(0.149 mm)	10.5
No. 200	(0.074 mm)	6.5

Table 3. Aggregate Properties.

Laboratory test	Method	Specification		Result
		Min	Max	
Coarse Agregate				
Bulk specific gravity (Gsb)	AASHTO T84/T85	2.5	–	2.67
Surface Saturated Dry gravity		2.5	–	2.69
Apparent specific gravity (SG)		2.5	–	2.73
Absorption		–	3	0.89
Flakiness Index	SNI 03-4137-1996	–	25%	13.15%
Los Angeles Abration	AASHTO T96	–	40%	38%
Medium Agregate				
Bulk specific gravity (Gsb)	AASHTO T84/T85	2.5	–	2.67
Surface Saturated Dry gravity		2.5	–	2.69
Apparent specific gravity (SG)		2.5	–	2.74
Absorption		–	3	0.92
Flakiness Index	SNI 03-4137-1996	–	25%	22.25%
Fine Agregate				
Bulk specific gravity (Gsb)	AASHTO T84/T85	2.5	–	2.67
Surface Saturated Dry gravity		2.5	–	2.69
Apparent specific gravity (SG)		2.5	–	2.73
Absorption		–	3	0.89
Clay lumps	SNI 03-4428-1997	–	8%	3%

The natural rock asphalt deposits in Indonesia was found on Buton Island known as Asbuton or BNA (Buton Natural Asphalt). There is a huge reserve of natural rock asphalt with containing an asphalt content of 5-30%. The size of deposits has been estimated at up to 677 million tonnes [8]. BNA still have a great chance to be utilized as partial or whole of oil asphalt substitutes in asphalt mixtures. BNA is produced by extracting Buton rock asphalt and more widely used as an modifier or additive because its very low penetration [9]. BNA-R (Buton Natural Asphalt – Rubber) is a product which consist of Buton rock asphalt extraction with crumb rubber. BNA-R consist of 60% bitumen semi extraction and 40% crumb rubber and reported has several advantages such as high adhesiveness, high softening point, high stiff modulus, high fracture resistance, long life and economical [10-11].

The content of BNA-R of 5%, 10% and 15% of total bitumen weight was chosen to modify the properties of base bitumen. The pulverised BNA-R into the smaller particle size then mixed with the base bitumen at the temperature of 140°C for 30 minutes and stir with the speed about 2,000 rpm to get the homogen mixture. For reducing the mixing temperature in the warm mix asphalt (WMA) the additive from material local was used.

The natural zeolit is one of natural deposit of Indonesia was chosen after activated first. The 2% of zeolit powder added to the aggregate as a substitution of filler before hot bitumen blended. With this additive the mixing temperature and compaction temperature can be reached 15 degree lower than conventional mix asphalt (hot mix asphalt).

The dynamic shear theometer (DSR) test was performed on the bitumen before and after modify with BNA-R using 10 rad/sec (1.519 Hz) frequency sweeps at the temperature between 52°C and 70°C with the increment of 6°C and prepared with two samples condition, viz un-aged and RTFO aged. RTFO condition is simulate the condition of bitumen in the filed when making the asphalt mixture where bitumen has been aged or called by short term aging.

The DSR test is generally used for evaluating the mechanical rheology of bitumen where the main parameter that will be analyzed from the test results are complex shear modulus (G*) and phase angle (δ). The total resistance of a material to deformation when subjected to a sinusoidal shear stress was expressed by G* (complex shear modulus). G* consist of recoverable (elastic) and non-recoverable (viscous) components. The phase angle (δ), is an indicator of proportion of viscous and elastic components. The temperature and frequency of loading are very influencing the values of G* and δ. Bitumen behave like viscous fluids with little capacity for recovering at high temperature and conversely at the low temperature, bitumen behave like elastic solids that completely recovery from deformation [12].

3 Results and discussions

3.1 Bitumen Rheology

The addition BNA-R to the base bitumen decreases the penetration grade of bitumen, increasing softening point and viscosity as shown in Table 4. This corresponds to the characteristics of Asbuton that have very low penetration [8], [9]. Generally with modify the base bitumen with BNA-R, the bitumen became more stiff. The result of conventional rheology of bitumen showed that with the addition of BNA- R increase the softening point. The increase in softening point is very advantages since bitumen with higher softening point may be less susceptible to permanent deformation (rutting). This is relevant with the decreasing penetration grade of bitumen after modifying with BNA-R. The viscosity values increase with the addition of BNA-R.

Table 4. Properties of bitumen modified.

Laboratory Test	Content of BNA-R			
	0%	5%	10%	15%
Penetration, 25°C	64.20	55.60	52.40	50.80
Softening Point, °C	51.50	52	53	53.75
Viscosity Kinematic at 135°C, cST	459.60	472.08	484.58	503.67

The mechanical rheology of bitumen can be evaluate from the result of DSR test. The changing of complex shear modulus (G*) and phase angle (δ) of both base bitumen and modified bitumen at frequency level of 1.519 Hz and at five different temperature are presented in Fig.1 and Fig. 2 respectively.

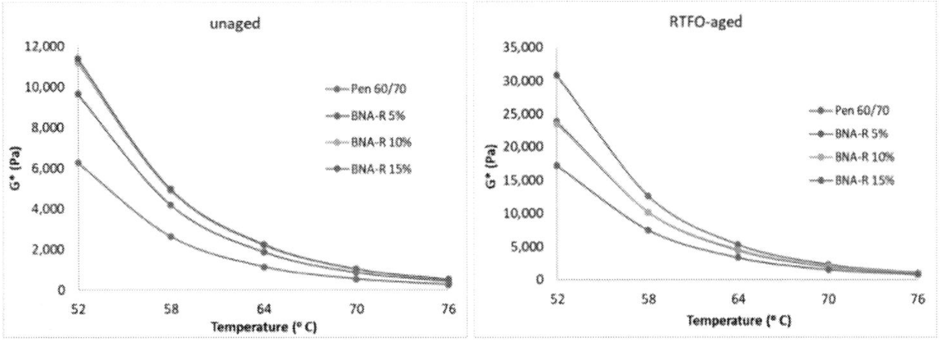

Fig.1. The changing of complex shear modulus (G*) at unaged and RTFO-aged.

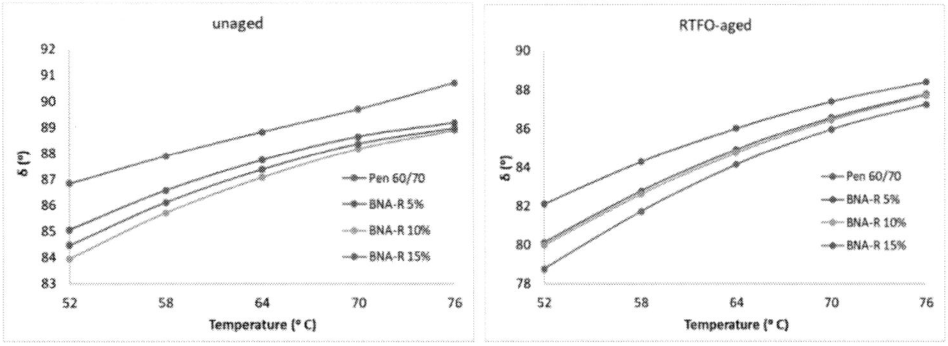

Fig.2. The changing of phase angle (δ) at unaged and RTFO-aged.

As depicted in Fig. 1, as temperature increases, complex shear modulus decreases. A decreasing in G* indicates lower elastic part. The phase angle increases as the test temperature increases as depicted in Fig. 2. The rheology of bitumen can be evaluated by measuring two viscoelastic parameters including complex shear modulus (G*, stiffness) and phase angle (δ, viscous or elastic indicator) throughout various test temperatures and frequencies so that the behaviour of asphalt pavement by various conditions can be predicted. From the Fig. 1 and Fig. 2, the addition of BNA-R as a modifier to the base bitumen influence the value of G* and δ. Generally, the higher content of modifier, the higher value of G* and lower value of δ. It indicate that the BNA-R make the bitumen more stiff and elastic.

The capability of pavement to withstand the deformation or rutting is related to both binder properties and aggregate properties. The G*/sin δ value is considered as a numerical indicator of rutting resistance, while higher G*/sin δ values are favorable for permanent deformation resistance [13, 14]. A high complex shear modulus G* and low phase angle δ are both desirable for rutting resistance. Fig. 3 shows the anti rutting factor G*/sin δ for both unaged and short term aging (RTFO-aged). At the low of test temperature, the addtion of BNA-R give the significant increment of G* and G*/sin δ, but not at the high ones.

3.2 Permanent Deformation Performance

The samples of asphalt mixture was prepared base on the optimum bitumen content at the each content of modifier (BNA-R) from the Marshall test. In order to make the warm mix asphalt where the mixing and compacting temperature lower than conventional, the natural zeolit powder was used at the content of 2% as a substitution of filler of the asphalt

 https://doi.org/10.1051/e3sconf/20183401032

mixture. The decreasing temperature with using natural zeolit reach about 15°C than hot mix asphalt.

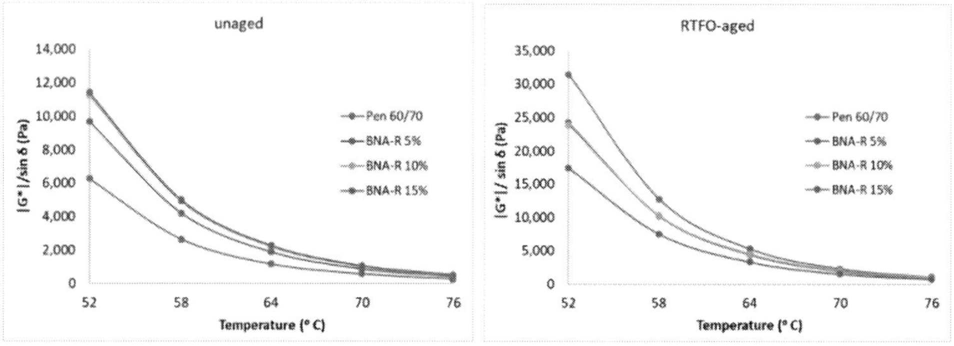

Fig.3. The changing of anti rutting factor (G*/sinδ) at unaged and RTFO-aged.

To evaluate the performance of asphalt mixture to the permanent deformation criteria, the wheel tracking machine test with the extended cycles until 3 hours (3,780 trajectories) was used. The conventional asphalt mixture was prepared too as a comparison. Fig. 4 and Fig. 5 shows the result of the extended wheel tracking test at the test temperature of 40°C for hot mix asphalt and warm mix asphalt respectively.

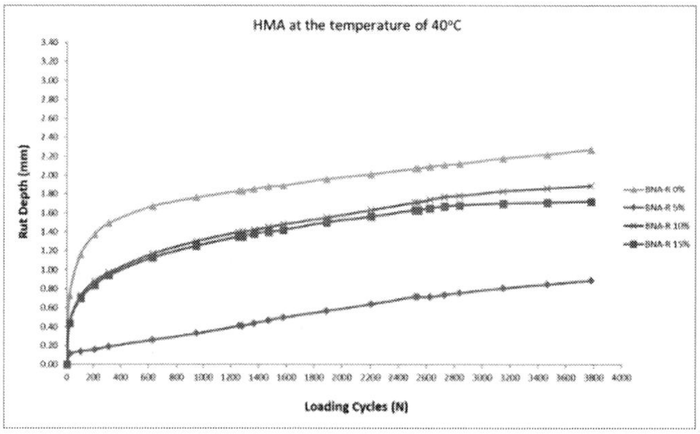

Fig. 4. The deformation of hot mix asphalt.

As depicted at the Fig. 4 and Fig. 5, the deformation (rut depth) of asphalt mixture with unmodified bitumen higher than mixture with modified bitumen both of hot mix and warm mix asphalt. Because both mixtures was prepared at the same gradation of aggregate, the performance of mixtures deformation very dependant on the bitumen properties. As mentioned above, the rheology of bitumen changed because of the addition of BNA-R especially showed in anti rutting factor (G*/sinδ), this influencing the resistance of mixtures to the deformation overall.

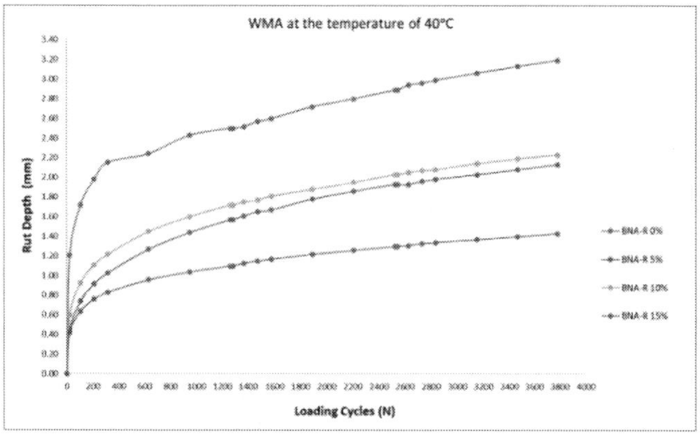

Fig.5. The deformation of warm mix asphalt.

Comparing the performance of hot mix asphalt and warm mix asphalt in the deformation, looked that the low mixing and compacting temperature (135°C), the appeareance of rutting of warm mix asphalt higher than hot mix. As Fig.d in Fig. 4 and Fig. 5, the maximum rutting of hot mix asphalt of about 2.2 mm but for warm mix asphalt reach of about 3.3 mm, for mixture without modified bitumen. The reduction in rutting depth is quite significant when the mixture use modified bitumen, both of hot mixture or warm mixture.

4 Conclusions

Based on the tests, the conclusion of the study is the changes of the bitumen rheology can affect the performance of asphalt mixture. The decreasing of mixing and compacting temperature in the warm mix asphalt tend to lowering the capability to the resistance of permanent deformation. BNA-R as a local material from Indonesia can be used as bitumen modifier to enhance the bitumen resistance to the deformation. The composition of BNA-R and natural zeolit as an additive for reducing mixing temperature in WMA must be suitable for keeping the ability of asphalt mixture in resist the deformation permanent.

This research was sponsored by the grant Penelitian Disertasi Doktor (PDD) from Ministry of Research, Technology and Higher Education Republic of Indonesia year 2017 with the contract number of 009/SP2H/LT/DRPM/IV/2017. The laboratory work was completed in the Material and Structure Laboratory Universitas Indonesia, Laboratory of Highway engineering and Traffic Institut Teknologi Bandung and Research Centre and Development for Road and Bridge Laboratory - Ministry of Public works Republic of Indonesia. The paper was prepared with the support from *Workshop dan Klinik Peningkatan Kualitas Hasil Penelitian Program Peningkatan Kapasitas Riset* was held by DRPM Ministry of Research, Technology and Higher Education Indonesia collaboration with Universitas Katolik Parahyangan.

References

1. Y. H. Huang, Pavement Analysis and Design (2003)
2. J. Read and D. Whiteoak, *The Shell bitumen handbook* (2003)
3. S. D. Capitalo, L. G. Picado-Santos, F. Martinho, Constr. Build. Mater. **36**, 1016–1024 (2012)
4. A. Topal, B. Sengoz, B.V. Kok, M. Yilmaz, P.A. Dokandari, J. Oner, D. Kaya, Constr. Build. Mater. **57**, 38–44 (2014)

5. A. Chowdhury, J. Button, Texas Transportation Institute. Texas A&M University System. **7**(2), 75 (2008)
6. X. Huang, Y. Zhang, Road Mater. Pavement Des. **11**(3), 969–991 (2010)
7. S. P. Hadiwardoyo, R. Jachrizal Sumabrata, R. H. Aryapijati, N. Wahjuningsih, A. Kurnia, R. Abrar, ARPN J. Eng. Appl. Sci. **11**, 24 (2016)
8. N. Suaryana, Int. J. Pavement Res. Technol. **9**(5), 387–392 (2016)
9. S. P. Hadiwardoyo, E. S. Sinaga, H. Fikri, Constr. Build. Mater. **42**, 5–10 (2013)
10. N. Anagi, S. P. Hadiwardoyo, R. J. Sumabrata, N. Wahjuningsih, *AIP Conference Proceedings*, **1855** (2017)
11. N. Wahjuningsih, S. P. Hadiwardoyo, R. J. Sumabrata, *AIP Conference Proceedings*, **1855** (2017)
12. B. Sengoz, L. Bagayogo, J. Oner, A. Topal, Constr. Build. Mater. **154**, 1105–1111 (2017)
13. Y. Li, S. Liu, Z. Xue, W. Cao, Constr. Build. Mater. **54**, 533–540 (2014)
14. A. Behl, S. Chandra, A. M. Asce, J. Mater. Civ. Eng. **29** (2012), 1–7 (2017)

Mechanical and Thermal Analysis of Classical Functionally Graded Coated Beam

Abdolreza Toudehdehghan[1,2*], *Md.* Mujibur Rahman[2], and *Faris* Tarlochan[3]

[1]Department of Mechanical Engineering, INTI International University, Jalan BBN 12/1, Bandar Baru Nilai, 71800 Nilai, Negeri Sembilan, Malaysia
[2]Department of Mechanical Engineering, Universiti Tenaga Nasional, km7, Jalan Kajang-Puchong, 43009 Kajang, Selangor, Malaysia
[3]Mechanical and Industrial Engineering, College of Engineering, Qatar University, P.O. Box 2713, Doha, Qatar

Abstract. The governing equation of a classical rectangular coated beam made of two layers subjected to thermal and uniformly distributed mechanical loads are derived by using the principle of virtual displacements and based on Euler-Bernoulli deformation beam theory (EBT). The aim of this paper was to analyze the static behavior of clamped-clamped thin coated beam under thermo-mechanical load using MATLAB. Two models were considered for composite coated. The first model was consisting of ceramic layer as a coated and substrate which was metal (HC model). The second model was consisting of Functionally Graded Material (FGM) as a coated layer and metal substrate (FGC model). From the result it was apparent that the superiority of the FGC composite against conventional coated composite has been demonstrated. From the analysis, the stress level throughout the thickness at the interface of the coated beam for the FGC was reduced. Yet, the deflection in return was observed to increase. Therefore, this could cater to various new engineering applications where warrant the utilization of material that has properties that are well-beyond the capabilities of the conventional or yesteryears materials.

1 Introduction

Functionally Graded Materials (FGMs) were initially used as a thermal barrier by Japanese scientists in 1984. Since then, this kind of materials has been expanded and developed as a heat-resistant material. As an example, the surface temperature of the space shuttle is approximately 21000 K, that the temperature difference between the outside to the inside reaches 16000 K. Therefore, conventional composites cannot resist this temperature difference [1]. Unlike the conventional composites, in FGMs, a combination of materials is changed continuously from high resistance temperature materials such as ceramics on one side to the other side, which has strong mechanical toughness materials, such as metals. FGM is used not only as a thermal barrier but also as a corrosion and wear coating resistant

* Corresponding author: Abdolreza.toudehdehghan@newinti.edu.my

[2]. Javaheri and Eslami [3] analyzed functionally graded rectangular plate under thermal loads based on higher order shear deformation theory and power law function for FGMs. Most of the researcher, in order to investigate on FGM structures, are used power law function [4-6]. Ho Chi and Chung [7] used power-law, exponential and sigmoid function on the classical plate theory and compared the results with applied the finite element method.

Generally, the surface of structures or elements is weaker than the inside of them. Therefore, the coating has been used for decades to increase the resistance of the surface of structures, reduce the stress concentration on the surface and stop cracks, which have been generated. In the conventional surface coating method, one layer as a coated will be covered the surface of the body and the thickness range is from a micrometer to several millimeters. The coating thickness usually is depended on the type of material which is used as a substrate layer [8]. For example, in Thermal Barrier Coating (TBC) one-layer ceramic is bonded to the substrate layer. The disadvantage of this type of coating is that due to the use of two completely different materials (ceramic-metal), the concentration of stress between the two layers' increases, and thus the two layers are separated [9]. To improve the weakness, FGMs have lately been suggested to modify the conventional coating [10]. Due to the characteristic of FGM composites, the properties of materials in the composite change very gradually between two different materials. This property makes the separation between the composite layers not present. Mainly, under a high-temperature atmosphere like a nozzle of the shuttle, nuclear fusion reactor, internal combustion engine and so on at the interface of two layers due to different thermal expansion, the mismatch will be happened [7]. Coating structures with FGMs are called FGC. FGC beams under mechanical, thermal and thermos-mechanical loads are studied by several of researches [11-17].

The present article, a static analysis of FGC beam is studied by using a series displacement as a linear combination of know function which satisfies the boundary conditions and unknown parameters. The governing equation of FGC beam based on classical beam displacement theory and principle of virtual is derived. The material properties of FG layer through the thickness is considered an exponential function. In order to demonstrate the behavioural difference of a conventional composite with a modern composite which is consisting of an FGM layer. Two composite models are considered. Lastly, the effect of the rectangular clamped-clamped beam under thermal and mechanical loads are investigated.

2 Extracting governing equations

The displacement field components in Euler-Bernoulli's beam are considered in the form below [18]:

$$\begin{aligned} u(x,y,z) &= u_0(x,y) - z\frac{dw_0(x)}{dx} \\ V(x,y,z) &= 0 \\ w(x,y,z) &= w_0(x) \end{aligned} \tag{1}$$

Thus, the axial strain component in mid-plane direction of beam is

$$\varepsilon_{xx} = \frac{du}{dx} = \frac{du_0}{dx} - z\frac{d^2 w_0}{dx^2} \tag{2}$$

2.1 FGMs material properties

The continuous composition of FGMs can be described by using several mathematical models to show the material properties of the composite. The material properties of FGMs are inhomogeneous microscopically and change with changing temperature, therefore the material properties of FGMs depend on position and temperature. One of the mathematical models of material properties was used in a lot of research in order to study fracture mechanics, crack propagation, vibration, and bending, is exponential function [7, 11, 19-26].

Material properties with exponential function of FGMs are:

$$\begin{aligned} E_{fe}(z) &= E_c e^{\left(\frac{1}{h}\ln\left(\frac{E_m}{E_c}\right)\left(z+\frac{h}{2}\right)\right)} \\ \alpha_{fe}(z) &= \alpha_c e^{\left(\frac{1}{h}\ln\left(\frac{\alpha_m}{\alpha_c}\right)\left(z+\frac{h}{2}\right)\right)} \\ k_{fe}(z) &= k_c e^{\left(\frac{1}{h}\ln\left(\frac{k_m}{k_c}\right)\left(z+\frac{h}{2}\right)\right)} \end{aligned} \tag{3}$$

2.2 Governing equation of beam

To derive the governing equations of the beam, we use the principle of virtual work defined as [27]:

$$\delta W = \delta W_I + \delta W_E = 0 \tag{4}$$

where δW_I and δW_E are virtual work cause of internal forces and virtual work cause of external forces respectively. The boundary conditions of clamped-clamped beam are

$$w(0) = w(a) = \frac{dw}{dx}(0) = \frac{dw}{dx}(a) = 0 \tag{5}$$

Also, the variation form of the beam should satisfy the equation (5):

$$\delta w(0) = \delta w(a) = 0 \qquad \frac{d\delta w}{dx}(0) = \frac{d\delta w}{dx}(a) = 0 \tag{6}$$

The stress-strain relation in the mid-plane direction under thermal and mechanical loading is

$$\sigma_{xx} = E(z)\varepsilon_{xx} - E(z)\alpha(z)T(z) \tag{7}$$

The total internal virtual work done is

$$\delta W_I = \int_0^a \int_A \sigma_{xx} \delta \varepsilon_{xx} dA dx \tag{8}$$

The virtual work done by external distributed load by using the virtual displacement is

$$\delta W_E = -\int_0^a b P_0 . \delta w dx \tag{9}$$

By using equations (8) and (9) into equation (4) and with consideration the boundary conditions (equations (5) and (6)), it can be obtained the governing equation of beam, taking into account Euler-Bernoulli theory, according to the following:

$$\int_0^a D_{xx} \frac{d^2 w}{dx^2} \frac{d^2 \delta w}{dx^2} dx - \int_0^a P^*(x) \delta w dx - \int_0^a b P_0 \delta w dx = 0 \tag{10}$$

where $D_{xx} = IE_{fe}(z)$

2.3 Equations of the temperature change cross the thickness of functionally graded beam

Two models coated beam were considered. The first is conventional coated beam (HC model). In this model the coated beam has two different layers. The homogeneous ceramic layer (h_3) as a thermal barrier coating and substrate as a homogeneous metal layer (h_1) which are located from top to bottom of the model in thickness direction respectively. The second model is FGC beam (FGC model). FGC model has FGM layer (h_2) on top and substrate metal homogeneous (h_1).
By considering the temperature variation is happened just in the thickness direction of the structures and steady-state heat condition. The heat conduction equation through the thickness of the beam is as following [28]:

$$-\frac{d}{dz}\left[k(z)\frac{dT}{dz}\right] = 0 \tag{11}$$

2.3.1 Temperature equations of homogeneous layers of beam

To find the temperature function across the thickness of homogeneous single layer has to use thermal conductivity coefficient of homogeneous material and boundary condition across the thickness of the layer into the equation (11). Thermal equation across the thickness of the homogeneous ceramic coated layer and homogeneous metal substrate respectively will be as below:

$$T_c(z) = \frac{1}{h_2}\left((T_m - T_u)z + \left(\frac{h_1 + h_2}{2}\right)T_m + \left(\frac{h_2 - h_1}{2}\right)T_u\right) \tag{12}$$

$$T_s(z) = \frac{1}{h_1}\left((T_L - T_m)z - \frac{h_2 - h_1}{2}T_L + \frac{h_2 + h_1}{2}T_m\right) \tag{13}$$

2.3.2 Temperature equation of Functionally Graded layer of beam

Nonlinear thermal equation across the thickness of FGM layer based on exponential function is as follows:

$$T_{fe}(z) = T_u + (T_m - T_u)\eta_E(z)$$

$$\eta_E(z) = \frac{1}{C_E}\left[\left(\frac{z+\frac{h}{2}}{h_2}\right) - \frac{1}{2!}\ln\left(\frac{k_m}{k_c}\right)\left(\frac{z+\frac{h}{2}}{h_2}\right)^2 + \frac{1}{3!}\left(\ln\left(\frac{k_m}{k_c}\right)\right)^2\left(\frac{z+\frac{h}{2}}{h_2}\right)^3\right.$$

$$\left.-\frac{1}{4!}\left(\ln\left(\frac{k_m}{k_c}\right)\right)^3\left(\frac{z+\frac{h}{2}}{h_2}\right)^4 + \frac{1}{5!}\left(\ln\left(\frac{k_m}{k_c}\right)\right)^4\left(\frac{z+\frac{h}{2}}{h_2}\right)^5 - \frac{1}{6!}\left(\ln\left(\frac{k_m}{k_c}\right)\right)^5\left(\frac{z+\frac{h}{2}}{h_2}\right)^6\right] \quad (14)$$

$$C_E = 1 - \frac{1}{2!}\ln\left(\frac{k_m}{k_c}\right) + \frac{1}{3!}\left(\ln\left(\frac{k_m}{k_c}\right)\right)^2 - \frac{1}{4!}\left(\ln\left(\frac{k_m}{k_c}\right)\right)^3 + \frac{1}{5!}\left(\ln\left(\frac{k_m}{k_c}\right)\right)^4 - \frac{1}{6!}\left(\ln\left(\frac{k_m}{k_c}\right)\right)^5$$

2.4 Solution to rectangular coated beam

An approximate solution, the deflection of the beam will be obtained based on boundary conditions (equation (6)) and geometry which is occupying the space defined by

$$0 \le x \le a \qquad -\frac{h}{2} \le z \le \frac{h}{2} \quad (15)$$

Where a and h are the length of the beam and the total thickness of beam included the two layers respectively.
To find the answer to equation (10), based on boundary condition and geometry of beam, an approximate solution is obtained as:

$$w = \sum_{i=1}^{m} A_i \sin^2\left(\frac{i\pi x}{a}\right) \quad (16)$$

The thermal load $P^*(x)$ in equation (10) by using Fourier series and some simplification will be as:

$$P_x^*(x) = \sum_{m=1}^{\infty}\frac{4m\pi}{a^2}M_T \sin\left(\frac{m\pi x}{a}\right) \quad (17)$$

$$m = 1,3,5,...$$

By using the solution equation, variation form of solution and thermal load in equation (10) we get

$$R_{(\gamma)(i)}A_i - F_\gamma = 0 \quad (18)$$

2.4 Numerical calculations

As mentioned in section 2.3, HC and FGC models have two layers with the same thickness. Therefore, the thickness of the coated layer and metal substrate are 4mm and 6mm respectively. In order to find the material properties which are defined as the exponential functions in equation (3). The modulus of elasticity, the thermal expansion coefficient, and thermal conductivity were assumed as follow:

E_c=100 GPa, E_m=10 GPa, $\alpha_c = 1.4\times10^{-5}\left({}^{o}K^{-1}\right)$, $\alpha_m = 0.9\times10^{-5}\left({}^{o}K^{-1}\right)$
k_c = 40 W/(m.°K), k_m=90 W/(m.°K) (19)
The Poisson's ratio was assumed constant and equal to $\nu = 0.3$.

2.4.1 Static analysis

The dimensions of the coated beam in both models were taken as a = 50cm, b = 5cm, and h = 1cm. On top of the coated beam with clamped-clamped boundary conditions (equation (5)), the uniformly distributed load and temperature are assumed to be $P_0 = 22000kPa$ and $T_u = 1000^o K$ respectively. The coordinate axes were located in the middle of the volume of the rectangular coated beam. Only bending behaviour of the coated beam was considered, thus displacement of mid-plane along x-axis equal to zero. The temperature of the bottom surface of the coated beam was assumed $T_L = 300^0 K$. Also in equation (18), γ and i were considered equal to two.

Fig. 1 shows deflection of the centre of the coated beam for both models. The deflection of FGC model much bigger than HC model, because of the ceramic percentage in FGC model is less than HC model. Fig. 2 shows the stress variation throughout the thickness of the coated beam for both models. The graph of the HC model clearly demonstrates the stress singularity at the interface of the metal substrate and homogeneous ceramic coated layer. Therefore, this phenomenon causes to create a de-bonding at the interface of two layers. Vice versa for FGC mode, the graph shows the stress gradually changed at the interface of two layers. Thus stress singularity and concentration will decrease.

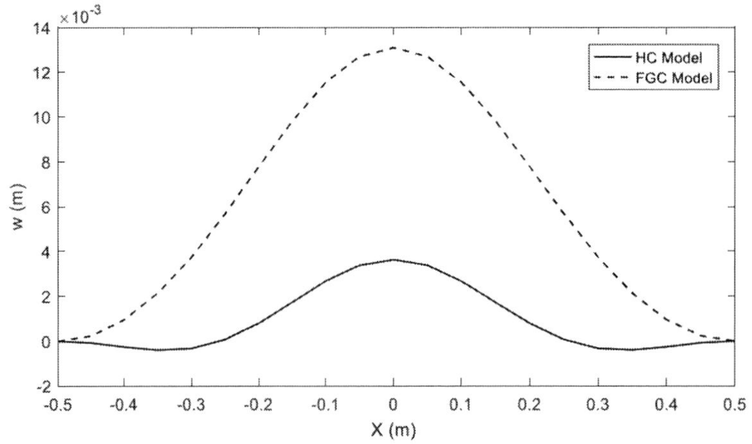

Fig. 1. The deflection (w) along the length of the coated beam in HC and FGC model.

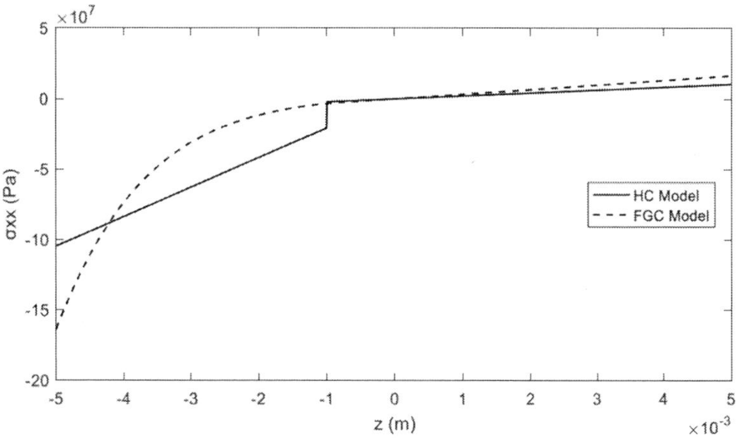

Fig. 2. The stress (σ_{xx}) throughout the thickness of centre of the coated beam in HC and FGC model.

3 Conclusions

The static analysis of conventional coated and FGC rectangular beam under the transverse distributed load and thermal excitation due to the thermal environment. The assumption of position dependent material properties was considered as an exponential function. The governing equation is derived based on the Euler-Bernoulli theory and exponential function for FGM layer by using the virtual work principle. Transverse displacement and stress throughout the thickness of the beam are considered. With obtained an approximation function which is supported clamped-clamped boundary conditions, the displacement or behaviour of the beam structure and stress throughout the thickness under thermal and mechanical loads are found. As noted initially, the HC model will experience delamination upon the introduction of stress to the model. Hence, this study has successfully demonstrated the superiority of FGC composite against conventional coated composite. From the analysis, the stress level throughout the thickness at the interface of the coated beam for the FGC was reduced. Yet, the deflection in return was observed to increase. Therefore, this could cater to various new engineering applications where warrant the utilization of material that has properties that are well-beyond the capabilities of the conventional or yesteryears materials.

References

1. M. Koizumi, Composites Part B: Engineering **28** 1-4, (1997)
2. S. Suresh, A. Mortensen, *Fundamentals of functionally graded materials* (The Institut of Materials, 1998)
3. R. Javaheri, M. Eslami, Journal of thermal stresses **25**, 603-625 (2002)
4. W. Zhang, Y. Hao, X. Guo, L. Chen, Meccanica **47**, 985-1014 (2012)
5. Y. Hao, W. Zhang, J. Yang, S. Li, Journal of Mechanical Science and Technology **25**, 1637-1646 (2011)
6. R. Javaheri, M. Eslami, AIAA Journal **40**, 162-184 (2002)
7. S.H. Chi, Y.L. Chung, International Journal of Solids and Structures **43**, 3657-3674 (2006)
8. J. Mencik, *Mechanics of components with treated or coated surfaces* **42** (Springer Science & Business Media, 2013)

9. Y.L. Chung, C.F. Pon, International Journal of Solids and Structures **38**, 75-90 (2001)
10. M. Kashtalyan, M. Menshykova, International Journal of Solids and Structures **44**, 5272-5288 (2007)
11. B. Sankar, Composites Science and Technology **61**, 689-696 (2001)
12. R. Shahsiah, K. Nikbin, M. Eslami, Iranian Journal of Mechanical Engineering **10**, 64-80 (2009)
13. A. Davoodinik, G. Rahimi, Acta Mechanica Sinica **27**, 767-777 (2011)
14. G. Rahimi, A. Davoodinik, IUST International Journal of Engineering Science **19**, 105-113 (2008)
15. S. Esfahani, Y. Kiani, M. Eslami, International Journal of Mechanical Sciences **69**, 10-20 (2013)
16. B. V. Sankar, J. T. Tzeng, AIAA journal **40**, 1228-1243 (2002)
17. L. Ma and D. Lee, Composite Structures **93**, 831-842 (2011)
18. C. Wang, J. N. Reddy, K. Lee, Elsevier (2000)
19. F. Erdogan, Composites Engineering **5**, 753-770 (1995)
20. J. Yang, Y. Chen, Composite Structures **83**, 48-60 (2008)
21. F. Erdogan, A. C. Kaya, P. F. Joseph, Journal of Applied Mechanics, **58**, 410-418, (1991)
22. W. B H, Journal of applied Mechanics **64**, 449 (1997)
23. H. J. Choi, Journal of Applied Mechanics **63**, 479-486 (1996)
24. Z.H. Jin, N. Noda, Journal of Applied Mechanics **61**, 738-740 (1994)
25. N. Noda, Z.H. Jin, JSME international journal. Ser. A, Mechanics and material engineering **38**, 364-369 (1995)
26. B. Wang, J. Han, S. Du, Journal of applied mechanics **67**, 87-95 (2000)
27. J. N. Reddy, *Energy principles and variational methods in applied mechanics* (John Wiley & Sons, 2017)
28. H.S. Shen, *Functionally graded materials: nonlinear analysis of plates and shells* (CRC press, 2016)

Peat Soil Stabilization using Lime and Cement

Nadhirah Mohd Zambri[1,*], and *Zuhayr* Md. Ghazaly[1]

[1]School of Environmental Engineering, University Malaysia Perlis, Kompleks Pusat Pengajian Jejawi 3, 02600 Arau, Perlis, Malaysia

Abstract. This paper presents a study of the comparison between two additive Lime and Cement for treating peat soil in term of stabilization. Peat and organic soils are commonly known for their high compressibility, extremely soft, and low strength. The aim of this paper is to determine the drained shear strength of treated peat soil from Perlis for comparison purposes. Direct Shear Box Test was conducted to obtain the shear strength for all the disturbed peat soil samples. The quick lime and cement was mixed with peat soil in proportions of 10% and 20% of the dry weight peat soil. The experiment results showed that the addition of additives had improved the strength characteristics of peat soil by 14% increment in shear strength. In addition, the mixture of lime with peat soil yield higher result in shear strength compared to cement by 14.07% and 13.5% respectively. These findings indicate that the lime and cement is a good stabilizer for peat soil, which often experienced high amount of moisture content.

1 Introduction

Soils are normally used as construction for dams, embankment, and other earth structure. Ground improvement is a rapidly developing field because the good sites for construction are becoming limited day by day [1]. Therefore, soil is a key of component in construction industry.

Soft soils possess high water content and low workability allows an increase in the cost and duration of projects on soft soil [2]. Soft soils are known to have low strength, high compressibility and the water contents are also known to be high at almost to its liquid limits[3]. They are often unstable under structure and they are susceptible to problems from changes in moisture content [2]

Peat soil is a representative material of soft soils and classified as highly organic [4][1]. Peat deforms greatly under applied loading on account of its exceptionally high water content and organic nature. Peat also has remarkably high strength compared with mineral soils at similar water content. In order to construct a safe, stable and serviceable road, road engineers have to overcome these engineering problems and find suitable solutions to construct roads on peat soil [5]. Peat is subjected to problems of instability such as local sinking and development of slip failure. It is also subjected to very large primary and long term settlement under an even moderate increase in load [6]. The peat soil is problematic

* Corresponding author: nadhirahazeela@gmail.com

soil that will become more complex for construction process such as the construction of building and road [5]. The pile foundation is often recommended for the buildings on peat. However, the ground may still settle around the building. Also, sometimes the construction line such as road embankment is not only just subjected to only localized sink and development of the slip failure, but also large magnitude of primary and long term settlement [6][2]. There have been are many construction failures related to the inability of capturing this behaviour and so the proper model for compression and settlement prediction becomes a priority issue [7].

All materials deformed when subjected to loads. For soil, the load deformation relationship is usually complex, varying widely with different soils and particularly in the plastic range of cohesive soils, where time plays a major role [8][5]. The settlement increases in magnitude with an increase in load, although not linearly.

2 Chemical Stabilizer

Chemical stabilization is a major category of soil stabilization involving the use of chemical agents for initiating reactions within the soil for modification of its geotechnical properties [9]. The soil modification, stabilization, or both can be classified in group of soil improvement. Soil modification is an addition an active additives such as lime and cement for soil to change its index properties, while soil stabilization is the treatment of soils to improve the texture, increase strength, increase the CBR value and reduce shrink-swell characteristics such that they become totally suitable for construction for a long term [3].

A number of construction options that can be applied to peat and organic soils, namely excavation-displacement and reinforcement to enhance soil strength and stiffness, such as by stage construction and preloading, stone columns, piles, thermal precompression and preload piers, or by reducing driving forces by light- weight fill and chemical admixture such as cement and lime. These chemical admixtures can be applied either as deep in situ mixing method (lime-cement columns) or as surface stabilizer [10].

2.1 Lime

The additional of lime has an influence in reducing the optimum water content and increasing the maximum dry unit weight of dry density of the peat soil [11]. Lime has actually been used as a soil stabilizing agent since Roman time [11-12]. Lime was also used widely to stabilize weak fly ash to reduce settlement and to increase bearing capacity [3]. The hydrated lime [$Ca(OH_2)$] was used as an additive for peat soil to improve some engineering properties of the soil. Quick lime is more efficient to effect change in soil strength compared with hydrated lime [11]. Lime soil reactions can be broadly classified into short term and long term reactions. The short term reactions include ion exchange, flocculation [13] and aggregation, carbonation [9,14] and pozzolanic [11].

2.2 Cement

Cement stabilization is the most common stabilization methods adopted for soil treatment. A Portland cement particle is a heterogeneous substance, containing minute tri-calcium silicate (C^3S) dicalicum (C^2S), tricalcium (C^3A), and solid solution described as tetra calcium alumino-ferrite (C^4A) [15][16]. When the pore water of the soil encounters with cement, hydration of the cement occurs rapidly and the major hydration (primary cementitious) produces hydrated calcium silicates (C^2SH^x,C^4AH^x), and hydrated lime $Ca(OH)^2$ [15][4]. The addition of cement decrease the soils liquid limit [11]. The previous study state that addition of the chemical admixture, cement and lime, can improve the engineering properties of tropical peat soils [11]. Cement stabilization results in good compressive strength and is preferred for cohesionless to moderately cohesive soil but loses effectiveness in the case of highly plastic soil [9].

3 Soil Stabilization in Peat Soil

Peat is subjected to problems of instability such as local sinking and development of slip failure. Peat soil achieved the lowest shear strength with extremely soft and high organic content. The construction cost will reduce and effective stabilization by using lime and cement as the additive to stabilize the soil. During rainy seasons, weather related delays can be reduced by stabilizing the soil in order to continue site work. This can impact on the construction schedule in a positive way and save the cost of having to wait for good weather to continue work [17].

4 Methodology

4.1 Peat Sample

Peat samples were taken from paddy filed Jejawi, Perlis in the disturbed condition 0.3-0.6m from the ground surface. All the samples were obtained using Mud Auger with two opening cylinder design and sharp edge at the end of the auger with 0.45m of diameter and 1.56m height and then they were placed into the plastic container sealed with aluminium foil to maintain the properties of peat soil like moisture content. The sample getting dried up in the oven with temperatures 105°C to 110°C in duration 24 hours (BS 377: Part 2: 1990). The dried up soil was then broken into pieces by using rubber hammer and were sieved passing through the sieve diameter 1.18mm.

4.2 Direct Shear Box Test

All the disturbed samples were tested under direct shear test with drained condition sample based on BS standard 1377:1990 (Part 7). The specimen was in square with the size 60mm × 60mm × 25mm. Normal stress is applied are 136.2 kN/m^2, 272.5 kN/m^2 and 408.8 kN/m^2. The treated sample was prepared and tested with shear test. Each of a sample contained 10%, 20% of lime, 10% and 20% of cement.

5 Results and discussions

5.1 Peat Properties

The moisture content of Perlis peat soil is 327.14%. High moisture content in peat is the main factor contributing to the characterization of peat with low bearing capacity and bulk density [18].

5.2 Direct Shear Test

By conducting these direct shear test, given that direct shear are considered as one of the most accurate testing techniques there is in geotechnical studies, it is possible to obtain the shear strength parameters of the stabilized soil, namely the friction angle, ϕ and the cohesion of soil, c. To ensure the accuracy of parameters calculations, all the tests were performed under three different pressures. In this research work, three normal stresses, 136.2, 272.5 and 408.8 kN/m^2 was applied to determine shear properties of untreated and treated peat soil. The friction angle and cohesion of soil was determined from Mohr-Coulomb.

5.2.1 *Untreated Peat Soil*

Fig. 1 illustrates the shear displacement curve of the untreated and treated peat soil with different amount of additives. The load applied is 136.2 kN/m^2. The internal friction angle of control sample was found to be 40.78°. Based on [19] the average effective angle of friction is 53° for the peats and clearly above the average angle of 41° for the organic soils. The relatively high value of friction angle at control sample can be anticipated due to the angular features of the sand particles. In the past found that the fibrous peat is in range 27°-32° under normal pressure 3-50 kN/m^2. Besides that, in certain cases the friction angle is 50° for amorphous granular peat and for fibrous peat is in range of 53°-57°. The high friction angle does not represent the strength of the peat. However, it departs from the existence of fiber content in peat soil [20].

5.2.2 *Treated Peat Soil (Lime and Cement)*

Based on Fig. 1, 2, and 3 the experiment results show that the addition of lime and cement improve the strength properties of soil. The high of shear strength of peat will not actually reflect high of friction angle. This is because, [19] state that the fact the fibers are not always solid and may be filled with water and gas. From the all figure it can be observed that peat achieve the high shear strength with addition of lime compare to cement. The stabilizer such as lime and cement, depend on their chemical reaction with the soil elements in the presence of water and percent of the amount of stabilizer agent. The addition of lime (Ca^{+2}) strengthening the bond between the soil particle. The soil particle becomes closer to each other, causing the soil texture to change. The factors affecting lime treated soil are lime content, short period reaction and soil mineralogy. The cement reactions are time dependent and require long periods of time (years) because such reaction are functions of temperature, calcium quantity, pH value and percentage of silica in soil mineral[21].

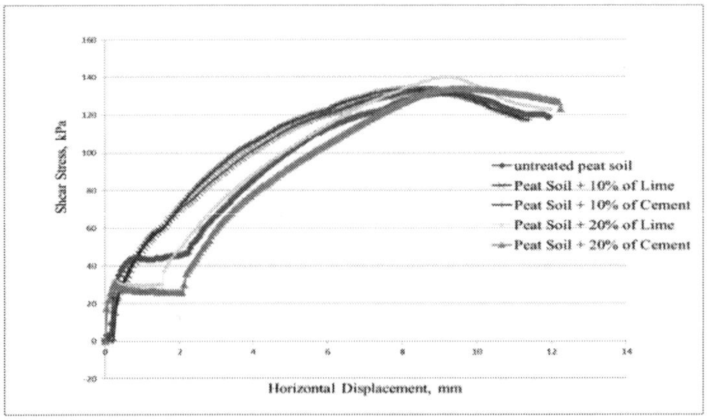

Fig. 1. Graph shear stress vs displacement from direct shear box test for untreated and treated peat soil with normal stress applied 136.2 kN/m^2.

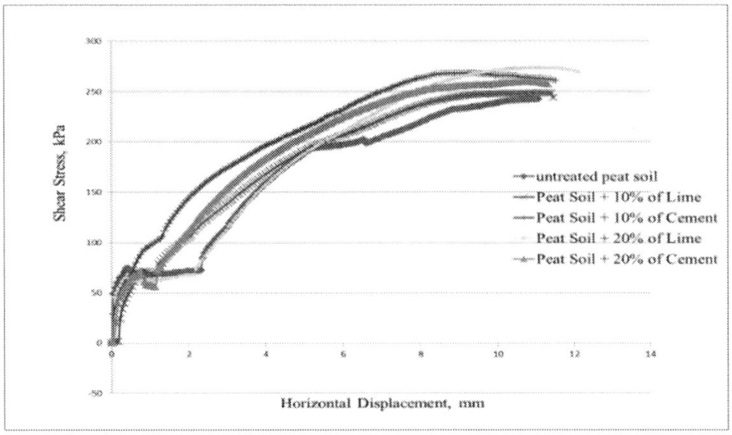

Fig. 2. Graph shear stress vs displacement from direct shear box test for untreated and treated peat soil with normal stress applied 272.5 kN/m^2.

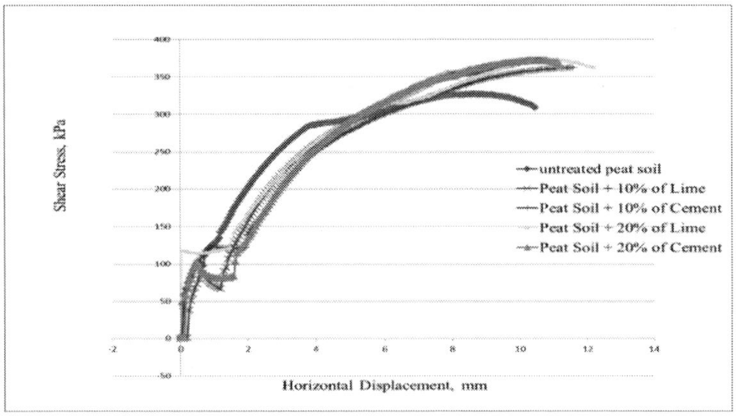

Fig. 3. Graph shear stress vs displacement from direct shear box test for untreated and treated peat soil with normal stress applied 408.8 kN/m^2.

Table 1. Summary result of direct shear box test for Perlis peat soil for untreated and treated soil.

Figure Number	1			2			3			4			5		
Sample Description	Control sample			Peat soil + 10% of lime			Peat soil + 10% of cement			Peat soil + 20% of lime			Peat soil + 20% of cement		
Normal Stress (kN/m^2)	136	273	409	136	273	409	136	273	409	136	273	409	136	273	409
Peak Shear Stress (kN/m^2)	134	246	327	133	248	372	138	268	362	140	274	373	138	260	371
Angle of Shear Resistance	40.78°			41.24°			40.05°			34.44°			41.05°		
Cohesion (kN/m^2)	27.72			21.79			28.40			50.09			17.66		

Based on the experiment results, the addition of lime and cement with percentages of mixing between 10% and 20% with peat give result in improving the strength characteristics of peat soil. The peak shear stress is also found to increase and the displacement with increase the normal stress. By treating the peat with lime and cement some of water in soil is replaced so that the density of soil will tends to increase on stabilization and those chemicals that increase the pH make an appropriate environment for stabilization process. This experiment result may be vary with others researcher as well known that the difference in fibrous content in peat. In addition, the deeper of peat soil the more decomposed is more consolidated.

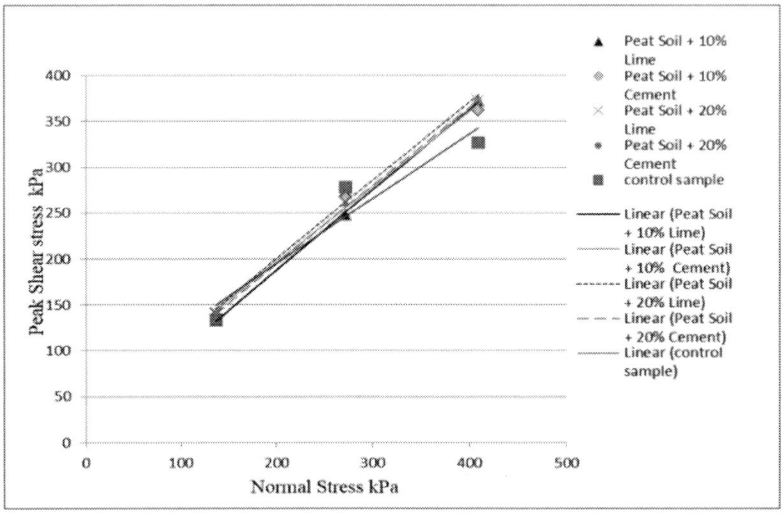

Fig. 4. Direct shear box test result for Perlis peat soil treated with 20% of cement with different of normal stress.

Fig. 4 shows the effect of additive to peat result from direct shear test for different amount of additive, with Mohr-Culomb lines fitted. In each case, cohesion, C, is given by the y-intercept and angle of internal friction, ϕ, by the line gradient. The additional of lime and cement may increase the strength for drained condition on peat. This is because the shear strength of peat is increased while the water content is decreased. Based on the figure, the specimens treated with the lime content reached the highest shear strength with the cohesion 50.09 kN/m^2, resulting in a strength increase with the percentage of additive increase in comparison with the untreated peat strength while the peat treated with cement, the result shows decreased. It an be noticed from the Fig. 4 the cohesion and friction angle of peat treated with cement supposedly show the increased trend but in this experiment, the value is decreased than the raw sample. Based on that, the result may be increased if the percentage of cement will increase to 20% and above. On the other and, for others experiment sample that increase the percent of additives, the result of peat strength were even higher.

6 Conclusions

From the result of this study it can be concluded that addition of lime and cement will improve the strength characteristics of peat soil. The angle of shear resistance and cohesion of peat is found to increase significantly with increase in lime and cement content. The performance of lime is better than the cement in term of shear strength of peat soil. It can be prove by the result of shear stress for treated peat (20% of lime) increased 14.07% from untreated soil compare the increment of shear stress for treated peat (20% of cement) increased 13.5% from untreated soil. The relationship between shear strength and peat soil stabilized by different stabilized agent and amount is less significant strength gain unless cement is added to the soil in large dosage. Stabilization of organic soil with lime and cement will reduce the damage and destruction of the superstructure. Furthermore, the strength of peat can increase in term of stabilization, improve the deformation properties and cost saving.

This work was financially supported by Ministry of Higher Education Malaysia (MOHE) under Research Acculturation Grant Scheme (RAGS) 2015, No. 9018-00088. This support is gratefully acknowledged.

References

1. S. Boobathiraja, P. Balamurugan, M. Dhansheer, A. Adhikari, Int J Civ Eng Res **5**, 4, 431–438, (2014)
2. A. I. M. Ismail, Z. L. Belal, Nat. Sci. **12**, 3, 73 - 78, (2014)
3. N. Khalid, M. Mukri, F. Kamarudin, N. Sidek, M. F. Arshad, *InCIEC 2013*, (Springer Singapore, 2014)
4. S. Deboucha, R. Hashim, and A. Alwi, J. Geotech. Eng-EJGE. **13**, 1–9, (2008)
5. S. N. M. Razali, I. Bakar, A. Zainorabidin, Procedia Eng. **53**, 145–155, (2013)
6. S. Kazemian, A. Prasad, B. B. K. Huat, M. Barghchi, Int. J. Phys. Sci. **6**, 8, 1974–1981, (2011)
7. E. Papadaki, *Modelling of Peat Compressed under Sand Bodies: Experimental and Numerical Approach*, (Doctoral dissertation, Master thesis at Technical University Delft, 2013)
8. J. N. Cernica, *Geotechnical engineering: soil mechanics*, (Wiley, 1995)
9. J. James, P. K. Pandian, Adv. Civ. Eng. **2016**, 2016.
10. T. B. Edil, *Proc. 2nd Int. Conf. Soft Soil Eng. Technol.* (Putrajaya, Malaysia, 2003)
11. B. B. K. Huat, S. Maail, T. A. Mohamed, Am. J. Appl. Sci. **2**, 7, 1113–1120, (2005)
12. P. F. F. Lancaster-Jones, E. McKeand, F. G. Bell, *Foundation engineering in difficult ground*, (London, Boston, Newnes-Butterworth, 1978)
13. S. Bhuvaneswari, T. Thyagaraj, R. G. Robinson, S. R. Gandhi, Indian Geotechnical Conference–2010, GEOtrendz, 609–612, (2010)
14. Z. Metelková, J. Boháč, R. Přikryl, I. Sedlářová, Appl. Clay Sci. **61**, 37–43, (2012)
15. D. T. Bergado, *Proc. of the seminar on ground improvement application to Indonesian soft soils,* (Indonesia: Jakarta, 1996)
16. R. Hashim, M. S. Islam, Electron. J. Geotech. Eng. **13**, 1–9, (2008)
17. G. R. Otoko, Int. J. Eng. Technol. Res. **2**, 5, 1–6, (2014)
18. N. M. Sa'don, A. R. A. Karim, W. Jaol, W. H. W. Lili, UNIMAS e-Journal Civ. Eng. **5**, 6–12, (2015)
19. B. B. K. Huat, A. Prasad, A. Asadi, S. Kazemian, *Geotechnics of organic soils and peat.* (CRC Press, 2014)
20. H. Moayedi, R. Nazir, Geotech. Geol. Eng. 1–11, (2017)
21. I. T. Jawad, M. R. Taha, Z. H. Majeed, T. A. Khan, Res. J. Appl. Sci. Eng. Technol. **8**, 4, 510–520, (2014)

Physical and rheological properties of Titanium Dioxide modified asphalt

Rosnawati Buhari[1,*], *Mohd Ezree* Abdullah[1], *Mohd Khairul* Ahmad[2], *Ai Ling* Chong[1], *Rosli* Haini[3], and *Siti Khatijah* Abu Bakar[1]

[1]Faculty of Civil and Environmental Engineering, Universiti Tun Hussein Onn Malaysia, 83000 Batu Pahat, Johor Malaysia
[2]MINT-SRC, UniversitiTun Hussein Onn Malaysia, 86400, Parit Raja, BatuPahat Johor, Malaysia
[3]Faculty of Civil Engineering, Universiti Teknologi Malayia, Skudai , Johor, Malaysia

Abstract. Titanium Dioxide (TiO_2) has been known as a useful photocatalytic material that is attributed to the several characteristics includes high photocatalytic activity compared with other metal oxide photocatalysts, compatible with traditional construction materials without changing any original performance. This study investigates the physical and rheological properties of modified asphalt with TiO_2. Five samples of asphalt with different concentration of TiO_2 were studied, namely asphalt 2%, 4%, 6% 8% and 10% TiO_2. The tests includes are penetration, softening point, ductility, rotational viscosity and dynamic shear rheometer (DSR) test. From the results of this study, it is noted that addition of TiO_2 has significant effect on the physical properties of asphalt. The viscosity tests revealed that asphalt 10% TiO_2 has good workability among with reducing approximately 15°C compared to base asphalt. Based on the results from DSR measurements, asphalt 10% TiO_2 has reduced temperature susceptibility and increase stiffness and elastic behaviour in comparison to base asphalt. As a result, TiO_2 can be considered to be an additive to modify the properties of asphalt.

1 Introduction

Asphalt is the binder in the asphaltic concrete mixture for the construction of road surface. Asphalt is a liquid at higher temperature and is inelastic at lower temperature. These behaviors can result at higher temperature and cracking at low temperature. Therefore, modifying asphalt is now a common alternative to enhance the performance of asphalt by addition of various modifiers based on the aims and objectives of the purposes to be using for. In recent years, nanomaterials are commonly used as additives for hot mix asphalt or warm mix asphalt [1-2]. Several nanomaterials used includes nanoclay, nano hydrated lime, titanium dioxide nanoparticle, carbon nanoparticles etc. In general, Nanotechnology will produce benefits in two ways by making existing products and processes more cost effective, durable and efficient and by creating entirely new products. It also improve the storage stability in polymer modified asphalt, increase the resistance to UV aging, reduce

* Corresponding author: rosna@uthm.edu.my

the moisture susceptibility under water, snow and deicers, improve the properties of asphalt mixtures at low temperature, improve the durability of asphalt pavements, save energy and cost, decreasemaintenance requirements [3]. Titanium Dioxide is naturally occurred oxide that easy to find material in the earth's crust that is 0.6%. Its chemical formula is TiO_2 and it belongs to the family of transition metal oxides. The bulk material of TiO_2 is well known to have three crystal structures: anatase, rutile and brookite [4]. The fundamental structure unit in three TiO_2 crystals is (TiO_6) octahedra, but their modes of arrangement and links are different. Brookite and anatase are metastable and transform exothermally and irreversibly to rutile over a range of temperatures [5-6]. The composition of TiO_2 is temperature dependant. TiO_2 normally used has nanoparticles with a great specific surface (area to volume ratio) and diameter inferior to 100 nm. FESEM and Transmission Electron Microscopy (TEM) is the apparatus that is frequently used to determine the characteristic of TiO_2 through its morphology and to observe the characteristic of the TiO_2 growth, composition etc. The TEM study is to confirm the FESEM study regarding the diameter of the sample.

2 Experiment

2.1 Material and sample preparation

The base asphalt was of 80/100 penetration grade. Titanium dioxide white powder, procured from industry was used as a modifier for asphalt. The physical properties of the base asphalt and modifier TiO_2 are listed in Table 1 and the morphology of both TiO_2 powder as shown in Fig.1. The particle size ranges from 40 to 70 nm was observed and the Energy Dispersive X-ray Spectroscop (EDS) tests conform that the powder is Titanium dioxide has crystal structures of Anatase.

Table 1. Physical properties of asphalt and TiO_2 modifier.

Material	Properties	Unit	Limits	Value
Asphalt 80/100	Specific gravity	-	1.00-1.05	1.03
	Penetration @ 25°C,	0.1mm	80-100	70
	Softening point	°C	42-52	47
	Viscosity @ 35°C	(Pa.s)	≥ 3	0.5
TiO2	Ductility @ 25°C	cm	≥ 100	> 100
	Size	nm	-	40-60

2.2 Experimental Procedure

In the process of modifying asphalt, type of the modifying agent, preparation temperature and modification process will influence the performance of modified asphalt. Considering the different kinds of base asphalt, polymers and nanomaterial the preparation process of modified asphalt can vary dramatically. In addition to these factors, control of the preparation process parameters such as shearing speed, temperature and time to prepare the modified asphalt with good compatibility and thermal storage stability are also areas of work for future research in nanomaterial modified asphalt. To reduce the viscosity in preparing the modified asphalt it is necessary to raise the temperature to facilitate nanoparticle diffusion (Brownian motion). Since asphalt is easily aged with oxygen during the composite preparation, the temperature of the mixture should not be too high and the shearing time should not be too long which presents a paradox [7]. In this study, the TiO_2 modifier was used for the preparation of five differently modified asphalt samples at concentration of 2%, 4%, 6%, 8% and 10%. The asphalt was melted and a Silverson high shear mixer was used for the mixing at a constant temperature of 155°C and speed of 3500 rpm for 45 min to produce homogeneous mixtures. The homogeneity of mixtures evaluated using storage stability test for every single mixture.

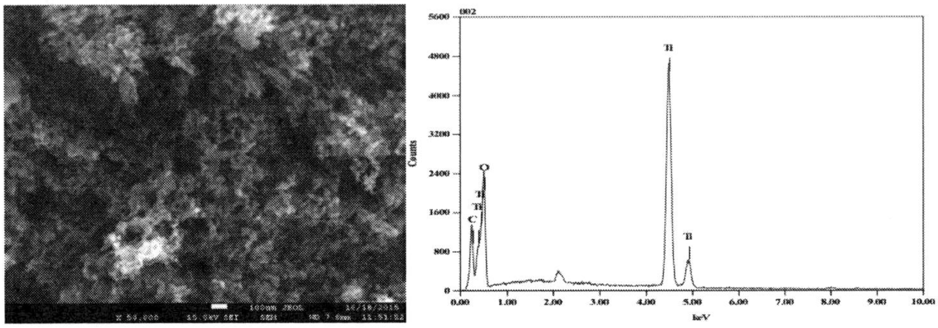

(a) (b)

Fig. 1. (a) Fe-SEM image of TiO_2 and (b) EDS image.

For physical properties penetration test (ASTM D5), softening point (ASTM D36) and ductility (ASTM D113) were run to ensure reproducibility and to evaluate the properties changes of modified asphalt in comparison to the original asphalt. The penetration, softening point and ductility are the test to measure stiffness, temperature at the asphalt starts to become a fluid and the maximum distance of asphalt to elongate at specific rate and temperature respectively. Others, the viscosity of the asphalt were measured by rotational viscometer as per ASTM D4402. The original and modified asphalt were tested at various temperature that are 120°C, 135°C, 150°C, 165°C, and 180°C. The dynamic shear rheometer (DSR) is used to characterise the viscous and elastic behaviour of asphalt at high and intermediate temperature. Other than that, DSR measures the complex modulus G^* and phase angle, δ at the desired temperature and frequency of loading as per AASHTO T315. The failure temperature are considered for all asphalt blends according to Superpave specification for $G^*/\sin \delta$ values less than 1.0 kPa in case of unaged asphalt and less than 2.2 kPa after rotational thin film oven test (RTFOT).

3 Results and discussions

3.1 Physical Properties

The influence of various concentration of TiO_2 on the properties of asphalt in the value of penetration, softening point and ductility are given in Table 2. All TiO_2 modified asphalts shows pronounced decrease in penetration with the strongest difference is asphalt 10% TiO_2 for up to approximately 36% from the original asphalt. Conversely, the softening point values were slightly decreasing for 0.4% for asphalt 2% $TiO_2$and increasing for up 7.8% for other percentage of TiO_2 blend. The decrease in penetration and increase in softening point denote an increasing in stiffness of TiO_2. The Penetration Index (PI) value in the table also assured the stiffness of the blend, where increasing PI value denoted the rising the hardness and also leads to improved temperature susceptibility of TiO_2 blend and vice versa. It is noted that asphalt 10% TiO_2 is the most hardness among asphalts blend and strongly comparable with the original asphalt. Meanwhile the viscosity of the asphalt 8% and 10% TiO_2 are decreasing leads to improve its workability. It also noted that ductility value for all TiO_2 blends are all more than 100 mm, which passed the standard requirements.

Table 2. Physical properties of asphalt and TiO_2 blends.

Control & TiO_2 Modify asphalt	Penetration @25C	Softening Point	PI	Viscosity@135°C, Pa.s	Ductility
Control 80/100	89.1	46.4	-0.71	0.5	>100
TiO_2 2 %	88.2	46.2	-0.80	0.6	>100
TiO_2 4%	75.3	46.6	-1.07	0.6	>100
TiO_2 6 %	67.1	50.0	-0.5	0.6	>100
TiO_2 8 %	60.8	48.6	-1.1	0.5	>100
TiO_2 10 %	56.4	49.2	-2.85	0.5	>100

3.2 Viscosity

The viscosity value at 135°C shows increasing for up to 4% TiO_2 and same as asphalt control for asphalt 6%, asphalt 8% $TiO_2$and 10% TiO_2. It is proved that asphalt TiO_2 passed the pumped and handled ability as it lower than 3Pa.s. Moreover, the mixing and compaction temperature of the mixture also obtained from the viscosity value for several temperatures where determined based on the reference viscosity values of 0.17-0.2 Pa.s and 0.28-0.3 Pa.s. It can be observed from Fig. 2 that the viscosity values decrease immediately with increasing test temperature for all concentration of TiO_2. It is also shows that the concentration of 10% TiO_2 blends has the lowest viscosity value results lowest mixing and compaction temperature. The mixing and compaction temperature were reduced significantly to approximately about 5°C and 15°C respectively.

3.3 Rheological properties of TiO_2 blends

3.3.1 Isochronal plots

Isochronal plots in Fig. 3 and Fig.4 plot a viscoelastic variable, complex modulus and phase angle, δ versus temperature (°C) at constant frequency of 1 rad (approximately 0.1592 Hz) and 10 rad (approximately 1.592 Hz). The isochronal plot is correspondence to the expected response of the asphalt binder under actual traffic conditions. It is noted that, at 1 rad, G* values asphalt 10% TiO_2 is higher that original asphalt thus leading to improved temperature susceptibility among all investigated asphalt blending. Moreover, the same pattern obtained at 10 rad where complex modulus of asphalt 10% TiO_2 is higher and the phase angle is lower than original asphalt. Other than that, all asphalt blending show pronounce increasing in G* values at low frequency and temperature, as well as at high frequency and temperature which represents an improvement in terms of the temperature susceptibility [8-9]. Moreover, it is noted that the asphalt 10% has a slightly lower phase angle than the other which means it has better elastic recovery performance.

https://doi.org/10.1051/e3sconf/20183401035

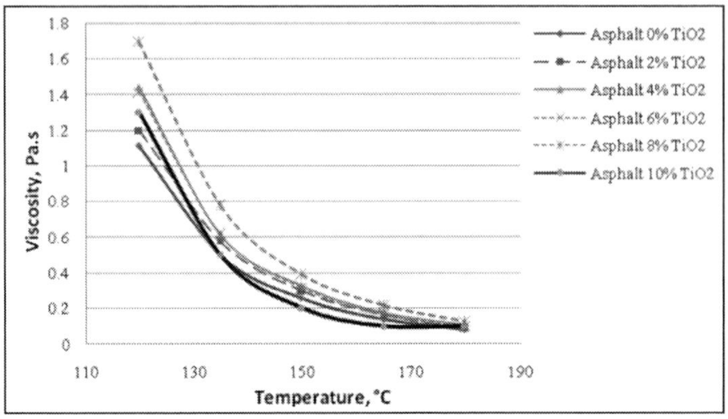

Fig. 2. Rotational viscosity of asphalt 80/100 and TiO_2 blends.

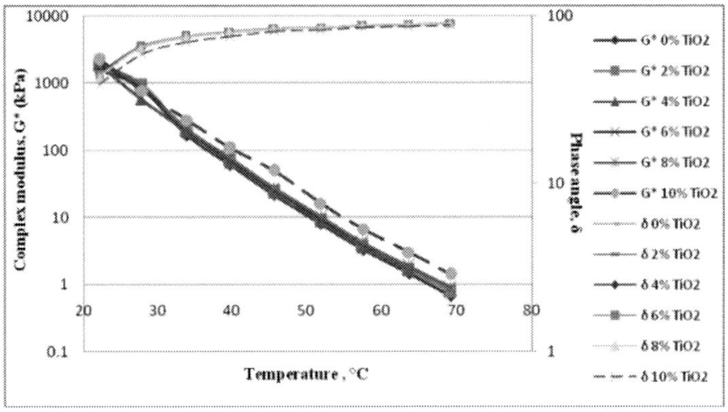

Fig. 3. Isochronal plots of the Complex Modulus G* at 1 rad.

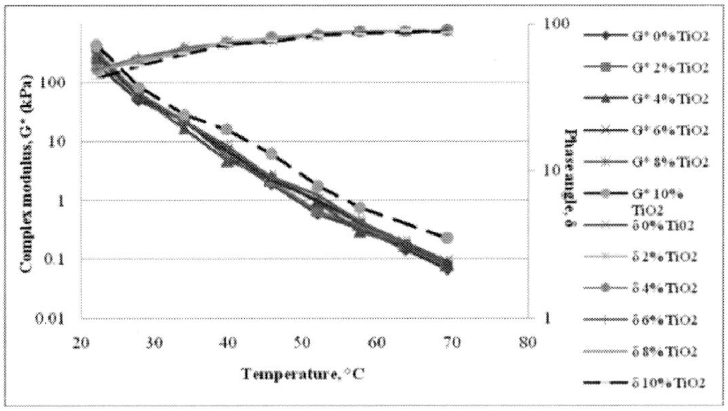

Fig. 4. Isochronal plots of the Complex Modulus G* at 10 rad.

3.3.3 Rheological master curve

Master curve are used to representation of rheological measurements, such as the complex modulus or phase angle at varies frequencies and temperature. In the present case, the master curve reflects the time dependency of bitumen over a wide range of loading times. It start with the selection of a reference temperature is selected and then the data at all other temperatures are shifted horizontally with respect to time to produce a single smooth curve. The frequency dependence of G* for original asphalt and asphalt 10% TiO_2 were assessed in Fig. 5 by producing rheological master curves at a reference temperature of 40°C. The shifting was done by using the shift factor, which varies for each test temperature. Other asphalts with lower percentage of TiO_2 are not included due only slightly comparable with the asphalt original. The asphalt 10% TiO_2 are shown to be the most comparable among other percentage. It is noted that increases in the complex modulus upon increasing TiO_2 are observed. In general, the complex modulus of the asphalt 10% TiO_2 is increasing at low and high load frequency as the phase angle slightly lower than original asphalt (refer Fig. 3 and Fig. 4) that means it has better elastic recovery performance [8].

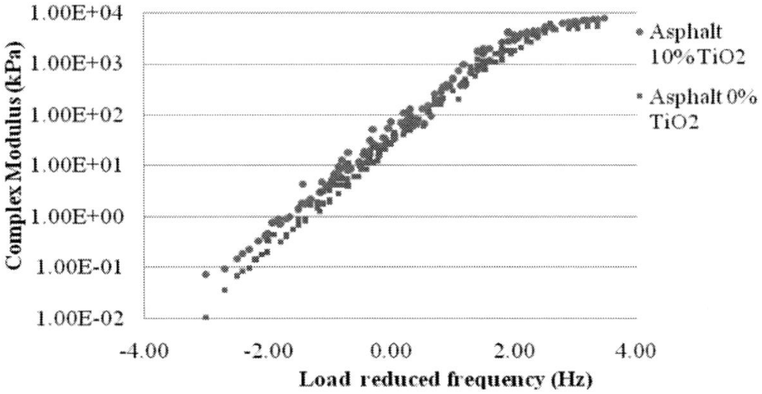

Fig. 5. Complex modulus master curve for original asphalt and asphalt 10% TiO_2 at the reference temperature of 40°C.

4 Conclusions

In this study, the physical and rheological characteristic of the asphalt TiO_2 were investigated using conventional and DSR test. The following findings and conclusions can be drawn:

- Base on penetration, softening point, viscosity and ductility test, it was approved that the hardness of asphalt increased, thus resulting improvement in temperature susceptibility.
- The isochronal plot reveals an increasing in stiffness (complex modulus) as a result of the improvement temperature susceptibility for asphalt 10% TiO_2, while other blends are slightly difference from original asphalt.
- The asphalt 10% TiO_2 elastic behaviour (phase angle) was slightly increase compared with the base asphalt.

The authors wish to thank the Universiti Tun Hussein Onn Malaysia. This study was funding by Ministry of Higher Learning Malaysia under FRGS Grant, vot number1458.

References

1. M.E. Abdullah, K.A. Zamhari, R. Buhari, N. Hidayah, M. Kamaruddin, N. Nayan, R.P. Jaya, Jurnal Teknologi **4**, 69–76 (2015)
2. G. Shafabakhsh, S.M. Mirabdolazimi, M. Sadeghnejad, Constr. Build. Mater. **54,** 566–571 (2014)
3. J. Yang, S. Tighe, Procedia - Social and Behavioral Sciences **96,** 1269–1276,

https://doi.org/10.1051/e3sconf/20183401035

(2013)
4. X. Chen, S.S. Mao,. *Titanium Dioxide Nanomaterials: Synthesis, Properties, Modifications, and Applications* (2007)
5. J.T. Yates, Surface Science **603,** (10–12), 1605–1612, (2009)
6. A. Fujishima, T.N. Rao, D.A. Tryk, Titanium dioxide photocatalysis **1**, 1–21 (2000)
7. C. Fang, R. Yu, S.Liu, Y. Li, J. Mater. Sci. Technol **29** (7), 589–594, (2013)
8. S.I.A. Ali, A. Ismail, N.I.M. Yusoff, M.R. Karim, R.A. Al-Mansob, D.I. Alhamali, Constr. Build. Mater. **93**, 326–334 (2015)
9. R.A. Al-Mansob, A. Ismail, A.N. Alduri, C.H. Azhari, M.R. Karim, N.I.M. Yusoff,. Constr. Build. Mater. **63**, 242–248, (2014)

E3S Web of Conferences **34**, 01036 (2018) https://doi.org/10.1051/e3sconf/20183401036
CENVIRON 2017

Psychological distance of pedestrian at the bus terminal area

Mohd Firdaus Mohamad Ali[1], *Muhamad Salleh* Abustan[1,*], *Siti Hidayah* Abu Talib[1], *Ismail* Abustan[2], *Noorhazlinda* Abd Rahman[2], *Hitoshi* Gotoh[3]

[1]Faculty of Civil and Environmental Engineering, Universiti Tun Hussein Onn Malaysia, 86400 Parit Raja, Batu Pahat, Johor, Malaysia
[2]School of Civil Engineering, Universiti Sains Malaysia, 14300 Nibong Tebal, Penang, Malaysia
[3]Department of Civil and Earth Engineering, Kyoto University, Japan

Abstract. Walking is a part of transportation modes that is effective for pedestrian in either short or long trips. All people are classified as pedestrian because people do walk every day and the higher number of people walking will lead to crowd conditions and that is the reason of the importance to study about the behaviour of pedestrian specifically the psychological distance in both indoor and outdoor. Nowadays, the number of studies of crowd dynamics among pedestrian have increased due to the concern about the safety issues primarily related to the emergency cases such as fire, earthquake, festival and etc. An observation of pedestrian was conducted at one of the main bus terminals in Kuala Lumpur with the main objective to obtain pedestrian psychological distance and it took place for 45 minutes by using a camcorder that was set up by using a tripod on the upper floor from the area of observation at the main lobby and the trapped area was approximately 100 m^2. The analysis was focused on obtaining the gap between pedestrian based on two different categories, which are; (a) Pedestrian with relationship, and (b) Pedestrian without relationship. In total, 1,766 data were obtained during the analysis in which 561 data were obtained for 'Pedestrian with relationship' and 1,205 data were obtained for "Pedestrian without relationship". Based on the obtained results, "Pedestrian without relationship" had shown a slightly higher average value of psychological distance between them compare to "Pedestrian with relationship" with the results of 1.6360m and 1.5909m respectively. In gender case, "Pedestrian without relationship" had higher mean of psychological distance in all three categories as well. Therefore, it can be concluded that pedestrian without relationship tend to have longer distance when walking in crowds.

1 Introduction

Walking is one of the effective modes of transportation for either short or long trips and that are depending on the pedestrian preferences. Every single person can be classified as pedestrian because people do walk every single day and due to that particular reason, more

* Corresponding author: sallehb@uthm.edu.my

people tend to walk rather than using vehicles that will cost a lot of consumptions of gas, time and as well as the money. Pedestrianization plays an important part in the sustainable urbanization that can be advantageous in reducing air pollution that caused by fuel consumption, sound pollution that caused by vehicles on the street and also death that can be triggered by many factors like car crash. In fact, the increasing number of pedestrian will lead to crowds and that is the logic of the importance to study about this issue both indoor (e.g.: shopping mall, bus terminal complex, and airport) and outdoor (e.g.: recreation park and pedestrian sidewalks).

The studies of human crowds and pedestrian behaviour since three decades ago had attracted more attentions from many researchers and it is now one of the important researches field not only in engineering but also in other fields such as computer science, mathematics, physics, psychology, urban planning, architecture and humanities [1, 3, 7, 10, 13, 14]. Pedestrian traffic flow simulation models were invented due to the needs of predicting the pedestrian movement behaviour in various kinds of environment in order to help decision makers to identify any hazardous situations and therefore may increase the safety level of pedestrian from accidents by implementing the evaluation and analysis of data that will be obtained during the experiments. Furthermore, the number of studies of crowd dynamics among pedestrian nowadays have increased due to the concern about the safety issues primarily related to the emergency cases such as fire, earthquake, festival and etc. [4]. These issues had attracted many researchers to implement their experiments for learning the causes and impacts of the importance of safety management issues of crowds and also in strengthening the need to have more refined models of pedestrian dynamics. It is necessary to understand the local interaction laws underlying collective crowd dynamics in order to develop reliable prediction models for the design of urban infrastructures, traffic management, and crowds' safety during mass events or evacuation processes. Therefore, a deep understanding of pedestrians' behaviour is vital in order to predict their motion in a complex collective characteristics emerged. Although crowds represent a group of people, it is important to study about the pedestrian behaviour individually. However, current models of crowd dynamics specifically for validating the efficiency of individual-level are in dire need. It is significant to study about the psychological distance of pedestrian and for that specific reason, the author had decided to conduct an experiment stochastically at a bus terminal area due to the high number of pedestrian that can be easily observed by making a video recording and also to understand more about the psychological distance between pedestrian at that location.

2 Methods and data analysis

An observation of pedestrian was conducted at one of the main bus terminals in Kuala Lumpur and it took place for 45 minutes by using a camcorder. The camera was set using a tripod on the upper floor from the site of the observation where the trapped area in the video recording was approximately 100 m^2 and the recorded video was converted into image sequences using DVD VideoSoft software. Hence, data analysis was performed by using a software named Human Behaviour Simulator (HBS) that was plugged into Autodesk Maya and all of the image sequences were imported into the latter software to track the gap or also can be explained as psychological radius and psychological distance between pedestrians in the video. Figure 1 shows an example of how the gap between two walking female pedestrian was obtained in data analysis stage and the process of obtaining gap between pedestrians in the video was just the same.

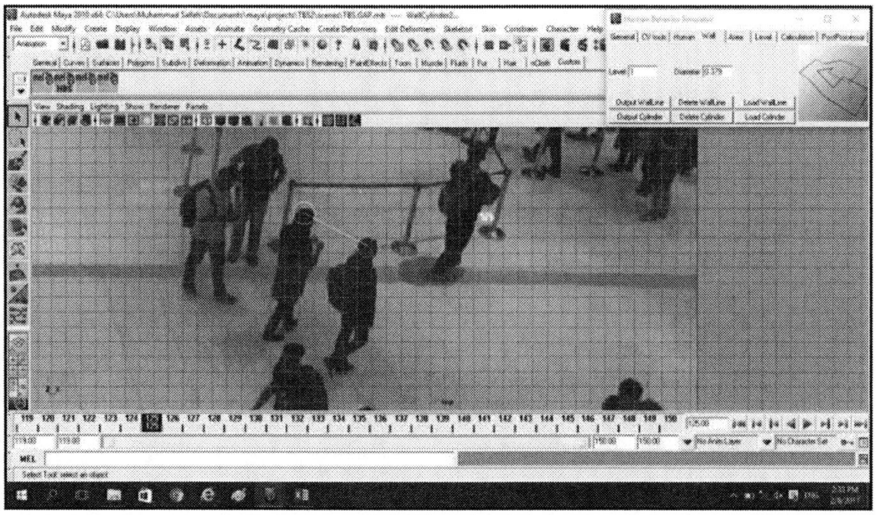

Fig. 1. An example of the step of pointing out the coordinate between two female pedestrian.

After pointing out the coordinates between the two pedestrians, the coordinates of x-axis and y-axis were generated in notepad format as shown in Figure 2 in which the first column represents x-axis (x1 and x2) while the second column represents y-axis (y1 and y2). Thus, the mathematical operation for calculating the length between two coordinates was applied to find the gap between two pedestrians.

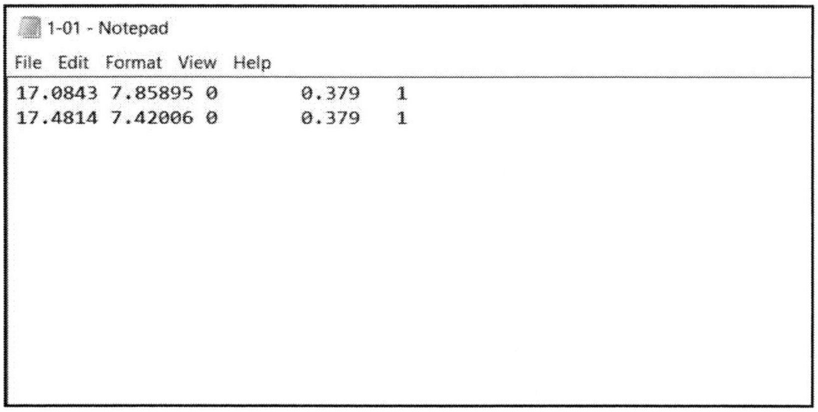

Fig. 2. Sample of the generated coordinates between two pedestrians using HBS software.

The data analysis was concentrated on obtaining the gap between pedestrian based on two different categories which are; (a) Pedestrian with relationship, and (b) Pedestrian without relationship and there are three criteria under each category; (i) Man to woman, (ii) Man to man, and (iii) Woman to woman. Wallace et al., (2016) [15] had quoted that Hall et al., (1968) invented the term proxemics when studying the role of interpersonal distances in human interactions, and defined zones of interaction ranging from intimate (less than 0.5m), to personal (0.5m to 1.2m), social (1.2m to 3.6m), and public (3.6m to 8m). Hence, based on these criteria, the range of psychological distance of pedestrian and also the gender classifications were designed as shown in Table 1 in which the gap between pedestrian were divided into six groups. After categorizing all of the obtained gaps, a statistical analysis was done by using T-test in SPSS software.

Table 1. Data classification based on range group and age categories.

Gender	Range (m)
Man to woman	Category 1: < 0.5999
Man to man	Category 2: 0.6000 - 1.0999
Woman to woman	Category 3: 1.1000 - 1.5999
	Category 4: 1.6000 - 2.0999
	Category 5: 2.1000 - 2.5999
	Category 6: > 2.6000

3 Results and discussions

In total, 1,766 data were obtained during the analysis in which 561 data were obtained for 'pedestrian with relationship' and 1,205 data were obtained for 'pedestrian without relationship'. The results are presented by category as aforementioned in Table 2 until Table 5 where total distance in all tables is referred to the sum of psychological distance obtained accumulatively from the number of data by each category.

Table 2. Psychological distance between pedestrian with relationship based on gender.

Category	Total distance (m)	Number of data	Standard deviation (m)	Psychological distance (m)		
				Lower	Upper	Mean
Man to Woman	26.883	216	0.4943	0.8378	0.9704	0.9041
Man to Man	23.192	121	0.5200	1.0028	1.1900	1.0964
Woman to Woman	36.538	224	0.3401	0.7854	0.8750	0.8302

Table 3. Psychological distance between pedestrian without relationship based on gender.

Category	Total distance (m)	Number of data	Standard deviation (m)	Psychological distance (m)		
				Lower	Upper	Mean
Man to Woman	43.528	508	0.6825	1.2586	1.3776	1.3181
Man to Man	47.458	458	0.6164	1.3103	1.4235	1.3669
Woman to Woman	32.318	239	0.5246	1.0299	1.1636	1.0967

Table 2 shows the results of psychological distance between pedestrian based on gender differences under 'Pedestrian with relationship' category. The most differences observed were among 'Man to Woman' with 216 number of data but the highest average value of distance was obtained by 'Man to Man' category with the mean of 1.0964m gap length while the lowest was obtained from 'Woman to Woman' category with only 0.8302m gap length and 'Man to Woman' category comes in the middle of the two categories with 0.9041m gap length. On the contrary, 'Pedestrian without relationship' category shows higher number of data based on the three classifications as shown in Table 3 where 'Man to Man' still has the highest mean of distance with 1.3669m gap length while the lowest is still among 'Woman to Woman' category with only 1.0967m gap length. Similar to Table 2, 'Man to Woman' shows the average of 1.3181m gap length as the second highest.

Logically, man is known as an aggressive type of people and they walk fast compare to woman and this leads to the higher value of distance between them. However, most women walk slower and they smaller in size compare to men in Malaysia also in other countries and as the results, the distance exists between them are not long compare to distance between men. Hence, Table 4 and Table 5 represent the results of distance between pedestrian based on range due to the consideration of the maximum and the minimum length of distance that were obtained during the analysis.

Table 4. Psychological distance between pedestrian with relationship based on range.

Range (m)	Total distance (m)	Number of data	Standard deviation (m)	Psychological distance (m)		
				Lower	Upper	Mean
Category 1 < 0.5999	61.116	137	0.0882	0.4455	0.4753	0.4604
Category 2 0.6000 - 1.0999	96.426	272	0.1393	0.7978	0.8310	0.8144
Category 3 1.1000 - 1.5999	98.575	103	0.1325	1.2611	1.3129	1.2870
Category 4 1.6000 - 2.0999	71.418	36	0.1492	1.7258	1.8268	1.7763
Category 5 2.1000 - 2.5999	40.875	8	0.1570	2.1370	2.3995	2.2683
Category 6 > 2.6000	26.634	5	0.2467	2.6326	3.2454	2.9390
Mean average of psychological distance between pedestrian (m)						**1.5909**

Table 5. Psychological distance between pedestrian without relationship based on range.

Range (m)	Total distance (m)	Number of data	Standard deviation (m)	Psychological distance (m)		
				Lower	Upper	Mean
Category 1 < 0.5999	57.057	126	0.0927	0.4547	0.4874	0.4711
Category 2 0.6000 - 1.0999	124.416	438	0.1438	0.8415	0.8685	0.8550
Category 3 1.1000 - 1.5999	158.219	296	0.1457	1.3230	1.3564	1.3397
Category 4 1.6000 - 2.0999	171.302	210	0.1544	1.8040	1.8461	1.8251
Category 5 2.1000 - 2.5999	148.129	90	0.1486	2.2894	2.3516	2.3205
Category 6 > 2.6000	47.331	45	0.4258	2.8765	3.1324	3.0044
Mean average of psychological distance between pedestrian (m)						**1.6360**

In Table 4, the most obtained average of psychological distance occurred among pedestrian was in the range of Category 2 with the number of 272 data and it is the same as shown in Table 5 with 438 number of data and the measurement of psychological distance between these two tables in Category 2 are only slightly different with 0.8144m and 0.8550m psychological distance respectively. This has proven that most pedestrian will keep their distance in the aforementioned range. Based on Malaysian ecological system, the body height of the people is not as tall as the people in other countries that are mostly tall such as in the United States, Japan or in the United Kingdom and due to that particular reason, range of gap in Category 2 should be reasonable for Malaysian people among normal body size. Range of more than 2.600 m shows the least data obtained during the analysis for both cases of pedestrian with and without relationship with only 5 and 45 number of data respectively. In this case, some people prefer or tend to keep their distance in a wider range due to their larger body size that requires them to have more spaces for their movement. Also, some people have prejudice attitude and this keeps them even further from the people that they find strange in term of physical appearance. But this does not classify a pedestrian has negative thoughts towards others but it is for the sake of safety issue due to many crime cases happened nowadays such as pickpocketing or even worse kidnapping. Another difference that can be seen is Table 5 shows that the top three range of distance among strangers are starting from Category 2 until category 4. On the other way around, Table 4 shows that the top three range among people that know each other are in Category 1, Category 2 and Category 3. Based on logic, it is reasonable for strangers to have wider range of distance between them compare to pedestrian that have relationships and both Table 4 and 5 show strong evidence of the situation.

Psychological distance between pedestrian is an important thing to be investigated in order to have better understanding about pedestrian walking behaviour. This field of studies may contribute to the improvement of current pedestrian simulation model where these results can be applied for making a simulation of pedestrian movement such as evacuation process during emergency cases (e.g. fire, earthquake and storm). Also, these results of pedestrian psychological distance can be useful in determining the optimum design of the building area so that a better evacuation planning can be arranged effectively and for an example, the standard width of a pathway based on [2] is 2.25m and in making a simulation of an evacuation process, this width and the results in this research studies can be applied in order to understand of how many people can be possibly evacuated in certain cases. Moreover, future researchers can also apply the results in this research in improving the individual-level simulation model that are varied in every country due to the difference in some aspects such as lifestyle as reviewed by [5], people that live in Tokyo, Japan with a very busy lifestyle walk faster with the speed of 1.56 m/s than Malaysian with the walking speed of 1.16 m/s as obtained by [9] and the obtained Japanese speed of walking may contribute to the difference of psychological distance between pedestrian in that country compare to Malaysian as studied in this research. Furthermore, the difference of body size in term of height may also contribute to the difference of pedestrian distance where taller people with longer legs usually walk even faster.

4 Conclusions

A bus terminal is an easy area for obtaining the video of walking people for the purpose of studying the behaviour of pedestrian. As successfully obtained in this research, there were 1,766 data of psychological distance among the free walking people that was observed for 45 minutes on a normal weekday. From the analysis, the pedestrian were categorized based on pedestrian with and without relationship, also based on range categories that the author found reasonable in order to have better understanding about the criteria that may contribute to various psychological distance among pedestrian. Based on the obtained results, higher measurements of psychological distance were obtained from pedestrian without relationship in line of both gender and gap range. Pedestrian without relationship tend to keep their average distance for about 1.6360m in length while pedestrian with relationship had their average distance among each other for about 1.5909m and the results between these two categories were only 2.70% different. In gender cases, the highest length of average psychological distance for both cases was 'Man to Man' category where pedestrian without relationship had shown higher value of distance than pedestrian that has relationship in which the results were 1.3669m and 1.0964m respectively. Same goes to both 'Man to Woman' and 'Woman to Woman' categories, pedestrian without relationship had greater psychological distance than pedestrian with relationship with the difference of 31.41% and 24.30% respectively. In future, the author would like to propose more research studies on psychological distance among pedestrian in different places because the results may vary due to different functions of buildings or zones. So for that reason, a simulation model of evacuation process that can predict the behaviour of pedestrian for both individual-level and group cases can be developed and suitable for various case of studies.

This research is funded by the Ministry of Higher Education (MOHE) of Malaysia under Fundamental Research Grant Scheme (FRGS) Vot 1526.

References

1. A. Abdelghany, K. Abdelghany, H. Mahmassani, Transportation Research Part A **86**, 159–176 (2016)
2. L.R. Board, *Uniform Building By-Laws 1984 [G.N. 5178/85]. Malaysia: International Law Book Services* (2013)
3. D. Helbing, L. Buzna, A. Johansson, A., T. Werner, *Transportation Science* **39(**1), 1–24 (2005)
4. F. Huo, W. Song, L. Chen, C. Liu, K.M. Liew, *Safety Science* **86**, 165–173 (2016)
5. K.K. Laxman, R. Rastogi, S. Chandra, **136**(1), 23–33 (2010)
6. B. Li, Transportation Research Part B: Methodological **65**, (2014)
7. D. Li, B. Han, **80**, 41–55 (2015)
8. Z. Li, M. Tang, D. Liang, Z. Zhao, *Procedia Engineering* **135**, 615–620 (2016)
9. M.S. Abustan, E. Harada, and H. Gotoh, *In Proceeding of Coastal Engineering, JSCE*, **Vol. 3**, pp. 56-60, (2012).
10. S.I. Park, F. Quek, Y. Cao, **5**, 155–164 (2013)
11. D.S. Pawar, G.R. Patil, Journal of Safety Research **52** (2015)
12. D.S. Pawar, G.R. Patil, **86**, 295–303 (2016)
13. N. Perozo, S. Garnier, D. Helbing, G. Theraulaz, **5**(4), 1–7 (2010)
14. G. Vizzari, S. Bandini, *IEEE Intelligent Systems*, **28**(5), 56–60 (2013)
15. J.R. Wallace, N. Iskander, (2016)

Evaluating the Rheological Properties of Waste Natural Rubber Latex Modified Binder

Siti Khatijah Abu Bakar[1], *Mohd Ezree* Abdulah[1,*], *Mazlina* Mustafa Kamal[2], *Raha* Abd Rahman[1], *Kamarul Arifin* Hadithon[2], *Rosnawati* Buhari[1] and *Saiful Azhar* Ahmad Tajudin[1]

[1]Faculty of Civil and Environmental Engineering, Universiti Tun Hussein Onn Malaysia, 86400 Parit Raja, Batu Pahat, Johor, Malaysia
[2]Malaysian Rubber Board, Technology Centre, RRIM Research Station Sg. Buloh, 47000 Selangor, Malaysia

Abstract. Road surface is designed to be the durable surface material to sustain the traffic loading. However, due to physical and mechanical stress, pavement deterioration is accelerated. Thus, modifying conventional bitumen by improving its properties is seen as the best method to prolong pavement in-service life. The purpose of this paper is to study the effect of waste natural rubber (NR) latex on rheological properties of bitumen. Conventional bitumen PEN 80/100 was modified with different content of waste NR latex using a high shear mixer at temperature of 150°C. The modified binder properties were characterized by conducting physical test (i.e. softening point, penetration and penetration index) and rheological test (i.e. dynamic shear rheometer, DSR). Results showed that, the addition of waste NR latex improved the rheology properties, which indicates by improving of rutting factor ($G^*/\sin\delta$). This properties improvement has also shows a potential to resist deformation on road surface despite of high traffic loading.

1 Introduction

Bitumen is viscoelastic material and its rheological properties are very sensitive to temperature variation as well as rate of loading [1]. The sensitivity of asphalt pavement to temperature and traffic loading cause high temperature permanent deformation, fatigue cracking and low temperature cracking [2–5]. Efforts have been made in improving bitumen properties by selecting the proper starting crude, tailoring the refinery processes and modify the bitumen [4].

Among the existing solutions, modifying bitumen with polymer is seems to compensate the bitumen deficiency. The most common polymers used in bitumen modification are SBR, SBS, EVA, polyethylene and rubber [3,5] . Since polymers are rather expensive, it is necessary to seek cost-effective ways to mitigate and prevent asphalt pavement premature failure [6]. In this sense, using waste rubber is viable alternative to modify bitumen due to economic and environment preservation [4,7].

* Corresponding author: ezree@uthm.edu.my

Basically, rubber industry waste is disposed of through incineration and landfill disposal [8]. Unfortunately, the abundance and improper handled of waste rubber could cause serious environmental impact [7]. Researches on modifying bitumen with waste rubber were initiated when granular and powder forms of rubber were developed in 1930 [9]. Since then, extensive studies have been conducted on various type of waste rubber: crumb rubber (from waste tires), rubber powder (from rejected gloves) and natural rubber latex [9–14].

One of the most well known, widely used as bitumen modifier is crumb rubber, which is scrapped from waste tires. Basically, crumb rubber contains of natural rubber which helps to improve the elasticity behaviour; carbon black and synthetic rubber which are able to improve the thermal stability of the modified binder [15]. Ongoing research and practices demonstrated that crumb rubber can be effectively incorporated into road surfacing [16]. In some cases, crumb rubber is difficult to disperse throughout the bitumen due to the rubbers have been used in a vulcanized state. Ineffective dispersion requires high temperatures and long mixing times and can yield a heterogeneous binder [15]. Rubber dispersion, nevertheless, can be improved by using chemical solvent (e.g. Toluene) to dissolve rubber, without changing the rheological properties of base bitumen [17].

Contrary to crumb rubber, natural rubber latex can provide more homogeneous binder as the rubber particles are dispersed in water phase, which ensure the rubber fuses with bitumen easily. Since the concentrated natural rubber latex contains almost 40 percent of water, mixing natural rubber latex with bitumen resulting in bitumen foaming and frothing due to the water evaporates at high mixing temperature [12-13,18].

Previous research found that modifying bitumen with rubber improves the physical properties of bitumen (i.e. penetration, softening point and viscosity). Moreover, interaction between rubber particles and bitumen helps in increasing elasticity and stiffness behaviour at high temperature which indicates by the increasing of complex shear modulus, G* and phase angle, δ [1]. The improvement of these rheological properties results in enhancing fatigue and permanent deformation resistance [12,19]. Tuntiworawoit et al. [12] suggested that natural rubber latex as the best alternative as the natural rubber improves the flexibility and stability of asphalt pavement and bringing greater service life expectancy [12].

The main objective of this work is to study the rheological properties of waste latex modified bitumen. To this end, rheological properties of unmodified and modified bitumen are compared.

2 Materials and methods

Bitumen Penetration Grade (PEN) 80/100 was used in this study. In order to ensure the quality of the bitumen, fundamental physical testing as specified in Standard Specification for Road Work (JKR/SPJ/2008-S4) were performed. Since all the values comply with the standard, PEN 80/100 was used as base bitumen and modified with waste NR latex, which was supplied by Malaysian Rubber Board Standard.

2.1 Preparation of modified bitumen

The preparation of waste NR latex modified binder involves wet mixing process by means of high shear mixer. Typical mixing procedure began by heating the base bitumen at 150°C. As soon as the desired temperature was reached, waste NR latex (5-15% by weight) was added gradually into the base bitumen under stirring condition at 1500 rpm for 45 min. Caution should be taken as foam could occur when the waste latex was added.

In order to deal with the rheological behaviour at in-service and handling temperature, the modified bitumen and base bitumen were exposed to high temperature of 163°C for 85 minutes, to artificially aged the bitumen. Ageing process of bitumen was performed as accordance in AASHTO T 240. Meanwhile, rheological properties of the bitumen were determined using dynamic shear rheometer (DSR) as specified in AASHTO T 315. DSR is equipped with parallel plates, one that fixed, and another is oscillating. Rheological parameters: complex shear modulus (G*) and phase angle (δ) for each modified sample were measured and used to determine the rutting resistance and failure temperature of the modified binder.

3 Results and discussions

3.1 Physical testing

The addition of waste NR latex improves bitumen's properties by lowering penetration values and increasing the softening point. Softening point and penetration index (PI) are the important physical parameters in enhancing bitumen's properties. Generally, higher softening point and PI are preferable in tropical countries, as they represent the bitumen response to temperature variation. Basic physical bitumen testing of modified binder is tabulated in Table 1.

Table 1. Physical properties of modified binder.

Properties	80/100	5NR	10NR	15NR
Penetration at 25°C, 100g 5 sec, 0.1mm (AASHTO T 49)	89	76.3	80.2	75.4
Softening Point, °C (AASHTO T 53)	46	47.5	52.4	48.8
Penetration Index (PI)	-0.32	-0.83	-0.33	0.44

3.2 Failure temperature

High service temperature or failure temperature of binders is determined at which G*/sin δ value is greater than 1.0 kPa for unaged and 2.2 kPa for short term aged binder. Generally, bitumen is rated based on its performance at different temperatures which known as Performance Grade system. Typical bitumen Penetration Grade (PEN) 80/100 is equivalent to Performance Grade (PG) 64 [20] which indicate the bitumen meets the high temperature physical properties up to 64°C. Whereas, failure temperature of a binder is obtained through the interpolation of temperatures [21].

Table 2 showed the high failure temperature for both unaged and short term aged condition for waste NR latex modified binder. High failure temperature changes with the addition of waste NR latex content. 15NR recorded the highest failure temperature meanwhile 5NR recorded the lowest failure temperature for both conditions. Difference of failure temperatures between 15NR and base bitumen were 3.72°C for unaged and 3.97°C for short term aged condition. This result revealed that modifying base bitumen with waste NR latex could expand the failure temperature, which later indicates better rutting resistance.

Table 2. Failure temperature of waste latex modified bitumen.

Binder type	Failure temperature (°C)	
	Unaged (G*/sin δ > 1.0 kPa)	Short term aged (G*/sin δ > 2.2 kPa)
80/100	64.91	64.53
5NR	67.65	65.41
10NR	68.18	67.4
15NR	69.63	68.5

3.3 Rutting resistance

Theoretically, rutting resistance parameter at selected test temperature was measured by G*/sin δ. Figure 1 depicted G*/sin δ values of waste NR latex modified binder for each condition: unaged and short term aged. Result showed that, waste NR latex modified binder displayed a good trend as G*/sin δ values increased with the increasing of waste latex content until the maximum value at 15%. The improvement of rutting resistance reaching up to 6% reflected to the addition of waste NR latex when compared to base bitumen. Although modifying bitumen with waste NR latex remained the same performance grade (PG64), rutting resistance recorded slightly well than the base bitumen. Table 3 presented the rutting resistance values for each waste latex content at 64°C.

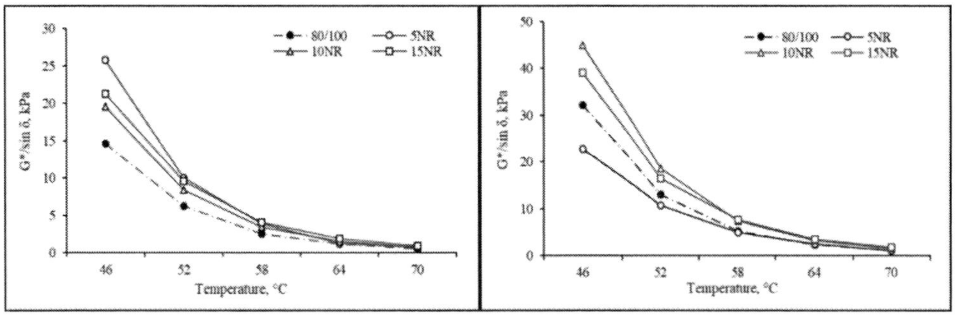

(a) (b)

Fig. 1. G*/sin δ of (a) unaged condition, (b) short term aged condition.

Table 3. Rutting resistance measured at 64°C.

Binder type	G*/sin δ (kPa) at 64°C	
	Unaged	Short term aged
Control	1.187	2.301
5NRL	1.319	2.534
10NRL	1.578	3.214
15NRL	1.933	3.575

4 Conclusions

Waste NR latex was successfully used to modify bitumen. Result showed that, waste NR latex slightly changed the bitumen's physical properties. Furthermore, incorporating waste NR latex into bitumen had positively affected the bitumen response to temperature changes which was represented by the increasing of softening point and penetration index (PI). Further investigation on rheology properties found that modifying waste NR latex into PEN

80/100 improved bitumen's rutting resistance. However, it was noted that the modified bitumen performance grade (PG) remains the same PG with the base bitumen.

The author would like to acknowledge the financial support by Ministry of Higher Education Malaysia (MOHE), under MyBrain15 programme, and IGSP U669, Universiti Tun Hussein Onn Malaysia.

References

1. N. S. Mashaan, A. H. Ali, M. R. Karim, and M. Abdelaziz, Int. J. Phys. Sci. **6**, 684–690, (2011)
2. K. Soudani, V. Cerezo, and S. Haddadi, Constr. Build. Mater. **104**, 126–133, (2016)
3. F. Xiao, S. Amirkhanian, H. Wang, and P. Hao, Constr. Build. Mater. **64**, 316–323, (2014)
4. Z. N. Kalantar, M. R. Karim, and A. Mahrez, Constr. Build. Mater. **33**, 55–62, (2012)
5. Y. Yildirim. Polymer modified asphalt binders, Constr. Build. Mater. **21**, 66–72, (2007)
6. F. Moreno, M. Sol, J. Martín, M. Pérez, and M. C. Rubio, Mater. Des. **47**, 274–280, (2013)
7. T. U. Ganiron, Int. J. Serv. Sci. Technol. **7**, 181–194, (2014)
8. T. Clark, *T. Int. Lat. Conf.*, 1-21, (2013)
9. M. H. Harun and R. Razali, J. Jab. Kerja Raya **1**, 58–70, (2013)
10. D. Dong, X. Huang, X. Li, and L. Zhang, Constr. Build. Mater. **29**, 316–322, (2012)
11. N. Vichitcholchai and N. Panmai, Jaratsri, Na-Ranong, Rubber Thai J. **39**, 32–39, (2012)
12. N. Tuntiworawit, D. Lavansiri, and C. Phromsorn, *Proc. East Asia Soc. Transp. Stud.* **5**, 679–694, (2005)
13. C. S. Ruggles. The efficient use of environmentally-friendly natural rubber latex in road construction – Past, present and future, www.revertex.com.my, (2004)
14. F. J. Navarro, P. Partal, F. Martínez-Boza, C. Valencia, and C. Gallegos, *Chem. Eng. J.* **89**, 53–61, (2002)
15. R. A. Al-Mansob, A. Ismail, A. N. Alduri, C. H. Azhari, M. R. Karim, and N. I. M. Yusoff, *Constr. Build. Mater.* **63**, 242–248, (2014)
16. J. P. Wu, P. R. Herrington, and K. Neaylon, *NZ Transport Agency Research Report 578*, (2015)
17. A. Ismail, R. . Al-Mansob, N. I. B. . Yusoff, and M. . Karim, *Aust. J. Basic Appl. Sci.* **6**, 97–101, (2012)
18. N. M. Azahar, N. A. Hassan, R. P. Jaya, M. A. A. Kadir, N. Z. M. Yunus, and M. Z. H. Mahmud, *Int. J. Agric. For. Plant.* **2**, 212–218, (2016)
19. M. R. Ibrahim, H. Y. Katman, M. R. Karim, S. Koting, and N. S. Mashaan, *Adv. Mater. Sci. Eng.* 1–8, (2013)
20. J. Ahmad, N. I. M. Yusoff, M. R. Hainin, M. Y. A. Rahman, and M. Hossain, *Constr. Build. Mater.* **50**, 567–576, (2014)
21. M. E. Abdullah. Performance of warm mix asphalt (WMA) mixture using nanoclay modified asphalt binder, *PhD thesis, Universiti Tun Hussein Onn Malaysia*, (2014)

Properties of cold-bonded lightweight artificial aggregate containing bottom ash with different curing regime

Norlia Mohamad Ibrahim[1,*], *Khairul Nizar* Ismail[2] , *Roshazita* Che Amat[1] and *Mohamad* Iqbal Mohamad Ghazali[1]

[1]School of Environmental Engineering, Universiti Malaysia Perlis, 01000 Kangar, Malaysia

Abstract. Cold-bonded pelletizing technique is frequently used in aggregate manufacturing process as it can minimise the energy consumption. It has contributed to both economical and environmental advantages because it helps to reduce the gas emissions problems. Bottom ash collected from municipal solid waste incineration (MSWI) plant was selected as raw material in this study and was utilised as a partial replacement for cement for artificial aggregate production. Several percentage of ash replacement was selected ranged from 10 to 50%. Aggregate pellets were subjected to different types of curing condition which is room-water (RW), room-room (RR), oven-room (OR) and oven-water (OW) condition. Properties of aggregate pellets were examined to obtain its density, water absorption, aggregate impact value (AIV) and specific gravity (SG). The results indicated that the most efficient curing regime is by exposing the aggregate in RW condition. The optimum aggregate was selected at 20% where it has satisfied the required density of 739.5kg/m^3, and classified as strong aggregate with AIV 14. However, the water absorption of aggregate increased proportionately with the increment of ash content.

1 Introduction

The amount of solid waste dumped into landfill nowadays is beyond our expectation and imagination. It was estimated that currently, over 23,000 tonnes of waste is produced each day in Malaysia. However, this amount is expected to rise to 30,000 tonnes by the year 2020. This solid waste is either being dumped in sanitary landfill or disposed by means of incineration process. Volume of solid waste can be reduced up to 80 – 95 % depending on the composition and the degree of recovery materials [1]. Nevertheless, the combustion of solid waste will produce two by-product ashes, which is fly ash and bottom ash. Many research attempts have been done by previous researchers to utilise fly ash in aggregate and concrete due to its pozzolanic characteristic that can contribute to the strength of materials [2-3]. Studies have been made to explore the effect of fly ash volume to the compressive strength of concrete [4-6]. However, bottom ash from incineration process was left

[*] Corresponding author: norlia@unimap.edu.my

untreated and considered as trash and dumped in the landfill. Most of the time, bottom ash is used in the field of road construction and pavement [7]. Seldom efforts were taken place to reuse and recover the potential use of bottom ash especially MSWI bottom ash as mineral admixtures in concrete. Therefore, it is important to investigate the properties of bottom ash so that full potential of bottom ash can be utilised. Recent studies shows that bottom ash either with or without prior treatment can be incorporated in concrete manufacturing and gives significant improvement in the quality of concrete [3], [8–12]. Several studies have been conducted to use bottom ash in the production of artificial aggregate.

Aggregate pellets can be produced using two main techniques i.e. sintering process and cold-bonding pelletization process. It is well known that in sintering process, it involves high-energy consumption during pelletization, which will lead to increasing cost of production. Pellets were produced by a bloating process and the temperature usually can reach up to 1200 °C [12-13]. Meanwhile, more economic technique has been proposed whichis known as cold-bonding pelletization method. In this method, the agglomeration of fine materials happened at ambient temperature with the aid of wetting agent, usually water, to become bigger solid particles. The spherical and rounded shaped of cold-bonded aggregate facilitated the workability of concrete and reduce the water cement ratio.

Curing process of aggregate is as important as the curing process of concrete. It will significantly influence the strength of structural element that used these aggregate pellets. Conventional curing method is the most preferred way to cure samples whereby they were immersed in water for 28 days. However, the presence of optimum heat and moisture are also important. Method of curing are normally depending on several factors such as availability of curing materials, size, shape, age of concrete, production facilities either in place or on plant, esthetic appearance and economic. A study from Ibrahim et al. (2015) shows that aggregate that have been cured in the oven at 100 ± 5 °C possessed comparable strength to the aggregate cured in water. Therefore, in this study four (4) different types of curing regime were employed to examine the properties aggregate produced. The curing conditions were set to room-water (RW), room-room (RR), oven-room (OR) and oven-water (OW) conditions.

The main objective of this study is to study the potential use of bottom ash in lightweight aggregate production and to investigate the physical and mechanical properties of the aggregate. The effect of different curing regime towards the strength of aggregate was also determined.

2 Materials and methods

2.1 Raw materials preparation

The main materials involved in this study were prepared prior the commencement of any experiment. Type 1 ordinary Portland cement (OPC) and bottom ash from MSWI plant were used in this study. The chemical composition of the cement and bottom ash is presented in Table 1 after they were determined using X-Ray Fluorescence (XRF). The bottom ash used was collected from an incineration plant located in Cameron Highland, Malaysia. It comes in wet and black solid form with different particles size and shape. To remove excessive moisture content, it was dried in the oven at 100 ± 5 °C for at least 24 hours and sieved to obtain required size (300 μm) before it was used as partial replacement for cement. It is important to distinguish between irrelevant ash and low-grade ash that contained high carbon compound, which will affect the strength of aggregate.

Table 1. Chemical Compositions for OPC and bottom ash.

	Chemical Compositions (%)	
	OPC	Bottom Ash
Silicon dioxide (SiO_2)	13.40	31.5
Aluminium oxide (Al_2O_3)	2.60	14.4
Ferric Oxide (Fe_2O_3)	4.43	26.48
Calcium Oxide (CaO)	72.90	17.9
Magnesium Oxide (MgO)	-	0.28
Sodium Oxide (Na_2O)	-	-
Pottasium Oxide (K_2O)	1.54	2.89
Sulfur trioxide (SO_3)	3.73	1.04
Others	1.00	3.87

The lightweight artificial aggregate (LWA) in this study was produced using cold-bonded pelletization method. Lightweight property of the aggregate was obtained by introducing foam into aggregate mixture. The foam will create pores that subsequently will produce less dense material. A protein-based foaming agent was used with the dilution ratio of 1: 33 where one portion of foaming agent was mixed with 33 portion of water.

The required weight of the foam was calculated by multiplying the foam density by the foam volume to be added. It was added immediately into the dry mixture and mixing it until uniform paste was achieved. Then, pelletization process will take place where spherical shaped aggregate was produced. Five different percentage of dry mixture containing only bottom ash and OPC were prepared. The percentage of bottom ash is 10, 20, 30, 40 and 50 %. After pelletization, the aggregate was subjected to different curing regime to study its physical and mechanical properties. Four different curing regimes were set namely oven-room (OR), oven-water (OW), room-room (RR) and room-water (RW). In OR and OW curing condition, the pellets were put in the oven for 24 hours at temperature 100 ± 5 °C. Next, it will be either left for curing at ambient temperature or immersed in water for another 14 days. As for RR and RW, the pellets were left at curing room first before it was subjected to next curing regime, which is immersion in water. The pellets in RR were left in curing room for continuous 14 days. The different curing regime was proposed in order to study the effects of different curing conditions toward the strength of aggregate. The samples were examined based from their types of curing regime and percentage of ash replacement for example one sample was denoted as BAOR10 which refers to aggregate that was cured in oven-room condition with 10 % bottom ash replacement in the mixture.

2.2 Testing Procedures

2.2.1 Density and specific gravity

Physical and mechanical properties of lightweight aggregate are interrelated with one another.better conclusions cannot be made if each and every aspects of produced aggregate is not well-analysed. Therefore, the mos timportant criteria for aggregate was determined including its specific gravity, water absorption, AIV and density. The density of aggregate was determined by following standard method which was stipulated in ASTM 330 [15]. The oven dried density (ODD) of lightweight aggregate must not exceed 1040 kg/m^3. It is necessary to wash aggregate with tap water to remove impurities in lightweight aggregate.

Lightweight aggregate were wiped and dried using water absorbing towel. Then, it was weighted and the mass was recorded as B. After that, lightweight aggregate were immediately immersed in water, weighted, C, and dried in the oven for 24 hours with 110 °C ± 5 °C. The mass was recorded as A. ODD can be calculated using equation 1 as follows:

$$ODD = A / (B - C) x 997.5 \quad (1)$$

where A = oven-dried mass of bulk aggregate

B = saturated mass of aggregate

C = surface dry mass of aggregate

On the other hand, specific gravity test was also conducted based on the saturated surface dry condition of the aggregate. Specific gravity of aggregate is defined as the ratio of the mass of solid in a given volume of sample to the mass of an equal volume of water at the same temperature. The specific gravity was determined using Electronic Densimeter Model MD-300S. The results were indicated in SG and V for volume. Fig. 2 shows the apparatus to obtain specific gravity measurement. This machine adopts Archimedes' Principal and the determination of relative density value based on the density of 4 °C : 1 g/m^3.

2.2.1 Water Absorption

The second test is to obtain water absorption of aggregate. Water absorption was generally known as increment in mass of aggregate due to penetration of water into pores of particle during a certain time but not including water adhering to the outside surface of particle ASTM C127, 2010 [16] . Water absorption of aggregate can be determined using equation 2 as follows:

$$Water\ absorption, \% = (M_A - M_S) / M_A \times 100\% \quad (2)$$

where M_A = mass of dry moist sample

M_S = mass of dry oven sample

2.2.3 AIV

The aggregate impact value is needed to classify the strength of aggregate produced in this study. An apparatus used in this study is as shown in Fig. 1 and the test method was strictly followed testing method as stipulated in BS 812 (1975). The value of AIV will represents the strength of aggregate produced. It should be noted that, AIV for aggregate should not exceed 25 % for heavy-duty concrete application. AIV can be determined using equation 3.

$$Aggregate\ impact\ value\ (AIV) = (B/A) \times 100 \quad (3)$$

where A = net weight of aggregate (g)

B = weight of crushed aggregate passing 2.36 mm sieve (g)

Fig. 1. AIV testing machine.

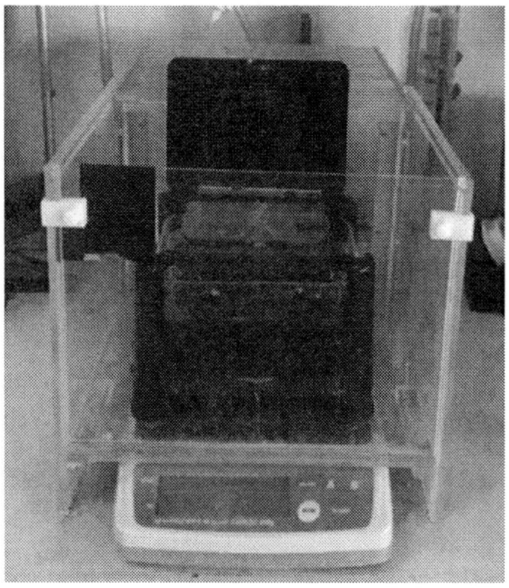

Fig. 2. An Electronic Densimeter Model MD-300S.

3 Results and discussions

3.1 Density and specific gravity

Density of lightweight aggregate will be used to classify the types of aggregate. According to BS EN 13055-1 (2002), concrete can be identified as lightweight if the density obtained was below than 2000 kg/m^3. Meanwhile, aggregate can be classified as lightweight aggregate if its unit weight was less than 880 kg/m^3 as stated in ASTM C330, [15]. It can be noted from results shown in Table 2, 90 % of aggregate produced in this study can be classified as LWA regardless their curing regime. Only BARR10 and BARW10 obtained slightly higher density from the limit stated in the standard which increased by 1.14 % and 1.59 % respectively. Light weight criteria of aggregate was contributed by the existance of pores structures inside the mixture It was found that maximum value of density was obtained from BARW10 with 894 kg/m^3 and the lowest BAOR50 and BAOW50 which is 570 kg/m^3. Greater percentage replacement of bottom ash will produce lower density of aggregate.

Specific gravity of aggregate also plays an important role in reducing the density of aggregate. From the test that has been done, 100 % of aggregate samples had less than 2 (SG) with the lowest SG are 1.28 for 50 % bottom ash replacement. Meanwhile, highest SG was identified at BARW10. Value of SG acquired in this study is interrelated with the value of density. It was due the fact that bottom ash itself has lower SG compared with OPC.

The relationship of density and SG also can be discussed based on their curing regimes. From the same table, we can see that the value of density is higher when LWA undergoes water curing compared to air curing process. For example, 20% LWA with RW condition exhibits higher density compared to RR condition, which is 740 and 725 kg/m^3. This is because LWA with air curing at ambient temperature for 14 days has lesser moisture content compared to LWA with water curing regime. However, excessive moisture content will resulted in lower compressive strength [17].

Table 2. Density, water absorption, AIV and specific gravity of LWA.

Sample	Density (kg/m^3)	Water absorption (%)	AIV (%)	Specific gravity
BARR10	890	22.1	17.2	1.72
BARW10	894	21.1	15.4	1.83
BAOR10	843	26.5	25.4	1.71
BAOW10	861	24.5	22.1	1.67
BARR20	725	23.6	16.9	1.51
BARW20	740	20.0	13.9	1.52
BAOR20	719	29.5	22.4	1.35
BAOW20	721	27.3	20.4	1.49
BARR30	684	33.7	29.7	1.37
BARW30	698	30.2	28.6	1.40
BAOR30	678	39.3	30.8	1.35
BAOW30	668	35.7	31.3	1.35
BARR40	600	37.1	36.0	1.34
BARW40	609	35.7	38.6	1.38
BAOR40	589	40.6	42.4	1.34
BAOW40	593	36.7	39.8	1.31
BARR50	574	39.8	57.9	1.28
BARW50	588	39.0	55.7	1.31
BAOR50	570	41.3	61.0	1.28
BAOW50	570	39.5	58.5	1.30

3.2 Water absorption

In all samples, water absorption in aggregate increased proportionately with the increased of bottom ash volume inside aggregate mixture. Previous research suggested that cold-bonded LWA has higher water absorption than that sintered fly ash. The moisture inside aggregate pellet was evaporated during the burning process of aggregate. Table 2 also shows result of water absorption test for 20 samples of LWA in five different percentages, which is 10, 20, 30, 40 and 50%, and at different curing regime. Water absorption test was carried to measure the amount of water absorbed by LWA. It can be seen that the value of water absorption are increasing according to the percentage of bottom ash used in LWA production. For example, it can be observed that the result of 10% until 50% of LWA with RR curing condition is increasing from 22.6, 23.6, 33.7, 37.6 and 39.8 %. Same trend was also found in OW curing condition where the water absorption increased from 24.5 to 39.5 % for 10% until 50% bottom ash replacement. Higher water absorption will adversely affect the strength of concrete produced using LWA. It can be concluded that, the higher the amount of bottom ash used in LWA production, the higher value of water absorption of LWA and reduced the strength of LWA. Bottom ash in aggregate mixture tends to absorb more water and affect the strength of aggregate. The existence of internal pores created by foams was also contributed to higher value of water absorption. The water absorption capacities are depending on the raw materials composition, curing parameters and also hydration kinetics.

3.3 AIV

The AIV test is carried out to determine the strength of aggregate and it was tested by using compaction technique. Based on Table 2, it can be observed that aggregate impact value increasing proportionally with the increase of percentage bottom ash used in LWA production. For example, we can see that the result of 10% until 50% LWA is increasing from 17.2 % for BARR10 until 58.5% for BAOW50. Padhi, 2014 [18] in his study stated that, any aggregate that has AIV greater than 20 % cannot be used in concrete production and, if higher than 30 %, it can be considered as very weak aggregate. This finding also tallied with the requirement stated in BS 812 (1975) that is, AIV for aggregate should not exceed 25 % for heavy-duty concrete application. The results obtained in this study shows that, the higher the percentage of bottom ash used in LWA production resulted in lower AIV. The AIV also can be discussed based on their respective curing condition. All BARW10, BARW20, BARW30, BARW40, BARW50, BAOW10, BAOW20, BAOW30, BAOW40 and BAOW50 that undergoes at least once water-curing regime tend to have lower aggregate impact value compared to air curing but will increase if the volume of bottom ash increased.

4 Conclusions

As a conclusion, sample denoted with BARW20 can be selected as the optimum LWA among the entire samples because it has the lowest AIV which 13.9 %. According to the result, the AIV obtained can be classified as strong aggregate. In addition, it also met another criteria based on water absorption test which is the water absorption value is in the range of 5 until 25 % referring to ASTM C127 [16] and lowest among all samples with 20% water absorption value. Lastly, it was also categorised as LWA because it has density 740 kg/m^3.

References

1. C. Hwang, L. A. Bui, K. Lin, and C. Lo, Cem. Concr. Compos. **34**(10) 1159–1166 (2012)
2. F. Colangelo, F. Messina, and R. Cioffi, J. Hazard. Mater. **299**, 181–191 (2015)

3. J. R. Pan, C. Huang, J. Kuo, and S. Lin, "Recycling MSWI bottom and fly ash as raw materials for Portland cement," **28**, 1113–1118 (2008)
4. S. Mengxiao, W. Qiang, and Z. Zhikai, Constr. Build. Mater. **98**, 649–655 (2015)
5. E. P. Kearsley and P. J. Wainwright, Cem. Concr. Res. **31**, 105–112 (2001)
6. H. K. Kim, J. H. Jeon, and H. K. Lee, Constr. Build. Mater. **29**, 193–200 (2012)
7. J. G. Roessler, T. G. Townsend, and C. C. Ferraro, J. Hazard. Mater. **300**, 830–837 (2015)
8. U. Müller and K. Rübner, Cement and Concrete Research **36**(8), 1434–1443 (2006)
9. R. Cioffi, F. Colangelo, F. Montagnaro, and L. Santoro, Waste Manag. **31**(2), 281–288 (2011)
10. G. Pecqueur, C. Crignon, and B. Que, Waste Manag. **21**(3), 229–233 (2001)
11. J. M. Chimenos, M. Segarra, M. A. Fernandez, and F. Espiell, J. Hazard. Mater. **64**(3), 211-212 (1999)
12. M. Han, D. Han, and J. Shin, Constr. Build. Mater. **99**, 192–199 (2015)
13. C. R. Cheeseman, A. Makinde, and S. Bethanis, Resources, Conservation and Recycling **43**(2), 147–162 (2005)
14. N. M. Ibrahim, L. Q. Wen, M. A. Rahim, K. N. Ismail, R. C. Amat, S. Salehuddin, and N. L. Rahim, Applied Mechanics and Materials **755**, 348–353 (2015)
15. S. Specification, *Standard Specification for Lightweight Aggregates for Structural Concrete 1*, 1–4 (2015)
16. C. Ag-, B. Statements, and C. Aggregate, *Standard Test Method for Density , Relative Density (Specific Gravity), and Absorption*, 1–6 (2013)
17. C. Hwang and V. Tran, Constr. Build. Mater., **87**, 78–85 (2015)
18. Padhi, *Aggregate Impact Test-10+ Important Notes To Remember*, *http://civilblog.org/2014/07/16/aggregate-impact-value-test-important-note-to-remember/*, 2014

Microstructural characterization of catalysis product of nanocement based materials: A review

Norsuzailina Mohamed Sutan[1,*] , *Nur Izaitul Akma* Ideris[1], *Siti Noor Linda* Taib[1] ,*Delsye Teo Ching* Lee[1] , *Alsidqi* Hassan[1] , *Siti Kudnie* Sahari[2] , *Khairul Anwar* Mohamad Said[3], and *Habibur Rahman* Sobuz [4]

[1]Civil Engineering Department, Universiti Malaysia Sarawak, 94300 Kota Samarahan, Sarawak, Malaysia
[2]Electrical and Electronic Engineering Department, Universiti Malaysia Sarawak, 94300 Kota Samarahan, Sarawak, Malaysia
[3]Chemical Engineering and Energy Sustainability Department, Universiti Malaysia Sarawak, 94300 Kota Samarahan, Sarawak
[4]Department of Building Engineering and Construction Management, Khulna University of Engineering and Technology (KUET), Khulna, Bangladesh

Abstract. Cement as an essential element for cement-based products contributed to negative environmental issues due to its high energy consumption and carbon dioxide emission during its production. These issues create the need to find alternative materials as partial cement replacement where studies on the potential of utilizing silica based materials as partial cement replacement come into picture. This review highlights the effectiveness of microstructural characterization techniques that have been used in the studies that focus on characterization of calcium hydroxide (CH) and calcium silicate hydrate (C-S-H) formation during hydration process of cement-based product incorporating nano reactive silica based materials as partial cement replacement. Understanding the effect of these materials as cement replacement in cement based product focusing on the microstructural development will lead to a higher confidence in the use of industrial waste as a new non- conventional material in construction industry that can catalyse rapid and innovative advances in green technology.

1 Introduction

In facing the new era, issues of sustainability and durability of construction industry are often emphasized on the pursuit of finding materials as partial cement replacement to decrease the high energy consumption and carbon dioxide emission during cement production [1]. The candidates that can be used as cement replacement must have two characteristics: high reactive silica content and very fine particle size. Any material that has these characteristics can produce pozzolanic reaction that can improve cement based

* Corresponding author: msnorsuzailina@unimas.my

product qualities. The scope of this review is on two materials namely spent fluid cracking catalyst (SFCC and nanosilica (nS).The spent fluid cracking catalyst (SFCC) which is produced from the catalytic cracking in petroleum industry is an inorganic waste material [2-3] and its contribution in cement-based production can bring environmental and economic benefits, thus reducing the final cost of cement-based products and the environmental impacts related to cement production and can be used as partial cement or sand replacement[4-5]. The material can be classified as non- hazardous material [6]for cement replacement because it was found that low concentration of heavy metal that can leach out of the SFCC which gives higher confidence in its possible use as a construction material.Other researchers [7] stated that nano-materials have received wide attention for use in the constructions sector and the increasing in the performance has been discovered for these materials in terms of smart functions and sustainable characteristics [8]. Many studies have shown that the contribution of nano silica (NS) in cementitious systems especially in construction of cement-based products is by improving the characteristics of the plastic and hardened material[9]. Silica in the nano-particles size has a filler effect by filling up the voids between the cement grains [10]. Referring to the previous study conducted, [11] stated that silica nano-particles are environmental friendly material and has no impact on health issues. It can help to create something new and sustainable to reduce the use of cement and reduce the cost of final construction works.

In the construction industry, cement replacement with pozzolanic materials, produce cement-based products that help in the positive use of waste materials, saving the environment, reduce carbon dioxide emissions and improve their performance [12-14]. Various methods have been implemented to improve the characteristics of cement-based products in the past [15]. The aim of this paper is to review the possible techniques to characterize the cement hydration products. The possibilities of both materials possessing cementing characteristics are to be studied according to their pozzolanic reaction and pore filling effect. Both materials are classified as pozzolanic where the term pozzolan is defined by previous researchers [16].

As a siliceous or aluminous material, either natural or artificial, which react chemically with calcium hydroxide (CH) or with materials that can release calcium hydroxide in the presence of water to form compounds that possess cementing characteristics.In order to understand the microstructural development caused by pozzolanic reaction from SFCC and NS as partial cement replacement, specialized microstructural characterization techniques were utilized by many researchers. This review will highlight their effectiveness in characterizing calcium hydroxide (CH) and calcium silicate hydrate (C-S-H) formation during hydration process.

2 Pozzolanic Reactions and Pore Filling Effect of SFCC and NS Materials in Cement Hydration Process

The pozzolanic reaction is where the reactive siliceous and aluminous phases react with calcium hydroxide (CH) to form new calcium silicate hydrate (C-S-H) in the presence of water at room temperature, hence possessing cementitious characteristic. Pore filling effect can be described as the ability of materials to fill in the voids, finally produce dense, compact microstructure and enhance strength of the cement-based products. The hypothesis of cement hydration process, which is influence by partial replacement of both SFCC and NS materials, is described in Fig. 1 and 2:

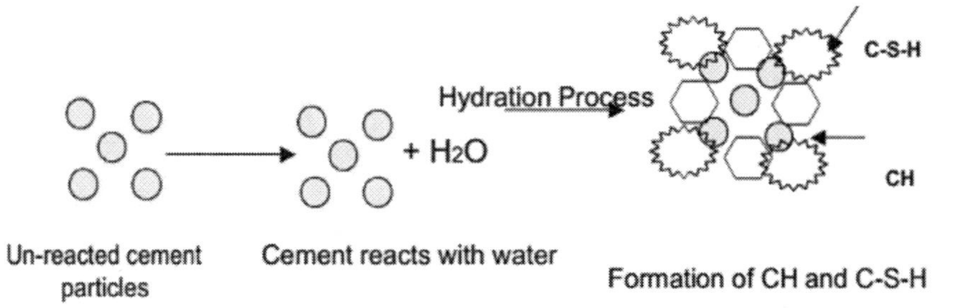

Fig. 1. Hydration Process of Ordinary Portland Cement.

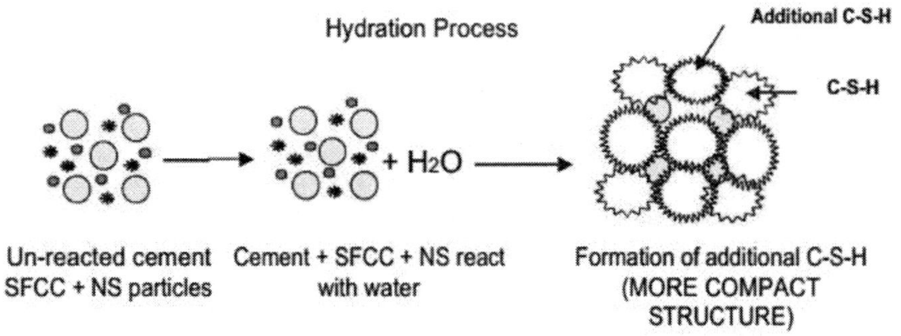

Fig. 2. Hydration Process of Partial Replacement of Cement by SFCC and NS.

In cement hydration process, there are some amounts of CH remaining in the cement structure. Those remaining CH can be replaced by consuming them with silica oxide to form additional C-S-H which is the main contribution of cement strength [17]. This silica oxide can be found through the pozzolanic material known as pozzolan where this study used SFCC and NS materials. The chemical reaction occurs in cement hydration process can be described as:

Chemical Reaction in Ordinary Portland Cement (OPC) Hydration Process:

$$\begin{aligned}
&\text{Cement } (C_3S) + H_2O \longrightarrow Ca^{2+} + H_2SiO_4^{2-} + OH^- \\
&Ca^{2+} + H_2SiO_4^{2-} \longrightarrow \text{C-S-H} \\
&H_2SiO_4^{2-} + OH^- \longrightarrow Ca(OH)_2 \textit{(Remaining CH)}
\end{aligned}$$

Chemical Reaction in Pozzolan Replacement Cement Hydration Process:

$$\left.\begin{aligned}
&SiO_2 + H_2O \longrightarrow H_2SiO_4^{2-} \\
&Ca(OH)_2 \textit{ (Remaining CH)} + H_2O \longrightarrow Ca^{2+} + OH^- \\
&H_2SiO_4^{2-} + Ca^{2+} \longrightarrow \text{C-S-H } \textit{(Additional)}
\end{aligned}\right\} \text{Pozzolanic Reaction}$$

SFCC waste was proven to behave as an excellent pozzolan for early curing age by improving both durability and mechanical behaviour in the cementitious system [18]. According to literature in previous study, SFCC which consists primarily of active silica (SiO_2) and alumina (Al_2O_3) [19] shows pozzolanic activity and contributes to the acceleration of the hydration and setting processes [20]. It has been found that, in the compositions with Portland cement, the SFCC waste reacts with calcium hydroxide which is a product of cement hydration as a pozzolanic material [21]. SFCC in a nano-particles size can reduce significantly the porosity [22] of the cement-based products by acting as filler to the voids causing denser microstructures of interfacial transition zone (ITZ) [23].

Pozzolanic reaction of NS with CH, which occurs during the hydration of cement, produces additional C–S–H which is the main constituent for strength and density in the hardened cementitious system [24]. At the same time, CH which was not contributes to strength development is consumed. Cement-based products with well dispersed NS particles inside it has a dense microstructure even if the NS particles are added in small quantity, but if NS particles are not dispersed properly, it may results in voids and weak zones formation [25]. NS particles act as nucleation sites to accelerate the hydration [26]. Due to high surface area and spherical shape, NS can enhance performance of cement-based material due to pozzolanic reaction of silica with calcium hydroxide, CH to form additional calcium silicate hydrate, C-S-H which represent the strength of cement-based product [27].As the most widely investigated of nano material in cement and concrete research, NS was found to be effective in enhancing the hydration process [28], the mechanical properties[29], and the durability [30] of cement-based material to a great extent.

3 Characterization Techniques Used to Confirm Pozzolanic Reaction of SFCC and NS Materials as Partial Cement Replacement.

3.1. Scanning Electron Microscopy (SEM)

SEM is used to study the morphology characteristic in order to confirm the presence of C-S-H products from the cement hydration process in the form of image analysis. In this study approach, the formation of the additional C-S-H occurs from the pozzolanic reaction is to be confirmed by the image obtained from certain phases of curing ages. The results of those reactions can be seen through the compactness of microstructure image as influence from pore filling effect of the material used. The study on SFCC by previous study [23] indicated that the microstructure of cement-based products appeared quite dense and compact with great modification or improving for interfacial transition zone (ITZ) between cement paste and aggregate. This proved that the pozzolanic reaction occurred in the cement-based product samples where amorphous silica from SFCC reacted with CH from cement hydration was able to produce additional C-S-H which the composition of those additional C-S-H found to be similar to those obtained in ordinary portland cement hydration. The SEM image indicated that SFCC can enhance the hydration products and improve the microstructure. SEM results reported in other previous study [31] shows that the microstructure of cement-based product partially replaced with NS was denser than cement-based product without the NS replacement because of the pozzolanic reaction by consumption of CH, and formation of additional C-S-H which fill the pores system and causing densification effect which improved the microstructure of cement system. A more compact microstructure image caused by pozzolanic reaction in cement-based products with replacement of nano silica was reported in the study done by another researcher [8].

3.2 Fourier Transformation Infrared Spectroscopy (FTIR)

FTIR is used to identify the functional group of the hydration products. The CH and C-S-H amount due to pozzolanic reaction in cement-based products can be measured using this technique. The amount of the chemical compound presence in the cement hydration process might affect the filling ability of those hydration products to fulfil the capillary pores in cement system. The FTIR analysis for SFCC studied in previous [32] indicated the formation of hydrated calcium silicate of the C- S-H gel type as caused by nature alkaline activation of SFCC as the main reaction product. This result showed that the pozzolanic reaction has occurred in the cement- based sample and gave pore filling effect to the cement microstructure as more C-S- H gel produced in the reaction during cement hydration process. The FTIR test result of functional group identification shown in Figure 3[33], a study of characteristics of blended cements containing NS, indicated that NS particles promoted the pozzolanic reaction by transforming CH into C–S–H gel. The existence of C–S–H was expected to be in the cement system reaction where the FTIR spectra of hydrated cement paste curing up to 90 days showed the band that appeared in the range of 950-980 cm-1 which is assigned to the tobermorite like phase, C-S-H. Meanwhile, in [34] studies using FTIR to characterize CH and C-S-H formation in the nanopozzolan modified cement using small percentage of NS gave the strongest C-S-H peaks at 970 cm-1 which confirm the pozzolanic reaction in the samples.

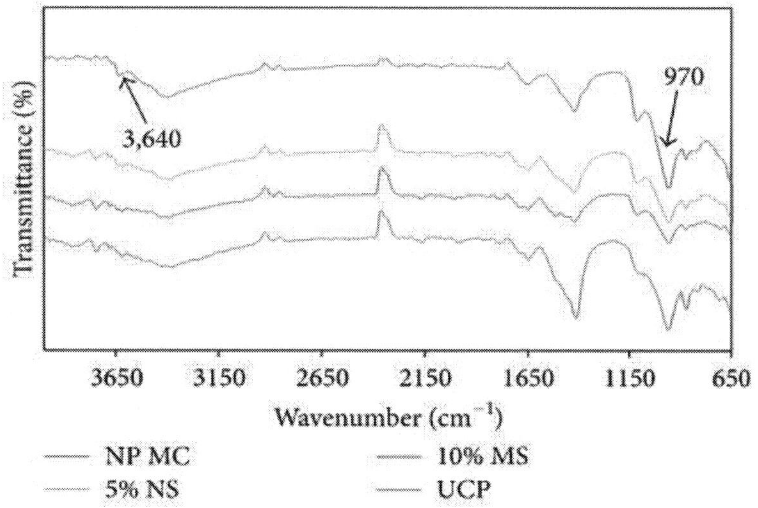

Fig. 3. FTIR spectra for NPMC, 5% NS, 10% MS, and UCP on day 28 [33].

3.3 X-Ray Diffraction (XRD)

XRD phase analysis is used to characterize and identify the hydration products. Pozzolanic reaction of those materials can be examined in a crystalline structure presenting those CH and C-S-H formation. The formation of additional C-S-H will influence cement strength and produce denser microstructure of cement-based products.The previous research study[20]using XRD indicated the presence of hydrated calcium silicate, C-S-H at a broad diffuse halo 2θ angle of about 26° and non-hydrated calcium silicate, CS and portlandite, CH and silica phases. The presence of those active silica which participated in cement hydration process will enhance pozzolanic reaction, thus chemically convert the remaining CH during cement hydration into C-S-H which is the strength responsible phase. Author [35]observed in their study an increasing of CH amount due to the increasing of the C-S-H formation in the 3 days of curing period. The amount of crystalline CH seems to be decrease in the 28 days of curing period. This was indicated by the shorter length of CH peaks observed in the longer period of curing. This again showed the pozzolanic reaction of NS with CH to form additional C-S-H has occurred. Previous researchers [36],also cited that NS was beneficial to promote the pozzolanic reaction as early as the hydration process in cement-based product. It slightly changed the amount of hydration products in 28 days of curing period where the content of the CH was consumed.

4 Conclusions

This paper discusses the possibilities of cement replaced by SFCC and NS materials and their pozzolanic reaction in cement hydration process through characterization techniques analysis in order to investigate the ability of both waste and nano material works together for partial cement replacement. Those catalysis hydration products (additional C-S-H) from the pozzolanic reaction during cement hydration process consider havingpositive influence to improve the characteristics of cement system. The hydration mechanism of cement-based products is still not completely understood. Slow reaction rate of pozzolanic material have left a significant amount of calcium hydroxide (CH) remaining in the cement-matrix during hydration process making the evolution of the pozzolanic reaction should be observed in the long term.

The authors wish to acknowledge Ministry of Higher Education and University Malaysia Sarawak for supporting this work under FRGS/TK04(02)/1138/2014(05)

References

1. M. S. Norsuzailina, I. Yakub, H. Jau, Applied Mechanics and Materials **695**, 643–646 (2014)
2. Y. Huang, X. Chen, Y. Deng, D. Zhou, L. Wang, Chemical Engineering Journal **269**, 434–443 (2015)
3. S. V. Payá, Jordi, Jose M. Monzo, Maria V. Borrachero, Ad Cem Res **25**, 1–7 (2013)
4. B. Pacewska, M. Nowacka, I. Wilińska, W. Kubissa, V. Antonovich, J Therm Anal Calorim **105**(1), 129–140 (2011)
5. K. Alshamsi, M. Baawain, K. Aljabri, R. Taha, Z. Al-kamyani, Procd Soc Behv **53**, 326–334 (2012)
6. E. D. Rodríguez, S. A. Bernal, J. L. Provis, J. D. Gehman, J. M. Monzó, J.Payá, M. V. Borrachero, Fuel **109**, 493–502 (2013)
7. S. A. E. Mohamed, A. E. Ragab, Int J Eng Res Technol **3**(7), 339–358 (2014)
8. H. Biricik and N. Sarier, Mater Res **17**(3), 570–582 (2014)
9. P. Hou, S. Kawashima, D. Kong, D. J. Corr, J. Qian, S. P. Shah, Composites Part B: Engineering **45**(1), 440– 448 (2013)
10. A.Lazaro Garcia, G. Quercia, H. J. H. Brouwers, J. W. Geus, World Journal of Nano Science and Engineering **3**, 41–51 (2013)
11. E. Horszczaruk, E. Mijowska, K. Cendrowski, S. Mijowska, P. Sikora, Construction and Building Materials **66**, 418–421 (2014)

12. H. Gurdián, E. García-Alcocel, F. Baeza-Brotons, P. Garcés, E. Zornoza, Materials **7**(4), 3176–3197 (2014)
13. C. Costa, M. S. Ribeiro, N. Brito, Materials Sciences and Applications **5**, 905–914 (2014)
14. T. Castellanos, T. Agredo, M. De Gutiérrez, Ingeniería E, Investigacion **33**(1), 18–22 (2013)
15. S. Velázquez, J. Monzó, M. Borrachero, J. Payá, Materials **7**(4), 2561–2576 (2014)
16. M. S. Amin, S. M. A. El-Gamal, S. A. Abo-El-Enein, F. I. El-Hosiny, M. Ramadan, HBRC Journal, 1687–4048 (2014)
17. V. K. Vadivel Rajendiran, Stalin, Advanced Nanomaterials and Emerging Technologies, 706–712 (2013)
18. L. Soriano, J. Monzó, M. Bonilla, M. M. Tashima, J. Payá, M. V.Borrachero, Cement and Concrete Comp **42**, 41–48 (2013)
19. R. Taha, Z. Al-kamyani, K. Al-jabri, M. Baawain, K. Al-shamsi, J Hazard Mater **229–230**, 122–127 (2012)
20. Y. Morozov, a. S. Castela, a. P. S. Dias, M. F. Montemor, Cement Concrete Res **47**, 1–7 (2013)
21. M. Aleknevic, V. Antonovic, I. Pundien, J Therm Anal Calorim **109**, 537–544 (2012)
22. B. Pacewska, M. Nowacka, M. Aleknevičius, V. Antonovič, Procedia Engineering **57**, 844– 850 (2013)
23. K. Al-Jabri, H. Shoukry, Construction and Building Materials **73**, 636–644 (2014)
24. U. S., S. A. L. P. Singh, S. K. Agarwal, S. K. Bhattacharyya, Nanomater. nanotechnol **1**(1), 44– 51 (2011)
25. K. S. S., Maheswaran, Bhuvaneshwari B, Palani G.S, Nagesh R Iyer, Research Journal of Recent Sciences **2**, 17–24 (2013)
26. G. L., D. Stephan, J Mater Sci **47**, 1011–1017 (2012)
27. N. M. Sutan, I. Yakub, S. Hamdan, Concrete Research Letters **5**, 760–767 (2014)
28. M. Stefanidou, I. Papayianni, Composites Part B: Engineering **43**(6), 2706– 2710 (2012)
29. S. R., R. F. Sharma Parveen, Journal of Nanomaterials 1–19 (2013)
30. A.B., P. Hosseini, Concrete Research Letters **2**(1), 167–180 (2011)
31. S. K. Sayed Abd El-Baky, Sameh Yehia, Nanocon 2013, 5 (2013)
32. C. Kae-Long Lin, Chao-Lung Hwang, Je-Lueng Shie, AIChE - Environmental Progress & Sustainable Energy Journal, 1–7 (2013)
33. M. Heikal, S. Abd El Aleem, W. M. Morsi, HBRC Journal **9**(3), 243–255 (2013)
34. N. M. Sutan, I. Yakub, M. S. Jaafar, K. A. Matori, S. K. Sahari, Journal of Nanomaterials, 1–7 (2015)
35. N.M. Sutan, S. Hamdan, H.R. Sobuz, V.Laja, M.S. Islam, Concrete Research Letters **2**(4), 326-33 (2012)

Lignin recovery and it effects quality of anaerobic treated palm oil mill effluent (AT-POME)

Abdul Haqi Ibrahim[1,2*], *Muhammad Fahmi* Ridwan[1,2], *Che Zulzikrami Azner* Abidin[1,2], *Soon Ann* Ong[1,2], *Yee Shian* Wong[1,2], *Ayu Wazira* Azhari[1], and *Siti Norruhaidawati* Ozir[2]

[1]WAREG, School of Environmental Engineering, Universiti Malaysia Perlis (UniMAP), 01000, Perlis, Malaysia.
[2]School of Environmental Engineering, Universiti Malaysia Perlis (UniMAP), 01000, Perlis, Malaysia.

Abstract. Lignin is one of the main structural polymers present in plant tissue. It can also be found as an isolated product of the pulp and paper industry. Palm oil mill effluent (POME) has been known as high strength industrial wastewater that is difficult to treat due to its large variety of inorganic and organic contents. The main purpose of this study is to recover soluble lignin from anaerobically treated palm oil mill effluent (AT-POME) and indirectly improves the quality of AT-POME. AT-POME was adjusted to different pH using different type of acids. Response Surface Methodology (RSM) was utilized to obtain the optimum operating parameters as well as to analyse the interaction between them. Model shows that 74.67 % of lignin can be recovered from AT-POME after 5 minutes reaction time using sulfuric acid (H_2SO_4) at pH 5. Hence from the experiment, it was proved that simple pH adjustment could precipitate the soluble lignin from AT-POME.

1 Introduction

Lignin is one of the important compounds of the cell wall plant. Lignin is nature's most aromatic polymer, and the second renewable natural resource (after cellulose) and it is traditionally viewed as waste material or low value by product.

The application of lignin is depending on the possibilities of either degrading it into low molecular weight aromatics or to take advantage of lignin as a multi-functional macromolecule [1]. Because of its polyaromatic nature and great availability, lignin has big potential to serve as future "green" source for aromatic chemicals, especially phenols [2].

In addition, various possible value-added applications for lignin without conversion such as blending with polyolefins giving UV stabilization [3-4] or blending with PL giving 100% renewable resins [5] or take advantage of its antimicrobial and antioxidant action [6-8].

*Corresponding author: abdulhaqi@unimap.edu.my

Palm oil mill effluent (POME) is one of the major pollutants in Malaysia. This is due to mass production of palm oil in the country. Malaysia is rank as world's second largest producer of palm oil. It is estimated that about 60 million tonnes of POME was produced yearly by the palm oil industries in Malaysia. Anaerobic digestion is widely employed biotechnologies for POME because it is capable to digests the high organic contents of present in POME.

However, lignin (part in lignocellulosic material) is very difficult to biodegrade in an anaerobic environment. As it is hardly/non-biodegradable, soluble lignin imparts black colour (aesthetically inconvenient) and produce bad odour to anaerobically treated POME (AT-POME). Hence, precipitation and recovery of soluble lignin from AT-POME could treat the solution faster. In addition, the recovered lignin can be modified for other uses/purposes.

2 Methodology

2.1 Materials and chemicals

Samples of AT-POME were collected from Malpom Sdn. Bhd., Nibong Tebal, Penang. Sulfuric acid (H_2SO_4) and hydrochloric acid (HCl) from R&M Chemicals were used to precipitate lignin from AT-POME. Lignin content in AT-POME was determined by the Tyrosine Method using Tannin-Lignin reagent set produced by HACH.

2.2 Lignin removal and recovery

Raw AT-POME was centrifuged at 4000 rpm for 10 minutes to separate any coarse particles present in AT-POME sample. Centrifuged sample was tested for its initial COD and lignin content. Next, 350 ml of AT-POME was poured into conical flask and the pH was adjusted accordingly. Samples were shaken using mechanical shaken for duration 5, 10 and 15 minutes. After the shaking process, AT-POME was centrifuged once more to separate the precipitated lignin and the supernatant. The precipitated lignin was collected and washed then dried at 105° C until constant weight. The supernatants were also tested for its remaining COD and lignin content.

2.3 Experimental design

Experimental design and statistical analysis were performed using Design Expert software (version 7.1). The Box-Behnken Design (BBD) and Response Surface Methods (RSM) were used to set up the experimental protocols and optimize the independent process variables for lignin precipitation. The variables and their levels are summarized in Table 1. The chemical oxygen demand (COD) and lignin recovery efficiency was selected as the dependent response variable for this process.

Table 1. Experimental setup for Box-Behnken Design.

Factor	Variable	Range	
		+1	-1
A	pH	5	2
B	Reaction time (min)	5	15
C	Acid Type	H_2SO_4	HCl

3 Results and discussions

3.1 Lignin recovery

Raw and centrifuged AT-POME contains 1900 mg/l of COD and 220 mg/l soluble lignin. Fig. 1 shows the result of the lignin recovery by acid precipitation. The results were varying between all experiments. Highest lignin removal was recorded in Experiment 1 (79.94%) by using acid H_2SO_4 and was adjust using pH 3.50 at 2.93 minutes. While lowest lignin removal was in Experiment 8 that using pH 1.38 for 10.00 minutes.

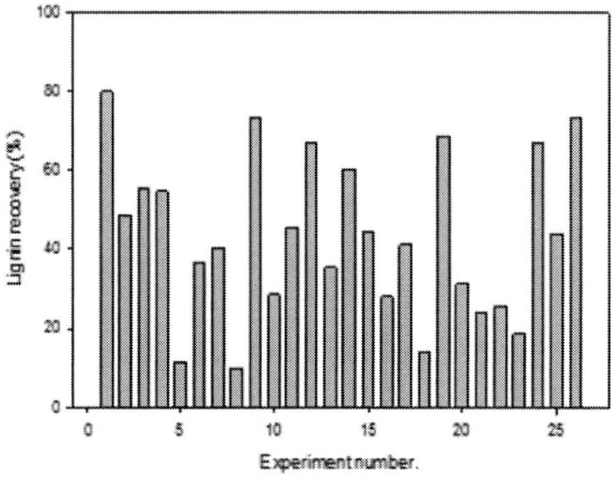

Fig. 1. Lignin recovery from AT-POME.

3.2 ANOVA analysis of experiment results

The analysis of variance (ANOVA) of lignin recovery process is shown in Table 2. It suggests that quadratic model was the significant model for precipitation of lignin from AT-POME by acids since the F-value was 1.59. F-value for two factors and cubic was the smallest which 1.51 and 0.36. Hence, quadratic model still the most significant model for this experiment. The suggested equation by the RSM as in Equation 1 where A is pH, B is reaction time and C is acid type.

$$\textit{Lignin recovery (\%)} = +45.08 + 10.13A - 3.57B - 5.50C - 15.06AB + 2.72AC + 1.49BC - 7.33A^2 + 3.89B^2 \quad (1)$$

Table 2. Statically analysis data for lignin recovery from AT-POME.

Model	Sum of Square	Degree of Freedom	Mean Square	F-Value
Two-Factor interaction	1623.52	3	541.17	1.51
Quadratic	1076.01	2	538.01	1.59
Cubic	745.7	5	149.14	0.36
Linear	2038.78	3	679.59	1.77

3.3 Adequacy of fitting model

The "Model F-value" of 1.75 implies the model is not important to noise. There is a 15.77 % chance that a "Model F-value" this large could occur due to noise. Values of "Prob > F" less than 0.0500 indicate model terms are significant. In this case A, AB are significant model terms. Values greater than 0.1000 indicate the model terms are not important. The "Lack of Fit F-value" of 1.21 implies the Lack of Fit is not significant relative to the pure error. There is a 38.70% chance that a "Lack of Fit F-value" this large could occur due to noise.

Table 3. Analysis of Variance (ANOVA) for the quadratic model.

Response	Source	Sum of square	Degree of freedom	Mean square	F-value	P-value (Prob>F
Lignin Removal (%)	Model	4738.32	8	592.29	1.75	0.1577
	Residual	5748.87	17	338.17	-	-
	Lack of Fit	2983.01	8	372.88	1.21	0.387
	Pure Error	2765.86	9	307.32	-	-
	Total	10487.19	25	-	-	-
	R^2=	0.4518				

Fig. 2 shows the graph for normal plot of residual, which is almost in straight line which supporting the condition that the error terms are normally distributed. While in Fig. 3, it shows the plot of residual versus predicted values. There is no definite increase of residual with predicted level as all plots are in the range (between –3.00 and +3.00). Actual and predicted results are plotted and shown in Fig. 4. Results show that the model is not fit very well as some points are away from the fitted line.

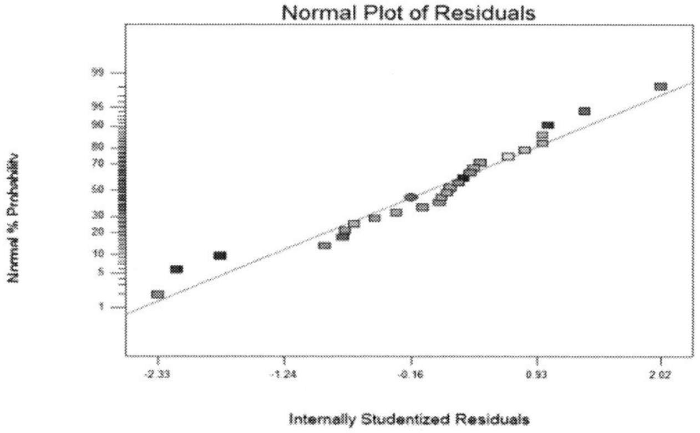

Fig. 2. Normal plot of residue.

https://doi.org/10.1051/e3sconf/20183402001

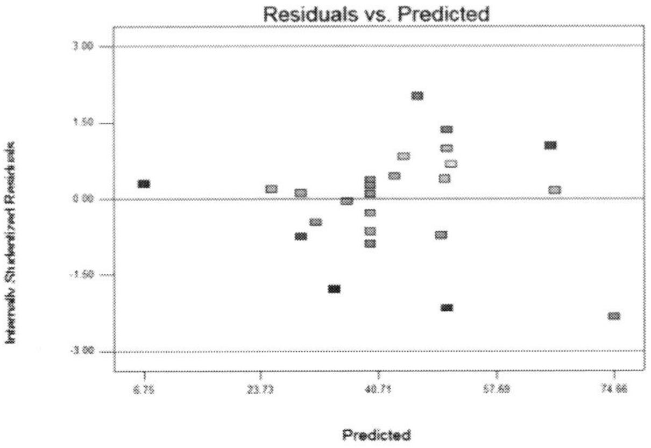

Fig. 3. Residual versus predicted value for percentage of lignin removal.

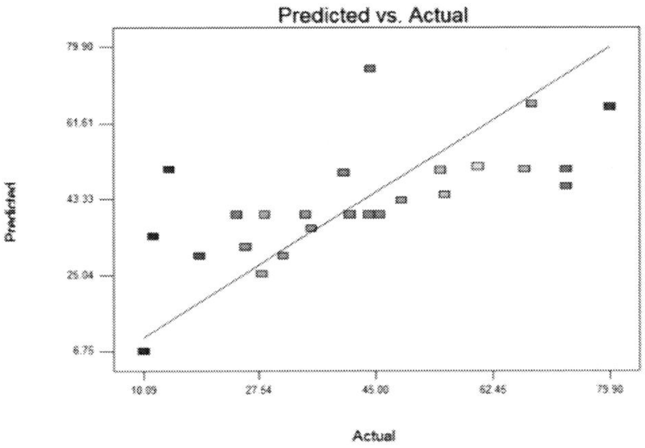

Fig. 4. Predicted versus actual values for percentage of lignin removal.

3.4 RSM for lignin removal

Lignin recovery from AT-POME was very much dependent on pH and reaction time. Fig. 5 and Fig. 6 show the effect of both parameters on lignin recovery from AT-POME using HCl and H_2SO_4 respectively. It was found that the lowest lignin recovery was 10 % at pH 2 with the reaction time of 5 minutes using HCl. The recovery was improved to almost 69% after the pH was increased to pH 5. However, lignin recovery from AT-POME can be further improved when H_2SO_4 is used to adjust the pH. When H_2SO_4 was used, the lowest and highest recovery approximately 27% and 75% respectively. In addition, the effect of reaction time on lignin recovery becomes less when H_2SO_4 is used to adjust the pH as shown in Fig. 6.

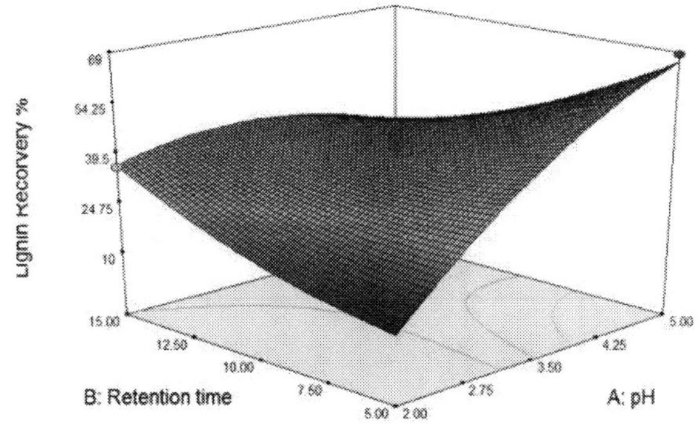

Fig. 5. Three-dimensional surface graph describing the effect of pH and reaction time on lignin removal (%); HCl.

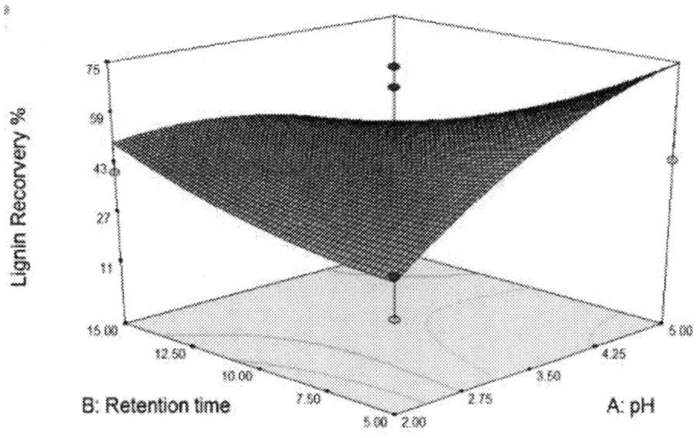

Fig. 6. Three-dimensional surface graph describing effect of pH and reaction time on lignin removal (%); H_2SO_4.

4 Conclusions

Acid precipitation is a feasible and simple method to recover lignin from AT-POME. Result shows that sulfuric acid works better compared to hydrochloric acid and pH of sample solution has been identified as main factor that contribute to the success of lignin recovery from AT-POME.

Authors would like to thank the Ministry of Higher Education for funding this research under the Fundamental Research Grant Scheme (FRGS 9003-00477).

References

1. D. Stewart, Ind Crops Products **27,** 202-207 (2008).
2. C. Amem-Chen, H. Pakdel, C. Roy, Bioresource Technol **79**, 277-299 (2001).
3. E. Scott, H.J. De Beukelar. Patent EP0980888, (1999).
4. B. Kosikova, V. Demianova, M. Kacurakova. J Applied Polymer Science **47**, 1065-1073 (1993).
5. Q.X. Hou, X.S. Chai, R. Yang, T. Elder, A.J. Ragauska, J Applied Polymer Science **99,** 1346-1349 (2006).

6. X. Dong, M. Dong, Y. Lu, A. Turley, T. Jin, C. Wu, Ind Crops Products **34,** 1629-1634 (2011).
7. C. Pouteau, P. Dole, B. Cathala, L. Averous, N. Boquillon, Polym Degradation Stab **81**, 9-18 (2003).
8. M.P. Vinnardell, V. Ugartondo, M. Mitjans, Ind Crops Products **27,** 220-223 (2008).

Temporal Variation of Ambient PM_{10} Concentration within an Urban-Industrial Environment

Yoon-Keaw Wong[1], *Norazian* Mohamed Noor[1*], and *Nur Izzah* Mohamad Hashim[1]

[1]School of Environmental Engineering, Universiti Malaysia Perlis, Kompleks Pengajian Jejawi 3, 02600 Arau, Perlis, Malaysia.

Abstract. PM10 concentration in the ambient air has been reported to be the main pollutant affecting human health, particularly in the urban areas. This research is conducted to study the variation of PM10 concentration at the three urban-industrial areas in Malaysia, namely Shah Alam, Kuala Terengganu and Melaka. In addition, the association and correlation between PM10 concentration and other air pollutants will be distinguished. Five years interval dataset (2008-2012) consisting of PM10, SOX, NOX and O3 concentrations and other weather parameters such as wind speed, humidity and temperature were obtained from Department of Environment, Malaysia. Shah Alam shows the highest average of PM10 concentration with the value of 62.76 μg/m³ in June, whereas for Kuala Terengganu was 59.29 μg/m³ in February and 46.61 μg/m³ in August for Melaka. Two peaks were observed from the time series plot using the averaged monthly PM10 concentration. First peak occurs when PM10 concentration rises from January to February and the second peak is reached in June and remain high for the next two consecutive months for Shah Alam and Kuala Terengganu. Meanwhile the second peak for Melaka is only achieved in August as a result of the transboundary of smoke from forest fires in the Sumatra region during dry season from May to September. Both of the pollutants can be sourced from rapid industrial activities at Shah Alam. PM10 concentration is strongly correlated with carbon monoxide concentration in Kuala Terengganu and Melaka with value of r2 = 0.1725 and 0.2744 respectively. High carbon monoxide and PM10 concentration are associated with burning of fossil fuel from increased number of vehicles at these areas.

1 Introduction

In Malaysia, PM_{10} is one of the major air pollutants and is decisive in the determination of Malaysian Air Pollution Index (API) [1]. It is believed to have an effect on the human respiratory system which in turn may result in chronic obstructive pulmonary disease and asthma. Located in the centre of South East Asia, Malaysia is experiencing rapid urban growth and is affected by local and regional air pollution [1]. The main sources of air

* Corresponding author: norazian@unimap.edu.my

pollutants in Malaysia are mobile sources particularly motor vehicles, stationary and transboundary emission.

Air pollution can only be controlled through the emission sources pathways. Alteration of both indoor and outdoor air quality may happens when chemical, biological or physical agents are present in the original air composition. In fact, particulate matter in air pollution has been reported to be the imperceptible health deteriorating issue in Malaysia among these environmental pollution problems as indication in the Air Pollution Index (API) [1]. About 6.5 million people were reported death due to deteriorate indoor and urban air quality in 2012 [2]. This research is conducted for the purpose to statistically describe characteristic of particulate matter, identifying correlation of PM_{10} with other air pollutants using Pearson Correlation. The areas of interest are within an urban-industrial environment that are Shah Alam, Kuala Terengganu and Melaka with the claim that urbanisation with industrial development can speeds up the rate of vehicle exhaust emission.

2 Methodology

2.1 Study Areas

Three urban-industrial areas were chosen in this study, namely Shah Alam, Kuala Terengganu and Melaka. Selangor is one of the 13 states that are well recognized to be the largest industrial region located at the west coast of Peninsular Malaysia. Air quality monitoring station of Shah Alam is geographically located at latitude N 03° 06' 17" and longitude E 101° 33' 22" at Sekolah Menengah Kebangsaan TTDI Jaya, Shah Alam. Second study area is Kuala Terengganu, which is the capital city of Terengganu that is located at the eastern coastal area of the peninsular Malaysia. The air quality monitoring station is situated at latitude N 05° 18' 455" and longitude E 103° 07' 213" at SK Pusat Chabang Tiga. Third air quality monitoring station is in Melaka which is geographically located at latitude N 02° 15'510" and longitude E 102° 10' 364" at Sekolah Menengah Kebangsaan Bukit Rambai, Melaka The capital city of Melaka covers a total area of 1664 km^2 and is bordered by Strait of Malacca. During the study period, the population has reached 842,500 in 2012 [3].

Besides PM_{10} concentration, other air pollutants (i.e. carbon monoxide, ground level ozone, nitrogen dioxide, sulfur dioxide, non-methane hydrocarbon concentration) and meteorological parameters (i.e. relative humidity, ambient temperature, wind speed, UVB) were also used for studying the variation of PM_{10} in these areas.

2.2 Association of PM_{10} concentration with the trace gases and weather parameters

Pearson correlation measures the degree of the linear relationship between two variables. By linear relationship, the relationship can be well-characterized by a straight line. Correlation ranges from -1.0 to +1.0, denoted by the letter r in Pearson correlation. There are three types of relationships: i) positive relationship (higher scores on other air pollutants concentration are associated with higher scores on PM_{10} concentration), negative (higher scores on other air pollutants concentration are associated with lower scores on PM_{10} concentration) and no relationship (there is no linear relationship between other air pollutants concentration and PM_{10} concentration).

3 Results and Discussions

3.1 Characterisation of PM_{10} Concentration

Fig. 1 illustrates the monthly concentration of PM_{10} in the study areas. The results show that there were two peaks observed in all the three study areas, which occurred in February and June (Shah Alam and Kuala Terengganu) while for Melaka, the second peak is in August. This is due to the southwest monsoon that occurred in Malaysia where the dry season started. In addition, the effect of transboundary particulate pollution from Sumatras will also make the peaks more significant [4]. The first peak is achieved in earlier month of the year at the three study areas where PM_{10} concentration rises from January to February as shown in Fig. 1 (a), (b) and (c). This happened due to minimum rainfall during the southwest monsoon. PM_{10} concentration for the three stations decreased in March before started to increase gradually to reach the second peak in June at Shah Alam, Kuala Terengganu and in August in Melaka. PM_{10} concentration remained high for the next two consecutive months as the ambient temperature rises during the dry monsoon (southwest monsoon) season [5]. Next, PM_{10} concentration decreased to the lowest concentration throughout the end of a year in November and increased slightly in December. The low PM_{10} concentration in November and December occurred due to high rainfall rate over the southwest and east coastal areas of peninsular Malaysia, where these areas experienced the northeast monsoon. The particulate matter that suspended in the atmosphere can be washed away by the rain precipitation and deposition [6].

The overall comparison of the average monthly PM_{10} concentration for Shah Alam, Kuala Terengganu and Melaka is indicated in Fig. 1(d). Overall, the highest PM_{10} concentrations were recorded in Shah Alam followed by Kuala Terengganu and Melaka. Most of the time, the concentration of PM_{10} in Shah Alam and Kuala Terengganu did not exceed the recommended value of Malaysia Ambient Air Quality Guideline (MAAQG) that is 150 $\mu g/m^3$. The PM_{10} concentration in Kuala Terengganu is the highest in February and in July due to high population with variety of commercial activities including airport station results in air pollution at this study area as the city center of Terengganu state. This situation can also be explained by the evaporation of seawater, which form salt particles in the atmosphere [7]. On the other hand, the PM_{10} concentration in Shah Alam recorded higher concentration than Kuala Terengganu from May until October. This may be due to the location of the station that is surrounded by the busy road in a mixed residential and commercial area. Besides that, the location of Shah Alam at the southwest coastal which near Indonesia also brought hot wind during southwest monsoon. Melaka shows the lowest PM_{10} concentration compared to the other two stations. The PM_{10} concentration exceeded Kuala Terengganu in the August and September due to southwest monsoon.

Table 1 indicates the tabulated values of the total number of unhealthy hours and unhealthy days recorded in Shah Alam, Kuala Terengganu and Melaka during the 5 years of the study period. The unhealthy hours and days were calculated based on the number of exceedances of PM_{10} concentration that were exceeded 150$\mu g/m^3$ according to Malaysia Ambient Air Quality Guidelines (MAAQG).

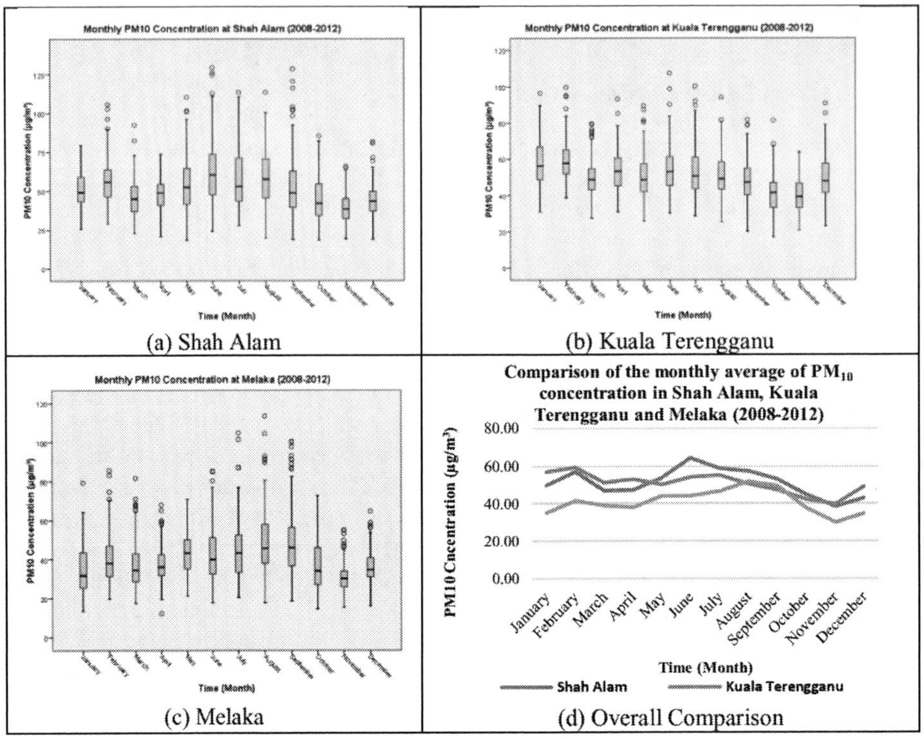

(a) Shah Alam (b) Kuala Terengganu

(c) Melaka (d) Overall Comparison

Fig. 1. Monthly concentration of PM_{10} in the study areas.

Table 1. Exceedances of PM_{10} Concentration.

Stations	**Exceedance**									
	Total Number of Hours*					**Total Number of Days****				
	2008	**2009**	**2010**	**2011**	**2012**	**2008**	**2009**	**2010**	**2011**	**2012**
Shah Alam	24	57	3	34	78	22	23	3	11	14
Kuala Terengganu	32	59	25	34	19	21	34	18	23	16
Melaka	11	116	75	33	15	10	54	44	22	12

Remark :
* represents number of hours that the PM_{10} concentration exceeds 150 µg/m³ according to Malaysia Ambient Air Quality Standard
** represents number of days account for at least 1-hour PM_{10} concentration exceeds 150 µg/m³ according to Malaysia Ambient Air Quality Standard

The highest numbers of unhealthy hours due to high PM_{10} concentration for Shah Alam is 78 hours in 2012 as haze attributed mostly due to the transboundary air pollution caused by the fire burning on the Indonesia island of Sumatras, where the strong wind brings the fumes across the Melaka Strait to Malaysia during the dry monsoon season [8]. For Kuala Terengganu and Melaka, both recorded the highest numbers of unhealthy hours for 59 hours and 116 hours in 2009. On the other hand, the highest numbers of unhealthy day are observed to be in 2009 for 23 days, 34 days and 54 days at the three study areas respectively. During that year, all the three studied areas recorded the highest number of unhealthy days that were 54 µg/m³, 34 µg/m³ and 23 µg/m³ in Melaka, Kuala Terengganu and Shah Alam respectively. This situation happens due to the effect of El Nino that occurred in June 2009 until May 2010. This is a periodical occurrence where the sea surface temperatures in the central and eastern Pacific Ocean become warmer than usual.

The consequence is significant throughout Malaysia especially in Melaka, which situated near the Pacific Ocean [9].

3.2 Correlation of PM_{10} concentration with the trace gases and meteorological parameters

Table 2 shows the result of Pearson analysis for Shah Alam, Kuala Terengganu and Melaka. The correlation between the dependent (PM_{10}) and the independent variables (CO, O_3, NO_2, SO_2, relative humidity, ambient temperature and wind speed) is as bolded. There is a significant positive relationship between the non-methane hydrocarbon (NmHc) concentration and PM_{10} concentration at Shah Alam with r = 0.534. High PM_{10} concentration associated with non-methane hydrocarbon because of the emission of acetylene, alkenes from combustion of fossil fuel, release of i-butane and i-pentane from paint and solvent utilization and volatilization of benzene and toluene from chemical cleaning process [10].

Table 2. Pearson correlation matrix of PM_{10} concentration with the trace gases and weather parameter

Shah Alam										
	PM_{10}	CO	O_3	NO_2	SO_2	NmHC	H	UVB	T	WS
PM_{10}	1									
CO	.498**	1								
O_3	.206**	-.050*	1							
NO_2	.503**	.681**	.030	1						
SO_2	.303**	.192**	.084**	.273**	1					
NmHC	**.534****	.688**	-.020	.598**	.376**	1				
H	-.264**	.113**	-.355**	.094**	-.172**	.115**	1			
UVB	-.026	-.097	.285**	-.175**	.038	-.040	-.343**	1		
T	.303**	-.016	.405**	-.014	.205**	-.003	-.838**	-.382**	1	
WS	-.221**	-.509**	.109**	-.486**	-.115**	-.526**	-.377**	.251**	.281**	1
Kuala Terengganu										
	PM_{10}	CO	O_3	NO_2	SO_2	H	T	WS		
PM_{10}	1									
CO	**.416****	1								
O_3	.308**	-.008	1							
NO_2	.364**	.351**	.144**	1						
SO_2	.109**	.085**	-.104**	.082**	1					
H	-.381**	.019	.303**	.038	-.090**	1				
T	.230**	.028	.094**	.031	.085**	-.603**	1			
WS	.041	-.241**	.455**	-.385**	-.166**	-.208**	-.082**	1		
Melaka										
	PM_{10}	CO	O_3	NO_2	SO_2	H	T	WS		
PM_{10}	1	.547**	.375**	.444**	.283**	-.186**	.275**	.010		
CO	.547**	1	-.094**	.601**	.145**	.093**	.029	-.233**		
O_3	.375**	-.094**	1	-.054*	.216**	-.432**	.378**	.143**		
NO_2	.444**	.601**	-.054*	1	.352**	.245**	.071**	-.367**		
SO_2	.283**	.145**	.216**	.352**	1	-.081**	.223**	.047		
H	-.186**	.093**	-.432**	.245**	-.081**	1	-.684**	-.263**		
T	.275**	.029	.378**	.071**	.223**	-.684**	1	.088**		
WS	.010	-.233**	.143**	-.367**	.047	-.263**	.088**	1		

For Kuala Terengganu, the most significant positive relationship was found to be between PM_{10} concentration and carbon monoxide concentration that was released from the combustion of fossil fuel in automobiles with the r = 0.416 [11].

The air pollutant that has the most significant positive correlation with PM_{10} concentration in Melaka was carbon monoxide concentration with r = 0.547. Apart from transportation, a significant concentration of carbon monoxide are also due to combustion of paddy field in agricultural activities [1].

4 Conclusions

The variations of PM_{10} concentration in the three urban-industrial areas i.e. Shah Alam, Kuala Terengganu and Melaka were analysed in this study. From the monthly variation, two peaks were observed in February and June. The first peak started from November to March while the second peak occurred in June for Shah Alam and Kuala Terengganu and August for Melaka. This is due to the southwest monsoon that occurred in Malaysia where the dry season starts from June to September. Other external contributing factor is the effect of transboundary particulate by forest fires from Sumatras. The exceedances of the PM_{10} concentration according to Malaysia Ambient Air Quality Guidelines (MAAQG) at $150\mu g/m^3$ were done to observe the total hours and total days of PM_{10} concentration exceeding the recommended value. The highest occurrence of unhealthy hours and unhealthy days are recorded in 2009. This can be explained that extreme high ambient temperature caused by El Nino effect, which leads to high particulate event in Shah Alam, Kuala Terengganu and Melaka. From Pearson correlation, the factor that has the most significant relationship with the PM_{10} concentration at Shah Alam is non-methane hydrocarbon (NmHc) whereas for Kuala Terengganu and Melaka, high concentration of PM_{10} is associated with the high carbon monoxide concentration. These relationships can be explained due to the surrounding of the monitoring stations that were located at the busy road and nearby the commercial areas.

The author want to thank Ministry of Higher Education for the Fundamental Research Grant Scheme (FRGS 9003 00508) and Department of Environment, Malaysia for the air pollutants dataset.

References

1. D. Dominick, H. Juahir, M.T. Latif, S.M. Zain, A.Z. Aris, Atmos. Environ. **60**, 172-181 (2012).
2. M. Khor, The Star Online, Malaysia (2016).
3. Department of Statistics, Malaysia, Department of Statistic Malaysia, (2016).
4. E.J. Hyer, B.N. Chew, Atmos. Environ **44** (11), 1422-1427, (2010).
5. L. Juneng, M.T. Latif, F. Tangang, Atmos. Environ. **45** (26), 4370-4378, (2011).
6. M. Demuzere, R.M. Trigo, J. Vila-Guerau de Arellano, N.P.M. Van Lipzig, Atmos. Chem. Phys. **9** (8), 2695-2714, (2009).
7. J.H. Seinfeld, S.N. Pandis, K. Noone. *Atmospheric chemistry and physics: from air pollution to climate change* (Wiley, 1998).
8. C.Y. Lin, H.M. Hsu, Y.H. Lee, C.H. Kuo, Y.F. Sheng, D.A. Chu, Atmos. Chem. Phys. **9** (20), (2009).
9. F. Tangang, M.T. Latif, L. Juneng, National University of Malaysia, (2010).
10. M. Wang, M. Shao, M.W. Chen, S. Lu, Y. Liu, B. Yuan, M. Hu, Atmos. Chem. Phys. **15** (3), 1489-1502 (2015).
11. N.A. Ghazali, N.A. Ramli, A.S. Yahaya. Eur. J. Sci. Res. **37**(2), 192-205 (2009).

Incorporation of layered double nanomaterials in thin film nanocomposite nanofiltration membrane for magnesium sulphate removal

Muhammad Hanis Tajuddin[1,2], *Norhaniza* Yusof[1,2,*], *Wan Norharyati* Wan Salleh[1,2], *Ahmad Fauzi* Ismail[1,2], *Nur Hanis Hayati* Hairom[3], and *Nurasyikin* Misdan[3]

[1]Advanced Membrane Technology Research Centre (AMTEC), Universiti Teknologi Malaysia, 81310 Skudai, Johor Bahru, Malaysia
[2]Faculty of Chemical and Energy Engineering (FCEE), Universiti Teknologi Malaysia, 81310 Skudai, Johor Bahru, Malaysia
[3]Faculty of Engineering Technology, Universiti Tun Hussein Onn Malaysia, 86400 Parit Raja, Johor Malaysia

Abstract. Thin film nanocomposite (TFN) membrane with copper-aluminium layered double hydroxides (LDH) incorporated into polyamide (PA) selective layer has been prepared for magnesium sulphate salt removal. 0, 0.05, 0.1, 0.15, 0.2 wt% of LDH were dispersed in the trimesoyl chloride (TMC) in n-hexane as organic solution and embedded into PA layer during interfacial polymerization with piperazine. The fabricated membranes were further characterized to evaluate its morphological structure and membrane surface hydrophilicity. The TFN membranes performance were evaluated with divalent salt magnesium sulphate ($MgSO_4$) removal and compared with thin film composite (TFC). The morphological structures of TFN membranes were altered and the surface hydrophilicity were enhanced with addition of LDH. Incorporation of LDH has improved the permeate water flux by 82.5% compared to that of TFC membrane with satisfactory rejection of $MgSO_4$. This study has experimentally validated the potential of LDH to improve the divalent salt separation performance for TFN membranes.

1 Introduction

Significant growth in population and economy have increased the demand of clean water which led to water shortage specifically in water-stressed region areas. To improve the current technologies, the search for potential advanced materials has gained significant attraction among researchers to cope with clean water demand.

The current commercial of nanofiltration membrane are dominated by polyamide thin film composite (TFC) membrane. With the performance and characteristics pores lies between ultrafiltration and reverse osmosis, nanofiltration (NF) process have high retention

* Corresponding author: norhaniza@petroleum.utm.my

flux, relatively low operating pressure and high rejection towards multivalent salts [1-3]. However one of the major drawbacks of TFC membrane is susceptibility towards fouling thus rendering their overall performances.

Thin film nanocomposite (TFN) membrane incorporated with nanomaterials is gaining significant attention since its first introduction in 2007 by Lee Hoek [4]. It is believed that incorporation of nanomaterials would play an important role in improving membrane properties such as hydrophilicity, selectivity and water permeability [4-5]. In recent years many nanoparticles have been utilized in fabrication of TFN membranes, including zeolites [7, 8, 9] TiO_2 [10], Ag, CNTs and silica [5]. Furthermore, incorporation of nanomaterials into thin film structures has significantly improved the performances of available composite nanofiltration membrane and reverse osmosis membranes.

In recent years, layered materials as illustrated in Fig. 1 such as layered double hydroxides has gained significant attention especially in membrane fabrication for desalination process [10-11]. With their ability to intercalated in interlayer regions and tunable ratio composition they were found prominently in many applications such as photo-catalyst, adsorbents and medicines. Example of layered compounds are layered transition mixed metal oxides, layered hydroxide salts and layered double hydroxides. LDHs have general formula of $[M(II)_{1-x}M(III)_x(OH)_2]^{x+}[(A^{n-})_{x/n}.yH_2O]$, where M(II) and M(III) represents divalent (Mg, Zn, Co, Cu, Ni) and trivalent cations (Al, Fe, Ga). A^{n-} is an interlayer anion such as Cl^-, SO_4^{2-} or CO_3^{2-} etc. and the x value is between 0.2-0.33 can be describe as molar ratio between M(II)/ [M(III) + M(III)] [13].

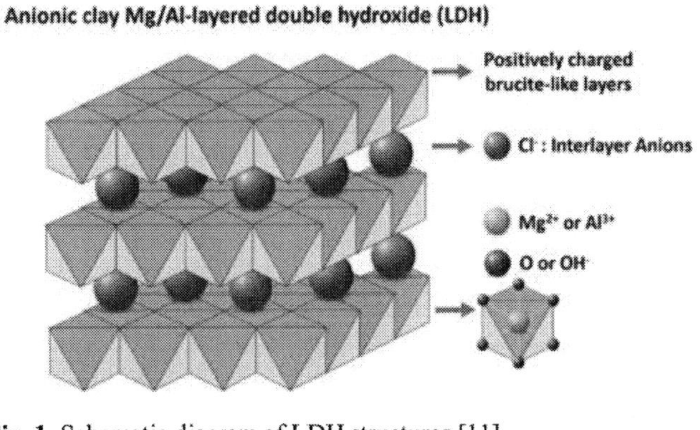

Fig. 1. Schematic diagram of LDH structures [11].

Therefore, this study aimed to synthesize a new type of thin film nanocomposite by incorporating layered double hydroxides into polyamide selective layer which was formed on top of polysulfone substrate. The physicochemical properties of as prepared Cu-Al LDHs were evaluated in term of its morphological structures. The influences of LDHs were systematically investigated through its effect in membrane morphology, surface hydrophilicity and separation performances of $MgSO_4$ salts.

2 Materials and methods

2.1 Materials

Aluminium nitrate nonahydrate $Al(NO_3)_2.9H_2O$ and copper (II) nitrate trihydrate $Cu(NO_3)_2.3H_2O$ powder were purchased from Sigma Aldrich and Merck respectively. Sodium hydroxide (NaOH) and sodium carbonate (Na_2CO_3) was obtained from from

Sigma Aldrich and Qrec respectively. All the chemicals were of analytical grade and used without further purification. Commercial PSf pellets purchased from BASF SE Germany is the main component in membrane fabrication. Polyvinlypyrolidone (MW= 40, 000 g/mol) supplied by Sigma Aldrich as pore forming agent. For the preparation of ultra-thin PA layer, piperazine (PIP) and 1,3,5 - benzene tricarboxylic acid chloride or trimesoyl chloride (TMC) were obtained from Sigma Aldrich. N-methyl-2-pyrollidone (NMP) (Purity > 99.5%) obtained from Sigma Aldrich and used as a solvent for the preparation of substrate membrane. *n*-hexane (purity > 99.5%) will be used as organic solvent for TMC monomer, were purchased from Merck. For membrane performance, $MgSO_4$ by Sigma Aldrich was used to prepare aqueous salt solution for membrane flux and rejection determination.

2.2 Synthesis of LDH

Firstly, $Cu(OH)_2$ was precipitated by adding 100 ml of 0.2 mol/L NaOH into 100 ml of 0.1 mol/L $Cu(NO_3)_2$ solution. Then, the precipitate of $Cu(OH)_2$ was added into 100 ml of 0.05 mol/L $Al(NO_3)_2$ solution under vigorous stir for 60 min. The pH was adjusted until pH 12 by using 1M of NaOH solution. The precipitates were aged at 6h and separated by centrifuge at 7000 rpm and washed with deionized water several times. Finally, the obtained precipitates was dried at 60°C for 24 h.

2.3 Preparation of thin film nanocomposite membrane

An appropriate amount of Cu-Al LDH was added in 20 ml of 0.1 (w/v)% TMC in n-hexane and dispersed via ultrasonicating for 1 hour. Firstly, the substrate was clamped in between glass plate and silicone rubber. Then, 20 ml of 2 (w/v)% PIP aqueous solution was poured on top of PSf substrate for 2 min. After that, the excess PIP solution on the surface of substrate membrane was removed by rubber roller. Then, 20 ml of 0.1 (w/v)% TMC in n-hexane solution was poured onto substrate for 1 min and polyamide layer was instantaneously form. Subsequently, the membrane was stored in oven at 60°C for 5 min, the resultant membrane rinsed with DI water and stored in deionized water before used. The protocol of preparation composite membrane was illustrated in Figure 3.2.The amount of Cu-Al LDH concentration were varied from 0, 0.05, 0.1, 0.15 to 0.2 (w/v)%. The resulting membrane was denoted as TFC, TFN 0.05, TFN 0.1, TFN 0.15 and TFN 0.2.TFC membrane without Cu-Al LDH was prepared to serve as a control in this experiment.

2.4 Characterization of Cu-Al LDH nanoparticles, TFC and TFN membranes

Different SEM micrographs with various magnifications were obtained from cross section, skin layer thickness and surface of TFC/TFN membrane and LDH nanoparticles by using scanning electron microscopy (Model: TM 3000, Hitachi). LDH nanofillers were characterized by Transmission Electron Microscopy (TEM) (Orius SC 1000A) to observe their particle properties. Meanwhile, surface hydrophicility of TFC and TFN was analysed by contact angle meter (Dataphysics OCA 15pro).

2.5 Membrane separation performances

The flux and rejection of fabricated TFC and TFN membranes was assessed by using dead-end filtration system (Sterlitech^{TM} HP4750 Stirred Cell) under nitrogen atmosphere. The effective surface are of the membrane is 14.6 cm^2. The TFC and TFN membranes were compacted at a pressure 8 bar for about 30 min to achieve flux steady state condition. The experiments will be performed by using 1000 ppm of $MgSO_4$ solution at operating pressure and temperature set at 7 bar and 25 C respectively. Membrane water flux *(F)* will be subsequently measured by using the following equation:

$$F = \frac{V}{t \times A} \tag{1}$$

where V is the permeate volume in litre (L), A is the membrane area (m^2) and t is the experimental time to obtain V (h).The salt concentration in the feed and permeate solutions will be measured by bench conductivity meter (Jenway 4520). The membrane salt rejection will be determined by using the following equation:

$$R(\%) = (1 - \frac{C_p}{C_f}) \times 100 \tag{2}$$

where C_p is the permeate concentration (ppm) and C_f is the feed concentration (ppm) respectively.

3 Results and discussions

3.1 LDH morphological structures

From the morphological structures showed in TEM images in Fig. 2 a) and b), LDH nanofillers displayed a layer structure materials as indicated in square red line. This characteristics properties were found as similar as shown by research reported from Dong and co-workers [11]. Besides that, the SEM images of LDH nanofillers shows that the Cu-Al LDH particles cluster all together as shown in red circle in figure 2 c) into a large aggregates mainly caused by high surface charge and surface tension [11,13].

3.2 Effects of LDH on membrane properties

The top surface of TFN membrane morphologies are presented in Fig. 3. Fig. 3 (a) show typical poly(piperazine amide) with a rough surface and firmly grab globules. These traits can be seen similarly found in research reported by [15]. However after addition of LDH nanofillers, the surface become smoother compared with pristine TFC and red circle in Fig. 4 (b), (c), (d), (e) indicating the presence of LDH in selective layer. Generally, incorporation of hydrophilic materials reduce surface roughness of membrane with similar behaviour was observed when Zarrabi et al. [16] embedded amine functionalized MWCNT.

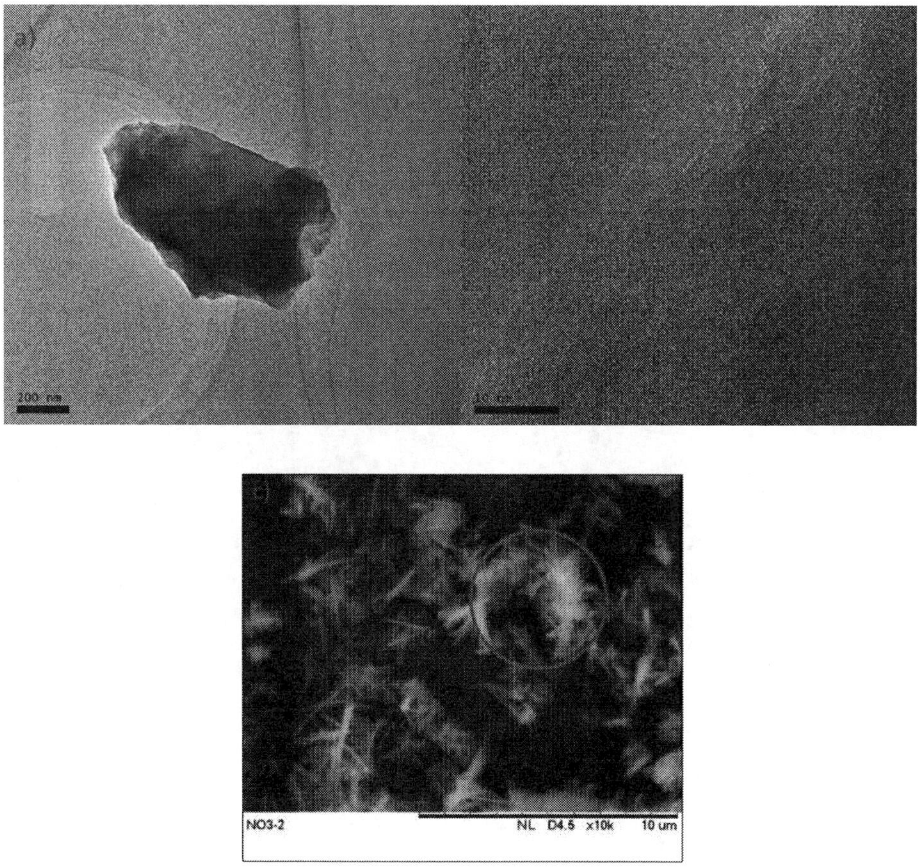

Fig. 2. a) and b) TEM displayed the irregular shape of LDH c) SEM images of synthesized LDH nanofillers.

Fig. 4 presented cross section image of poly(piperazine amide) thin film membrane. According to the cross section images, the red line in Fig. 4 (b) indicating the formation of dense selective layer in thin film composite membranes. These traits were further supported by Fig. 4 (c) shows the thickness of poly (piperazine amide) layer with measurement of averagely ranging from 246 nm to 294 nm.The same features also displayed by Wang and co-researchers when they were fabricating thin film composite nanofiltration membranes [17].

The water contact angle of TFC and TFN membranes are presented in Fig. 5. From the Fig. 5, it can be observed that pristine TFC displayed high contact angle of 53.24°. Moreover, after the incorporation of nanofillers into polyamide layer, the contact angle values decreased as the amount of LDH loadings increased from 0.05 wt% (48.64°) to 0.2 (37.50) wt% indicated the improvement in membrane hydrophilicity. This improvement of membrane hydrophilicity could be ascribed by hydrophilic property and presence of hydroxyls group in LDH nanofillers. Another reported research by Xia et al [18] show the same improvement in membrane hydrophilicity as they incorporated nanofillers in polyamide membrane. Furthermore, the hydrophilic surface membrane are favourable in desalination process because it could produce better water permeability and fouling resistance [10,17].

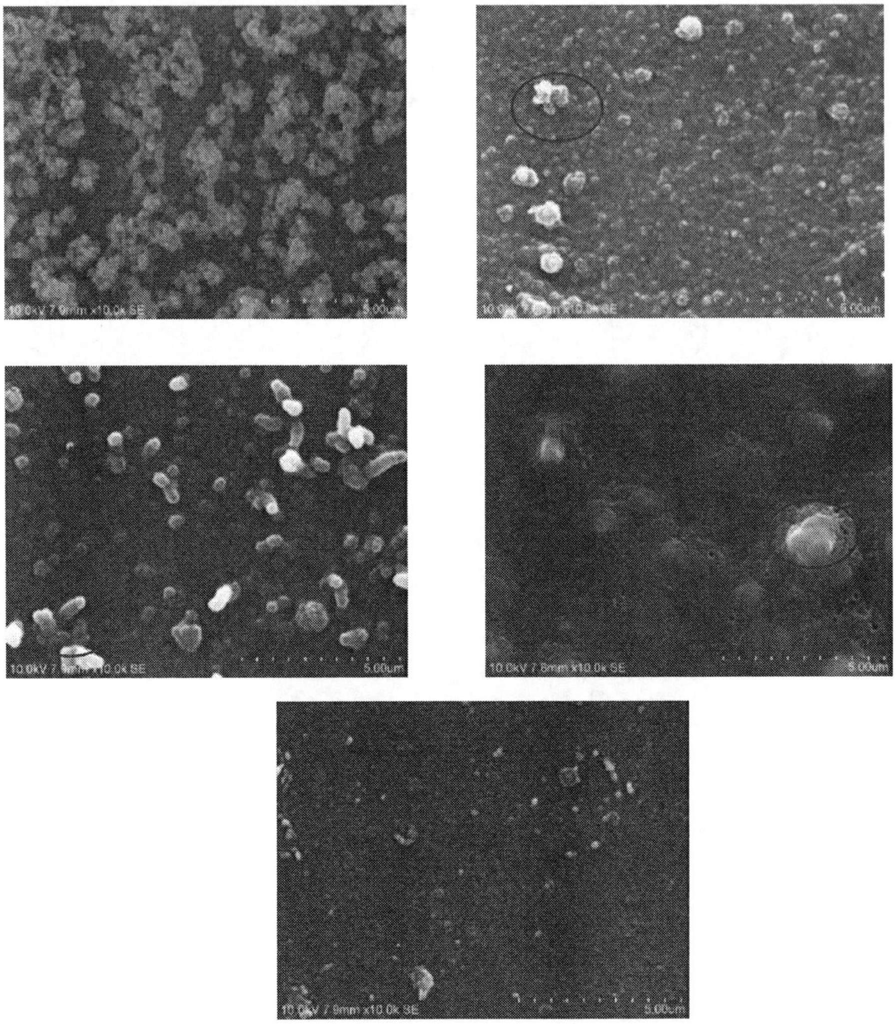

Fig. 3. SEM images of a) TFC b) TFN 0.05 c) TFN 0.1 d) TFN 0.15 and e) TFN 0.2 membrane shows the appearance of LDH on top of the polyamide layer.

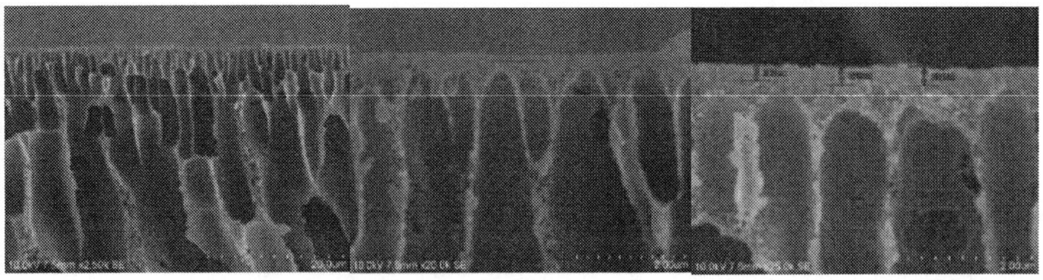

Fig. 4. Cross section image of poly(piperazine amide) membrane a) 2500k magnification b) 20000k magnification c) 25000k magnification.

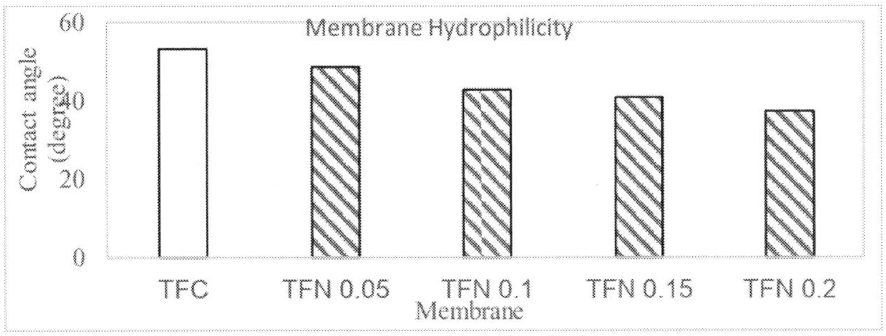

Fig. 5. Membrane surface hydrophilicity.

3.3 Membrane separation performances

The separation performance of TFC and TFN membranes were evaluated by using $MgSO_4$ a representative of divalent salts. The applied pressure of 7 bar was applied to TFC and TFN membrane and were examined in term of water permeability and $MgSO_4$ rejection as shown in Fig. 6.

The water permeability of TFC, TFN 0.05, TFN 0.1, TFN 0.1 and TFN 0.2 membrane under the studied condition of 7 bar were recorded at 1.32, 1.75, 3.23, 5.57 and 7.50 LMH/Bar respectively. The significant improvement of water permeability from TFC to TFN probably caused by incorporation of LDH nanoparticles into polyamide layer. Moreover, the notably increased of water permeability are correlated to membrane hydrophilicity as stated in figure 5. with significant increment by 82.4 %. As the values of contact angle measurement was decreased the hydrophilicity of membrane improved due to existence of layered structure of LDH nanofillers [11-12].

Meanwhile the MgSO4 salt rejection of LDH-polyamide membrane for TFC, TFN 0.05, TFN 0.1, TFN 0.1 and TFN 0.2 were recorded at 91.23%, 93.63%, 95.41%, 94.01% and 90.57%. At the 0.1 wt% of LDH, the TFN membrane displayed the best separation performance with consideration of high water permeability (3.23 L/m^2.h.bar) and salt rejection (95.41%) with slight improvement of 4.18% compared with pristine TFC membrane. However, the decrease of salt rejection was observed with increment of loading at 0.15 wt% and 0.2 wt% LDH into polyamide layer. It could be mainly attributed to agglomeration of LDH in the selective layer thus decreasing the rejection as represented in SEM image in Fig. 3.

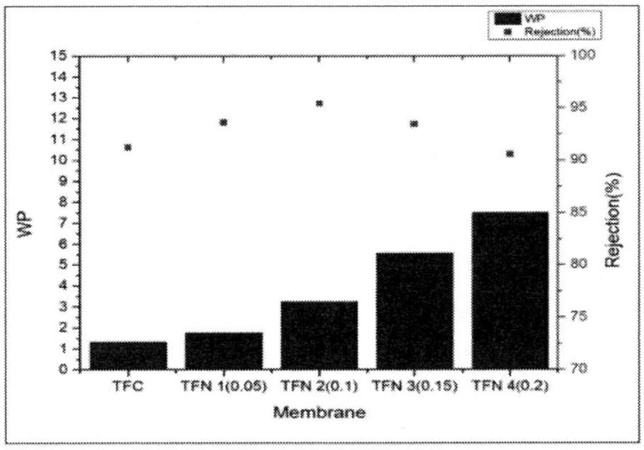

Fig. 6. Water permeability and $MgSO_4$ rejection.

4 Conclusions

TFN membrane incorporated with LDH have been developed for salt removal. The presence of LDH in polyamide layer has altered the properties of membrane and increased the membrane hydrophilicity. Moreover, the water permeability of TFN membrane has significantly improved by 82.4% compared with pristine TFC. Furthermore, the optimum condition of $MgSO_4$ rejection is displayed by 0.1 wt% of LDH in selective layer. Despite having higher rejection at 95.41% as the loadings were increased mainly cause by the agglomeration of nanofillers in selective layer of membrane. Since the modification of pristine membrane enhanced the water permeability and selectivity, we believed this facile approach can offer another potential materials for improvement in desalination process.

The authors would like to acknowledge the financial support from the Ministry of Higher Education Malaysia and Universiti Teknologi Malaysia under Fundamental Research Grant Scheme (R.J130000.7846.4F929), GUP grant (Q.J130000.2546.16H29), and Higher Institution Centre of Excellence (HiCOE) grant (R.J090301.7846.4J180) and (R.J090301.7846.4J179). The authors would also like to acknowledge the technical and management support from Research Management Centre (RMC), Universiti Teknologi Malaysia.

References

1. S. Pourjafar, A. Rahimpour, M. Jahanshahi, J. Ind. Eng. Chem. **18,** 1398–1405 (2012).
2. M. Liu, Y. Zheng, S. Shuai, Q. Zhou, S. Yu, C. Gao, Desalination **288**, 98–107 (2012).
3. B. Rajaeian, A. Rahimpour, M.O. Tade, S. Liu, Desalination. **313,** 176–188 (2013).
4. J. Yin, B. Deng, J. Memb. Sci. **479,** 256–275 (2015).
5. S. Al Aani, C.J. Wright, M.A. Atieh, N. Hilal, (2016) (to be published).
6. M. Fathizadeh, A. Aroujalian, A. Raisi, J. Memb. Sci. **375,** 88–95 (2011).
7. N. Ma, J. Wei, R. Liao, C.Y. Tang, J. Memb. Sci. **405–406,** 149–157 (2012).
8. L.X. Dong, X.C. Huang, Z. Wang, Z. Yang, X.M. Wang, C.Y. Tang, Sep. Purif. Technol. **166**, 230–239 (2016).
9. E. Bet-moushoul, Y. Mansourpanah, K. Farhadi, M. Tabatabaei, Chem. Eng. J. **283**, 29–46 (2015).
10. H. Dong, L. Wu, L. Zhang, H. Chen, C. Gao, J. Memb. Sci. **494,** 92–103 (2015).
11. P. Lu, S. Liang, L. Qiu, Y. Gao, Q. Wang, J. Memb. Sci. **504**, 196–205 (2016).
12. M.A. Djebbi, M. Braiek, P. Namour, A. Ben Haj Amara, N. Jaffrezic-Renault, Appl. Surf. Sci. **386,** 352–363 (2016).
13. N.S. Ahmed, R. Menzel, Y. Wang, A. Garcia-Gallastegui, S.M. Bawaked, A.Y. Obaid, S.N. Basahel, M. Mokhtar, J. Solid State Chem. **246**, 130–137 (2017).
14. P. Wen, Y. Chen, X. Hu, B. Cheng, D. Liu, Y. Zhang, S. Nair, J. Memb. Sci. **535,** 208–220 (2017).
15. H. Zarrabi, M.E. Yekavalangi, V. Vatanpour, A. Shockravi, M. Safarpour, Desalination. **394,** 83–90 (2016).
16. C. Wang, Z. Li, J. Chen, Z. Li, Y. Yin, L. Cao, Y. Zhong, H. Wu, J. Memb. Sci. **523,** 273–281 (2017).
17. S. Xia, L. Yao, Y. Zhao, N. Li, Y. Zheng, Chem. Eng. J. **280,** 720–727 (2015).
18. E.M. Vrijenhoek, S. Hong, M. J. Memb. Sci. **188,** 115–128 (2001).

E3S Web of Conferences **34**, 02004 (2018) https://doi.org/10.1051/e3sconf/20183402004
CENVIRON 2017

Biodiesel Derive Bio-oil of *Hermetia illucens* Pre-pupae Catalysed by Sulphonated Biochar

Siew Yoong Leong[1*], *Soo Shin* Chong[1], and *Kah Seng* Chin[1]

[1]Department of Petrochemical Engineering, Faculty of Engineering and Green Technology, University of Tunku Abdul Rahman, Jalan Universiti, Bandar Barat, 31900 Kampar, Perak

Abstract. This study investigates the development of biochar catalyst from bamboo applied for biodiesel synthesis. A non-conventional biodiesel feedstock was used in the in-situ transesterification reaction. This non-conventional feedstock is obtained from an insect's fly, the *Hermetia illucens* fly. Biochar derived from bamboo has been investigated as a promising catalyst for biodiesel synthesis. The biochar acid catalysts were prepared by sulphonation via impregnation with concentrated sulphuric acid. The prepared catalysts were investigated for their performance to catalyse in-situ transesterification via ultra-sonication of *Hermetia illucens* bio-oil. The effects of carbonisation time (1 hour and 2 hour) and temperature (400°C, 500°C and 600°C) as well as catalyst loading (5-20 wt% on oil basis) on the transesterification yield were studied. Result showed that the highest yield of FAME obtained was 95.6% with catalyst loading of 15 wt% carbonized at 500°C for 2 hours. Sharp band of methyl ester functional groups were observed in the FTIR spectra at 1735-1750cm^{-1}. The composition of this methyl ester was further deduced using gas chromatography and the fatty acid was predominantly lauric acid.

1 Introduction

Chemically, biodiesel is also known as fatty acids of mono alkyl ester [1-3]. Biodiesel is produced via transesterification reaction with a suitable catalyst either homogeneous [3] or heterogeneous [4]. The choice of catalyst selection is influenced by the amount of free fatty acids (FFA) presence in the feedstock. Homogeneous catalysts have always been the choice of catalyst for biodiesel production in large-scale operation. However, there are some shortcomings of using homogeneous catalyst due to non-recovery of the catalyst; poor thermal stability and require excessive of purification [5].

On the contrary, heterogeneous catalysts are capable to overcome the problems faced by the former ones. However, they were mostly derived from non-renewable resources and are expensive [6-7]. Recently, carbon based catalysts derived from agriculture waste have gained the attention from various researchers [6,8,10]. This type of catalysts offers numerous advantages, including renewable resources, non-toxic, reusable and reproducible, high catalytic activity, and stability in both acidic and basic conditions [2].

* Corresponding author: leongsy@utar.edu.my

Carbon based catalyst were derived from various sources such as from palm waste [8], carbohydrates [9], coconut shell [10] and coffee grounds [11]. Past study showed that, sulfonated biochar catalyst has been used to for free fatty acid (FFA) esterification and transesterification of vegetable oil due to higher catalytic activity [6,9,12-13]. Lou [9] reported that, carbohydrate-derived catalysts for both esterification and transesterification have markedly enhanced the yield of methyl esters in converting waste cooking oils containing 27.8 wt% high free fatty acids (FFAs) into biodiesel. Study by Kastner [12], shows that, the conversion from esterification of the fatty acids was attained approximately at ~90–100% conversion within 30–60 min at 55–60°C in excess of methanol blends with vegetable oil and animal fat (5–15 wt.% FFA). Buasri [13] also carried out similar study on transesterification of waste frying oil (WFO) with methanol using potassium hydroxide catalyst supported on *Jatropha curcas* fruit shell activated carbon (KOH/JS). Approximately, 86.7% of the conversion yield was achieved with optimum conditions at reaction temperature of 60°C for 2 h with methanol/oil molar ratio of 16 and catalyst bed height of 250 mm.

To the best of our knowledge, the utilization of carbon-based solid acid catalyst to synthesize biodiesel is mainly dedicated on the esterification reaction of FFA [12], edible oil, non-edible oil [14] and waste vegetable oil [8, 13]. Based on our literature findings, the use of bamboo biochar catalyst to prepare biodiesel from the insect's bio-oils has not been reported so far. Therefore, the gap of this study has to be filled in order to develop a better understanding on the effect of the biochar catalyst towards biodiesel yield from our non-conventional feedstock (*Hermetia illucens* pre-pupae). In this study, bamboo biochar is chosen as the catalyst support due to chemically inert, inexpensive and naturally porous. Moreover, the morphological structure and physicochemical properties of biochar can be customised by altering the nature of biomass via pyrolytic conditions and/or treating biomass biochar [15]. The objective of this study is to develop low-cost, carbon-based acid catalyst from bamboo and to investigate the potential application of the catalyst for biodiesel production from *Hermetia illucens* pre-pupae bio-oil.

2 Methodology

2.1 Catalyst preparation

First phase is to develop the biochar catalyst using bamboo as heterogeneous catalyst. The biochar catalysts were prepared in a series of carbonization time (1 and 2 h) and temperature (400, 500 and 600°C). Subsequently, the biochars were sulphonated with concentrated sulphuric acid for 24 h, filtered and washed with distilled water. The sulphonated biochars were dried in an oven for 24 h at 80°C. The sulphonated biochar is denoted as SBx_y_z (e.g: SB400_1_5), where x is carbonisation temperature (°C), y is carbonisation time (min) and z is catalyst loading (wt%).

2.2 In-situ transesterification via ultrasonication

Second phase is the in-situ transesterification of fatty acid methyl ester (FAME) via ultrasonication using the prepared catalyst. Approximately, 5 g of dried *Hermetia illucens* pre-pupae was dispensed into a reaction vessel containing methanol, biochar catalyst and n-hexane. Reaction parameters were fixed at 55°C for 248 min using 8:1 methanol to mass ratio at different catalyst loading of 5, 10, 15 and 20 wt% carbonized at different time and temperature. Prior analysis, sample FAME obtained was separated via centrifugation and was washed with distilled water at 70°C. The experiments were conducted in triplicates.

2.3 Characterisation of biochar catalyst and FAME

The morphologies of the sulfonated biochars were carried out using SEM with accelerating voltage of 2.0 kV, working distance of 6.4 mm and magnification of X1000. To qualitative assess, FTIR analysis was performed on the unsulphonate and sulphonated biochar catalyst. The biochar sample was finely grinded along with KBr IR grade using an agate mortar and pastel. The grinded sample was pressed into a disk and was placed at sample holder. The determination of FAME functional group was carried out using Fourier transform infrared (FTIR) equipped with attenuated total reflectance (ATR). Spectra were scanned over a wavenumber range of 4000 – 650 cm^{-1} at a resolution of 4 cm^{-1}.

Analysis of FAME compositions were done in triplicates and determined using gas chromatography (GC) equipped with a capillary column, BPX 70. The compositions of FAMEs were compared with a standard, Supelco® 37 Component FAME Mix purchased from Sigma Aldrich. The yield of FAME obtained was calculated according to Eq. (1).

$$\text{Yield of FAME, \%} = \left(\frac{\text{masss of FAME (g)}}{\text{mass of oil (g)}} \times 100\% \right) \quad (1)$$

The total acid density of the sulphonated catalysts was determined using the standard acid-base back-titration method. The back titration was done by titrating 0.02 M of HCl on 0.10 g of catalyst suspended into 60 mL of 0.01M NaOH.

3 Results and discussions

3.1 Effect of carbonization time & temperature on the sulphonated biochar morphology

Morphology of the sulphonated biochar catalysts is illustrated in Fig. 1 and Fig. 2. As observed, the overall morphologies exhibit a porous structure, indicating the porosity was developed along with the increasing carbonization time and temperature. Furthermore, thicker cell walls were observed when the biochar was carbonised at 400 and 500°C (Fig. 1 and 2). The development of the pore size increases when the biochars were carbonised at 600°C as compared to those carbonised at 400 and 500°C (Fig. 1 and 2). However, as the carbonisation temperature and time increases, the pores structure showed greater destruction (Fig. 1c). This phenomenon is due to prolong heat treatment that leads to decomposition of bamboo mass and subsequently weakens the matrix structure in bamboo vascular.

The morphology shown in Fig. 1(c) exhibits rougher surface and irregular pores patterns as compared to the morphologies in Fig. 1 (a) and (b). This phenomenon is due to the devolatilisation of volatile matter occurs with increase in temperature causes the destruction of globular structure that presence inside the biochar. Similar phenomenon was also reported by Huang et al. [16].

Pore cracks were observed in Fig. 2(a) due to the damaged of the adjacent wall between the micropores to form mesopores and macropores at high temperature. Thangarajan et al., (2015) [17], also reported a similar finding. Moreover, pore cracks are an indication to a reduction in pore volume. Furthermore, mixtures of large pore and small pores were observed when the carbonisation temperature increases (Fig. 2 c).

 https://doi.org/10.1051/e3sconf/20183402004

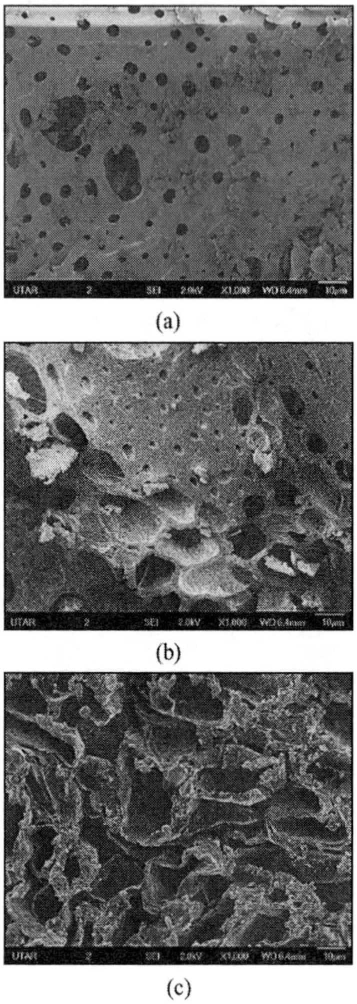

Fig. 1. SEM images for biochar carbonised at (a) 400°C, (b) 500°C and (c) 600°C for 1h.

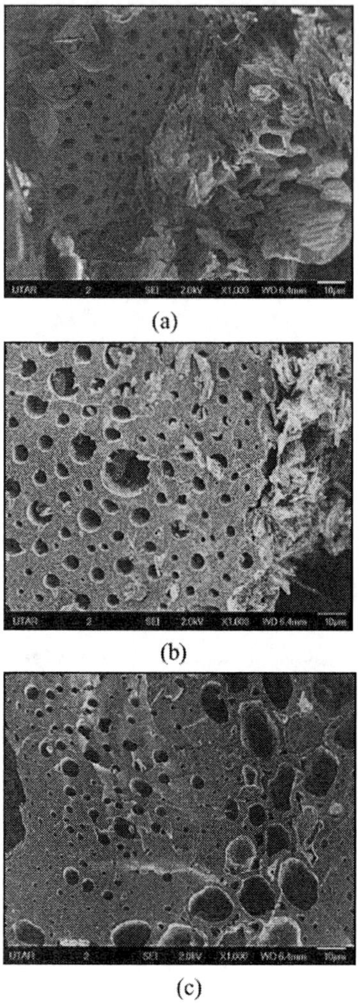

Fig. 2. SEM images for biochar carbonised at (a) 400°C, (b) 500°C and (c) 600°C for 2h.

3.2 Effect of catalytic activity on FAME yield

In this study, in-situ transesterification of FME was carried out using the *Hermetia illucens* pre-pupae biomass. The reaction conditions were fixed at 55°C with 5.8:1 of methanol to mass ratio for 248 mins. The effects of the catalytic activity were studied for the carbonization time and temperatures as well as catalyst loading. Fig. 3 illustrates the yield of FAME obtained with various parameters.

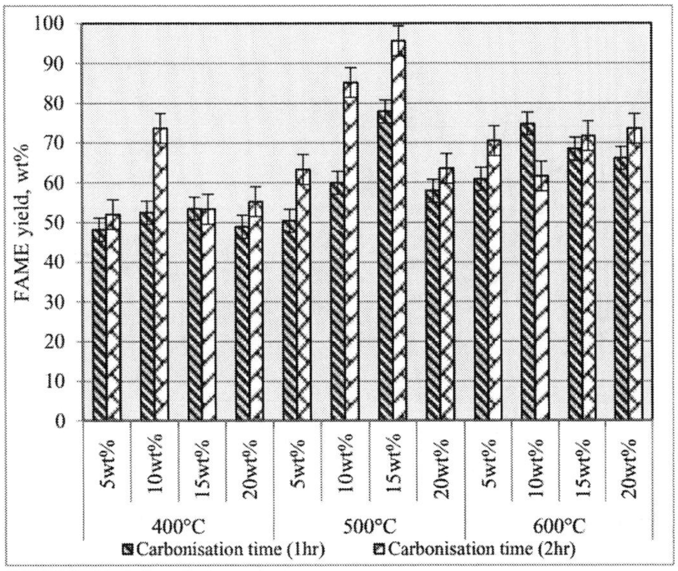

Fig. 3. The yield of FAME obtained with various catalyst loading and carbonisation parameters.

The results obtained indicated that the catalyst loading plays an important role in catalytic performance for FAME production. In general, the yield of FAME increases along with the increase of catalyst loading from 5 to 15 wt%. Further increase of catalyst loading from 15 wt% up to 20 wt% will lead to a decreases in yield. This trend was observed for catalysts carbonised at 1 hour for 400, 500 and 600°C, respectively. Furthermore, catalytic performances of FAME yield were also affected by carbonisation time and temperature. This is evident for carbonization at 400 and 500°C, showing a rising trend in FAME yield from 1 to 2 hour, respectively. However, the catalytic activity decreases with further increased in carbonisation temperature to 600°C. The performances of the catalyst at 600°C did not show much improvement on the yield production (60.84 – 74.82 wt%) with effect of time and catalyst loading. Conversions of FAME at this temperature had negatively influenced the FAME yield. The catalytic activities at this point were drastically decreasing than that of 500°C.

The catalytic performance showed higher yield with longer carbonization duration for 400 and 500°C. Highest catalytic performances were better at 2 h of carbonisation than that of 1h. The highest yield obtained was attained at 95.6 wt% for reaction with SB500_2_15 catalyst with 4.007 mmol/g total -SO_3H acid density (Fig. 4). Increase in yield is attributed to the morphological structure that is better pore development and therefore increases the probabilities of the acid group attachment. The increasing time and temperature during carbonisation have led to the decrease in carbon sheets and destruction of pores structure thus resulting lower FAME yield. Subsequently, lessens SO_3H actives site catalysts attachment on the biochar. The poor performance is evidence from Fig. 4 depicting the low acid site ranging from 2.01 – 2.21 mmol/g. Hence, this explains the poor catalytic performances at 600°C. Moreover, high catalyst loading effect negatively on the yield due to ether formation generated from the excess of acid [18]. Besides that, emulsifications occurred when excess of catalyst loading is applied and in the long run affects the separation between FAME and by product. Hence, this is also another reason leading to lower FAME yield.

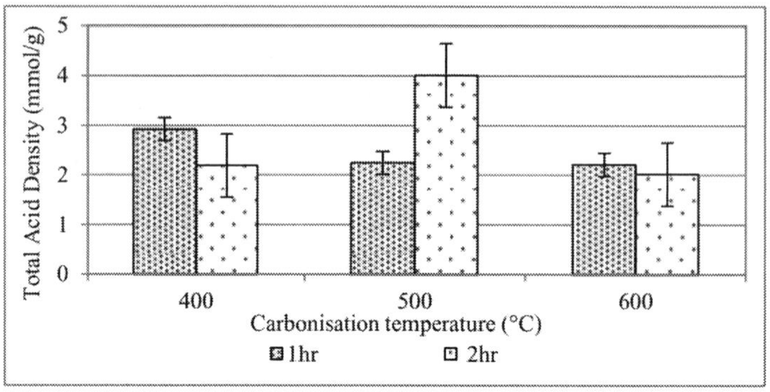

Fig. 4. Total acid density (-SO_3H) of the sulphonated biochar catalyst.

3.2 Analysis of functional group for FAME and biochar catalyst

Fig. 5 compares the characteristics of functional group between the sample and standard FAME. As observed, both FTIR spectra shows the presence of strong stretching carbonyl group (C=O) at 1746.9 cm^{-1} and that confirms the presence of mono alkyl esters. Furthermore, bending vibrations of CH_2 and CH_3 aliphatic groups was detected in both FAME and the standard at absorption band of 1449.36 cm^{-1}. Absorption bands observed at 2921.68 and 2855.45 cm^{-1} were attributed by asymmetric and symmetric vibration of aliphatic –CH_2 fatty acids hydrocarbon chains, respectively. The bands at 721 and 862.65 cm^{-1} showed the CH_2 rocking out-of-plane vibration. The FTIR spectrum in Figure 5 confirms that the identified functional groups in the FAME were also perceived in the standard.

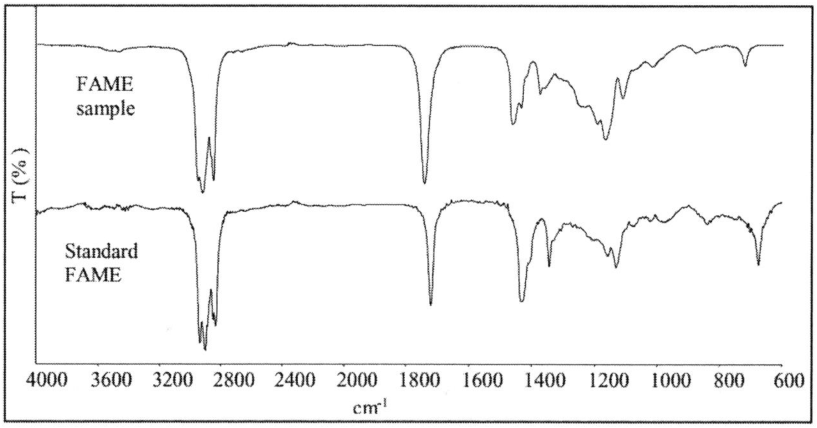

Fig. 5. FTIR spectra of FAME.

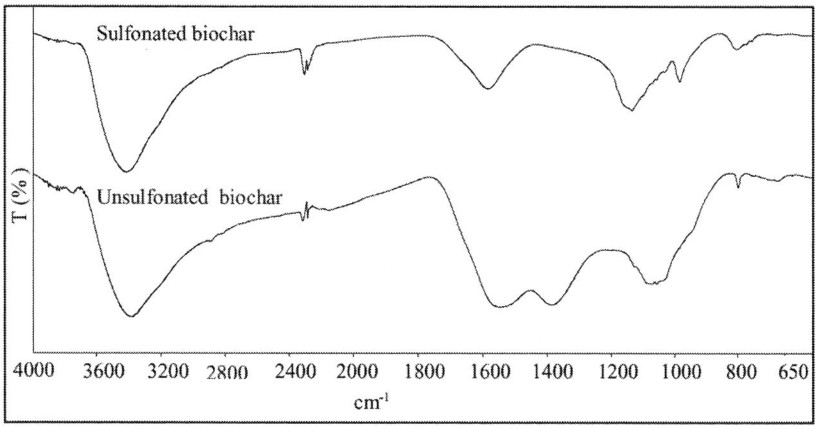

Fig. 6. FTIR for (a) unsulphonated and (b) sulphonated bamboo biochar.

Fig. 6 compares surface functional groups between unsulphonated and sulphonated biochar catalyst. The FTIR spectrum of sulphonated biochar shows a strong C–H bending at 880 cm^{-1} that attributed to 1,2,4-trisubstituted ring of aromatic and heteroatomic compounds. The band 1442 cm^{-1} is attributable to medium stretch of C–C aromatic compounds found in the unsulphonated biochar. This band disappeared after sulphonation indicating cellulose chain continues to convert into polycyclic aromatic hydrocarbon structure during sulphonation. The absorption bands between 780 – 850 cm^{-1} indicates the out of plane bending in the unsulphonated biochar. The band at 1200 and 1055 cm^{-1} specifies a strong S=O stretching of sulphate and sulphoxide functional groups. However, these two bands were not observed in the unsulphonated biochar. This indicates the presence of SO_3H groups, by this means confirming the incorporation of sulphonic groups onto the biochar matrix. Both samples in 1650 – 1566 cm^{-1} were a group of C=C medium stretching vibration peak, which may attribute to cyclic alkene, monosubstituted and disubstituted (cis) alkene [19]. All samples also indicate a broad O–H bending between 3550 – 3200 cm^{-1} indicating the presence of oxygen containing functional groups such as phenolic hydroxyl and carboxylic groups in polycyclic aromatic skeleton.

4 Conclusions

In conclusion, the development of sulphonated biochar catalyst has successfully shown the catalytic ability to produce FAME. Significant effect on the catalytic activity of the sulphonated biochar catalyst was displayed by SB500_2_15. This catalyst showed the highest catalytic performance as evidence from the yield of FAME which attained at 95.6 wt%. However, the catalytic activity did not give rise to FAME yield when the catalyst was carbonised at 600°C. This is due to the decrease in carbon sheets and destruction of pores structure at 600°C. Hence, makes the attachment of sulfonic group more difficult and eventually results in lower FAME yield. The catalyst derived from bamboo in this study has shown a promising application as a solid acid catalyst for biodiesel production.

The authors would like to thank the University of Tunku Abdul Rahman (UTAR) for the laboratory facilities.

References

1. L.J. Konwar, J. Boro, D. Deka, Renew. & Sus.t Energ. Rev. **29**, 546–564 (2014)

2. A.P.S. Chouhan, A.K. Sarma, Renew. & Sust. Energ. Rev. **15**, 4378–4399 (2011)
3. A. Demirbas. Int. J. of Green Energ. **4**, 15–26 (2007)
4. R. Ormsby, J.R. Kastner, J. Miller, Catal. Today **190**, 89–97 (2012).
5. Y. Zhang, M.A. Dube, D.D. McLean, M. Kates, Bioresour. Technol. **89**, 1–16 (2003)
6. A.M. Dehkhoda et al. Appl. Catal. A: General **382**, 197–204 (2010)
7. D.E. López, J. Goodwin, D.A. Bruce, E. Lotero, Appl. Catal. B **295**, 97–105 (2005)
8. Q. Shu, J. Gao, Z. Nawaz, Y. Liao, D. Wang, J. Wang, Appl. Energ. **87**, 2589–2596 (2010)
9. W.Y. Lou, M. H Zong, Z.Q Duan, Bioresour. Technol **99**, 8752–8758 (2008)
10. Y.M. Sani, A.O. Raji, P.A. Alaba, Bioresour. **10**, 3393–3408 (2015)
11. D.R. Vardon, B.R. Moser, W. Zheng, K. Witkin, R.L. Evangelista, ACS Sustainable Chem. Eng. **1**, 1286−1294 (2013)
12. J.R. Kastner, J. Miller, D.P. Geller, J. Locklin, L.H. Keith, T. Johnson, Catal. Today **90**, 122 – 132 (2012)
13. A. Buasri, N. Chaiyut, V. Loryuenyong, C. Rodklum, T. Chaikwan, N. Kumphan, Appl. Sci. **2**, 641 – 653 (2012)
14. F. Pua, Z. Fang, S. Zakaria, F. Guo, C. Chia, Biotech. for Biofuels **4**, 1 – 8 (2011)
15. A.U. Rajapaksha., S.S. Chen., D.C.W. Tsang., M. Zhang., M. Vithanage., S. Mandal., B. Gao., N.S. Bolan., Y.S. Ok. Chemosphere **148**, 276–291 (2016)
16. P. Huang, J. Jhan, Y. Cheng, H. Cheng. The Scientific World Journal; 1 – 8 (2014)
17. R. Thangarajan, N. Bolan, S. Mandal, A. Kunhikrishnan, G. Choppala, et al (2015). *Biochar for inorganic contaminant management in soil* (Chapter 5): *'Biochar: Production, Characterization and Applications'* (Eds. Ok Yong Sik, Uchimiya Minori, Chang Scott and Nanthi Bolan). Pages 99-138; FL, USA: CRC Press, ISBN-1482242303, 9781482242300.
18. J. Emrani and A. Shahbazi, Journal of Biotech. & Biomat. **2**, 124 (2012)
19. S.Y. Ouyang, K. Xiangming, X. Qiong, Y. Dulin, Journal of Mat. Sci. & Chem. Eng. **2**, 4 – 8 (2014)

Study on The Effectiveness of Egg Tray and Coir Fibre as A Sound Absorber

Masiri Kaamin[1*], *Nor Farah Atiqah* Ahmad[1], *Norhayati* Ngadiman[1], *Aslila* Abdul Kadir[1], *Siti Nooraiin* Mohd Razali[1], *Mardiha* Mokhtar[1], and *Suhaila* Sahat[1]

[1]Centre for Diploma Studies, Universiti Tun Hussein Onn Malaysia, Batu Pahat, Johor, Malaysia.

Abstract. Sound or noise pollution has become one major issues to the community especially those who lived in the urban areas. It does affect the activity of human life. This excessive noise is mainly caused by machines, traffic, motor vehicles and also any unwanted sounds that coming from outside and even from the inside of the building. Such as a loud music. Therefore, the installation of sound absorption panel is one way to reduce the noise pollution inside a building. The selected material must be a porous and hollow in order to absorb high frequency sound. This study was conducted to evaluate the potential of egg tray and coir fibre as a sound absorption panel. The coir fibre has a good coefficient value which make it suitable as a sound absorption material and can replace the traditional material; syntactic and wooden material. The combination of pyramid shape of egg tray can provide a large surface for uniform sound reflection. This study was conducted by using a panel with size 1 m x 1 m with a thickness of 6 mm. This panel consist of egg tray layer, coir fibre layer and a fabric as a wrapping for the aesthetic value. Room reverberation test has been carried to find the loss of reverberation time (RT). Result shows that, a reverberation time reading is on low frequency, which is 125 Hz to 1600 Hz. Within these frequencies, this panel can shorten the reverberation time of 5.63s to 3.60s. Hence, from this study, it can be concluded that the selected materials have the potential as a good sound absorption panel. The comparison is made with the previous research that used egg tray and kapok as a sound absorption panel.

1 Introduction

Noise also known as the unwanted sound has been recognized as part of environmental nuisance. A person can face progressive loss of hearing with an increase in the threshold of hearing sensitivity due to the continuous exposed to the noise that is 86-90 dBA over a lifetime in industrial settings [1]. Excessive noise can cause the negative effects such as hearing impair, sleep disturbance, communication interface and health effect [2, 3]. In recent years, the demand for noise barrier has increased due to its importance as part of noise pollution control management.

* Corresponding author: masiri@uthm.edu.my

Along with the development of technology, noise control management has become one of the major requirements to improve the quality of living environment. One of the best methods that can control to reduce these unwanted sounds is by using sounds absorption material. The best practice applied is to install sound absorption panels at desirable area. A sound absorption panel has been installed at certain part in the building to reduce the sound level. However, sound absorption panels that are available in the market nowadays used fiberglass as their primary material [4], which has negative impact not only towards the environment but to the human health as well. Although synthetic fibres sound absorption panels performed well, it is costly and not sustainable [5]. These days, natural fibre materials from agricultural such as coconut, rice husk, coir fibre and oil palm frond fibre have widely used as a substitute to synthetic fibres and wood–based material for acoustic absorption purpose [6]. Hence, researchers are more interested in reviewing the application of wood [7], rice straw [8], rice husk [9], wheat straw [10], and bamboo [11] as absorption panels' materials. These sustainable materials are easy to obtain and safe to be use in the environment. This study was conducted by emphasis the characteristics of coir fibre that potentially be a good sound absorber. The coir fibre, which has porous and hollow space, has the ability to absorb the sounds well.

The orientation of sound-absorbing material plays an important role in the effectiveness as a sound absorber. Effect of geometric orientation in the rate of sound absorption has been investigated based on the triangular orientation, semi-circular, rectangular and plate [12]. Result shows that the triangle or pyramid structure can be a good sound absorber for high frequency. This type of structure can be seen on the surface of egg tray and indirectly become the main factor of material selections for this study. The egg tray structure is capable to absorb and reflect the sound waves [13]. Hence, the objective of this study is to identify the characteristics of sound absorption panel based on the natural and recycling materials, which are egg tray and coir fibre as a sound absorbing panel. This study was focused on the use of environmentally sound absorption panel in the reverberation room with a low frequency such as classroom and lecture hall.

2 Sound Absorption Panels

All construction materials have acoustic criteria where it does absorb, rebound or transmit sound. Generally, acoustic materials are designed to be used for sound absorption. In acoustic architecture and manual prepared by the manufacturers, these commercial acoustic materials are characterized by sound absorption coefficient where the mean value is taken at 250 Hz, 500 Hz, 1000 Hz and 2000 Hz frequency. These values are useful in comparing the overall absorption level of commercial acoustic materials for reducing sound level.

This study focused on effectiveness characteristic of coir fibre as natural fibre sound absorption panels based on its porous composition. According to [2], an excellent sound absorption coefficient at low frequency is when the room is covered with hollow plats. The increasing of the absorption coefficient can be done by increase the surface density of the materials and increase the airflow obstacles.

The material orientation for sound absorption panels also plays an important role to ensure its effectiveness. A study on the effect of the materials orientation has been done by [12] that based on basic orientation such as triangle, hemisphere, rectangular and plate and the result concluded triangle/pyramid orientation gives the best absorption for high level frequency. This pyramid orientation can be seen and easily obtained from egg cartons. As supported by [13], the orientation of egg cartons is capable of dispersing and rebound sound wave.

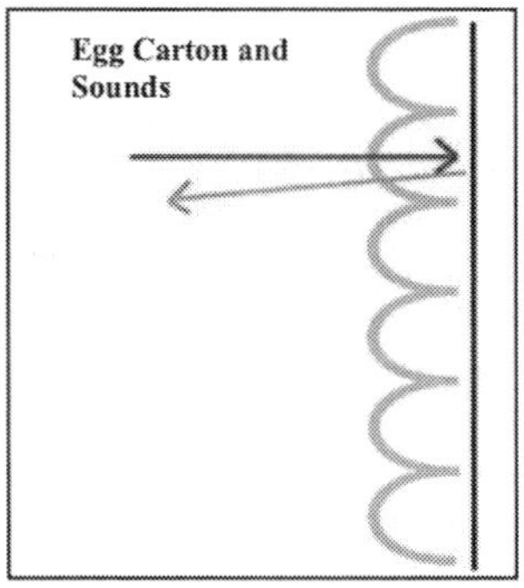

Fig. 1. Sound Wave when hit the egg tray surface [13].

3 Methodology

Fig. 2 shows the main materials used for this study; coir fibre, egg cartons, plywood. The combination of these materials helped in strengthening and improving sound absorption characteristics.

(a) (b) (c)

Fig. 2. Basic Materials for sound absorption panels; (a) Coir fibre, (b) Egg cartons, (c) Plywood.

3.1 Panels Fabrication

An impedance tests were carried out to determine the sound absorption coefficient (SAC) before both design could be made. For the fabrication of the composite egg-carton cores, the plywood was used. The design of the panel is illustrated as in Fig. 3. By using stapler gun, each of egg-carton cores were then filled with coir fibre before it been wrapped by black cloth as a layer of finishing for the panels.

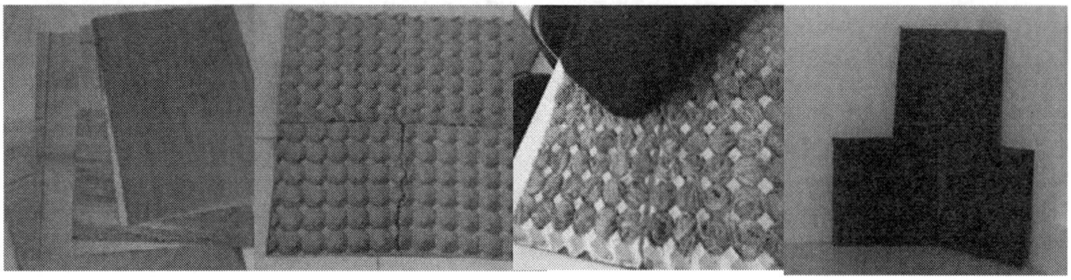

1m x1m plywood	Egg tray been put together	Coir fibre filled the egg tray hollow space	Black cloth as final finishes

Fig. 3. Absorption Panels Design Stage.

Thirteen panels were made for the study and were hanging in a room with dimensions as described in Fig. 4. According to [14], the testing room must be filled at least 15 to 20 percent with the absorption panels. Due to that, 15% of the corresponding testing room were filled with absorption panels. The characteristics of the design panels are listed in Table 1.

Table 1. Details sound absorption panels

Parameter	Unit	Panels
Structure	-	Sound Absorption Panel
Material	-	Egg-carton and coir fibre
Thickness	mm	45.0
Density	kg/m^3	0.42
Area	m^2	0.0036

3.2 Reverberant Field Method

Fig. 4 shows the equipment setup for the tests and layout of the echo chamber located at Acoustic Laboratory, FKAAS, UTHM. This room was designed represented a real situation to investigate maximum effect of noise from various angles. Absorption panels have been tested in echo chamber concedes to the ISO 354 standards. The required parameter that needed is an echo time (seconds) for the empty room and room with the absorption panels. This echo time were collected at two different positions of microphone; Microphone 1 and Microphone 2.

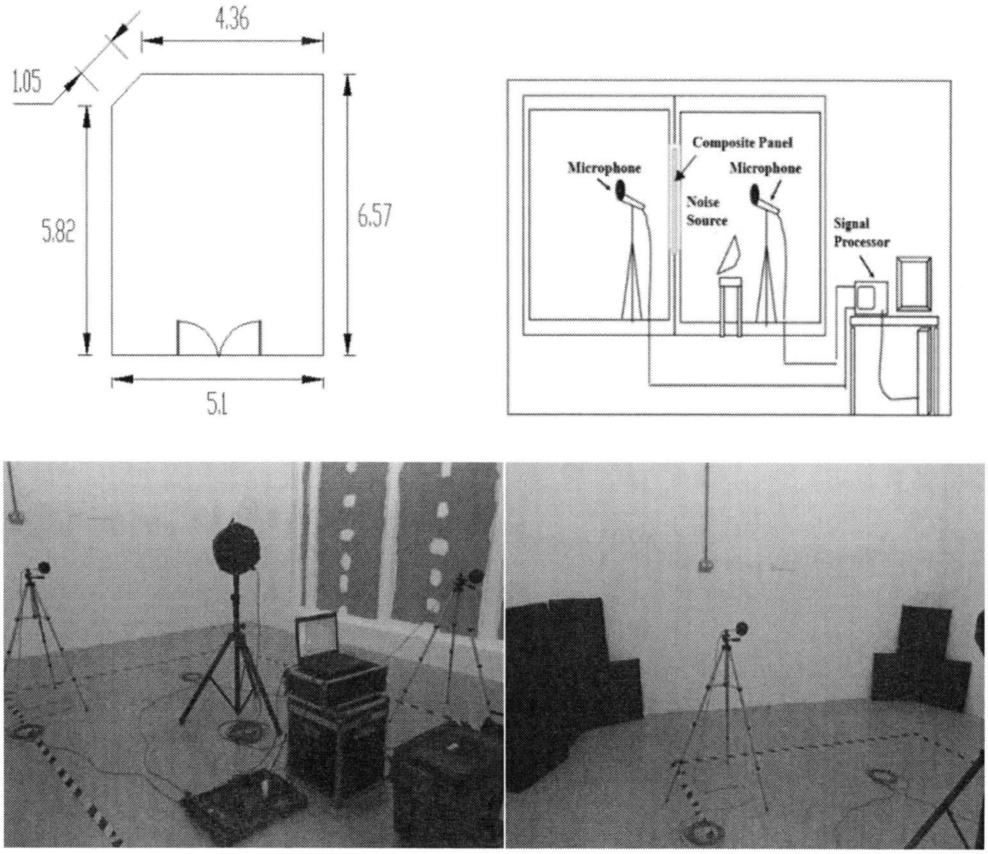

Fig. 4. Equipment setup and layout of the echo chamber.

4 Results and discussions

Echo that produced due to the hitting of the sound waves to any obstacles is different depending on the obstacle material. It can be heard in small space with hard walls such as in a cave. But, if these echoes meet a soft surface, the sound wave will be absorbed. In this study, there is difference of echoes time obtained for the room with and without absorption panels. Fig. 5 and Fig. 6 both shows the results for Microphone 1 and Microphone 2 respectively. The result shows an obvious different of echoes time loss at low frequency between 31.5Hz – 1600Hz compared to high frequency. Generally, the echoes time at both microphones is shorter in a room filled with absorption panel than the empty room. The maximum echoes time loss for the empty room is 5.63s and 5.74s for Microphone 1 and Microphone 2 respectively. While for room filled with absorption panels, the echoes time losses are 3.60s for Microphone 1 and 3.73s for Microphone 2. Due to that matter, it is safe to conclude that this absorption panel that are made from egg-cartons and coir fibre is an excellent absorption panel for low frequency. This can be related with the size of coir fibre itself. As highlighted in [12] study, the fine fibre can increase the coefficient of sound absorption. Since this material used in the original size, this panel is more suitable for low frequency. So, it's proven that agricultural waste is in an absorbing material group [10].

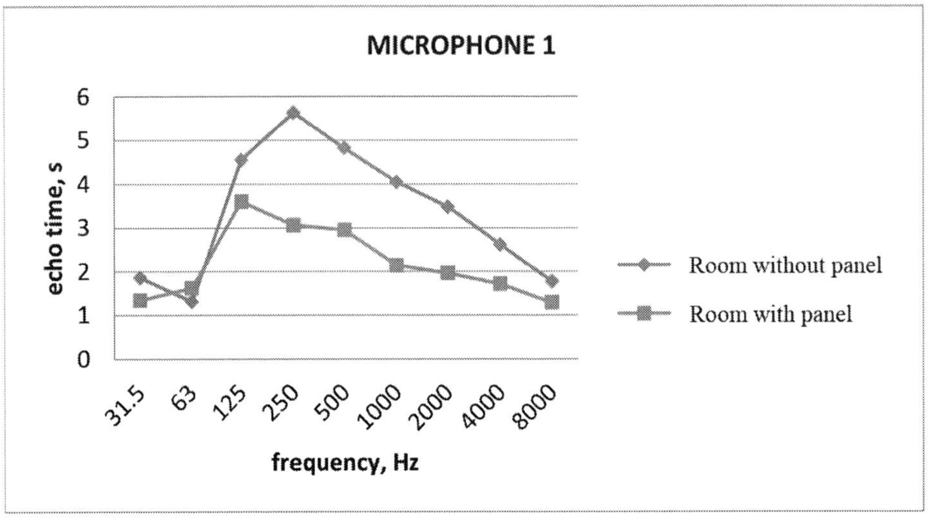

Fig. 5. Echoes time for Microphone 1.

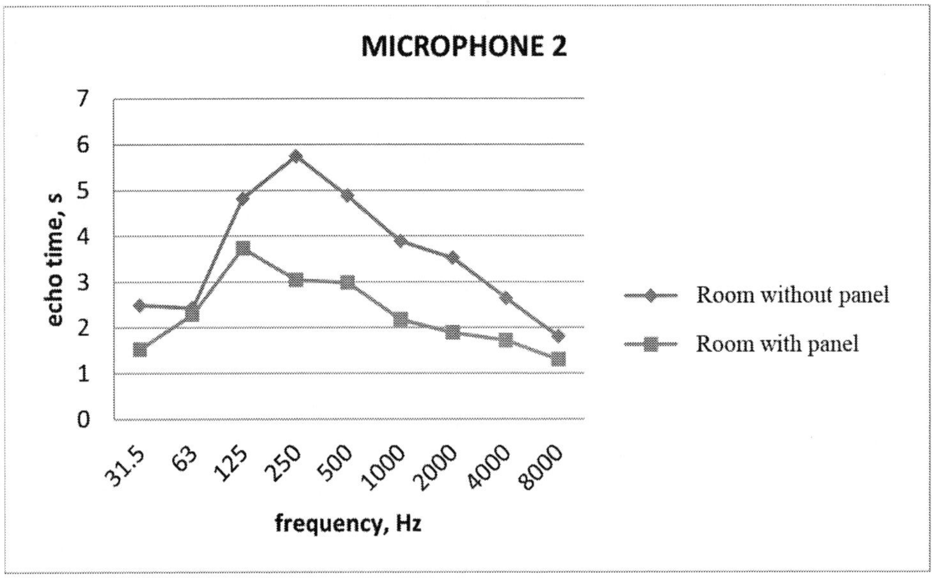

Fig. 6. Echoes time for Microphone 2.

5 Conclusions

There are many other ways in utilizing recycled materials like egg-cartons and coir fibre. Stepping up to the challenge, this study has proven that the useable egg-cartons and coir fibre have the criteria of performing an effective sound absorption pane that is more environmental friendly. For future study, other natural fibers can be used to replace the egg-cartons.

Authors would like to extend his gratitude to all participants who helped directly and indirectly in this study. The authors also thanked Office for Research, Innovation, Commercialization and Consultancy

Management (ORICC), UTHM, Johor for their support and also to Siti Nur Ezianie Mohamad A'Ashri, Suria Fathri Mohd Rethuan and Mai Zatul Amira Mior Ramlee for their efforts.

References

1. S.A. Stansfeld, M. P. Matheson, British M.B **68** (1), 243-257, (2003)
2. R. Zulkifli, Zulkarnain, M. J. M. Nor, American Journal of Applied Sciences **7** (2), 260-264 (2010)
3. N. Singh, S.C. Devar, J. Hum, Ecol. **16** (3), 181-187 (2004)
4. M. Kaamin, N. S. M. Mahir, A. A. Kadir, N. B. Hamid, M. Mokhtar, N. Ngadiman, Sound absorption study on acoustic panel from kapok fibre and egg tray. *AIP Conference Proceedings* **1901**, 130012 *(2017)*
5. A. Nick, U. Becker, W.Thoma, Journal of Polymers and The Environment **10** (3),115-118, (2002)
6. M.J. M. Nor, N. Jamaluddin, F.M Tamiri, Electronic Journal-Technical Acoustics **3** (8), 1-8 (2004)
7. C. Wassilief, *Applied Acoustic* **48**, 339-356, (2003)
8. H. S. Yang, D. J. Kim, H. J. Kim, Bioresource Technol **86**, 117-21 (2003)
9. V. Ajiwe, C.A. Okeke, S.C. Ekwuozor, Uba IC., Bioresource Technol **66**, 41-43 (1998)
10. M. Saadatnia, G. Ebrahimi, M. Tajvidi, *Proceedings of the 17th World Conference on Nondestructive Testing*, 25-28 October 2008, Shanghai, (2008)
11. T. Koizumi, N. Tsujiuchi, A. Adachi, *High Performance Structures and Composite Book* (WTT Press, Southampton, UK., 2002)
12. Y. E. Lee, W. J. Chen, AUTEX Research Journal **3**(2), 78-84 (2003)
13. M. Boughan, Egg Cartons Get Egg on Their Face in Acoustics Test. Para, 3. Retrieved from https://acousticalsolutions.com (2012)
14. F. Simon, J. Pfretzschner, Noise and Vibration Worldwide **35** (1) (2004)

Preparation, Characterization and Adsorption Study of Granular Activated Carbon/Iron oxide composite for the Removal of Boron and Organics from Wastewater

Augustine Chioma Affam[1*], *Chee Chung* Wong[1], *Mohammed A. B.* Seyam[1], *Chelsea Ann Anak* Frederick Matt[1], *Josephine Lantan Anak* Sumbai[1], and *Abdullahi Mohammed* Evuti[2]

[1]Civil Engineering Department, School of Engineering and Technology, University College of Technology Sarawak, 96000 Sibu, Sarawak, Malaysia
[2]Chemical Engineering Department, University of Abuja, P.M.B. 117, Abuja, Nigeria

Abstract. Boron and organics maybe in high concentration during production of oil and gas, fertilizers, glass, and detergents. In addition, boron added to these industrial processes may require to be removed by the wastewater treatment plant. The preparation, characterization and application of iron oxide-activated carbon composite for removal of boron and COD was studied. The one variable at a time (OVAT) method was implemented to obtain desirable operating conditions (adsorbent dosage 5 g/L, reaction time 2 h, agitation speed 100 rpm, pH 5 for COD removal and pH 9 for boron removal). It was found that boron and organics present in a sample wastewater may require to be treated separately to remove the contaminants. The study achieved 97 and 70% for boron and COD removal, respectively. Adsorption as an alternative cheap source of treatment and its practicability for small communities is recommended as effective in removal of contaminants from river water.

1 Introduction

The presence of boron and high organic concentration in wastewater is particularly of interest because of its toxic nature and at the same time beneficial use in some other situations. Its presence in surface water occurs frequently as a consequence of the discharge of treated sewage effluents. Boron and organics are found in nature in the underground aquifers and this basically is dependent on the locality and its environmental geological features. World Health Organization (WHO) has provided a guideline for the boron concentration in drinking water to be less than 0.5 mg/L. In irrigation water, boron below 1 mg/L could be beneficial and above 1 mg/L is termed toxic [1]. In terms of being beneficial, boron available as boric acid could be absorbed for development of chemical and biological structures of plants. Its deficiency may cause a reduction in its growth ability

* Corresponding author: author@ucts.edu.my or affamskii@yahoo.com

and reproduction [2]. Boron can be easily leached and requires to be replaced in plant nutrition and development [3]. Organics maybe in high concentration during production of fertilizers, glass, detergents etc. In addition, boron added to these industrial processes may require to be removed by the wastewater treatment plant. Treatment of this type of industrial wastewater using various methods have been reported [4-10]. However, applying cheap and easy to handle treatment processes may still require more studies. Removal of boron and organics simultaneously may be necessary and has not been reported in literature. In this study, the focus is to examine the contaminants in a sample wastewater and to treat same for removal of boron and organics.

This study addresses two main issues. The first is to characterize the granular activated carbon/iron oxide composite and the wastewater composition, and to eventually treat the wastewater in order to remove boron and organics present in the sample using activated carbon/iron composite adsorbent through adsorption process. The application of activated carbon/iron oxide for sorption of phenol and other pollutants from water has been reported [11]. The availability of micropores and large surface area are characteristics of activated carbon. Modification of the surface usually increases the adsorption capacity and efficiency of the composite material. Iron oxide has been found to perform well when combined with activated carbon as against other oxides e.g. aluminum oxide, titanium oxide [11], and can be reused after each application, thus making it cheaper and easier to handle. This research investigated the performance of the combined activated carbon/iron oxide composite for adsorption of boron and organics contained in a sample wastewater. It is pertinent to obtain desirable operating conditions that will remove the highlighted contaminants in the selected wastewater.

2 Materials and methods

2.1 Chemicals

Iron (III) oxide was purchased from Sigma Aldrich, UK, BoroVer 3 Boron Reagent Powder Pillows and COD regent was obtained from HACH, USA.

2.2 Analytical Methods

Chemical oxygen demand (COD) was determined according to the Standard Method 5220 [12]. Boron concentration was determined according to Carmine Method 8015 [12]. The pH measurements were done using a pH meter SENION MM374 Bench Top Multi Parameter Meter HACH Company, USA. Biodegradability was measured by 5-day biochemical oxygen demand (BOD_5) test according to the Standard Methods [12]. Scanning electron Microscope JSM-6010PLUS/LV, JEOL, USA was used to obtain the morphology of the composite. Brunauer–Emmett–Teller (BET) surface area apparatus (ASAP 2000, Micrometrics, USA) was used for obtaining the adsorbent morphology.

2.3 Wastewater Composition

The wastewater was collected from the Rajang River in Sibu downstream area. The composition of the wastewater is shown in Table 1. The wastewater sample was analysed generally following the Standard Methods for Examination of Wastewater [12].

2.4 Preparation of Granular Activated Carbon/Iron Oxide Composite

The preparation of the adsorbent was done according to the method adopted by Abussaud et al. [13] with a little modification. This is as shown in Fig. 1.

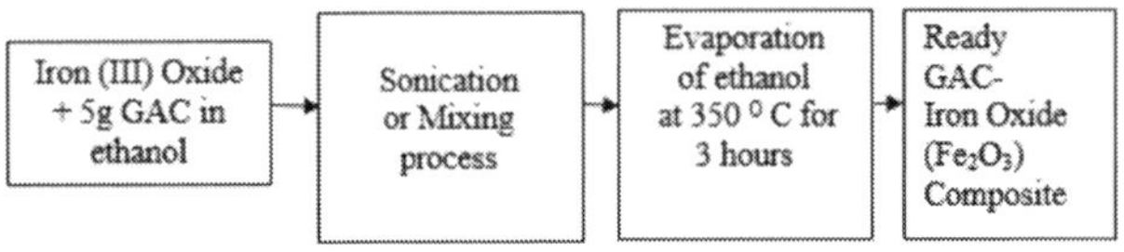

Fig. 1: Procedure for adsorbent preparation.

Table 1. Composition of Wastewater.

Parameter	Concentration
Boron	80 mg/L
pH	7.26
COD	548 mg/L
BOD	9.6 mg/L
UV_{254}	0.653 cm^{-1}
Turbidity	92 FAU/NTU
EC	14.68 μs/cm
TDS	9.84 mg/L
Colour	98 pt-co
TSS (mg/l)	2030

2.5 Experimental procedure

The wastewater collected from Rajang River was sent to the Environmental Laboratory of University College of Technology Sarawak and stored at 4°C until needed. Where sample pH required adjustment, 2 to 3 drops of 1 M H_2SO_4 or NaOH was used. Adsorbent concentration, time, speed of agitation, pH were varied while the wastewater concentration remained same. A 250 ml Erlenmeyer flask containing100 ml of the wastewater was placed on a desk orbital shaker allowing varying time intervals before removal. This was according to the one variable at a time (OVAT) design of experiment. Samples were centrifuged at 5000 rpm using a bench centrifuge for thorough separation of supernatants for analysis. Results were obtained in duplicate analysis for consistency.

3 Results and discussions

3.1 Effect of pH

Understanding the effect of pH is a very vital component in wastewater treatment in order to determine the existing charges on the surfaces of adsorbents [14]. Fig. 2 shows COD removal efficiency at various pH values from 2 to 9. Other operating conditions of the system was 100 rpm, 30 to 120 min and 5 g/L adsorbent dosage. It was observed that the removal efficiency was strongly dependent on the pH of the solution. Based on the obtained result, pH 5 has the highest removal of 54 % for COD. At pH 9, up to 97 % of boron was removed as shown in Fig. 3. It was reported that the removal of boron when optimised occurs around pH 9 – 9.5 [15]. High boron removal of more than 70% was achieved above pH 11 [16].

 https://doi.org/10.1051/e3sconf/20183402006

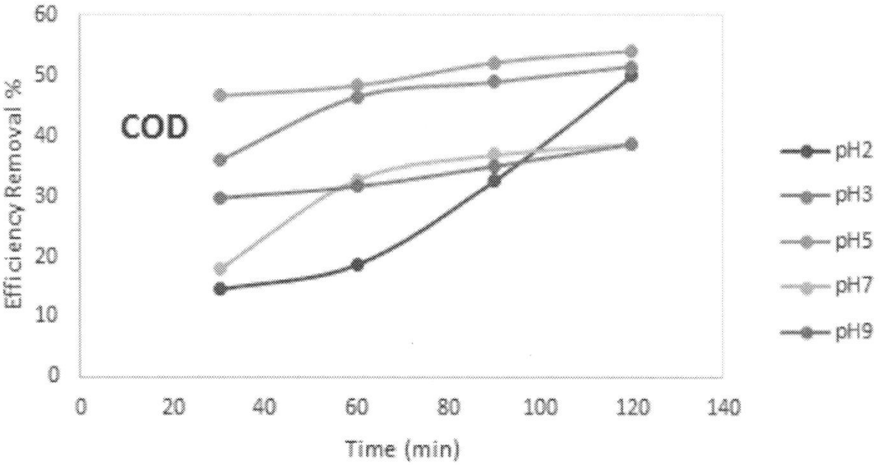

Fig. 2. COD removal efficiency from pH 2 to 9.

3.2 Effect of Agitation Speed

Optimum speed of agitation is essentially needed to maximize the interactions between wastewater sample and adsorption sites of adsorbents in a solution. In this study, the effect of agitation speed (100 to 300 rpm) on adsorption of the contaminants was investigated. The percentage removal of COD and boron under selected operating conditions (30 to 120 min and 5 g/L adsorbent dosage) are shown in Fig. 4 and 5. It was found that COD and Boron removal efficiencies increased with an increase from 100 to 300 rpm. This increase can be explained by the fact that for a relatively increase in speed, the system is completely mixed with probably more energy used. With further increase in agitation speed rate from 100 to 300 rpm, the effect on the adsorption became comparatively negligible; since the system was well mixed under a comparatively high shaking rate [17]. As a consequence, a shaking rate of 100 rpm was selected as desirable for subsequent studies because it was obvious that there was no significant increase or effect on the process. A research reported that, higher mass transfer rate and removal efficiency were obtained when the solution was agitated at higher mechanical speed [18].

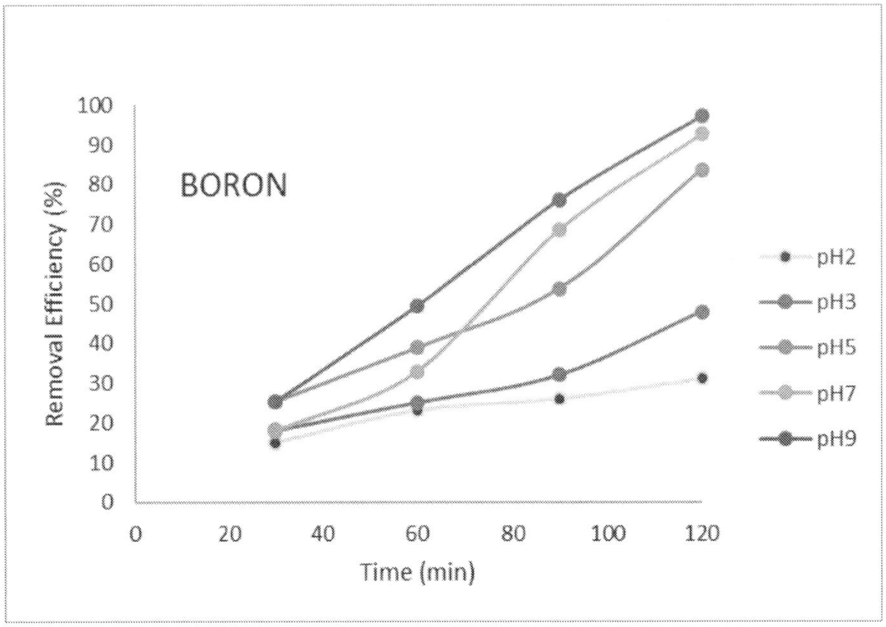

Fig. 3. Boron removal efficiency from pH 2 to 9.

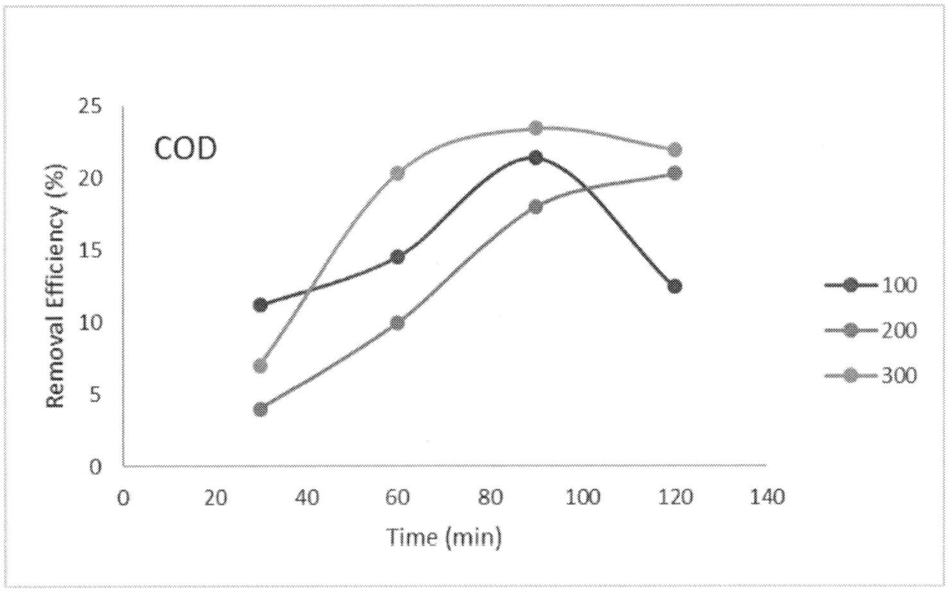

Fig. 4. COD removal efficiency from 100 to 300 rotor speed per minute.

https://doi.org/10.1051/e3sconf/20183402006

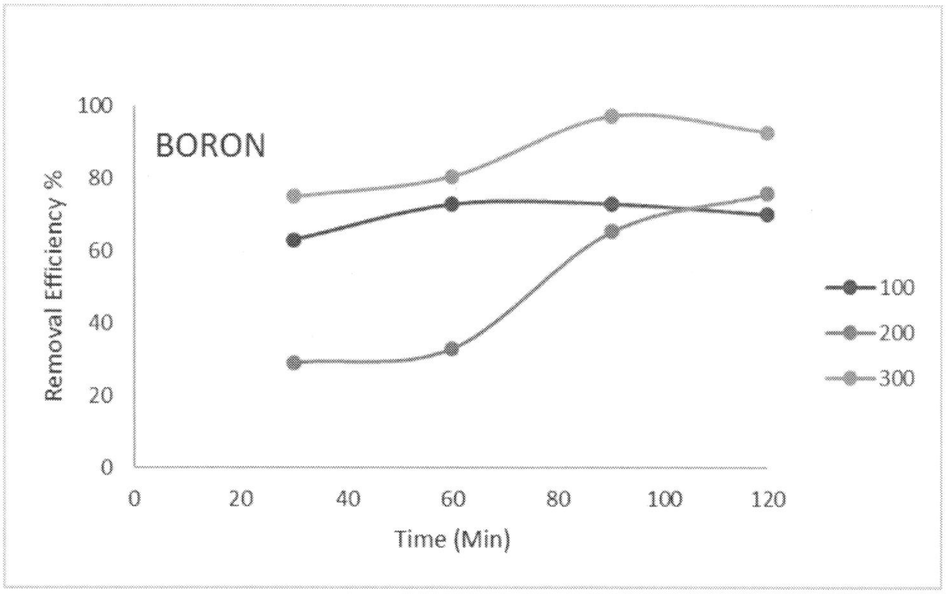

Fig. 5. Boron removal efficiency from 100 to 300 rotor speed per minute.

3.3 Effect of Contact Time

The COD and boron removal were studied at different contact time duration. Fig. 6 shows the effect of time on the removal of organics and boron at different time intervals from 1 to 4 h. Other operating conditions of the system were 100 rpm, 1 to 4 h and 5 g/L adsorbent dosage. The removal efficiency increased from 1 until 2 h for COD and boron, thereafter a drop in performance was observed from the 3 until 4 h. This suggests that the adsorbent can easily and rapidly adsorb contaminants due to high specific surface area at the initial phase and gradually become fully saturated when it reaches the equilibrium phase in a known time interval. From the results obtained from the experiment, contact time of 2 h removed COD and boron of 70 and 97%, respectively (Fig. 6).

3.4 Isotherm Model

In order to analyse the removal of the pollutants in terms of COD and boron, isotherm studies were conducted. Langmuir and Freundlich isotherms represents the equilibrium distribution of metal ions between the solid and liquid phases, defining the correlation with the amount of adsorption and liquid phase concentration. The Langmuir isotherm is based on the assumption that the adsorption will take place only at specific localized sites on the surface and the saturation coverage corresponds to complete occupancy of these sites, and each site can accommodate only one molecule and the surface is energetically homogeneous [19]. The adsorption isotherm for heterogeneous surface is most times described by the Freundlich adsorption isotherm, whose main assumption is that the metal ions can be applied for multilayer sorption [20]. To determine which model describes the removal of pollutants, the data from this experiment was fitted to the models separately.

The Langmuir isotherm can expressed in the equation 1.

$$C_e/q_e = 1/bq_{max} + C_e/q_{max} \quad (1)$$

Where the qmax (mg/g) is the maximum adsorption capacity, the b is the Langmuir constant (L/mg). The q_{max} and b are obtained from the linear regression plot of C_e/q_e Versus C_e.

The Fruendlich is expressed by the equation 2.

$$\ln q_e = \ln K_F + 1/n \ln C_e \quad (2)$$

where K_F and n are constants from Fruendlich equation. K_F represents the capacity of the adsorbent for the adsorbate, and 1/n is the reciprocal of reaction order. A linear regression plot of ln q_e versus ln C_e gives the K_F and n values required [20].

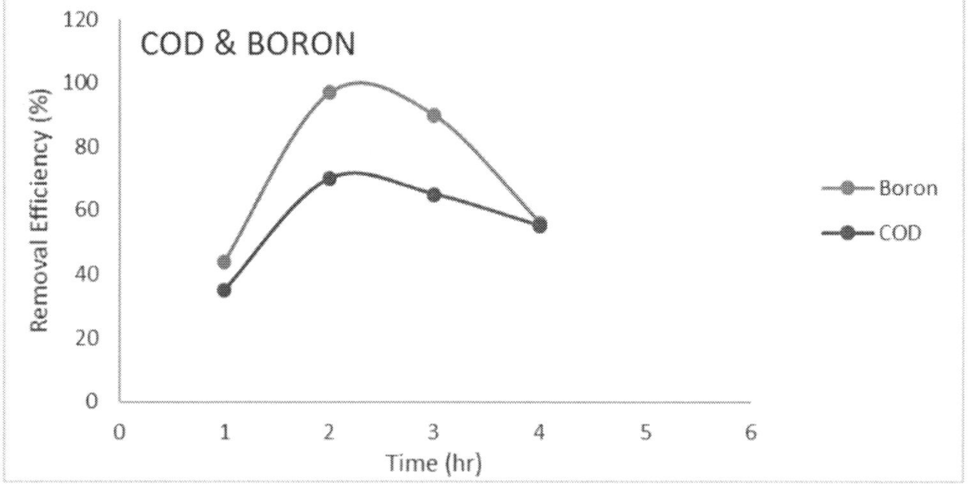

Fig. 6. COD and Boron removal efficiency at 1 to 4 h contact time.

In this study, the obtained isotherm model results show that both Langmuir and (R^2 = 0.9997) and Fruendlich (R^2 = 0.9997) for COD removal and Langmuir (R^2 = 0.8451) and Fruendlich (R^2 = 0.9049) for boron removal. It was observed that in both cases the data fits well for the adsorption of COD and boron onto the GAC/iron oxide, respectively. Thus, it shows that the adsorption process was occurring on both homogeneous and heterogeneous surfaces of the adsorbent composite. Masindi and Gitari [21] reported similar findings in their work.

3.5 Scanning Electron Microscope

The analysis by scanning electron microscope (SEM) of the adsorbent demonstrates that the mechanism of non-diffusive sorption is involved since the iron oxide was adsorbed on the porous matrix of the activated carbon. The SEM obtained the morphology of the adsorbent before and after preparation as seen in Fig. 7 and 8 for commercial GAC and GAC/iron oxide, respectively. The figures show that the activated carbon had an irregular and porous surface indicating relatively high surface areas.

Fig. 7. GAC.

Fig. 8. GAC/Iron oxide.

3.6 Brunauer-Emmett-Teller (BET) Characteristics

The BET analysis gives the GAC/iron oxide composite surface area, which is the sum of micropore area and external surface area. The large pore diameter of composite could permit the pollutants (organics and boron) to diffuse into the mesoporous material pore channels. Thus, pollutants can be adsorbed onto the inner and outer structures of the adsorbent composite. Initial BET surface area for the GAC was 995.56 m^2/g whereas the GAC/iron oxide was 1474.36 m^2/g. The increase in the surface area with the modification using iron oxide was very obvious. In order to improve the adsorbent particle size, it was initially further reduced using a blender when it was supplied. An increase in surface area will correspondingly increase the adsorption capacity of an adsorbent.

4 Conclusions

The study was able to investigate the possibility of simultaneous removal of boron and organics from a sample wastewater collected from a primary source of raw water namely the Rajang river. Under the desirable operating condition obtained, it was found the maximum COD removal efficiency was 70% at pH 5 and 97% for boron removal at pH 9. This implies that both pollutants may require to be treated separately to remove them efficiently. In addition, when pH is in the acidic region it will be more favourable for the removal of COD compared to removing boron. However, the study shows that granular activated carbon/iron oxide composite can be used for removal of low concentration of boron and organics that maybe in wastewater.

The authors are grateful to the management of University College of Technology Sarawak (UCTS) for providing the internal grant No. UCTS/RESEARCH/1/2016/10> (01) to support this work and also for the provision of facilities to conduct this research.

References

1. R.O. Nable, G.S. Banuelos, J.G. Paull, Plant and Soil **193**, 181 (1997)
2. A.E. Yilmaz, R. Boncukcuoglu, M.M. Kocakerim, Desalination **230**, 288 (2008)
3. M. Tarig, C.J.B. Mott, Asian J. Plant Sci. **5**, 963 (2006)

4. K.L. Benko, J.E. Drewes, Environ. Eng. Sci. **25**, 239 (2008)
5. D.L. Shaffer, L.H. Arias Chavez, M. Ben-Sasson, S. Romero-Vargas Castrillón, N.Y. Yip, M. Elimelech, Environ. Sci. Technol. **47**, 9569 (2013)
6. F.R. Ahmadun, A. Pendashteh, L.C. Abdullah, D.R.A. Biak, S.S. Madaeni, Z.Z. Abidin, J. Hazard. Mater. **170**, 530 (2009)
7. J.Q. Jiang, Y. Xu, K. Quill, J. Simon, K. Shettle, Environ. Chem. **3**, 350 (2006)
8. G. Sayiner, F. Kandemirli, A. Dimoglo, Desalination **230**, 205 (2008)
9. M.H. Isa, E.H. Ezechi, Z. Ahmed, S.F. Magram, S.R.M. Kutty, Wat. Res. **51**, 113 (2013)
10. N. Esmaeilirad, K. Carlson, P.O. Ozbek, J. Hazard. Mater. **283**, 721 (2015)
11. B. Abussaud, H. A. Asmaly, Ihsanullah, T. A. Saleh, V. K. Gupta, T. laoui, M. A. Atieh. J. Mol. Liq. **213**, 351 (2016)
12. APHA, AWWA, WPCF, Washington, DC, USA (2005)
13. H.A. Asmaly, Ihsanullah, B. Abussaud, T.A. Saleh, T. Laoui, V.K. Gupta, M.A. Atieh, Desalin. Water Treat. **57** (2015)
14. M. Ahmaruzzaman, V. K. Gupta, Ind. Eng. Chem. Res. **50**, 13589 (2011)
15. A. E. Yilmaz, R. Boncukcuoğlu, M. M. Kocakerm, E. Kocadağistan, Desalination, **230**, 288 (2008)
16. L. A. Richards, M. Vuachère, A. I. Schäfer, Desalination, **261**, 331 (2010)
17. M. V. Karkare, Z. V. P. Murthy, J. Ind. Eng. Chem.**18**, 1301 (2012)
18. M. Shams, M. H. Dehghani, R. Nabizadeh, A. Mesdaghinia, M. Alimohammadi, A. A. Najafpoor, J. Mol. Liq. **224**,151 (2016)
19. R. Sime, *The Langmuir Adsorption Isotherm*, **7** (2000)
20. A. Dada, A. Olalekan, A. Olatunya, O. Dada, IOSR J. Appl. Chem. **3**, 38 (2012)
21. V. Masindi, M. W. Gitari, 2017. Removal of boron from aqueous system using magnesite. J. Wat. Reuse Desalin. **7**, 205 (2017)

Heavy metal recovery from electric arc furnace steel slag by using hydrochloric acid leaching

Lim Jin Wei[1, 2*] , *Ong* Teng Haan[1], *Thomas* Choong Shean Yaw[2], *Luqman* Chuah Abdullah[2], *Mus'ab* Abdul Razak[2], *Tezara* Cionita[1], and *Abdolreza* Toudehdehghan[1]

[1]Department of Mechanical Engineering, INTI International University, Jalan BBN 12/1, Bandar Baru Nilai, 71800 Nilai, Negeri Sembilan
[2]Department of Chemical and Environmental Engineering, Universiti Putra Malaysia, 43400 Selangor, Malaysia

Abstract. Electric Arc Furnace steel slag (EAFS) is the waste produced in steelmaking industry. Environmental problem such as pollution will occur when dumping the steel slag waste into the landfill. These steel slags have properties that are suitable for various applications such as water treatment and wastewater. The objective of this study is to develop efficient and economical chlorination route for EAFS extraction by using leaching process. Various parameters such as concentration of hydrochloric acid, particle size of steel slag, reaction time and reaction temperature are investigated to determine the optimum conditions. As a result, the dissolution rate can be determined by changing the parameters, such as concentration of hydrochloric acid, particle size of steel slag, reaction time and reaction temperature. The optimum conditions for dissolution rates for the leaching process is at 3.0 M hydrochloric acid, particle size of 1.18 mm, reaction time of 2.5 hour and the temperature of 90°C.

1 Introduction

Steelmaking is the process of manufacturing steel by using iron ore and other elements. Two commonly used steelmaking process are basic oxygen and electric arc furnaces (EAF). BOF uses iron ore from the mining as the raw material for the process while the main raw material for EAF is steel scrap [1]. In Malaysia, EAF is widely used in steel mills. EAF slags are formed from the manufacturing of molten steel [2]. Ferric Chlorides could be extracted from EAF steel slags that are used to remove phosphate from wastewater which cause the eutrophication. When excessing phosphate enters the water, it can cause the microorganisms to grow faster because phosphate is a type of nutrients. With the increase in microorganisms such as algae, the oxygen composition in the water will decrease and therefore killing fish and reducing water quality [3].

$$Fe^{3+} + H_nPO_4^{3-n} \leftrightarrow FePO_4 + nH^+ \quad (1)$$

Another uses of $FeCl_3$ are to absorb moisture from the air when in crystalline form and it is usually supplied as a solution. $FeCl_3$ was sold economically in water treatment

* Corresponding author: ljwei84@yahoo.com

industries. It is used as a coagulant for the turbidity removal and the removal of organic contaminants in drinking water. $FeCl_3$ as a drinking water coagulant is inexpensive if maintain at a pH values around neutral pH [4]. Research shows that the presence of Ferric oxides, 33.34% have the highest weight composition in EAF steel slag followed by Calcium oxides, 27.22% and silicon oxides, 20.80% by using X-ray fluorescence (XRF) technique [5]. One of the effective ways to extract iron from EAF slag is leaching process by using hydrochloric acid.

$$Fe_2O_3 + 6HCl \rightarrow 2FeCl_3 + 3H_2O \tag{2}$$

2 Materials and methods

2.1 Materials

EAF steel slags are collected from Southern Steel Berhad, which is a steel manufacturing company in Malaysia. The samples were crushed and sieved into different size (0.3 mm, 0.5 mm, 0.6 mm, 1.0 mm and 1.18 mm). Sieved EAFS were washed with distilled water to remove impurities. The, EAFS were dried in oven at 110°C for 24 hrs and stored in heavy duty plastic bags until used [6]. For the chemical reagents preparation, distilled water was used to dilute and prepare all the hydrochloric acid solutions.

2.2 Methods

2.2.1 Steel slag leaching with using Hydrochloric Acid

In order to determine the effect of Hydrochloric acid concentration on steel slag leaching, each sample of weight 1.0 gram is mixed with hydrochloric acid with the concentrations of 0.5 M, 1.0 M, 2.0 M, 3.0 M and 4.0 M for 24 hours. About different sample particle sizes, the sizes of the particles of 0.3 mm, 0.5 mm, 0.6 mm, 1.0 mm and 1.18 mm in 1 gram were tested using hydrochloric acid concentration of 4.0 M for 24 hours. Particle size of 1.0 mm in 1 gram is tested using hydrochloric acid concentration of 4.0 M with different heating temperatures of 60°C, 70°C, 80°C and 90°C for 1 hour each. Particle size of 1.0 mm in 1 gram was tested using hydrochloric acid concentration of 4.0 M with different reaction time of 0.5 hour, 1.0 hour, 1.5 hour, 2.0 hour and 2.5 hour. Therefore, all the sample solutions were placed on the orbital shakers for 2 hours to ensure the thorough mixture. Then, filter paper is used to filtrate the residue from each sample. Therefore, the residue is dried in oven for 60°C in 2 hours. The percentage of weight loss of EAF steel slag is measured and calculated. After that, dichromatic titration process was conducted to quantify the weight of ion Fe^{3+} presence after the leaching process.

2.2.2 Titration process

Potassium dichromate is chosen for the oxidation of Fe^{2+} ion in the titration process. The samples were heated until boil and then $SnCl_2$ solution is added into each of the sample until the yellow color disappears. After that, the solution is cooled and 10 ml of $HgCl_2$ was added immediately. A white precipitate should appear. Furthermore, 5 ml of concentrated H_2SO_4 and 7 ml of H_3PO_4 was added to pre-reduced the solution. The solution is then diluted with distilled water and 8 drops of diphenylamine sulfonate indicator was added. After that, the solution is titrated with $K_2Cr_2O_7$ solution until it turns from a color of blue-green to violet. The volume of the titrated $K_2Cr_2O_7$ solution was recorded. The experiment is repeated two times to get the average value.

$$2FeCl_3 + SnCl_2 \rightarrow 2FeCl_2 + SnCl_4 \quad (3)$$

$$SnCl_2 + 2HgCl_2 \rightarrow SnCl_4 + Hg_2Cl_2 \quad (4)$$

$$Cr_2O_7^{2-} + 14H^+ + 6e^- \rightarrow 2Cr^{3+} + 7H_2O \quad (5)$$

$$Fe^{2+} \rightarrow Fe^{3+} + e^- \quad (6)$$

$$Cr_2O_7^{2-} + 6Fe^{2+} + 14H^+ \rightarrow 2Cr^{3+} + 6Fe^{3+} + 7H_2 \quad (7)$$

$$\mathbf{Fe^{3+} = Total\ Fe - Fe^{2+}} \quad (8)$$

3 Results and discussions

3.1 Material Characterization

3.1.1 Scanning Electron Microscopy (with Energy Dispersive X-ray Spectroscopy)

SEM is the sample scanning method that using focused beam of electron with high amount of energy in order to observe the topographical, morphological and compositional information of material in solid form [7]. A test on sample of EAF slag was conducted by using a scanning electron microscope (model S -3400N, HITACHI, Japan) to collect the morphology, crystalline structure and orientation of based material creating the EAF slag sample at an accelerating voltage of 20.0 kV, beam current of 50 mA and working distance of 9.0 mm – 10.0 mm with utilization of backscattered electrons (BSE) set objectively with aperture of 3 [7]. Images at low magnification (300×) and high magnification (1000×) were captured and the micrographics of the non-reacted EAF slag are presented in Fig. 1. Areas with lighter color contain a higher proportion of Fe, Mg, Mn and Al while the shady areas contain a higher proportion of Ca and Si [8].

EAF steel slag of 2 samples was analysed by using energy-dispersive x-ray spectroscopy. The primary elements found in the sample A included 29.93% CaO_2, 22.01% Fe_2O_3 and 21.41% SiO_2. On the other hand, sample B included 23.29% Fe_2O_3, 22.76% SiO_2 and 21.1% CaO_2. It is concluded that the main element in EAF steel slag is Ca, Fe and Si.

3.1.2 X-Ray Diffraction

The studies on mineralogical composition on steel slag were conducted by several researchers. Many overlapping peaks appear in XRD patterns indicates the presence of crystalline phases in the steel slag [9]. The more crystalline structured compound has its XRD peaks more easily to be identified [10]. The common mineral phases present in steel slag include two- and three-component compositions of the type CaO-SiO_2, CaO-Fe_2O_3, CaO-MgO-SiO_2, MgO-FeO-SiO_2 and CaO-Al_2O_3-SiO_2 [9,11], and the most highly represented minerals in slag are dicalcium and tricalcium silicates, while different aluminates and silicates are likely to appear as well [12].

https://doi.org/10.1051/e3sconf/20183402007

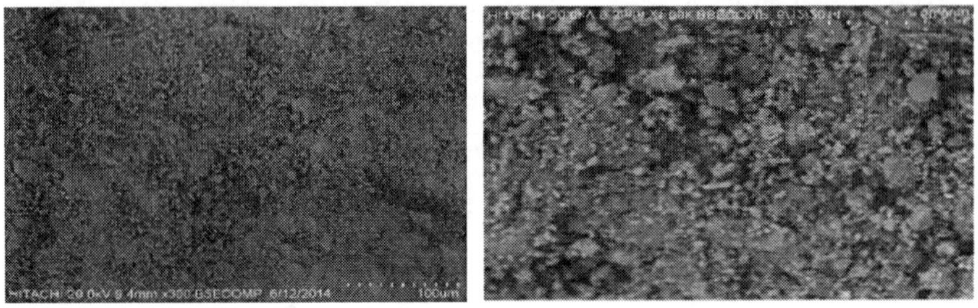

(a) (b)

Fig 1. SEM image on unreacted EAFS at magnification of (a) 300x and (b) 1000x.

The XRD analysis of unreacted EAFS is shown in Fig. 2. The diffraction pattern is heterogeneous, consisting of a mixture of crystalline phases [13]. The analysis identifies calcium oxide (CaO), silicon oxide (SiO_2), hematite (Fe_2O_3), manganese (III) oxide (Mn_2O_3), magnesium oxide (MgO), aluminium oxide (Al_2O_3), phosphorus oxide (P_2O_5), metallic iron (Fe), srebrodolskite ($Ca_2Fe_2O_5$), dicalcium silicate (Ca_2SiO_4) and pyroxene ($CaAl_2SiO_6$).

3.2 Effect of hydrochloric acid concentration

The effects of HCl concentrations on the dissolution EAF slag were studied at varying concentration of HCl. It is observed that the dissolution of EAF slag increased with increasing concentration of HCl at constant temperature and time. When the concentration of HCl increased from 0.5 M to 4.0 M, the dissolution increased 17.06 %. The optimum concentration of HCl for the leaching process is found to be 3.0 M which 77.41 % was leached. The dissolution rate of EAF slag can be calculated by equation (9).

$$\alpha\,(0.5\text{ M}) = \frac{\Delta mass\ of\ EAFS\ (gram)}{initial\ mass\ of\ EAFS\ (gram)} \qquad (9)$$

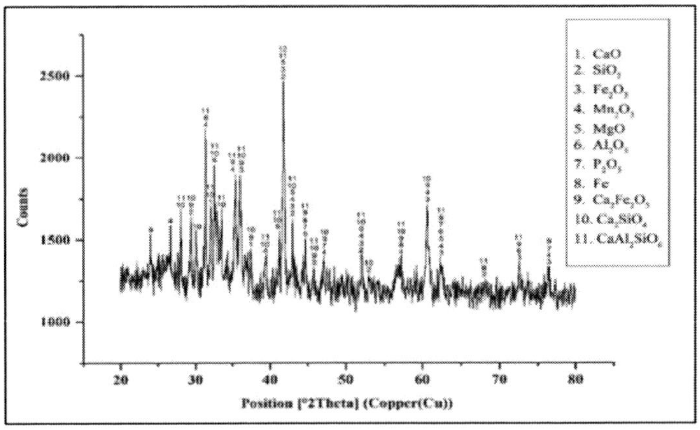

Fig. 2. XRD pattern on unreacted EAFS.

Table 1. Dissolution percentage and weight of Fe^{3+} on varying HCl concentrations.

Concentration of HCl (M)	Dissolution (%)	Weight of Fe^{3+} (gram)
0.5	60.35	0.14559
1.0	64.58	0.16886
2.0	69.74	0.15975
3.0	77.41	0.20674
4.0	77.22	0.18837

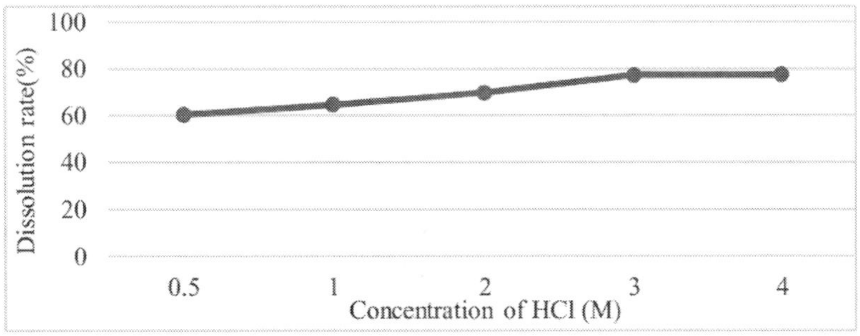

Fig. 3. Dissolution rate against concentration of HCl.

3.3 Effect of particle sizes

The effect of particle sizes was studied at varying particle sizes. It is observed that the dissolution of EAF slag increases with increasing particle size at constant HCl concentration, which is 3M. When the particle size increased from 0.3 mm to 1.18 mm, the dissolution increased 11.87%. The optimum particle size for the leaching process is found to be 1.18 mm which 75.12 % was leached.

Table 2. Dissolution percentage and weight of Fe^{3+} on varying particle size.

Particle size (mm)	Dissolution (%)	Weight of Fe^{3+} (gram)
0.3	63.25	0.13213
0.5	64.01	0.13255
0.6	66.43	0.12091
1.0	74.98	0.13684
1.18	75.12	0.14045

 https://doi.org/10.1051/e3sconf/20183402007

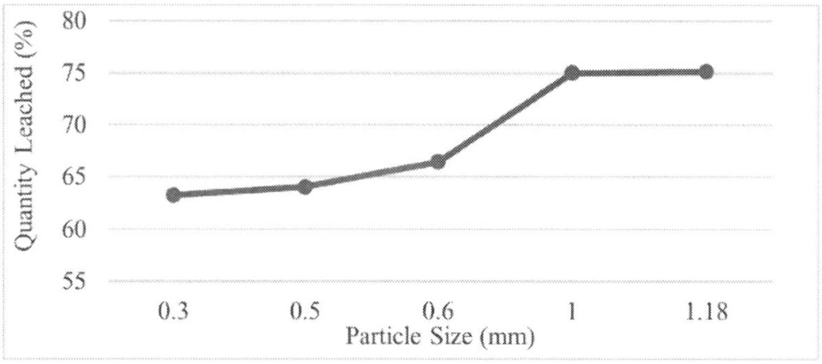

Fig. 4. Percentage of steel slag leached against particle sizes.

3.4 Effect of reaction time

The effect of reaction time was studied at varying reaction time. It is observed that the dissolution of EAF slag increases with increasing reaction time at constant HCl concentration, which is 3M. When the reaction time increased from 0.5 hour to 2.5 hour, the dissolution increased 22.95%. The optimum reaction time for the leaching process is found to be 2.5 hour which 55.45 % was leached.

Table 3. Dissolution percentage and weight of Fe^{3+} on varying reaction time.

Reaction time (hour)	Dissolution (%)	Weight of Fe^{3+} (gram)
0.5	32.50	0.05726
1.0	42.35	0.0704
1.5	43.48	0.0963
2.0	51.02	0.0698
2.5	55.45	0.0730

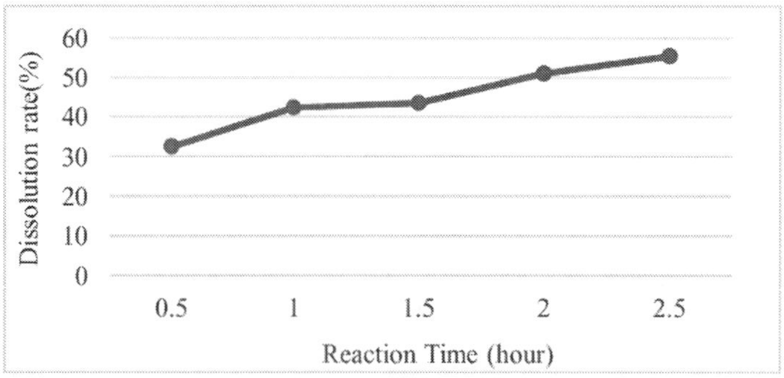

Fig. 5. Dissolution rate against reaction time.

3.5 Effect of reaction temperature

The effect of reaction temperature was studied at varying reaction temperature. It is observed that the dissolution of EAF slag increases with increasing reaction temperature at constant HCl concentration, which is 3M. When the reaction temperature increased from

60°C to 90°C, the dissolution increased 20.84%. The optimum reaction temperature for the leaching process is found to be 90°C which 42.34 % was leached.

4 Conclusions

By using material characterization methods, it shows that EAF steel slag containing a suitable amount of Fe ions. It is proven that the leaching of EAF steel slag contains a suitable amount of $FeCl_3$ for further usage such as wastewater treatment. In this experiment, different variables that affect the concentration of Fe^{3+} are chosen to be the focus when conducting the EAF slag leaching process. The highest dissolution percentage of the leaching process are 77.14%, 75.12%, 55.45% and 42.34% with HCl concentration, particle size, reaction time and reaction temperature respectively. The highest weights of Fe3+ leached are 0.20674 gram, 0.14045 gram, 0.073 gram and 0.07141 gram for HCl concentration, particle size, reaction time and reaction temperature respectively. It can be observed that the optimum conditions for the leaching process is at 4.0 M HCl, particle size of 1.18 mm, reaction time of 2.5 hour and reaction temperature of 90°C.

Table 4: Dissolution percentage and weight of Fe^{3+}on varying reaction temperature.

Reaction temperature (°C)	Dissolution (%)	Weight of Fe^{3+} (gram)
60	21.50	0.0243
70	33.48	0.02841
80	37.70	0.05714
90	42.34	0.07141

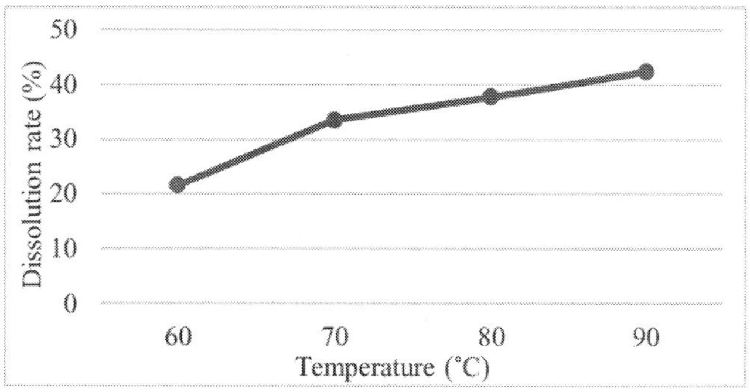

Fig 6. Dissolution rate against reaction temperature.

The authors would like to gratefully acknowledge Ministry of Higher Education (MOHE), and Universiti Putra Malaysia (UPM) for the financial support of this work.

References

1. Neil, S.D. *Thermal Remediation of stainless steel Electrical Arc Furnace (EAF) Dust*, Master Thesis, McGill University (1999).
2. Pao Ter Teo, Anasyida Abu Seman, Projjal Basu and Nurulakmal Mohd Sharif, Procedia Chemistry **19**, 842-846 (2016)
3. Raymond DeVries IV, *Ferric Chloride in Wastewater Treatment*, Carthage College (2011).

4. Peter Leopold and Sue D Freese, Water Research Commision (2009).
5. Huang Yi, Guoping Xu, Huigao Cheng, Junshi Wang, Yinfeng Wan and Hui Chen Procedia Environmental Sciences **16**, 791-801 (2012)
6. P. D. E. Worch, *Adsorption Technology in Water Treatment: Fundamentals, Processes, and Modeling, Dresden: Walter de Gruyter GmbH & Co.* (2012)
7. J. W. Lim, L. H. Chew, T. S. Y. Choong, C. Tezara, M. H. Yazdi, IOP Conference Series: Earth and Environmental Science **36**(1), 012067 (2016)
8. Bankole L., Rezan S. and Sharif N. *IOP Conf. Series: Earth Environ. Sci.* **19**, 012004 (2014)
9. Yildirim, I. and Prezzi, M. Advances In Civil Engineering, 1-13 (2011)
10. Nwaubani, S. and Muntasser, T. Journal Of Civil Engineering And Construction Technology, **3**(11), 291-301 (2012)
11. Rađenović, A., Malina, J. and Sofilić, T. Advances In Materials Science And Engineering, 1-6. (2013)
12. Selih, J., Ducman, V., Mladenovič, a., Sever, S.A., Pavšič, P. Makarovič, M., Legat, A. Mater.Techno. Ljubljana, **38**, 79-86. (2004)
13. Lim, J.W., Chew, L.H., Choong, T.S., Tezara, C. and Yazdi, M.H., *MATEC Web of Conferences* **74**, 00026 (2016)

Generalization of Water Pricing Model in Agriculture and Domestic Groundwater for Water Sustainability and Conservation

Tan Kim Hek[1], *Mohammad* Fadzli Ramli[1*], *Iryanto*[2], *Siti* Rohana Goh[1], and *Mohd* Faiz M. Zaki[3]

[1]Institut Matematik Kejuruteraan, Universiti Malaysia Perlis, Kampus Pauh Putra, 02600 Arau, Perlis, Malaysia
[2]Departement Matematika, Universitas Sumatera Utara, Jln. Perpustakan, No. 1, Medan, Indonesia
[3]Pusat Pengajian Kejuruteraan Alam Sekitar, Kompleks Pengajian Jejawi, Universiti Malaysia Perlis, 02600 Jejawi, Perlis, Malaysia

Abstract. The water requirement greatly increased due to population growth, increased agricultural areas and industrial development, thus causing high water demand. The complex problems facing by country is water pricing is not designed optimally as a staple of human needs and on the other hand also cannot guarantee the maintenance and distribution of water effectively. The cheap water pricing caused increase of water use and unmanageable water resource. Therefore, the more optimal water pricing as an effective control of water policy is needed for the sake of ensuring water resources conservation and sustainability. This paper presents the review on problems, issues and mathematical modelling of water pricing based on agriculture and domestic groundwater for water sustainability and conservation.

1 Introduction: Intrinsic of water value for the human needs

Access to water is a fundamental right for every person without discriminate. It makes the water is an intrinsic value that should be considered more than the value for the social, economic, religious and environmental [1].

As a fundamental right, the water needs must be fulfilling ± 500 liters per person per day, which is essential needs for a standard living in life. The higher standard living of life, then the greater usage of water. For example, in the city of Chicago and Los Angeles each required 800 and 640 liter of water respectively. In Paris, it requires 480 liter and Tokyo requires 530 liter per person per day. Meanwhile in poor countries such as Somalia only 9 liters acquires per person per day [2-3]. Whereas in developing countries like Indonesia, in a large city requires 200-400 liters per person per day while in rural areas only 60 liter per person per day covering 30 liter for bathing, 15 liter for drink (including cooking and beverages) and 15 liter for other purposes [4].

* Corresponding author: mfadzli@unimap.edu.my

According to [5], only in 18 districts are citizens enjoyed water services out of the 35 districts in Subang district of West Java. One of the obstacles faced by rural communities is poverty thus this people have limited access to fulfill the needs of clean water that is safe and feasible. Poverty and participatory type of project is a significant factor affecting the condition of water supply system [6]. It require the help of the government or donor agencies to build water projects. Providing clean water by utilizing ground water with two bore holes each with a capacity of 5 liters/sec. Water pumped from wells drilled into the two towers of water (volume respectively 4 and 18 m^3). From water tower, then it is supplied to houses. However, this method is ineffective because it requires a large cost.

In Surabaya, Indonesia, water problem is serious in terms of both quantity and quality. In 2025, it is predicted a deficit of water for a population of 3.04 million people and water need to reach 47.05 m^3 per second while water availability is only 39.62 m^3 per second. Therefore, it is water deficit of 7.43 m^3 per second. This happens because water pollution is severe in Surabaya as a supplier of drinking water for the channeling of water increasing the cost of prevention and the use of chemicals treatment is high. As a result, 85% of people/households seek other alternatives to buying drinking water and household behavior bring economic consequences in the form of additional expenditure [7].

In Java Island, availability of water per capita per year is only 1750 m^3, which is below water sufficiency standard of 2,000 m^3 per capita per year. Conditions are expected to worsen in 2020 where availability only 1200 m^3 per capita per year [8]. This is because the population is widespread and limited ability of local communities to obtain financial support in financing the investments needed to improve service facilities and infrastructures, as well as the investment of non-physical consisting of management, technical and human resource development in water management.

2 Optimal water price reflecting sustainability

Water occupies a very important position and strategic location in the life of the community and the nation [9]. Therefore the water should be available at a particular location and the right time and should be retrieved easily. For the protection of everyone's access to water, the water right becomes the highest priority in human rights. Therefore, the state must respect, protect and fulfill the needs for public survival. However, the government/state often ignores the rights evidenced by the many areas such in Medan is still a shortage of clean water availability and the price of water is relative expensive [10].

According to [9], in order to obtain optimal water price, it must consider the total water comes into the storage container (aquifer) and the total water that comes out because the total water affects estimates of evaprotranspiration. The amount of water that cans entry to the storage container is:

$$INFLOW = Q_{SWI} + Q_{GWI} + Q_{PPT} + Q_{TW} \quad (1)$$

where: Q_{SWI} = The flow of surface water entering aquifer
Q_{GWI} = Incoming groundwater flow
Q_{PPT} = Precipitation
Q_{TW} = Water entering aquifer from another source.

Total water that comes out is:

$$OUTFLOW = Q_{ET} + Q_{ND} + Q_{SWO} + Q_{GWO} \quad (2)$$

where: Q_{ET}: Evapotranspiration
Q_{ND}: Water requirements
Q_{SWO}: Surface water flow out
Q_{GWO}: Ground water flow out

In Chile, price of water is an economic policy instrument that may affect the environment, social and economic efficiency. Water scarcity has caused high water prices, causing consumers reduce water consumption and also worsen by less access to water services. The price of water is an economic policy instruments that are useful to affect the efficient use of water. A well designed water price will produce good objectives such as the promotion of economic efficiency, equity, environmental and water conservation measures in water scarce areas, public acceptance and politics [11]. The proposed water rate is based on the following idea:

$$R_{FB} + R_{SB} = CR_{UR} + ER_S + ER_{WS} \tag{3}$$

where R_{FB} and R_{SB} represent total perceived revenue from the first and second block respectively, CR_{UR} is the current revenue received under the uniform rate tariff structure, ERs represents the required extra revenue to cover subsidy costs and ER_{WS} is the additional revenue due to the increased tariff to signal water scarcity.

3 Mathematical modelling of sustainability and conservation of water resources in the future

Indeed, none of technology appropriate to prevent such degradation of water resources, even if the change in approach that offers a selection of technical, financial and institutional, but there will be changes to the environment, to the community needs to respond to the condition of the state with the prevention, protection and preservation sustainable water resources [12].

A study in [13], in Fendong District Fenhe Region, Shanxi Province, China an area of critical water resources, with irrigation efficiency water use is 65.8%, industrial (14.6%) and ecology (6.2%). The development of water resources is slower in comparison to the development of industrialization and urbanization. These are due to rapid population growth and determine the policy mistakes of the water so that the sustainability of water needs to be concerned. The optimization model for agricultural water transfer:

Objective:

$$\text{Max } Z^{\pm} = BA^{\pm} BS + BT - (AB^{`} + BS^{`} + BT^{`}) \tag{4}$$

The objective of the optimization model for agricultural water transfer is maximizing incremental benefit of water resources. The water amount transferred from agriculture to second industry and tertiary industry are chosen as decision variables.

A study by [14, 19-22] explored the behavior of water resources system imposed by climate change and population growth. They developed a dynamic model that stimulates changes in water governance through the integration process of supply, demand and asset management. Integration is a feedback loop that is used to express the relationship between the structure of rates, the level of demand and funding capacity which is designed in the modelling the scenario of various countries. This model shows an alternative rate structures to lower demand in the period of scarcity. The water demand sub-model estimates annual water demand, which is driven by three key variables: population, per capita water demand and climate impacts. These are calculated using the following equations:

$$AWD = P \; x \; WD_{PC} \tag{5}$$

$$P = \int (P_{ch}) \, dt + P_i \tag{6}$$

$$WD_{pc} = WU_{pc} \, x \, C_i \quad (7)$$

where *AWD*: Annual Water Demand; *P*: population; P_{ch} : population change; P_i: initial population; WD_{pc}: per capita water demand; WU_{pc}: per capita water use projection; C_i: climate impacts.

A study by [15] explains that increasing economic and population growth in developing countries led to problems of water scarcity and water pollution as the case in the area of Tian Jin, North China. The model adopted is a dynamic model that reflects the water management system which includes three sub-models: (i) socioeconomic, (ii) cycle springs, and (iii) the flow of pollutants. All three sub-models can react with each other to encourage and limit the development of each model. This model was used to analyze the potential of reclaimed water as waste water that can be recycled which is safe for a variety of uses. This study shows a good simulation of the socio-economic and environment arising from the use of reclaimed water.

The dryness caused by reduced groundwater recharge and depletion of groundwater, causing a threat to farmers in Iran [16]. The technique used in this study was engineered storage of water in the aquifer and is a technique that could minimizes the water loss due to high evaporation from dams and reduces soil depletion due to excessive exploitation of ground water that is causing a threat to drought. They developed ASR (Aquifer Storage Recovery) working to move surface water into aquifers through infiltration or injection of water in the dam then you have available can be used with conventional pumping and with the design of the dam it can reduce groundwater recharge. Model ASR profitable for farmers in these regions due to water depletion rate decreased significantly (from 14.5 meter per 40 years to 3 meters during the same period) and evaporation of the reservoir decreased by 50 million m^3 during the simulation period. During periodic simulation, aquifer storage, S_i is related to aquifer head using the relationship between storage and head in an unconfined aquifer:

$$S_t = (h_t - Z_{bed}).\Delta \; X^2 S_y \quad (8)$$

S_y = the specific yield of the aquifer;
Z_{bed} = the bedrock elevation (L).

A study by [17] extends that the water needed by each person depends on groundwater. The ground water is of growing importance in agricultural production, especially in the Middle East country. As the climate changes that cause global warming and low rainfall resulted in groundwater levels decreased in Hamedan-Bahar, Iran. Ground water has affected the public water suppliers in the extraction of the competition over the use thus causing more exploitation consequently causing damage to the order of the aquifer. The groundwater volume changes due to individual variables of the aquifer balance equation (such as precipitation or exploitation) the water table depth also changes according to the following equation:

$$\Delta S_i = S_t - S_{t-1} = \Delta V.\frac{1}{A.q} \quad (9)$$

S_t is the water table depth (height of water pumping) in year *t*; *A* is the area of aquifer (m^2), and *q* is the specific yield of the aquifer.

A study by [18] explained that in Medan it is frequent water shortages and interruption of water by distrbutor at certain hours. This happens because the population of Medan is large with 2,191,140 inhabitants of 502,735 households and the widespread diversification across various sectors of economic use of water for both irrigation and livestock farming. Pollution is high that approximately 2 million tons of waste every day which pollute the water, environment and catchment area increasingly destroys the water resources [1]. In addition, the management of the water supply is still weak for their cross-subsidy occurs between consumer usage of water (based on Regulation No. 23, 2006) that generate inequality in the consumption of water with same water quality. Therefore, it needs optimal water pricing to guarantee water sustainability. The dynamic model expressed in mathematical equation established equation of revenue, *R* from selling water is influenced by the size prescribed rate; the rate consists of a low rate, *LR*; the basic rate, *BR*; full rate, *FR*; special rate, *SR*; and the total cost, *TC* required for operation. The equation is proposed as follows:

$$R = \alpha_0 + \alpha_1 TC + \alpha_2 BR + \alpha_3 LR + \alpha_4 FR + \alpha_5 SR + \alpha_6 lag\ R + \xi_1 \quad (10)$$

R is called the overall demand or aggregate demand, *AD*. This revenue formed optimal water price:

$$P = \lambda_0 + \lambda_1 R + \xi_7 \quad (11)$$

where;

$$R = (\alpha_0 + \alpha_1 TC + \alpha_2 BR + \alpha_3 LR + \alpha_4 FR + \alpha_5 SR + \alpha_6 lag\ R + \xi_1) \quad (12)$$

where α_0, α_1, α_2, α_3, $\alpha_4, \alpha_5, \alpha_6$ are parameter estimation and ξ_1, ξ_7 is disturbances/error terms. The optimal price is influenced by revenue, *R* and the price is called the aggregate supply, *AS*. In the order for optimal water price can be achieved when *AD* = *AS* (reach equilibrium).

4 Conclusions and suggestions

Based on research that has been done a number of conclusions as follows. Water infrastructure must be well maintained to achieve sustainability thus improve the community welfare and future generations. Determination of the optimal price of water is necessary because water prices concerns about the value of water and the actual price of good water management, proper and reality set by channeling the water to provide services for the community. Based on research that has been done and scenarios of water pricing policies, some suggestions are: water resources is the main priority of infrastructure and basic needs of the community so that the correct water management and optimal pricing can prosper public life. The traditional presume that they thought the water resources as belonging to the public and controlled by the state so that the waste water usage occurs while this is now water has become economic goods that need to be guarded and maintained as well as the investment required to maintain sustainability.

References

1. Sanim Bunasor, *Sumberdaya Air Dan Kesejahteraan Publik Suatu Tinjauan Teoritis dan Kajian Praktis*, 25-37 (2011)
2. Widiyanti, Ni Luh Putu Manik, Ristiati Ni Putu, Ekologi Kesehatan **3**, 64 (2004)
3. Rohmania Prihatini, *Kualitas Air Isi Ulang Pada Depot Air Minum Di Wilayah Kabupaten Bogor Tahun 2008-2011*, Makalah Universitas Indonesia, Jakarta (2012).
4. Depkes, *Peraturan Menteri Kesehatan Republik Indonesia No.492/Menkes/Per/IV/2010, Tentang Kualitas Air Minum* (2010)
5. Saskya Sastavyana, *Pemodelan Sistem Penyediaan Air Minum Perdesaan yang Berkelanjutan dengan Menggunakan Analytic Network Process (Studi Kasus: Kabupaten Subang)*, **21**, 81 (2010).

6. Masduqi Ali, Noor Endah, Eddy Soedjono, *Sistem Penyediaan Air Bersih Perdesaan Berbasis Masyarakat* (2008)
7. Kusumawardani Deni, *Valuasi Ekonomi Air Bersih Di Surabaya (Studi Kasus Pada Air PDAM)* (2010)
8. Maryati Sri, *Keterkaitan Variabel Lingkungan Terhadap Biaya Penyediaan Air Minum*, (2009)
9. Nababan Ester. S. M, *Model Daya dukung Ketersediaan Air Dengan Koefisien Stabilitas (Studi Kasus pada Kawasan Industri Sei Semangkei. Kabupaten Simalungun)* (2012)
10. Ediyanto, *Kajian pengelolaan Air Indonesia ke Arah Pembangunan yang berkelanjutan* (2009)
11. Maria Molinos-Senante, Guillermo Donoso, Utilities Policy **43**, 107 (2016)
12. Pujawan, I. N., *Ekonomi Teknik*, 2nd Edition (2009)
13. Dongmei Zhang, Ping Guo, Agricultural Water Management **170**, 5 (2016)
14. Sahin O., Siems R. S., Stewart R. A., Michael G. Porter, Environmental Modelling & Software **75**, 348 (2016)
15. Nan Xiang, Jinghua Sha, Jingjing Yan, Feng Xu, Water **6**, 17 (2014)
16. Amir Niazi, Shiv O. Prasher, Jan Adamowski, Tom Gleeson, Water **6**, 2300 (2014)
17. Hamid Balali, Davide Viaggi, Water **7**, 5258 (2015)
18. Tan Kim Hek, Mohammad Fadzli Ramli, Iryanto, AIP Conference Proceedings **1847**, 020022 (2017)
19. Rob Dellink, Roy Brouwer, Vincent Linderhof, Karin Stone, Ecological Economics **71**, 63 (2011)
20. Kostas Bithas, Antonios Kollimenakis, Georgios Maroulis, Zafeiria Stylianidou, Procedia Economics and Finance **8**, 73 – 82, (2014)
21. Jaime J. Sadhwani, Jose M. Veza, Carmelo Santana, Desalination **185**, 1 (2005)
22. Vašak M., Banjac G., Baotie M., Matuško J., Procedia Engineering **89**, 1031 (2014)

Ozonation of return activated sludge for disintegration and solubilisation with synthesized titanium oxide as catalyst

S.F.Z. Mohd Sarif[*,1], *S.S.* Alias[1], *F.* Muhammad Ridwan[1], *K.S.* Ku Salim[1,2], *C.Z.A.* Abidin[1], and *U.F.* Md. Ali[1]

[1]School of Environmental Engineering, Kompleks Pusat Pengajian Jejawi 3, Universiti Malaysia Perlis, 02600 Arau Perlis, Malaysia
[2]Shorubber (M) SDN. BHD, No 2, Jalan Perindustrian Ringan 6, Kawasan Perindustrian Jejawi, Arau 02600 Perlis, Malaysia

Abstract. Ozonation of activated sludge in the present of titanium dioxide (TiO_2) as catalyst to enhance the production of hydroxyl radical was evaluated in comparison to the sole ozonation process. In this process, the catalytic ozontion showed improvement in increasing ozone consumption and improving activated sludge disintegration and solubilisation. The reduction of total suspended solid (TSS), volatile suspended solid (VSS) and soluble chemical oxygen demand (SCOD) solubilisation was better in the catalytic ozonation system. Initial pH 7 of activated sludge was found best to disintegrate and solubilise the sludge flocs. However upon additional of sodium hydroxide (NaOH) in pH adjustment enhanced the solubilisation of organic matter from the flocs and cells, making the initial pH 9 is the best condition for activated sludge solubilisation. Yet the initial pH 7 of activated sludge supernatant was the best condition to achieve SCOD solubilisation due to sludge floc disintegration, when it had stronger correlation between TSS reduction and SCOD solubilisation (R^2=0.961). Lower amount of catalyst of 100 mgTiO_2/gTSS was found to disintegrate and solubilise the activated sludge better with 30.4% TSS reduction and 25.2% SCOD solubilisation efficiency, compared to 200 mgTiO_2/gTSS with 21.9% and 17.1% TSS reduction and SCOD solubilisation, respectively.

1 Introduction

The activated sludge is a biological process involves microorganism in organic pollutant degradation in the wastewater [1]. Domestic and industrial wastewater treatment system adopted this method in organic pollutant degradation because of high efficiency. Despite the good performance, it produced a large volume of waste in the end of the process, which requires 65% from the total plant operation cost for treatment and disposal [2]. Therefore such research has been conducted involves lysis-cryptic growth to reduce the sludge production and increase organic matter solubilisation in water line, for example by means

[*] Corresponding author: msfatimahzahra@gmail.com

of biological sludge worm [3], chemical oxidation by potassium permanganate [4] and ultrasonic [5].

The lysis-cryptic growth occurs in biological system, however the cell lysis is hard to achieve and become the rate limiting step in the process. Therefore, the following process of cryptic growth and degradation of organic pollutant is retarded. Lysis is a process to break the sludge cell so that the intracellular organic matter solubilised in supernatant and became the food source for endogenous bacteria, while part of it will be oxidized into water and carbon dioxide. Meanwhile the cryptic growth is the growing of new cell from the soluble substrate [6]. The damage sludge structure and cell contribute to reduction volume of the activated sludge.

Ozone is a powerful oxidant with potential of 2.07 eV makes it popular in colour improvement, organic and inorganic removal from wastewater [7]. However, ozone is very selective and unable to oxidize certain organic compounds (e.g. natural organic compound) [8] and small molecular organic acids such as acetic acid, oxalic acid, and glyoxalic acid [9, 10]. The ozonation process in aqueous solution may occur in two routes: direct route is when the reaction by molecular ozone, and/or indirect route which the reaction of OH• involves. The efficiency of disintegration and solubilisation of activated sludge is expected to be higher due to the presence of more OH•, the powerful oxidant (2.33eV) [8].

The promotion to generate more OH• in oxidation process is called the advanced oxidation process [8]. One of a method to enhance the production of OH• by ozone decomposition is using solid metal as catalyst. The catalyst provides more active sites on the surface for the molecular ozone to adsorb and decompose into OH•. Such metal oxide has been used in ozonation process like TiO_2[11], MnO_2 [12] and ZnO [13]. Of all, TiO_2 has the advantages of higher stability, lower production cost [14] and harmless to living issue [15]. The good performance showed in removing organic contaminants from river water [16], degradation of nitrobenzene [17] and atrazine [18].

Up to this paper is written, the only study focusing on sludge reduction was conducted by Lee et al. using various metal ions, and Mn ion with ball mill for sewage sludge reduction [19, 20]. Utilizing metal ions (homogeneous catalyticozonation) caused the ions to stay in water and therefore, increase the secondary pollutant [21] and require additional treatment to remove, which is impractical [22]. Therefore utilizing solid metal as catalyst in ozonation process is preferable to degrade and mineralize the organic pollutant with high efficiency yet low negative effect [23].

Since there are lack of study embarking the activated sludge ozonation withmetaloxidecatalyst to enhance disintegration and solubilisation, this research was conducted to fill this gap. The efficiency of activated sludge disintegration and solubilisation in ozonation process with and without catalyst was compared. Synthesized catalyst namely TiO_2 was used for this purpose. The efficiency of catalytic ozonation (CO) is greatly depends on catalyst and solution pH [8], so this research was conducted to evaluate the performance of activated sludge ozonation with catalyst at different dose and initial pH.

2 Materials and methods

2.1 Sample collection

The activated sludge sample was collected from a rubber glove processing manufacturing in Arau, Perlis, Malaysia (Shorubber (M) SdnBhd) which adopted extended activated sludge process for the wastewater treatment. The activated sludge was taken from the return activated sludge (RAS) line using high density polyethylene (HDPE) bottle. Sample was stored in a refrigerator at 6°C ±1 until further usage. The initial characteristics of raw activated sludge were 7.78 ± 0.4 g/L, 6.27 ± 0.3 g/L, 60.5 ± 3.2 mg/L and 7565 ± 469 mg/L for TSS, VSS, SCOD and TCOD respectively. The SSV was 973 ± 2.02 mL/L and pH was 8.09 ± 0.2.

2.2 Catalyst preparation

The catalyst was prepared according to Karami [24] and Yang et al. [17]with a little modification. Tertabutylorthothitanate ($C_{16}H_{36}O_4Ti$) as precursor was mixed with ethanol and deionised water at room temperature with volumetric ratio of 1:4:0.5. The mixture was stirred to obtain dry-gel form. Then it was dried at 105 °C for 4 hours, before calcined at 500 °C for 6 hours to obtain dry powder. The pHpzc of catalyst was measured by Drift Methods, as according toBenhebal et al. [25]. Crystal size, lattice structure and phase were determined by XRD (Model Bruker D2 Phaser) in fast scan range from 10 - 80°, with Cu-Kα radiation (30 kV, 10 mA) and analyzed by DIFFRAC.EVA Software.

2.3 Ozonation experiment

The ozonation process was conducted in a capped cylindrical glass reactor with 80 mm diameter and 500 mm height. 1200 mL of working activated sludge was stirred with magnetic stirrer throughout the experiment for homogeneity. An ozone reactor (A2Z Model Z-3G) was used to generate ozone with 99% purified oxygen inflow of 2 L/min. A porous fritted glass diffuser was used to allow ozone gas to flow into the reactor. The ozonation was run for maximum of 170 minutes. 100 mL of activated sludge sample was withdrawn from the sampling port for analysis at different interval time, and another 100 mL of raw sample was return into the reactor to maintain its concentration and working volume [26]. Residual ozone coming out from the reactor was trapped with 2% potassium iodide. The oxygen was kept flowing after ozonation for 3 minutes to remove trapped ozone in the tubes and reactor. Initial pH of 5, 7 and 9 was used to study the effect of initial pH on catalytic ozonation performance. For effect of catalyst dosage, 100 and 200 mgTiO_2/gTSS was used and conducted at original pH of the sludge. Unless stated, the dosage for catalyst was 200 mg/gTSS for the experiments.

2.4 Sample analysis

TSS and VSS concentration was measured according to Standard Method [27]. Filtered supernatant was used to measure the SCOD concentration meanwhile the unfiltered sample for TCOD. The COD sample was prepared by Closed Reflux Method according to the Standard Method. The concentration was measured using a spectrophotometer (HACH 2800). HI 2211 pH meter (HANNA Instrument) was used to measure the pH of activated sludge. Iodometry Method was adopted to measure ozone concentrations in aqueous solution(APHA, 1995). The calculations were as per equations below, where 24 = conversion factor ((24 000 me/L)/(1 000 mL/L)), V_t = volume of titrated $Na_2S_2O_3$ (mL), N = normality of $Na_2S_2O_3$ (mg/me), V_{sludge} = sludge working volume (L) and SS_i = initial concentration of suspended solid.

$$\text{Ozone concentration (mg)} = 24 \times V_t \times N \quad (1)$$

$$\text{Ozone dose } (mgO_3/mgSS) = \frac{\text{Mass of O3 (mg)}}{\text{Vsludge x SSi}} \quad (2)$$

Sludge disintegration efficiency was measured by the difference of TSS before and after ozonation. Solubilisation efficiency was calculated by dividing the difference SCOD concentration with the initial TCOD [28].

3 Results and discussions

3.1 Catalyst characterization

The synthesized TiO_2 was estimated by the DIFFRAC.EVA Software to be in size of 64.4 nm and tetragonal lattice structure. There were 6 diffraction peaks matched with anatase and rutile phases. The anatase peaks were 25, 38, 48 and 62at plane (101), (004), (200) and (204) respectively. The rutile peaks found at 54 (105) and 55 (211) [29-32]. The planes and peaks as referred to Joint Committee on Powder Diffraction Standards (JCPDS) database, JCPDS 21-1272 and JCPDS 21-1276 for anatase and rutile respectively. The catalyst was in the mixture phase and predominately by anatase. The mixture of amorphous and rutile was found to perform efficiently in pollutant degradation [11, 16, 17, 18]. Yet, the mixture of rutile and anatase also found to degrade naproxen, carbamazepine [15] and clofibric acid [33]. The catalyst synthesized had the pHpzc of 6.7. This value for synthesized TiO_2 can bevaried between 6.4 to 7.3, which in the neutral condition [11, 15, 18].

3.2 Catalytic ozonation performance

The concept of sludge ozonation is to break the sludge floc into smaller particle, followed by lysis of unit cell of microorganism that formed the sludge matrix.Both processes involve the disintegration and solubilisation. The extrapolymeric substances (EPS) is a complex high-molecular-weight polymer that responsible in agglomerating the microorganisms to form a larger floc [34]. Therefore, by breaking the sludge flocs, smaller particle produced which resulted in reduction of the TSS and also increased the soluble organic material in the supernatant.TheEPS also secreted by the microorganisms due to cellular activities, product of cell lysis, and also formed by the adsorbed organic materials in the wastewater [35, 36]. The SCOD concentration was measured to evaluate the performance of activated sludge solubilisation.Since the EPS is a part of the sludge floc, the disintegration also can be evaluated by SCOD concentration.

The performance of sludge disintegration can be evaluated in the TSS reduction, as in Fig. 1(a) below.The same reduction of 4% at 30 minute of ozonation (29.6 $mgO_3/gTSS$) was achieved for both processes. The reason why they have the same performance is that less or no formation of OH• at this time to break the sludge flocs in CO. The OH•from ozone decomposition with aid of catalyst is believed to strong enough and start attacking the sludge flocsat dosage higher than 29.6 mg/gTSS.The sludge flocs may compose of recalcitrant materials and too compact to be destructed with ozone even in the presenceof catalyst. The reduction was slower at ozone dose higher than 94.7 mg/gTSS (110 minute) onwards, and 0.48% reduction difference between both processes achieved at the end of the process.

In CO, 94.7 $mgO_3/gTSS$ required to achieve 19% reduction of TSS. To obtain the same efficiency in non catalytic ozonation (NCO) system, it took higher ozone dose of 135.6 mg/gTSS (150 minute). Further ozonation period as to increase the ozone doseshowed no increment in efficiency. This may be caused by the remaining hard-to-degrade particles and cannot be disintegrated with the supply ozone and available OH• [37].

The SCOD resulted from solubilisation of EPS that bind the microorganism to form sludge floc, and from the inner cell [38]. Therefore disintegration of sludge floc and

 https://doi.org/10.1051/e3sconf/20183402009

breakage of sludge cell caused the SCOD concentration to increase. The intracellular material of the cell also contributed to SCOD concentration in supernatant once it was disrupted. The performance of both processes is presents in Fig. 1(b) below. The CO reached maximum solubilisation (19.38%) at 110 minute ozonation process and showed no increment after this period. Meanwhile at this contact time, the NCO achieved 10.14% and the maximum solubilisation was 13.98% at 150 minute. Higher solubilisation in CO system shows the present of more OH• aided the ozone in disintegration and solubilisation of sludge flocs and cell breakage.

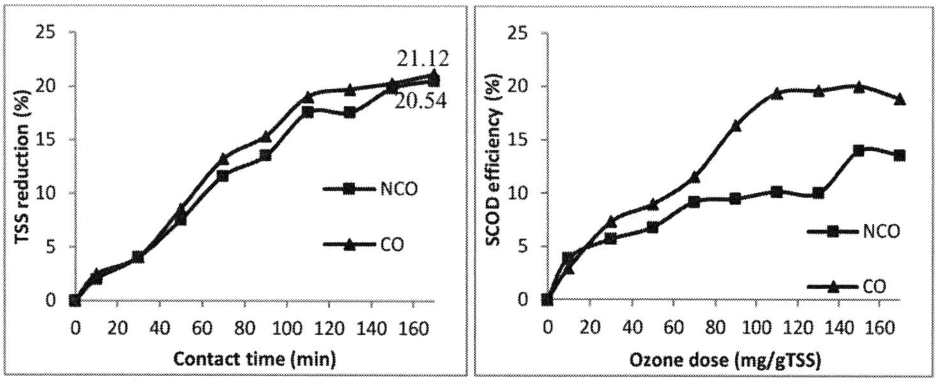

a) b)

Fig. 1. The TSS reduction (a) and SCOD solubilisation efficiency (b) for both processes.

During ozonation process, the EPS and sludge cell destruction released the deoxyribonucleic acid (DNA) and caused the sludge supernatant to be acidic. Production of fatty acid due to oxidation of higher to lower molecular weight organic matter also contributes to the reduction of supernatant pH [1]. Reduction of supernatant pH of 1.36 and 1.24 observed for CO and NCO system respectively. Chu et al.[39] and Dytczak et al.[40] also observed this changes in their processes.

3.3 Effect of initial pH

A graph of TSS reduction was plotted for the sludge disintegration efficiency observation, as in Fig. 2(a). At initial pH 7, the TSS reduced gradually and started to become slower at 110 minute to achieve maximum reduction of 21.12%. The maximum reduction achieved was the best among other pH. This is probably caused by the formation of OH• by neutral hydroxyl on the catalyst surface that more effective at pH close to its pH_{PZC}[11], and the synthesized TiO_2 had the pH_{PZC} of 6.7, which is almost neutral.

An interesting trend at initial pH 9, where a rapid reduction observed at the early process, no reduction percentage increment until the 70th minute, and the TSS reduced again to the maximum of 18% and no increment until the end of the process. Ozone decomposition occur faster at higher pH due to predominant of OH- as initiator [8]and caused the sludge disintegration to occur faster even at low ozone dose, due to more production of OH• in the alkaline supernatant. This explained the rapid reduction of TSS at the early process. Onwards, released of DNA and lower molecular organic compound dropped the pH to 7.64, cause the OH- to unprevail and therefore, lesser production of OH• to aid the disintegration efficiency, and reflected by the unchanged trend until the 70th minute of ozonation. The reduction efficiency increased again was believed due to molecular attack of accumulation ozone and the decomposed ozone into OH• on the sludge floc as the ozone dose increased.

No disintegration of sludge floc initially at pH 5 due to less or maybe no OH• present at lower pH [8]. The TSS reduction was observed after 30 minute of ozonation, when enough concentration of molecular ozone to disintegrate the sludge flocs at this contact time. The acidic solution made the catalyst surface to be positively charged, and the negative charge of sludge floc attracted and adsorbed on the catalyst.Instead of ozone decomposition reaction, the sludge floc is believed to oxidize on the catalyst surface by the attack of molecular ozone and OH• from the bulk solution.

The performance at initial pH 5 and 9 was found to unchanged at ozone dose higher than 82.6 $mgO_3/gTSS$ (90 minute). However it can be seen that the reduction of TSS as initial pH 7 still increasing until the end of the process, and concluded as the ideal pH for sludge disintegration at high ozone dose. However, in the point of economical view, it is not effective as more cost requires to supply high ozone dose for sludge floc disintegration. On the other side, additional of chemicals in the activated sludge for pH adjustment also need to be considered since additional of chemical may cause chaos in the microbiologic activities since the responsible bacteria was interrupted [41].

As for the solubilisation efficiency of organic matters, the performance is refers to Fig. 2(b) below. The solubilisation increased with ozone dose for all conditions of initial pH. The higher the initial pH, the higher the SCOD solubilised was observed. The efficiency achieved was 24.9%, 18.9% and 10.1% for initial pH of 9, 7 and 5 respectively.CO enhances the disintegration of sludge floc, release more organic matter as the result of EPS and sludge cells destruction. By additional of NaOH during pH adjustment, the cell wall was disrupted, released organic matter from the microorganism and thus increased the SCOD concentration [42, 43]. The SCOD efficiency in initial pH 9 increased throughout the process, although the supernatant pH dropped to 7.64 after 10 minute of ozonation. This increment may be due to the breakage of sludge flocs and cells by NaOH during initial pH adjustment, followed by OH• when the CO started. Therefore, smaller particles and organic materials was able to be continuously attacked by molecular ozone and OH• from ozone decomposition throughout the process.

Between the three initial pH experiments, the initial pH 7 was found to have the strongest correlation of SCOD solubilisation with TSS reduction when the R^2 value is closer to 1 compared to other (R^2=0.961). Therefore, the SCOD solubilisation due to sludge floc disintegration was best occurred at initial pH 7.

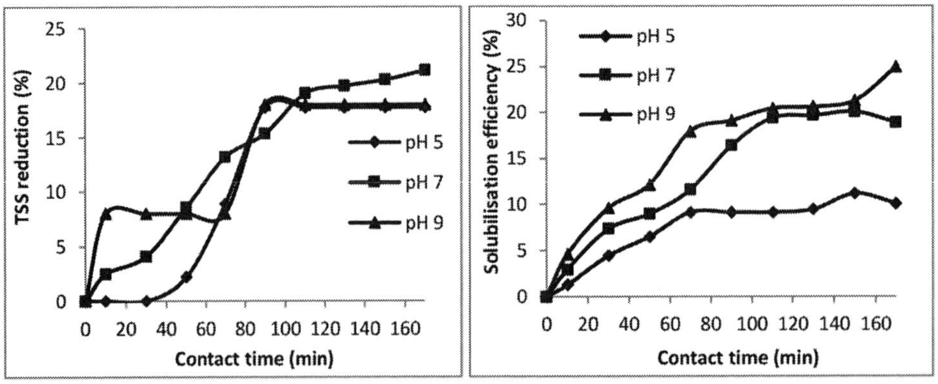

a) b)

Fig. 2. The TSS reduction (a) and SCOD solubilisation efficiency (b) for CO at different initial pH.

3.4 Effect of catalyst dosage

Catalyst dosage plays a role in the efficiency of a catalytic ozonation process. Theoretically, by increasing the catalyst dosage, the efficiency of a process also increased accordingly due to the availability of active sites for reaction to occur in decomposition of ozone into OH•[44, 45]. Therefore, more OH• would present to faster the reaction rate.

In investigating the effect of catalyst dosage, two different catalyst doses (100 and 200 mgTiO_2/gTSS) was manipulated to evaluate the performance efficiency. Using 100 mg/g catalyst found to be more efficient in activated sludge solubilisation compared to that 200 mg/g, with 25.2 and 17.1% respectively. Sludge floc disintegration was found to happen after 10 minute of the process, however ended with higher reduction of TSS (30.4%). Oxidizing of suspended organic materials also occurred after 30 minute of ozonation process and oxidized up to 64% of the VSS. Although the higher dosage of 200 mg/g catalyst offers great ability to disintegrate sludge flocs and oxidize the organic matter initially, the efficiency was lower than that ozonation at lower catalyst dosage, with 21.9 and 33% of TSS and VSS reduction respectively. It is believed that in the initial process, more active sites available offers at higher TiO_2 dosage. Even though more molecule ozone presents in bulk solution as increase of ozone dose, the amount of active sites on catalyst surface for ozone to be adsorbed and decomposed into OH• only offered by a minimum amount of catalyst. Therefore, providing excess active sites by increasing the catalyst dosage do not enhance the process efficiency [17]. An ongoing experimental work using 150 mg is conducted to observe if it is able to disintegrate and solubilise the activated sludge better than that the 100 and 200 mg of TiO_2.

4 Conclusions

The efficiency of sludge ozonation with synthesized TiO_2 as catalyst was evaluated in this paper. With additional of TiO_2 as catalyst, the ozone consumption in catalytic system was doubled than that the sole ozonation. From this research, some points can be summarised as follow:

1) The catalytic ozonation performed better in sludgesolubilistion. It was observed that the reduction of TSS was 21.12% and 20.54% for CO and NCO system, respectively. This small difference in TSS disintegration suggest that the present of catalyst does not improve much in disintegration, yet better in solubilisation of SCOD with 19.38%and 10.14% in CO and NCO respectively. The rubber processing activated sludge is a strong matrix that composed of recalcitrant compound.
2) Initial pH 7 was the best condition to disintegrate and solubilise the activated sludge when the correlation of TSS reduction and SCOD solubilisation obtained was nearest to 1 (R^2=0.961). Although at initial pH 9 was able to solubilised SCOD more, this better efficiency also may be attributed by the role of NaOH in pH adjustment that helped in activated sludge disintegration and solubilisation.
3) Lower catalyst dose of 100 mg/g disintegrate and solubilise the activated sludge better than 200 mg/g TiO_2. Although faster reaction in TSS and VSS reduction using 200 mg/g TiO_2observed initially, the efficiency of these parameters and the SCOD was better at lower dose of catalyst at the end of the process.

From this research, it was suggested that the activated sludge ozonation with present of TiO_2 as catalyst had positive effect in solubilisation. Further research to study the biodegradation on ozonated activated sludge is ongoing.

The authors fully acknowledged the Malaysian Ministry of Higher Education (MOHE) for the research grant (FRGS 9003-00379),School of Environmental Engineering (UniMAP) for the facilities, Sustainable Engineering Research Cluster (UniMAP) for the XRDandShorubber(M) Sdn Bhd.

References

1. G. Zhang, J. Yang, H. Kiu, J. Zhang, Bioresour. Technol. **100**, 1505–1509 (2009)
2. Y. Liu, Chemosphere **50**, 1-7 (2003)
3. X. S. Guo, J. X. Liu, Y. S. Wei, L. Li, J. Environ. Sci. **19**, 257-263 (2007)
4. C. Wu, G. Zhang, P. Zhang, C. C. Chang, Chem. Eng. J **240**, 420-425 (2014)
5. P. Zhang, G. Zhang, W. Wang, Bioresour. Tech. **98**, 207-210 (2007)
6. L. Chu, S. Yan, X. H. Xing, X. Sun, B. Jurcik, Water Res. **43**, 1811-1822 (2009)
7. L. M. Dorfman, G. E. Adams, *NSRDS-NBS*-**46** (1973)
8. B. Kasprzyk-Horden, M. Ziolek, J. Nawrocki, Appl. Catal., B: Env. **46**, 639-669 (2003)
9. K. He, Y. M. Dong, Z. Li, L. Yin, A. M. Zhang, Y. C. Zheng, J. Hazard. Mater. **159**, 587-592 (2008)
10. T. S. Ping, L. W. Hua, Z. J. Qing, C. C. Nan, Ozone Sci. Eng. **24**, 117-122 (2002)
11. Y. Yang, J. Ma, Q. Qin, X. Zhai, J. Mol. Catal A: Chem. **267**, 41–48 (2007)
12. S. P. Tong, W. P. Liu, W. H. Leng, Q. Q. Zhang, Chemosphere **50**, 1359–1364 (2003)
13. Y. Dong, H. Zhao, Z. Wang, G. Wang, A. He, P. Jiang, B. Korean Chem. Soc., **33**, 215-220 (2012)
14. K. J. Kim, K. D. Benkstein, J. Van De Lagemaat, A. J. Frank, Chem. Mater. **14**, 1042-1047 (2002)
15. R. Rosal, A. Rodriquez, M. S. Gonzalo, E. Garcio-Calva, Appl. Catal. B: Env. **84**, 48–57 (2008)
16. Y. X. Yang, J. Ma, S. J. Wang, J. Zhang, *7th International Conference on Applied Electrostatic (ICAES)* (2012)
17. Y. X. Yang, J. Ma, J. Zhang, S. J. Wang, Q. D. Qin, Ozone, Sci. Eng. **31**, 45–52 (2009)
18. Y. Yang, H. Cao, P. Peng, H. Bo, J. Hazard. Mater. **279**, 444–451 (2014)
19. M. J. Lee, Y. S. Kim, C. K. Yoo, J. H. Song, S. .J. Hwang, Environ. Technol. **31**, 7-14 (2010)
20. M. J. Lee, T. H. Kim, G. Y. Yoo, B. K. Min, S. J. Hwang, KSCE J. of Civ. Eng. **14**, 693-697 (2010)
21. J. Xiao, Y. Xie, H. Cao, Chemosphere, **121**, 1–17 (2015)
22. J. Nawrocki, B. Kasprzyk-Hordern, Appl. Catal., B: Env. **99**, 27-42 (2010)
23. N. Pugazthenthiran, P. Sathishkumar, S. Murugesan, S. Anandan, Chem. Eng. J. **168**, 1227-1233 (2011)
24. A. Karami, JICS, **7**, S154-S160 (2010)
25. H. Benhebal, M. Chain, T. Salmon, J. Geens, A. Leonard, S. D. Lambert, M. Crine, B. Heinrichs, Alex. Eng. J., **52**, 517-523 (2013)
26. J. Zhang, J. Zhang, Y. Tian, N. Li, L. Kong, L. Sun, M. Yu, W. Zuo, Chem. Eng. J. (2016)
27. American Public Health Association. *Standard Method for Examination of Water and Wastewater* (1995)
28. S. T. Yan, L. B. Chu, X. H. Xing, A. F. Yu, X. L. Sun, B. Jurcik, Water Res. **43**, 195-203 (2009)
29. S. Dain, Y. Wu, T. Sakai, Z. Du, H. Sakai, M. Abe, Nanoscale Res. Lett., **5**, 1829-1835 (2010)
30. G. S. Devi, K. A. S. Kumar, K. S. Reddy, Part. Sci. Technol. **33**, 219-223 (2015)

31. E. Hosone, S. Fujihara, K. Kakiuchi, H. Imai, JACS **126**, 7790-7791 (2004)
32. K. Thamaphat, P. Limsuwan, B. Ngotawornchai, Nat. Sci. **42**, 357–361(2008)
33. R. Rosal, M. S. Gonzalo, A. Rodriguez, E. Garcia-Calvo, J. Hazard. Mater. **169**, 411-418 (2009)
34. J. Wingender, T. R. Neu, H. C. Flemming, *Microbial extracellular polymeric substances: characterization, structures and function*, 1-18 (1999)
35. Y. Liu, H. H. P. Fang, Crit. Rev. Env. Sci. Technol. **33**, 237–273 (2003)
36. P. H. Nielsen, A. Jahn, *Microbial extracellular polymeric substances: Characterization, structure and function*, 49-72 (1999)
37. W. Saktaywin, H. Tsuno, H. Nagare, T. Soyama, J.. Weerapakkaroon, Water Res. **39**, 902–910 (2005)
38. G. P. Sheng, H. Q. Yu, X. Y. Li, Biotechnol. Adv. **28**, 882–894 (2010)
39. L. B. Chu, S. T. Yan, X. H. Xing, A. F. Yu, X. L. Sun, B. Jurcik, Chemosphere. **72**, 205-215 (2008)
40. M. A. Dytczak, K. L. Londry, H. Siegrist, J. A. Oleszkiewicz, Water Res. **41**, 543-550 (2007)
41. S. I. Perez-Elvira, P. Nieto Diez, F. Fdz-Polanco, Rev. Environ. Sci Bio. **5**, 375-398 (2006)
42. M. J. Brown, J. N. Lester, Appl. Environ. Microbiol. **40**, 179-185 (1980)
43. G. P. Sheng, H. Q. Yu, Z. Yu, Appl. Microbiol. Biotechnol. 67, 125– 130 (2005)
44. M. Muruganandham, J. J. Wu, Catal. Comm. **8**, 668-672 (2007)
45. J. Park, H. Choi, J. Cho, Water Res. **38**, 2285-2292 (2004)

Analysis of Multi-Criteria Evaluation Method of Landfill Site Selection for Municipal Solid Waste Management

Habiba Ibrahim Mohammed[1*], *Zulkepli* Majid[1], *Norhakim* Bin Yusof[1], *and Yamusa* Bello Yamusa[2]

[1] Faculty of Geoinformation and Real Estate, Universiti Teknologi Malaysia, 81310 Skudai, Johor, Malaysia
[2] Department of Civil Engineering, Nuhu Bamalli Polytechnic, 810001 Zaria, Nigeria

Abstract. Landfilling remains the most common systematic technique of solid waste disposal in most of the developed and developing countries. Finding a suitable site for landfill is a very challenging task. Landfill site selection process aims to provide suitable areas that will protect the environment and public health from pollution and hazards. Therefore, various factors such as environmental, physical, socio-economic, and geological criteria must be considered before siting any landfill. This makes the site selection process vigorous and tedious because it involves the processing of large amount of spatial data, rules and regulations from different agencies and also policy from decision makers. This allows the incorporation of conflicting objectives and decision maker preferences into spatial decision models. This paper particularly analyzes the multi-criteria evaluation (MCE) method of landfill site selection for solid waste management by means of literature reviews and surveys. The study will help the decision makers and waste management authorities to choose the most effective method when considering landfill site selection.

1 Introduction

Identifying a suitable site for the disposal of municipal solid waste is a very demanding task requiring a lot hard work for the municipality. There are different factors and criteria that needs to be given careful consideration before siting a landfill [1]. The selection process is very demanding because of its effects on the economy, environments and health are numerous. The developing countries are facing difficulty in waste management due to the population growth, urban migration, inappropriate planning, and lack of sufficient resources [2]. Getting rid of unwanted materials through landfills is an unavoidable event in solid waste management practice.

Waste disposal of residual proportion produced on land is an inevitable aspect of solid waste management even if all the measures of reduction, reuse and recycling are being employed in the environment. The sanitary landfill has become the most well-known and

* Corresponding author: mydearhabiba@yahoo.com

popular method of solid waste disposal in many developed and developing countries [3]. Furthermore, proper landfill site selection is an essential step in solid waste disposal practice and towards ensuring sustainable environment, good health, and quality of life for the future. To avoid the undesirable long-term effects of the landfill to the environment, it is important to implement a proper landfill siting. Therefore, careful consideration of siting criteria (constraints and regulations) set up by environmental agencies or local authorities should be employed for a suitable landfill site [4].

One of the most important tool used in land use suitability analysis recently is the Geographical information systems (GIS), also this same tool can be applied for landfill siting [5]. The benefit of the GIS-based method in site selection study include time and energy saving, cost reduction as well as provision of digital data for the site monitoring for long-term [6]. The reason is because GIS has the storage capacity, easy management and maintenance of the analyzed geographical data, and it is well designed for spatial and descriptive dependency data [7]. The combination of GIS features and multi criteria decision analysis (MCDA) methods gives rise to a convenient method for spatial multi-criteria decision analysis (SMCDA) for optimal landfill siting. This integration method of both techniques has been broadly used to a great extent in different landfill site selection studies [5]. In landfill siting decisions, various factors have to be incorporated and for this type of study, and geographic information systems (GIS) are the best ideal tools to carry out such study because of its ability in spatial data management especially in large volume [8]. The GIS is very important in siting process because it efficiently display and manage data from different type of sources, also, the cost and time of the siting process is reduced [9, 10]. Recently, over the past few years GIS developed into a very useful strong tool that can perform different type of analysis using spatial data. This is s also due to the technological advancement in computing technology and the availability of various kind of spatial data. Environmental models and GIS function with a broad range of related geospatial data used for a variety of applications and spatial analysis at different levels. Data evaluation and organization into a meaningful form produces a reliable information, which then gives rise to suitable analysis and modeling [11].

Literature review regarding landfill site selection and its methodology was conducted from various studies in different part of the world. This review paper specifically considers in detail the Geospatial based strategies of landfill site selection for municipal solid waste particularly the Multi-Criteria Evaluation techniques (MCE) by means of related literature and surveys. The study will help the decision makers and waste management authorities to know which method is to be considered most effective.

2 Literature Review Related to Landfill Siting

A lot of literature related to landfill site suitability analysis has been conducted over the years with the integration of GIS and MCDA. These methodologies have been widely used by many researchers in different countries in order to achieve suitable landfill site for waste disposal e.g. [1, 3, 12]. Fig. 1 describes the process and steps to be taken when siting a new landfill considering environmental, socio-economic, design, areas availability and technical criteria.

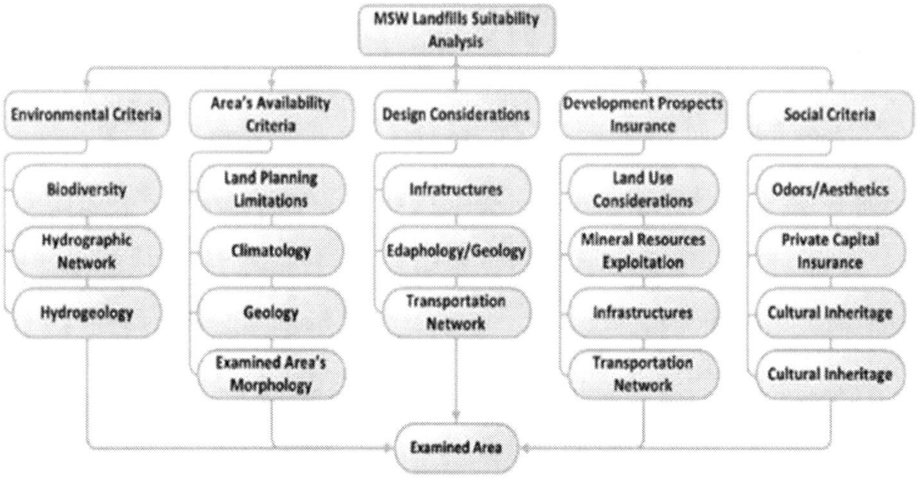

Fig. 1. Landfill site selection process [13].

Furthermore, several studies discussed the integration of GIS and multi-criteria decision analysis and its efficiency in solving the site selection problems. Landfill site selection process involves multiple criteria and conflicting objectives in decision- making therefore, it considers not only scientific analysis and data-mining, but also political factors and public acceptance [14]. This is because the site selection for landfill is very difficult as it requires large number of criteria both economic, environmental, social, political, and technical as shown in Fig. 1. In order to identify a landfill site, an evaluation process is required for optimal and best site selection [15]. This site selection process is increasingly becoming more complex and difficult, and it must be in accordance with rules and regulations governs by the authority [16].

3 Methods of Multi Criteria Decision Analysis

Multi-Criteria Decision-Making method support decision-making involved with conflicting objectives and multiple criteria [17]. The MCDA methods assist the decision makers in coming out with the best solutions or alternatives [18]. The MCDM is able to deal with large number of criteria, which makes it to be considered as an efficient method for decision support tool. Fig. 2 shows the structure of framework for spatial multicriteria decision analysis.

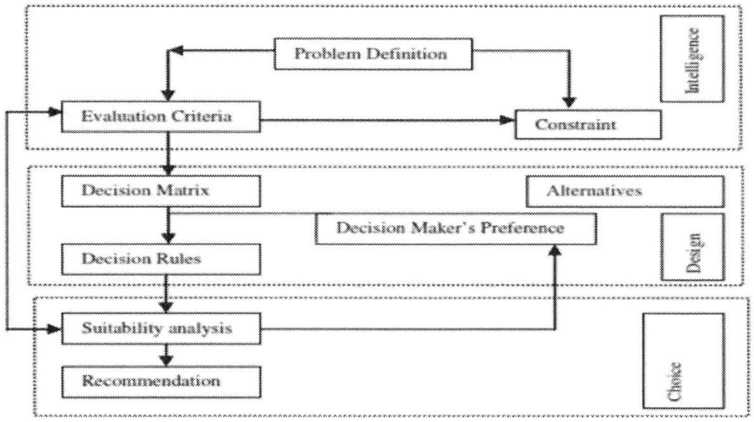

Fig. 2. Structure of spatial multi-criteria decision analysis [19].

3.1 Fuzzy Logic

This method was first innovated by Zadeh (1965) to identify and describe the uncertainties related with criteria and sub-criteria assessment. The selection criteria are weighted to represent the importance and contribution of each factor [19]. With Fuzzy Logic, criteria standardization is very accurate with the use of fuzzy membership function.

3.2 Weighted Linear Combination

This method is applied when dealing with Multi Attribute Decision Making (MADM). It is an analytical technique, which is used when there is more than one attribute to be considered based on the content of weighting average. The attributes are known as criteria and each of these criteria are assigned value (weight) based on its importance where the favorable outweigh the unfavorable. The outcome of it is multi-attribute features with a standardized score allocated to each. From the scores, the areas with higher scores indicate the level of adequacy. Example of this weighted linear combination method include land suitability analysis [20]. One of the most commonly used MCE methods in GIS-based studies is Weighted Linear Combination especially in land suitability analysis, site selection and evaluation analysis [21]. This method can be performed by GIS and overlapping capacities of this system. Use of this method is practical in both raster and vector format [22].

3.3 Analytic Hierarchy Process (AHP)

The most popular and most commonly used MCDA method is Analytic Hierarchy Process (AHP) which was introduced and established by Saaty [23]. It is a decision-making tool used in solving multiple criteria problems because of its ability in the determination of relative weight of multiple criteria, which are expressed in numerical order of 1 to 9. Pair wise comparison method is carried out among the criteria through the scores and weights assigned to each criterion [24].

AHP is complicated when there is large number of criteria to be the considered in the decision-making process. Saaty [25] explain the scale values and their definition as shown in Table 1.

Table 1. Pair wise comparison scale.

Intensity of importance	Definition
1	Extremely less important
2	Very strongly less important
3	Strongly less important
4	Moderately less important
5	Equally important
6	Moderate important
7	Strongly important
8	Demonstrate important
9	Extreme important

3.4 Ordered Weighted Averaging (OWA)

Ordered weighted average is a recent MCDA integration technique which is used to rank criteria. OWA is able to identify specific uncertainties related to criteria ranking. There is some trade-off among the criteria which are controlled by the ordered weights thus, giving rise to the best output in planning process [26]. It is also flexible to use because the level of importance of criteria can be changed based on expert decision [27]. Table 2 shows some outline of the advantages and disadvantages of the MCDA methods.

4 Multi-Criteria Decision Analysis using GIS

Presently, integration of GIS and MCDA plays an important role in selecting site for landfill. GIS-based approach has the capability of providing the spatial data in digital format which can be used for long term monitoring of the site, also it is able to save time and cost in the landfill siting process [10]. Some researchers [28] provided a method describing the process for municipal landfill sites identification in urban areas in Sierra Leone using a multi-criteria and GIS approach. The weighted linear combination (WLC) and ordered weighted averaging (OWA) was used as the MCE technique. From their studies, it was revealed that 83% of the study areas were unsuitable for the landfill siting.

A study conducted by [12] applied both GIS-based and multi-criteria evaluation approach for characterizing and assessing potential favorable landfill sites in the Polog Region in Macedonia. The environmental and economic factors were considered as multi-criteria decision framework, which were standardized by fuzzy membership functions after which they were integrated with analytical hierarchy process (AHP) and ordered weighted average (OWA) techniques. The AHP methodology was used to evaluate the importance of criteria and generates the global weights, which were used in conjunction with the local weights in OWA procedure for producing the decision alternatives.

Furthermore, [29] developed an integrated fuzzy logic and analytic network process (F-ANP) with the aim of locating a suitable site for disposing municipal solid wastes generated in Kahak Town, Qom, Iran. Their findings shows that the application of fuzzy logic and analytical neural process can be implemented in landfill site selection in similar study areas.

In addition, [22] used different models of the analytic multi-criteria decision-making process, such as an analytical hierarchy process (AHP), weighted linear combination (WLC), and Boolean logic, to manage layers to establish specific databases for urban waste landfills site location. Their results indicated that AHP and WLC methods in the early stages had better decision-making powers for locating landfill sites when compared to Boolean logic.

Table 2. Comparison of some MCDA methods.

S/No	MCDA Methods	Advantages	Disadvantages
1	AHP	- it is easy to handle complex conflicting objectives in decision-making process. - It enables the computation of consistency ratio. - It is straightforward and simple to use by using pairwise comparison.	- Subjective evaluation i.e. conversion from verbal to numeric scale. - Limited to the use of 1-9 scale. - Very tedious in terms of calculations of pairwise comparison matrix, normalization, and consistency ratio. - Inconsistent judgement by decision makers
2	WLC	-It has distinct advantages in preference modeling, weight elicitation, and aggregation performance. -It can be applied both in raster and vector format.	-Absence of unit of measurement will lead to error.
3	FUZZY Logic	- It is able to identify and describe the uncertainties related with criteria assessment. - It enables criteria weighting and standardization. - Ability to provide sufficient mapping of the decision-making vagueness and Imprecise knowledge.	- Inconsistencies cannot be processed in multiple criteria decision problems such as criteria ranking.

5 Conclusions

An attempt has been made to review the recent existing literature regarding GIS-based and MCDA application in landfill site selection process. From this paper, it has been proved that they are both efficient tools to be used in this type of site selection analysis. The main aim of the MCDA is to provide aid to the decision-makers in choosing the best candidate site from the variety possible choice alternatives under the presence of multiple priorities. Also, it has been noticed that from the previous decades, the integration of geospatial based strategies coupled with multi-criteria decision analysis has increased rapidly, this can be because of its accuracy, efficiency, and less time consuming compared to the conventional methods. Furthermore, the incorporation of the GIS-based technology and MCDA in landfill site selection and other land suitability analysis has been found useful and thus, aids in sustainable development. The study will help the waste management authorities and decision-makers to know which method is to be considered most effective when siting a new landfill.

References

1. Z. G. Rahmat, M. V. Niri, N. Alavi, G. Goudarzi, A. A. Babae, Z. Baboli, M. Hosseinzadeh, KSCE J Civ Eng **21**(1), 111-118 (2016)
2. A. Mornya, R. Majid, L. Yola, *Identification of Landfill Sites by Using GIS and Multi-Criteria Method in Batam, Indonesia*, in *3rd International Graduate Conference on Engineering, Science and Humanities, University Tekologi Malaysia. amazonaws. com/academia. edu.* (2010)
3. N. Alavi, G. Goudarzi, A. A. Babaei, N. Jaafarzadeh, M. Hosseinzadeh, Waste Manag Res **31**(1), 98-105 (2013)
4. S. Z. Ahmad, S. M. S. Ahmad, H. Wan, *Comparative site selection process based on different policies and guidelines for municipal solid waste landfill site, 10th International symposium & exhibition on geoinformation 2011 (ISG 2011), Selangor, Malaysia,* 27–29 September (2011)
5. M. Eskandari, M. Homaee, S. Mahmodi, Waste Manag **32**(8), 1528-38 Aug (2012)
6. V. R. Sumathi, U. Natesan, and C. Sarkar, Waste Manag **28**(11), 2146-60 Nov (2008)
7. A. Elahi and H. Samadyar, Middle East J Sci Res **22**(9), 1294-1307 (2014)
8. B. Şener, M. L. Süzen, and V. Doyuran, Environ. Geol **49**(3), 376-388 (2005)
9. O. B. Delgado, M. Mendoza, E. L. Granados, D. Geneletti, Waste manag **28**, (7), 1137-1146 (2008).
10. M. Moeinaddini, N. Khorasani, A. Danehkar, A. A. Darvishsefat, M. Zienalyan, Waste Manag **30**(5), 912-20 May (2010)
11. M. A. Alanbari, N. Al-Ansari, H. K. Jasim, Natural Sci **06**(05), 282-304 (2014)
12. P. V. Gorsevski, K. R. Donevska, C. D. Mitrovski, J. P. Frizado, Waste Manag **32**(2), 287-96, Feb (2012)
13. O. Demesouka, A. Vavatsikos, K. Anagnostopoulos, Waste Managt & Res **32**(4), 270-296 (2014)
14. A. Beskese, H. H. Demir, H. K. Ozcan, H. E. Okten, Environ. Earth Sci **73**, (7), 3513-3521 (2014)
15. M. Eskandari, M. Homaee, S. Mahmoodi, E. Pazira, M. T. Van Genuchten, Environ Sci Pollut Res Int **22**(10), 7754-65 (2015)
16. N. B. Chang, G. Parvathinathan, J. B. Breeden, J Environ Manage **87**(1), 139-53 (2008)
17. H. Voogd, *Multicriteria evaluation for urban and regional planning*. Pion London, (1983)
18. M. Doumpos and C. Zopounidis, "An overview of multiple criteria decision aid," in *Multicriteria Analysis in Finance*: Springer, 11-21 (2014)
19. D. Khan and S. R. Samadder, Waste Manag Res **32**(11), 1049-62 (2014)
20. R. Al-Adamat, A. Diabat, G. Shatnawi, J. Arid Environ **74**(11), 1471-1477 (2010)
21. J. Malczewski, Progress in planning **62**(1), 3-65 (2004)
22. H. Shahabi, S. Keihanfard, B. B. Ahmad, M. J. T. Amiri, Environ. Earth Sci **71**(9), 4221-4233 (2013)
23. T. L. Saaty, *What is the analytic hierarchy process?* in *Mathematical models for decision support*: Springer, 109-121, (1988)
24. J. Aerts, M. Van Herwijnen, R. Janssen, T. Stewart, J Environ Plann Man **48**(1), 121-142 (2005)
25. T. Saaty, *The analytic hierarchy and analytic network processes for the measurement of intangible criteria and for decision-making, Multiple criteria decision analysis: state of the art surveys,* 345-405, (2005)
26. C. Makropoulos and D. Butler, Environ. Model. Softw. **21**(1), 69-84 (2006).

27. Trupti I. Lokhande, Sachin J. Mane, S. T. Mali, Int J Res in Eng, Sci and Tech **03** (2017)
28. S. P. Gbanie, P. B. Tengbe, J. S. Momoh, J. Medo, V. T. S. Kabba, Appl Geogr **36**. 3-12 (2013)
29. A. A. Isalou, V. Zamani, B. Shahmoradi, H. Alizadeh, Environ. Earth Sci **68**(6), 1745-1755 (2012)

Performance of a Retrofitted Multicyclone for $PM_{2.5}$ Emission Control

Dewika, M[1], *Rashid,* M[2*], and *Ammar,* M.R[3]

[1] Centre of American Education, Sunway University, 47500 Subang Jaya, Selangor, Malaysia
[2] Air Resources Research Laboratory, MJIIT, 54100 UTM Kuala Lumpur, Malaysia
[3] AMR Environmental Sdn Bhd, 81110 Johor Bahru, Malaysia

Abstract. This paper presents on the performance of a retrofitted multicyclone system, which aims to increase the collection efficiency of $PM_{2.5}$ (i.e. particulate size fraction $\leq$ 2.5 μm) emission. The multicyclone was retrofitted by extracting 15% and 20% of the total volumetric air flow rate at the dust hopper of the unit using an additional Induced Draft Fan. The total collection efficiency with and without the extraction was measured at various air volumetric flow rates and particulate mass inlet concentration. The results showed that there was a reduction of 12% to 54% depending on the inlet concentration of $PM_{2.5}$ emission in the stack with compared to without extraction increasing the collection efficiency of the retrofitted multicyclone. The finding suggests that a simple technique of applying gas extraction at the dust hopper of a multicyclone as reported in this study able to increase the overall performance in fine particulate collection.

1 Introduction

Cyclones are one of the most popular air pollution control device used in separating particulate matter (hereinafter PM) from flue gas in industries. Generally, it is a preferred choice due to geometric simplicity, safety, low maintenance and operating costs, and wide range of operating conditions [1,2]. In industries, since the volumetric flow of contaminated flue gas is large, a multicyclone is used instead of single cyclone whereby it is made up of a number of small-diameter cyclones which are arranged in parallel and have a common gas inlet and outlet [3,4]. These multicyclones remove PM from the contaminated flue gas before emitting to the air, but unfortunately its more efficient in collecting coarse PM rather than fine. The fine PM fractions particularly smaller than 2.5 μm or PM2.5 are harmful to human health especially in relation to respiratory and cardiovascular disease [5,6] hence the emission needs to be addressed earnestly. The conventional multicyclone has limitations of low collection efficiency on fine particulate, therefore it is merely used as pre cleaner rather than as a main air pollution control system in industry [7]. Hence emerging a motivation in increasing its performance especially

* Corresponding author: mrmy99@gmail.com

collection of particulate with size fraction of 2.5 μm and below (hereinafter $PM_{2.5}$), the existing conventional multicyclone system was retrofitted at its dust hopper.

2 Methodology

2.1 The retrofitted pilot plant

Fig. 1 presents the experimental setup of the pilot plant scale multicyclone system, which showed the retrofitted section (images inside the dotted box) of an additional extraction stream attached to the dust hopper. The conventional multicyclone system consists of four identical miniature cyclones in a single compartment equipped with a dust feeder and a pressure gauge to determine multicyclone's differential pressure, dust hopper, multicyclone ID fan (designated as ID multicyclone or IDC) and multicyclone stack (CS). The conventional multicyclone was based on the designed done in the previous study [8]. The retrofitted extraction stream is equipped with a valve and an ID Extraction fan or IDS to regulate the air volumetric flow rate of the extraction stream. Similarly, the airflow rate of the IDC is manipulated separately through a control panel attached to the pilot plant. The experiment was performed without and with extraction flow rate created at the dust hopper via the extraction stream by the IDS.

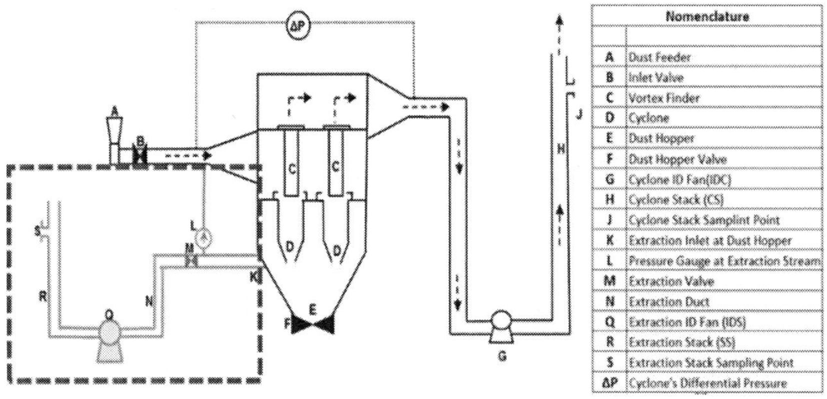

Fig.1. The pilot plant multicyclone system with the retrofitted stream in dotted region.

2.2 The particulate sample

The palm oil mill boiler fly ash (POMBFA) used in the experiments were obtained from two separate mills from the dust hopper of their respective air pollution control system. Each sample was dried in an oven at temperature 105±5°C for 24hr [9] and then the samples were thoroughly mixed and sieved through a 100um sieving tray (Endecotts Laboratory Test Sieve BS 410-1). The sample passing through the 100μm tray was used in the experiment.

2.3 The experiments on non-retrofitted multicyclone

Initially, the experiments were conducted with and without the multicyclone unit at air volumetric flow rate of $0.19m^3/s$ using two different inlet dusts loading of 0.4 and 0.6 mg/m^3, where the emission concentration of $PM_{2.5}$ was measured at the cyclone stack. Later, the experiments were performed with the multicyclone at two different air volumetric flow rates of 0.11 and 0.19 m^3/s, which corresponds to inlet velocity of 6.0 and 11.0 m/s respectively. A similar inlet dust loading of 0.4 and 0.6 g/m^3 was applied in this case. The emission concentration of $PM_{2.5}$ was measured isokinetically using Dekati® PM_{10} Impactor following the USEPA Method 17: Determination of Particulate Emissions From Stationary Sources -In-Stack Filtration Method. The $PM_{2.5}$ was collected on a filter media placed in the impactor where its particulate mass emission concentration was calculated using Equation 1.

$$C = \frac{M_f - M_i}{Q \times t} \quad (1)$$

where,
C = emission concentration, g/m^3
M_f = mass of filter paper after experiment, g
M_i = mass of filter paper before experiment, g
t = time of sampling, s
Q = isokinetic flowrate of sampling pump, m^3/s

2.4 The Experiments of Retrofitted Multicyclone

The experiments of the retrofitted multicyclone was performed at flow rate of 0.11 m^3/s (corresponding to inlet velocity of 6.0 m/s) at two different inlet dust loading of 0.4 and 0.6 g/m^3 with the air extraction of 15% and 20% of the total air volumetric flow rate of the unit. In this experiment, $PM_{2.5}$ emission concentration was measured at both cyclone and extraction stack sampling point. The emission concentration was obtained from with and without extraction and the results were compared in this study.

3 Results and Discussion

3.1 Performance on the non-retrofitted multicyclone

Fig. 2 presents the $PM_{2.5}$ emission concentration of the pilot plant with and without the presence of multicyclone units at a flowrate 0.19 m^3/s at inlet dust loading of 0.4 and 0.6 g/m^3 which showed that $PM_{2.5}$ emission concentration was reduced at range of 48% – 51% compared to without the multicyclone unit. In addition, the finding suggest that $PM_{2.5}$ constitutes 30-32% of the inlet dust loading.

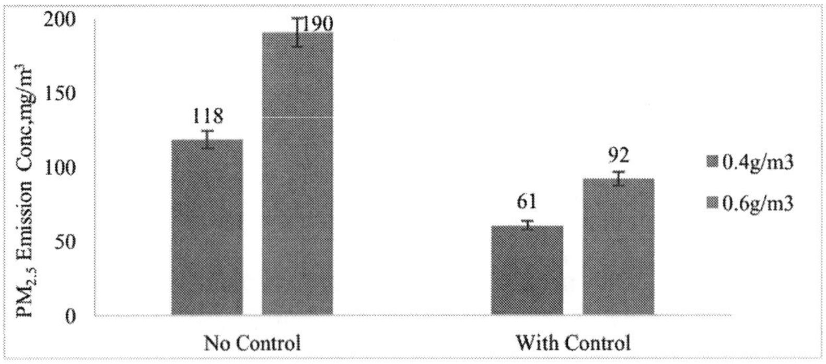

Fig. 2. $PM_{2.5}$ emission concentration with and without multicyclone unit at flow rate of 0.19 m^3/s at the respective inlet dust loading.

Fig. 3 presents the emission concentration of the plant with cyclone at inlet velocities of 6.0 and 11.0 m/s at two different inlet dust loadings of 0.4 and 0.6 g/m^3 which showed that $PM_{2.5}$ emitted from the cyclone increases with inlet dust loading and velocity as per observed in Fig. 2.

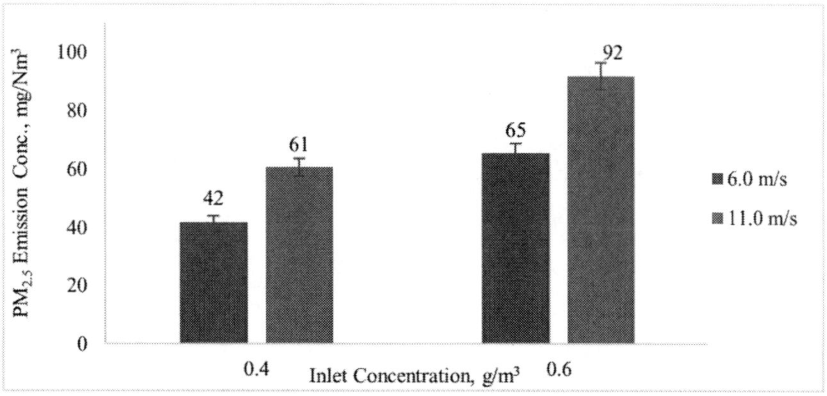

Fig. 3. $PM_{2.5}$ emission concentration of the conventional multicyclone with different inlet velocities and inlet concentration.

Similarly as observed in Fig. 3, a higher inlet concentration contains more of $PM_{2.5}$ therefore as expected more $PM_{2.5}$ would be emitted out from the multicyclone due to the limitations of collecting the overloaded $PM_{2.5}$. It seemed that the cyclone loses some of its swirl intensity in its vortex at high inlet concentration causing a dampening in its gas flow turbulence, which increases the emission concentration. Derksen *et al.* (2005) performed three-dimensional, time-dependent Eulerian–Lagrangian simulations of the turbulent gas–solid flow in a cyclone where through large-eddy representation of the gas flow, solid particles with different sizes were tracked and found that the collection efficiency gets affected in opposite senses: negatively by the loss-of-swirl, positively by the reduced turbulence in a cyclone[10].

In addition at higher inlet velocity of 11.0 m/s, the emission concentration of $PM_{2.5}$ is more compared to lower inlet velocity due to $PM_{2.5}$ re-entrainment where finer particle would easily escape from the cyclone due to the centrifuging out of the particles in tangential velocity field [11].

3.2 Performance on the retrofitted multicyclone

Fig. 4 presents the $PM_{2.5}$ emission concentration of conventional compared to retrofitted multicyclone at various inlet dust loading at constant inlet velocity of 6.0 m/s under 15% and 20% extraction ratio of extraction flowrate to multicyclone flowrate which showed a reduction in $PM_{2.5}$ emission upon applying the retrofit. This proved that the retrofit managed to selectively extract $PM_{2.5}$ out of the multicyclone's dust hopper simultaneously reducing the $PM_{2.5}$ emission concentration by 12 to 54% depending on the inlet concentration. As in Fig. 3, the collection efficiency which shows at extraction 15% for inlet concentration 0.4 g/m^3 the efficiencies are 89.5% and 93.5% and for 0.6 g/m^3 are 89.1% and 95% for conventional and retrofitted multicyclone respectively. On the contrary, the increase in extraction ratio at constant inlet velocity does not seem to contribute the reduction of $PM_{2.5}$ emitted which requires further investigations.

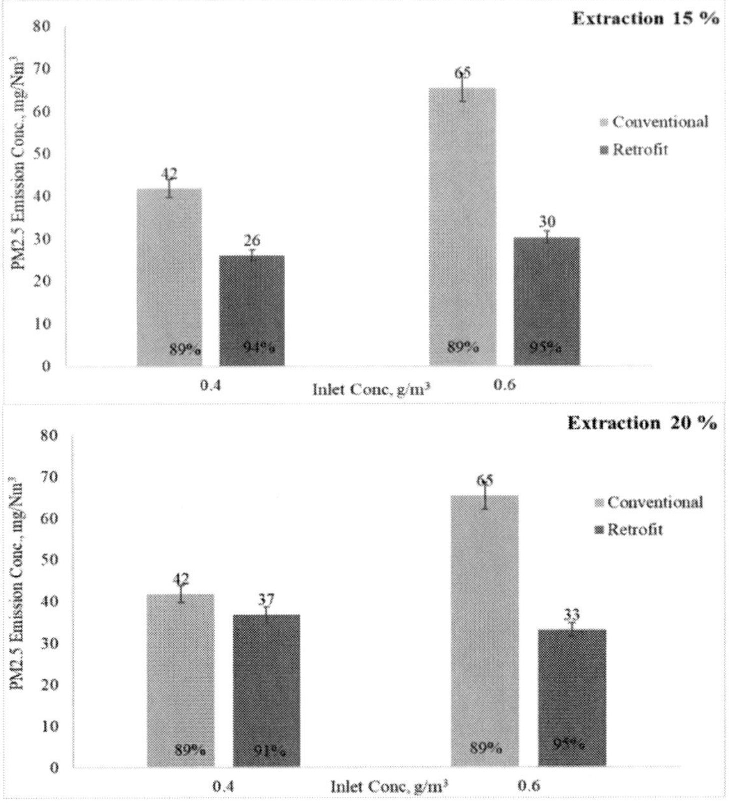

Fig. 4. $PM_{2.5}$ emission concentration at various inlet concentration at constant inlet velocity of 6.0m/s at 15% and 20% extraction ratio of extraction flow rate to multicyclone flow rate.

Based on Obemair *et al.* (2003), separation process in a gas multicyclone can be divided in three steps[12], where one of it is actually happens at the dust hopper therefore by introducing an extraction stream which is operated higher than the terminal falling velocity of $PM_{2.5}$, (0.425 mm/s) would cause more of $PM_{2.5}$ to be extracted out from dust hopper. Table 1 further validates that the extraction stream has selectivity towards $PM_{2.5}$ compared to the bigger size fraction because the ratio of $PM_{2.5}$ to PM_{10} is consistently at 90%. This phenomena is due to the inner vortex of the multicyclone is being extended under the influence of extraction causing fine particulate to be carried along with the extraction stream as modelled by CFD [13].

Table 1. Ratio of $PM_{2.5}$ to PM_{10} in the extraction stream.

Ratio $PM_{2.5}/PM_{10}$	15%	20%
0.4 g/m³	0.9	0.9
0.6 g/m³	0.9	0.9

4 Conclusions

The performance of the multicyclone on collection of $PM_{2.5}$ was evaluated by experiments before and upon installing the extraction stream at the dust hopper. Presence of extraction stream at the dust hopper of the multicyclone divides the airflow rates into two where a portion of $PM_{2.5}$ has been selectively extracted out from the dust hopper causing a reduction in $PM_{2.5}$ emitted from the multicyclone. The performance of retrofitted multicyclone in collecting $PM_{2.5}$ increases by at least 12% to 54% depending on the inlet concentration particularly in the collection of $PM_{2.5}$.

The first author is a doctoral candidate at the Malaysia-Japan International Institute of Technology (MJIIT), Universiti Teknologi Malaysia, sponsored by the Ministry of Higher Education Malaysia under the MYBrain15 research fellowship.

References

1. A. N. Huang. Sep. Purif. Tech. **174**, 408–416 (2017)
2. T.C. Hsiao, S.H. Huang, C.W. Hsu, C.C. Chen, P.K. Chang, J. Aerosol Sci. **86**, 1–12 (2015)
3. N. Hasyimah, M. Rashid, H. Norelyza, C. M. Hasrizam, *Int. Conf. Flu. & Chem. Eng.* 25–27 (2015)
4. N. Hasyimah, M. Rashid, H. Norelyza, Adv. Mater. Res. **1125**, 317–321 (2015)
5. K. Darcovich, K. A. Jonasson, C. E. Capes, Adv. Powder Technol. **8,** 179–215 (1997)
6. A. Seaton, D. Godden, W. MacNee, K. Donaldson, Lancet, **345**, 176–178 (1995)
7. K. S. Lim, H. S. Kim, K. W. Lee, J. Aerosol Sci. **35**, 103–116 (2004)
8. H. Norelyza, M. Rashid, J. Environ. Res. Dev. **7,** 1392–1398 (2013)
9. Y. Zarina, A. M. Mustafa, H. Kamarudin, I. K. Nizar, A. R. Rafiza, Rev. Adv. Mater. Sci. **34**, 37–43 (2013)
10. J. J. Derksen, Eng. Turb. Model. **163**, 929–938 (2005)
11. Y. Jo, C. Tien, M. B. Ray, Pow. Tech. **113**, 97–108 (2000)
12. S. Obermair, J. Woisetschläger, G. Staudinger, Pow. Tech. **138**, 239–251 (2003)
13. M. Dewika, M. Rashid , C.M. Hasrizam, *Int. Conf. Air Quality & Env. Sus.* (2017)

Acid Pretreatment of Sago Wastewater for Biohydrogen Production

Noor Illi Mohamad Puad[1,*], *Nurainin Farhan* Abd Rahim[2], and *Azlin Suhaida* Azmi[1]

[1]Bioprocess and Molecular Engineering Research Unit, Department of Biotechnology Engineering, Kulliyyah of Engineering, International Islamic University Malaysia, P.O Box 10, 50728 Kuala Lumpur, Malaysia

Abstract. Biohydrogen has been recognized to be one of the future renewable energy sources and has the potential in solving the greenhouse effects. In this study, *Enterobacter aerogenes* (*E. aerogenes*) was used as the biohydrogen producer via dark fermentation process using sago wastewater as the substrate. However, pretreatment of sago wastewater is required since it consists of complex sugars that cannot be utilized directly by the bacteria. This study aimed to use acid pretreatment method to produce high amount of glucose from sago wastewater. Three different types of acid: sulfuric acid (H_2SO_4); hydrochloric acid (HCl) and nitric acid (HNO_3) were screened for the best acid in producing a maximum amount of glucose. H_2SO_4 gave the highest amount of glucose which was 9.406 g/L. Design of experiment was done using Face-centred Central Composite Design (FCCCD) tool under Response Surface Methodology (RSM) in Design Expert 9 software. The maximum glucose (9.138 g/L) was recorded using 1 M H_2SO_4 at 100 °C for 60 min. A batch dark fermentation using *E. aerogenes* was carried out and it was found that pretreated sago wastewater gave a higher hydrogen concentration (1700 ppm) compared to the raw wastewater (410 ppm).

1 Introduction

Hydrogen gas has the potential as a viable alternative fuel and energy carrier of the future since it is a clean fuel with no CO_2 emissions and can be used for generation of electricity. Hydrogen has an enthalpy energy of 122 kJ/g which is 2.75 times greater than hydrocarbon fuels [1]. Hydrogen can be produced chemically and biologically. Indirect biophotolysis, photofermentation, dark fermentation and hybrid system are the examples of biohydrogen process [2]. Dark fermentation is the production of biohydrogen from microorganisms using agricultural waste as the substrate [3]. *E. aerogenes* is a facultative anaerobe hydrogen producing bacteria that can alter the oxygen condition to anoxic condition that recovered the activity of Fe-hydrogenase to produce hydrogen via dark fermentation [4].

In this study, sago wastewater was used instead of sago fibrous pith residues as the alternative substrate for source of glucose production since it contains a high amount of starch [5]. It was reported that about 2.2 mol H_2/mol glucose was produced by *Escherichia*

* Corresponding author: illi@iium.edu.my

coli using sago wastewater as a substrate [6] which showed. Since sago wastewater has complex structure sugars, it needs to be broken down into the simpler fermentable sugar to be utilized by *E. aerogenes*. Two common hydrolysis methods in sago wastewater pretreatment are acid and enzymatic pretreatment. In acid hydrolysis, acid acts as a catalyst that breaks the glycosidic bond into maltose, maltotriose, glucose and dextrin depending on which chain is attacked [7]. In the enzymatic hydrolysis, amylase is used to breakdown the starch into fermentable sugars [8]. However, those reported pretreatments are specifically for sago sago fibrous pith residues and not sago wastewater.

This study was carried out to determine the optimum conditions for the acid pretreatment of sago wastewater in order to obtain maximum fermentable sugar to be used as the substrate by *E. aerogenes* for biohydrogen production.

2 Method

2.1 Collection, preparation and charaterization of sago wastewater

Sago wastewater was collected from a sago mill in Batu Pahat, Johor. The sago wastewater was in brownish color with approximately composed of 60 % water and 40 % sediment. Characterization of sago wastewater was conducted using the standard method for examination of wastewater [9]. Upon use, the wastewater was kept in the cold room to prevent the growth of bacteria while its sediment was sun-dried for 9 hours and then blended into the powder form.

2.2 Screening of acid solution for sago wastewater pretreatment

Three acid solutions that have been chosen for the screening purpose were H_2SO_4, HCl and HNO_3. The aim of the screening was to select the acid that produced the highest amount of glucose from the sago wastewater. Thus, only type of acid solution was varied, while the value of other parameters which were acid concentration (0.5 M), incubation temperature (90 °C) and incubation time (40 min) were kept constant. Sago wastewater samples were used with the composition of 100 mL of sago liquid and 2 g of dried sediment. About 150 mL of 0.5 M of each acid was added into 250 mL conical flask and the solution was stirred well for 1-2 minutes on the hotplate magnetic stirrer at the room temperature. The solution was then placed into the water bath at 90 °C for 40 minutes.

After that, samples were cooled down in the ice water bath and filtered using vacuum filter in order to separate any solid particles from the aqueous solution. NaOH (2 M) solution was added into the filtered solutions for neutralization. The solution was then submerged in the ice water bath to absorb the heat of neutralization and to avoid further decomposition of fermentable sugars [7]. Lastly, glucose concentration was analyzed using dinitrosalicylic acid (DNS) method [10].

2.3 Optimization of pretreatment process parameters

Three parameters of acid pretreatment were chosen to be optimized which are acid concentration (M), incubation temperature (°C) and incubation time (min) with the selected ranges as shown in Table 1. This selection of ranges was varied respectively based on the previous reports [11-13]. Then, the experiment was designed using FCCCD of RSM in the Design Expert 9 software for the analysis of optimum condition for acid pretreatment of sago wastewater. The response was the concentration of glucose (g/L) and all experiments were conducted in triplicates.

Table 1. Experimental range and level of three parameters used in the study.

Factor / Level	Incubation temperature (°C)	Incubation time (min)	Acid concentration (M)
1	100	90	1.5
0	85	60	1.0
-1	70	30	0.5

2.4 Fermentation of *E. aerogenes* for biohydrogen production

2.4.1 Preparation of E. aerogenes inoculum

The inoculum was prepared aerobically by adding 10 mL of *E. aerogenes* broth in a 250 mL conical flask with 90 mL LB media. The flask was placed in an incubator shaker at 180 rpm and 37°C for 18 hours. The bacteria were actively grown at the exponential phase after 4 hours of inoculation and it was then used as the inoculums or stock culture [14]. Then, the OD of the inoculum was measured at the wavelength of 600 nm using a spectrophotometer.

2.4.2 Media preparation

Two different media that contains the pretreated and raw (non-treated) sago wastewater, respectively were prepared. The raw sago wastewater was prepared by mixing 2 g of sediment with 100 mL of sago liquid. The pretreated sago wastewater was prepared by mixing 2 g sediment with 100 mL sago liquid and the pretreatment process was conducted according to the suggested parameters from the RSM model in the Design Expert 9.0 which were for 88.01 min at 99.9 °C using 1.31 M H_2SO_4 concentration and it was then followed by filtration and pH neutralization as mentioned in Section 2.2. The initial glucose concentration for both media were then measured using DNS method.

2.4.3 Experimental set-up and batch fermentation process

Serum bottles with a capacity of 125 mL were used for the dark fermentation of *E. aerogenes* which was air-sealed and covered with a silicon stopper and aluminium seal cap [15]. An amount of 4 mL inoculum was added to each 76 mL raw and pretreated sago wastewater to make a final working volume of 80 mL. The fermentation was conducted in duplicates at 37 °C with 180 rpm for 48 hours [14]. After 48 hours of incubation, the glass syringe was used to collect the hydrogen gas accumulated in the serum bottle and then the hydrogen analysis was performed. The final glucose concentration in each serum bottle was measured using DNS analysis to observe the glucose conversion after acid pretreatment.

2.4.4 Hydrogen gas analysis

Hydrogen gas was accumulated in the serum bottle after 48 hours. Hydrogen gas concentration for each sample was analyzed using a portable hydrogen gas analyzer and expressed in the unit of part per millions (ppm). If the concentration was more than 1125 ppm, gas dilution was done by taking 10 mL of biogas and mixing it in the 125 mL of empty serum bottle. After that, the concentration of hydrogen was measured again. During gas collection using the glass syringe, the gas pushed up the glass syringe. Thus, the total

 https://doi.org/10.1051/e3sconf/20183402012

volume of biogas equals to the total volume of gas in the serum bottle (45 mL) and gas in the glass syringe. The biogas in the serum bottles consist of carbon dioxide, hydrogen, methane and other gases.

3 Results and Discussions

3.1 Screening of acid solution

It was observed that pretreatment using H_2SO_4 produced the highest glucose concentration which was 9.406 g/L compared to HCl and HNO_3 (Table 2). This result is in agreement with [16] which reported that H_2SO_4 was the best solution for lignocellulosic biomass pretreatment in producing simple sugars. This is due to the ability of diluted H_2SO_4 to effectively remove and recover most of the hemicellulose as dissolved sugars. It was also concluded that H_2SO_4 reached about 72 % dissolution of polysaccharides compared to HCl which was only 41 % dissolution [17]. HNO_3 has also recorded a good conversion of starch to glucose, but its higher price would result in less cost effective in acid recovery for hydrolysis [11]. Based on these reasons, H_2SO_4 was used as the acid solution for optimizing sago wastewater pretreatment parameters.

Table 2. Comparison of glucose concentration after sago wastewater pretreatment using different types of acid at 0.5 M.

Type of acid (0.5 M)	Glucose concentration (g/L)
HCl	6.902
H_2SO_4	9.406
HNO_3	7.615

3.2 Optimization of process parameters for sago wastewater acid pretreatment

Based on the experimental design using FCCCD in Design Expert 9.0, a total of 20 runs were carried out for different ranges of incubation temperature (70, 80 and 100 °C), incubation time (30, 60 and 90 min) and acid concentration (0.5, 1 and 1.5 M). The initial glucose concentration in the raw sago wastewater was measured and its concentration was less than 0.1 g/L. Hence, it can be considered that before pretreatment, sago wastewater contained insignificant amount of glucose. Table 3 lists out glucose concentration (g/L) for all 20 runs that haven conducted. Acid pretreatment at 100 °C for 60 minutes with 1 M H_2SO_4 gave the highest amount of glucose which was 9.138 g/L (Run 13).

Further analysis on the glucose concentration using ANOVA was carried out and the result is summarized in Table 4. It shows that the model given was significant since (Prob>F-value<0.05) and the lack of fit was not significant (Prob>F value>0.1). In addition, the most significant factor affecting the maximum glucose concentration is the incubation temperature where its value of "Probability>F" was 0.0001 which was less than 0.05. It was then followed by acid concentration and incubation time which were less significant. This result can be supported by the previous findings which concluded that at the mild temperature (~120 °C), there was a significant recovery of sugar produced while a higher temperature of pretreatment would cause in the sugar degradation and thus lead to the formation of inhibitors [18].

Table 3. Glucose concentration in sago wastewater after acid pretreatment as a response during optimization using FCCCD.

Run	Incubation temperature (°C)	Incubation time (min)	Acid concentration (M)	Glucose concentration (g/L)
1	85	30	1	2.54
2	85	60	0.5	3.47
3	85	60	1	3.41
4	85	60	1	4.47
5	85	60	1	4.02
6	85	60	1	5.14
7	85	60	1	8.93
8	85	60	1.5	4.68
9	85	90	1	5.84
10	70	90	0.5	0.69
11	100	90	1.5	9.08
12	100	90	0.5	7.00
13	100	60	1	9.13
14	100	30	0.5	4.77
15	100	30	1.5	8.67
16	70	30	0.5	0.48
17	70	60	1	0
18	70	30	1.5	0.33
19	70	90	1.5	1.45
20	85	60	1	6.39

Since incubation temperature is a crucial factor that affects directly to the production of glucose compared to other factors, choosing the appropriate temperature is essential depending on the types of agricultural wastes used in the pretreatment. The "Lack of Fit F-value" of 0.37 implies that the lack of fit condition is not significant and it is a 92.17% chance of the value could occur due to noise. This non-significance lack of fit is good since the model is required to be fitted to the linear regression. The final equation in terms of actual factors is shown in Equation 1.

$$\textit{Glucose concentration} = -18.71970 + 0.23806 \times \textit{Incubation temperature} + 0.02424 \times \textit{Incubation time} + 1.5592 \times \textit{Acid concentration} \quad (1)$$

Although the highest glucose amount has been achieved from acid pretreatment of sago wastewater in Run 13 (Table 3), its process parameters have yet to be optimized. This is based on the 3D response surface model shown in Fig. 1. It depicts that a higher temperature (>100°C) should be used in order to obtain an optimized pretreatment temperature. Monavari *et al.* (2009) [19] proposed that acid pretreatment should be carried out using H_2SO_4 with the concentration range of 0.5 to 1.5 % (w/w) at the temperature of 121-160 °C. From the pretreatment conditions, the yield of sugar from hemicelluloes achieved was 70 % - 95 %.

Table 4. ANOVA Analysis Regression coefficient of predicted models for the response of glucose concentration.

Source	Sum of square	DF	Mean square	F-value	Probability> F	
Model	138.88	3	46.29	20.20	<0.0001	**Significant**
Incubation temperature (°C)	127.51	1	127.51	55.64	<0.0001	
Incubation time (min)	5.29	1	5.29	2.31	0.1483	
Acid concentration (M)	6.08	1	6.08	2.65	0.1230	
Residual	36.67	16	2.29			
Lack of fit	16.44	11	1.49	0.37	0.9217	**Not significant**
Pure error	20.23	5	4.05			
Cor total	175.55	19				

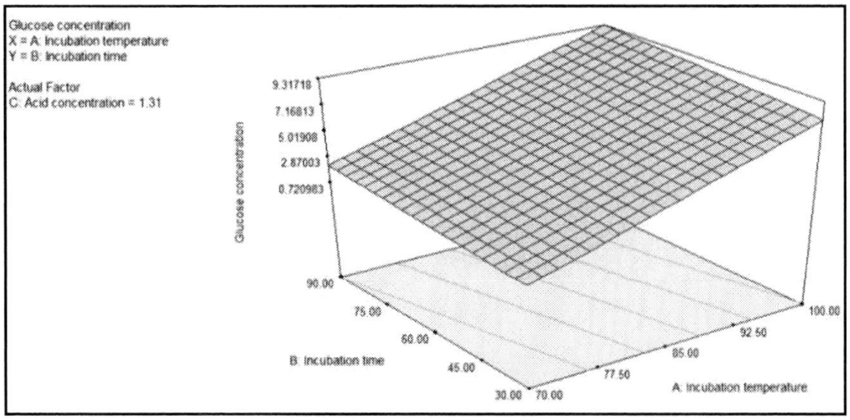

Fig. 1. 3D response surface for model graph.

3.3 Fermentation of *E. aerogenes* for biohydrogen production using pretreated sago wastewater

A batch dark fermentation experiment was carried out to examine the feasibility of pretreated sago wastewater as the substrate for biohydrogen production using *E. aerogenes*. Raw sago wastewater was used as the control. The results shown in Table 5 demonstrates that the pretreated sago wastewater is a potential substrate for *E.aerogenes* to produce biohydrogen. The final concentration of biohydrogen achieved in the media supplemented with pretreated sago wastewater was 1700 ppm (3.59 µmol) which is about 4 times higher compared to the raw sago wastewater. It shows that pretreatment is essential in order to break down the complex sugar into the simplest one so that *E. aerogenes* can easily utilize the sugar to carry out their metabolic activities for biohydrogen production. Previously, it was reported that *E. aerogenes* produced 1.0 mol of hydrogen from 1.0 mol of glucose [20].

Table 5. The final concentration of biohydrogen produced by *E. aerogenes* after 48 hours of incubation.

Substrate	Final concentration of hydrogen (ppm)	Concentration of hydrogen (µmol)
Raw sago wastewater	410	0.81
Pretreated sago wastewater	1700	3.59

4 Conclusions

H_2SO_4 was used in the sago wastewater acid pretreatment since the glucose concentration based on its absorbance showed the highest value compared to HCl and HNO_3. The optimization study of pretreatment process parameters for maximum glucose production using FCCCD was yet to achieve due to the range of temperature used but the highest glucose produced (9.138 g/L) was achieved from run 13 (1 M H_2SO_4 at 100 °C for 60 minutes) with a significant model based the ANOVA analysis. For hydrogen analysis after fermentation, the pretreated sago wastewater achieved was about 4 times higher of hydrogen concentration compared to the raw sago wastewater.

Overall, acid pretreatment sago wastewater is an important step before any agricultural waste can be used as the substrate for fermentation in producing biofuel and biohydrogen. Optimization of important pretreatment parameters such as incubation temperature and concentration of acid is important to improve the yield of glucose from the complex sugars.

References

1. Y. Nalakath, C. M. Malikudy, J. Jose, International Journal of Engineering Research & Technology **6**, (2017)
2. K. Nath, & Das, D. , Journal of Scientific & Industrial Research **63**, (2004)
3. T. A. Ulhiza, N. I. M. Puad, A. S. Azmi, Journal of Advanced Research in Materials Science **38**, 21-31 (2017)
4. N. I. M. Puad, N. A. Mamat, A. S. Azmi, *International Proceedings of Chemical, Biological & Environmental Engineering* (IPCBEE) **93**, 55-61 (2016)
5. K. Sabariswaran, Papitha, P., Selvakumar, S., & Raj, S. P. , Journal of Biodiversity and Environmental Sciences **3**, 17-23 (2013)
6. N. I. M. Puad, S. Sulaiman, A. S. Azmi, Y. Shamsudin, M. Mel, Journal of Engineering Science and Technology, 11-21 (2015)
7. E. C. Kumoro, A. C. Ngoh, G. C. Hasan, M. Ong, C. H. Teoh, Asian Journal of Scientific Research **1**, 412-420 (2008)
8. R. Sunaryanto, Handayani, B.H., & Safitri, R., Microbiology Indonesia **7**, 68-74 (2013).
9. A. P. H. *Association, in Work.* Assoc. Environ. Fed. (1998), pp. 552.
10. G. L. Miller, Analytical Chemistry **31**, 426-428 (1959)
11. M. Tutt, Kikas, T., & Olt, J. , *Biosystem Engineering Special Issue*, 269-276 (2012).
12. K. J. Dussán, Silva, D. D. V, Moraes, E. J. C., Priscila, V., & Felipe, M. G. A., Chemical Engineering Transactions **38**, 433-438 (2014)
13. M. Idrees, A. Adnan, S. Sheikh, F. A. Qureshic, EXCLI Journal **12**, 30-40 (2013)
14. P. Mishra, D. Das, International Journal of Hydrogen Energy **39**, 7496-7507 (2014)
15. J. Procházka, J. Mrázek, L. Štrosová, K. Fliegerová, J. Zábranská, M. Dohányos, Engineering in Life Sciences **12**, 343-351 (2012)

16. L. P. Ramos, Química Nova **26**, 863-871 (2003).
17. S. Koti, S. P. Govumoni, J. Gentela, L. Venkateswar Rao,. SpringerPlus **5**, 1545 (2016)
18. B. Yang, C. E. Wyman, Biofuels, Bioproducts and Biorefining **2**, 26-40 (2008)
19. S. Monavari, M. Galbe, G. Zacchi, Biotechnology for Biofuels **2**, 6 (2009)
20. S. Tanisho, N. Kamiya, N. Wakao, Bioenergetics **973**, 1-6 (1989)

Reusability Performance of Zinc Oxide Nanoparticles for Photocatalytic Degradation of POME

Nur Zarifah Zainuri[1], *Nur Hanis Hayati* Hairom[1,*], *Dilaelyana Abu Bakar* Sidik[1,2], *Nurasyikin* Misdan[1], *Norhaniza* Yusof[3,4], and *Abdul Wahab* Mohammad[5]

[1]Department of Chemical Engineering Technology, Faculty of Engineering Technology, Universiti Tun Hussein Onn Malaysia, 86400 Parit Raja, Batu Pahat, Johor, Malaysia
[2]Department of Science and Mathematics, Center of Diploma Studies, Universiti Tun Hussein Onn Malaysia, 86400 Parit Raja, Batu Pahat, Johor, Malaysia.
[3]Faculty of Chemical and Energy Engineering, Universiti Teknologi Malaysia, 81310 Skudai, Johor Bahru, Malaysia
[4]Advanced Membrane Research Technology Centre (AMTEC), 81310 Skudai, Johor Bahru, Malaysia
[5]Centre for Sustainable Process Technology (CESPRO), Faculty of Engineering and Built Environment, Universiti Kebangsaan Malaysia, 43600 UKM Bangi, Selangor, Malaysia

Abstract. Performance and reusability of different zinc oxide nanoparticles (ZnO-PVP and ZnO-PEG) for photocatalytic degradation of palm-mill oil effluent (POME) has been studied. The nanoparticles properties were characterised with fourier transform infrared spectroscopy (FTIR), X-ray diffraction (XRD) and transmission electron microscopy (TEM). The TEM results show that ZnO-PEG nanoparticles exhibit the smaller size than ZnO-PVP with less agglomeration. It was found that ZnO-PEG shows better effectiveness than ZnO-PVP in reducing turbidity, colour and increasing the dissolved oxygen (DO). By using two types of reusability methods: (a) oven drying (b) hot water rinsing, the oven drying method portrayed the most efficient route for POME treatment. This research would be a solution to the palm oil industry for photocatalyst recovering as well as reduction of the chemical usage in order to meet the development of advanced and greener technologies.

1 Introduction

While Malaysia relishes the profitable benefit from the massive production of palm oil, the impact of the production of a large volume of waste has also turn out to be a distressing consequent. Ominously, other than printing and dyeing wastewater, major water pollutants are contributing from the oil palm plantation. POME can be described as a thick brownish effluent which is rich in organic contaminants and discharged at the temperature range 80-90°C[1]. Some approach had been done in order to minimize the environmental threat, which is by using biological treatment. Nonetheless, the brownish colour of the POME still

* Corresponding author: nurhanishayati@gmail.com

cannot be diminished. Therefore, in order to protecting the environment, there were several regulatory controls over by the palm oil mills before they discharge the POME as in-scripted in the Environmental Quality (Prescribed Premises) (Crude Palm Oil) Regulations, 1977 promulgated under the Environmental Quality Act (EQA) 1974 and fallen under the jurisdiction of the Department of Environmental (DOE)[2]. Table 1 showed the standard discharged limit set by DOE and EQA[3].

Table 1. The standard discharged limit set by DOE and EQA.

Parameters	DOE discharged limit (1986 onwards)	Environmental Quality Act
pH	5.0	5.0-9.0
Temperature (°C)	45	45
BOD_3 (mg/L)	50	100
COD (mg/L)	1000	1000
Total solids (mg/L)	1500	1500
Suspended solids (mg/L)	400	400
Oil and grease (mg/L)	50	50
Ammoniacal nitrogen (mg/L)	100	150
Total nitrogen (mg/L)	200	200

One of the promising approaches that can be implemented in order to degrade the colour of POME is by using photocatalytic degradation process [4]. During photocatalytic degradation process, hazardous organic chemical will be mineralized to become water, carbon dioxide and simple mineral acids [5]. Previously, Mondal & Sharma claimed that photocatalytic degradation is one of the promising methods due to their capability for fully mineralising the organic contaminants under mild conditions such as ambient temperature and pressure [6]. Other than that, Beydoun *et al.,* also reported that the advantage of the photocatalyst used in photocatalysis processes is they can be reused or recycled by their self-generated [7]. In addition, amongst all the methods for the photocatalytic degradation of dyes, ultraviolet (UV) irradiation had been acknowledged for its effectiveness in remove the pollutants and the environmental purification[8].

The performance of the photocatalyst such as ZnO nanoparticles in the photocatalysis process is undeniable especially in the utilization of coupling photocatalytic reactor and membrane. Recently, Desa *et al.,* and Chiranont *et al.,* reported that ZnO-PVP and ZnO-PEG nanoparticles have a great potential as a photocatalyst for industrial dye wastewater treatment [10-11]. Along with its steady wurtzite structure and wide bandgap (3.4 eV) compound semiconductor, ZnO nanoparticles has been generally used as a catalyst for dye wastewater treatment [11] By using ZnO as photocatalyst, Hairom *et al.,* managed to achieve 100% colour removal, 100% turbidity reduction, 100% suspended solid rejection by coupling the photocatalysis and the membrane filtration [12].

There were few researches concerning the reusability of photocatalyst in photocatalysis process. Most of them were succeed to reutilize the photocatalyst back in the remediation of the wastewater with a slightly reduced performance after a few cycle of consecutive run. For instance, Velmurugan *et al.,* had been undergo three consecutive run in order to study the photocatalyst reusability which is the ZnO nanocrystals. The photocatalyst did shown a remarkably photostability by degraded around 85.8% of dye after the third run [13]. Other than that, Gomez *et al.,* had been reutilize the photocatalyst by filtering, washed, and dried in air oven under 70°C. The result did not portray a large different in term of degradation percentage nor the mineralization for the 1st and the 8th cycle [14]. It is very important for the industries as they can lower their chemical usage besides reducing their cost.

Hence, the main objective of this study is to elucidate the performance and reusability of zinc oxide nanoparticles in photocatalytic degradation of POME. The output from this study will help to provide the fundamental understanding of photodegradation process if they are applied within a membrane photocatalytic reactor for industrial wastewater treatment.

2 Materials and chemicals

2.1 Synthesis of ZnO nanoparticles

ZnO nanoparticles were synthesized by precipitation method in a manner corresponding to Hairom *et al.*,[12]. They were prepared by added 0.15 M of oxalic acid dehydrate solution (obtained from R&M Marketing, Essex, UK) slowly into 0.1 M zinc acetate dehydrate solution (obtained from R&M Marketing, Essex, UK) with the molar ratio 1.5. They were conducted in the room temperature (25°C) under two conditions:

(i) vigorous stirring in the presence of polyethylene glycol (ZnO-PEG)
(ii) vigorous stirring in the presence of polyvinyl pyrrolidone (ZnO-PVP)

For the preparation of ZnO-PVP, 0.015g/L of PVP and 0.025 g/L of PEG (both obtained from R&M Marketing, Essex, UK) were transferred into the mixture respectively after 5 min of reaction. Then, the mixture was stirred for 12 hours under room temperature. The reaction yields precipitate which was filtered and dried at 100°C in oven for 1 hour to remove the excessive water. Subsequently, the precipitate was calcined in furnace under 550°C for 3 hours to remove all the impurities.

2.2 Characterization of ZnO nanoparticles

Chemical bond of the ZnO nanoparticles was characterized by using Fourier transform infrared spectroscopy (FTIR) spectrometer. FTIR spectra was recorded in transmittance mode in an Agilent Technologies Cary 600 series FTIR spectrometer at 4cm-1 of spectral resolution in the range 400-4000cm-1 with 16 scans. The FTIR spectra were obtained directly on the powdery sample repeating the acquisition for 2 times. Next, the purity and crystallinity of the ZnO-PVP and ZnO-PEG samples were analysed with X-ray diffractometer (XRD) (Bruker AXS GmbH model) by using the radiation of Cu Kα with λ = 1.5418Å at 40 kV and 40 mA. The data were collected at room temperature in the range of 2Θ = 10-80°, continuously with the scan rate of 0.025°. High-resolution images for the structural properties of ZnO nanoparticles was analysed by the high resolution transmission electron microscopy (TEM), 200 kV (JEOL JEM-ARM 200F). It offers the ultimate stability for imaging and analysis at the sub-nanometer scale.

2.3 POME sampling and preservation

Raw POME sample wastewater was directly collected from the effluent point of the ponding treatment system of palm oil mill located in Johor. The initial POME characteristics such as pH, dissolved oxygen (DO) turbidity and colour were analysed by using standard methods.

2.4 Photocatalytic degradation of POME

A UV lamp, predominately emit at 365 nm with the definite power of 15 W was employed as the light source. The schematic diagram of the laboratory-scale of this study is shown in Fig. 1. ZnO nanoparticles were added into the POME wastewater. Before starting the photocatalysis process, the mixture was well agitated at 300 rpm for 30 min in the dark to reach adsorption-desorption equilibrium of photocatalyst. The operation temperature was kept constant at room temperature (25°C) by immerse it inside the water bath. The treated POME wastewater was sampled by 5ml at the interval of 5 minutes and the mixed solution (POME and ZnO nanoparticles) was separated by using centrifuge Gyrozen 1736R model for 20 min. The treated wastewater was analysed in term of DO, turbidity and colour intensity. The color intensity was determined with DR 5000 UV-Vis Laboratory Spectrophotometer (Hach). DO was measured by using DO meter (Hach, H280G). The turbidity was analysed using turbidity meter respectively. The percentage of dye removal and DO increment were calculated according to Equation (1) and (2), respectively: C_o and C_t denote the colour intensity at time zero and at time t, respectively whereby DO_o and DO_t represent the dissolve oxygen at time zero and at time t, respectively.

$$Percentage\ of\ dye\ removal = C_o\text{-}C_t\ /\ C_o \times 100 \quad (1)$$

$$Percentage\ of\ DO\ increment = DO_o\text{-}DO_t\ /\ DO_o\ 100 \quad (2)$$

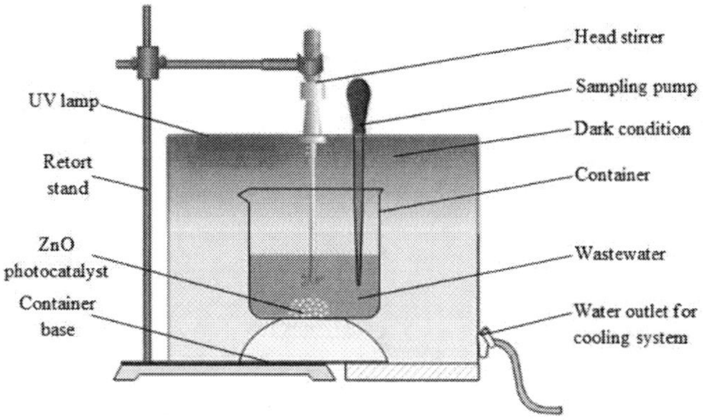

Fig. 1. The schematic diagram of photocatalysis process.

2.5 Reusability method of the ZnO nanoparticles

2.5.1 Oven drying method at 100°C

This method was employed by referring Subash et al. [15]. The photocatalyst was recovered after the photocatalysis process and was run with water. After that, it was dried at 100°C in air oven for 5h. Then, the photocatalyst was ready to use in the next run.

3.2.1 Hot water rinsing method

This method had been studied by Fathy et al. [16]. The catalyst was reutilized by washed it thoroughly using hot distilled water and dried at 80°C overnight for every cycle. Then, the photocatalyst was ready to use in the next run.

3 Results and discussions

3.1 Characterisation of ZnO nanoparticles

3.1.1 FTIR analysis of ZnO nanoparticles

The purity, nature and chemicals bonds of ZnO-PEG nanoparticles were shown in Fig. 2. The first figure, which is Fig. 2(a) indicates the FTIR pattern of ZnO-PEG. Prominent absorption peaks for ZnO-PEG were obtained at 3414.14, 1509, 877.55 and 632 cm^{-1}. The peaks at 3414.14 cm^{-1} and 877.55 cm^{-1} indicates the O-H stretching and deformation due to the adsorption of moisture from surrounding during synthesizing the photocatalyst. The peak observed at 632 cm^{-1} are correspond to Zn-O stretching and deformation vibration while the peak at 1509 cm^{-1} indicates the vibration and stretching of C=O bond. The second figure, which is Fig. 2(b) designates the FTIR pattern of ZnO-PVP. As can be seen in the figures, there were several characteristics absorption bands appearing in the range of 600-4000 cm^{-1} due to covalent bonds of PVP [17]. The peaks that appeared from 1068 and 612 cm^{-1} are attributed as the stretching of C-C rings of breathing of pyrrolidone. The absorption bands at 1634 cm^{-1} due to ZnO stretching and deformation vibration respectively [18].

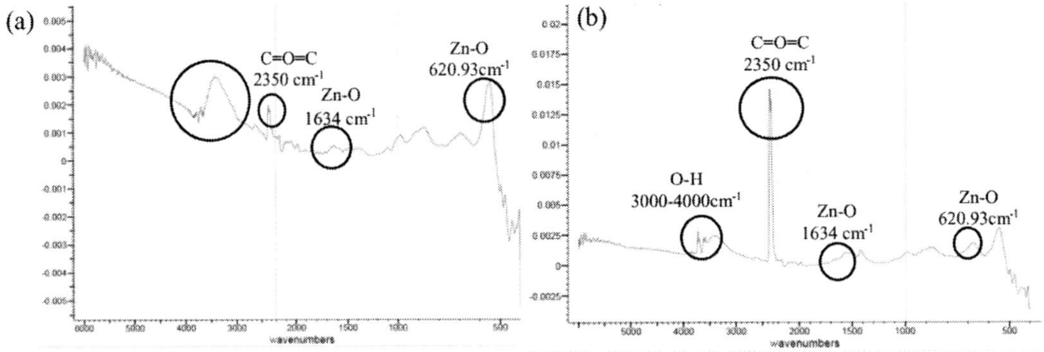

Fig. 2. IR spectra of: (a) 0.015g/L ZnO-PEG and (b) 0.025g/L ZnO-PVP.

The results showed that C=O bond existed in the range from 2327 until 2420 cm^{-1} in the sample as stated by Singh *et al.*, [19]. They also revealed that C=O group carboxylic derivatives present due to residue of zinc acetate used in reaction. It might ensue because of the precursor material and reaction product. The peaks located from 3000 until 4000 cm^{-1} represents as the vibration mode of O–H group of the moisture which make the bond arises and proved that ZnO-PVP samples has absorbed the existence water from surroundings [17].

3.1.2 XRD analysis of ZnO nanoparticles

The crystallinity of the ZnO nanoparticles had been shown by the XRD pattern in Fig. 3. The crystalline regions of the ZnO nanoparticles had been clearly observed at the range 30° to 40° (2θ values). The results of the XRD patterns corresponded well with the standard wurzite structure at planes ⟨1 0 0⟩, ⟨0 0 2⟩ and ⟨1 0 1⟩ [20]. Furthermore, it was found that the less intense peak at the range 40° to 80°, which indicates the high crystallinity of the different forms of ZnO.

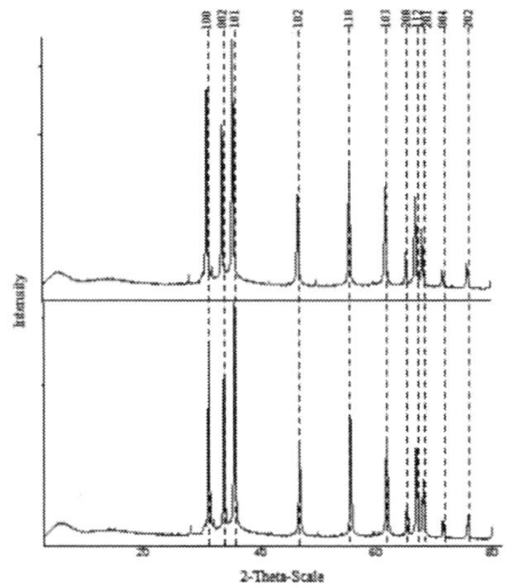

Fig. 3. XRD patterns of: (a) ZnO-PEG and (b) ZnO-PVP.

3.1.3 TEM analysis of ZnO nanoparticles

The morphology and size of the ZnO-PEG and ZnO-PVP nanoparticle were studied using TEM as shown in Fig. 4 The result showed a less agglomerated particle between two types of the ZnO nanoparticle. However, it was observed that the particle size of ZnO-PEG provided the smaller size compared to ZnO-PVP nanoparticles. The particle size of ZnO-PEG ranged from 60 to 150 nm and particle size of ZnO-PVP ranged from 80 to 150 nm respectively. According the result, ZnO-PEG give a smallest particle due to the high performance of the PEG as capping agent compared to PVP during the synthesis of the ZnO nanoparticle. The smaller size of the ZnO-PEG is due to the PEG capped performance during the synthesize of the zinc oxide nanoparticles as proven by Tshabalala *et al.,* and Rahim *et al.,* [22-23]

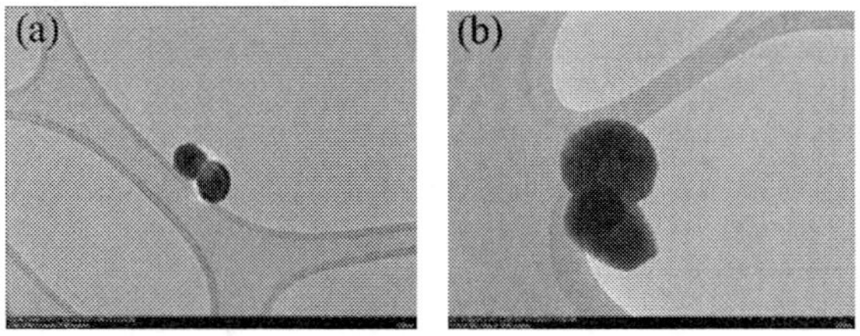

Fig.4. TEM image of: (a) ZnO-PEG and (b) ZnO-PVP.

3.2 Reusability performance of ZnO nanoparticles for photocatalytic degradation of POME

3.2.1 DO increment percentage analysis

From Fig. 5, the graph represented the maximum DO increment percentage of photocatalysis by ZnO nanoparticles. Based on the graph, it can be clearly observed that, there were enormous different between DO percentage for the 1st cycle and 2nd cycle. For the reusability methods, the oven drying method succeeded to demonstrate a higher performance compared than hot water rinsing method for both type of ZnO nanoparticles. This is might be the hot water could damage the chemical structure of the ZnO while reutilizing it back. Despite to the fact that the performance of the ZnO nanoparticle had been reduced for the second time usage, they manage to increase the DO to the standard, which can be legally discharge to the environment, which is at the range 0.9-1.5 mg/L. Therefore, the treated wastewater could be discharged to environment and safe to aquatic life. From the analysed results, it could be deduced that the treated wastewater is safe to discharge to environment since their DO value is in the acceptable range [24].

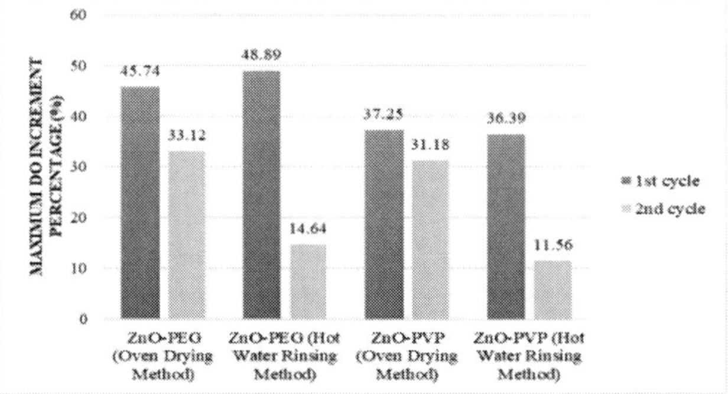

Fig. 5. Maximum DO increment percentage of photocatalysis by ZnO nanoparticles.

3.2.2 Turbidity reduction percentage analysis

The turbidity of treated wastewater is very relative to the percentage of dye degradation. Fig. 6 showed the maximum percentage of turbidity reduction on two different types of ZnO nanoparticles and two different types of reusability methods. It can be deduced that, by using oven drying method, it can provide a better performance in reducing turbidity of the wastewater for the ZnO-PEG and ZnO-PVP. In 2012, Carre *et al,* also claimed that UV photons excite ZnO, which then reacts with water molecules or hydroxide ion to produce hydroxyl radicals [25]. Then, colours were absorbed at the surface of ZnO; which then be degraded. Consequently, the degradation leads the breaking at least one of the conjugated bonds since the colour of wastewater formed by conjugated molecules. However, the turbidity might increase if the photocatalysts use beyond the optimum dosage resulting in fewer catalyst active sites available for photon absorption to generate □OH and O_2^- radicals [27–29].

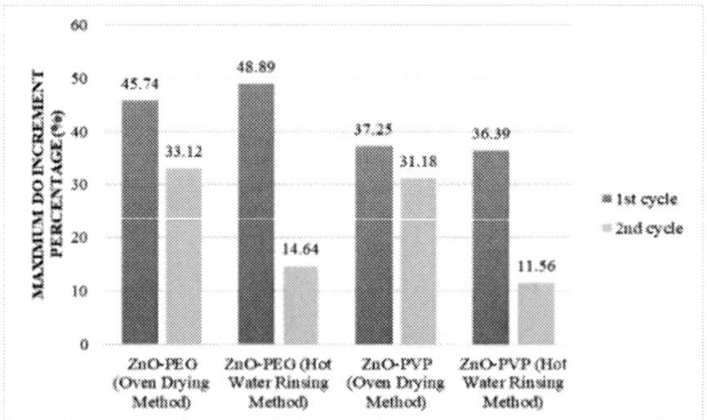

Fig. 6. Maximum turbidity reduction percentage of photocatalysis by ZnO nanoparticles.

3.2.3 Colour removal percentage analysis

Fig. 7 portrayed the comparison of experimental data of the colour removal percentage of POME by using two different types of ZnO nanoparticles by using two different types of reusability methods. The color removal results for both nanoparticles were exhibited the oven drying method could be applied as a method to reutilize back the nanoparticles. Even though the performance was reduced, it still managed to remove the colour almost 40%. Moreover, the method for the oven drying method was much simpler compared to the hot water rinsing method. The results portray the ability of the both photocatalyst to recover event though there might be slightly different of their performance. In addition, The smaller size of the ZnO-PEG really affected its capability in degrading the colour of POME due to higher effective surface area [11].

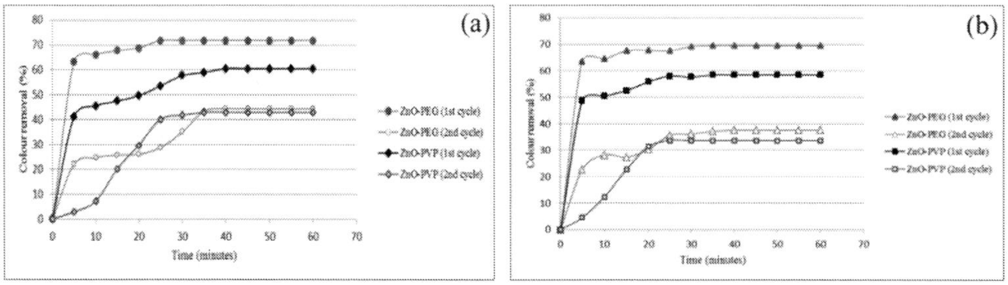

Fig. 7. Colour removal percentage by: (a) Oven drying method at 100°C and (b) Hot water rinsing method.

4 Conclusions

Both ZnO-PEG and ZnO-PVP were successfully synthesized and characterized. Their morphology, crystallinity, purity and chemical bond has been revealed by using Fourier transform infrared spectroscopy (FTIR), X-ray diffraction (XRD) and transmission electron microscopy (TEM). ZnO-PEG exhibit smaller size of nanoparticles compared to ZnO-PVP at range 60-150nm. And the presence of capping agents; PEG and PVP obviously reduced the agglomerations and could be seen in their performance of reducing colour of POME. Comparing for both type of ZnO-nanoparticles, both of them have capability to be reused in the next run. However, ZnO-PEG still showing its effectiveness after the second cycle compared to ZnO-PVP. For the reusability method, it can be concluded that the most suitable method that can be apply for the reutilize the ZnO-nanoparticle is by using oven drying method at 100°C.

This work was supported by Ministry of Higher Education Malaysia (MOHE) through a Grant (FRGS) Vot 1616 and Short Term Grant U659 Universiti Tun Hussein Onn Malaysia. The authors would like to acknowledge Department of Chemical Engineering Technology, Universiti Tun Hussein Onn Malaysia for providing necessary facilities.

References

1. A. Kongnoo, T. Suksaroj, P. Intharapat, T. Promtong, C. Suksaroj, Environ. Eng. Sci. **29**, 855–859 (2012)
2. C. K. Cheng, M. Rizauddin Derahman, M. R. Khan, J. Environ. Chem. Eng. **3**, 261–270, (2015)
3. N. S. Shahrifun A, N. Ab'lah, H. Hussain, A. Aris, Q. Omar, N. Ahmad, Malaysian J.

Civ. Eng. **27**,144–151 (2015)
4. M. H. Alhaji et al., Process Saf. Environ. Prot. **102**, 673–686 (2016)
5. U. G. Akpan, B. H. Hameed, J. Hazard. Mater. **170**, 520–529 (2009)
6. K. Mondal, A. Sharma, Indian Inst. Tecnol. 36–72 (2016)
7. D. Beydoun, R. Amal, G. Low, S. McEvoy, J. Nanoparticle Res. **1**(4), 439–458 (1999)
8. N. Tripathy, R. Ahmad, J. Eun Song, H. Ah Ko, Y. B. Hahn, G. Khang, Mater. Lett. **136**, 171–174 (2014)
9 A. L. B. Desa, *Optimization of Photocatalytic Degradation of Industrial Dye Wastewater in Presence of ZnO-PVP Nanoparticles*, **January** (2017)
10. M. C. A. K. Chiranont, *Optimization of Industrial Dye Wastewater via Photocatalysis Process in Presence of ZnO-PEG Nanoparticles*, Universiti Tun Hussein Onn Malaysia, (2017)
11. N. H. H. Hairom, A. W. Mohammad, A. A. H. Kadhum, Chem. Eng. J. **260**, 907–915 (2015)
12. N. H. H. Hairom, A. W. Mohammad, L. Y. Ng, A. A. H. Kadhum, Desalin. Water Treat. **54**, 944–955 (2015)
13. R. Velmurugan, K. Selvam, B. Krishnakumar, M. Swaminathan, Sep. Purif. Technol. **80**(1), 119–124 (2011)
14. S. Gomez, C. L. Marchena, M. S. Renzini, L. Pizzio, L. Pierella, Appl. Catal. B Environ. **162**, 167–173 (2015)
15. B. Subash, B. Krishnakumar, R. Velmurugan, M. Swaminathan, Sep. Purif. Technol. **101**, 98–106 (2012)
16. N. A. Fathy, S. M. El-Khouly, N. A. Hassan, R. M. S. Awad, J. Water Process Eng., **16**, 21–27 (2017)
17. P. Lee, E. Saion, N. Al-Hada, N. Soltani, Metals (Basel). **5**, 2383–2392 (2015)
18. N. Goswami, D. K. Sharma, Phys. E Low-Dimensional Syst. Nanostructures **42**, 1675–1682 (2010)
19. A. Singh, R. Kumar, N. Malhotra, A. Professor, E. Dept, Int. J. Sci. Emerg. Technol. with Latest Trends **4**, 49–53 (2012)
20. S. A. M. Lima, F. A. Sigoli, J. Jafelicci M., M. R. Davolos, Int. J. Inorg. Mater. **3**, 749–754 (2001)
21. M. A. Tshabalala, B. F. Dejene, H. C. Swart, Phys. B Condens. Matter **407**, 1668–1671 (2012)
22. S. Rahim, M. Sasani Ghamsari, S. Radiman, "Surface modification of titanium oxide nanocrystals with PEG," Sci. Iran. **19**, 948–95 (2012)
23. Y. Peng, J. Ji, X. Zhao, H. Wan, D. Chen, Powder Technol., **233**, 325–330 (2013)
24. A. Tamime, *Cleaning-in-Place: Dairy, Food and Beverage Operations: Third Edition.* (2009)
25. P. Huber, B. Carré, BioResources **7**, 1366–1382 (2012)
26. F. C. F. Low, T. Y. Wu, C. Y. Teh, J. C. Juan, N. Balasubramanian, Color. Technol.,**128**, 44–50 (2012)
27. M. Hassan, Y. Zhao, B. Xie, Chemical Engineering Journal **285**, 264–275 (2016)
28. S. Mohammadzadeh, M. E. Olya, A. M. Arabi, A. Shariati, M. R. Khosravi Nikou, J. Environ. Sci. (China) **35**, 194–207 (2015)

E3S Web of Conferences **34**, 02014 (2018) https://doi.org/10.1051/e3sconf/20183402014
CENVIRON 2017

Improving the performance of streamflow forecasting model using data-preprocessing technique in Dungun River Basin

Ervin Shan Khai Tiu[1*], *Yuk Feng* Huang[1], and *Lloyd* Ling[1]

[1]Lee Kong Chian Faculty of Engineering and Science, Universiti Tunku Abdul Rahman, Sungai Long, Selangor, Malaysia

Abstract. An accurate streamflow forecasting model is important for the development of flood mitigation plan as to ensure sustainable development for a river basin. This study adopted Variational Mode Decomposition (VMD) data-preprocessing technique to process and denoise the rainfall data before putting into the Support Vector Machine (SVM) streamflow forecasting model in order to improve the performance of the selected model. Rainfall data and river water level data for the period of 1996-2016 were used for this purpose. Homogeneity tests (Standard Normal Homogeneity Test, the Buishand Range Test, the Pettitt Test and the Von Neumann Ratio Test) and normality tests (Shapiro-Wilk Test, Anderson-Darling Test, Lilliefors Test and Jarque-Bera Test) had been carried out on the rainfall series. Homogenous and non-normally distributed data were found in all the stations, respectively. From the recorded rainfall data, it was observed that Dungun River Basin possessed higher monthly rainfall from November to February, which was during the Northeast Monsoon. Thus, the monthly and seasonal rainfall series of this monsoon would be the main focus for this research as floods usually happen during the Northeast Monsoon period. The predicted water levels from SVM model were assessed with the observed water level using non-parametric statistical tests (Biased Method, Kendall's Tau B Test and Spearman's Rho Test).

1 Introduction

Dungun River Basin is chosen as the study area for this research study. It is one of the flood prone areas in Peninsular Malaysia and had attracted the attention of a group number of researchers. The phenomenon of the flood could result in a lot of destructions to the livelihood and the local populations within the affected area. In the year-end of 2016, the National Disaster Management Agency (NADMA) in Malaysia had reported that heavy rain has caused flooding in the Terengganu state and consequently forcing hundreds to evacuate. Dungun was one of the affected areas.

* Corresponding author: ervintiu92@gmail.com

There were some research studies done towards the Dungun River Basin, like flood forecasting and early warning system [1] and hydrological extreme flood event [2]. For data-preprocessing technique, VMD which is more robust to sampling and noise was introduced lately, as an alternative to Empirical Mode Decomposition (EMD) model [3]. For streamflow forecasting model, Artificial Intelligence (AI) has starting to be used widely in engineering and science problems since middle of the 20th century. AI has performed significantly in forecasting and modelling non-linear hydrological applications. SVM approach has been selected as the model in streamflow forecasting for this research study.

As stated by Dragomiretskiy and Zosso [3], the multi-resolution VMD model was newly introduced lately as an alternative to an adaptive technique called EMD to overcome its limitations. According to Aneesh et al. [4], the VMD decomposed the signal into various modes or Intrinsic Mode Functions (IMFs) using calculus of variation. The VMD is able to extract the modes concurrently, properly balancing errors between them and separate tones of similar frequencies as contrary to the EMD [3]. VMD was found to be contributed widely and possessed an outstanding performance in different field of studies such as biomedical signal denoising [5-6], analysis of international stock markets [7].

For streamflow forecasting model, SVM had been classified into classification and regression analysis, which developed by Vapnik [8]. SVR has been further improved from time to time and successfully applied in different problems and prediction, and not to mention SVR has also contributed widely in hydrological field [9-15]. However, SVR model has its own limitations regarding the highly non-stationary of original hydrological data that vary over a range of scales although it performed quite well in flexibility of hydrological time series forecasting [16-17].

In this paper, the aim of this study is to investigate the improvement of selected streamflow forecasting model using data-preprocessing techniques for Dungun River Basin, Terengganu. Some of the specific objectives of this study are to process observed rainfall data using data-preprocessing technique as the input for the streamflow forecasting model, calibrate and validate the streamflow forecasting model using processed rainfall data from the data-preprocessing technique and the observed rainfall data, generate water level based on the processed rainfall data from the data-preprocessing technique and the observed rainfall data, and assess the performance of streamflow forecasting model with and without of data-preprocessing technique using appropriate performance evaluation.

2 Methodology

2.1 Workflow

The streamflow forecasting using chosen time series model began with the rainfall and water level data collection from the Department of Irrigation and Drainage (DID) Malaysia [18]. The collected observed rainfall data were sorted using homogeneity test to produce a more homogenous input rainfall data. Not to mention, the normality test was conducted onto the collected observed rainfall data to decide the method of statistics to be employed. The selected data-preprocessing technique is VMD whereas the chosen streamflow forecasting model is SVM. Before the SVM procedure, the observed rainfall data were analysed using the selected data-preprocessing technique. Last but not least, the predicted water levels, which generated from SVM were assessed with the observed water level using suitable method of quantitative standard statistical performance evaluation measures.

2.2 Study area

Dungun River Basin, which is one of the seven districts in Terengganu state, Malaysia has been chosen as the study area for this research study. The location of Dungun River Basin in Peninsular Malaysia and distribution of the rainfall stations are clearly shown in Fig. 1.

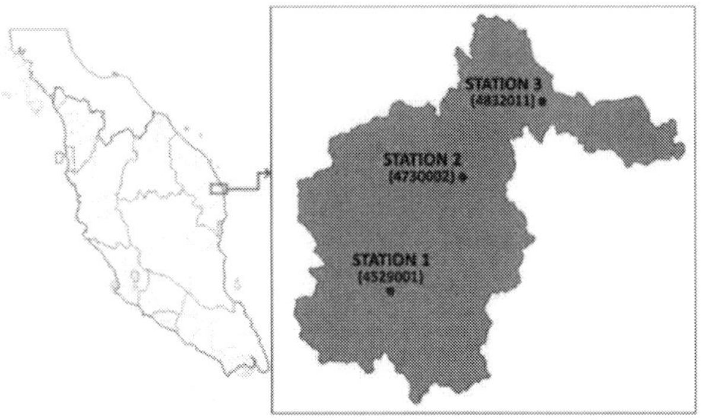

Fig. 1. Location of dungun river basin in Peninsular Malaysia.

All the rainfall data, water level data and rating curve were obtained from the DID Malaysia for 6 rainfall stations within the Dungun River Basin in Terengganu state [18]. However, only 3 rainfall stations within the catchment have a continuous good record. A common series length of 20 years (1996-2016) of that 3 rainfall stations has been determined. The details of the rainfall stations are clearly shown in Table 1.

Table 1. The name, study period and coordinates of the selected stations.

Station No.	Station Code	Station Name	Latitude	Longitude
1	4529001	Rumah Pam Paya Kempian at Pasir Raja	4°34'05"N	102°58'45"E
2	4730002	Kg. Surau at Kuala Jengai	4°44'05"N	103°05'15"E
3	4832011	Jambatan Jerangau at Terengganu	4°50'35"N	103°12'15"E

A mean rainfall data series within the catchment (3 rainfall stations) was computed using Thiessen Polygon method. The area weighted for the 3 rainfall stations are clearly shown in Fig. 2 and Table 2.

2.3 Homogeneity test

Homogeneity tests are implemented for rainfall data of every rainfall stations before further investigating to obtain a more homogenous input data. Four homogeneity tests which applied to the rainfall data, namely Standard Normal Homogeneity Test (SNHT), the Buishand Range test (BR), the Pettitt test (PET) and the Von Neumann Ratio test (VNR).

In frequentist statistics statistical hypothesis testing, the data are tested against the null hypothesis which the data is homogenous. If the significance level, $\alpha > 0.05$, means that the data are homogenous; if significance level, $\alpha < 0.05$, means that the data are non-homogenous. The criteria that used in this research study is similar to the one that proposed by Wijingaard et al. [19] which used to determine the homogeneity of data series. If at least three out of four tests reject the homogeneity, the time series is categorized as "suspect". If two out of four tests reject the homogeneity, then the time series is categorized as "doubtful". And, the time series is known to be "useful" when only one or zero test rejects the homogeneity.

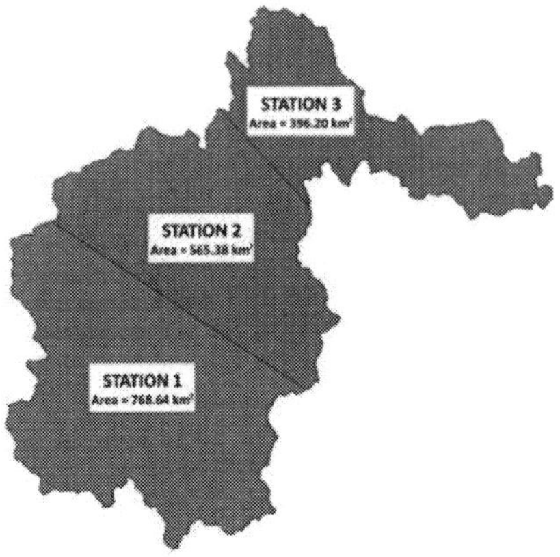

Fig. 2. Map of area weighted of 3 rainfall stations using Thiessen Polygon method.

Table 2. Area weighted of 3 rainfall stations using Thiessen Polygon method.

Station No.	Station Code	Station Name	Area, A_i (km^2)	Area Weighted (A_i/A_T)
1	4529001	Rumah Pam Paya Kempian at Pasir Raja	768.64	0.4442
2	4730002	Kg. Surau at Kuala Jengai	565.38	0.3268
3	4832011	Jambatan Jerangau at Terengganu	396.20	0.2290
		Total (A_T) =	1730.22	1.0000

2.4 Normality test

For normal distribution of data, parametric statistic tests had been implemented to evaluate the data. If the data are non-normally distributed, non-parametric statistic tests might be preferred. There are several tests that used to assess the normality of input data set. For normal distributed of data, the input data must pass all the stated normality tests. The normality tests are Shapiro-Wilk Test, Anderson-Darling Test, Lilliefors Test and Jarque-Bera Test. In normality test, the data are tested against the null hypothesis that it is normally distributed. If the significance level, $\alpha > 0.05$, mean the data are normal; if significance level, $\alpha < 0.05$, mean the data are not normal. In order to pass the normality test, it is necessary to turn the green light for all the normality tests.

2.5 Variational Mode Decomposition (VMD) for data-preprocessing

VMD method which proposed by Dragomiretskiy and Zosso [3] decomposes the incoming signal into k discrete number of sub-signals (modes) where each mode possesses the limited bandwidth in spectral domain. Therefore, each of the mode k is required to be compacted the most around a centre pulsation, ω_k which determined along the decomposition process. The VMD will find out the central frequencies and IMFs which located at the centre of the frequencies concurrently using an optimization method namely Alternate Direction Method of Multipliers (ADMM). The original formulation of the optimization problem is continuous in time domain.

The constrained variational problem which given by Dragomiretskiy and Zosso [3] is stated in the following equations:

$$\underset{u_k, \omega_k}{min} = \left\{ \sum_k \left\| \partial_t \left[\left(\delta(t) + \frac{j}{\pi t} \right) * u_k(t) \right] e^{-j\omega_k t} \right\|_2 \right\} \tag{1}$$

subject to,

$$\sum_k u_k = f \tag{2}$$

where f is the signal, u is the mode, ω is the frequency, δ is the Dirac distribution, t is the time script, k is the number of modes and $*$ is the convolution. The mode u with high-order k represents the low frequency components.

2.6 Support Vector Machine (SVM) for streamflow forecasting

SVM which proposed by Vapnik [8], is known as classification and extended to regression. The SVR is derived from the SVM which used to solve the regression problems with SVM. The regression function of SVM that relates the input vector x to the output $\hat{y}$ is formulated as equation stated below.

$$\hat{y} = \mathrm{f}(\mathrm{x}) = \omega^T . \phi(x) + b \tag{3}$$

where ω^T is a weight vector, b is a bias and ϕ is a nonlinear transfer function that maps the input vectors into a high-dimensional feature space in which theoretically a simple linear regression can matched with the complex nonlinear regression of the input space.

The final regression function can be rewritten as the follow equation:

$$f(\mathrm{x}) = \sum_{k-1}^{N_{sv}} \alpha_k K(x_k, x) + b \tag{4}$$

where x_k is the kth support vector and N_{sv} is the number of the support vectors.

2.7 Model performance evaluation

2.7.1 Parametric statistical tests

For parametric statistical tests, the input data must be normally distributed. Here, the accuracy of the predicted water levels to the observed water level has been evaluated. In this research study, the models' performance is evaluated by using three chosen standard performance measures, namely Root Mean Square Error (RMSE), Mean Absolute Percentage Error (MAPE) and Coefficient of Efficiency (CE).

2.7.2 Non-parametric statistical tests

If the input data is non-normally distributed, non-parametric statistical tests need to be implemented. As similar to parametric statistical tests, the accuracy of the predicted water levels to the observed water level has been evaluated. In this research study, the models' performance is evaluated by using four standard performance measures, namely Biased Method, Kendall's Tau B Test and Spearman's Rho Test.

3 Results and discussions

3.1 Homogeneity test

The homogeneity tests were implemented on three rainfall stations to obtain a more homogenous input data. If the significance level, α > 0.05, means that the data are homogenous. It was shown that either one or zero tests reject the homogeneity for every certain month. Hence, the homogeneity of observed rainfall data series is useful. The results of homogeneity tests for three rainfall stations are clearly shown in Table 3, Table 4 and Table 5. The significance level which highlighted in orange colour stated that the respective homogeneity test rejects the homogeneity for that certain month.

Table 3. Results of homogeneity tests for Station 1 (4529001).

Month	Homogeneity Test (Significance Level, α)				Homogeneity
	SNHT	BR	PET	VNR	
Jan	0.376	0.278	0.619	0.682	Useful
Feb	0.419	0.788	0.156	0.668	Useful
Mar	0.618	0.847	0.407	0.407	Useful
Apr	0.532	0.356	0.556	0.114	Useful
May	0.371	0.418	0.119	0.187	Useful
Jun	0.839	0.737	0.630	0.372	Useful
Jul	0.422	0.215	0.313	0.296	Useful
Aug	0.634	0.385	0.548	0.032	Useful
Sep	0.192	0.122	0.205	0.142	Useful
Oct	0.890	0.660	0.189	0.119	Useful
Nov	0.144	0.112	0.487	0.147	Useful
Dec	0.142	0.161	0.484	0.120	Useful

Table 4. Results of homogeneity tests for Station 2 (4730002).

Month	Homogeneity Test (Significance Level, α)				Homogeneity
	SNHT	BR	PET	VNR	
Jan	0.206	0.119	0.159	0.728	Useful
Feb	0.494	0.741	0.974	0.740	Useful
Mar	0.325	0.220	0.683	0.024	Useful
Apr	0.194	0.078	0.302	0.015	Useful
May	0.459	0.273	0.307	0.521	Useful
Jun	0.815	0.582	0.695	0.581	Useful
Jul	0.940	0.749	0.802	0.075	Useful
Aug	0.065	0.029	0.074	0.066	Useful
Sep	0.179	0.111	0.239	0.006	Useful
Oct	0.789	0.811	0.166	0.043	Useful
Nov	0.154	0.139	0.226	0.432	Useful
Dec	0.215	0.127	0.484	0.382	Useful

Table 5. Results of homogeneity tests for Station 3 (4832011).

Month	Homogeneity Test (Significance Level, α)				Homogeneity
	SNHT	BR	PET	VNR	
Jan	0.277	0.149	0.232	0.490	Useful
Feb	0.518	0.780	0.045	0.717	Useful
Mar	0.817	0.708	0.677	0.362	Useful
Apr	0.068	0.041	0.145	0.170	Useful
May	0.698	0.770	0.597	0.746	Useful
Jun	0.497	0.296	0.869	0.404	Useful
Jul	0.140	0.085	0.208	0.127	Useful
Aug	0.239	0.148	0.353	0.049	Useful
Sep	0.640	0.864	0.249	0.672	Useful
Oct	0.836	0.588	0.882	0.395	Useful
Nov	0.411	0.433	0.739	0.766	Useful
Dec	0.130	0.083	0.327	0.140	Useful

3.2 Normality test

The normality tests such as Shapiro-Wilk Test, Anderson-Darling Test, Lilliefors Test and Jarque-Bera Test are applied on the observed rainfall data. The significance level, α value was shown to be lesser than 0.05 for all normality tests. Hence, the input data are non-normally distributed. The results of normality tests for three rainfall stations are clearly shown in Table 6.

Table 6. Results of normality tests for observed rainfall data.

Station No.	Station Code	Normality Test (Significance Level, α)			
		Shapiro-Wilk Test	Anderson-Darling Test	Lilliefors Test	Jarque-Bera Test
1	4529001	< 0.0001	< 0.0001	< 0.0001	< 0.0001
2	4730002	< 0.0001	< 0.0001	< 0.0001	< 0.0001
3	4832011	< 0.0001	< 0.0001	< 0.0001	< 0.0001

3.3 Application of data-preprocessing technique

VMD as data-preprocessing technique was carried out on the monthly and seasonal rainfall series of Northeast Monsoon (November to February) only. The training and validation sets of rainfall data are separated into 18-year and 2-year, respectively. For the period of 1996-2016, VMD method was first used to process the early 18 years of rainfall data which act as the training set of rainfall data, followed by processing the last 2 years of rainfall data which act as the validation set of rainfall data.

3.4 Application of streamflow forecasting model

SVR was used to predict water level based on the processed and observed rainfall data. At this stage, the first 18 years and last 2 years of processed and observed rainfall data were trained and validated, respectively. The predicted water level from both processed and observed rainfall data were assessed with the observed water level using certain statistical measures. The hydrographs of observed water level and predicted water level using processed and observed rainfall data for validation sets (2 years of rainfall data) are shown in Fig. 3.

(a)

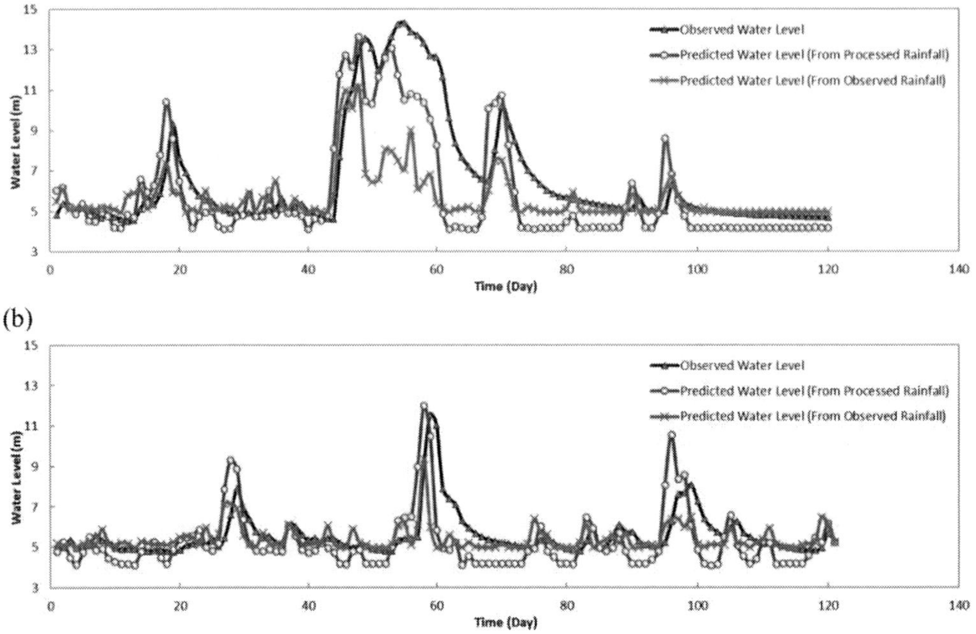

(b)

Fig. 3. Hydrograph of observed water level and predicted water level using processed and observed rainfall data for validation sets of (a) year 2014-2015 and (b) year 2015-2016.

3.5 Model performance evaluation

As shown in the normality test, the input data are non-normally distributed. Non-parametric statistical tests were implemented to evaluate the models' performance. The goodness-of-fit of every statistical test which used to evaluate the models' performance are shown in Table 7, for both training and validation sets of predicted water level from both processed and observed rainfall data.

Table 7. Performance evaluation of the predicted water levels from both processed and observed rainfall data.

Period	Time (Year)	Predicted Water Level	Goodness-of-fit		
			Biased Method	Kendall's Tau B Test	Spearman's Rho Test
Training	18	From processed rainfall	-1616.86	0.272	0.395
		From observed rainfall	-3075.72	0.268	0.379
Validation	2	From processed rainfall	-133.38	0.319	0.442
		From observed rainfall	-134.44	0.292	0.405

It can be observed that accuracy of predicted water level was shown to be improved when the rainfall data was processed with VMD. In the validation period, the predicted water level from processed rainfall data showing an improvement which obtained the better biased, correlation coefficients of Kendall's Tau B test and Spearman's Rho test statistics of -133.38, 0.319 and 0.442, respectively as compared to that of predicted water level from observed rainfall data which obtained the biased, correlation coefficients of Kendall's Tau B test and Spearman's Rho test statistics of -134.44, 0.292 and 0.405, respectively.

4 Conclusions

The application of VMD has been shown as a robust data-preprocessing technique to denoise the rainfall data before putting into the respective streamflow forecasting model. The accuracy of streamflow forecasting model was analysed via the non-parametric statistical tests. The accuracy of streamflow forecasting model was shown to be improved when the model was coupled with VMD. In overall, the application of VMD as data-preprocessing technique is able to improve the performance of the streamflow forecasting model.

References

1. I. Hafiz, N.D.M. Nor, L.M. Sidek, H. Basri, K. Fukami, M.N. Hanapi, L. Livia, IOP Conf. Series: Earth and Environmental Science **16**, 012129 (2013)
2. M.S. Lariyah,, *Conference: 13th International Conference on Urban Drainage, Sarawak, Malaysia* (2014)
3. K. Dragomiretskiy, D. Zosso, IEEE Transactions on Signal Processing, **62**, 531–544 (2014)
4. C. Aneesh, K. Sachin, P.M. Hisham, K., P., Soman, Procedia Computer Science, **46**, 372-380 (2015)
5. S. Lahmiri, M. Boukadoum, IEEE BIOCAS, 340–343 (2014c)
6. S. Lahmiri, M. Boukadoum, IEEE ISCAS, 806–809 (2015c)
7. S. Lahmiri, Physica A, **437**, 130–138 (2015d)
8. V. Vapnik, *The Nature of Statistical Learning Theory. Springer, New York* (1995)
9. H. Moradkhani, K.-L. Hsu, H.V. Gupta, et al., J. Hydrol. **295**, 246–262 (2004)
10. P.S. Yu, S.T. Chen, I.F. Chang, J. Hydrol. **328**, 704–716 (2006)
11. J.Y. Lin, C.T. Cheng, K.W. Chau, Hydrolog. Sci. J. **51**, 599–612 (2006)
12. C.L. Wu, K.W. Chau, Y.S. Li, J. Hydrol. **358**, 96–111 (2008)
13. G.F. Lin, G.R. Chen, P.Y. Huang, et al., J. Hydrol. **372**, 17–29 (2009)
14. S.T. Chen, P.S. Yu, Y.H. Tang, J. Hydrol. **385**, 13–22 (2010)
15. H. Yoon, S.C. Jun, Y. Hyun, et al., J. Hydrol. **396**, 128–138 (2011)
16. B. Cannas, A. Fanni, L. See, G. Sias, Phys. Chem. Earth. **31** (18), 1164–1171 (2006)
17. J. Adamowski, H.F. Chan, J. Hydrol. **407**, 28–40 (2011)
18. Department of Irrigation and Drainage (DID) Malaysia, *Kuala Lumpur: Department of Irrigation and Drainage Malaysia* (2016)
19. J.B. Wijingaard, T. Klein, A.M.G., G.P. Können, International Journal of Climatology, **23**, 679-692 (2003)

Spatial and Temporal Variations of Water Quality and Trophic Status in Sembrong Reservoir, Johor

Syarifah Intan Najla Syed Hashim[1], *Siti Hidayah Abu* Talib[1*], and *Muhammad Salleh* Abustan[1]

[1]Faculty of Civil and Environmental Engineering, Universiti Tun Hussien Onn, Malaysia

Abstract. A study of spatial and temporal variations on water quality and trophic status was conducted to determine the temporal (average reading by month) and spatial variations of water quality in Sembrong reservoir and to evaluate the trophic status of the reservoir. Water samples were collected once a month from November 2016 to June 2017 in seventeen (17) sampling stations at Sembrong Reservoir. Results obtained on the concentration of dissolved oxygen (DO), water temperature, pH and secchi depth had no significant differences compared to Total Phosphorus (TP) and chlorophyll-a. The water level has significantly decreased the value of the water temperature, pH and TP. The water quality of Sembrong reservoir is classified in Class II which is suitable for recreational uses and required conventional treatment while TSI indicates that sembrong reservoir was in lower boundary of classical eutrophic (TSI > 50).

1 Introduction

Reservoir is an artificial lake where water is stored. The importance of reservoir to society was expected to increase over time as population, irrigation demand and economic activity grows. Main function of reservoir was to mitigate floods in downstream plain and provide water supply. Even though modern hydraulic systems consist of many elements to appropriate both surface and groundwater supplies, in many regions reservoirs are the most important components for water supply. The quality of water resources is declining from both natural (changes in precipitation; erosion) and anthropogenic (industrial and agricultural activities) factors [1]. In addition, all reservoirs are subjected to periodic fluctuations in water levels, due to rain, hydrological regimes that are influenced by irrigation of agricultural lands and temperature changes [2]. In Sembrong reservoir, human activities from agriculture and development, pollution from septic systems and sewers, sewage sludge spreading, and other human-related activities increase the flow of both inorganic nutrients and organic substances into ecosystems thus inducing the rapid growth of plants and algae which leads to eutrophication. Thus this study will determine the

* Corresponding author: hidayaht@uthm.edu.my

temporal (average reading by month) and spatial variations of water quality in Sembrong reservoir and to evaluate the trophic status of the reservoir.

2 Materials and Methods

Sembrong dam is located about 10 km from Ayer Hitam, Johor. The dam lies at the coordinate of 2°01'35"n - 1°58'29"n latitude and 103°09'32"e - 103°12'57"n longitude. The reservoir surface area is 8.5 km^2, with capacity of 18 million m^3 along with a catchment area of 130 km^2 and dam height of 11 m. Sembrong dam was completed in 1984 and formed part of the western johor integrated agricultural development project in the state of johor. The dam was primarily for flood mitigation but the stored water (up to rl 8.5m) is tapped by syarikat air johor for water supply. Currently, they are abstracting approximately 2 million liters per day (mld) directly from the dam for domestic water demand. The reservoir is surrounded by oil palm plantation. Other minor land use are swamp forest, modern agriculture and husbandry. There are two major tributaries flowing into the reservoir, namely Sembrong River and Merpoh River. Due to indiscriminate farming, agricultural activities and the planting of oil palm trees which covers at least 87% of the dam's 130 km2 catchment area, this water reservoir is slowly endangered due to eutrophication or better known as algae bloom. Three sampling zones (Z1, Z2 and Z3) were established within the reservoir area (Fig. 1). All descriptions and locations are described in Table 1.

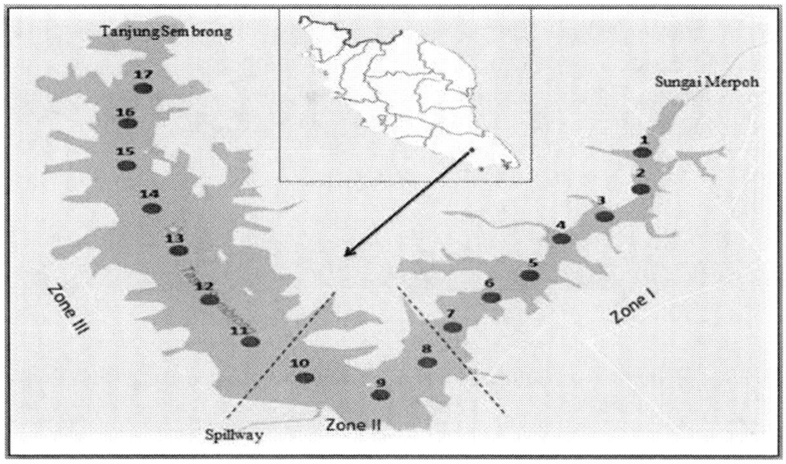

Fig. 1. Sampling Location (Z1, Z2, Z3) in Sembrong Reservoir.

2.1 Nutrient Samples Collection and Analysis

The water was collected about 10cm below the water surface from the boat using grab sample method and stored in 1L polyethylene bottles. All samples were kept in an icebox and transported back to the laboratory for analysis. Nutrient samples were delivered to the laboratory and stored in a refrigerator within the 48 h. Laboratory analysis for total phosphorous (TP), nitrate (NO_3^-), ammonia nitrogen (NH_4^+) and chlorophyll-a were tested using DR 6000 Spectrophotometer and the method were outlined in the Standard Method for the Examination of Water and Wastewater (APHA, 1999).

Table 1. General Characteristics of Sampling stations in Sembrong Reservoir.

Zone	Station	Coordinates of sampling location	Description of station
Zone I	1	N 02° 00' 01.1" E 103° 12' 16.6"	This station is located near to the inlet of Sungai Sembrong. The stations are surrounded by farming and agricultural activities. Riparian vegetation consists of *Eichhornia crassipes* and grass.
	2	N 01° 59' 50.4" E 103° 12' 08.0"	
	3	N 01° 59' 41.8" E 103° 12' 01.4"	
	4	N 01° 59' 33.1" E 103° 11' 54.1"	
	5	N 01° 59' 24.4" E 103° 11' 45.6"	
	6	N 01° 59' 12.8" E 103° 11' 36.7"	
Zone II	7	N 01° 58' 58.5" E 103° 11' 24.7"	This station located near to the spillway of reservoir.
	8	N 01° 58' 49.2" E 103° 11' 19.4"	
	9	N 01° 58' 52.8" E 103° 11' 13.1"	
	10	N 01° 58' 53.4" E 103° 11' 06.5"	
	11	N 01° 59' 03.5" E 103° 10' 44.7"	
Zone III	12	N 01° 59' 24.7" E 103° 10' 32.3"	This station is located near to the inlet of Sungai Merpoh. There were many stump and dead trees at this area.
	13	N 01° 59' 38.6" E 103° 10' 26.0"	
	14	N 01° 59' 55.8" E 103° 10' 18.5"	
	15	N 02° 00' 11.3" E 103° 10' 11.5"	
	16	N 02° 00' 22.0" E 103° 10' 05.7"	
	17	N 02° 00' 34.8" E 103° 10' 01.4"	

2.2 Carlson's Trophic State Index (TSI)

The level of eutrophication can be determined by the trophic state index (TSI). Transparency (measured by secchi depth), total phosphorous (TP) and chlorophyll-a concentrations are used in estimating trophic status and one of the most commonly used is Carlson's Trophic State Index (TSI) (Carlson 1977) with the formula:

$$TSI\ (SD) = 60 - 14.41\ ln\ (SD) \quad (1)$$
$$TSI\ (Chl\text{-}a) = 9.81\ ln\ (Chl\text{-}a) + 30.6 \quad (2)$$
$$TSI\ (TP) = 14.42\ ln\ (TP) + 4.15 \quad (3)$$
$$TSI = \frac{[TSI\ (TP) + TSI(CA) + TSI(SD)]}{3} \quad (4)$$

where SD = Secchi depth (m), TP = total phosphorus(ppb), and CA = chlorophyll-a (ppb).

3 Results and discussions

3.1 Hydrological Data

Fig. 2 shows the fluctuations of rainfall data and mean water level in Sembrong reservoir. The highest rainfall amount was recorded in January 2017 (650 mm) while the lowest amount was recorded in March 2017 (200 mm). Mean water level (m) recorded at Sembrong reservoir has increased from November 2016 to February 2017 and started to decreased until June 2017. The Northeast Monsoon (Monsun Timur Laut) began on November 2016 and is expected to end in March 2017 where Malaysia is expected to experience a series of heavy rainfall. The reservoir had the lowest mean water level with the value of 2.45 m (November 2016) and the highest water level was recorded in February 2017 (3.50 m).

https://doi.org/10.1051/e3sconf/20183402015

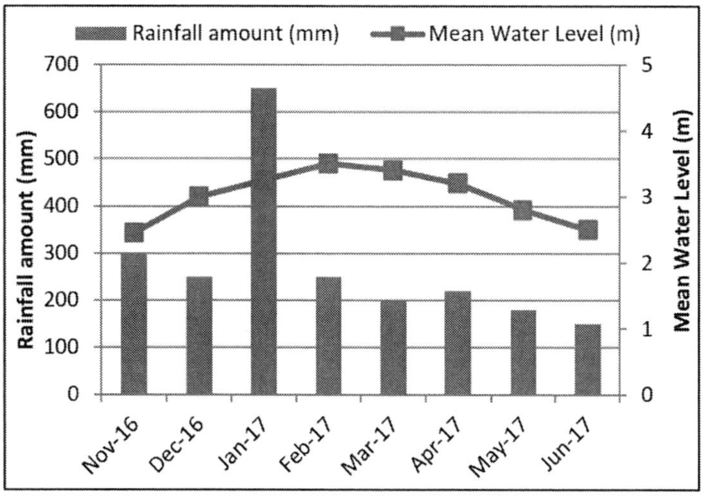

Fig. 2. Temporal variations of actual rainfall data with mean water lever in Sembrong Reservoir.

3.2 Temporal Variations of Physiochemical Parameters

Fig. 3 shows the average readings of all sampling stations by month from November 2016 to June 2017 of physiochemical parameters and TSI of Sembrong reservoir. Almost all lowland reservoirs are irrigation reservoirs and their hydrological regimes are significantly influenced by irrigation of agricultural lands [3]. The Sembrong reservoir is subjected to periodic fluctuations in water levels, due to rain, temperature changes and how they are operated for their functional role.

Mean concentrations of DO were ranged from 4.88 to 6.61, which the values decreases started from January 2017 to June 2017 (Fig. 3). High values of DO in water are not unusual and mainly attributed to the direct contact of surface water with air and to the relatively higher photosynthetic activity compared to that in deeper depth of water. The DO concentration in Sembrong reservoir was regulated by the process of diffusion of oxygen from the atmosphere, photosynthetic activity, respiration and decomposition of aquatic organisms [4]. There was no significant difference in the concentration of DO among sampling stations because of the same water bodies and geographical area.

Mean water temperature had no significant difference as the values were consistent for every month at the range of 30 °C (Fig. 3). The water temperature in Sembrong reservoir was consistent throughout the year (November 2016 to June 2017) and between stations, although a drop in the temperature was observed in certain months. Cloudy and rainfall that occurred during the sampling influenced the reading of the water temperature [5]. Similar water bodies and geographical area are the reason of less variation in the water temperature among sampling stations [6].

Mean values of pH had less significant difference and ranged from 6.4 to 6.87. The pH was decreasing every month from November 2016 to June 2017 (Fig. 3). The values of pH for both spatial and temporal were consistent throughout the study, fell around six to seven (neutral) hence suitable for most aquatic organisms. Generally, pH is almost the same in every layer of Sembrong reservoir due to the shallowness of the reservoir [7,8]. pH is a very important factor of water body for the fish culture as it controls the amount of soluble ions in the water [4]. An acidic pH of water reduces the growth rate, metabolic rate and other physiological activities of fishes [9], thus suggested that pH values varying from 6.5 to 9.0 as suitable for the normal growth of fishes.

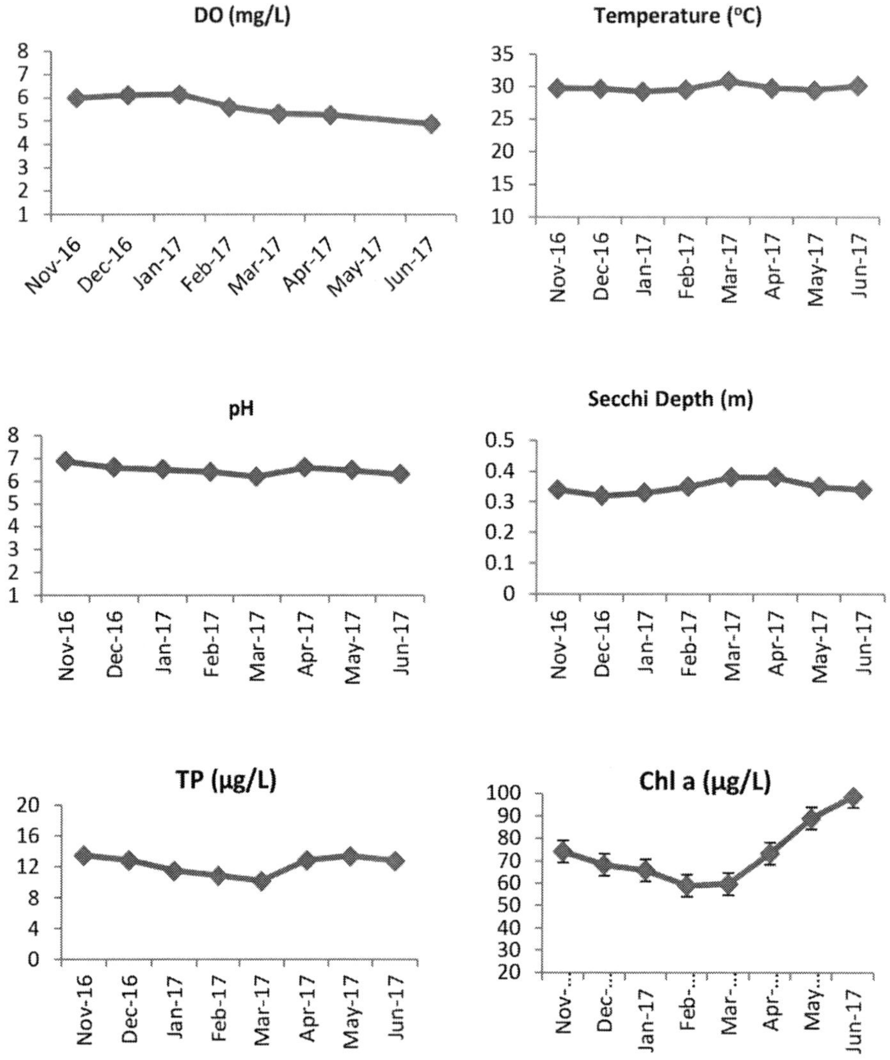

Fig. 3. The average reading of all sampling stations by month of physiochemical parameters such as DO, temperature, pH, secchi depth, TP and Chlorophyll-a in Sembrong Reservoir.

Water clarity based on secchi depth value was consistent every month with the lowest light penetration was recorded in December 2016 (0.32 m) and the deepest light penetration was recorded in March 2017 (0.38 m). High concentrations of dissolved and particulate substances (eroded soils and nutrients) that were transported from the surrounding catchment into the reservoir after rainfall has been reducing light transmission and water clarity [10].

There was a significant difference in the mean concentration of TP between November 2016 and January 2017 (0.0185 and 0.0915 respectively). Human activities (human and animal waste, fertilizers and erosions) were the main factors increasing the concentration of the phosphorus in the water bodies [11]. The decreasing of TP was probably due to the decreased of microbial activity [12]. The secchi depth values are associated with most of aquatic plants because the chlorophyll content inside them are depending upon light availability, temperature and nutrients [13,14].

3.3 Trophic Status Analyses

Fig. 4 shows the TSI from November 2016 to June 2017. The TSI were less varied and belong to the same trophic state due to the location of the same water bodies and geographical areas. The TSI showed that Sembrong reservoir was in Eutrophic (TSI > 50). The result indicated that Sembrong reservoir was in lower boundary of classical eutrophy: Decreased transparency, anoxic hypolimnia during the summer, macrophyte problems evident, warm-water fisheries only and dominance of blue-green algae, algal scums probable, extensive macrophyte problems.

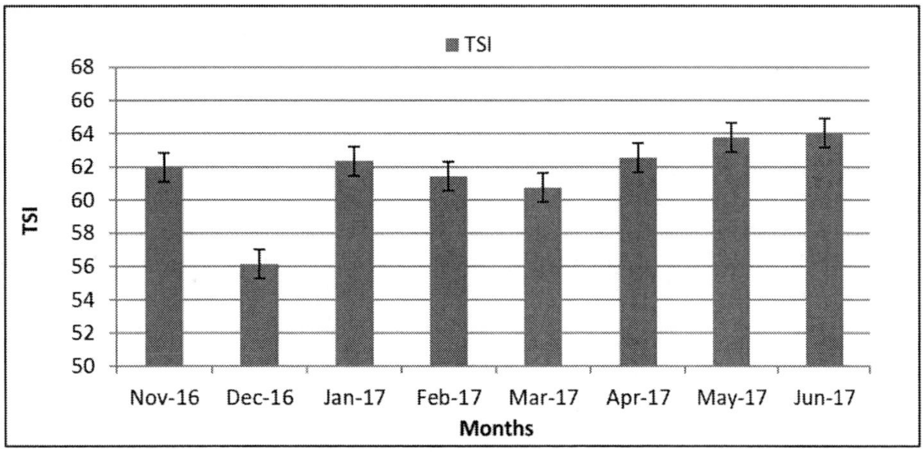

Fig. 4. Trophic Status Index (TSI) of all sampling stations by month in Sembrong Reservoir.

TSI	Chl-a (µg/L)	SD (m)	TP (µg/L)	Attributes
<30	<0.95	>8	<6	Oligotrophy: Clear water, oxygen throughout the year in hypolimnion
30-40	0.95-2.6	8-4	6-12	Hypolimnia of shallower lakes may become anoxic
40-50	2.6-7.3	4-2	12-24	Mesotrophy: Water moderately clear; increasing probability of hypolimnetic anoxia
50-60	7.3-20	2-1	24-48	Eutrophy: Anoxic hypolimnia, macrophyte problem possible
60-70	20-56	0.5-1	48-96	Blue-green algae dominate, algal scums and macrophyte problem
70-80	56-155	0.25-0.5	96-192	Hypereutrophy: (light limited productivity). Dense algae and macrophyte
>80	>155	<0.25	192-384	Algal scums, few macrophytes

4 Conclusions

The concentration of dissolved oxygen (DO), water temperature, pH and secchi depth had variations temporally. Meanwhile, the TP and chlorophyll-a had variations in mean reading among the sampling stations. The water quality of Sembrong reservoir is classified in Class II and TSI indicates that Sembrong reservoir was in lower boundary of classical eutrophic (TSI > 50).

We thank the Batu Pahat Department of Irrigation and Drainage for their hospitality, Universiti Tun Hussein Onn for funding this project under the Research University Grant (Geran Penyelidikan Kontrak Khas 2015) vot no. U422 and Research, Innovation, Commercialization, Consultancy Office UTHM (ORICC) for Postgraduate Research Grant (GPPS).

References

1. A. M. Sheela, J. Letha, S. Joseph, M. Chacko, S. P. Sanal Kumar & J. Thomas. Lakes & Reservoirs: Research and Management. *Water quality assessment of a tropical coastal lake system using multivariate cluster, principal component and factor analysis* **17**, 143-159 (2012)
2. A. Duncan, J. Kubečka. Hydrobiologia **303**, 11-30 (1995)
3. D. Sahoo, P. K. Smith. Journal of Hydrology **367**(3), 217-227 (2009)
4. M. F. A. Mollah, M. Moniruzzaman, M. M. Rahman. Journal of the Bangladesh Agricultural University **9**(2), 327-338 (2011)
5. F. J. D. M. Pedrazzi, F. T. D. Conceição, D. D. S. Sardinha, V. Moschini- Carlos, M. Pompêi. Journal of Water Resource and Protection **5**, 64-71 (2013)
6. A. Sreenivasan. Internationale Revue der gesamten Hydrobiologie und Hydrographie **59**(3), 327-342 (1974)
7. Department of Fisheries Malaysia (DOF). *Compilation of Research Studies and Reports on Rivers, Lakes and Reservoirs in Peninsular Malaysia. Report on Limnological Study of Reservoirs in Malaysia for Inland Fisheries Development*. 51-117 (1993)
8. A. S. Sarnita. *Symposium of Indo Pacific Fishery Commision, Colombo, Sri Lanka. Some aspects of fisheries and their development in Man-made Lakes in Indonesia with special reference to Lake Jatiluhur, West Java* **17**, 1-21 (1976)
9. H. S. Swingle. *Proceedings of the 9th Pacific Science Congress. Relationships of pH of pond water to their suitability for fish culture* **10**, 72-75 (1967)
10. C. O. Akinbile, M. S. Yusoff, S. H. A. Talib, Z. A. Hasan, W. R. Ismail, U. Sansudin. Water Science & Technology: Water Supply. *Qualitative analysis and classification of surface water in Bukit Merah Reservoir in Malaysia* **13.4**, 1138-1145 (2013)
11. S. C. Chapra. *Surface Water-Quality Modeling. UK.* (1997)
12. W. R. Ismail & S. A. M. Najib. Lakes & Reservoirs: Research and Management **16**, 179-184 (2011)
13. W. T. Edmondson. Limnology and Oceanography **25**(2), 378-379 (1980)
14. R. G. Wetzel, G. E. Likens. *Limnological Analyses* (2000)

Adequacy of Flood Relief Shelters: A Case Study in Perak, Malaysia

Nur Zawani Zahari[1]*, and *Ahmad Mustafa* Hashim[1]

[1]Department of Civil and Environmental Engineering, Universiti Teknologi PETRONAS, 32610 Bandar Sri Iskandar, Perak, Malaysia

Abstract. The recent flood event occurred in 2014 had caused disastrous effects in Peninsular Malaysia in states of Kelantan, Pahang, Terengganu, Perak, Johor and Perlis. Perak state was reported with 12,115 victims from 2,896 families registered at 77 relief shelters. There are several issues encountered by the victims and related agencies which caused inconveniences and interruptions during the flooding period. Besides, the usage of public buildings as relief shelters contributes to deterioration of the infrastructures whereby their suitability, convenient, capacity and safety might not be optimum for longer period of time. This paper focuses on the assessment of relief shelters established in Perak Tengah district, Perak. Standards and guidelines for relief shelters were reviewed according to the most relevant agreed principles for humanitarian response. Data and information in this study were obtained from survey activities, interview sessions and observations. In Perak Tengah, more than 50% of the previous relief shelters were public buildings with low capacity areas. Strategic location of shelters with proper design standards should be established to ensure safe and healthy environment for the victims. Findings from this paper provide important outcomes to serve as better preparation in handling future disaster.

1 Introduction

Flood is the most devastating and costly natural hazards in the world that leads to significant economic and social damages [1-3]. Malaysia had experienced several major floods in the last few decades due to monsoon rainfall throughout the year [4]. Average rainfall of approximately 2,500 mm a year for all states in Malaysia making it one of the countries with the heaviest rainfall in the world [5]. In 1926, a worst memory of flood event caused severe damages at several states in Malaysia. Another disastrous floods back in 1967 and 1971 surged across the states in Malaysia that left huge impacts in the flood history [1, 6]. Heavy monsoon rainfall in the year of 2006-2008 had also led to flooding in some of the affected states in Peninsular Malaysia [1, 6]. The recent major flood event in Peninsular Malaysia in December 2014 to January 2015 had caused tremendous damage to several states including Kelantan, Pahang, Terengganu, Perak, Johor and Perlis [4, 7].

* Corresponding author : wanie.zahari@gmail.com

Causative factors of floods are interrelated between various weather factors such as heavy rain, storm surges, inadequate drainage systems and structural failures of dams [8]. Nevertheless, people living within the flood plain areas or in close proximity to geographic regions such as coastal areas and river basins are vulnerable to flooding [8, 9]. Relief shelters play a vital role as part of disaster response and recovery in providing optimum security, personal safety, protection from climate and improve resistance to ill health and diseases for victims who lost their accommodations as a result from the disaster itself [10-11]. When a disaster happened, most of the evacuees particularly the elderly and low income communities rely on the relief shelters for their temporary house or shelters [12]. However, some of the evacuees were either stayed in hotels, relatives' houses, or even reluctant to leave their own houses in the times of flooding. Living in sheltered places with unfamiliar people may lead to invasion of personal space and privacy of the evacuees [7].

Relief shelters are not only function as immediate and short-term shelter for the victims but also help them to recover from trauma of the disaster as well as provide a base to start rehabilitation process [10]. Victims from disasters are not only require financial support due to losses and property damages, but some of them may need physical and emotional support after experiencing unexpected situations [13]. Since relief shelters may be used for several weeks or even months after a disaster period, numerous factors should be taken into account in planning and designing the shelters such as physical location and needs of likely users [10]. As a long-term flood disaster management strategy, a series of dedicated and well-designed relief shelters should be seriously considered. This paper presents the general standards of relief shelters and assessment of relief shelters established in Perak Tengah district, Perak.

2 Relief shelters standards and guidelines

The handbook of Humanitarian Charter and Minimum Standards in Disaster Response as compiled by The Sphere Project and Handbook for Emergencies by The United Nations High Commissioner for Refugee (UNHCR) had been classified as the most relevant agreed principles for response [14].

2.1 The Sphere Project - Humanitarian Charter and Minimum Standards in Disaster Response

2.1.1 Description

The Sphere Project was initiated by a group of humanitarian non-governmental organizations (NGOs) and the International Red Cross and Red Crescent Movement to develop a set of universal minimum standards for humanitarian response during the disaster. The handbook is the product of collective experiences of many people and agencies based on two core beliefs, first is that those affected by disaster or conflict have a right to live with dignity and receive assistance, second is that all possible steps should be taken to alleviate human suffering arising out of disaster or conflict [11]. A set of minimum standards are identified which describe conditions that must be achieved for disaster-affected populations to survive and recover in stable conditions and dignity [11].

2.1.2 Minimum standard for relief shelter

Table 1 shows the minimum standards in humanitarian response according to the standards specified in The Sphere Project handbook [11].

Table 1. Minimum standards in humanitarian response [11].

Minimum Standard	Description
Type of Shelter	Tents, plastic sheeting, prefabricated buildings, temporary or transitional shelters, permanent public buildings.
Living Area	Covered area of 3.5 m^2 per person.
Sanitation	Family latrines are preferred. Start with 50 persons per latrine and lower to maximum of 20 persons per latrine (segregated by sex).
Water Availability	Basic water needs are 7.5-15 litres/person/day. Maximum distance to nearest water point is 500 metres.
Health Facilities	Essential health services including control of communicable diseases, child health, sexual and reproductive health, injury, mental health and non-communicable diseases. One health centre/50,000 people with qualified health workers (medical doctors, nurses, midwifes).
Food and Nutrition	Minimum nutritional requirements of 2,100 kcal/person/day consists of 10 % protein, 17 % fat and adequate micronutrient intake.
Non-Food Items	Rapid market analysis can determine whether goods can be sourced locally or in neighbouring, non-affected areas.
Accessibility	Access road should consider proximity to transport hubs for the supply of relief assistance, seasonal constraints, hazards and security risks. Escape routes should avoid isolated or screened areas that could pose a threat to users' safety.
Transportation	Primary storage and food distribution accessible by heavy trucks from an all-weather road. Other facilities accessible by light vehicles.
Climate and Environmental Condition	Shelter should provide protection from seasonal climate. Need to minimize adverse impact on the local natural environment resources when providing shelter to the affected population.
Education and Activities	Encourage community to organize structured and supportive educational activities for children. Access to community-based social support and self-help activities that promote psychosocial well-being.

2.2 The United Nations High Commissioner for Refugee (UNHCR) – Handbook for Emergencies

2.2.1 Description

The United Nations High Commissioner for Refugee (UNHCR) is mandated to lead and coordinate the international action for protection of refugees and the resolution of refugee problems [15]. The handbook defined refugees in any situation in which their life or well-being will be threatened unless immediate and appropriate action is taken. UNHCR's primary purpose is to safeguard the rights and well-being of refugees and others irrespective of their race, religion, political opinion or gender. General guidance on emergency assistance included in the handbook aimed to provide protection and ensure necessary needs of the refuges are met effectively and appropriately [15].

2.2.2 Minimum standard for relief shelter

Table 2 shows the minimum standards in emergency management according to the standards specified in the Handbook for Emergencies by UNHCR [15].

Table 2. Minimum standards in emergency management [15].

Minimum Standard	Description
Type of Shelter	Plastic sheeting, light weight emergency tents (LWET), prefabricated shelters.
Living Area	Floor space of 3.5 m^2 per person.
Sanitation	One latrine per family or communal latrines that not exceeding 20 persons per latrine facility (segregation of sexes).
Water Availability	Minimum provision of 7 litres/person/day. Maximum distance to water distribution point should not more than 100 metres.
Health Facilities	Priority should be placed on health programmes of measles immunization, nutritional support, control of communicable diseases and epidemics, reproductive health problem and public health surveillance. One health centre per facility (20,000 persons).
Food and Nutrition	General ration of 2,100 kcal/person/day (meet essential vitamin and mineral requirements).
Non-Food Items	Assess goods that are readily available on the local market.
Accessibility	Access roads should be all-weather connecting to sources of necessary supplies, above flood level and have adequate drainage.
Transportation	Number of light and heavy vehicles should be determined. Ensure sufficient trained drivers, fuel, lubricants, spare parts, tyres, maintenance personnel and facilities.
Climate and Environmental Condition	Free from major environmental health hazards and suitable year-round climatic condition.
Education and Activities	Provide learning activities in safe and child-friendly spaces. Non-formal education included with full participation of the community.

3 Methodology

Field surveys were done by collecting data at the flood-affected areas and interviewing the victims to access the issues encountered during the 2014 flood event. The list of relief shelters that had been established were also identified which can act as temporary accommodation for the affected victims during a disaster. Collection of data involved combination or integration of qualitative and quantitative research methods. Qualitative data usually includes open-ended with emerging responses while quantitative data tends to be closed-ended with predetermined responses.

3.1 Data Survey

Field surveys were conducted at areas affected by 2014 flood event in Perak Tengah district, Perak. Survey activities were administered at 75 villages from 10 sub-districts in Perak Tengah which were Kota Setia, Bandar, Pasir Salak, Bota, Pulau Tiga, Lambor, Pasir Panjang Hulu, Layang-Layang, Kampung Gajah and Belanja. A number of 449 respondents were selected randomly from local people who were also the flood victims at the respective areas. Respondents were comprised of 69 % female and 31 % male ranged from teenagers to senior citizens. The age of respondents was majority ranged from 41 to 60 years old. Structured questionnaires were developed and used as a guidance while conducting the survey. The elements in the questionnaires consist of respondents' personal details, flood-related information, experiences in facing the flood, condition in relief shelters, losses estimation, health conditions and psychological aspects.

3.2 Interview

Interview sessions were carried out with the local authorities, related agencies and public community. The topics discussed during the interviews included flood management procedures, agencies involved in handling the flood, early warning and preparation, evacuation process, emergency relief shelters for the victims, distribution of supplies and assistance and mitigation measures for future flood disaster. Acquisition of data involved meeting and discussion with Perak Tengah District Council, District and Land Office of Perak Tengah, Department of Social Welfare of Perak Tengah, Department of Irrigation and Drainage (DID) Perak Tengah, Department of Civil Defence of Perak Tengah, Department of Public Works of Perak Tengah, Perak Tengah District Youth Council, district headman, villagers' head and flood victims.

3.3 Observation

Site observations were also done during the survey activities. Information related to flooding was recorded based from the observations at the surrounding areas. Facilities used as relief shelters during flooding were also visited as part of data collection activities. From the observations, relief shelters in Perak Tengah used existing permanent buildings to accommodate the flood victims. In most cases, the evacuees were located at the nearest shelters from their houses to facilitate the evacuation process during the flooding period.

4 Results and discussions

4.1 Relief shelters in Perak Tengah district and its classification

After disastrous impacts from 2014 major flood event, the authorities aimed to improve the disaster management system including early preparation and warning, risk reduction measures, immediate recovery and rehabilitation of affected communities. One of the initiatives in handling future flood disaster was provision of relief shelter locations that can accommodate the victims during the disaster period.

Table 3. Relief shelters distribution in Perak Tengah district.

Sub-District	Number of Relief Shelter	Capacity
Belanja	15	2,770
Layang-Layang	6	1,350
Lambor	4	1,900
Bandar	2	800
Bota	16	6,000
Kampung Gajah	5	3,250
Pasir Salak	1	1,000
Pulau Tiga	2	800
Kota Setia	1	500
Total	**52**	**18,370**

Distributions of relief shelters in Perak Tengah district are listed in Table 3. The information was obtained from the Department of Social Welfare of Perak Tengah whereby the shelters can accommodate a maximum capacity of 18,370 persons at one time. A number of 52 relief shelters were established in Perak Tengah district to act as temporary accommodations for the community affected by a disaster. Shelters were chosen by related authorities as early preparation in evacuating and placing the affected communities in a well-managed situation during disaster period in the future. All of the shelters are permanent buildings whereby Fig. 1 illustrated the classification of the shelters according to their types including schools, public halls and mosque or *surau*. Majority of the shelters are schools (63 %), followed by public halls (29 %) and mosques or *surau* (8 %). The schools included were secondary and primary schools, religious schools, and training institutes. Most of the shelters were selected as the buildings were equipped with basic facilities including toilets, separate areas and suitable for a wide range of population.

From the collected data, relief shelters were categorized based on their maximum capacity. In Perak Tengah, the highest number of relief shelter is for the capacity of 200 and below where it contributes to 35 % from the total number of relief shelters (Fig. 2). It is followed by 200 to 400 capacity (31 %), 500 to 700 capacity (21 %) and 800 to 1,000 capacity (13 %). It can be said that the existing shelters can accommodate certain number of people at one time due to the size of their buildings and surrounding areas. Low capacity shelters were public halls, mosque or *surau* and schools located at the rural areas. In addition, relatively there were also higher capacity temporary shelters that were assigned to Secondary School Sultan Muzaffar Shah, Mosque of Izzuddin Shah Bota, District Council's Public Hall, Institute of Agriculture Titi Gantong, Training Institute UITM Sg Galah, Primary School Dato' Sagor and MARA Junior Science College (MJSC) Pasir Salak.

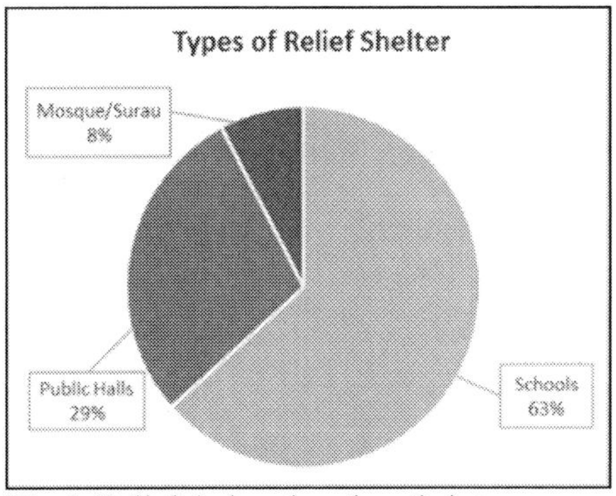

Fig. 1. Relief shelters based on their types.

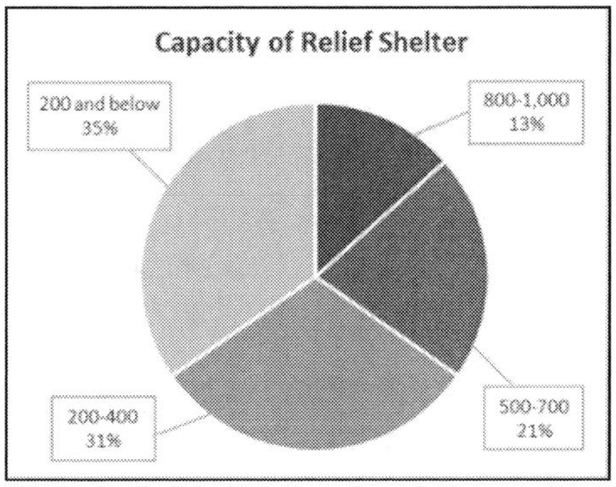

Fig. 2. Relief shelters based on their capacity.

4.2 Issues during flooding and setbacks

The survey conducted revealed several pertinent issues during the 2014 flood event. The issues highlighted in this study were related to situations or problems encountered by the victims during flooding and agencies involved in handling the disaster. Most areas in sub-districts of Kampung Gajah, Pasir Salak, Pulau Tiga, Kota Setia and Bandar were inundated for more than two weeks with some areas had been disconnected and normal access by land was temporarily hindered. As a result, Department of Civil Defence of Perak Tengah that acts as the rescue team stated that the distribution of food supplies, daily necessities and assistance to the shelters were disrupted.

In times of flooding, victims were evacuated to relief shelters as temporary accommodation until their actual houses are safe from any harmful conditions or risks of danger. During the peak of flooding, some of the relief shelters were also inundated resulted in relocation of victims to other shelters. This had caused difficulties to the affected victims in terms of relocation process, evacuation transport mode, sudden need to move and risk of danger due to rising of flood water level. Other than that, increasing number of flood victims also led to issues of crowded condition in the relief shelters. In case of emergency, victims were evacuated to the nearest centres. Instances where a high number of victims were placed in a relief shelter resulted in uncomfortable environment for the victims.

The prolonged flood inundation resulted in damages to building structures and disruptions to services and operation of the affected premises. Public buildings such as schools should be only used as short-term accommodation before providing more suitable shelters. Consequently, school sessions had been suspended due to inundation at the school areas, maintenance and repair works, recovery period and preparation for learning tools and materials. Besides, the infrastructure of the building including water, electricity and sanitation will deteriorate quickly from concentrated use to the extent that living conditions can become relatively unhealthy. Hence, the buildings may damage rapidly because they are unsuited to large number of occupants in a long period of time. In addition, low sense of responsibility by the users also contributes to the deterioration of the buildings.

In Malaysia, there is no implementation of standards or guidelines for relief shelters as temporary accommodation for the victims during a disaster. Properly designed and strategically located shelters based on acceptable standards are crucial to ensure practical, conducive and safe shelters for the victims. Therefore, relief shelters in Malaysia should be improved progressively in terms of strategic locations, optimum covered area for victims, availability of health facilities and sanitation, sufficient source of food and water, transportation, secure accessibility and more importantly conducting community-based activities to promote a balanced psychosocial well-being for the victims.

5 Conclusions

From the highlighted issues encountered during the 2014 flooding, appropriate planning and management strategies should be implemented to substantially reduce the inconveniences and risks to the victims during disaster. Besides, the use of public buildings as relief shelters may not only affect the formal usage of such premises but it may also not able to offer the convenience and appropriate facilities, especially for much longer stay. Capacity of the relief shelters reflects the maximum occupants that should be placed in the buildings at one time to provide safe and conducive environment for the affected victims. Furthermore, it is important to ensure regular operational maintenance of the buildings in order to maintain good facilities throughout the usage period. Establishment of dedicated relief shelters should be considered for future preparedness so that healthy environment can be maintained as well as to reduce the damage effects from the disaster.

The initial part of the research was partially funded by FRGS grant (FRGS/1/2015/TK02/UTP/02/6). The authors would also to thank the Centre for Graduate Studies and Department of Civil and Environmental Engineering, Universiti Teknologi PETRONAS for their support and encouragement.

References

1. B. Pradhan, Spatial Hydrology **9**, 1-18 (2009)
2. M. Haq, M. Akhtar, S. Muhammad, S. Paras, J. Rahmatullah, The Egyptian Journal of Remote Sensing and Space Sciences 135-141 (2012)
3. M. Saharia, P.E. Kirstetter, H. Vergara, J.J. Gourley, Y. Hong, J. Hydrol **548**, 524–535 (2017)
4. Malaysian Medical Relief Society (MERCY), *Malaysian Flood Emergency Response Donor Report* (2014)

5. M.S. Khalid, S. Shafiai, International Journal of Social Science and Humanity **5**, 398-402 (2015)
6. M.M.A. Khan, N.A. Shaari, A.M.A. Bahar, M.A. Baten, D.A.B. Nazaruddin, World Applied Sciences Journal 32, **4**, 626-634 (2014)
7. A.Z. Othman, A. Dahlan, S.N. Borhani, H. Rusdi, Procedia - Social and Behavioral Sciences **234**, 125 – 134 (2016)
8. M.N. Halgamuge, A. Nirmalathas, Int. J. Disast. Risk Re. **24**, 1-11 (2017)
9. C.S. Mustaffaa, N.A. Marzukia, M.T. Ariffina, N.A. Salleha, N.H. Rahamana, Procedia - Social and Behavioral Sciences **155**, 197 – 202 (2014)
10. A. Bashawria, S. Garritya, K. Moodleya, Economics and Finance **18**, 924-931 (2014)
11. *Humanitarian Charter and Minimum Standards in Disaster Response* (The Sphere Project, United Kingdom, 2011)
12. B. Kar, M.E. Hodgson, Transactions in GIS **12**, 227–248 (2008)
13. M.R. Rateau, Archives of Psychiatric Nursing **31**, 561–565 (2017)
14. The International Federation of Red Cross and Red Crescent Societies (IFRC), *Shelter and Settlements Standards*. Retrieved January 5, 2016 from http://www.ifrc.org/en/what-we-do/disaster-management/responding/services-for-the-disaster-affected/shelter-and-settlement/shelter-library/standards/
15. *Handbook for Emergencies*, (United Nations High Commissioner for Refugees, UNHCR, Geneva, 2007)

Dam pre-release as an important operation strategy in reducing flood impact in Malaysia

Nurul Hidayah Ishak[1*], and *Ahmad Mustafa Hashim*[1]

[1]Department of Civil and Environmental Engineering, Faculty of Engineering, Universiti Teknologi PETRONAS, 32610 Bandar Seri Iskandar, Perak, Malaysia

Abstract. The 2014 flood was reported to be one of the worst natural disaster has ever affected several states in the northern part of Peninsular Malaysia. Overwhelming rainfall was noted as one of the main factors causing such impact, which was claimed to be unprecedented to some extent. The state of Perak, which is blessed with four cascading dams had also experienced flood damage at a scale that was considered the worst in history. The rainfall received had caused the dam to reach danger level that necessitated additional discharge to be released. Safety of the dams was of great importance and such unavoidable additional discharge was allowed to avoid catastrophic failure of the dam structures. This paper discusses the dam pre-release as a significant dam management strategy in reducing flood impact. An important balance between required dam storage to be maintained and the risk element that can be afforded is the crucial factor in such enhanced operation strategy. While further possibility in developing a carefully engineered dam pre-release strategy can be explored for dam operation in Malaysia, this has already been introduced in some developed countries. Australia and South Africa are examples where pre-release has been practiced and proven to reduce flood risk. The concept involves controlling the dam lake level throughout the year, in reference to the rainfall data and the hydrological properties for the catchment area of the dams. Plentiful data analysis need to be done in contemplation of producing the optimal pre-release model. The amount of heavy rainfalls received is beyond human control but the distribution of the discharge from the dams can be further managed with the appropriate pre-release strategy.

1 Introduction

Flood is one of the main disasters faced by countries worldwide. Some of the natural disaster faced worldwide includes floods, volcanic eruptions, earthquakes, tsunamis and other geologic processes. Since Malaysia is not located within the active volcanic plates and tornado alley, flood is one of the main disasters that keep the government concerns almost every year. According to Emergency Events Database (EM-DAT), a global database on natural and technological disasters, flood is the most frequent disaster that happened in

* Corresponding author: nurulhidayahishak@gmail.com

Malaysia, which constitutes 48% of total disaster occurrences [1]. Fig. 1 shows the frequency of natural disasters that occurred in Malaysia, from 1965 to 2016.

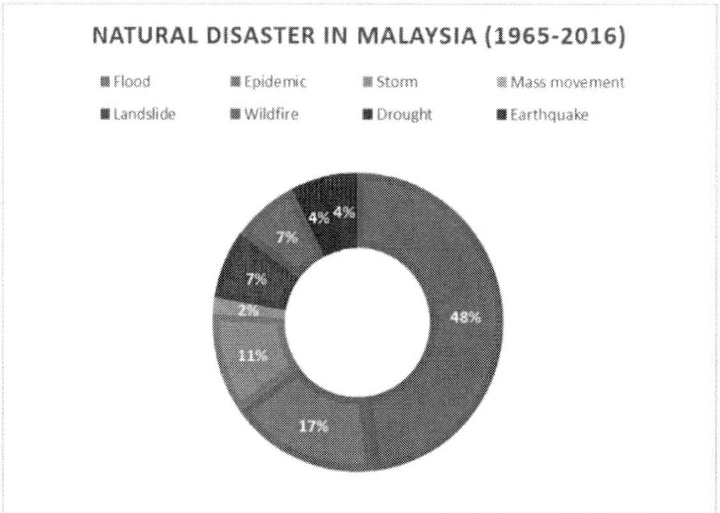

Fig. 1. Frequency of natural disasters that happened in Malaysia from 1965 to 2016 [1].

On December 2014, some parts of Malaysia had been hit with extreme flood, and in some places, it was even recorded as the worst event ever occurred. Perak, a Northern state in Malaysia, was one of the states that were badly affected by flood. In this study, Perak is chosen as the main area for assessment as it involves a unique case – four cascading hydroelectric dams. Among the areas in Perak that had been flooded were Kampung Gajah, Lenggong, Kerian, Bota, Parit, Kuala Kangsar and Sungai Siput.

Generally, the 2014 flooding event in Perak is associated with the overwhelming rainfall affecting the dams along Perak River, which are Temenggor dam, Kenering dam, Chenderoh dam and Bersia dam. During heavy rainfalls event on December 2014, an additional discharge was observed to be released from the dams due to over capacity of water level in the reservoir. Such action was understandably taken to avoid the dams from collapsing which could results in much more severe damage towards the downstream areas. Flooding and destruction happened in Perak because houses and properties were mostly located near river and within flood prone areas. Besides that, the long inundation period of the flood, which was up to one month in some areas, has caused a higher rate of destruction.

In some places, the villagers claimed they never experience flooding before and 2014 flood event was the first flood disaster they have ever encountered. Among the villages mentioned were Taman Murni and Kampung Tersusun Sejagop in Perak Tengah district, Kampung Lintang, Kampung Pulau Mentimun, PPRT Kampung Senggang, Kampung Pulai Ara Panjang and Kampung Biong Sauk in Kuala Kangsar.

2 Dam pre-release

This paper further discusses on pre-release concept and how the implementation will help to reduce the flood impact environmentally and economically. Pre-release in this context is defined as the strategic release of water to actively manage the water level in a dam prior to forecast rainfall. There are some countries, which already practiced this pre-release concept, such as Australia and South Africa. They even have pre-release and surcharge policy. Most of the dams are mainly for water storage; to supply potable water. Generally, pre-release is introduced to flood control dam. The principles of flood control dam consist of flood absorption and pre-release [2].

2.1 Melbourne Water, Australia

The purpose of the Melbourne Water's Dam Pre-Release and Surcharge Policy is to outline the company's practices on the dams pre-release and surcharge. The policy is practical to all the dams, which are under their management, including harvesting dams with or without catchments, seasonable storages and obsolete dams. It covers the Melbourne Water's practices, which are related to the pre-release of water and water level surcharge from the dams.

Melbourne Water is in charge of 16 dams, where each of them has been designed to function as water storage for drinking water source and in certain cases for drainage. All of them have uncontrolled spillways which would allow the dam water level to be maintained below the crest of the dam in case of intense flood situation. However, if the water level is above the spillway, the flow is uncontrollable thus additional discharges will naturally occur during flood events. Despite that, in regular 'non flood' operational situations, Melbourne Water does manage the discharge amount from their dams, as it is part of the environmental flow responsibilities. This is where they take into account the volume of rainfalls received so that the discharge could be well-planned. In term of safety, Melbourne Water always ensures that the stability of the dams are continually observed and they always deliver discharge report to related emergency organizations during flood events.

In managing their dams, Melbourne Water initiates reserved discharges of water from its dams when it is believed compulsory for dam safety purposes. Their dam operation aims to amplify water intake while diminishing hazard to the neighbourhood and assuring conformity with its Bulk Water Entitlements and environmental flow duties [3].

2.2 South Gippsland Water, Australia

The South Gippsland Water (SGW)'s Dam Pre-Release and Surcharge Policy is to delineate the operating rules and relevant measures related to pre-release and extra discharge of dams during heavy rainfall events.

SGW maintains their dams by ensuring that measured discharges of water by means of scour passages from the dam will commence when it is considered needed for dam safety purposes. They will amplify water intake while diminishing any possible safety or flood threat to the public and ensuring conformity with environmental flow duties.

Their management policy also specifies measures in reducing possible flood effects. The dams are not designed to discharge vast amount of water, before or during any overflowing. The spillway will make to function when the water level is above FSL where it will allow safe passageway of the overflow. SGW dams which have unrestrained spillways do not have allied flood mitigation guidebooks and therefore it is impossible to control the water release during normal or flood occasions. In utmost flood occasions, the spillways will ensure that the water level is kept below crest of the dam to safeguard the stability of the structure. To establish agreement with environmental flow responsibilities, regulated small discharge from their storages which are prearranged and actively accomplished are frequently made in order to make sure that their action does not commit to flood or riverbank overflows [4].

2.3 Goulburn-Murray Water, Australia

Implying to predicted inflows, pre-release will be contemplated when the dams have exposure to failing. In determining the amount of water to be pre-released, their utmost policy requires that any volume of water discharged should be replenished with a high degree of reliance. The emphasize is to ensure that the pre-release would not cause any man-made flooding and there is no previous flooding downstream that would be impacted by the pre-release unprepared. Generally, their dam management also focuses so that downstream flood peak can be decreased [5].

2.4 Orange-Fish-Sundays River Basin, South Africa

In operating the reservoir for the Orange-Fish-Sundays River Basin in South Africa, MIKE II was used to model and simulate the reservoir operations and river flows. Real-time data was used and optimized for sideways inflow, river water level, salinity and reservoir procedure to make sure the fulfilment of the multiple objectives of the simulation-optimization which includes flood control, hydropower generation and irrigation. The simulation-optimization framework is very flexible because it can be applied to any control parameters and objectives of interest [6].

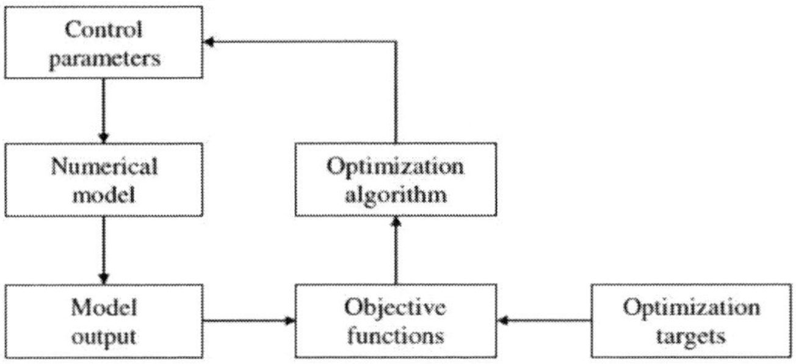

Fig. 2. Simulation-Optimization framework for Orange River Basin, South Africa [6].

As indicated in earlier study, the main objective was to minimise the water loss from the Orange-Fish-Sundays River System and ensure that the irrigation demands were fulfilled [6]. This had also included maintaining the water level in the dam between Minimum Operating Level (MOL) and Full Supply Level (FSL). Under normal operating system, the water level will be kept closed to MOL but under flood events condition, if the forecasted rainfalls would exceed the FSL, pre-release would be executed. From the simulation in the late 2015 and early 2016 on that river basin, it is distinguished that pre-release was executed when the water level reached FSL of the reservoir.

3 Pre-release concept

As mentioned above, pre-release has been practiced by some of the countries, but none of them are hydroelectric dam. In Perak, there are four cascading hydroelectric dams, as shown in Fig. 3, comprise of Temenggor Dam, Bersia Dam, Kenering Dam and Chenderoh dam, which are managed by Tenaga Nasional Berhad (TNB). It is naturally expected that TNB's procedure for releasing water is based on electricity demand; as their main service is providing electricity for industrials and also domestic uses. Thus, for hydroelectric dam, among the vital criteria that need to be considered in optimizing this concept are the dam level and dam discharge. For hydroelectric dam, these two parameters are crucial because every drop of water that the dam release is crucial for their operation. The dam cannot simply discharge water without proper planning ahead. Fig. 4 shows the basic concept of how the pre-release would work.

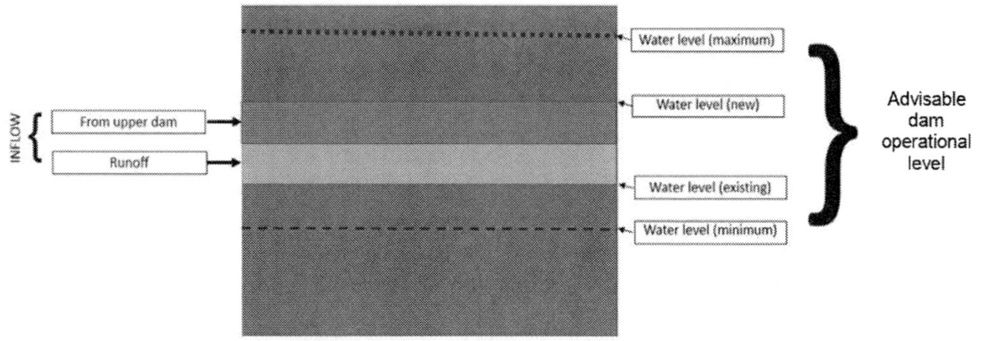

Fig. 3. Cascading dams in Perak (Reproduced from [8]).

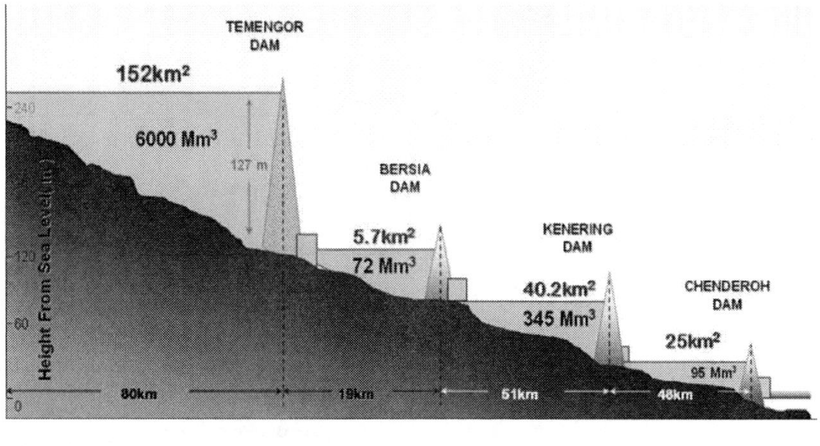

Fig. 4. Advisable dam operational level.

Fig. 4 shows the schematic illustration on dam cross section. The basic concept of pre-release is based on controlling the dam level in between minimum water level and maximum water level. The proposed scheme for volume of water that need to be discharged is:

$$\textit{Volume of water need to be discharged} = \textit{new storage volume} - (\textit{forecast inflow from upper dam} + \textit{forecast runoff}) \quad (1)$$

It is important to ensure that the water level in the dam is maintained at least the minimum required level so as not to affect the operational requirement of the electricity generation.

3.1 Pros & cons of pre-release

By introducing pre-release into the dam management systems, it also means to allow some risks to the dam. The benefits of implementing pre-release would include the dam level is maintained and monitored throughout the year; so that no excessive water will be discharged during peak rainfall season. Other than that, it will prevent unnatural flooding; flooding due to extra discharge from the dams thus reducing the flood impact to the downstream areas; environmentally and economically (if flood are to happen during monsoon seasons)

4 Conclusions

Pre-release is not a new method in the dam management. However, pre-release done by hydroelectric dam would somehow be difficult to handle. In Perak's unique case, the dams are cascading; which the flow from the upper dam will accordingly affect the lower dam. Thus, proper management need to be carried out. Since pre-release is human-controlled, a sound and fail-proof policy would be needed beforehand and sufficient studies and simulation would be required before the implementation, especially when involving hydroelectric dam. Regardless of that, to apply this approach to the cascading hydroelectric dams in Malaysia remains as a challenging task.

The authors also wish to express special thanks to the Centre of Post Graduate Studies and Department of Civil and Environmental Engineering, Universiti Teknologi PETRONAS for their support and encouragement.

References

1. EM-DAT - The International Disaster Database. Retrieved February 15th 2017 from http://www.emdat.be/ [online]
2. J. Jordaan, A. Bell. EOLSS Publications. (2009)
3. "Dam Pre-Release and Surcharge Policy". Retrieved March 14th 2017 from https://www.melbournewater.com.au/aboutus/whoweare/Legislationandpolicies/policies-and-procedures/Documents/Dam%20pre-release%20and%20surcharge%20policy.pdf [online]
4. "Dam Pre-Release and Surcharge Policy". Retrieved March 14th 2017 from http://www.sgwater.com.au/wp-content/uploads/2014/04/Dam-pre-release-and-surcharge-policy.pdf [online]
5. "Policy PRE-RELEASE FROM DAMS". Retrieved March 14th 2017 from http://www.g-mwater.com.au/downloads/gmw/Storages/Pre-Release_from_Dams_Policy.pdf [online]
6. C. B. Pedersen, H. Madsen, C. Skotner. *13th SANCIAHS Symposium* (2007)
7. F. Chou, C. Wu, Journal of Hydrology **497**, 1-14. (2013)
8. M.I. Najid, L.M. Sidek, B. Hidayah, Z.A. Roseli, *International Conference on Advances in Renewable Energy and Technologies (ICARET 2016)* (2016)

The Study of Heavy Metals on Sediment Quality of Kuala Perlis Coastal Area

A.A. Ahmad Zubir[1], *F.N.* Mohd Saad[1*], and *F.A.* Dahalan[1]

[1]School of Environmental Engineering, Universiti Malaysia Perlis, 02600 Arau, Perlis, Malaysia

Abstract. The contamination of heavy metals gives bad implications to the aquatic environment. Thus, a study was conducted to assess the sediment quality by using different contamination indices such as Enrichment Factor (EF), Geo-accumulation Index (I_{geo}), and Pollution Load Index (PLI). Each sediment sample was collected at the surface (0-15cm) at 5 locations based on the land use activity; jetty port (A), seaside restaurant (B), roadside area (C), power plant (D) and residential area (E). All samples were undergoes acid digestion and analyzed with AAS. Four elements identified from the sediment samples which are Cr, Cu, Pb and Zn were used to calculate the respective indices. Results show that, the highest EF value of Pb which categorized as very severe enrichment was at point E. Meanwhile minor enrichment was detected at point B for Cu while Cr and Zn at point A. Based on I_{geo} value the sediment quality along Kuala Perlis was in the unpolluted-moderately polluted condition (class 1). As for PLI it shows that the sediment of the coastal area is unpolluted (PLI<1). Therefore, this study revealed that, the main contributor of heavy metals in this area is Pb while the sediment quality of Kuala Perlis was in minor pollution condition.

1 Introduction

The increasing of industrialization and urbanization activities in Kuala Perlis, more waste waters that containing various types of toxic contaminants such as heavy metals were release to the rivers, surface runoff and finally reach the coastal waters [1]. The transportation of the toxic contaminants to the coastal water ways finally deposited in the marine sediment over the time exposure [2].The heavy metals contaminated the aquatic environment, through the bioaccumulation in the sediments thus toxicity of the metals that altered the environmental background conditions. This resulting the long term bad implications to the marine habitat and human health along the food chain in the ecosystem[3]. In order to overcome this problem, a pollution study must be conducted to assess the sediment quality in the coastal area. This assessment is very significance to give an understanding the degree of sediment pollution as well as basic information for future supervision such as monitoring and the mitigation of pollution control. Therefore, this study is aimed to identify the type of heavy metals that polluting the sediment of coastal area

* Corresponding author: farahnaemah@unimap.edu.my

 https://doi.org/10.1051/e3sconf/20183402018

based on the land use activities as well as to assess the pollution degree by using different contamination indices such as Enrichment Factor (EF), Geo-accumulation Index (I_{geo}), and Pollution Load Index (PLI). The enrichment factor (EF) of metals is an indicator reflecting the status and degree of environmental contamination of that metal while Geo-accumulation Index (I_{geo}) shows the various degree of enrichment above the background value ranging from unpolluted to very polluted sediment quality [4]. Pollution Load Index (PLI) is the nth root of the product of n contamination factors (CF) values and also one of the indices to assess the pollution in the sediment [4].

2 Materials and methods

2.1 Sampling location

Five sediment samples pointed as A, B, C, D and E were taken along the coastal area of Kuala Perlis as shown in Fig. 1. The distances between each point approximately 1 km-3 km and these 5 locations were selected based on different land use activity. The sample on each location was collected at the subsurface (0-15cm) by using hand mud auger before taken to laboratory for analysis.

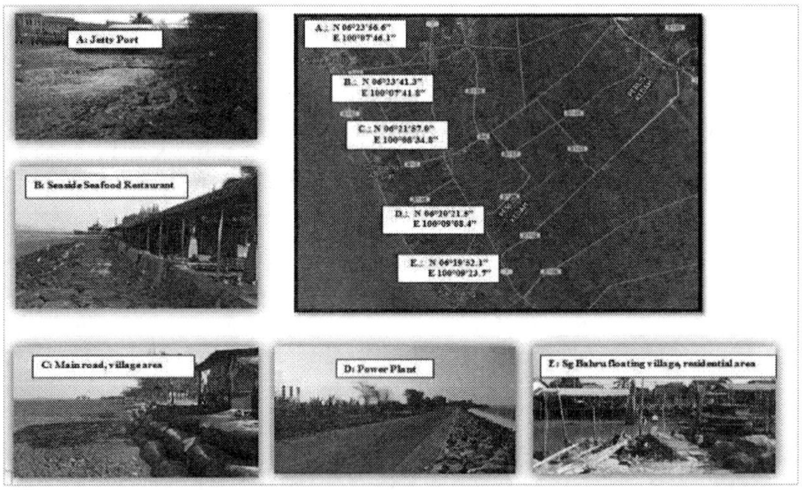

Fig. 1. Location of each sampling point based on land use activity.

2.2 Experimental methods

The sediment samples were oven dried at 110°C for 24 hours. The moisture content of the sediment was measured and the pH value was determined by pH meter. Then, the dried samples were sieved to obtain the particle size distribution of the sediment (ASTM D422-63). Some of the dried samples were also sieved into 60 μm to identify the types of heavy metal by using XRF (X-Ray Fluorescence) analyser [1]. Then, by following aqua regia method [5], 2g of sample was diluted with 25ml deionized water (wetting), and digested with H_2O_2, HNO_3, $HClO_4$, and H_2SO_4. The digested samples were filtered into 50ml of volumetric flask [5,6] and analysed by AAS (Atomic Absorption Spectrometer) to determine the element's concentration.

The quality of sediments in this study was assessed through three ways which are EF, I_{geo} and PLI. In this study, Fe is used as a reference element [10] while average shale concentrations given by Wedepold and Turekian [11] are used as the background concentration for the element tested. Therefore, the equation 1, 2 and 3 are used to calculate EF, I_{geo} and PLI.

$$EF = \frac{(\frac{Cn}{Cref})sample}{(\frac{Bn}{Bref})} \quad (1)$$

where *Cn* (sample) is concentration of the examined element in examined environment, *Cref* (sample) is the concentration of the examined element in reference environment, *Bn* is the concentration of the reference element in examined environment and *Bref* is the concentration of the reference element in reference environment.

$$I\text{geo}=\text{Log2}\,(\frac{Cn}{1.5Bn}) \quad (2)$$

where C*n* is the concentration of metal in examined environment, and B*n* is the concentration of metal in reference/background environment.

$$PLI = (CF1 \times CF2 \times CF3 \times \ldots \times CFn)^{\frac{1}{N}} \quad (3)$$

where CF is the contamination factor ($C_{sample}/C_{background}$) of the element while N is the number of metals.

3 Results and discussions

3.1 Sediment characteristics

Sediment moisture content, sediment pH and sediment classification of each sample collected at 5 locations are determined in this study. Table 1 shows the moisture content of each sediment collected at different location. It shows that Point A and B recorded high moisture content with more than 60%. This condition is due to the watery sediment condition. On the other hand, sediment sample taken from Point C, D and E have lower percentage of moisture content ranging from 46% to 51%. Based on the observation, it was found that the sediment in Point C,D and E was a little bit dry thus the water content is lesser than sample in Point A and B.

Table 1. Moisture content and pH value of sediment sample.

Parameter	Point A	Point B	Point C	Point D	Point E
Moisture content (%)	65.7	63.5	46.1	47.5	50.9
pH	7.48	7.51	7.99	8.17	8.01

As stated in Table 1, the sediment pH value in point A, B, C, D and E are 7.48, 7.51, 7.99, 8.17 and 8.01 respectively. Normally, the pH value of marine water is closely to 8.2 since most of the water has some capacity to resists the changes of pH value due to the effects of carbonate-buffer system [7]. Bicarbonate ions which are the main component in the system, were produce from the weathering of silicate or carbonate minerals as rainwater passes through the soil zone [7]. During the hydrolysis of bicarbonate ion, hydroxyl ions are produce to neutralise hydrogen ions thus maintain the pH near the constant level [7].

However, the presence of contaminants carried by the contaminated waterways altering the normal pH conditions. In this study, pH value in point A and B are in the range of 7.1-

7.5 which shows the environment is in slightly alkaline condition. The pH condition was quite similar to the study done by Idris and Ahmad [8] which shows that Juru river sediment condition were also in slightly acidic to slightly alkaline condition which are due to the presence of heavy metals (nickel and chromium) in the sediment. Meanwhile, point C, D and E were exists in moderately alkaline environment since the pH range between 7.6-8.3 which the pH where not mainly altered by the waterways contaminants. According to Singare et al.[9] the sediment can be considered as acidic and hazardous if the pH<6, while it can be also harm the aquatic life if the pH>10 which contaminated with strong base such as NaOH and $Ca(OH)_2$[9]. Therefore, in this study, since the pH is in the range of 7.4 to 8.1 thus, the sediment in each location can be considered as not in hazardous condition.

Particle size distribution is also studied in this study to classify the sediment. Table 2 shows the sieve analysis of the sediment in each point along Kuala Perlis coastal area. According to Unified Soil Classification System (USCS)[10], point A, B, C, D and E gives a similar soil type which is silt and clay. Silt and clay is the smallest grain in soil classification. According to Seyedeh Belin [11] the concentration of heavy metal enriched more in fine sediment particle compared to sand particle. This is due to the adsorption of metal from water to deposit at the bottom sediment are more significant. Besides, the capacity of fine grained particles are high to retain the heavy metals within the sediment compared to other sand fraction types[11-13].

Table 2. Sieve analysis.

Soil type (%)	Sampling Point				
	A	B	C	D	E
Sand	38.5	41.7	24.5	36.9	32.9
Silt & Clay	61.5	58.3	75.5	63.1	67.1

3.2 Pollution assessment

3.2.1 Sediment Quality Guidelines (SQG)

Sediment Quality Guidelines (SQG) has been developed by MacDonald[14] in order to predict the adverse effects in polluted sediment [15]. In this study, the concentration of metals were compared with SQG of PEL: Probable Effect Level, SEL: Severe Effect Level, and TET: Toxic Effect Threshold [14].

Table 3 shows the metal concentration in subsurface sediment of this study area. The results reveal that, the concentration of Cu, Cr, Pb and Zn contain in the sediment sample does not exceed the probable effect level, severe effect level, and toxic effect threshold of the sediment quality guidelines. Thus, this indicates that, the sediment of Kuala Perlis coastal area can be considered as a good quality and will not give the harmful effect to the marine environment.

3.2.2 Contamination Indices : Enrichment Factor (EF), Geo-accumulation Index (I_{geo}), Contamination Factor (CF) and Pollution Load Index (PLI)

Table 4 shows the types of contamination indices and the terminologies[16]. The assessment of sediment contamination in this study was carried out by using the Enrichment Factor (EF), Geo-accumulation Index (I_{geo}), Contamination Factor (CF) and Pollution Load Index (PLI). Enrichment factor (EF) of metals (Cu, Cr, Pb and Zn) on each study point were tabulated in Table 5.

Table 3. Comparison of metals concentration and Sediment Quality Guidelines (SQG).

Point	Heavy metals concentration (mg/kg)			
	Cu	Cr	Pb	Zn
A	0.538	1.182	5.119	1.469
B	0.315	0.367	3.088	0.687
C	0.165	0.588	3.722	0.514
D	0.266	0.414	4.218	0.655
E	0.314	0.058	4.428	0.545
	USEPA : Dry weight (mg/kg)			
PEL	197	90	91.3	315
SEL	110	110	250	820
TET	86	100	170	540

Among the metals studied, Pb gives the highest enrichment in the sediment collected at Point A and E since the EF values are ranging between 25 to 50 which categorized as very severe enrichment. Meanwhile, point C shows that Pb is in the severe enrichment since EF value is in the range of 10-25. The minor enrichment of Zn ($1<EF<3$) was shown in point A,B,D and E while there are no Zn enrichment in point C. There is also minor enrichment of Cr in point A and no enrichment on the other points. For Cu, there are minor enrichment detected at point A, B, E and no enrichment in point C and D. Therefore, it can be seen that there is enrichment of elements in Point A, B, and E which are the Jetty Port, Seaside restaurant and Sg. Bahru Estuary respectively.

Since Pb gives the highest enrichment factor so, it can be considered as the possible contamination from these land use. The shipping and ferries activities that operates daily may incidentally expelled the discharged of fuel and oil spillage directly to the coastal waters [2]. This is because in the marine environment there are two main factors of metal leaching into the waters such as from the anthropogenic activities and the earth natural processes from minerals and the physical characteristics of the sediments [17].

Previous studies proved that, the combustion of fuels by the vehicles and boat activities resulted in the high enrichment of Pb in the sediment [18]. Besides, Ong and Kamaruzzaman[19] also reported that the sediment of eastern coast of Johor resulted from high Pb enrichment due to the natural erosion of earth's surface.

Table 6 shows the results of Geo-accumulation Index (I_{geo}), Contamination Factor (CF) and Pollution Load Index (PLI) of the sediment in 5 locations. For I_{geo} assessment, it shows that Point A gives the highest I_{geo} value followed by Point E, D, C and B. According to the mean of I_{geo} value, all the sediment samples in the study area are in unpolluted- moderately polluted condition and categorized as class 1. As for CF assessment, the result reveals that Point A gives the highest CF value while the lowest of CF value is identified at point B. Overall, it can be concluded that low elements contaminations were detected in the sediments of study area since the CF value are less than 1. As the overall assessment, the Pollution Load Index (PLI) shows the highest value in point A while the lowest was in point D. Since the PLI values in all the sampling points were <1, this has shown that the sediments of Kuala Perlis coastal area are in unpolluted condition.

Table 4. Contamination indices categories and terminologies.

EF		I_{geo}			CF		PLI	
Value	Enrichment condition	Value	Class	I_{geo} condition	Value	Condition	Value	Condition
EF<1	No enrichment	≥0	0	Unpolluted	CF<1	Low CF	>1	Polluted
EF=1-3	Minor enrichment	0-1	1	Unpolluted-moderately polluted	1≤CF<3	Moderate CF		
EF=3-5	Moderate enrichment	1-2	2	Moderately polluted	3≤CF<6	Considerable CF		
EF=5-10	Moderately severe enrichment	2-3	3	Moderately-strongly polluted	CF≥6	Very high CF		
EF=10-25	Severe enrichment	3-4	4	Strongly polluted			<1	Unpolluted
EF=25-50	Very severe enrichment	4-5	5	Strongly-extremely polluted				
EF>50	Extremely severe enrichment	>5	6	Extremely polluted				

Table 5. Enrichment Factor (EF) of metals.

Point	EF value of each metal			
	Cu	Cr	Pb	Zn
A	1.56	1.71	33.31	2.01
B	1.58	0.92	34.79	1.63
C	0.41	0.72	20.56	0.6
D	0.9	0.7	32.06	1.05
E	1.22	0.11	38.86	1.01

Through observation it is believed that the source of pollution in point A and point E are mainly form the shipping and boat activities. Since the boat are operates daily, the coastal water are contaminated with the effluent from the boats such as the oil, fuel and other toxic agent for boat antifouling paints[20]. Point A gives the highest sediment moisture content so, the overlying waters carries the pollutants [21] and finally deposited and settled on the bottom of the sediment. Besides, both point A and E are located at nearby estuary which all the municipal and domestic wastes created by the residents in the area are flown in the waterway before ended up to the sea. On the other hand, the existence of metals in point C are from the surface runoff of the road or the effect of paddy field activities which surrounding the area.

As mentioned by Wahidatul [1]the present of Pb is mainly from the using fertilizers and pesticides in the paddy fields. In point D, the present of Cr, Cu, Zn and Pb might be from the power plant that release the vapour effluent containing metals contaminants to the air. Thus, the metal contains in the air will cause the acid rain at nearby area. These acid rain will effects the costal water quality and finally deposited in the sediment [22].

Table 6. Geo-accumulation Index (Igeo), Contamination Factor (CF) and Pollution Load Index (PLI) of sediments taken from 5 locations.

Point	I_{geo} value of each metal					Cf value of each metal					PLI
	Cu	Cr	Pb	Zn	mean	Cu	Cr	Pb	Zn	mean	
A	0.0023	0.0026	0.0514	0.0031	0.015	0.012	0.013	0.256	0.015	0.074	0.028
B	0.0014	0.0008	0.031	0.0014	0.008	0.007	0.004	0.154	0.007	0.043	0.013
C	0.0007	0.0013	0.0373	0.0011	0.010	0.003	0.006	0.186	0.005	0.050	0.012
D	0.0012	0.0009	0.0423	0.0013	0.011	0.006	0.004	0.211	0.006	0.056	0.011
E	0.0014	0.0001	0.0444	0.0012	0.012	0.007	0.001	0.221	0.005	0.058	0.014

4 Conclusions

This study has shown that the sediment of Kuala Perlis coastal area is currently contain with four elements which are copper (Cu), chromium (Cr), zinc (Zn) and lead (Pb). The result reveal that the highest enrichment of element in the sediment is Pb, thus indicating it as the most contributors of heavy metals in the sediment of Kuala Perlis. The high enrichment of Pb is effected by the land use resulting from the anthropogenic activities that are operates daily in the area. Point A and E were identified to have high metals enrichment since each location has lots of activities that contribute to the release of these metals to the environment. However, according to SQG, the sediments are in good quality since the metal concentration does not exceed the Probable Effect Level (PEL), Severe Effect Level (SEL), and Toxic Effect Threshold (TET). Based on the geo-accumulation index the sediment were consider as class 1 which is unpolluted- moderately polluted condition. While PLI assessment shows that the coastal area is unpolluted sediment.

The author's gratitude goes to the support of School of Environmental Engineering and School of Material Engineering Universiti Malaysia Perlis (UniMAP) for providing the laboratory facilities in conducting this research. In addition, we are also thankful to the Kuala Perlis Maritim Department for providing the schedule and useful knowledge regarding the sea water tidal in Kuala Perlis which are related to the sampling activities.

References

1. N. Wahidatul, A. Zainon, S. A. Mohammed, S. H. Ismail, W. Amiza, Int. J. Civ. Environ. Eng. **12**(6), 28–33 (2012)
2. K. Lias, T. Jamil, S. N. Aliaa, J. Appl. Chem. **4**(1), 48–54 (2013)
3. P. Saikia, S. Das, R. K. Shah, S. Islam, Int. J. Adv. Biol. Res. **5**(2), 150–154 (2015)
4. S. M. Praveena, M. Radojevic, M. H. Abdullah, Int. J. Environ. Sci. Educ. **2**(3), 60–68 (2007)
5. Z. Y. Hseu, Z. S. Chen, C. C. Tsai, C. C. Tsui, S. F. Cheng, C. I. Liu, H. L. Ta, N, Water Air Soil Pollut. **141**,189–205 (2002)
6. E. Geana, A. M. Iordache, C.Voica, M.Culea, R.Ionete, Asian J. Chem. **23**(12), 5213–5216 (2011)
7. L. Radke, http://www.ozcoasts.gov.au/indicators/ph_coastal_waterways.jsp. (2015)
8. A. A. Idriss, A.K. Ahmad, J. Environ. Prot. **4**, 1245–1250 (2013)
9. P. U. Singare, M. P. Trivedi, R. M. Mishra, Mar. Sci. **1**(1), 22–29 (2011)
10. U. Lowell, http://www.dphu.org/uploads/attachements/books/books_2789_0.pdf. (2008)

11. S. Belin, T. Sany, A. Salleh, W. Port, N. Port, S. Port, Water Soil Pollut. **224**, 1476, 2–18 (2013)
12. G. M. S. Abrahim , R. J. Parker, Environ. Monit. Assess. **136**, 227–238 (2008)
13. E. P. Nobi, E. Dilipan, T. Thangaradjou, K. Sivakumar, L. Kannan, Estuar. Coast. Shelf Sci. **87**(2), 253–264 (2010)
14. D. D. MacDonald, C. G. Ingersoll, T. A. Berge, T. A. Berger, Arch. Environ. Contam. Toxicol. **39**(1), 20–31(2000)
15. A. Naji , A. Ismali, Environ. Asia **4**(1), 30–38 (2011)
16. A. S. Likuku, K. B. Mmolawa, G. K. Gaboutloeloe, Environ. Ecol. Res. **1**, 2, 32–402 (2013)
17. B. M. Mwashote, West. Indian Ocean J. Mar. Sci. **2**, 1 25–34 (2003)
18. S. Sabri, M. I. M. Said, S. Azman, Malaysian J. Anal. Sci. **18**(1), 37–42 (2014)
19. M. C. Ong, B.Y. Kamaruzzaman, Am. J. Appl. Sci. **6**(7), 1418–1423 (2009)
20. E. J.A.McCoy, L. T. Johnson, http://ucanr.edu/sites/coast/files/59476.pdf. (2010)
21. M. E. Goher, H. I. Farhat, M. H. Abdo, S. G. Salem, Egypt. J. Aquat. Res. **40**(3), 213–224 (2014)
22. A. Singh, M. Agrawal, J. Environ. Biol. **29**(1), 15–24 (2008)

Effect of pH, temperature and moisture content during composting of rice straw burning at different temperature with food waste and effective microorganisms

Irnis Azura Zakarya[1,2*], *Siti Noor Baya* Khalib[2], and *Norhasykin* Mohd Ramzi[2]

[1]Centre of Excellence for Biomass Utilization, Universiti Malaysia Perlis
[2]School of Environmental Engineering, Universiti Malaysia Perlis, Kompleks Pusat Pengajian Jejawi 3, 02600 Arau, Perlis

Abstract. Rice straw is considered as one of the most important agricultural residues and represented as one of the major by-products from rice production process. Normally, rice straw that produced after harvesting season been directly burned on-farm. Conversion of rice straw into value added compost will improve the productivity of plant, reduction of pollution towards environment and reduction of local pollution due to open burning activity. The objective of this study was to evaluate the performance of composting rice straw ash (RSA) with food waste (FW) and effective microorganisms (EM) in term of the compost quality (pH, temperature, moisture content). RSA was prepared by burning the raw rice straw at three different temperature of 300°C, 400°C and 500°C for one hour. EM used during the composting process was prepared by mixing of brown sugar, 'tempe' and water that can be used after one week of fermentation process. There are four treatments of RSA-compost; RSA (300°C), RSA (400°C), RSA (500°C) and control (raw rice straw) with the same amount of compost medium; 1kg black soil, 0.5kg RSA, 3L EM and 1kg FW. The composting process happens for 30 days. During the composting process, all the parameters of RSA-compost obtained in a range like; pH value 8-10, temperature 20-50°C and moisture content 40-60%. The result showed that all compost quality of rice straw ash compost obtained in an acceptable range for final compost to establish.

1 Introduction

One of the main agricultural productions in Asia is rice. Hence the production of rice straw in Asia is higher. Statistic shows that, each kilogram of grains harvested, rice straw generated are between 1.0-1.5 kg [1]. Normally rice straw produced after harvesting season been directly burned. Conversion of rice straw into value added compost will improve the productivity of plant, reduction of pollution towards environment and reduction of local pollution due to open burning activity [2].

* Corresponding author: irnis@unimap.edu.my

On the other hand, in 2011 at least 1184371.6 tonnes of rice straw has been produced in Northern Region of Malaysia [3]. Usually, farmers will manage the residue by on site burning or use it as household domestic use such as for cooking. In Malaysia, straw burning method was use due to crop rotation that needs to be done. Decomposing of rice straw naturally cannot be done due to limitation of time. Rice straw burning will cause haze. Based on a study that has been done in a journal, they stated that by burning a tones of rice straw, 3 kg of particulate matter, 1460 kg of carbon dioxide, 60 kg of carbon monoxide, 2 kg of sulphur dioxide and 199 kg of dust will be produced [4].

According to National Solid Waste Management Department also stated that between 31% until 45% of 36000 tonnes of garbage is food waste. Most of the food waste that been produced will be sent to landfill while the incineration and food waste treatment will consume large cost. Malaysian also did not realize that roting food waste is one of the main factors of greenhouse gas due to production of methane gas [5]. Existing of methane gas will increase 23 times of global warming compare to carbon dioxide. Based on a research that has been done by Environment Protection Agency (EPA)'s statistics, it show that less than 3% of food waste been recycle or recovered in 2010 showing that 97% of the food waste will be dumped into landfill area. In the same article written in The Star Online, based on statement from Waste and Resource Action Programme (WRAP) in 2015, about 7% of greenhouse gas emission that equal to 3.3 billion tones in a year has been produced through food waste wasted. They added that 1 billion tones per year of greenhouse gas can be reducing by reducing the food waste amount [5]. One of the way to reduce the amount of food waste end up to landfill, we can reuse or recycle them as composting material.

From the issue that has been stated, rice straw and food waste can be combining together to be used as composting material for organic fertilizer production. The content of fertilizer that will be produced in this research must at least as same as commercial's fertilizer content. So that the global warming issue due to rice straw's open burning especially by farmers will be reduce through composting of food waste and rice straw ash. At the same time, utilization of landfill towards food waste disposal will be minimized. In this research, rice straw ash was produced by using different burning temperature in order to determine the quality of compost possessed by each temperature during composting process.

2 Materials and methods

2.1 Raw Sample Preparation

Rice straw been collected from Wang Ulu's paddy field in Perlis for about 10 kg. Then the collected rice straw allowed drying under open air for a week. After a week, the dried rice straw was grinded using Plastic Shredder Machine into small pieces which were between 1-5mm. Rice straw was burning at different temperature of 300°C, 400°C and 500°C using CF1100 muffle furnace. The amount of rice straw ash that should be produced at each temperature was 0.5 kg.

Then the process continue by production of 12 litre fermented liquid using 1 kg of brown sugar, 4 pieces of *tempe* and 12L of water. The mixture was left for a week in the dark place to allow the fermentation process to occur.

Food waste was collected from Dragon's food stall. Food waste had been rinse using tap water before blended. The amount of food waste use for each compost medium including control was 1kg.

2.2 Preparation of Compost Containers

The type of containers use in this project were baskets that contain holes along it side to allow aeration process occurred. The dimension of compost container was shown in Fig. 1. The diameter of the container was 31 cm while height of container was 30 cm.

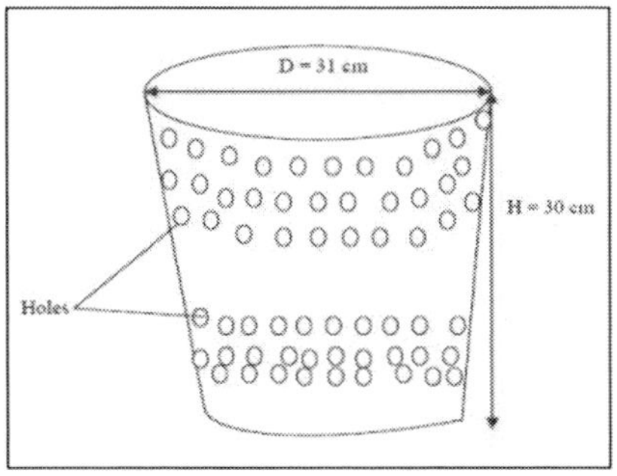

Fig. 1. Dimension of basket use as compost container.

2.3 Preparation of Compost Medium

The compost medium was prepared by mixing 1kg of black soil, 0.5 kg of rice straw ash, 3L of fermented liquid and 1 kg of food waste. The process was repeated for each temperature of rice straw ash. While for the control, raw rice straw without burning was used. The illustration on composition of compost medium prepared was showed in Fig. 2 and Fig. 3.

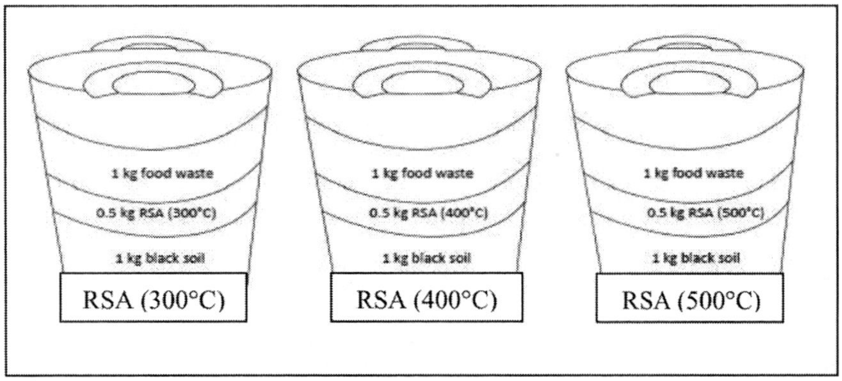

Fig. 2. Cross-section of compost medium using different temperature of rice straw ash.

2.4 Parameters Analysis

Parameter that been analyse during compost period were pH, temperature and moisture content. Temperature been analyse by using mercury thermometer by placed at certain point for 60 seconds. There were 3 different point should be test for temperature [6]. pH that measure by dissolving 1g of compost sample into 10 ml of distilled water. The ratio of sample dissolved should be in 1/10 [7]. Then the pH was taken using pH meter after 1 hour. Moisture content of sample was determined by usng ASTM 02974 oven-dry method where 20 g of sample been placed into oven for 24 hours at temperature of 105°C [8].

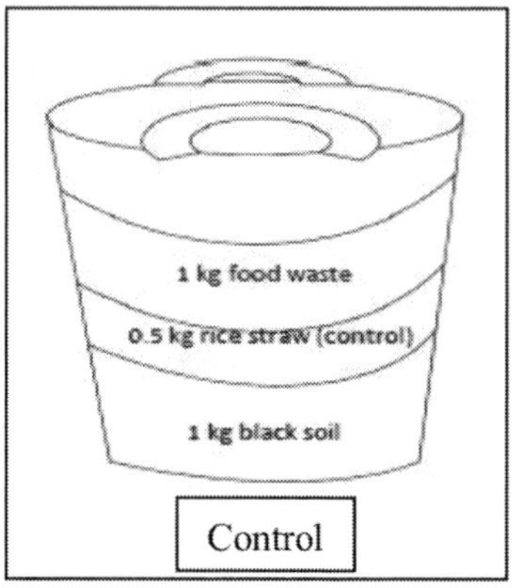

Fig. 3. Cross-section of compost medium using rice straw as control.

3 Results and discussions

3.1 Effect of pH during the composting process

Based on a study conducted, the pH value during compost process was varied within the pH of 4.9-8.3. While the optimum pH should exist between range of 7-8 when a test on pH dependency of microbial activities using liquid medium of proteins and glucose was conducted [9].

Based on the Fig. 4, the pH value is increasing from day 1 until day 12. The increase in pH value indicates that the decomposition of organic matter inside compost medium was occurred and ammonium was formed [10]. pH pattern for RSA(300°C) was slightly different from control and RSA (400°C) and RSA(500°C) where the value begin to decrease until reaching slightly neutral condition as reaching day 21 with 7.9 value. As for control, the pH become neutral on day 18 with value 7.6 which was 3 days earlier compared to RSA (300°C). Besides, the pattern of pH for RSA (300°C) show the longer decline period compared to other medium where the pH value start to decline from day 15 until day 21 while other medium pH decrease up until day 18. Decrease in pH value might caused by formation of carbon dioxide gas and organic acid during organic matter decomposition [10, 11].

After day 21, the pH value in RSA(300°C), start to change from neutral to alkaline condition on day 24 where the value was 8.6 and start to decline back on day 27 until day 30 from 8.3 to 8.1. Changes on pH value from 7.9 to 8.6 for RSA(300°C) and 7.6 to 8.1 for rice straw(control) may affected by the frequency of turning during composting period [12].

As for RSA (400°C) and RSA (500°C), the pH increase from day 18 until day 30. As for RSA (400°C) the pH increase from 8.4 on day 18 until reach value of 9.6 on last day of composting period. While for RSA (500°C), the pH increase from 8.2 on day 18 to 9.8 as it reaching day 30. The changes may cause by released of protein that comes from the mixture of vegetables remains that use along composting process [12]. Hence, from the result obtained based on pH value, the best quality of compost was RSA at 300°C since the pH value were almost similar with the optimum pH compared to other.

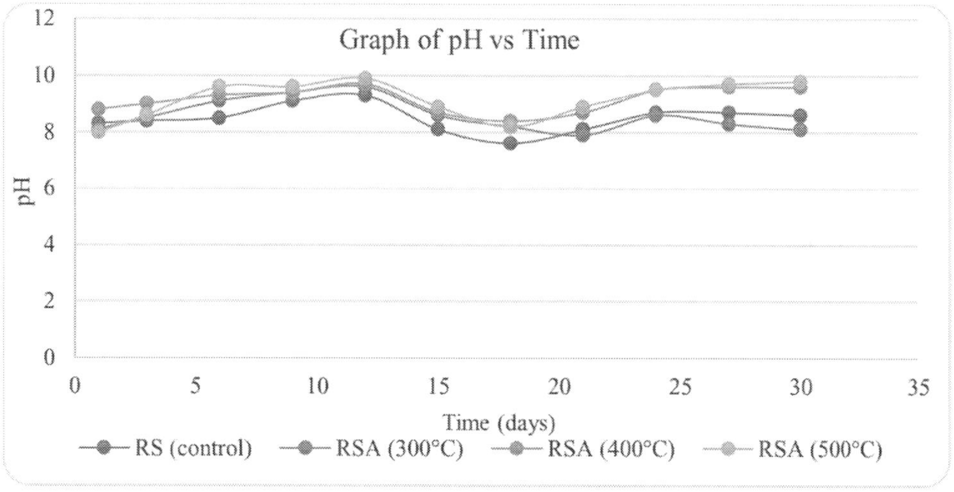

Fig. 4. Graph of pH of compost medium.

3.2 Effect of temperature during composting process

Based on the graph in Fig. 5, the highest temperature in control medium is at 50°C that recorded on day 6 before it falling to temperature 44°C on day 9. As for rice straw ash (RSA), the compost medium that has the highest temperature is RSA (500°C) which possessed the same temperature value as control on day 6 which was 50°C before decrease to 42°C on day 9. Increase in temperature was lead to the existing of thermophilic stage that mainly influence by the level of microbial activity that cause significant changes in psychochemical properties of composted material which also allow the decomposition of hemicelluloses, cellulose and lignin. By comparing the result of RSA(300°C) and RSA (400°C) with RSA(500°C), the microbial activity inside the RSA(300°C) and RSA (400°C) is lower compared to RSA (500°C). Besides, the changes in compost temperature was also influences by mixing frequency that can affect the oxygen supply towards compost medium [2].

After that the temperature of compost continues to decrease. From the result obtained, on day 24 the temperature of compost including control was 26°C and continue to decline until reaching day 30 with 22°C. The decrease in temperature indicates that the microbial activities started to slow down due to the low amount of organic material available. Hence the microbial activity stopped and reached the cooling stages [12].

Rice straw that been burned at temperature 500°C was seen as optimum temperature for rice straw ash composting since it achieved the highest value of temperature which was 50°C compared to other temperature of rice straw burning.

3.3 Effect of moisture content on composting process

The moisture content value obtained in this study showed in Fig. 6. The range of moisture content in this study was within 40 % to 59 %. Based on a journal, the range of an ideal moisture content that must be possessed by compost medium was within 40-60% [13]. While in other study, it stated that the optimum moisture content to ensure the condition in an active phase is within range of 45% to 50% [14].

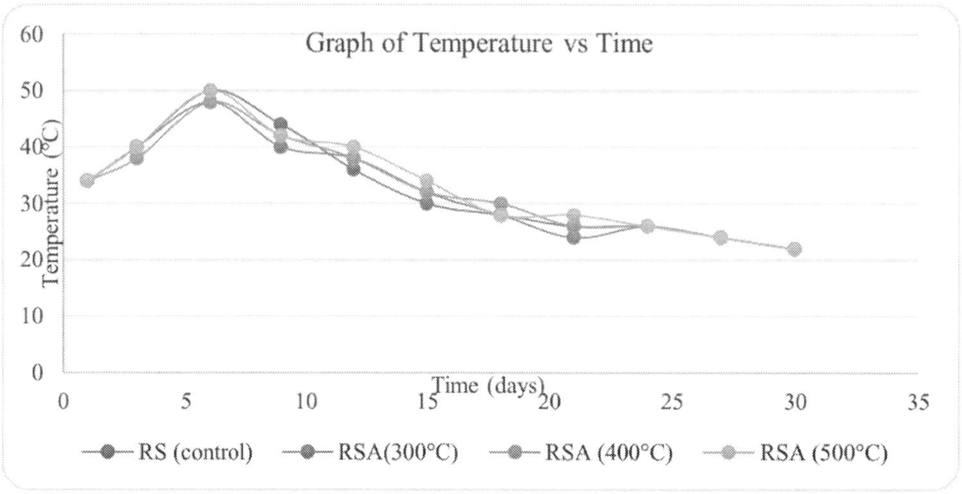

Fig. 5. Graph of temperature compost medium.

Roughly, based on the graph in Fig. 6, the amount of moisture content in each of the compost medium was showing a fluctuation. As for control, the highest moisture content is on day 21 while the lowest moisture content is on day 12 which were 59 % and 40 % respectively. As for RSA (300°C), RSA (400°C) and RSA (500°C), the highest moisture content was on day 1 which were 58% respectively. While the lowest moisture content for RSA (300°C) was 40 % on day 6. Lastly the lowest moisture content for RSA (400°C) and RSA (500°C) similar which were 45 % on day 30.

Moisture content affect by temperature distribution of compost medium. The main factor of moisture content loss is due to the rising in compost temperature [12]. However in this study, the temperature of compost did not give obvious effect towards moisture value since the amount of food waste inserted much higher than the rice straw ash and rice straw (control) as bulking agent along the compost period. High moisture content inside food waste cannot been absorbed by bulking agent efficiently causing the moisture content become unsteady [15].

The most good quality sample for this parameter was RSA(400°C) and RSA(500°C) since the range minimum moisture content laid within the optimum moisture content range which is 45% compared to RSA(300°C).

4 Conclusions

Composting using rice straw ash and food waste achieved the maturity of compost in 30 days. The pH possessed by control were within 7.6 to 9.3 compared to RSA (300°C), RSA (400°C) and RSA (500°C) which were between 7.9 to 9.6, 8.4 to 9.6 and 8.0 to 9.8 respectively. The range of temperature obtained from this study, control sample and RSA (500°C) temperature value were within the range of 22°C to 50°C. While for RSA (300°C), RSA (400°C) range of temperature were between 22°C to 48°C. Then from data analysis on moisture content, control value varied between 49% to 59% while for RSA (300°C), the moisture lied within 40% to 54%. As for RSA (400°C) and RSA (500°C), the value was in the range of 45% to 57% and 45% to 56% respectively.

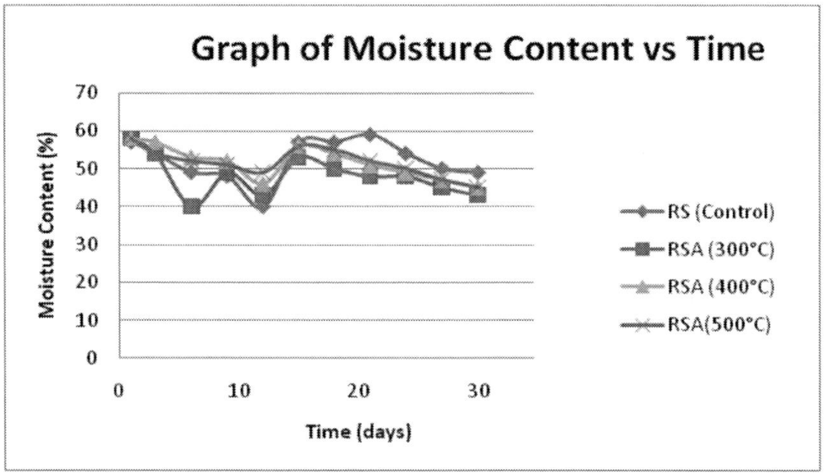

Fig. 6. Graph of moisture content of compost medium.

In this research, the classification of best composted was determine based on temperature. Hence based on result obtained by this research, among RSA (300°C), RSA (400°C) and RSA (500°C) the best compost medium was RSA (500°C) which it temperature at thermophilic phase was the highest which was 50°C. During composting process, the fermented liquid that made from *tempe* and brown sugar was used. Usage of fermented liquid considered as good since it was able to stimulate the soil microbial growth and activity. Besides, usage of brown sugar can be served as food towards the microbes either from *tempe* or compost medium itself. Optimum moisture content must be maintained to avoid foul odour production.

The authors gratefully acknowledge the supporting staff in the Laboratory School of Environmental Engineering, UniMAP for the good support in this research. This research was financially supported by the Centre of Excellence for Biomass Utilization.

References

1. A. Abraham, A. Mathew, R. Sindhu, A. Pandey, P. Binod. Bioresource Technol. **215** 1-3, (2016)
2. S. Hosseini, H. Abdul Aziz. Bioresource Technol. 240-241, (2013)
3. S. Shafie, H. Masjuki, T. Mahlia. Energy **70**, 404, (2014)
4. M. Rosmiza, W. Davies, C. Rosniza Aznie, M. Jabil. Geografia **10 (5)**, 31, (2014)
5. The Star Online http://www.thestar.com.my/news/nation/2016/05/24/malaysians-waste 15000-tonnes-of-food-daily (2016)
6. C. Zhou, Z. Liu, Z. Huang, M. Dong, X. Yu, P. Ning. Waste Manage. **40**, 38-41, (2015)
7. Y. Wei, Y. Zhao, H. Wang, Q. Lu, Z. Cao, H. Cui, et al., Bioresource Technol. **221**, 140, (2016)
8. A. International. ASTM Standard D2974, *Standard Test Methods for Moisture, Ash and Organic Matter of Pest and Other Organic Soils*, (West Conshohocken PA, United States, 2003)
9. Z. Li, H. Lu, L. Ren, L. He. Chemosphere **93(7)**, 1248-1250, (2013)
10. L. Meng, W. Li, S. Zhang, C. Wu, W. Jiang, C. Sha. Ecol. Eng. **94**, 240, (2016)
11. S. Kharrazi, H. Younesi, J. Abedini-Torghabeh. Int. Biodeter. Biodegr. **92**, 44 (2014)

12. I. Zakarya, S. Khalib, T. Tengku Izhar, S. Yusuf. Engineering Research and Technology **4(8)**, 181-184, (2015).
13. H. Johnson. *A Pre-feasibility study for the establishment of an organic fertilizer plant in Haiti,* 7, (2012).
14. E. Bazrafshan, A. Zarei, F. Mostafapour, N. Poormollae, S. Mahmodi, M. Zazouli, Health Scope **5(1)**, 3, (2016).
15. C. Wu, W. Li, K. Wang, Y. Li. Bioresource Technol. **516**, (2015).

Analysis of rainfall distribution in Kelantan river basin, Malaysia

Faizah Che Ros[1*] and *Hiroyuki* Tosaka[2]

[1]Department of Environmental Engineering & Green Technology, Malaysia-Japan International Institute of Technology, Universiti Teknologi Malaysia Kuala Lumpur, 54100 Kuala Lumpur, Malaysia
[2]Geosphere Environmental Technology Ltd. (GET) NCO Kanda-Awajicho 3rd Floor, 101-0063 2-1 Awajicho, Chiyoda-ku, Tokyo, Japan

Abstract. Using rainfall gauge on its own as input carries great uncertainties regarding runoff estimation, especially when the area is large and the rainfall is measured and recorded at irregular spaced gauging stations. Hence spatial interpolation is the key to obtain continuous and orderly rainfall distribution at unknown points to be the input to the rainfall runoff processes for distributed and semi-distributed numerical modelling. It is crucial to study and predict the behaviour of rainfall and river runoff to reduce flood damages of the affected area along the Kelantan river. Thus, a good knowledge on rainfall distribution is essential in early flood prediction studies. Forty six rainfall stations and their daily time-series were used to interpolate gridded rainfall surfaces using inverse-distance weighting (IDW), inverse-distance and elevation weighting (IDEW) methods and average rainfall distribution. Sensitivity analysis for distance and elevation parameters were conducted to see the variation produced. The accuracy of these interpolated datasets was examined using cross-validation assessment.

1 Introduction

Assessing rainfall variability is the key element to develop any conceptual and predictive models in many fields such as water resource management and flood disaster prediction [1]. Using rainfall gauge on its own as input carries great uncertainties regarding runoff estimation, especially when the watershed area is large and the rainfall is measured and recorded at irregular spaced gauging stations [2-3]. Therefore, spatial interpolation is the key to obtain continuous and orderly rainfall distribution for the input of distributed and semi-distributed numerical modelling. Few studies [4-6] indicated that using mean areal rainfall instead of spatially distributed rainfall inclines to undervalue the volumes and peak runoffs. However, the impact of distributed rainfall on runoff estimation is complex as it can be dependent on the rainfall's nature, catchment characteristics and the spatial scale used [7-8].

* Corresponding author: crfaizah@utm.my

The estimation of rainfall distribution can be grouped into two; indirect and direct method. Indirect rainfall estimation via satellite products, especially from radar remote sensing is increasingly used as covariates since they offer spatially detailed data information [3, 9]. The direct ground-based method has a wide range of interpolation techniques ranging from simple to more complex calculations, which can be divided into two main groups; deterministic and geostatistical. The mostly known deterministic approaches are Thiessen polygon and inverse distance weighting (IDW) whereas the geostatistical methods constitutes a discipline of mathematics and earth sciences, known as kriging method.

This study aims to investigate the IDW method in estimating rainfall distribution of Kelantan watershed by considering the spatial effect (large vs small scale) and by integrating elevation data into IDW, via inverse distance elevation weightage (IDEW) method. Sensitivity analysis for distance and elevation parameters were conducted to see the variation produced. The accuracy of these interpolated datasets will be examined by cross-validation assessment.

2 Materials and methods

2.1 Study area

In this study, the large Kelantan watershed was chosen as the main research area. Che Ros et al. [10] has conducted a homogeneity study on fifty rainfall stations. Out of fifty gauges, forty six were found homogenous. Thus, a total of forty six rainfall stations and their daily time-series were used to interpolate gridded rainfall surfaces using IDW and IDEW. The rainfall stations are managed by Department of Irrigation and Drainage Malaysia (DID). The IDW and IDEW were tried twice each by: a) applying on whole watershed and b) sub-watershed scale. So, a total of 4 rainfalls distribution gridded surface was produced in one space-time rainfall series for year 2007.

2.2 Methods

In this study, the Kelantan watershed was delineated into eight sub-watersheds as shown in Fig. 1 to develop the sub-areal based of IDW and IDEW, namely, Areal IDW (AIDW) and Areal IDEW (AIDEW). The delineation was based on the hydrological characteristics such as stream network and waterlines, and the manner of rainfall based on reliable long-term records.

2.2.1 Inverse Distance Weighting Method (IDW)

Inverse distance weighting method (IDW) is a famous conventional deterministic approach for spatial interpolation based on similarity or smoothness within a research area other than Thiessen method. The assigned values to unknown points are calculated as a weighted average of the values available at the known points [11]. Weight increases as the distance decreases from the known points to unknown points [2, 12]. The mathematical form of the IDW is given below:

$$p_e = \sum_{i=0}^{N} \lambda_i p_i \tag{1}$$

where p_e is the interpolated rainfall for a grid cell; N is the number of gauges that are used in the interpolation for the current grid cell; λ_i is the corresponding weight of each known points where it reduces as the distance between known points and predicted points increases; p_i is the rainfall value in mm/day of the ith gauge station. The weighting factors is called Shepard method [13] and determined by:

$$\lambda_i = \frac{d_{i0}^{-\alpha}}{\sum_{i=1}^{N} d_{i0}^{-\alpha}} \tag{2}$$

$$\sum_{i=1}^{N} \lambda_i = 1 \tag{3}$$

where the power of exponent α controls the influence of the distance among the sample points or gauge stations on the interpolations results. A low α leads to a greater weight towards a grid point value of rainfall from remote rain gauges. As the α decreases toward zero, the method approximates the Thiessen method [14-17]. Dirks et al. [15] had concluded, to minimize the interpolation errors, α of 2 is used for daily and monthly steps, 3 for hourly and 1 for yearly. However, most study [2, 12] set α to 2 where inverse square distances were used in their estimation. During the process of prediction, the weights of gauge station are proportional and the sum of these weights is equal to 1.

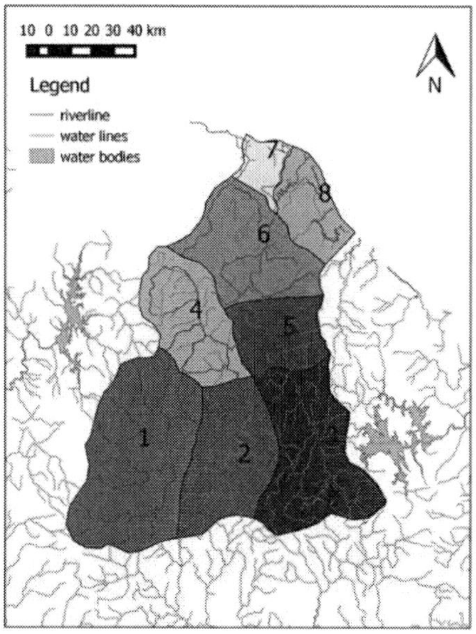

Fig. 1. Kelantan watershed divided into 8 sub-watersheds.

2.2.2 Inverse Distance and Elevation Weighting Method (IDEW)

The IDW interpolation only incorporates a single influence factor, which is horizontal distance. Therefore, it is not suitable to be applied in an area with abrupt changes in elevation, which could create uncertainty in estimating unknown information [27]. Inverse distance and elevation weighting (IDEW) technique provides more suitable results for mountainous regions where topographic impacts on rainfall are important [26,29]. IDEW algorithm is part of hydrological data processing software called HyKit developed by UNESCO-IHE [18]. However, in this study, C programming was used to do the interpolation by adopting the equations as shown below:

$$p_e = W_D \sum_{i=1}^{N} \frac{1}{D} w(d)_i p_i + W_Z \sum_{i=1}^{N} \frac{1}{Z} w(z)_i p_i \quad (4)$$

where W_D and W_Z are the importance factors for distances and elevations. Typical values for the importance factor are 0.8 for distance and 0.2 for elevation, respectively [19]. Similarly, $w(d)_i$ and $w(z)_i$ are the individual gauge weighting factors for distance and elevation respectively. Whereas D and Z are the normalization quantities given by the sum of individual weighting factors $w(d)$ and $w(z)$, respectively for all the gauges used in the grid interpolation. The weighting factors of $w(d)_i$ and $w(z)_i$ can be determined by:

$$w(d) = 1/d^{\alpha} \ for \ d > 0 \quad (5)$$

$$w(z) = \begin{cases} 1/z_{min}^{\beta} \ for \ z \le z_{min} \\ 1/z^{\beta} \ for \ z_{min} < z < z_{max} \\ 0 \ for \ z \ge z_{max} \end{cases} \quad (6)$$

where d is the distance in kilometer between current grid and the gauge station used for interpolation, z is the absolute elevation difference in meter between the current grid ceall and the gauge station used for interpolation, α and β are exponent factors for distance and elevation weightings where the distance weighting exponent is 2 and 1, respectively [19]. z_{min} and z_{max} are the minimum and maximum limiting values of elevation differences for computing elevation weightings [18, 19]. Limitations on z_{min} and z_{max} help to avoid dominance of the stations having very small elevation difference and enables data point inclusion to be restricted to a local elevation range [18]. Typical values of z_{min} are 100 to 300 meters while z_{max} from 500 to 2,500 m depending on the study area [18].

2.2.3 Cross-Validation & Performance Assessment

The performance of the interpolation methods was assessed and compared by cross-validation procedure. The procedure consists of temporarily discarding one observation value at a time from the data sets and re-estimating the discarded value from the remaining sampled point using each interpolation method. Nash-Sutcliffe efficiency (NSE) [20] and root mean square error (RMSE) [21] were adopted to evaluate the performance of interpolation methods. NSE criterion is a form of normalized least squares where perfect agreement between the observed and estimated values yields an efficiency of 1, whereas a negative efficiency represents lack of agreement where the observed mean is a better predictor than the estimation value. RMSE indicates the deviation between estimated values and observed one hence, smaller RMSE contributes to better estimation capability.

3 Results and discussions

3.1 Cross-validation and performance assessment

The results of four interpolation methods were validated and ranked via RMSE and NSE as shown in Table 1. The RMSE validation indicates that each interpolation method has relatively similar deviations between estimated and observed values with a mean residual of 10.175 ± 0.233. It is also observed that the range of deviation values is significantly large for all interpolation methods, ranging from 0.05 to 78.5. This shows that despite similar deviation, the confidence level between estimation and observation value is not as high as the external factors need to be considered, such as the quality of the rainfall data. Hence, RMSE is not able to determine the best interpolation method in this case as illustrated in Fig. 2.

Table 1. Cross-validation assessment performance and ranking of different interpolation methods based on root mean square error (RMSE) and Nash-Sutcliffe efficiency (NSE).

Interpolation Methods	Range RMSE	Mean RMSE	Rank of Mean RMSE	Range NSE	Mean NSE	Rank of Mean NSE
IDW	0.05-75.42	9.84	2	-0.87-0.76	0.09	4
IDEW	0.07-75.08	9.82	1	-0.28-0.99	0.93	2
AIDW	0.06-78.46	10.60	4	0.6-0.99	0.98	1
AIDEW	0.07-75.37	10.44	3	-0.97-0.99	0.92	3

On the other hand, the NSE validation shows that the IDW interpolation has the lowest mean value of NSE close to zero, while the rest of interpolation methods demonstrate a high mean value of NSE close to 1. Fig. 3 illustrates the NSE cloud distributions for all interpolation methods and it can be clearly seen that the NSE value for IDW is much scattered within a wide range, both positive and negative efficiency, as compared with others. This shows that the quality of rainfall estimation in Kelantan can be improved by incorporating elevation and smaller sub-watershed in the interpolation. This is obvious since the geophysical and topography of Kelantan is mixed with mountainous terrain and valley. It is also worth to note that integrating elevation in a smaller sub-watershed, i.e. AIDEW, did not significantly change the NSE value of AIDW. This could be due to the smaller elevation change that exists within the sub-watershed; hence elevation gives little impact in estimating the rainfall in smaller watershed. Overall, AIDW shows the best interpolation method to estimate the rainfall in entire Kelantan watershed according to NSE.

3.2 Sensitivity analysis of rainfall distribution parameter

Sensitivity of rainfall distribution parameters was investigated to see variations in the pattern and amount of rainfalls produced by changing the α and z_{min} and z_{max}. Sensitivity analysis of α parameter as shown in Fig. 4 shows that the greater of α will result in greater influence on area or grid that close to the interpolated points, whereas smaller α will result in interpolated values will be dominated by other points or rainfall gauges that are far away. This can be seen in day 341, year 2007 where when α equals to 1.5, weights of rainfall gauge that has high rainfall records in the east part was lesser compared with α equals 2 and 3. Same goes to day 346, year 2007 where when α equals to 3, both high rainfall and zero rainfall area gave greater influence to surrounding nearby grid compared with 2 and 1.5. Greater α value also results in greater amount of rainfall distributed over Kelantan watershed even though the number of rainfall gauges that have high rainfall value is small. For IDEW elevation parameter, only z_{min} was adjusted since z_{max} for Kelantan watershed is known (around 1800m above sea level). Masih et al [29] suggested the range of z_{min} is around 100 to 300 m. Fig. 6 illustrates the difference by applying z_{min} of 100 and 200 m. This shows that larger range of Δz leads to higher amount of rainfall across the watershed.

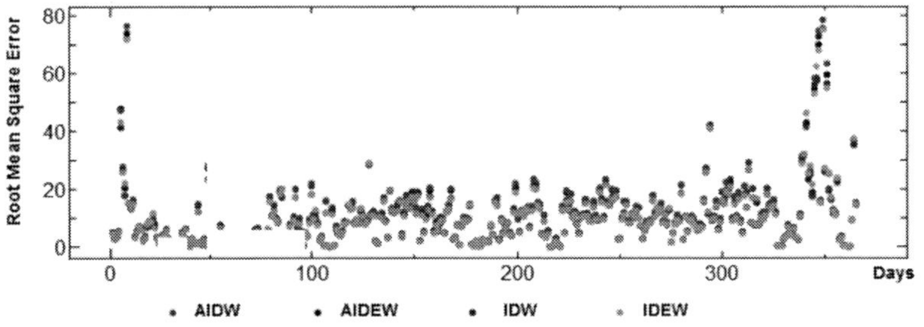

Fig. 2. RMSE clouds from 1st January to 31st December 2007.

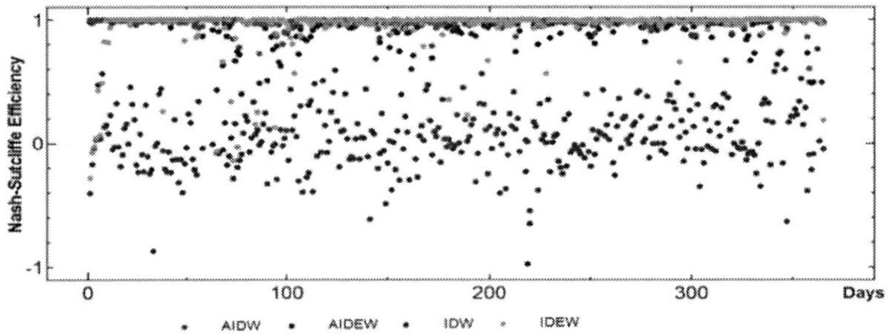

Fig. 3. NSE clouds from 1st January to 31st December 2007.

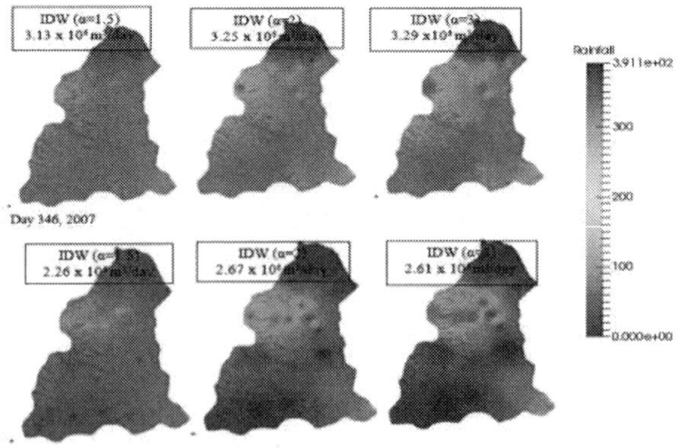

Fig. 4. Sensitivity Analysis of IDW parameter α.

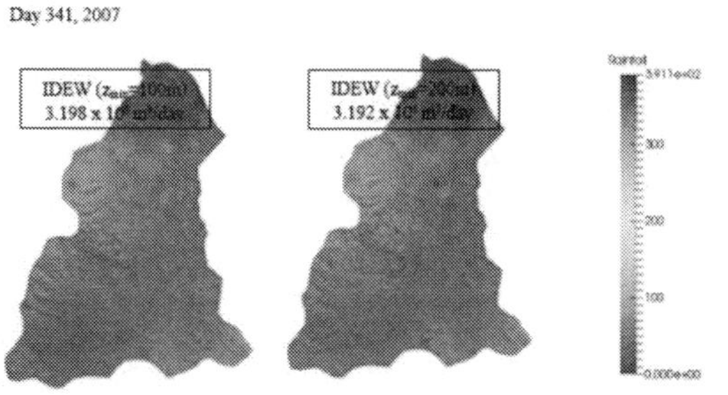

Fig. 5. Sensitivity Analysis of IDEW parameter z_{min} ($\alpha = 2$ and $z_{max} = 1800$m).

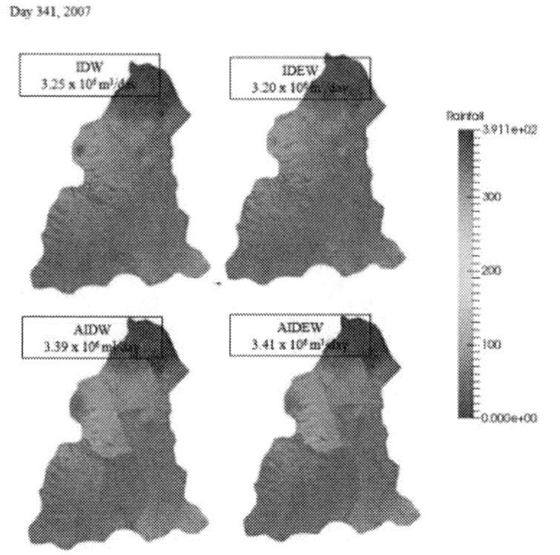

Fig. 6. Sensitivity analysis of rainfall distribution types. ($\alpha = 2$, $z_{min} = 200$ and $z_{max} = 1800$m).

4 Conclusions

Rainfall distributions over the large Kelantan watershed were developed on daily basis. Up to forty six rainfall stations and their daily time-series were used to interpolate gridded rainfall surfaces using inverse distance weighting (IDW), inverse distance and elevation weighting (IDEW). The IDW and IDEW were tried twice each by: a) applying on whole watershed and b) sub-watershed scale. So, a total of five rainfall distribution gridded surface was produced in one space-time rainfall series together with average rainfall distribution. The sensitivity analysis of rainfall distribution parameters was done to see variations in rainfall pattern and amount of rainfall produced by changing α and z_{min}. From the analysis, a low α leads to a greater weight towards a grid point value of rainfall from remote rain gauges. As α decreases toward zero, the method approximates the Thiessen method, whereas larger range of elevation leads to higher amount of rainfall produced across the watershed. For elevation parameter, larger range of Δz leads to higher amount of rainfall across the watershed.

The authors acknowledge the cooperation of the Department of Drainage and Irrigation Malaysia (DID) for providing the data required for this study.

References

1. D. R Dawdy, J. M. Bergman,Water Resour. Res. **5**, 958-966 (1969)
2. S. Ly, C. Charles, A. Degre, Hydrol. Earth Syst. Sci. **15,** 2259-2274 (2011)
3. P. D. Wagner, P. Fiener, F. Wilken, S. Kumar, K. Schneider, J. Hydrol. **464**, 338-400 (2012)
4. P. Arnaud, C. Bouvier, L. Cisner, R. Dominguez, J. Hydrol. **260**, 216-230 (2002)
5. J. M. Schuurmans, M. F. P. Bierkens, Hydrol. Earth Syst. Sci. **11**,677-693 (2007)
6. Y. Tramblay, C. Bouvier, P.-A., Ayral, A. Marchandise, Nat. Hazards Earth Syst. Sci. **11**, 157-170 (2011).
7. M. L. Segond, H. S Wheater, C. Onof, J. Hydrol. **347**, 116-131 (2007)
8. T. Tao, B. Chocat, S. Liu, K. Xin, J. Env. Protection **1**, 50-58 (2009)
9. A. Verworn, U. Haberlandt, Hydrol. Earth Syst. Sci. **15,** 569-584 (2011)
10. F. Che Ros, H. Tosaka, L. M. Sidek, H. Basri, Int. J. River Basin **14**, 151-163 (2016)
11. R. Franke, Math Computations, **33**, 181-200 (1982).
12. W. Xu. Y. Zou, G. Zhang, M. Linderman, Int. J. Climatol. **35,** 2898-2907 (2014)
13. D. Shepard, *ACM Annual Conference/Annual Meeting*, 517-524 (1968)
14. A. Mair, A. Fares, J. Hydrol. Eng. **16**, 371-383 (2011)
15. K. N. Dirks, J. E. Hay, C. D. Stow, D. Harris, J. Hydrol. **208**, 187-193, (1998)
16. M. Ashraf, L. C. Loftis, K. G. Hubbard, Agric. For Meteorol. **84**, 255-271 (1997)
17. S. Ly, C. Charles, A. Degre, Biotechnol. Agron. Soc. Environ **17**, 392-406 (2013)
18. I. Masih, S. Maskey, S. Uhlenbrook, V. Smakhtin, J. Am. Water Resour. Assoc. **47**, 179-195 (2011)
19. C. Daly, W. P. Gibson, G. H. Taylor, L. G. Johnson, P. Pasteris, Climate Research **22**, 99-113 (2002)
20. J. E Nash, J. V. Sutcliffe, J. Hydrol **10**, 282-290 (1970)
21. S. H. Hwang, D. H. Ham, J. H. Kim, Hydrol. Sc. J. **57,** 1257-1274 (2012)

Cultivating Composting Culture Activities among Citizens and Its Beneficial to Prolong the Landfill Lifespan

Irnis Azura Zakarya[1, 2*], *Khairul Azri* Jamial[1, 2], and *Norazlinda* Mat Tanda[1, 2]

[1]Centre of Excellence for Biomass Utilization, Universiti Malaysia Perlis
[1]School of Environmental Engineering, Kompleks Pengajian Jejawi 3, Universiti Malaysia Perlis, 02600, Arau, Perlis, Malaysia.

Abstract. Currently, the Ministry of Housing and Local Government manage solid waste in Malaysia, with the participation of the private sector. Food waste represents almost 60% of the total municipal solid waste disposed in the landfill. Material valorisation of food waste usually conducted by biological processes such as composting. Compost, an organic amendment, is the final product of the composting process. These processes are efficient, low cost and environmentally friendly alternative for managing food waste and are used extensively worldwide. Therefore, organic solid waste management practices program for the communities in Perlis was conducted. The main objective of this program was to instilling environment awareness especially among Perlis citizens. This study was investigated the impact of food waste or kitchen waste composting to the citizens in Perlis State and the beneficial of compost fertilizer to our environment especially in plant growth. Composting method was taught to the food premises owner, individuals, teachers, and students and their responses to the composting practices were then summarized. In future, we can prolong our landfill lifespan by practicing organic waste composting and can preserving our environment.

1 Introduction

Our country is struggling to reduce amount of municipal waste from dump onto landfill and food waste is one of main types of the municipal solid waste [1]. There were 30,000 kilogram of waste collected and dump onto the landfill daily recorded. More than 51% of wastes were dumped organic waste such as kitchen waste, food waste, garden waste and so on. On the other hand, inorganic wastes such as recyclable waste like plastics, papers, metals, electronic waste or e-waste and much more. This result brought us on shortens life span of landfill at every state in Malaysia. More than 100 landfills situated all over Malaysia and this is not included open dumping site. From 100 landfills, not all were sanitary landfill, which provided more technology and more efficient waste treatment or by-product treatment such as leachate.

* Corresponding author: irnis@unimap.edu.my

There are three landfills site in Perlis, two were closed and one currently active landfill. The active landfill located at Padang Besar District. There are various methods in managing solid waste such as incinerator, Material Recovery Facility (MRF), landfill and also composting. Composting also known as an organic fertilizer producer through biological decomposition of organic waste that happened for centuries [2]. Composting is an environmentally friendly technology and can be an effectively waste management strategy especially for food waste [3]. Studied claimed that composting is a biological process which organic matter can be utilized by aerobic thermophilic and mesophilic microorganisms as substrate and mainly converted into mineralized products (CO_2, H^2O, NH^{+4}) or stabilized organic matter (mostly as humic substances) [4] and also can adds nutrients to the soil surrounding [5].

2 Materials and Methods

Several stages were introduced during the program**.** Fig. 1 show preparation for food waste composting process process in this program.

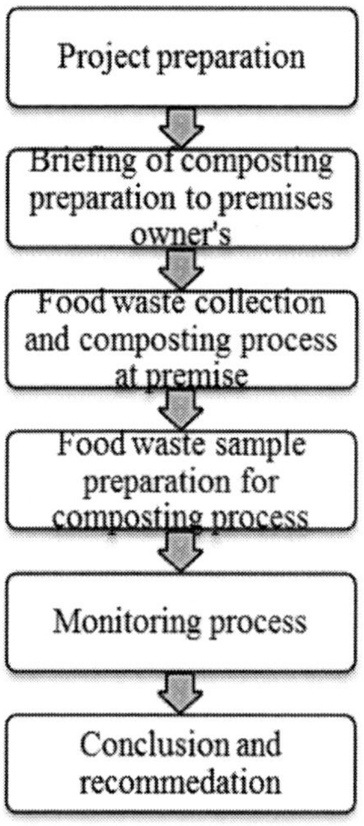

Fig. 1. Preparation for food waste composting at the premises.

2.1 Compost material

Fermented solution was prepared before it's given to the premises. Materials that needed for compost preparation are 'tempe', brown sugar, rice husk and black soil. All the material were given to the premises and showed to the owner of premises.

2.2 Composting process (Takakura method)

Takakura Method was used in this program since it is easier and takes less space to do. It took three stages on making the Best regards,. For the first stage, fermentation solution (Effective Microorganisms (EM)) was made to culture bacteria. This fungal and bacteria cultures were added to accelerate the degradation process in the compost [2]. Two kilograms of brown sugar were mixed with 20 litres of water. Then four pieces of fermented soybean were put into the brown sugar water and the container was closed for a week for fermentation process to be done.

For the Stage 2, bucket with soil and rice husk were prepared as compost medium. Rice husk was used as bulking agent to adjust the moisture content and Carbon Nitrogen ratio [3]. Rice husk also act as oxygen access agent for bacteria to breath from the compacted soil. The soil and rice husk were mixed with the prepared EM and then leaved it for a week until white mold appeared. When the white mold has appeared, food waste can now be put inside the compost for the bacteria to degrad into smaller size.

Temperature is the most important parameter for affecting the compost process. To avoid pathogens and weeds, the temperature should be remain in thermophilic condition which is range 50 – 60°C [6] the compost must experience the temperature exceeding 55°C for at least three (3) days so that pathogens and weeds can be eliminated [7].

During composting, microorganism needs moisture to transporting the dissolved nutrients required for physiological and metabolic activities [8]. Compost was monitored to make sure not exceeding 60% of moisture content. If there is too much water droplet, the compost might be too wet since food waste consumed almost 65 – 80% of moisture [9, 10].

2.3 Food premises and program organized

From January 2016 until July 2017, there were 25 food premises and schools get involved in this program. All participants were given all the compost materials and tools provided by the PPKAS and SWCorp for the first time basis. They were taught on how to prepare compost from food waste and will be monitored every week to overcome incoming problems or any advice that needed.

On 18 April 2017, School of Environmental Engineering has organised a program to educate citizens regarding food waste composting. There were almost 50 participants joining this one-day program that happened at Multipurpose Room, School of Environmental Engineering.

2.4 Parameter analysis

Compost samples from selected premises in this study were tested for physical and chemical parameters. Every samples were tested by three (3) times and average value were recorded. In physical parameter, the pH value and temperature was measured. The pH values of the compost were determined by mixing the compost into distilled water. 10 gram of waste was weighed and 25 mL distilled water was measured and mixed in a beaker. The samples were stirred for 10 minutes. After that, when the sample have settled after 3 minutes, the sample were tested with pH meter to get the reading. The temperature was taken for 7 days continuously to record the changes in temperature during composting process. Three times of reading were taken to have the average value of compost. The chemical parameters were carried out according to APHA (2005) standard method. The chemical parameter including Nitrogen (N), Phosphorus (P), and Potassium (K) were conducted in this study.

3 Results and discussion

3.1 Food premises and other related program

3.1.1 Food premises

Until November 2017, total of premises that still joining this project are 25. Table 1 has shown the list and location of the premises that take their part in this project. .

Table 1. List of food premises participated simplified by area in Perlis.

No.	Premise Name	Area
1.	Ziezan Station	
2.	Nasi Ayam Tip Top	
3.	Ummi Corner	
4.	Restoran Warisan Masyhur	
5.	Restoran D' Sinar	
6.	Iriss Café	
7.	Nasi Kerabu Abang Wan	
8.	Nasi Arab Abou Arabic and Western Cuisine	Kangar
9.	SK IPGM Perlis	
10.	SK Putra	
11	SMK Sanglang	Sanglang
12.	Kedai Roti dan Kek	
13.	TM Café	
14.	Aneka Nasi Berlauk	
15.	Kedai Minuman (Air Buah)	
16.	Nasi Campur (LOT 01)	
17.	Gerai Nasi Campur (Lot10)	
18.	Masakan Panas/Ala Carte (LOT 03)	UNIMAP Pauh
19.	Makanan Arab (Lot 06)	
20.	Kedai Makan Kak Yah Soto	
21	Kedai Nasi Kandar Zakri	
22.	Gerai Wan Azmi	
23.	Gerai Zainon Fitri	Arau
24.	Gerai Che Amah	
25.	SK Oran	

Based from the Fig. 2, it shows that the highest food waste generated were from schools canteen. It found out the low generation of food waste at the premises because of lack of interest to handle the compost and prefer the conventional way. Besides, the participants complaining about the smell occur during the compost process. The investigation has been done, and found out the bad smell occurred because of high moisture content. It was because the location of the compost bucket was exposed to the rain and some put under the leaking pipe. High moisture content tends to lower down the amount of oxygen access for the bacteria to grow and causes H^2S and NH_3 gases to be released [11].

Fig .2. The average weight of food waste produce per day at every premise.

3.1.2 Workshop on the Art of Composting

The Art of Composting was successfully held on 18 April 2017. All participants enjoyed and gained knowledge regarding on food waste composting and it's beneficial to the environment. This program held in two-way communication when speaker and participants kept changing opinions and knowledge on managing the waste produced. The participants were from various levels of citizens, staffs, university students, primary schools students and Kolej Vokasional Kangar students. Fig. 3 and Fig. 4 show the picture of workshop held.

Fig. 3. Briefing session.

Fig. 4. Practical session.

At the end of the program, participants agreed that in order to keep our environment clean and to prolong our landfill lifespan, food waste composting is a good method and should be applied to all people in Malaysia. It looked difficult for the first time but once applied it, there should not be a problem anymore and no need to depend on chemical fertilizer anymore.

3.2 Comparison of Nitrogen (N), Phosphorus (P) and Potassium (K)

Table 2 shows the comparison of NPK value between the food waste compost, organic compost and chemical fertilizer. Food waste compost had low percentage on N, P and K compare to chemical fertilizer But NPK value of food waste compost is similar to cow dung. This happened because carbohydrate, protein and fat were assumed to be less and had gives the lowest value of NPK [13].

Table 2. NPK value between organic compost and chemical fertilizer.

Sample	N (%)	P (%)	K (%)
Food waste compost	1.39	0.47	0.07
Cow dung	1.3	0.58	2.15
Chemical fertilizer	20	40	25

4 Conclusions

In conclusion, as conjunction of Separation at Source Program initiated by the Government (conducted by SWCorp), food waste composting could be the best way in reducing waste to the landfill. It perhaps can help to reduce almost 20 to 30 percent of total organic waste that has been dump onto the landfill before. Unfortunately, in order to have this target to be achieved, more works to be done especially on educating people from youngsters till the elders. Government much looks more seriously on having specific regulations on composting at food premises because most of the premises were the most contributor of producing food waste and kitchen waste to the landfill.

This program was funded by Knowledge Transfer Program (KTP) Scheme under Ministry of Higher Education Malaysia and supported by the Centre of Excellence for Biomass Utilization. The authors also wish to thank to School of Environmental Engineering, Universiti Malaysia Perlis and Solid Waste Corporation (SWCorp) Malaysia in this project.

References

1. N. Mekjinda,R. J. Ritchie, Waste Manag. **35**, 199-206 (2015)
2. S. Sarkar, S. Pal, S. Chanda, Procedia Environ. Sci. **35**, 435–440 (2016)
3. X. Wang, S. Pan, Z. Zhang, X. Lin, Y. Zhang, S. Chen, Bioresour. Technol. **224**, 397–404 (2017)
4. X. Qian, G. Shen, Z. Wang, C. Guo, Y. Liu, Z. Lei, Z. Zhang, Waste Manag. **34**, 530–535 (2014)
5. M. K. Manu,R. Kumar,A. Garg, Technol. **234**, 167–177 (2017)
6. L. Zhang, X. Sun, Bioresour. Technol. **218**, 335-343 (2016)
7. B. Nakhshiniev, M. K. Biddinika, H. B. Gonzales, H. Sumida, K. Yoshikawa, Bioresour. Technol., ***151***, 306–313 (2014)
8. R. Guo, G. Li, T, Jiang, F. Schuchardt, T. Chen, Y. Zhao, Y. Shen, Technol., ***112***, 171–178 (2012)
9. J.I. Chang, T.E. Hsu, Bioresour. Technol. **99**, 8068-8074 (2008)
10. J. I. Chang, J. J. Tsai, K. H. Wu, Technol. **97**, 116–122 (2006)
11. F.N.,Saad, M. Nadrah, N. Zain, S. Ezlin, N. Basri, N. Sofia, Jurnal Teknologi **2**, 89–95 (2013)
12. America Public Health Association (APHA), *Standard Methods for Examination of Water and Wastewater*, 21st ed., Washington DC, USA. (2005)
13. I. Zakarya, S. Khalib, T. Tengku Izhar, S. Yusuf. *Engineering Research and Technology* **4**, 181-184 (2015)

Settling properties of aerobic granular sludge (AGS) and aerobic granular sludge molasses (AGSM)

Azlina Mat Saad[1], *Farrah Aini* Dahalan[1*], *Naimah* Ibrahim[1], *Sara Yasina* Yusuf[1], *Siti Aqlima* Ahmad[2], and *Khalilah* Abdul Khalil[3]

[1]School of Environmental Engineering, Universiti Malaysia Perlis, Kompleks Pengajian Jejawi 3, 02600 Arau, Perlis, Malaysia
[2]Department of Biochemistry, Faculty of Biotechnology and Biomolecular Sciences, University Putra Malaysia, UPM 43400 Serdang, Selangor, Malaysia
[3]Department of Biomolecular Science, Faculty of Applied Sciences, Universiti Teknologi MARA, Seksyen 2, 40450 Shah Alam, Selangor, Malaysia

Abstract. Aerobic granulation technology is applied to treat domestic and industrial wastewater. The Aerobic granular sludge (AGS) cultivated has strong properties that appears to be denser and compact in physiological structure compared to the conventional activated sludge. It offers rapid settling for solid:liquid separation in wastewater treatment. Aerobic granules were developed using sequencing batch reactor (SBR) with intermittent aerobic – anaerobic mode with 8 cycles in 24 hr. This study examined the settling velocity performance of cultivated aerobic granular sludge (AGS) and aerobic granular sludge molasses (AGSM). The elemental composition in both AGS and AGSM were determined using X-ray fluorescence (XRF). The results showed that AGSM has higher settling velocity 30.5 m/h compared to AGS.

1 Introduction

Technology on wastewater treatment has improved over time based on various advanced processes to treat municipal and industrial wastewaters. Since the operational performance of the conventional activated sludge processes is limited by the aptitude of a clarifier for the solid;liquid separation of the activated sludge from the treated wastewater, Aerobic granular sludge (AGS) technology is an attractive option due to its settleability advantages. It has become an alternative to the conventional wastewater treatment method specifically activated sludge due to the excellent settling performance and high capacity for biomass retention. With high biomass aggregation of various indigenous microorganisms in the wastewater, AGS exhibits high capability to withstand extensive organic and inorganic loads and demonstrated to have high potential to treat various types of wastewaters [1].

According to [2] aerobic granulation is a combination of two processes involving biological and chemical engineering to form a very compact structure of activated sludge

* Corresponding author: farrahaini@unimap.edu.my

that possess high density. Physical dynamism such as vigorous aeration which contributes to effective hydrodynamic shear force has also influenced the acceleration on the formation of AGS. Preferably, a relative high shear force was found to favour rapid granulation.

In general, aerobic granulation involves a highly complex ecosystem which integrates physical, chemical and biological components synergistically to serve for a hybrid compact wastewater treatment system. To achieve optimum process, sequencing batch reactor (SBR) system has been creditably used as an ideal system to develop AGS [3]. During SBR process, activated sludge which is used as seed sludge is subjected to optimized physicochemical growth condition. This is significant to boost vital aggregation of microorganism consortia facultatively to form AGS naturally via cell-to-cell biochemical mechanisms without supplement of any artificial carrier for attached growth [4]. Basically, the development of AGS beginning from the transformation of seed sludge to flocs, and followed by the establishment of sludge granules.

Owing to the constructed design of the aerobic granulation system, the AGS produced have shown perceptible features on the morphology and settling velocity performance surpasses the conventional activated sludge. The structure of the AGS produced are normally possess notable characteristics which appears to be regular, dense, strong, compact, smooth, and uniform in shape and sizes.

However, so far, information about the enhancement of AGS using molasses wastewater and the settling performance is still sparse. Therefore, this study focuses to investigate the settling performance of AGS and AGSM cultivated initially under favourable SBR configuration using synthetic wastewater prior to acclimatization and further granulation in molasses wastewater. In addition, X-ray fluorescence (XRF) analysis was also performed in order to examine the chemical compounds occurred in both AGS and AGSM.

2 Methodology

2.1 Research framework

AGS was cultivated using SBR in the laboratory. A three-litre cylindrical column reactor was used and operated for 24 hours continuously with intermittent aerobic-anaerobic mode. Fresh activated sludge was collected from local wastewater treatment plant and used as seed sludge to initiate the cultivation process of AGS in SBR.

The cultured AGS were then acclimatized and enriched with molasses wastewater in conical flask under aerobic condition. AGS enriched in molasses wastewater are then known as aerobic granular sludge molasses (AGSM). The characteristics of activated sludge were investigated using APHA Standard Method [5]. The experimental workflow of the development of AGS and AGSM is presented in Fig. 1.

The physical performance and chemical characteristics of the AGS and AGSM were investigated. In this study, settling properties of both AGS and AGSM were significantly studied base on the gravitational settling velocity in order to determine the settleability rate which depict the feasibility of the system in solid:liquid separation performance. The elemental composition of AGS and AGSM were investigated by using XRF spectrometry.

 https://doi.org/10.1051/e3sconf/20183402022

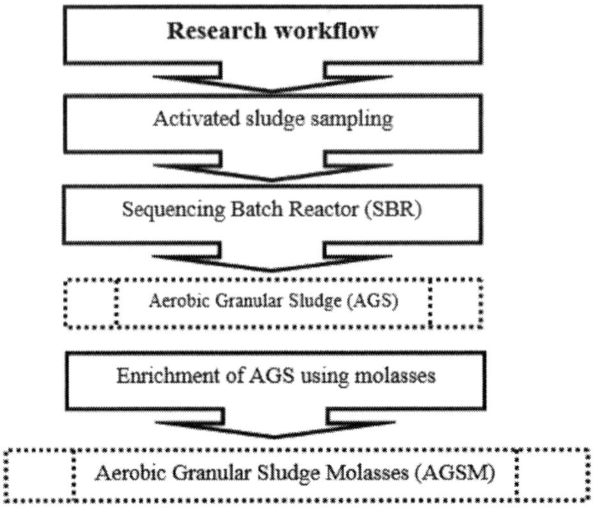

Fig. 1. Development of AGS and AGSM.

2.2 Bioreactor set-up and operation

The SBR used for cultivating aerobic granules was designed for a working volume of 2.4 L, with maximum height of 30.5 cm and 9 cm of internal diameter. The volumetric exchange ratio was fixed at 50%. Fig. 2 shows the schematic diagram of the SBR set-up used in this study.

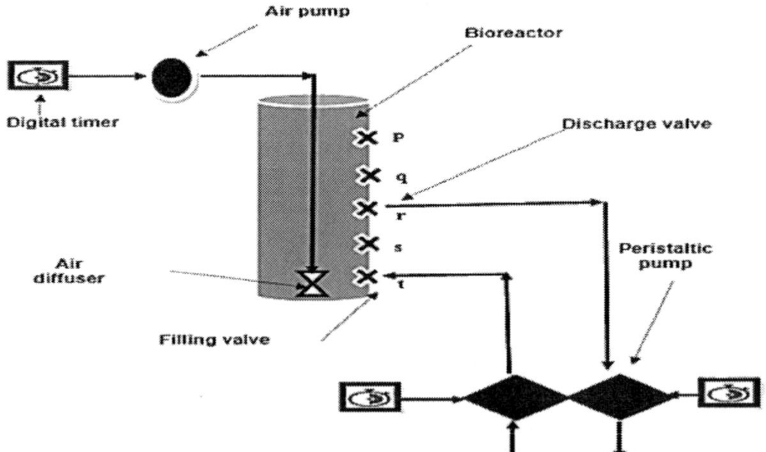

Fig. 2. Bioreactor Set-up.

The bioreactor was operated for 24 hours continuosly in complete cycle of 3 hours [6]. The bioreactor was operated in a fill-draw mode. The SBR configuration comprises of five treatment phases starting from influent filling, aeration, settling, effluent discharge to idle. Fig. 3 shows the cycle of SBR process in this study.

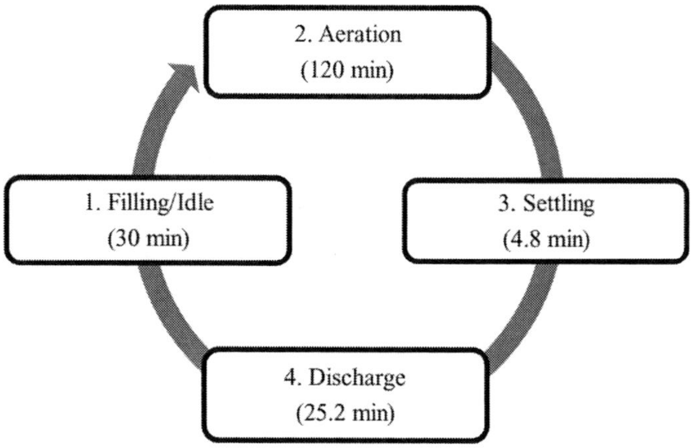

Fig. 3. SBR Operations.

The influent was introduced in the reactor through a valve positioned at the bottom of the SBR. The effluent was discharged through a discharged point at the middle of reactor height with 50% of height ratio. Air was supplied at the bottom of the reactor by a fine air bubble diffuser to give dispersed and optimum distribution of aeration. The SBR was operated at room temperature.

2.3 Seed sludge

Activated sludge was collected from a local wastewater treatment plant. About 900 mL of activated sludge was introduced initially into the bioreactor to start the process. Later, it was maintained at 50% volume exchange ratio. The chemical oxygen demand (COD), sludge volume index (SVI), concentration of mixed liquor suspended solid (MLSS) and mixed liquor volatile suspended solid (MLVSS) were determined using APHA Standard Method [5].

2.4 Synthetic wastewater

The reactors were fed with the acetate-containing synthetic wastewater. The composition of synthetic wastewater was based on [6] which comprise of medium X and medium Y. Medium X consists of NaAc, $MgSO_4.7H_2O$ and KCl. Medium Y consists of NH_4Cl, K_2HPO_4, KH_2PO_4. The trace element solution was also dosed.

2.5 Settling velocity

Settling velocity of the AGS and AGSM were determined using gravitational method. AGS and AGSM with different diameters ranging between 0.2 - 0.6 mm were prepared. An individual granule was selected and delicately placed in a cylinder filled with water to a height of 40cm, and then the settling velocity of individual granule was estimated base on teh time taken to reach the bottom. The materials and equipments needed are; (i) a measuring cylinder with 6.5 cm diameter and 40 cm depth fill with distilled water, (ii) a stopwatch (to record the settling time of granules), (iii) AGS, (iv) AGSM and (v) tweezers. Fig. 4 illustrates the experimental set-up to determine the settling velocity of AGS and AGSM in this study.

https://doi.org/10.1051/e3sconf/20183402022

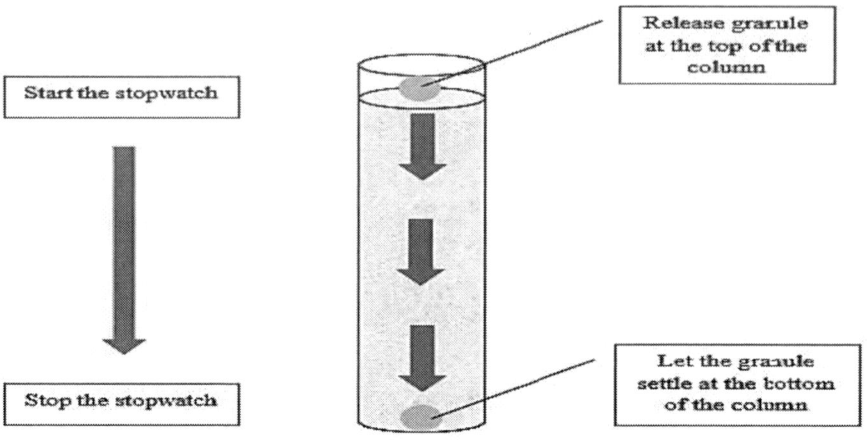

Fig. 4. Settling Velocity Procedure.

2.6 Elemental composition determination

The elemental composition of AGS and AGSM were determined using X-ray Fluorescence (XRF). Prior analysis, samples were dried at 110 °C for 6 hr in an oven and then followed by drying in a muffle furnace at 600 °C for 60 min. Samples (~2 g) was dried in a muffle furnace at 600 °C for 2 hr to determine the ratio of inorganic and organic matters.

3 Results and Discussions

3.1 Settling velocity of AGS and AGSM

Fig. 5 shows the settling velocity profile of AGS and AGSM for size >0.2<0.4 mm of each granules. During early stage (day-14 to day-19), the result shows that the AGS took a shorter time to settle at the bottom compared to AGSM. The ration of MLVSS to MLSS of inoculum was about 60%. This probably due to the adjustment and acclimatization of AGSM to the new enrichment medium and hence has not contributed to the the biomass weight and density.

On day-20, AGSM starts to increase their settling properties by having settling velocity of 23.1 m/h with MLSS of 12.6 mg/L and continued increasing up to day-30 with settling velocity of 30.5 m/h and MLSS of 15.7 mg/L. The increment is due to the accumulation of molasses into AGS and increase of biomass and density of AGSM after succesfully adapted to the new environment.

AGS showed good adaptability to molasses and and AGSM performed good settleability which favours solid-liquid separation in the SBR. [1-2] reported good settling behaviour and biomass retention of various aerobic granules. These maybe due to the inhabitation of different biomass with different conditions of bioreactor especially in nutrients and carbon sources consumptions.

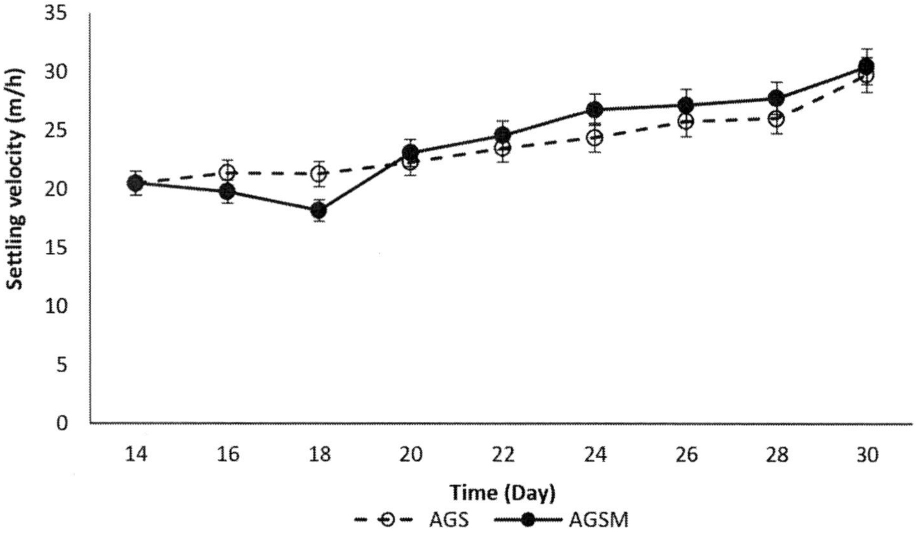

Fig. 5. Settling Velocity of AGS and AGSM.

3.3 Elemental compositions of AGS and AGSM

Fig. 6 shows the XRF results of AGS and AGSM in this study. The XRF results shows that main chemical composition of AGS is palladium (Pd) with 74%, followed by hafnium (Hf) with 13%, calcium (Ca) 4.7% and small amount of other elements like zinc (Zn), ferum (Fe), potassium (K) and rhenium (Re). The various chemicals were detected in AGS due to the samples which are collected from a biological tank from a wastewater treatment plant in a latex processing industry.

According to [7], there were various chemicals and materials used in industries such as in latex processing. Palladium and hafnium were used in numerous application including textiles manufacturing [8]. The XRF results indicate that the main chemical composition of AGSM is Hf with 74%. Other chemical compositions detected in AGSM are K with 16.4%, Re with 13%, Ca with 3.6% and small amount of other elements including Fe, Pd, Os and Zn. In another study, it was found that about seven types of elements present in molasses wastewater including Fe, Zn, manganese (Mn), copper (Cu), chromium (Cr), cadmium (Cd) and cobalt (Co). Each compounds were detected at a range from 0.08 mg/L to 348 mg/L in molasses wastewater samples [9].

Overall, seven chemical elements were detected in AGS, while eight composition elements are found in AGSM. Similar elements were detected in both the AGS and AGSM. Different elemental ratio shows that the acclimatization and enhancement of molasses wastewater on the cultivated AGS has influenced the chemical composition of granules formed depend on the nutrients available in the molasses wastewater. This study gives an insight on the potential of using AGS to treat molasses wastewater. Molasses wastewater contains high organic and inorganic compounds makes it difficult to achieve high solid:liquid separation and colour removal. With the the abundant nutrients present in molasses wastewater, the agglomeration and growth of microorganisms in AGS can be enriched. Hence, employing AGS in molasses wastewater could favour biomass accumulation and colour removal problem in molasses processing industry.

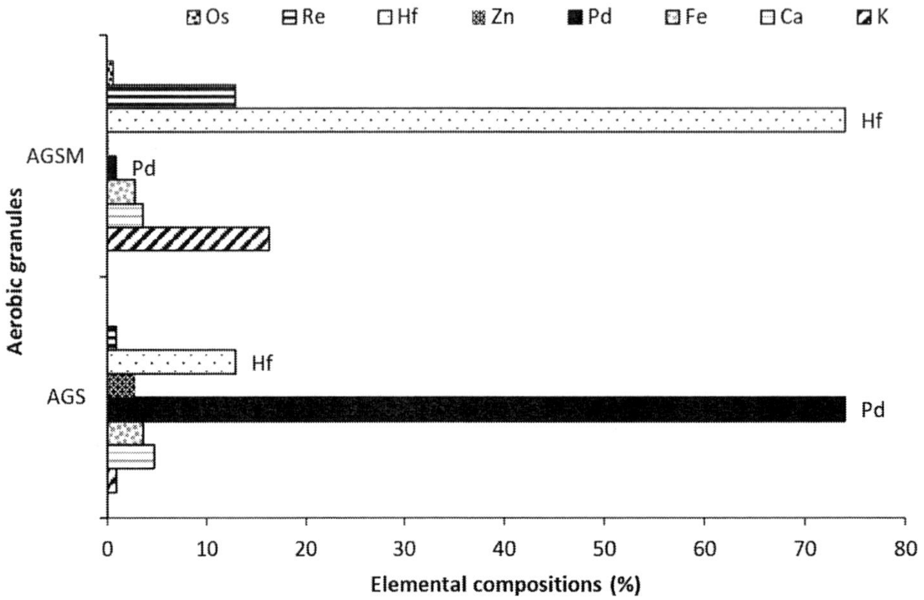

Fig. 6. Chemical Compositions in AGS and AGSM.

4 Conclusions

This study demonstrated that aerobic granular sludge could be cultivated in nutrient rich molasses wastewater. After 30-day operation, stable granules with greater settling properties were obtained with average size ~ 0.4 mm. With enhanced granulation in molasses wastewater, the settling performance of the granules formed gradually increased, as indication by a decreased sludge volume index and an increased settling velocity. The formed granules was regular, spherical and appear in black colour. The dark colour granules indicates the accumulation and enrichment of molasses in the granules. XRF analysis gives detailed various elemental compositions in the cultivated and enhanced granules. This suggests that the aerobic granular sludge served as a good starting inoculum for the colour removal treatment of molasses wastewater. Further studies will have to be resolved in a future investigation.

Authors wish to thank the Ministry of Higher Education Malaysia for the Fundamental Research Grant Scheme (FRGS) No. 9003-00386, RACE Grant (9017-0009), and Universiti Malaysia Perlis for the financial aid through Postgraduate Academic Activities Fund (PAAF).

References

1. A. Nor-Anuar., Z. Ujang, M.C.M. Van Loosdrecht, M.K. De Kreuk. Wat. Sci. and Tech. **56(**7), 55-63 (2007)
2. F.A. Dahalan, N. Abdullah, A. Yuzir, G. Olsson, M. Hamdzah, M.F.M. Din, S.A. Ahmad, K.A. Khalil, A.N. Anuar, Z.Z. Noor, Z. Ujang. Bio. Tech. **181**, 291-296 (2015)
3. P. P. Kalbar, S. Karmakar, S.R. Asolekar. Wat. Env. J. **27**(2), 261-268 (2013)

4. S.S. Adav, D.J. Lee, K.Y. Show, J.H. Tay. Bio. Adv. **26**(5), 411-423 (2008)
5. APHA (American Public Health Association). *Standard methods for the examination of water and wastewater.* (American Public Health Association, Washington DC, 2005)
6. A. Nor-Anuar. *Development of aerobic granular sludge technology for domestic wastewater treatment in hot climate.* Unpublished Ph.D dissertation. (Universiti Teknologi Malaysia, 2008)
7. E., Morgenroth, T. Sherden, M.C.M. Van Loosdrecht, J.J. Heijnen, P.A. Wilderer. Wat. Res. **31**(12), 3191-3194 (1997)
8. F.A. Dahalan, N. Abdullah, A. Yuzir, G. Olsson, M. Hamzah, M.F.M. Din, Ujang, Z. Bio. Tech. **181**, 291-296 (2015)
6. T. Akabane. Int. Poly. Sci. and Tech. **43**(5), T45 (2016)
7. U. Marga, S. Yanti, S. Made, P. Tita, I. Dian, K. Makuuchi, F. Yoshii. *Production of surgical gloves from low extractable protein.* RVNRL No. JAERI-CONF. 2001-005 (2001)
8. American Elements. Hafnium Nitrate Solution. Retrieved from https://www.americanelements.com/hafnium nitrate-solution-15509-05-4 (2018)
9. I.S. Thakur. *Industrial biotechnology: problems and remedies*. (IK International Pvt Ltd., New Delhi, 2010)

Noise Annoyance Produced by Commercial Vehicles Transit on Rumble Strips

Haron Zaiton[1,*], *Yahya* Khairulzan[1], *Darus* Nadirah[1], *Mashros* Nordiana[1], *Hezmi* Muhammad Azril[1], *Abu Samah* Rosiah[2], *Abdul Hameed* Abdul Mutalif[2], *Norudin* Wan Mohammad Asyraf[1], *Abd Halil* Muhammad Habab[1], *Jahya* Zanariah[1], and *Othman* Mohd Hanifi[3]

[1]Faculty of Civil Engineering, Universiti Teknologi Malaysia, 81310 Johor Bahru, Johor, Malaysia
[2]Pusat Kecemerlangan Kejuruteraan & Teknologi JKR (CReaTE), Jalan Kemus Simpang Empat, 7800 Alor Gajah Melaka
[3]Faculty of Civil and Environmental Engineering, Universiti Tun Hussein Onn Malaysia, 86400 Batu Pahat, Johor, Malaysia

Abstract. This paper reports on research examining the extent of noise annoyance affecting residents within the vicinity of installation of two types of transverse rumble strips (TRS), namely Middle Overlapped (MO) and Multilayer Overlapping (MLO). In order to assess the noise annoyance in the area, measurements were taken at 7.5m from centre of road with TRS installation using single vehicle test to determine the extent of changes of sound level indices and sound spectrum. Two light and two medium weight commercial vehicles were used. Indicators L_{Aeq}, L_{AFmax}, L_{AImax}, L_{AIeq}, and L_{ASmax} were used to determine impulsivity that led to noise annoyance. The results showed that, at 30 km/h, all commercial vehicles considered in this study that transited on MO produced impulsive noise, while only light commercial vehicles caused noise annoyance when they transited on MLO. The research also analysed the extent of low-frequency noise and found a significant low-frequency component, which indicated that noise annoyance might arise from the hitting of MO and MLO by the commercial vehicles. For night-time related annoyance, it was suggested that an additional weighting factor could be added to the average A-weighted value during night-time.

1 Introduction

Rumble strips can be defined as a set of yellow bars painted on the pavement with a specific thickness to alert road users through their colour, sound, and physical vibrations. TRS have a specific sectional profile and they are laid perpendicular to the vehicle's flow. In Malaysia, there are three common types of TRS profiles: Multilayer Overlapping (MLO), Rumbler Raised (RR), and Middle Overlapped (MO) [1, 2]. TRS are used to inform road users that there is a change in the road environment that requires drivers to be more cautious. They are placed at a critical point before reaching a junction, roundabout,

*Corresponding author: zaitonharon@utm.my

toll plaza, road hump, and pedestrian crossing. These are the alternatives used to reduce accident rate as they give drivers a warning to slow down.

TRS are mostly utilised as traffic calming measures implemented in the vicinity of schools, business areas, and neighbourhoods [2, 3]. Thus, one of the prime issues caused by the existence of TRS is noise annoyance. Night-time noise that causes sleep disturbance is the major source of concern, therefore, some TRS have to be replaced by other traffic calming devices due to the complaints by nearby residents. TRS' physical properties such as profile, thickness, number, and spacing between bar are selected without any proper references to standards or guidelines [4], and purely on the basis of the local traffic engineers' experience. Apart from that, vehicle speed on the road will also affect the level of noise during transit on the TRS. Furthermore, it was found that, as the vehicle speed increased the pass-by noise levels caused by the TRS also increased [1,5]. In addition to that, the vehicle's type and weight also influenced the noise level. Haron et al (2016) found that a 980 kg compact car caused a significant increase in pass-by noise level when it traversed the TRS and generated a significant impulsive sound that made it more annoying than a continuous noise [1]. According to An et al. [5] in general, the increase in noise was greater at 100 km/h than at 40 km/h and lower for a sedan than a truck.

Previous research suggested that the typical evaluations of noise annoyance based on the A-weighted equivalent level by the regulatory authorities were inadequate and can lead to incorrect decisions [6]. Annoyance caused by sound in the low-frequency region (20 Hz up to 100 Hz) cannot be assessed using the A-weighting but rather with the consideration of the C-weighting [7]. Moreover, it was suggested that low-frequency components may increase the adverse effects considerably, pose more detrimental impacts to public health, and become more annoying as claimed by individuals [8, 9]. Acoustically, noise annoyance is the result of interference with daily activities, feelings, thoughts, sleep, or rest, and may be accompanied by negative emotional responses, such as irritability, distress, exhaustion, a wish to escape the noise and other stress-related symptoms [10].

The current study investigated noise annoyance that arisen due to TRS installation by commercial vehicles with a particular emphasis on low frequency noise content. It focused on commercial vehicles because their percentage on the road is currently about 12% [11] and represented 10% of the total sales of vehicles in Malaysia as of March 2017. Furthermore, in traffic noise prediction, commercial vehicles are denoted as light, medium, and heavy vehicles. Therefore, the objectives of this research are to (1) determine the change in sound level indices associated with the installation of two TRS profiles, and (2) analyse noise annoyance objectively through sound level changes and low-frequency content.

2 Methodology

2.1 Measurement of sound level indices and frequency spectrum

The study focused on the change of sound level indices, impulsivity, and low-frequency content due to the transition of vehicles on the two types of TRS. Two road stretches in the state of Johor installed with MO (located at Skudai, width = 600 mm) and MLO (located at Kangkar Pulai, width = 400mm) were selected, with pavement and thermoplastic TRS profile were in good condition and have the same thickness (3 mm). Profile of both TRS is shown in Figure 1. The commercial vehicles adopted for the tests were 2 light vehicles (LV) (Toyota Hilux, with gross vehicle mass (GVW = 2780 kg), 1 HiAce van (GVW=2600 kg), and 2 medium vehicles (MV) namely a one ton lorry (GVW = 5000 kg) and a 40 seater bus (GVM = 7700 kg).

Measurements were carried out at mid-night to avoid disturbance from other noise sources. Also, wind speed and air temperature were below 5m/s and 5^0c to 40^0c, respectively for data validity [13]. The sound levels were measured using control pass-by (CPB) method in which microphones was placed at a defined distance from the vehicle path at the side of the roadway (Fig. 1). For measurement of noise level due to TS installation, sound level meter Pulsar Type 1 was placed at Point 1 at 7.5 m from the centre of the vehicle lane at a height of 1.2 m above the pavement [12]. Measurement for determination of sound level indices and frequency spectrum in this study were carried out separately. The noise indices recorded were L_{Aeq} , L_{AFmax}, L_{ASmax}, L_{AImax} and L_{AIeqT}. The L_{Aeq} is the constant noise level that expends the same amount of energy as a fluctuating level over the same time period [12]. L_{AFmax} is fast response maximum with the equivalent A-weighted sound pressure level while L_{AImax} is maximum A-weighted impulse response. Meanwhile, LAIeqT is the A-weighted impulse sound pressure levels averaged over the same time interval $_{\text{and}}$ L_{ASmax} is slow response maximum with the equivalent A-weighted sound pressure level. For determination of sound level indices and frequency spectrum, each vehicle was run using speed 30 km/h and 50 km/h as these speeds are the speed limit imposed on roadway at critical point and in each speed test were repeated 3 times. Vice versa point 2 was used to record the noise level when vehicles were run without TRS with the whole activities were repeated.

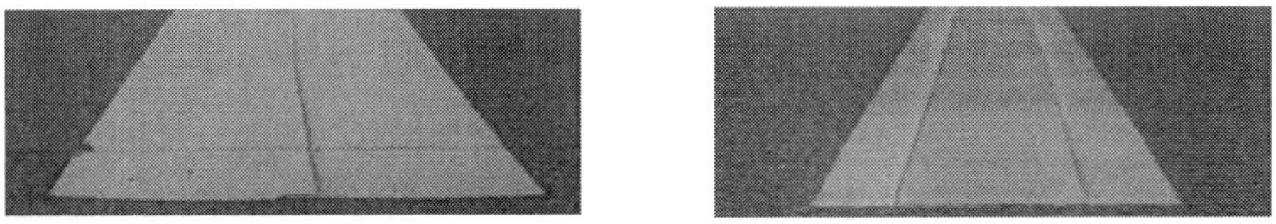

i) MO profile ii) MLO profile

(a) Rumble strips profiles

i) Hilux (2780 kg) ii)Hiace (2600kg) iii)Lorry (5000 kg) iv)Bus (7700 kg)

(b) Test vehicles types

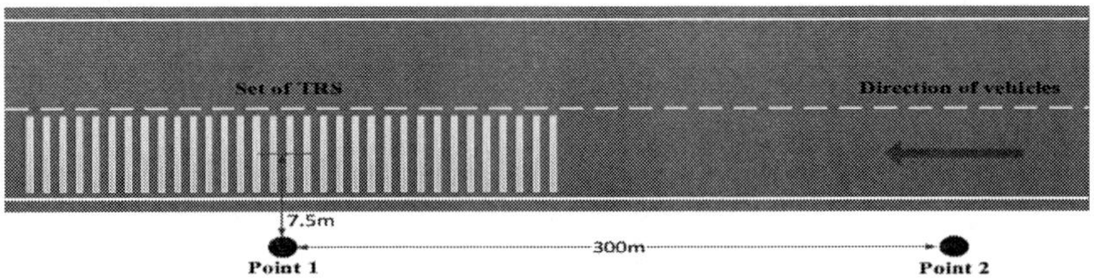

(c) Experimental set up

Fig. 1. Profiles of TRS, test vehicles and measurement layout.

2.2 Change in sound levels indices

The change of noise level, L_{Aeq} can be directly obtained from the disparities of values between the tests (with and without TRS). A 3-dB change in noise level is considered just discernible; a 5-dB change is clearly discernible; and a 10-dB change louder or softer is perceived as a doubling or halving of volume, respectively [17]. Further, in this study, the significance of impulsivity was determined using several criteria involving recorded noise indices as suggested in previous research. In order to objectify the annoyance, the difference in noise indices was compared with the reference value. If any of these reference values is exceeded, it is assumed that the annoyance/complaint is objectively attributable to the TRS source. The reference values of the guideline are shown in Table 1.

Table 1. Impulsive characteristic determination.

Different in noise indices	L_{AFmax}- L_{AImax}	L_{AFmax} -L_{Aeq}	L_{AIeqT} -L_{Aeq}	L_{AImax}- L_{ASmax}
References limit (dB (A))	2	10	2	6

2.3 Analysing the Low frequency noise for the annoyance

Data on frequency spectra were used to calculate dB(A) and dB(C). Furthermore, Kjellberg et al. [13] suggested that if the difference between A and C-weighted values exceeds 15dB then a low-frequency noise problem may exist. Then, Broner and Knight-Merz [14] method was considered in which they proposed simple criteria for the control of annoyance due to low-frequency noise. They suggested that if the noise level is fluctuating by 5dB(C), then a penalty of 5dB(C) should be added i.e., the criteria should be reduced. Consequently, an examination on the presence of a tone was carried out. Data on frequency spectra were used to identify the presence of a tone. The level in one 1/3rd octave band to the level in the two adjacent bands was compared according to ISO 1996-2 [15]. It was suggested that if the tone is 15 dB in the low-frequency one-third-octave band (25 Hz to 125 Hz), the annoyance might be due to low-frequency content. Then, a further procedure for the assessment of low frequency noise presented by Newman and McEwan [16] who referred a British Gas Corporation criterion for specifying noise control for gas turbines will be referred. The procedure involves a 60 dB limit in the 31.5 Hz octave band at the nearest dwelling. An additional weighting factor of 5 dB can be added to the average A-weighted if the limit is exceeded.

3 Results and discussions

3.1 Annoyance due to change in sound levels indices

The relationship between the L_{Aeq} produced with speed and MO and MLO is shown in Fig. 2 (a) and (b), respectively. The increasing trend in sound level with GVW was found to be in polynomial equation of second degree with very good relationship, where the GVW explained more than 90% of the variations. In general, both type of TRS generated higher sound level at higher speed (50 km/h) by maximum 8 dB(A) compared with speed 30 km/h only by 6 dB(A) when traversed by all tested commercial vehicles.

It can be seen that the MO, when transited by all types of commercial vehicle with speed of 50 km/h, produced increasing sound level from 1 to 2.2 dB(A) relative to the normal road, while at the lower speed, only the 40 seated bus (7700 kg) yielded sound level increment of 2.9 dB(A). On the other hand, the MLO seemed to absorb the sound level produced by the tyre and propulsion as the vehicle speed reached 50 km/h. The installation of MLO increased sound level at value of 2.4dB(A) only when the Hiace with GVW of 2600kg at 50 km/h traversed it and also from the one ton lorry with speed of 30 km/h and 3 dB(A) increase in sound level. Based on the increment of sound level, only the one ton lorry which travelled at 30km/h produced the increment of 3 dB(A). It is seen that the corresponding increase in sound level values for with and without do not exceed the 5 dB(A) [17], this indicates that there may not be a significant problem in the noise generated by both type TRS when vehicles transit on them.

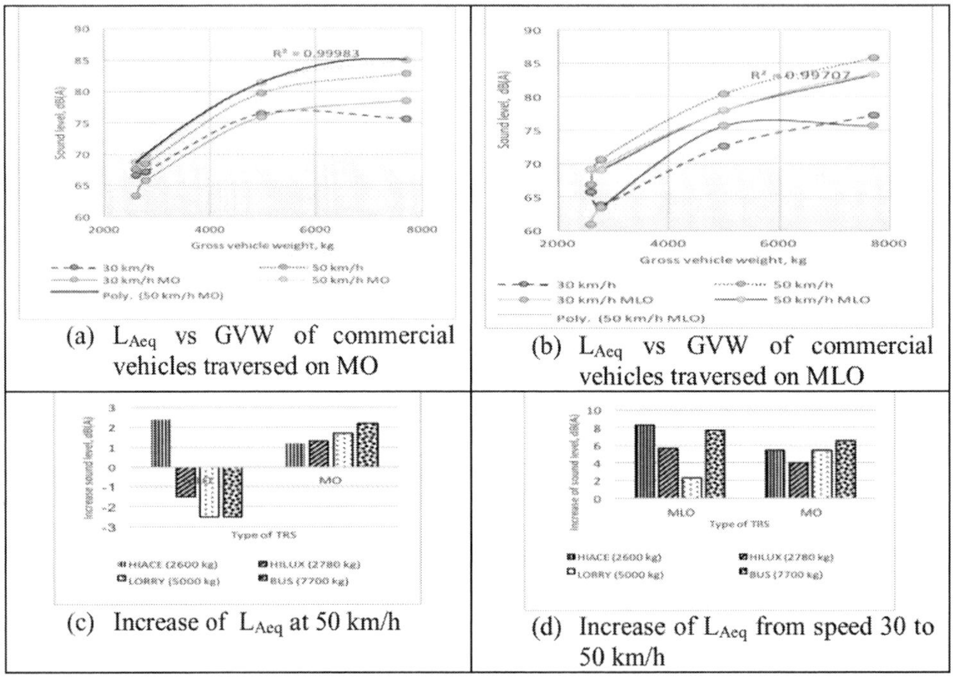

(a) L_{Aeq} vs GVW of commercial vehicles traversed on MO

(b) L_{Aeq} vs GVW of commercial vehicles traversed on MLO

(c) Increase of L_{Aeq} at 50 km/h

(d) Increase of L_{Aeq} from speed 30 to 50 km/h

Fig. 2. Changes in sound level.

3.1.1 Impulsive characteristics

Fig. 3 shows the comparison of impulsive chrateristics with reference limits. According to (LAIeqT -LAeq), it was found out that all vehicles do not poses impulsive noise (Figure 3c). However, by considering (LAFmax – LAImax), (LAFmax – LAeq) and (LAImax-LASmax), it was founded that the MO produced significant impulsive noise when both LV and MV transited at speed of 30 km/h. The significant impulsivity was obtained when GVW>5000 kg produced (LAFmax – LAImax) and (LAImax-LASmax) values greater than 10 (Fig. 3a) and 6 dB (Fig. 3d), respectively, while GVW< 5000 kg yielded (LAFmax-LAeq) value greater than 10 (Fig. 3b). For MLO, only LV produced significant impulsive noise when it traversed at speed of 30 km/h and this can be identified through (LAFmax – LAeq) value which exceeded 10 dB when GVW<4000 kg. Thus, MO and MLO will cause impulsive characteristics if the LV such as Hiace and Hilux transit on them at a lower speed. From the standpoint of annoyance, if the activity occurs at night-time the effect would be considerably more detrimental than day-time exposure.

3.2 Annoyance due to Low frequency noise

3.2.1 dB(C)-dB(A)

TRS profile also affects the frequency spectrum composition of the noise emitted in low, medium, and high frequencies. Fig. 4 shows sound level spectrum, A-weighted, and C-weighted sound level generated by a bus that travelled on MO and without MO. The average increase in sound level in low frequency fluctuated compared to the higher frequency. Overall, the increase of total dB(A) and total dB(C) from that of without MO were, 1.18 dB(A) and 1.39 dB(C), respectively. The results of the dB(C)-dB(A) values with and without TRS at different speeds highlighted that the corresponding values did not exceed the 15dB threshold as suggested by Kjellberg et al [13]. As mentioned previously, this situation indicated that there may not be a significant low-frequency problem in the noise generated by both types of TRS when the vehicles traversed them.

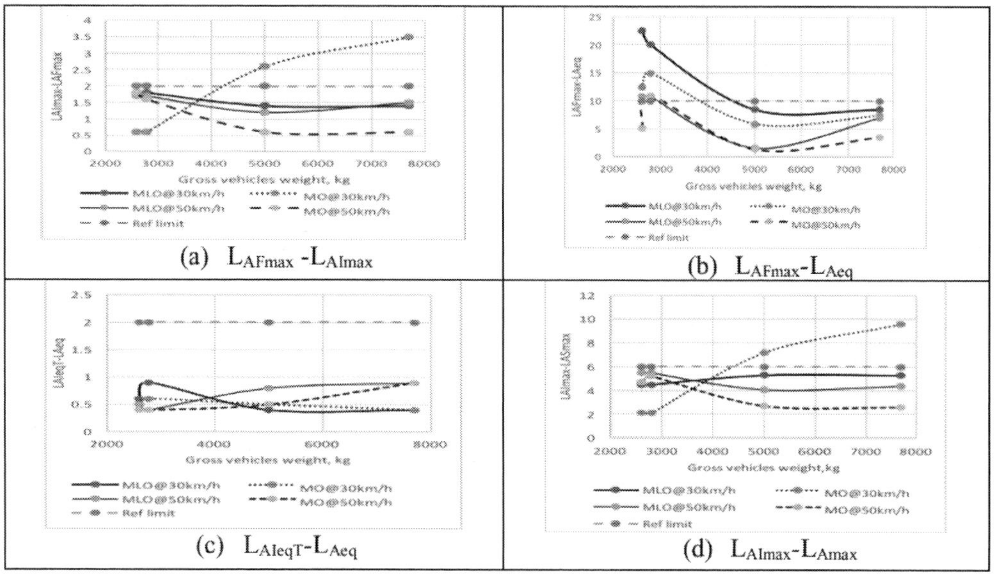

(a) L_{AFmax} -L_{AImax} (b) L_{AFmax}-L_{Aeq}

(c) L_{AIeqT}-L_{Aeq} (d) L_{AImax}-L_{Amax}

Fig. 3. Examination of impulsive characteristic.

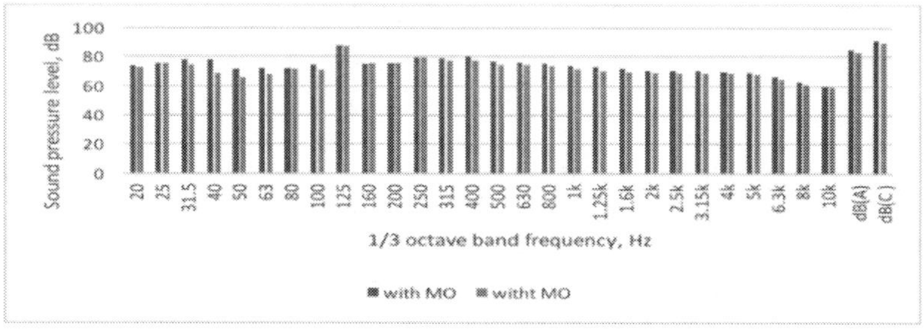

Fig. 4. Frequency spectra in dB, dB(A) and dB(C) when 40 set bus traverse on TRS at speed 50 km/h.

3.2.2 Fluctuation of dB(C)

Table 2 shows the fluctuation of dB(C) between with and without TRS and speeds of 30 km/h and 50 km/h. The noise level fluctuation was higher than 5 dB(C) when a lorry traversed the MLO, which indicated the presence of a low-frequency noise. Thus, a spectrum analysis was conducted to assess the extent of noise in the lower frequency bands, namely in the region between 20 Hz to 125 Hz.

3.2.3 Tonal examination in 20 Hz to 125 Hz

The investigation showed that there were no significant tones that can be identified using the methodology outlined in ISO 1996-2 when LV and MV, investigated in this study, traversed the TRS. Fig.5 shows the difference of level or tone produced between 20 Hz to 125 Hz when vehicles transit and do not transit the TRS. In general, the tone produced by the vehicles that transited on the TRS appeared to be reduced. The installation of MO reduced the annoyance from the Hiace, lorry, and bus travelling at speed of 50 km/h. Moreover, at 30 km/h, the MLO and and MO reduced the significant tone generated by the Hiace and lorry, respectively. Thus, it can be said that noise generated by the TRS produced a broadband source with a clear low-frequency content.

Table 2. Fluctuation of dB(C) with and without TRS.

Vehicles type	30 km/h		50 km/h	
	MO dB(C)	MLO dB(C)	MO dB(C)	MLO dB(C)
HiAce	0.72	-6.98	4.98	-15.04
Hilux	0.92	-4.32	3.67	-14.5
Lorry	-2.59	0.09	-5.49	5.59
Bus	0.93	-6.36	1.39	-3.95

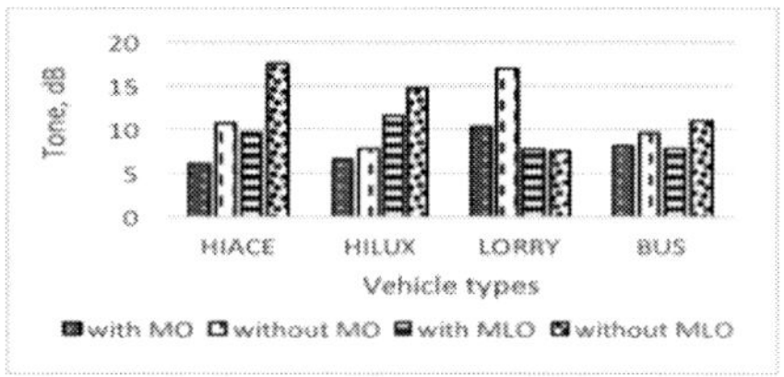

(a) speed 30 km/h

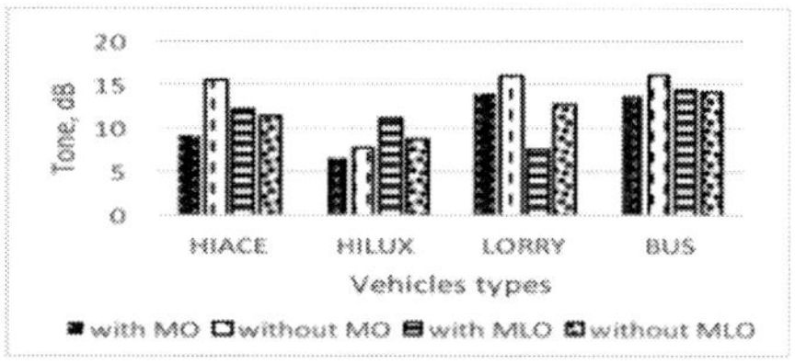

(b) speed 50 km/h

Fig. 5. Examination of tone existing between 25 to 125 Hz when vehicles traverse on TRS.

3.2.4 Examination of sound level in 31.5 Hz

Fig. 6 shows the logarithmic addition of the 25Hz, 31.5 Hz, and 40Hz third octave band levels comparison with the 31.5Hz octave band criterion of 60dB as set by British Gas Corporation [16]. It can be seen that total sound levels with the TRS was higher than that of without TRS and also the criteria were exceeded even though the MV transited and did not transit the TRS. The bus produced the highest sound levels when passing the road with MO at speeds of 30 and 50 km/h and exceeded by 15.7 dB and 19.9 dB, respectively. This indicated that a low-frequency noise was likely to be added to the overall night-time sleep disturbance and annoyance level at nearby residents. The results showed that the low-frequency noise was significant when the MV traversed the MO and MO, an additional weighting factor of 5 dB can be added to the average A-weighted value during night-time.

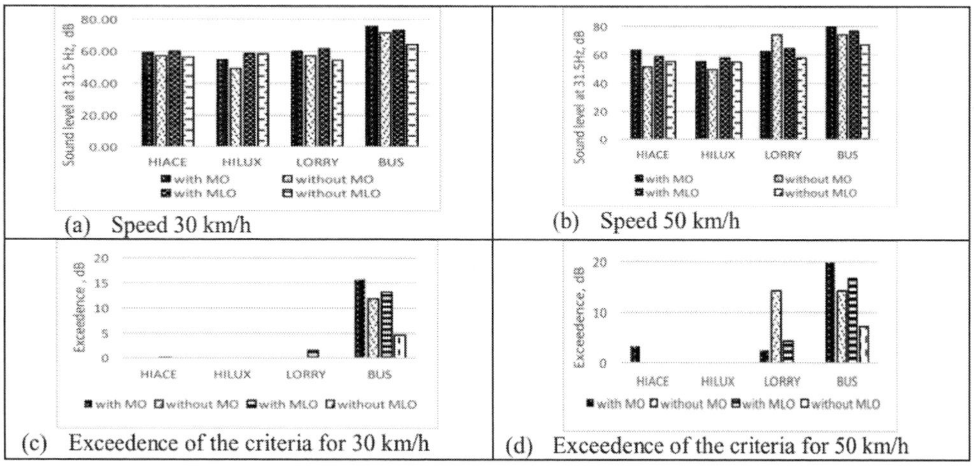

(a) Speed 30 km/h

(b) Speed 50 km/h

(c) Exceedence of the criteria for 30 km/h

(d) Exceedence of the criteria for 50 km/h

Fig. 6. Examination on sound level at 31.5 Hz.

4 Conclusions

The current study investigated noise annoyance that arised due to the MO and MLO installations as they were traversed by commercial vehicles. The following conclusions were obtained:

- Both types of TRS produced higher sound level at higher speed (50 km/h) compared to the lower speed (30 km/h) when traversed by all commercial vehicles.
- Annoyance did not arise due to the increase of sound level as it was below the limit of 5dB.
- Annoyance occurred when all commercial vehicles considered in this study transited the MO with speed of 30 km/h and only light commercial vehicles with the same speed of 30 km/h traversed the MLO.
- Annoyance also arised due to low frequency noise that resulted when light (Hiace) and medium weight commercial vehicles (bus and lorry) traversed both MO and MLO.
- MO reduced annoyance from Hiace, lorry, and bus that travelled at 50 km/h.

References

1. Z. Haron, M.H. Othman, M. H., K. Yahya, K., M.R. Hainin, K. Ambak, & N. Darus, *MATEC Web of Conferences*.(2017)
2. M. H. Othman, Z. Haron, M.R. Hainin, K. Yahya, H. Yaacob, M. E. Sanik: Jurnal Teknologi. **73** (2015)
3. M.M. Abd Manan, P.W.H, Alvin, Malaysian institute of Road Malaysian Research (2009)
4. Manual fasiliti keselamatan jalan, Cawangan Kejuruteraan Jalan dan Geoteknik, Jabatan Kerja Raya Malaysia (2014).
5. D. S. An, S.A. Kwon, J. Lee, Y.C. Suh, ASCE, J. Perf. Faci. ASCE (2016)
6. E. Murphy, E.A. King, Environment International **63** (2014)
7. K. Persson, M. Bjorkman, J Sound Vibration **127** (1988)
8. B. Berglund, T. Lindvall, D. Schwela, K.T. Goh, WHO (1999)
9. G.H. Leventhall, DEFRA (2003)

10. M.E. Beutel, C. Jünger, E.M. Klein, P.W.K Lackner, M. Blettner, H. Binder, M. Michal, J. Wiltink, E. Brähler, T. Münzel, https://doi.org/10.1371/journal.pone.0155357(2016)
11. http://www.maa.org.my/info_summary.htm, (assessed March 2017)
12. ISO (1997a). ISO 11819-1:2001
13. Kjellberg, M. Tesarz, K. Holmberg, U. *Landstrom (Environment International)* **23** (1997)
14. N. Broner, S. Knight-Merz, *Acoustics Australia* **39** (2011)
15. ISO 1996-2 (2007) ISO 1996-2:2007
16. J.R. Newman, K.I. McEwan K. I, Journal of Engineering for Power **102** (1980)
17. J.L. Darbyshire, J.D. Young, *Critical Care* **17** (2013)

Traffic Noise Assessment at Residential Areas in Skudai, Johor

F.S. Sulaiman[1], *N.* Darus[1], *N.* Mashros[2,*], *Z.* Haron[1], and *K.* Yahya[1]

[1]Department of Structure and Materials, Faculty of Civil Engineering, Universiti Teknologi Malaysia, 81310 Johor Bahru, Johor, Malaysia
[2]Department of Geotechnics and Transportation, Faculty of Civil Engineering, Universiti Teknologi Malaysia, 81310 Johor Bahru, Johor, Malaysia

Abstract. Vehicles passing by on roadways in residential areas may produce unpleasant traffic noise that affects the residents. This paper presents the traffic noise assessment of three selected residential areas located in Skudai, Johor. The objectives of this study are to evaluate traffic characteristics at selected residential areas, determine related noise indices, and assess impact of traffic noise. Traffic characteristics such as daily traffic volume and vehicle speed were evaluated using automatic traffic counter (ATC). Meanwhile, noise indices like equivalent continuous sound pressure level (L_{Aeq}), noise level exceeded 10% (L_{10}) and 90% (L_{90}) of measurement time were determined using sound level meter (SLM). Besides that, traffic noise index (T_{NI}) and noise pollution level (L_{NP}) were calculated based on the measured noise indices. The results showed an increase in noise level of 60 to 70 dBA maximum due to increase in traffic volume. There was also a significant change in noise level of more than 70 dBA even though average vehicle speed did not vary significantly. Nevertheless, L_{Aeq}, T_{NI}, and L_{NP} values for all sites during daytime were lower than the maximum recommended levels. Thus, residents in the three studied areas were not affected in terms of quality of life and health.

1 Introduction

Noise can cause stress, disturb sleep, disrupt activities, hinder work, and impede learning [1]. Therefore, noise pollution has always been a significant environmental problem for humans. It can be defined as unwanted sound that intrudes humans' daily activities. Sound pressure is a basic measurement of air vibration. Furthermore, the range of sound pressure that can be detected by human ears is very wide whereby these levels are measured in logarithmic scale in the unit of decibel (dB) [2]. Traffic is a dominant noise source in urban and rural environments. Noise sources include those from motor vehicles, aircrafts, and rail transport. This noise originates from the engine and frictional contact between the vehicle and the ground, and is affected by traffic flow rate that depends on vehicle's speed and road's surface condition [3].

* Corresponding author: mnordiana@utm.my

Nowadays, the living environment in many residential areas has been deteriorating due to the increase in traffic volume, which resulted in causing noise pollution that affected residents [4]. Environmental awareness and knowledge in impact of noise on health, performance, and well-being, are becoming increasingly important to curb noise pollution. The World Health Organization (WHO) has described noise as the principle environmental nuisance in industrial nations. Besides that, WHO also recognised it as a harmful pollution that causes adverse psychosocial and physiologic effects on human health. For example, a noise pollution study was conducted in Fulton County, Georgia. The research highlighted that many residents were exposed to high noise levels that put them at risk of annoyance or sleep disturbance, which can lead to serious health consequences [5].

Hence, determining traffic noise level in residential areas will help in understanding the actual condition regarding noise pollution in that environment. Then, the results will indicate whether precautions or certain actions are sufficient to handle the traffic noise issue among the affected neighbourhoods. Since traffic noise from roadways, especially those with higher traffic volume and vehicle speed, creates problems to nearby residents, this study intends to evaluate traffic noise at selected residential areas in Skudai, Johor, Malaysia. This study was conducted according to the objectives, namely, evaluating traffic characteristics at selected residential areas, determining related noise indices, and assessing impact of traffic noise on nearby residents.

2 Methodology

This study was conducted in three main stages: (1) evaluation of traffic characteristics; (2) determination of noise indices; and (3) assessment of traffic noise impact. The details of each stage are described in the following sections.

2.1 Evaluation of traffic characteristics

Three residential areas, Site 1-Taman Universiti, Site 2-Taman Sri Skudai, and Site 3-Bandar Baru Kangkar Pulai, located around Skudai, Johor were selected as sampling locations. These neighbourhoods were selected based on its high daily traffic volume, flat and straight roadways, single carriageways, and location of at least 70 m away from the nearest traffic light with the assumption that most vehicles in the traffic stream had already reached a steady speed [6]. To acquire the traffic characteristics of each sampling location, automatic traffic counter (ATC) was used in this study. Fig. 1 illustrates ATC's pneumatic road tubes configuration utilised on a single carriageway whereby two tubes are placed parallel to each other with 1 m spacing. The data recorder connected to the pneumatic road tubes was secured in a safety box, positioned at the roadside to avoid any loss or damage.

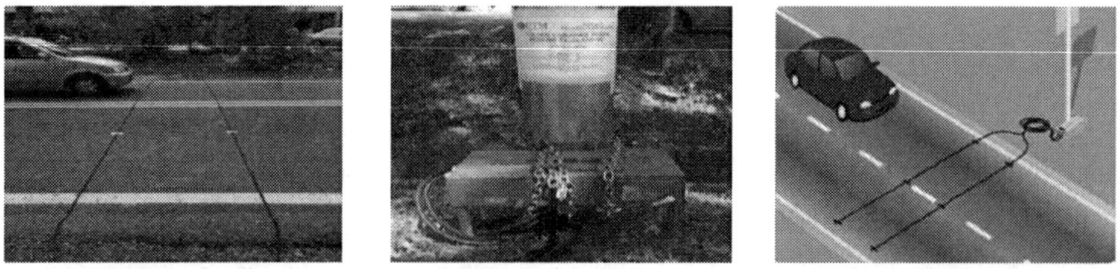

(a) Pneumatic tubes (b) Safety box (c) Installation layout

Fig. 1. Installation of ATC.

The traffic characteristics study was conducted for three days (Monday, Tuesday, and Wednesday) within a week. The start and end days of the week were excluded to avoid different traffic behaviours at sampling locations. The ATC was installed continuously for 24 hours to observe traffic characteristics at sampling locations. Furthermore, the peak period of measurement was determined because morning (7 a.m. to 8 a.m.), afternoon (12 p.m. to 1 p.m.), and evening (6 p.m. to 7 p.m.) peak hours contributed to high noise level. The parameters considered for this traffic study were traffic volume and vehicle speed. Data output that included traffic volume, vehicle speed, and composition were presented in the data recorder, Metro Count Software.

2.2 Determination of noise indices

The second stage in this study was conducted using a Pulsar Type 1 sound level meter (SLM) to measure related noise indices such as equivalent continuous sound pressure level (L_{Aeq}), and noise level exceeded 10% (L_{10}) and 90% (L_{90}) of measurement time. Fig. 2 shows the installation of SLM at sampling locations. The microphone must be positioned at a height of 1.5 m above the ground level, which corresponds to the ear level of an individual with average height [7]. It was calibrated prior and after the measurement to ensure a reading error within the permissible value of ±0.5 dBA. On the other hand, SLM was placed 7.5 m away from the middle of the nearest travel lane [8]. Noise measurement was conducted according to ISO 1996 guidelines [9–11]. It was carried out during daylight starting from 7 a.m. to 7 p.m. for three days under a good weather condition.

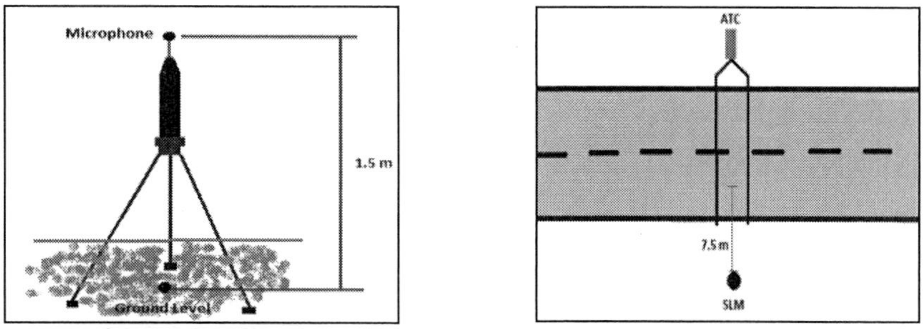

(a) Height of SLM (b) Distance of SLM

Fig. 2. Installation of SLM.

2.3 Assessment of traffic noise impact

Traffic noise index (T_{NI}) and noise pollution level (L_{NP}) were calculated based on the measured noise indices i.e. L_{Aeq}, L_{10}, and L_{90}. T_{NI} was derived by combining L_{10} and L_{90} values. In addition, T_{NI} attempts to make allowance for noise variability with respect to L_{10} level [12]. Moreover, T_{NI} values should not exceed 74 dBA to avoid dissatisfaction among nearby residents. Meanwhile, L_{NP} considers the variations in sound signal that serves as a better indicator of environment pollution [13]. T_{NI} and L_{NP} were computed using Equations 1 and 2, respectively [14]. Besides that, L_{Aeq} and L_{NP} values were compared with the standard guidelines from WHO and Department of Environment (DOE), Malaysia [15, 16]. Both guidelines stated that the tolerance limit due to traffic noise in urban areas should not exceed 55 dBA during daytime.

$$T_{NI} = 4\,(L_{10} - L_{90}) + L_{90} - 30 \quad (1)$$

$$L_{NP} = L_{eq} + (L_{10} - L_{90}) \quad (2)$$

3 Results and Discussions

3.1 Traffic volume

Table 1 displays the daily traffic volume for three days measured at selected road sections from three sampling sites. Based on the data, Site 3 recorded the highest total traffic volume with 60,153 vehicles. This indicates that Site 3 is the busiest road compared to other sites. Meanwhile, the total traffic volume at Sites 1 and 2 were not much different with 21,418 and 20,268 vehicles, respectively. Additionally, the daily traffic volume for Monday, Tuesday, and Wednesday recorded a similar value, for all sites.

Table 1. Daily traffic volume.

Day	Site 1	Site 2	Site 3
Monday	6,543	7,246	20,092
Tuesday	6,794	7,252	19,897
Wednesday	6,931	6,920	20,164
Total	20,268	21,418	60,153

Fig. 3 depicts the distribution of traffic volume within 24 hours for three days. It clearly shows that the highest traffic volume for each location occurred during morning (7 a.m. to 8 a.m.) and evening (6 p.m. to 7 p.m.) peak hours. These peak hours are typical rush hours at all sampling locations. This is due to the large number of private and public vehicles transporting passengers or goods through main traffic routes. On the other hand, the afternoon (12 p.m. to 1 p.m.) peak hour recorded a slightly lower volume in comparison to morning and evening peak hours.

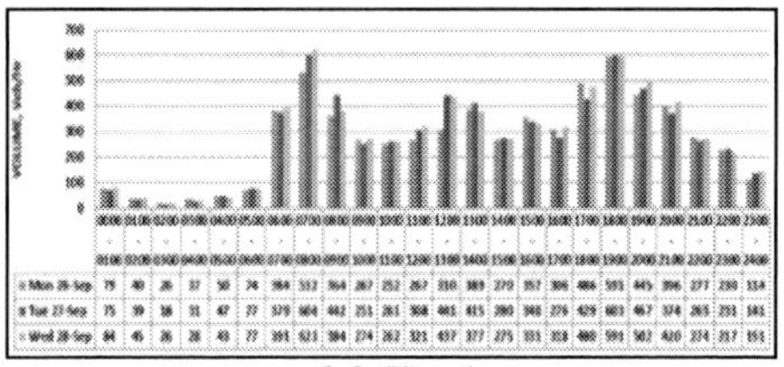
(a) Site 1

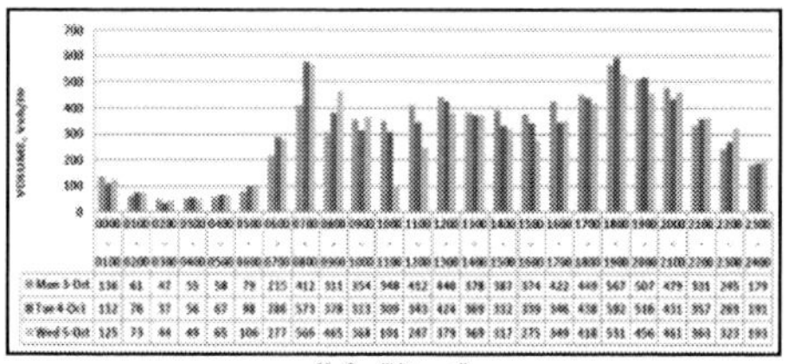
(b) Site 2

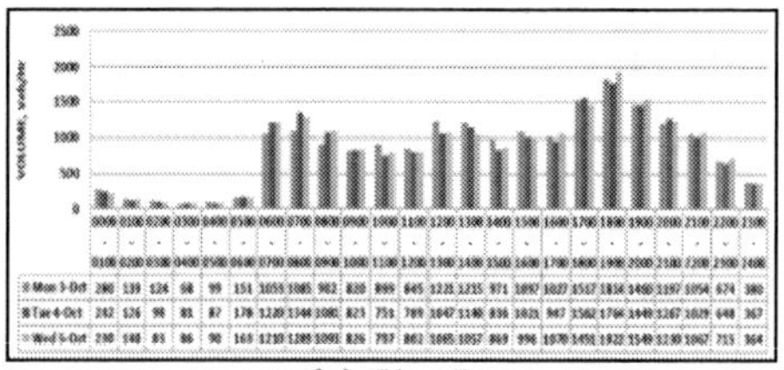
(c) Site 3

Fig. 3. Traffic volume at three sampling locations.

3.2 Traffic speed

Traffic speed distributions are presented in histograms with a class interval of 10 km/h. Fig. 4 displays the spot speed distributions of 300 vehicles during morning, afternoon, and evening peak hours at all sampling locations. Majority of spot speed at Sites 1 and 3 were within 41 to 50 km/h during all peak periods. Meanwhile, most vehicles' speed at Site 2 during the three peak periods was between 31 and 40 km/h.

During the evening peak hour at Site 3, vehicle speed was slightly higher and reached 80 km/h. This might be due to the different lane width at Site 3 compared to Sites 1 and 2. Besides that, all vehicular speed data collected during peak hours for three days were then converted into cumulative frequency distribution. The 85th percentile speed was determined to observe the majority speed selected by the motorists to reach their destination.

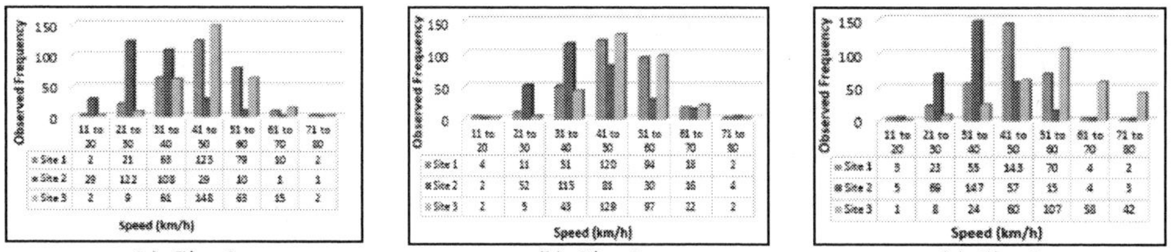

(a) Site 1 (b) Site 2 (c) Site 3

Fig. 4. Histogram of spot speed data for all peak hours.

Based on Fig. 5 to 7, the 85th percentile speed at Site 2 (35.14 km/h) was lower than Sites 1 and 3 (45.50 km/h and 41.45 km/h, respectively). Moreover, all sites recorded a slight difference, except for evening peak hour at Site 3, whereby the 85th percentile speed was 64.98 km/h. This value is higher in comparison to the other sites and could be due to higher traffic volume and wider lane width at Site 3, which allowed drivers to speed up their vehicles. Nonetheless, the traffic speed analysis showed that vehicles at all sites did not exceed the speed limits of 60 km/h for Sites 1 and 2 and 80 km/h for Site 3. Thus, it can be summarised that speed was higher during morning and evening peak hours, whereby most drivers were rushing to reach their workplace or other destinations.

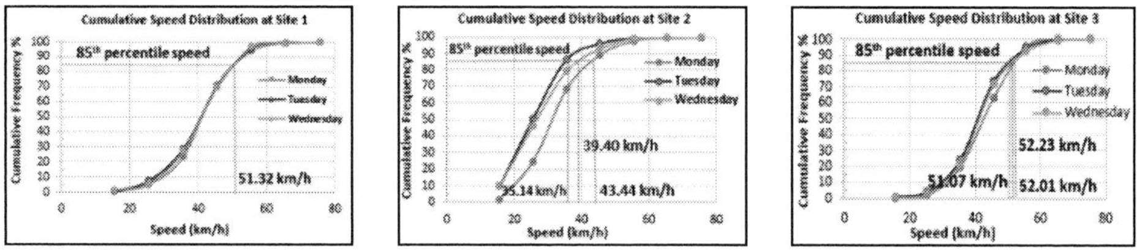

Fig. 5. Cumulative distribution of spot speed data during morning peak hour.

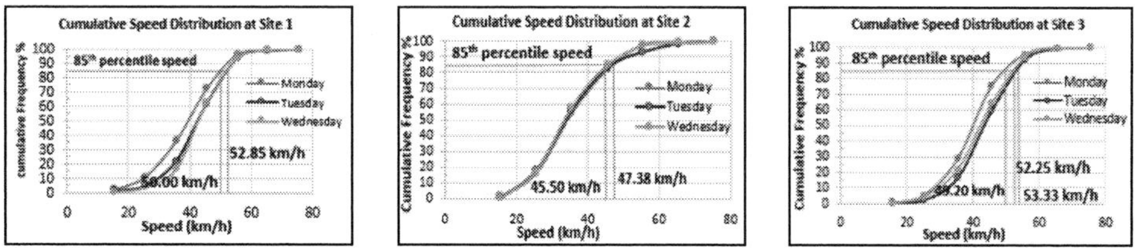

Fig. 6. Cumulative distribution of spot speed data during afternoon peak hour.

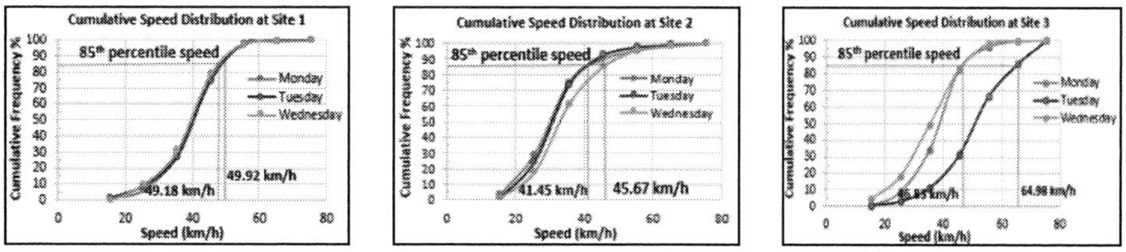

Fig. 7. Cumulative distribution of spot speed data during evening peak hour.

https://doi.org/10.1051/e3sconf/20183402024

3.3 Noise level

According to on-site noise measurement, background noise level at Sites 1 and 2 was 45 dBA whereas at Site 3 it was 48 dBA. Fig. 8 illustrates L_{Aeq} over a 12 hours period at three sampling locations in which many peak noise levels can be seen. Furthermore, most peak noise levels, which were between 60 and 75 dBA, occurred during morning and evening peak hours. Besides that, noise variation from 11 a.m. until 4 p.m. showed several peak noise levels. This scenario indicates that increase in noise level can be associated with certain driver behaviours such as honking, speeding up, and causing other sources of noise.

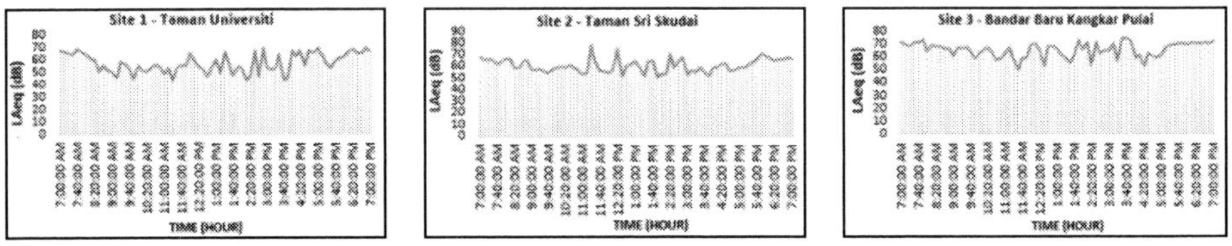

Fig. 8. L_{Aeq} over a 12 hours period at sampling locations.

3.4 Relationship between traffic volume and noise level

Fig. 9 illustrates the range of noise level, with Sites 1 and 2 scoring within 50–70 dBA while Site 3 recorded a higher noise level of 60–70 dBA. In general, the noise level produced was slightly higher during morning and evening peak hours. In addition, it was observed that noise level decreased about 5–10 dBA after morning peak hour between 9 a.m. to 4 p.m. Then, it continuously increased after 4 p.m. until the evening peak hour. Based on these results, noise level decreased as traffic volume decreased and vice versa. These results are in agreement with previous studies whereby loudness of traffic noise generally increased due to heavier traffic volumes. A total of 200 vehicles passing in one hour sound half as loud as 2000 vehicles. Based on previous study, traffic volume is a non-vehicle noise source and one of the major sources of traffic noise [17].

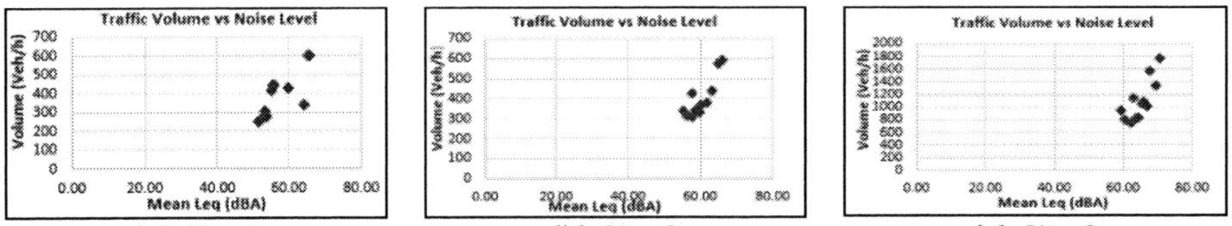

(a) Site 1 (b) Site 2 (c) Site 3

Fig. 9. Traffic volume vs. noise level.

3.5 Relationship between traffic speed and noise level

Fig. 10 shows that noise level increased as speed increased and vice versa. Furthermore, noise level for each sampling location increased during morning and evening peak hours. Noise due to engine transmission becomes a predominant influence at a lower speed. On the other hand, at a higher speed, the main influence would be interaction between vehicle tire and road surface. Moreover, noise level typically increases when a vehicle is accelerating. In this study, maximum vehicle speed for all sites was recorded during the evening peak hours, which were 52.3, 56.1, and 61.5 km/h for Sites 1, 2, and 3, respectively. In addition, highest noise level was observed at Site 3. Hence, this proves that traffic speed plays significant role in increasing traffic noise level. Reducing traffic noise is the most immediate way to reduce traffic noise level by enforcing suitable speed limit.

 https://doi.org/10.1051/e3sconf/20183402024

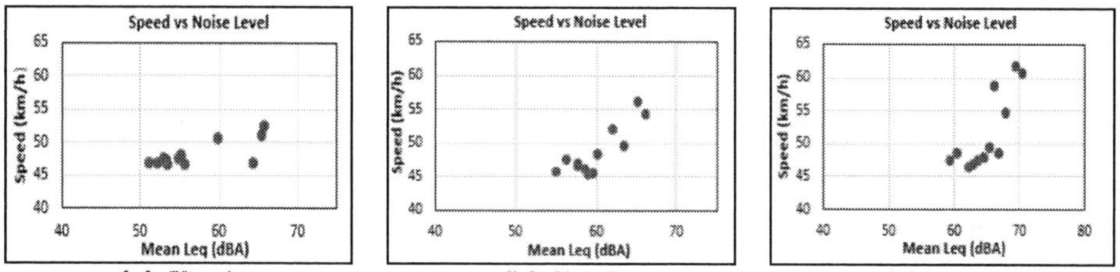

(a) Site 1 (b) Site 2 (c) Site 3

Fig. 10. Traffic speed vs. noise level.

3.5 Traffic noise impact

Table 2 presents the related noise indices and its comparison with DOE and WHO guidelines. L_{Aeq}, L_{NP}, and T_{NI} values at all sites recorded similar values while the highest values were found at Site 3. The L_{Aeq} and L_{NP} values were compared with DOE and WHO limits. Meanwhile, T_{NI} values were compared with the limit of 74 dBA. Based on the results obtained, L_{Aeq} and L_{NP} values for all sites were within the limit of 55 dBA as specified by DOE and WHO. For T_{NI}, all values were lower than 74 dBA, which is the limit for residential area. Therefore, traffic noise impact is not a concern at all sites since measured values were within permissible limits.

Table 2. Noise indices and comparison with guidelines

Time	Site	L_{10}, dBA	L_{90}, dBA	L_{Aeq}, dBA	L_{NP}, dBA	T_{NI}, dBA
				55 dBA		74 dBA
7 a.m. – 8 a.m. (1 hour)	1	66.9	56.4	65.8	76.3	68.4
	2	68.2	54.6	65.2	78.8	79.0
	3	72.5	54.4	69.4	87.5	96.8
8 a.m. – 6 p.m. (10 hours)	1	62.3	54.1	54.4	62.6	56.9
	2	61.5	52.3	57.3	66.5	59.1
	3	64.1	53.7	58.2	68.6	65.3
6 p.m. – 7 p.m. (1 hour)	1	70.4	55.4	65.8	80.8	85.4
	2	65.9	55.0	66.2	77.1	68.6
	3	71.6	59.4	70.4	82.6	78.2
15 hours (DOE)	1	50.7	43.5	45.0	52.2	42.2
	2	49.9	42.2	47.0	54.7	43.2
	3	52.3	43.4	48.1	57.1	49.2
16 hours (WHO)	1	47.5	40.8	42.2	48.9	37.7
	2	46.8	39.5	44.0	51.3	38.7
	3	49.1	40.7	45.1	53.5	44.3

4 Conclusions

This study has assessed traffic noise at three selected residential areas. Traffic characteristics and noise indices data were collected and then analysed according to the study's objectives. It was determined that traffic volume and vehicle speed are the main factors that influence increase of noise level. Site 3 recorded the highest traffic volume compared to Sites 1 and 2. Furthermore, traffic speed at all sites was in the range of 31 to 50 km/h. Other than that, noise level of 60 to 75 dBA was recorded during peak hours. Nonetheless, L_{Aeq}, L_{NP}, and T_{NI} values were lower than the maximum recommended noise limits. Thus, nearby residents' daily activities and health conditions are not affected.

The authors would like to express their deep gratitude to the Ministry of Higher Education, Malaysia and Universiti Teknologi Malaysia for providing financial assistance (Ref: FRGS/1/2016/TK08/UTM/02/2, Cost Centre No.: R.J130000.7822.4F867) and opportunity to carry out this research.

References

1. B. Thagesen, Tech. Uni. of Denmark, F & FN Spon, (1996)
2. D. Pal, D. Bhattacharya, Adv. in Fuzzy Sys **8** (2012)
3. M. J. Damavandi, M. Nowrouzi, Energy and Env. **266** (2012)
4. A. A. Kadar Hamsa, M. Miura, S. Inokuma, Y. Nishimura, J. of Asian Arch. and Build. Eng. **5** (2006)
5. K. Ki Hyun, J. Shamin Ara, K. Ehsanul, J. C. B. Richard, Env. Int **60** (2013)
6. G. Belojević, B. Jakovljević, O. Aleksić, Env. Int **23** (1997)
7. D. Ouis, J. Env. Psy **21** (2001)
8. H. K. Sy, P. P. Ong, S. H. Tang, K. L. Tan. App. Acoustics **18** (1985)
9. International Organization for Standardization, *ISO 1996-1*, Switzerland (1982)
10. International Organization Standardization, *ISO 1996-2*, Switzerland (1987)
11. International Organization Standardization, *ISO 1996-3*,Switzerland (1987)
12. P. D. Marathe, IJED **9** (2012)
13. B. K. Swain, S. Goawami, Int. J. E. Env **4** (2013)
14. L. Bengang, S. Tao, R. W. Dawson, App. Acoustics **63** (2002)
15. Department of Environment, Ministry of Natural Resources and Environment, Malaysia (2004)
16. World Health Organization, Congress on environmental health, OSLO (2000)
17. M. J. Crocker (Ed.) *Noise and Vibration Control.* (New Jersey, John Wiley & Sons, 2007)

E3S Web of Conferences **34**, 02025 (2018) https://doi.org/10.1051/e3sconf/20183402025
CENVIRON 2017

Characteristic of Noise-induced Hearing Loss among Workers in Construction Industries

Ain Naadia Mazlan[1], *Khairulzan* Yahya[1], *Zaiton* Haron[1,*], *Nik Amsharija* Mohamed [2], *Edrin Nazri* Abdul Rasib [2], *Nizam* Jamaludin [2], and *Nadirah* Darus[1]

[1]Faculty of Civil Engineering, Universiti Teknologi Malaysia, 81310 Johor Bahru, Johor, Malaysia
[2]Social Security Organisation, Menara PERKESO, No. 281, Jalan Ampang,50538 Kuala Lumpur

Abstract. Noise-induced hearing loss (NIHL) is among the most common occupational disease in industries. This paper investigates NIHL in construction related industries in Malaysia with particular emphasis on its relation with risk factors. The objectives of this research were to (1) quantify the prevalence of NIHL in construction related industries, and (2) assess the relationship between hearing loss and risk factors and it's characteristic. The study was conducted using 110 NIHL compensation record collected from Social Security Organisation (SOCSO), Malaysia. Risk factors namely area noise, age, temperature, smoking habit, hobby, diabetic and cardiovascular disease were identified and analysed. Results showed that there was no direct relationship between area noise with hearing impairment while there was only low relationship between age and hearing impairment. The range for area noise and age were between 70 to 140 dB(A) and 20 to 70 years, respectively. The other risk factors classified as categorical data and analysed using frequency method. Grade of impairment does not depend solely on area noise but also in combination with age and other risk factors. Characteristic of NIHL prevailed in construction related industries were presented using scatterplots and can serve as a references for future hazard control on site.

1 Introduction

Noise induced hearing loss (NIHL) is one of the most common occupational disease particularly in construction industries. NIHL remains a problem worldwide despite of step taken to reduce occupational noise exposure by the industries, strict standards for hearing protection and extensive public health awareness campaign [1-6]. In United States, Hong (4) found out that there were significant NIHL problem among operating engineers in construction industries in US. Over 60% of a total of 623 operating engineers showed hearing loss in the noise-sensitive higher frequencies of 4 and 6 kHz. Fig. 1 shows the prevalence of noise-induced hearing loss at selected construction trades at United States from 1996 to 2010 [2]. While in Malaysia, 1047 cases of NIHL have been reported with 499 cases involved in construction related industry from 2012 to 2014 [7].

* Corresponding author: zaitonharon@utm.my

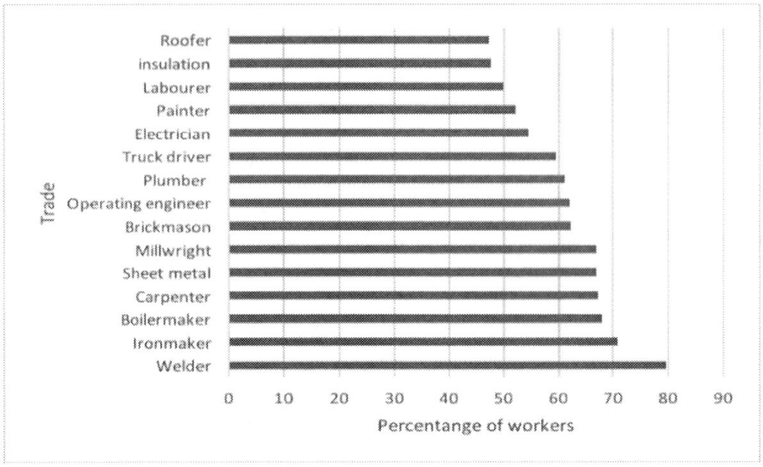

Fig. 1. Prevalence of noise-induced hearing loss at selected construction trades [2].

Noise exposure is an important and highly prevalent occupational hazard in the construction industry. Noise exposures for both ears are converted to binaural hearing impairment percentage [7]. Table 1 shows the grades of hearing impairment and, work performance and recommendation to workers according to binaural hearing impairment percentage. This table was used to obtain the grade of impairment [8]. The grade of impairment criteria are: slight (impairment less than 22.5%), moderate (impairment range between 22.6 to 52.5%), severe (impairment range of 52.6 to 82.5%) and profound impairment or deafness (impairment range more than 82.5%).

Table 1. Grades of hearing impairment [8].

Hearing impairment in %	Grade of impairment	Workers Performance	Recommendations to the workers
< 22.5	Slight	Able to hear and repeat words spoken in normal voice at 1 metre	Counselling. Hearing aids may be needed
22.6 - 52.5	Moderate	Able to hear and repeat words spoken in raised voice at 1 metre	Hearing aids usually recommended
52.6 - 82.5	Severe	Able to hear some words when shouted into better ear	Hearing aids needed. If no hearing aids available, lip-reading and signing should be taught
> 82.5	Profound /deafness	Unable to hear and understand even a shouted voice	Hearing aids may help understanding words. Additional rehabilitation needed. Lip-reading and sometimes signing essential

Beside loud noise exposure, there are other attribute that increases the likelihood of developing a known as risk factors. It includes age, cardiovascular disease, diabetes, smoking, chemical exposure, temperature and non-occupational activity or hobby. According to Factory and Machinery Act 1989 [9], a worker must not expose to noise level greater than 90 dB(A) for 8 hour in order to avoid the risk of developing hearing loss.

Moreover, a worker must not expose maximum instantaneous noise level of 135 dB(A) and 140 dB. Loud noise produce sound wave that destroys the tiny hair cell in the organ corti within the cochlear of the inner ear, and may cause trauma to ears, which results in chronic effect of hearing loss [10,11]. As the workers growing older, the risk of developing hearing loss increase as the inner hair cell deterioration are higher. In addition, if worker have cardiovascular problem the chances of getting hearing loss are more likely due to diminished circulation and narrowing of the blood vessels in the inner ear [12,13]. Cardiovascular disease is also associated with cochlear damage [10,14].

Further, diabetic factor show higher prevalence of hearing loss that is associated with a number of microvascular complication which commonly affecting the eyes and the kidneys. These microvascular abnormalities may also affect the ear and hearing ability. Moreover, diabetes and cardiovascular disease have strong correlated relationship and increase the chances of the risk of hearing loss is marginally higher [15,16]. In the case of workers with smoking habit, carbon monoxide, a common ingredient in cigarette smoke may cause a temporary threshold shift after long-term exposure by reducing the blood's capacity to deliver oxygen to the cochlea. This cause non-smoker shown lower risk of hearing loss compared to current smoker. Meanwhile ex-smoker presented lower risk of hearing loss compared to non-smoker but ironically passive smoker are facing high risk of hearing loss [10,12,14].

A mixed combination of chemical that have ototoxic properties agent such as toluene, styrene, lead, mercury and noise exposure can act in synergistic way. Chemicals in the bloodstream go through either the blood-labyrinth barrier into the cochlea or the blood–brain barrier to reach the eighth cranial nerve and the central nervous system. Ototoxic chemicals may affect the structures and/or the function of the inner ear (auditory and vestibular apparatus) and the connected neural pathways [18,19]. Likewise, aromatic solvents most likely affect hearing through chemical poisoning of hair cells, resulting in disorganization of their membranous structures. Higher/lower temperature in workplace can induce heat stress. Combination with high noise exposure and heat stress cause the rate of deterioration of hearing increase rapidly [20].

Any activity that is deemed not occupational or excluded from work scope description also contributed to total noise exposure. A 'buffer time' is required for the hair cells to return to its original positon in order to overcome the noise exposed in work place [21]. The study investigates NIHL among construction industry workers in Malaysia with particular emphasis on its relation with risk factors. The purpose of this study was to (1) determine the prevalence of noise-induced hearing loss among workers in construction industries, and (2) assess the relationship between hearing impairment percentage and risk factors.

2 Methodology

A retrospective analysis of record of successful claimants for NIHL compensation from Social Security Organisation (SOCSO) were analysed. These claimants' records were obtained from 2012 to 2014 for various type of job and salary less than RM3000.00. The claimants were 110 workers including machine operators, supervisors, mechanics, technicians, and general worker. Leesen et al. [22] used the same method for determination of exploring the relationship between hearing loss and noise intensity, noise exposure time and the use of hearing protection in the Dutch construction industry. In each record, data obtained were audiometric test for both ears, age, noise level or area noise with its temperature, smoking status, diabetic and cardiovascular status, and hobby. Audiometric test for both ears were then converted to binaural percentage of hearing impairment by using the SOCSO guideline [23]. Grade of impairment were obtained by using Table 1. The area noise level and age were classified as non-categorical data and its relationship with hearing impairment were plotted to detect if there were any significant patterns. Others risk factors were categorical data which represented by 'yes' and 'no' and its relatioship with hearing impairment were studied by using frequency analysis method.

3 Results and discussions

3.1 Prevalence of NIHL

Majority of the claimants came from blue-collar workers; operator (23%), supervisor (22 %), mechanic (16 %), technician (11 %), general worker (4 %). They represent 66% of the total sample population (N=110) as their nature of work that involved physical work duty and constantly exposed to noise. This result in agreement with previous research that discovered manual labour workers from construction has high risk of hearing loss [2,24]. Fig. 2 shows the distribution of worker with NIHL and its job types. Machine operator had highest percentage and this condition in good agreement with previous research by Haron et al. [24] that 68% of machine operators were exposed time-weighting average 85 dBA and 18% of them exceeded the permissible limit of 95 dBA. Legris and Pouline [25] also found that the machine or heavy equipment operator had highest noise exposure level.

Meanwhile, Fig. 3 shows the hearing loss obtained from the audiometric test of one of the workers. Analysed by using SOCSO [23] and Table 1, the percentage of hearing impairment and the statistical description of impairment grade is shown in Table 2. It was found that 60.91%, 20%, 17.27% and 1.82% of workers were categorised as slight, medium, severe and medium impairment respectively. The trend shows that the higher the grade of impairment the older the average age of worker but the trend between grade of impairment and average noise area was vice versa.

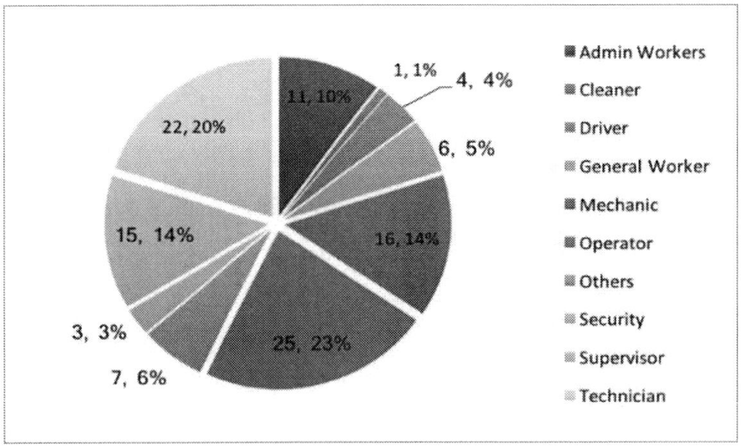

Fig. 2. Distribution of claimant according to type of job.

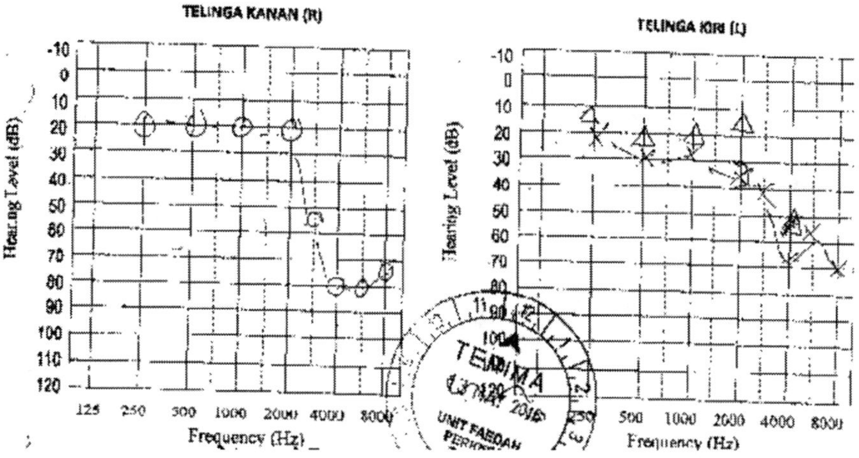

Fig. 3. Audiogram reading for right (R) and left ears (L) used for NIHL compensation claim.

Table 2. Statistical description for non-categorical risk factors.

Impairment grade	Number of workers (Percentage)	Area noise, dB(A)		Age of workers, years	
		Average	Std	Average	Std
Slight	67(60.91)	93.00	9.13	49.27	8.57
Medium	22(20.00)	92.54	9.50	53.45	6.69
Severe	19(17.27)	92.06	12.08	54.11	8.31
Profound/deaf	2(1.82)	89.10	2.12	59.50	7.78

Table 3 shows the percentage of workers with categorical risk factor in each impairment grade group. It is seen that in slight impairment grade there were 12 workers (17.91%) exposed to chemical environment, 19 workers (28.35%) exposed to high temperature, 13 workers had smoking habit (19.40%), 8 workers (11.94%) with diabetic, 4 workers (5.97%) with cardiovascular and 9 workers (13.43%) with risky/active life style as hobby. The data trend also show that sometimes a worker has more than one categorical factors. As the grade become worsen from medium to profound the number of risky factors appear become less. There were six risk factors for slight, five for medium and severe and none for the profound.

Table 3. Percentage of workers with categorical risk factors.

Impairment grade	**Chemical exposure**	**High temperature exposure**	**Smoking habit**	**Diabetic disease**	**Cardiovascular disease**	**Hobby**
Slight	12 (17.91%)	19 (28.35%)	13 (19.40%)	8 (11.94%)	4 (5.97%)	9 (13.43%)
Medium	4 (18.18%)	5 (22.72%)	3 (13.63%)	4 (18.18%)	1 (4.5%)	0
Severe	0	4 (21.05%)	4 (21.05%)	1 (5.26%)	1 (5.26%)	1 (5.26%)
Profound/deaf	0	0	0	0	0	0

3.2 Relationship between NIHL and risk factors

Fig. 4 shows the plot between hearing impairment and workplace area noise environment. The different zone in Y axis shows the grade of impairment criteria. Numbers shown in the plot are the ID number(ID#) for NIHL claimant. In general there were no specific pattern and the range of area noise exposed by the claimant was between 70 to 130 dB(A) but majority of the NIHL suferer are exposed in the range of 80 dB(A) to 100 dB(A) as denoted in circle. However, Figure 5 shows distribution of hearing impairment data according to age. Compared to Figure 4, there were a pattern of relation between percentage of hearing impairment and age of claimant with 12% of age variation defined the increase of hearing impairment (R^2=0.12). Data densely distributed around age 50s to 60s. This is consistent with previous study stating that older age is more prone to NIHL [26]. Case ID# 66 was outlier as he was a young worker that is already partial deaf (on left ear) and not included in regression.

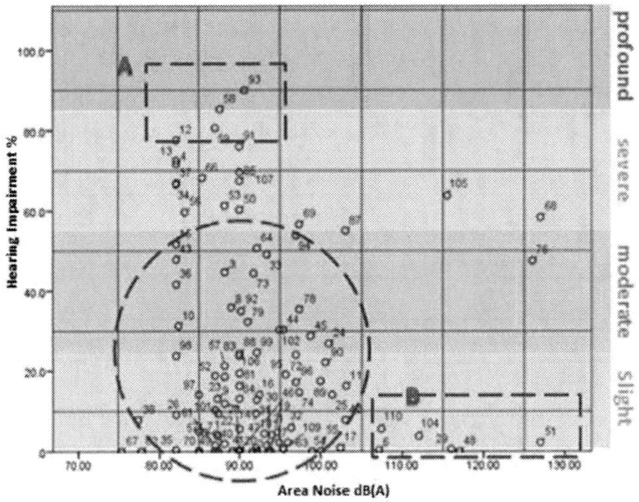

Fig. 4. Characteristic of NIHL: hearing impairment vs area noise.

As explained in Table 2, two sufferers of NIHL (in A zone) were diagnosed as having profound hearing loss who were a supervisor aged 65 (ID# 58) and a technician aged of 54 (ID# 93) and both of them has no hobby that can contribute to the extent of hearing loss. On the other hand, worker with ID# 58 a non-smoker who works for electrical manufacturing exposed to 86 to 93 dB(A) in high temperature, has diabetic and ear infection. Combination of risk factors of workplace condition (loud noise exposure, high temperature), age, and diabetic might be the reason for his high grade of hearing impairment and this was in agreement with previous research [10,12,14]. Meanwhile, ID# 93 worked at factory and exposed only with one risk factor of loud noise environment between 86 dB(A) to 93 dB(A). The same characteristics were observed by Dawes et al. [12]. There were two reasons that explaining these situations; first, NIHL is influence by age related hearing loss (presbycusis). Second, NIHL is a chronic disease that accumulates with time, continuous exposure manifested more profoundly among workers, as they are expose to noise in longer period of time. Table 4 shows the risk factors for selected claimants in profound and slight impairment grades.

 https://doi.org/10.1051/e3sconf/20183402025

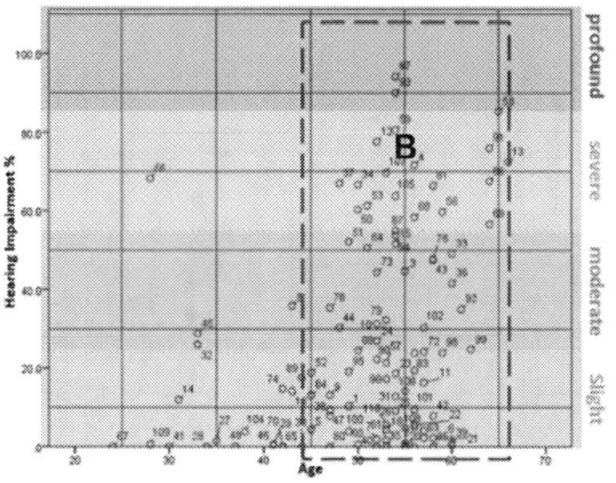

(a) Characteristic of NIHL: Hearing impairment vs age of workers.

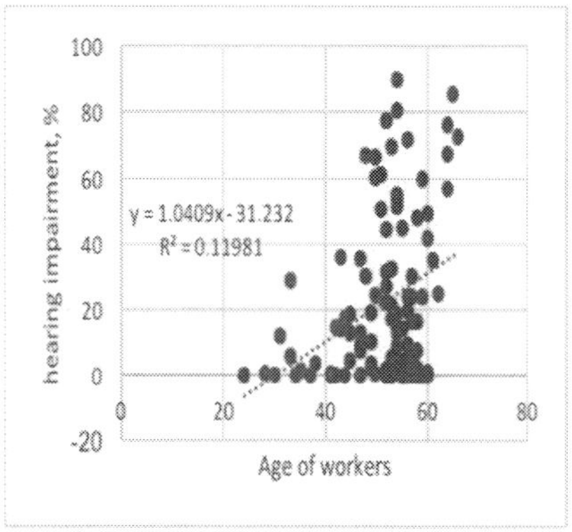

(b) Relation using linear regression.

Fig 5. Hearing impairment versus age.

Study has found 19 of the claimant were fall under the severe grade. The top three on the list were the workers with ID# 12, 59, and 91 (in A zone). No 12 is a driver in construction related company aged 50, had only exposed to average area of 82.1 dB(A). Similarly, worker no. 91 who worked in timber factory exposed only to area noise of 81 dB(A) without other risk factors is but slightly older of 64 years old. This case shows that while exposed to relatively lower noise, but in combination with age factors can lead to hearing deterioration as stated in [12,13]. While worker ID# 59 is also exposed to higher area noise 87.0 dB(A) and with high temperature and has cardiovascular diseases. This is in agreement with [20] that exposure to heat lead to stress and consequently can lead to cardiovascular disease and compounded the hearing loss induced by loud noise.

Further, for slight impairment category, 6 data cases that are of interest which include the workers with ID# 6, 28, 48, 51, 104, and 110 (in B zone) albeit were exposed to loud noise exceeded 100 dB(A). Workers ID# 29 (aged 41 years) worked in high temperature and noisy workplace but the slight impairment may be because he was still young. Worker ID# 48, a 37 years old and has active lifestyle as his hobby while worker ID# 110 , a 53 years old healthy fit but a smoker. Both of them were in agreement with Kolkhorst et al. [27] that there is moderate association of physical fitness and diminished temporary hearing loss experienced after noise exposure. In addition to that, no 110 have been using hearing protection and this is may be the reason for slight impairment grade he had. This implied that alterations in environmental, lifestyle, or other risks can influence severity of hearing loss regardless of age as suggested by [12,26]. Exception was worker ID# 6 who was surprisingly in this slight impairment category despite of his age of 60 years-old. As suggested by [27] NIHL severity depend much on individual susceptibility and so for worker ID# 6.

Thus, it was observed from the above that even for the identical area noise, hearing impairment grade can be different. Differences are more noticeable for cases in the range of 80 to 100 dB(A). In fact, profound grade are located in the 90 dB(A) instead of in high area noise such as at 110 dB or 140 dB(A). This shows the synergistic effect of the combination of the NIHL risk factors [10,28]. Meanwhile, NIHL shows a tendency toward older age despite several younger unconventional cases. Yet, this younger NIHL cases shows the significant effect of risk factors. Most of the high cases show a high influence combination of noise exposure, aging and other risk factors such as active life style or health background (diabetic). The scatterplot results provide simple presentation outlook on the situation occurring on the prevalent of NIHL among construction related industry workers. This scatterplot can be use in the future to help providing better awareness on the severity of NIHL risk factors and also law making occupational site regulation.

Table 4. Record of risk factors for selected workers.

Grade	ID#	Average area noise, dB(A)	Age	High temperature	Smoking	diabetic	Cardio-vascular	Chemical exposure	Hobby
Profound	50	90	65	Yes	No	Yes	No	No	No
	93	90.6	54	Yes	No	No	No	No	No
Severe	12	82.1	50	No	No	No	No	No	No
	59	87.0	54	Yes	No	No	Yes	No	No
	91	81	64	No	No	No	No	No	No
Slight	6	107.1	60	Yes	No	No	No	No	No
	29	116	41	Yes	No	No	No	No	No
	48	117	37	No	No	No	Yes	No	Yes
	51	127	49	No	No	No	No	No	No
	104	112	38	No	No	No	No	No	No
	110	107.4	53	No	Yes	No	No	No	Yes

4 Conclusions

The study was conducted using 110 NIHL record collected from SOCSO. Risk factors such as area noise, age, high temperature, chemical exposure, smoking, hobby, diabetic and cardiovascular are the list that experienced by the claimant of NIHL compensation. There was no direct relationship between area noise with hearing impairment but the range for area noise exposed by claimant were between 70 to 140 dB(A) with more data located in the 80 to 100 dB(A) zone. Meanwhile, there was low relationship between age and hearing impairment with the age range of claimant was between 20 to 70 years old with denser data between 45 to 65 years old. The profound impairment grade were consist of claimant with area noise of 90 dB(A) and age over 50s. There were also case where claimant had similar noise range and age but each case can have different hearing impairment level due to other NIHL risk factors. The scatterplots presented the characteristic of NIHL that prevailed in construction related industries and its can serve as a references for future hazard control on site.

References

1. M. Sliwinska-Kowalska, D. Adrian, Noise and Health **14**, 61 (2012).
2. CPWR,https://www.cpwr.com/sites/default/files/publications/CB%20page%2049.pdf (access, 12/1/18)
3. T. Robinson, J. Whittaker, A. Acharya, D. Singh, and M. Smith, Int J Occup Environ Health **21**, 1 (2015)
4. O. Hong, Int Arch. of Occup. and Environ. Health **78**, 7 (2005)
5. J. Dement, K. Ringen, L. Welch, E. Bingham, P. Quinn, https://dev.btmed.org/PortalDocs/Dement%20AJIM%20DOE%20Hearing%20%20Manuscript.pdf
6. N. S. Seixas, B. Goldman, L. Sheppard, R. Neitzel, S. Norton, S. G. Kujawa, Occup Environ Med **62** (2005)
7. SOCSO. *Noise Induced Hearing Loss Statistic*. Kuala Lumpur (2014)
8. A.N. Mazlan, *PhD Thesis*, UTM (2017)
9. *Factory Machinery Act*, (1989)

10. E. Daniel, Journal of School Health **77**, 5 (2007)
11. N.S. Seixas, R. Neitzel, B. Stover, L. Sheppard, P. Feeney, D. Mills, S. Kujawa, Occup and Environ Med, Oemed-2011 (2012).
12. P. Dawes, H. Fortnum, D.R. Moore, R. Emsley, P. Norman, K. Cruickshanks, K. Munro,. Ear Hear **35,** 3 (2014)
13. M. Seshamani, M.L. Kashima, *Head and Neck Surgery (5th Ed).* Philidaphia: PA: Elsevier Mosby (2010)
14. J. Shargorodsky, S.G. Curhan, R. Eavey, G.C. Curhan, the Laryngoscope **120** (2010)
15. O. Hong, J. Buss, E. Thomas, Disease-a-Month **59**, 4 (2013)
16. H. Sakuta, , T. Suzuki, H. Yasuda, T. Ito, Diabetes Research and Clinical Practise **75** (2007)
17. P. Dawes, K.J. Cruickshanks, D.R. Moore, M. Edmondson-jones, A. Mccormack, H. Fortnum, K. Munro, J of the Association for Research in Otolaryngology **15**, 4 (2014)
18. T.C. Morata, A.C. Johnson, P. Nylen, E.B. Svensson, J. Cheng, E.F. Krieg, J. Franks, Journal of Occupational and Environmental Medicine **44**, 9 (2002)
19. P. Campo, T.C. Morata, O. Hong, Disease-a-Month **59** (2013)
20. C. Chen, Y. Dai, Y. Sun, Y. Lin, Y. Juang, *Heat and Workload Exposure* (2007)
21. R. Neitzel, N. Seixas, B. Goldman, W. Daniell, The Annals of Occup Hyg **48**, 5 (2004)
22. M. C. J. Leensen, J. C. van Duivenbooden, W. A. Dreschler, Int Arch Occup Environ Health **84** (2011)
23. SOCSO, *Occup Diseases and Invalidity*. 2nd Edition. Kuala Lumpur. (2010)
24. Z. Haron, Z. A. Mardhiah, L. M. Han, K. Yahya, Z Jahya, K. Mohd Said, A. A. Saim, Advanced Materials Research, 838-841 (2014)
25. M. Legris & P. Poulin, American Industrial Hygiene Association Journal **59**, 11 (1998)
26. K. J. Cruickshanks, W. Zhan, W. Zhong, The Aging Auditory System (2010)
27. F.W. Kolkhorst, J.J. Smaldino, S.C. Wolf, L.R. Battani, B.L. Plakke, S. Huddleston, L. Hensley, Medicine and Science in Sports and Exercise **30**, 2 (1998)
28. A.L. Edwards, P.Van Coller, C. Badenhorst, Occup Health Southern Africa **16**, 4 (2010)

Impulsivity of Noise due to Single Lightweight Vehicles Transit on Transverse Rumble Strip

N. Darus[1,*], *Z.* Haron[1], *K.* Yahya[1], *M.H.* Abd Halil [1], *W.M.A.* Norudin [1], *M.H.* Othman [2], and *M.A.* Hezmi[1]

[1]Faculty of Civil Engineering, Universiti Teknologi Malaysia, 81310 Johor Bahru, Johor, Malaysia
[2]Faculty of Civil and Environmental Engineering, Universiti Tun Hussein Onn Malaysia, 86400 Parit Raja, Batu Pahat Johor, Malaysia

Abstract. Transverse Rumble Strips (TRS) acts as safety device that alert inattentive drivers from potential dangers. However, the noise produced due to TRS was reported as noise annoyance among the nearby residents lived adjacent to roadways. Thus, this paper investigates the impulsivity characteristic of noise due to single lightweight vehicles transit on TRS. The objectives of this study are to determine the increase of sound level and to evaluate the impulsivity of noise. Two TRS profiles namely middle overlapped (MO) and middle layer overlapped (MLO) were selected. Three types of single lightweight vehicles which include hatchback, sedan and multipurpose (MPV) were tested at speed of 30, 50 and 70km/h. The sound level was measured using sound level meter (SLM). Noise indices such as L_{Aeq}, L_{AIeqT}, L_{AImax}, L_{AFmax} and L_{ASmax} were obtained from the measurement. This study considered the differences of $L_{AImax} - L_{AFmax} >$ 2dBA, $L_{AFmax} - L_{Aeq} \geq$ 10dBA, $L_{AIeqT} - L_{Aeq} \geq$ 2dBA and $L_{AImax} - L_{ASmax} >$ 6dBA to evaluate the impulsivity of noise. It was found that TRS increased the sound level by at most of 6dBA. Furthermore, all single lightweight vehicles transit on TRS show significant impulsive characteristic. These results proved that TRS produce significant impact to the nearby residents.

1 Introduction

Transverse rumble strip (TRS) is one of the configurations of rumble strip installed on roadways apart from centreline rumble strip (CRS) and shoulder rumble strip (SRS). In Malaysia, TRS is commonly installed with various types of profiles such as middle overlapped, middle layer overlapped and raised rumble strip. It is usually installed at residential or urban area, which acts as one of the countermeasures in reducing traffic accident [1]. It alerts the inattentive and distracted drivers from potential dangers by transmitting sound and vibration through a vehicle. The sound and vibration will produce once the vehicle's tire interacts with the TRS surface. Installation of TRS helps to reduce accidents at pedestrian crosswalks, head-on collision, run off road crashes, fatal crashes and other potential accidents [2-6].

* Corresponding author: nddarus@yahoo.com

Several complaints were made by the nearby residents who lived adjacent to the roadways regarding the rumble strips noise annoyance. Thus, attempts had been made to find the extent of increase sound level, L_{Aeq} toward the noise annoyance incur by the rumble strip installations. In the case of CRS, a study reported that noise produced is a trade-off from the safety as it was 60dBA at distance of 60m away from roadway [7]. Furthermore, it was also found that noise produced is enough to be noticed by a normal human at a distance up to 45m away from the CRS [8]. On the other hand, TRS has increased noise level by 7dBA at a distance of 15m [9]. Besides that, installation of TRS at residential areas may generate noticeable noise that can be detected by the nearby residents as much as 4.5 dBA at 70km/h [10].

Beside the increase in sound level, most guidelines suggest the extent of impulsive characteristic of sound in order to examine the extent of noise annoyance. In Malaysia, Department of Environment (DOE) provides a penalty of 5dBA to the measured equivalent continuous A-weighted sound pressure level, L_{Aeq} for estimating annoyance [11]. However, impulsive characteristic of noise produced by the TRS application is very limited. Impulsive characteristic of noise annoys the nearby residents more than the normal continuous noise. It is also must be included in the guidelines by the local authority. Therefore, this study was conducted to investigate the impulsivity characteristic of noise due to single lightweight vehicles transit on TRS by determining the increase of sound level and evaluating the impulsivity of noise.

2 Methodology

2.1 Sound level due to single lightweight vehicles

In this study, middle overlapped (MO) and middle layer overlapped (MLO) were selected. The details of both TRS types were shown in Table 1. The TRS and road surfaces at the selected sites were in good and dry conditions. Both roadways at two sites are straight as required by the ISO 13325 [12]. Meanwhile, Table 2 shows the three types of selected test vehicle such as hatchback (Axia), sedan (Saga) and multipurpose (Alza). These vehicles were selected because they are typical cars used on the roadways in Malaysia. The same test vehicles were used throughout the measurement for both TRS types in order to avoid the effect of vehicle type and road condition [13]. All single vehicles were drove over both TRS types at running speeds of 30, 50 and 70km/h. These running speeds were tested as required by the PWD guideline [14].

A Type 1 Pulsar Model 33 sound level meter (SLM) was used to measure the sound level due these single lightweight vehicles transit on the TRS. The SLM was calibrated prior and after measurement using Model 105 acoustic calibrator in order to avoid reading error. It was set up at 1.5m above the ground level and 7.5m away from the middle of nearest travel lane as shown in Fig. 1. Each site has two measurement points where the distance between Point 1 and Point 2 is 300m which is sufficient for noise dissipation [15]. The measurement was repeated three times for each running speed in order to obtain the average value. The single vehicle test was conducted using controlled pass-by, a method derived from coast-by method [12]. This method can directly measure the tire-pavement noise due to single vehicle in free flowing and at constant vehicle speed.

https://doi.org/10.1051/e3sconf/20183402026

Table 1. TRS profile types.

Types of TRS profiles	Middle overlapped (MO) 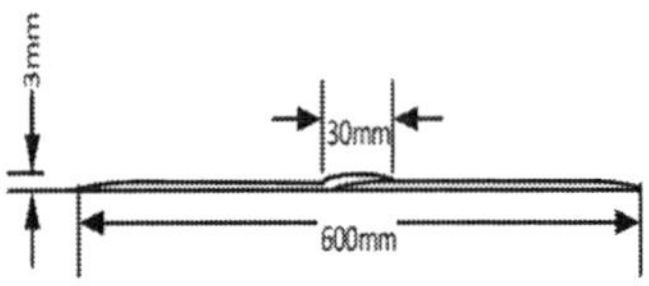	Middle layer overlapped (MLO) 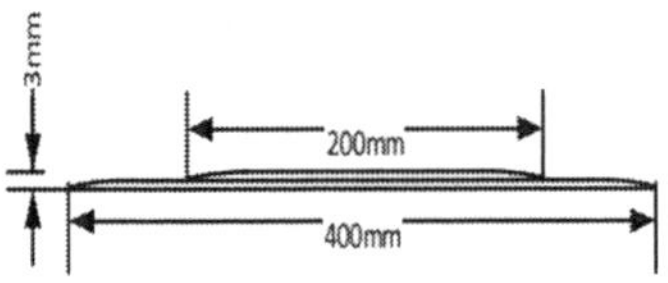
Thickness, mm	3	3
Width, mm	600	400
Spacing, mm	2350	2450
Length, mm	3350	2800
Number of strip	33	30

Table 2. Details of test vehicle.

Types of test vehicle	Hatchback 	Sedan 	Multipurpose 
Model	Perodua Axia	Proton Saga	Perodua Alza
Year	2016	2013	2007
Weight, kg	840	1017	1145
Displacement, cc	1.0	1.3	1.5
Tire Size	175/50 R14	195/55 R15	185/55 R15

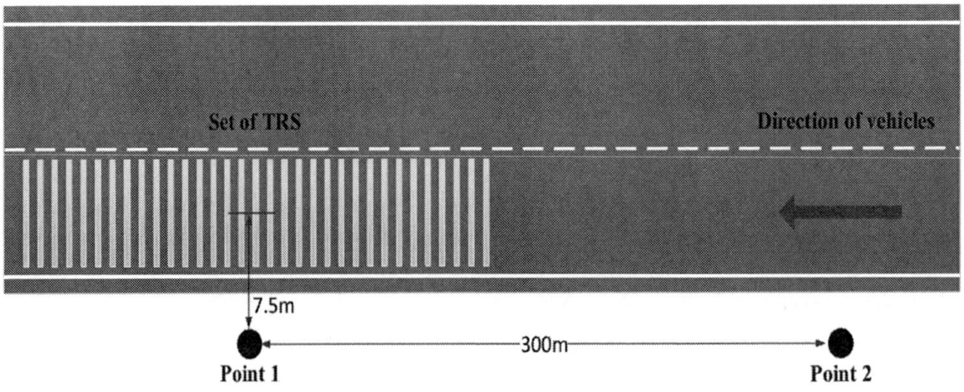

Fig. 1. Measurement layout.

2.2 Impulsivity due to single lightweight vehicles

The impulsive characteristic of noise was detected objectively using the following methods. The first one is by comparing the A-weighted maximum noise level between fast response (L_{AFmax}) with the equivalent A-weighted sound pressure level (L_{Aeq}), difference value of $\geq$10dB [16]. Besides that, the difference between A-weighted impulse sound pressure level, determined with time-weighting characteristic I (L_{AIeqT}), averaged over the same time interval and L_{Aeq} also can detect the impulsivity. A difference value of $\geq$2dB would indicate an impulsive characteristic [17]. Meanwhile, the third method is by comparing L_{AFmax} and A-weighted maximum impulse response (L_{AImax}) with >2dB difference value [18]. The last method is by comparing A-weighted, slow, maximum sound level (L_{ASmax}) with L_{AImax}. If the difference is $\geq$6dB, impulsive characteristic occurs [19].

3 Results and discussions

3.1 Sound level due to single lightweight vehicles

In this study, sound levels due to all types of single lightweight vehicles driven through both types of TRS were recorded by the SLM at Point 1. Fig. 2 to Fig. 7 shows the sound level (L_{Aeq}) versus time history of hatchback, sedan and MPV driven through MO and MLO and its differences with baseline (roadside sound level without TRS). The peak noise levels were shown at 6th second when the vehicle was directly in front of SLM at Point 1. The L_{Aeq} values produced due to single vehicles driven through roadways with and without TRS were presented using solid and dotted lines respectively. It can be observed that MO has significant lower sound level increment with peak level of 2.4dBA (hatchback) as compared to MLO with peak level of 6.6dBA (MPV). Besides that, hatchback and sedan which has lower engine power recorded lower sound level increment as compared to MPV.

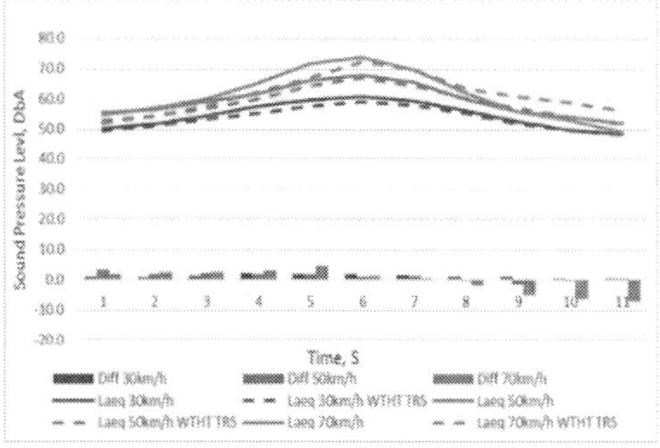

Fig. 2. L_{Aeq} vs. time using hatchback at MO.

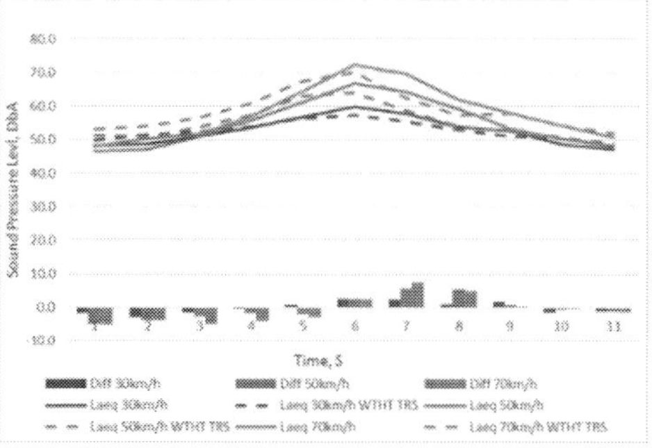

Fig. 3. L_{Aeq} vs. time using hatchback at MLO.

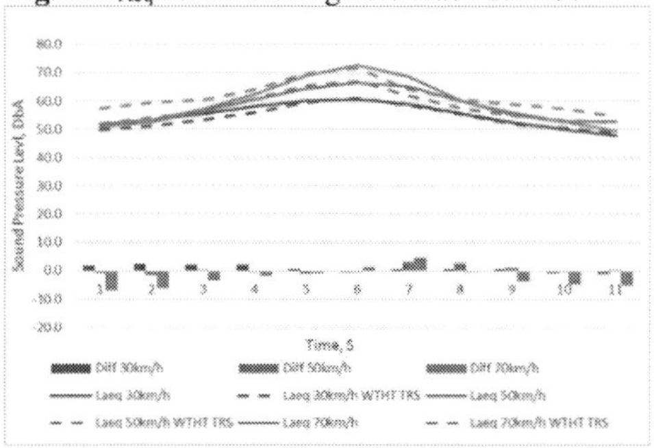

Fig. 4. L_{Aeq} vs. time using sedan at MO.

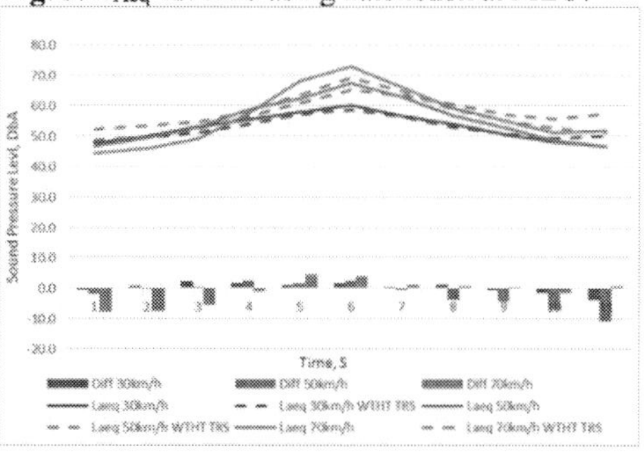

Fig. 5. L_{Aeq} vs. time using sedan at MLO.

 https://doi.org/10.1051/e3sconf/20183402026

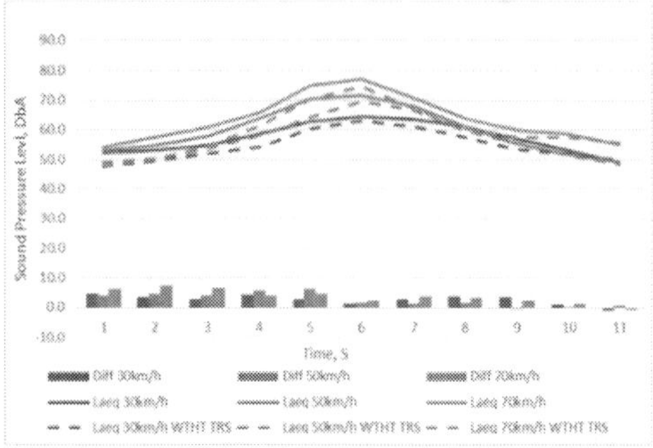

Fig. 6. L_{Aeq} vs. time using MPV at MO.

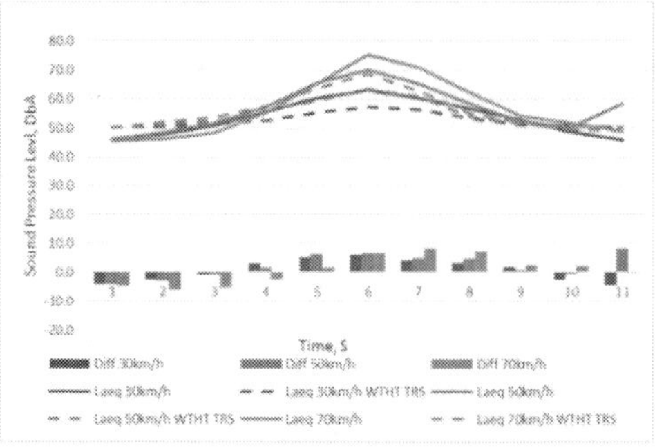

Fig. 7. L_{Aeq} vs. time using MPV at MLO.

Fig. 8 shows the relationship of average increment of L_{Aeq} and vehicle speeds for all vehicle types driven through MO and MLO. It is noted that the average increment of L_{Aeq} values for vehicles driven through MLO are higher than MO. It can be seen that only hatchback driven through both TRS types not recorded any increment for all vehicle speeds. At speeds of 30 and 50km/h, the increments are very low or the same where the values are below 3dBA. However, sedan at speed of 70 km/h and MPV at all three speeds driven through both TRS types recorded increment over 6dBA. There are no significant increments for low power engine vehicles such as hatchback and sedan which can be annoyance to the nearby residents. It shows that the overall average increment of L_{Aeq} values do not give significant impact when measured with L_{Aeq} only.

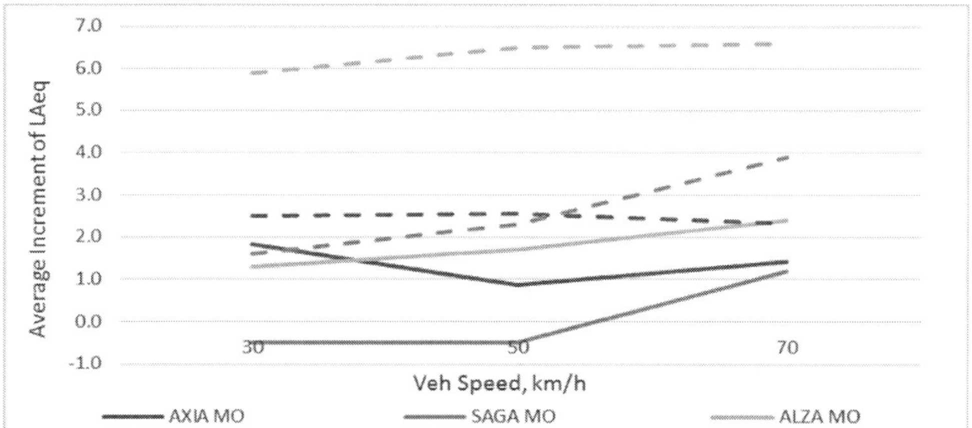

Fig. 8. Average increment of LAeq values versus vehicle speeds.

Other noise indices related to peak L_{Aeq} such as L_{AFmax}, L_{AImax}, L_{AIeqT} and L_{ASmax} were also evaluated. The values were compared with baseline of normal road (L_{AFmaxw}, L_{AImaxw}, L_{AIeqTw} and LA_{Smaxw}) as in Table 3. Some of the values are characterized by an increment for all vehicle speeds. However, most of the values shows the increment at speed of 30 to 50km/h whereas decrement at speed of 50 to 70km/h. This may be due to the values at higher speed were influenced by the sound produced by the power engine of the testing vehicles. Other than that, the values for all noise indices show decrement as compared to baseline at speed of 30km/h. There are also values decrement for hatchback and sedan at noise indices of L_{AFmax}, L_{AImax} and L_{ASmax} as compared to MPV. This can be assumed that the TRS act as a sound damper when vehicles are moving at low speed and power engines.

Table 3. Noise indices.

Vehicle types	TRS types	Speed, km/h	L_{Aeq}	L_{AFmax}	L_{AImax}	L_{AIeq}	L_{ASmax}	L_{Aeq}– L_{Aeqw}	L_{AFmax}– L_{AFmax}	L_{AImax}– L_{AImaxw}	L_{AIeq}– L_{AIeqw}	L_{ASmax}– L_{ASmax}
			dBA									
Hatchback	MO	30	59.9	81.8	84.2	61.4	74.6	2.5	-2.9	-1.0	3.2	-7.8
	MLO		61.0	73.6	77.4	61.8	68.7	1.8	-3.7	-3.5	1.9	-6.1
	MO	50	66.7	83.5	85.2	68.6	79.4	2.5	0.9	2.0	3.2	1.3
	MLO		68.1	86.0	87.3	69.1	83.9	0.9	7.3	6.4	1.2	9.6
	MO	70	72.3	80.2	81.8	74.2	75.1	2.3	4.9	5.3	3.1	2.8
	MLO		73.9	75.3	77.5	75.6	73.1	1.5	-9.3	-9.4	2.6	-5.1
Sedan	MO	30	60.3	81.8	84.2	61.1	74.6	1.6	-3.8	-2.5	1.9	-7.8
	MLO		60.6	74.7	78.6	61.4	69.0	-0.5	-11.6	-9.0	-0.7	-14.3
	MO	50	67.6	83.6	85.2	69.3	80.6	2.3	1.0	2.0	3.5	2.5
	MLO		66.6	86.0	87.3	67.5	83.9	-0.5	7.3	6.4	-1.1	9.6
	MO	70	73.1	82.3	85.5	75.6	75.3	3.9	4.3	6.2	5.9	0.7
	MLO		72.7	75.4	77.5	73.8	73.3	1.2	-9.2	-9.4	-0.3	-4.9
MPV	MO	30	63.1	77.1	79.7	64.3	71.8	5.9	-2.8	-0.9	6.6	-5.4
	MLO		64.5	83.6	84.4	65.3	81.4	1.3	14.3	13.5	1.5	14.4
	MO	50	70.0	77.7	81.3	71.5	69.7	6.5	1.1	3.1	7.5	-2.2
	MLO		71.5	78.6	80.3	72.6	76.3	1.7	5.1	6.2	2.1	4.5
	MO	70	75.2	76.9	78.6	76.6	74.0	6.5	-2.2	-2.1	7.2	0.5
	MLO		77.0	88.4	90.1	78.6	84.8	2.4	-8.1	-7.2	2.6	-8.0

3.2 Impulsivity due to single lightweight vehicles

Fig. 9 to Fig. 12 shows that impulsive characteristics exist for all types of vehicles driven through both TRS types at different running speeds. Based on Fig. 9, most of lightweight vehicles transit on both TRS at lower speed recorded difference values > 2dBA for L_{AImax} –

 https://doi.org/10.1051/e3sconf/20183402026

L_{AFmax}. Similar results were also recorded for $L_{AFmax} - L_{Aeq} \geq 10$dBA as in Fig. 10. Meanwhile, for $L_{AIeqT} - L_{Aeq}$, most of the vehicles recorded values of < 2dBA except for sedan driven through MLO at speed of 70km/h which show no significant impulsivity. However, for $L_{AImax} - L_{ASmax}$, most of the vehicles at different speed driven through both TRS types show impulsivity of > 6dBA. As compared to hatchback and sedan, MPV shows less impulsivity for both TRS types at higher speed. In overall, all single lightweight vehicles transit on TRS show significant impulsive characteristic.

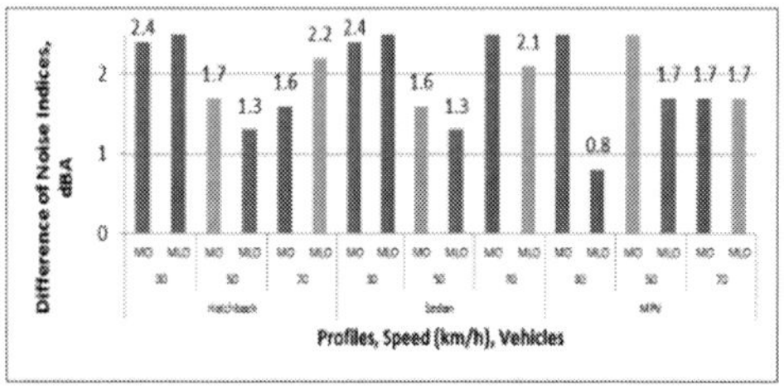

Fig. 9. $L_{AImax} - L_{AFmax}$

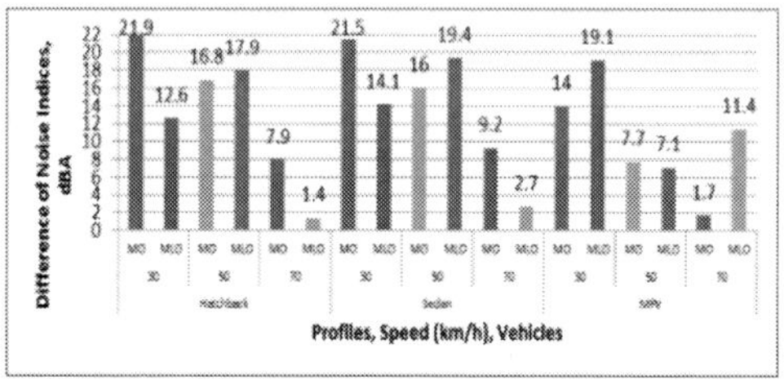

Fig. 10. $L_{AFmax} - L_{Aeq} \geq 10$dBA

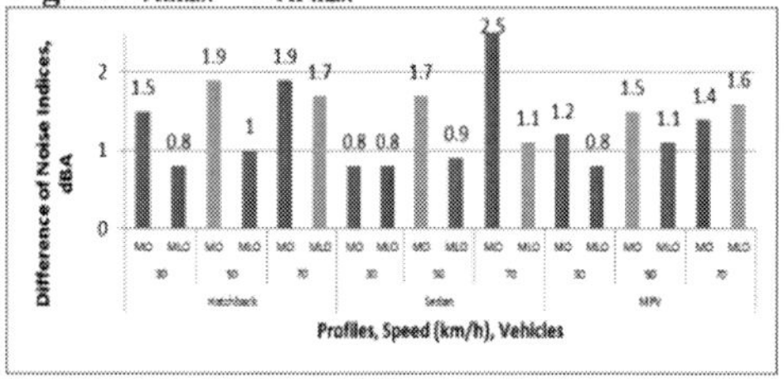

Fig. 11. $L_{AIeqT} - L_{Aeq}$

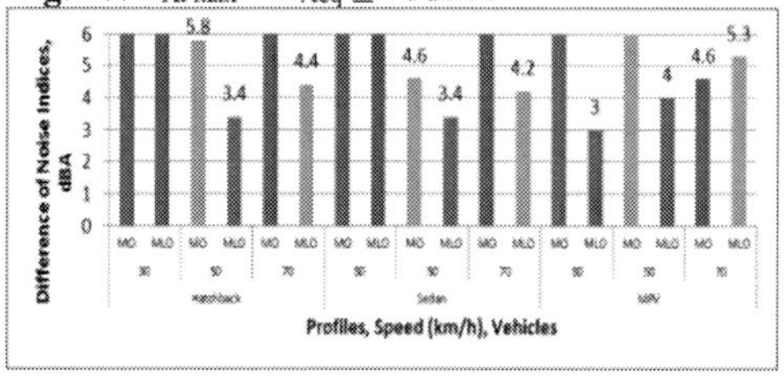

Fig. 12. $L_{AImax} - L_{ASmax}$

4 Conclusions

In conclusion, this study has investigated the impulsivity characteristic of noise due to hatchback, sedan and MPV transit on MO and MLO at speeds of 30, 50 and 70km/h. It was found that MO has significant lower sound level increment with peak level of 2.4dBA (hatchback) as compared to MLO with peak level of 6.6dBA (MPV). Besides that, hatchback and sedan has lower engine power recorded lower sound level increment as compared to MPV. Overall average increment of L_{Aeq} values does not give significant impact when measured with L_{Aeq} only. Based on other noise indices such as L_{AFmax}, L_{AImax}, L_{AIeqT} and L_{ASmax}, it can be assumed that the TRS act as a sound damper when vehicles are moving at low speed and power engines. Finally, all single lightweight vehicles transit on TRS show significant impulsive characteristic, which may affect the nearby residents.

The research was funded by GUP Q.J130000.2522.15H28 and FRGS R.J130000.7822.4F887, which was supported by RMC UTM. These supports are highly appreciated.

References

1. B. N. Persaud, R. A. Retting, C. A. Lyon, Accident Analysis and Prevention **36** (6), 1073 (2004)
2. P. Liu, J. Huang, W. Wang, C. Xu, Accident Analysis and Prevention **43** (6), 1947 (2011)

3. D. J. Torbic, J. M. Hutton, C. D. Bokenkroger, K. M. Bauer, D. W. Harwood, D. K. Gilmore, C. Lyon, *NCHRP Report 641-guidance for the design and application of shoulder and centreline rumble strips* (Transportation Research Board, Washington, D. C., 2009)
4. Federal Highway Administration, *Shoulder and edge line rumble strips* **39** (1), 1 (2011)
5. D. A. Noyce, V. V. Elango, *Safety evaluation of centreline rumble strips: a crash and driver behaviour analysis*, (Transportation Research Board, Washington, D. C., 2004)
6. T. D. Thompson, M. W. Burris, P. J. Carlson, J. of Trans. Research Board, 1 (2006)
7. D. E. Karkle, M. J. Rys, E. R. Russell, F. Asce, J. of Trans. Eng. **137** (5), 311 (2011)
8. R. Makarla, *Master Degree* (Kansas State University, 2009)
9. J. S. Higgins, W. Barbel, J. of the Trans. Research Board **27** (1984)
10. Z. Haron, M. H. Othman, L. M. Hee, K. Yahya, M. R. Hainin, N. Darus, M. S. S. Leong. Archives of Acoustics **42**(3), 449-457 (2017)
11. Department of Environment, *The planning guidelines for environmental noise limits and control* (Ministry of Natural Resources and Environment, Malaysia, 2004)
12. International Organization Standardization, *Tyres-coast-by methods for measurement of tyre-to-road sound emission: ISO 13325* (Switzerland, 2003)
13. U. Sandberg, J. A. Ejsmont, Tyre/road noise reference book (INFORMEX Ejsmont & Sandberg Handelbolaj, Kisa, Swesen, 2002)
14. Public Work Department, *Manual fasiliti keselamatan jalan raya* (Cawangan Kejureteraan Jalan & Geoteknik, Malaysia, 2014)
15. G. Bahar, T. Erwin, M. MacKay, A. Smiley, S. Tighe, *Best practice guidelines for the design and application of transverse rumble strips* (Transportation Association of Canada, Canada, 2005)
16. Environmental Protection Agency, *Guidance note for noise in relation to scheduled activities: 2nd edition* (Environmental Ireland, 2006)
17. International Organization Standardization, *Acoustics-guideline for the measurement and assessment of exposure to noise in a working environment: ISO 9612* (Switzerland, 1997)
18. Department of Environment and Resource Management, *Noise measurement manual: 3rd edition* (Queensland, Australia, 2000)
19. A. Sabato, C. Niezrecki, *Rumble strips noise emission effects on urban road traffic*, (Noise-Con, 2016)

Application of Geographic Information System (GIS) to Model the Hydrocarbon Migration: Case Study from North-East Malay Basin, Malaysia

Rudini[1,*], *Abd Nasir* Matori[1], *Jasmi* Ab Talib[2], and *Abdul-Lateef* Balogun[1]

[1]Civil & Environmental Engineering Department, University Technology of PETRONAS, Seri Iskandar, 32610 Tronoh, Perak, Malaysia
[2]Petroleum Geoscience Department, University Technology of PETRONAS, Seri Iskandar, 32610 Tronoh, Perak, Malaysia

Abstract. The purpose of this study is to model the migration of hydrocarbon using Geographic Information System (GIS). Understanding hydrocarbon migration is important since it can mean the difference between success and failure in oil and gas exploration project. The hydrocarbon migration modeling using geophysical method is still not accurate due to the limitations of available data. In recent years, GIS has emerged as a powerful tool for subsurface mapping and analysis. Recent studies have been carried out about the abilities of GIS to model hydrocarbon migration. Recent advances in GIS support the establishment and monitoring of prediction hydrocarbon migration. The concept, model, and calculation are based on the current geological situation. The spatial data of hydrocarbon reservoirs is determined by its geometry of lithology and geophysical attributes. Top of Group E horizon of north-east Malay basin was selected as the study area due to the occurrence of hydrocarbon migration. Spatial data and attributes data such as seismic data, wells log data and lithology were acquired and processed. Digital Elevation Model (DEM) was constructed from the selected horizon as a result of seismic interpretation using the Petrel software. Furthermore, DEM was processed in ArcGIS as a base map to shown hydrocarbon migration in north-east Malay Basin. Finally, all the data layers were overlaid to produce a map of hydrocarbon migration. A good data was imported to verify the model is correct.

1 Introduction

Geographic Information System (GIS) is a system for storing, managing, analyzing, and visual expressing geographic information [1]. GIS develops rapidly as computer information system to collect and manage the space data, then analyze them to make output of vision available [2].

The Oil and Gas industry is driven by an estimated 80% data that has a spatial component [3]. GIS is able to integrate all the spatial data in addition to manage, inquire,

*Corresponding author: rudini.rudi@gmail.com

analyze and visualize data [2]. The different data layers can be integrated statistically, based on data and/or knowledge-driven approaches, to map petroleum potential [4]. GIS can also produce more accurate output spatial and attribute information. Due to its impressive capabilities, GIS is able to meet the specified requirements in oil and gas industry [2].

Understanding hydrocarbon migration is important since it can mean the difference between success and failure in oil and gas exploration project. In recent years, the use of advanced geophysical methods has become standard in the reservoir characterization. The flow direction of hydrocarbon can be predicted and visualized with modeling hydrocarbon migration. However, many of existing methods are not suitable for large geographical areas of exploration. The hydrocarbon migration modeling using geophysical method is not particularly accurate because of the limitations of available data, since the seismic exploration has limited data covering.

With recent advances in GIS technology, it is possible to establish more accurate migration of hydrocarbon modeling. New studies have been carried out about the abilities of GIS to model migration of hydrocarbon. GIS-Based algorithm was successfully applied for modeling secondary migration of hydrocarbon in Songliao Basin, China [5]. Therefore, the purpose of this study is to model hydrocarbon migration using GIS in North-East part of Malay Basin. The objective of this study is to assess migration of hydrocarbon in North-East Malay Basin using GIS software.

2 Geological Setting

Malay basin is located in east peninsular Malaysia, between latitude 4° - 8° N and longitude 102° - 106° E. Its size is 500 km long and 200 km wide with more than 12 km sediment thickness in basin center [6]. It is located between Penyu and West Natuna Basin to the south and the Pattani Basin to the north as shown in Fig. 1.

The basin is formed during late Eocene-Oligocene to earliest Miocene as a result of tectonic activities in the basin. There are three main phases of tectonic activities in the basin; extensional synrift, thermal subsidence and compressional inversion (Fig. 2). The extensional synrift phase occurred during pre-Miocene era. This caused the basin to pull downward and created normal fault in the Malay Basin. The thermal subsidence occurred during early Miocene. At this phase, the basin is near to sea level, which is proven by the abundance of coal bearing strata interbedded in sand-shale deposition. The process of thermal subsidence is still on going. The last phase is basin inversion, which occurred in middle Miocene era and seems to have continued well in the Pliocene. The basin inversion has resulted in regional uplift and titling of southern part towards the northwest whereby the anticline structure is created from the downward extensional pulling and reactivated the major bonding faults [6].

Most of structural evolution of the southern Malay Basin has been relatively favorable to oil generation and entrapment [6]. Fault trending in Malay Basin can be divided into two parts. At the southeastern (SE) part, the faults are trending in EW and NW-SE directions while at the northwestern (NW) part the faults are trending in N-S direction as shown in Fig. 3.

Malay Basin comprises with two petroleum systems; there are Oligocene-Miocene Lacustrine Total Petroleum System with lacustrine shale source and reservoir and Oligocene-Miocene Lacustrine Total Petroleum System with coal and coaly shale source and fluvial, deltaic, near shore marine and offshore marine bar reservoirs and divided into two groups which are Group H and Group I. The Group I petroleum system is consist of mostly gas system which resulted by Group E and D when reached generation phase of the time of anticlinal traps were form. The major reservoir of Group I system are sands in the shallower unit. Vertical migration of hydrocarbon is the dominant mechanism for charging the relatively shallow reservoir since the Group I source rocks are structurally deeper in basin. Sub-vertical basement-linked fault acted as conduits for hydrocarbon migration. These gases are also like migrate into stratigraphically equivalent formation on the flank. The Group H Petroleum System derives its hydrocarbon from Group H source rock. The location of Group H oil kitchen indicated that the potential traps in Group D, E and B overlying it should be filled with oil. Group H oil is more likely to be preserved on the basin margins where it may migrate laterally via carrier beds. Hence, the main reservoirs for the Group H petroleum system would be in the fault-related traps at the basin margins [6].

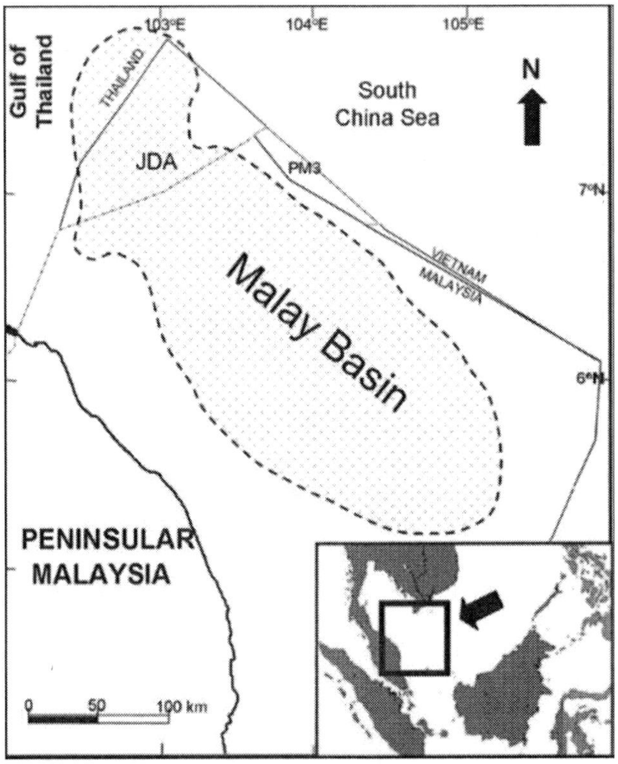

Fig. 1. Location of Malay Basin in South East Asia (inset) [6].

3 Methodology

The concepts, model and calculation of the development of a petroleum system are based on the current geological situation. The current geometry of the lithology and geophysical attributes are determining the current spatial distribution of the hydrocarbon reservoirs. For these reasons, accurate, correct and structure integration of all petroleum-related data and analysis as well as interpretation to model the petroleum environment is crucial to the result of hydrocarbon migration modeling.

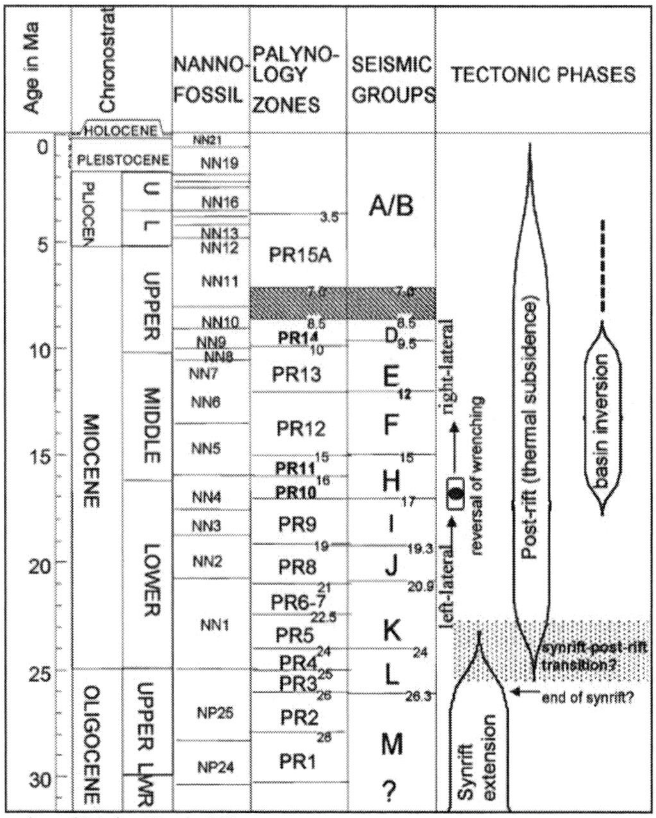

Fig. 2. Stratigraphy of Malay Basin [6].

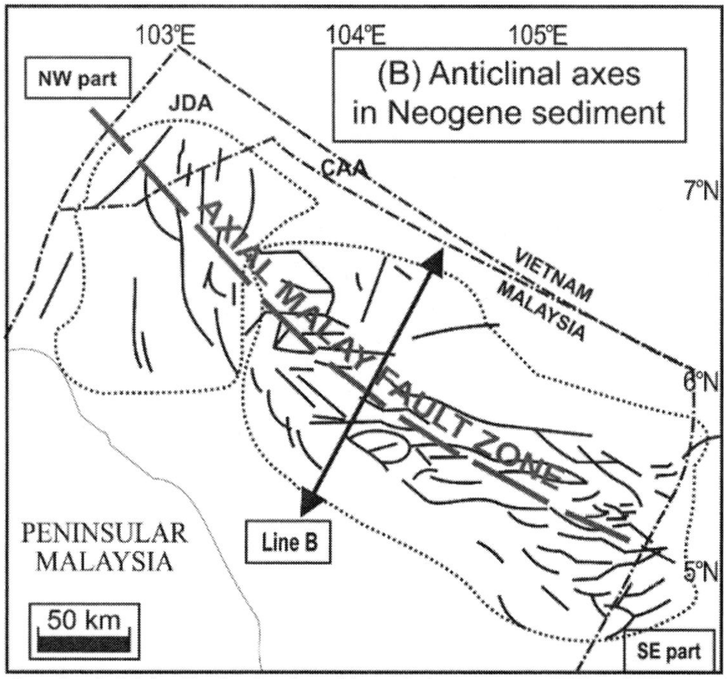

Fig. 3. Fault direction in Malay Basin; the dotted boundaries represent NW and SE part of the Basin [modified from 7].

https://doi.org/10.1051/e3sconf/20183402027

The methodology includes data preparation, data processing in petrel and ArcGIS and visualization. Seismic data, geological data and lithology data were collected for data preparation (Fig. 4). In petrel, Seismic data was interpreted and top of Group E horizon was selected as a basemap before it was exported to ArcGIS for advance processing. The process of searching for hydrocarbon migration is similar to the surface water flow concept. The difference lies in mechanisms and flow direction. The water flow direction is downward because it is driven by gravity force. Meanwhile, the flow direction of hydrocarbon migration is upward because it is driven by buoyancy and capillary pressure [5]. Based on this principle Digital Elevation Model (DEM) was constructed to stimulate the hydrocarbon migration modeling. DEM of Group E horizon was constructed by the result of seismic interpretation in ArcGIS software. Spatial analyst tool such as hydrology tool and 3D analyst tool were used in modeling hydrocarbon migration. Furthermore, 2-Dimensional Map of hydrocarbon migration was produced. In the end, well log data was imported to ArcGIS as an attribute data.

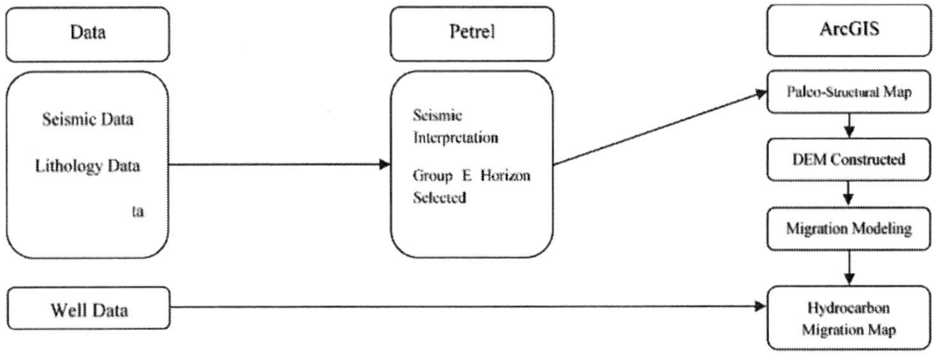

Fig. 4. Research workflow.

4 Results and discussions

Seismic data was interpreted in Petrel Software. Top of Group E horizon was selected as paleo-structural map and exported to ArcGIS software as a basemap for advance processing. In ArcGIS software DEM was constructed from paleo-structural map of Top of Group E horizon to start hydrocarbon migration modeling. Using spatial analyst tool and 3D analyst tool to create model of hydrocarbon migration (Fig. 5).

Fig. 6 shows the migration pathway in the northeast Malay basin, Malaysia from top of Group E formation horizon. The elevation of top of Group E horizon is between 1261 meters to 1692 meters below seafloor. The migration pattern indicates upward movement to highest elevation from lowest elevation. Hydrocarbon accumulated is accumulated in 1350 to 1400 meters below seafloor. Well data was imported to show the location of well and to verify the model is correct. Top of Group E horizon was chosen because the hydrocarbon migrated from lower group and accumulated in group E.

 https://doi.org/10.1051/e3sconf/20183402027

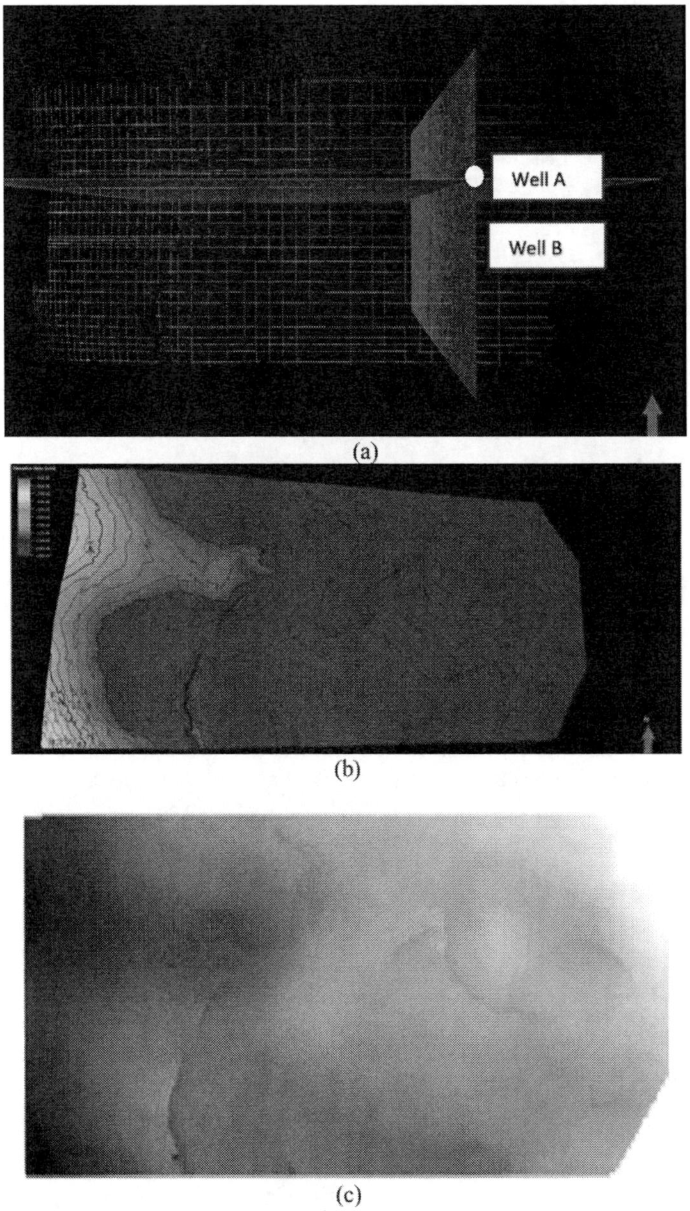

Fig. 5. Seismic data interpretation (a), Top of Group E horizon paleo-structural map (b) and Digital Elevation Model (DEM) of Top of Group E horizon (c).

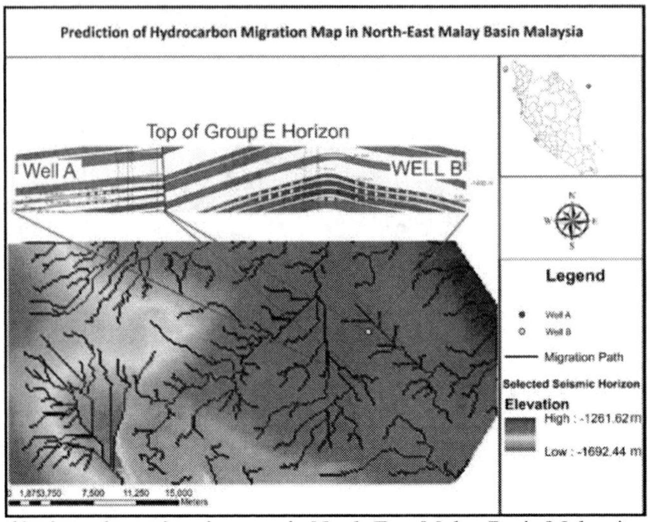

Fig. 6. Prediction of hydrocarbon migration map in North-East Malay Basin Malaysia.

5 Conclusions

GIS was successfully applied to modeling hydrocarbon migration in northeast part of Malay Basin, Malaysia. The hydrocarbon migration is matched with previous study about North-East part of Malay basin. The result shows that with current abilities and advantages of GIS, it is possible to successfully apply GIS to hydrocarbon migration modeling in oil and gas industry.

Thanks to University Technology of Petronas (UTP) to give me the opportunity to study this area.

References

1. D. J. Maguire, **1** (1991)
2. L. Wei, Qinyan, , ESIAT, (2010)
3. S. Deogawanka. https://www.gislounge.com/gis-in-oil-and-gas/ (2016)
4. E. J. M. Carranza, Delft University of Technology, (2002)
5. X. Liu, G. Zhong, J. Yin, Y. He, X. Li, Comp. Geosci. **34** (2008)
6. M. Madon, Y. Jiu-Shan, P. Abolins, A. Redzuan, M. Azmi, Z. S. Bahari, Geo. Soc. of Malay. Bull **49** (2006)
7. M. Madon, P. Abolins, M. J. Hoesni, M. Ahmad (1999)

Assessment of Three Flood Hazard Mapping Methods: A Case Study of Perlis

Nazirah Azizat[1], and *Wan Mohd Sabki* Wan Omar[1*]

[1]School of Environmental Engineering, Universiti Malaysia Perlis, 02600 Arau, Perlis, Malaysia

Abstract. Flood is a common natural disaster and also affect the all state in Malaysia. Regarding to Drainage and Irrigation Department (DID) in 2007, about 29, 270 km^2 or 9 percent of region of the country is prone to flooding. Flood can be such devastating catastrophic which can effected to people, economy and environment. Flood hazard mapping can be used is an important part in flood assessment to define those high risk area prone to flooding. The purposes of this study are to prepare a flood hazard mapping in Perlis and to evaluate flood hazard using frequency ratio, statistical index and Poisson method. The six factors affecting the occurrence of flood including elevation, distance from the drainage network, rainfall, soil texture, geology and erosion were created using ArcGIS 10.1 software. Flood location map in this study has been generated based on flooded area in year 2010 from DID. These parameters and flood location map were analysed to prepare flood hazard mapping in representing the probability of flood area. The results of the analysis were verified using flood location data in year 2013, 2014, 2015. The comparison result showed statistical index method is better in prediction of flood area rather than frequency ratio and Poisson method.

1 Introduction

Malaysia is located in the equatorial region being high temperature and high humidity throughout the year. The annual average of rainfall over Malaysia is 2500 millilitres and average temperature is 27 ^{0}C [1]. Most flood occurred resulting of cyclical monsoon with Southwest and Northwest monsoons. Heavy rainfall during monsoon season caused adversely impact to inundation area depends on magnitude of flood. Massive flooding have a devastating impact on economy and livelihood of people such as properties damage and loss of life. In Malaysia, an annual cost of loss is about US$ 274 million and claim lives of people [2]. Therefore, flood prediction is required to determine the area prone to flood and time of flood occurrence in terms of spatial and temporal. Spatial and temporal dynamic of parameter related to flood were analysed in prediction of where and when flood might occur [3].

Most part of floodplain area, flooding occurs repeatedly almost every year. The development of flood model is important to recognize the susceptible areas with different hazard level for decision makers [4]. Flood hazard mapping can be used to identify a prone

* Corresponding author: wansabki@unimap.edu.my

area to flood using statistical analysis in GIS software. In the previous study, many methods had been proposed to prepare flood hazard mapping primarily on hydrologic, meteorological and geomorphologic approaches [5]. Basically, method of flood mapping was chosen based on data availability and scale of study area. Selection of suitable method for preparation of flood mapping is very important in determination of accuracy of map.

2 Materials and Methods

2.1 Study area

Perlis is located in northern part of Peninsular Malaysia (Fig.1). The area of the state also surrounded by Thailand in the north, Kedah in the south and it western coastline is bordered by Straits of Malacca. Located near the equator, Perlis's weather is normally warm and wet throughout the year. It controlled by yearly fluctuation in position of Inter Tropical Convergence Zone resulting distinct rainy periods and prolonged drought period. Perlis was selected as study area on basis of frequent flooding and availability of data.

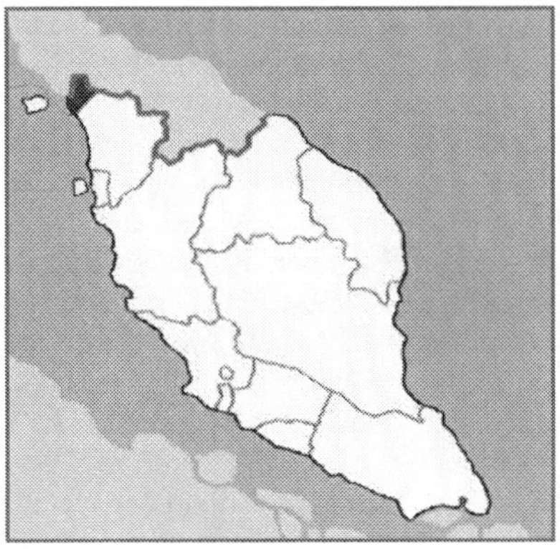

Fig. 1. Location of Study Area.

2.2 Data collection

Flood location map was produced in two set. One of flood location map in Perlis was prepared using November 2010 flood extent from DID in producing of flood hazard mapping (Fig. 1). In other hand, point of flood in year 2013, 2014 and 2015 were used in validation of flood mapping (Fig. 2). In analysis of flood hazard mapping, six spatial databases, which consider flood related factor were constructed including elevation, distance from the drainage network, rainfall, soil texture geology and erosion. Each factor was transformed into a grid spatial database by 30m x 30m cell size and grid of Perlis area was constructed by 929 columns and 1744 rows (1, 458, 158, 400 pixels; 1458 km^2).

A Digital Elevation Model which represent the land surface terrain of the study area was generated from Shuttle Radar Topography Mission (SRTM) which downloaded from United States Geological Survey (USGS) link: https://earthexplorer.usgs.gov/. Distance from drainage networks were calculated using multiple ring buffer in ArcGIS based on annual rainfall. A soil texture, geology and soil erosion were prepared based on field survey by Department of Agriculture.

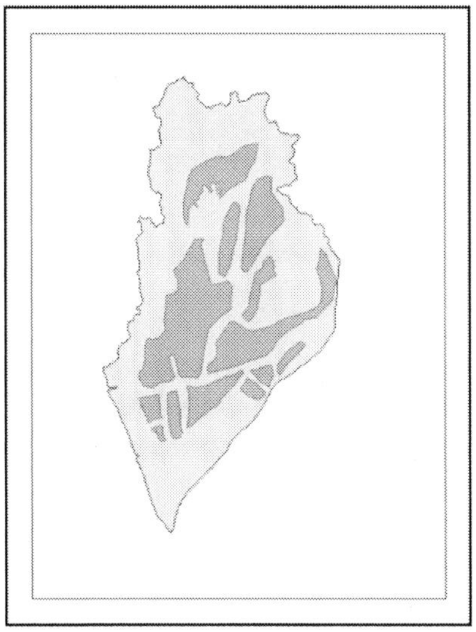

Fig. 2. Flood location map (Analysis).

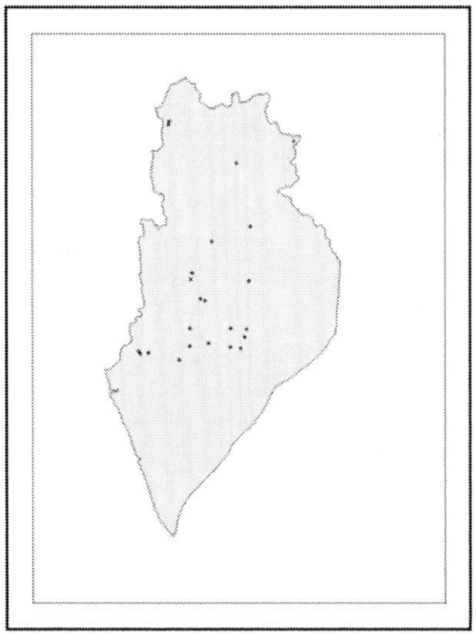

Fig. 3. Flood location map (Validation).

2.3 Frequency Ratio Model and its Application

Frequency ratio (FR) approached are based on relationship between location of flood area and each flood related factor to derived correlation between flood location and factors in the study area. Therefore, the frequency ratio of each factor's class were calculated from flood event. Analysis of relation is the ratio of area where flood occurred by total area. If value is greater than 1, its means a higher correlation and value below than 1 means lower correlation.

Frequency ratio method is used based on assumption that future flood will occurred under similar condition to the past flood using relationships between distribution of flood and each related factor. In calculation of Flood Hazard Index, values of each factor's frequency ratio were summed to produce flood area as in Equation 1:

$$Fr_i = \frac{N_{(Si)} / N_{pix(Ni)}}{\sum N_{pix(Si)} / \sum N_{pix(Ni)}} \quad (1)$$

$N_{no.(Si)}$: Number of pixel of flood area in a certain parameter class,
$N_{pix(Ni)}$: Number of pixel in a certain parameter class
$\Sigma N_{no.(Si)}$: Total number pixel of flood area in a certain parameter class
$\Sigma N_{pix(Ni)}$: Total number of pixel in a certain parameter class

To calculate the flood hazard index, each factor's frequency ratio values were summed using overlay tool in ArcGIS. Flood hazard value represents the relative hazard to flood occurrence resulting greater value is higher the hazard to flood and lower value means lower the hazard of flood. In order to facilitate the map, a map of flood hazard zonation is established by dividing flood hazard index into four different flood hazard zones.

2.4 Statistical index

Statistical index (SI) has close similarity to frequency ratio. In this method, weight of every class of each factor is defined as natural logarithm of the flood density in the class divided by total flood density using formula suggested by Van Westen (1997) [6].

$$\ln W_i = \ln\left(\frac{Densclass}{Densmap}\right) = \ln\left(\frac{N_{bil(Si)}/N_{pix(Ni)}}{\sum N_{bil(Si)}/\sum N_{pix(Ni)}}\right) \tag{2}$$

Wi value of each attribute was calculated. Flood hazard map was produced by overlaying all the layers. Flood hazard index for statistical index calculation is similar with frequency ratio.

2.5 Poisson method

In this method, the probability of flood is 1. Every class for each factor had been calculated in separately using formulae as below:

$$P[X] = \frac{\exp(-\theta)*\theta^X}{X!} \tag{3}$$

X = 1 (Probability of flood in every class)
θ = Average between number of flood in class and total number of flood

The calculation flood hazard index also similar with frequency ratio.

2.6 Validation method

For validation of flood hazard calculation methods, two basic assumptions are applied. The first assumption states that flood are related to spatial information such as elevation, distance from the drainage network, rainfall, soil texture geology and erosion. On the other hand, the second assumption is future flood will be triggered by specific factor such as rainfall. In this study, both are assumptions were satisfied when the flood occurrence is related to spatial information and triggered by heavy rainfall.

Validation was performed by comparing known flood location with flood hazard map. In this study area, 23 point location of flood in year 2013, 2014 and 2014 had been used in validation of three methods. Point of flood were extracted in every zone hazard of flood mapping to compare of accuracy of three methods.

3 Results and Discussions

The relationship between flood occurrence and flood related factors was analysed using frequency ratio, statistical index and Poisson method as shown in Table 1.

https://doi.org/10.1051/e3sconf/20183402028

Table 1. Frequency ratio, statistical index and Poisson method of flood occurrence.

Factor	Class	Number of pixel of flood in class	Total Number pixel in class	FR	SI	Poisson
Geology	Acid intrusives (undifferentiated)	0	5053	0.0000	0.0000	0.0000
	Limestone/marble	10568	137744	0.2459	-1.4026	0.0115
	Phyllite, slate, shale and sandstone: argillaceous rock are commonly carboniferous. Locally prominent development of limestone	138608	380704	1.1671	0.1545	0.1307
	Sandstone/metasandstone with subordinate siltstone, shale and minor conglomerate	0	219	0.0000	0.0000	0.0000
	Continental basin deposits of shale, sandstone, conglomerate and minor coal seams	7899	66061	0.3833	-0.9589	0.0086
	Clay, silt, sand and gravel - undifferentiated (Continental)	39272	51845	2.4283	0.8872	0.0413
	Clay and silt (marine)	87858	261201	1.0783	0.0754	0.0876
	Unclassified	0	8240	0.0000	0.0000	0.0000
Erosion	Moderate	126273	245011	1.6521	0.5021	0.1207
	High	1369	26316	0.1668	-1.7912	0.0015
	Moderately High	92577	292765	1.0137	0.0136	0.0918
	Very High	2294	124322	0.0592	-2.8277	0.0025
	Low	61035	221091	0.8850	-0.1222	0.0627
	Unclassified	657	1562	1.3484	0.2989	0.0007
Soil	Clay	14429	26395	1.7524	0.5610	0.0156
	Heavy Clay	11217	73597	0.4886	-0.7163	0.0122
	Mined Land	1354	6193	0.7009	-0.3554	0.0015
	Sandy Clay	9871	99091	0.3193	-1.1415	0.0107
	Sandy Clay Loam	88019	330590	0.8535	-0.1584	0.0877
	Sandy Loam	7493	18513	1.2975	0.2604	0.0082
	Silty Clay	146356	328542	1.4280	0.3563	0.3077
	Unclassified	3297	12461	0.8482	-0.1647	0.0115
	Lake & Pond	1512	14123	0.3432	-1.0695	0.0036
	Unclassifed	657	1562	1.3484	0.2989	0.0017

Table1 (Continued…)

Factor	Class	Number of pixel of flood in class	Total Number pixel in class	FR	SI	Poisson
Annual rainfall	Unclassified	19	123	0.4952	-0.7028	0.0000
	1418	19181	28197	2.1807	0.7796	0.0206
	1638	21303	87690	0.7788	-0.2500	0.0228
	1660	37017	79147	1.4993	0.4050	0.0390
	1696	18010	48859	1.1817	0.1669	0.0194
	1772	22007	40408	1.7459	0.5573	0.0236
	1780	1799	50475	0.1143	-2.1693	0.0020
	1801	9423	41641	0.7254	-0.3210	0.0102
	1803	35818	83990	1.3671	0.3127	0.0378
	1804	28751	40845	2.2565	0.8138	0.0306
	1810	37393	78848	1.5203	0.4189	0.0394
	1815	4400	96231	0.1466	-1.9202	0.0048
	1820	13745	58907	0.7480	-0.2904	0.0149
	1848	9356	61200	0.4901	-0.7132	0.0102
	1985	25981	72686	1.1458	0.1361	0.0277
	2015	0	41819	0.0000	0.0000	0.0000
Distance from drainage network	500	178031	548719	1.0400	0.0392	0.1607
	1000	39596	108844	1.1661	0.1536	0.0416
	1500	20213	58056	1.1160	0.1097	0.0217
	2000	9550	35172	0.8703	-0.1389	0.0104
	10000	36895	160444	0.7371	-0.3050	0.0389
DEM	-9 - 33	206672	505740	1.3097	0.2698	0.1808
	33 - 98	77106	287275	0.8602	-0.1506	0.0778
	98 - 206	480	59284	0.0260	-3.6516	0.0005
	206 - 354	0	38918	0.0000	0.0000	0.0000
	354 - 717	0	19825	0.0000	0.0000	0.0000

Flood hazard mapping was generated using FR method represent the summation of frequency for each classes of each factor (Fig. 4). Within FR analysis, the relationship between flood occurrence and classes of each factor were obtained. Regarding to the result from Table 1, geology has significant factor which influence of flood based on water behavior on the ground [7]. According to the result of geology, areas with clay, silt, sand and gravel are most influence on flooding in Perlis. Erosion is one of the huge contributors on flood disaster. Erosion is natural process that reshape of land and transporting soil. In the case of soil texture, most of flood occurred in soil types of clay. Rainfall is a main factor in triggering of flood in Perlis. The heavy rainfall makes the river overflow the bank river resulting temporarily covers the land by water [4]. The highest annual rainfall showed rainfall is significantly factor, which trigger of flood. Moreover, the areas with frequent flood occurrence were located near to the drainage networks. Urbanization aggravates flooding problem due to the increased runoff and sediment transporting capacity thus causing perturbations to river equilibrium [8]. The result of elevation showed the lower elevation was the most influential class on flooding. Flood hazard mapping was produced based on three methods are shown in Fig. 4, Fig. 4 and Fig. 6.

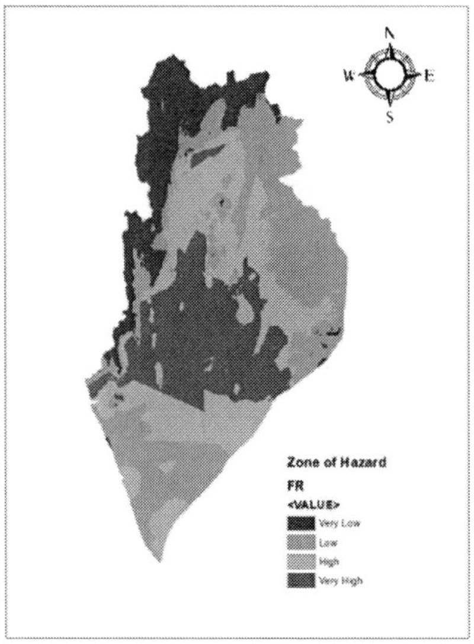

Fig. 4. Flood Hazard Mapping based on FR.

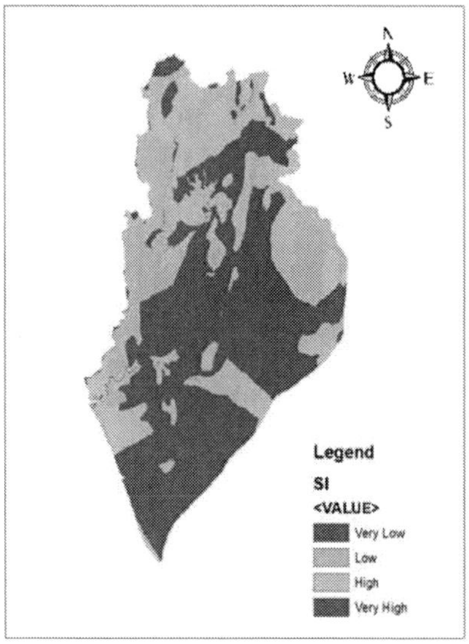

Fig. 5. Flood Hazard Mapping based on SI.

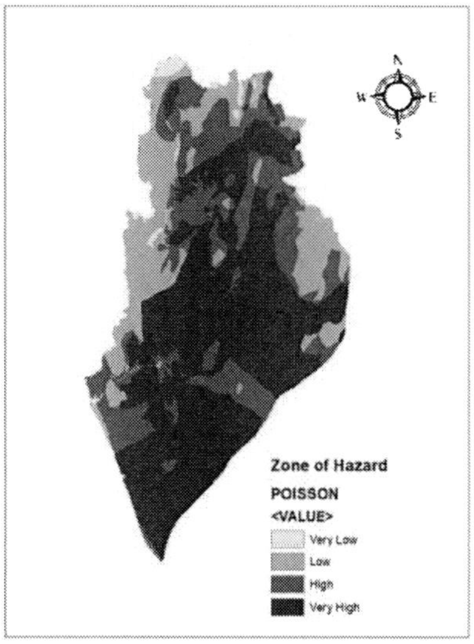

Fig. 6. Flood Hazard Mapping based on Poisson.

In this study, the validation result for both frequency ratio and statistical index method had shown prediction accuracy of 86.96 %. However, statistical index is better method than frequency ratio because 18 point location in very high of hazard zone meanwhile frequency ratio is 16 location.

Table 2. The accuracy of three methods.

Method	Number of flood in high and very high zone	% of flood in high and very high zone
Frequency Ratio	20	86.96
Statistical Index	20	86.96
Poisson Method	12	52.17

4 Conclusions

The comparison of three methods showed the statistical index method is best prediction accuracy in Flood Hazard Mapping. Therefore, the statistical method and frequency ratio in this study are valid for generalized planning and assessment purposes. Flood hazard mapping can be used as a great deal of importance for development of land. The findings of this study are intended to help all those involved in land use planning, development and management of land.

References

1. H. K. Hong, (2001)
2. N. W. Chan, 239-265 (2015)
3. Y. Hu, J. Huang, Y. Du , P. Han and W. Huang, Remote Sens. **7**, 494–520 (2015)
4. M. S. Tehrany, B. Pradhan and M. N. Jebur, J. Hydrol. **504**, 69–79 (2013)
5. L. T. K. Ho, M. Umitsu and Y. Yamaguchi, Remote Sens. Spat. Inf. Sci. **38**, 275–80 (2010)
6. K. L. K. Wern, T. L. Tien and H. Lateh, IEEE Int. Conf. Imaging Syst. Tech. 273–8 (2011)
7. M. S. Tehrany, B. Pradhan, S. Mansor and N. Ahmad Catena. **125**, 91–101 (2015)
8. P. Y. Julien, A. Ab. Ghani, N. A. Zakaria, R. Abdullah and C. K. Chang, J. Hydraul. Eng. **136**, 251–61 (2010)

Optimisation of Copper Oxide Impregnation on Carbonised Oil Palm Empty Fruit Bunch for Nitric Oxide Removal using Response Surface Methodology

Norhidayah Ahmad[1], *Sing Hung* Yong[1], *Naimah* Ibrahim[1*], *Umi Fazara* Md Ali[1], *Fahmi* Muhammad Ridwan[1] and *Razi* Ahmad[1]

[1]School of Environmental Engineering, Universiti Malaysia Perlis, 02600 Arau, Perlis, Malaysia

Abstract. Oil palm empty fruit bunch (EFB) was successfully modified with phosphoric acid hydration followed by impregnation with copper oxide (CuO) to synthesize CuO modified catalytic carbon (CuO/EFBC) for low-temperature removal of nitric oxide (NO) from gas streams. CuO impregnation was optimised through response surface methodology (RSM) using Box-Behnken Design (BBD) in terms of metal loading (5-20%), sintering temperature (200-800°C) and sintering time (2-6 hours). The model response for the variables was NO adsorption capacity, which was obtained from an up-flow column adsorption experiment with 100 mL/min flow of 500 ppm NO/He at different operating conditions. The optimum operating variables suggested by the model were 20% metal loading, 200°C sintering temperature and 6 hours sintering time. A good agreement (R^2 = 0.9625) was achieved between the experimental data and model prediction. ANOVA analysis indicated that the model terms (metal loading and sintering temperature) are significant (Prob.>F less than 0.05).

1 Introduction

Nitric oxides (NO_x) is one of the major air pollutants mainly contributed from fuel combustion process in stationary and mobile sources [1-2]. NO_x causes severe environmental problems such as photochemical smog, and acid rain which are harmful to both human and environment [3]. One of the widely used methods to control NO_x is selective catalytic reduction (SCR). SCR process involves the use of catalyst system like vanadium supported on titanium oxide with external reducing agents such as NH_3 or urea to convert NO_x into N_2 and H_2O. The catalytic reaction is effective at temperature higher than 350°C.

The use of carbonaceous materials derived from agricultural wastes in SCR of NO_x is an alternative technique that allows low temperature (100-300°C) application. Extensive study on carbonaceous materials for NO reduction is possibly due to the material abundance and cost. Besides, carbonaceous material derived from agriculture waste shows

* Corresponding author: naimah@unimap.edu.my

high porosity and surface area which are beneficial for removal of NO_x [4]. Such materials include olive stone [5], rice husk [6], orange skin [7], palm shell [2], plum stone [8], and coconut shell [9]. In Malaysia, oil palm biomass including empty fuit bunch (EFB) is available in abundance. Hence it is very useful and economical to utilize EFB as the carbon precursor in this study.

Impregnation of additives e.g. metal catalyst on the carbonaceous material's surface is another step to enhance NO_x reduction. Copper oxide (CuO) for example was used in many studies including our previous work as it is inexpensive and exhibits high reactivity towards NO reduction [10-12]. The variables chosen in the process of CuO impregnation onto the carbonised EFB (CuO/EFBC) may affect the effectiveness of NO_x reduction, therefore, this study aims to optimise CuO impregnation using response surface method, specifically using Box–Behnken design (BBD). In this independent quadratic design, the treatment combinations are at the midpoints of edges of the process space and at the centre. BBD is favourable because less number of experimental runs is required [4].

2 Experimental

2.1 Material preparation

Fibrous oil palm EFB sample was collected from United Oil Palm Mill Sdn. Bhd., Nibong Tebal, Malaysia. The collected EFB fibre was washed with water and dried in oven at 110°C for 24 h. The EFB was first carbonized and activated using one-stage chemical activation with phosphoric acid (H_3PO_4) at 1:1 (w/v) ratio and heated in the furnace for 24 h at 400°C to create pores, as described in previous work [10]. The heated EFB was then washed with distilled water to remove the remaining acid. Each 10 g of the carbonized material (EFBC) was further impregnated with CuO catalyst at varied loadings. Metal loading was as suggested by RSM based on minimum and maximum set of values obtained from the literature review. Then, 50 mL of distilled water was added to the mixture of EFBC and CuO and stirred with the magnetic stir bar for 4 hours. The mixture was then heated by double boiling method at 70°C until the liquid part evaporated. After dried, the samples were sintered in the furnace at temperature and time suggested by RSM based on fixed minimum and maximum set of values. Finally, the resulting sample of CuO/EFBC was weighed and sieved to gain average particles sizes in the range of 0.5 – 1 mm for further testing.

2.2 NO removal experiment

NO adsorption experiment was explained in detailed in the previous work [10], except that the operating parameters i.e. sample loading, gas concentration and reaction temperature were changed to 5 g of CuO/EFBC, 500 ppm NO/He and 200°C. The total gas flow rate was maintained at 100 mL/min, but passed through the packed bed reactor in an up-flow manner. Outlet NO concentration was measured by NO-NO_2 analyser (TESTO 340, Germany) at the end of the system for one hour. Differential amount of NO removed from gas stream and adsorbed by the CuO/EFBC catalyst system was measured in an experimental set-up shown in the same previous work. For the model response, NO removed from the gas streams or adsorbed by CuO/EFBC was calculated by the following equations (Eqs. 1 and 2). In this case, q was calculated using experimental NO concentration data from breakthrough curve at time equivalent, t_t, which can be obtained by integrating the fraction of NO removed ($1\text{-}C/C_o$) per minute using trapezoidal method (Eq.1) [2].

$$t_t = \int_0^\infty 1 - \frac{C}{C_o} \quad (1)$$

$$q = \frac{C_{NO} Q_f t_t y_f}{m_c} \quad (2)$$

where q is adsorption capacity of NO (mg/g), C_{NO} is NO concentration (mg/L), Q_f is gas flow rate (L/min), t_t is time equivalent (min) calculated using Eq. 1, y_f is mole fraction of NO, and m_c is mass of adsorbent (g).

2.3 Design of Experiment

2.3.1 Experimental design

CuO impregnation variables studied include (i) x_1, metal loading; (ii) x_2, sintering temperature and (iii) x_3, sintering time, while the performance indicator or response is the NO adsorption capacity. The minimum (-1), centre (0) and maximum (1) set of values for the three variables given in Table 1 were selected based on literature and preliminary studies. A total of seventeen (17) experimental runs were suggested by RSM. The sequence of experimental run was randomized in order to minimize errors due to uncontrollable factors.

Table 1. Independent variables and their coded levels for the Box Behnken Design.

Variables			Coded variables level		
	Code	Unit	-1	0	1
Metal loading	x_1	%	5	12.5	20
Sintering temperature	x_2	°C	200	400	800
Sintering time	x_3	hour	2	4	6

2.3.2 Model development

The response variables (i.e. NO adsorption capacity) obtained from the seventeen experimental runs were used to develop an empirical, quadratic model that correlates the response to the three factors, by using a second degree polynomial equation [13] given by Eq. 3.

$$Y = \beta_0 + \beta_1 x_1 + \beta_2 x_2 + \beta_3 x_3 + \beta_{12} x_1 x_2 + \beta_{13} x_1 x_3 + \beta_{23} x_2 x_3 + \beta_{11} x_1^2 + \beta_{22} x_2^2 + \beta_{33} x_3^2 + \varepsilon \quad (3)$$

where Y is the predicted response, β_0 is model constant, x_1, x_2, and x_3 are independent variables, β_1, β_2 and β_3 are linear coefficients, β_{12}, β_{13} and β_{23} are cross-product coefficients, β_{11}, β_{22} and β_{33} are the quadratic coefficients and ε is the statistical error term.

2.3.3 Statistical and graphical analysis

The correlation coefficient value (R^2), Fisher value (F-value) and probability (P-value) from analysis of variance (ANOVA) was used to evaluate the quality and significance of the model and model terms. Graphs were employed to analyse the combined effect of factors on the response using 3D plot and to analyse the predicted response variables versus experimental value.

2.3.4 Model optimisation and validation

Model optimisation was done to determine the optimum operating conditions for the process variables under consideration. To achieve this, goals were set based on the combined effect of factors on responses. Model validation was carried out by conducting NO removal experiment using CuO/EFBC, prepared based on optimum operating conditions suggested by the model. The experimental values (NO adsorption capacity) obtained were compared with the model predicted values.

3 Results and discussions

3.1 Development of model for optimisation of CuO impregnation

The response in term of NO adsorption capacity obtained from the seventeen experimental runs carried out according to operating variables suggested by RSM is presented in Table 2, together with the experimental design variables mentioned in Section 2.3.1.

Based on the data in Table 2, a quadratic model was developed using RSM. The final empirical model for NO adsorption was then obtained in terms of coded factors (parameters) as presented in Eq. 4.

$$Y_1 = 6.09672 \text{ x}10^{-3} - 1.64944 \text{ x}10^{-3} x_1 + 2.62639 \text{ x } 10^{-3} x_2 - 5.67000 \text{ x } 10^{-4} x_3 - 1.55556 \text{ x } 10^{-3} x_1x_2 + 1.01667 x_1x_3 - 4.79167\text{x } 10^{-7} x_2x_3 + 9.39556 \text{ x } 10^{-6} x_1^2 + 8.72222 \text{ x } 10^{-10} x_2^2 + 8.9 \text{ x } 10^{-5} x_3^2 \quad (4)$$

The fitness of the model developed was evaluated based on the correlation coefficient value (R^2) obtained from linear model extrapolation showed in Fig. 1. The R^2 value closer to unity is desirable as an indication that predicted values are closer to the actual values [14]. The R^2 value is relatively high at $R^2 = 0.9625$, indicating that there is good agreement between the adsorption capacity from experimental data and model prediction.

3.2 Statistical and graphical analysis of model

The adequacy of the model was further evaluated through analysis of variance (ANOVA). Table 3 shows the analysis done on the quadratic model for NO adsorption capacity (Y_1). Model terms, x_1, x_2, x_3 refer to metal loading, sintering temperature and sintering time.

Table 2. Experimental design matrix for optimisation of CuO impregnation onto EFBC.

Experimental Run	Variables			Response
	x_1 Metal loading (%)	x_2 Sintering temperature (°C)	x_3 Sintering time (hr)	Y Adsorption Capacity (mg/g)
1	5.0	200	4	0.00504
2	20.0	200	4	0.00614
3	5.0	800	4	0.00531
4	20.0	800	4	0.00501
5	5.0	500	2	0.00548
6	20.0	500	2	0.00576
7	5.0	500	6	0.00524
8	20.0	500	6	0.00613
9	12.5	200	2	0.00482
10	12.5	800	2	0.00539
11	12.5	200	6	0.00559
12	12.5	800	6	0.00501
13	12.5	500	4	0.00466
14	12.5	500	4	0.00479
15	12.5	500	4	0.00470
16	12.5	500	4	0.00475
17	12.5	500	4	0.00494

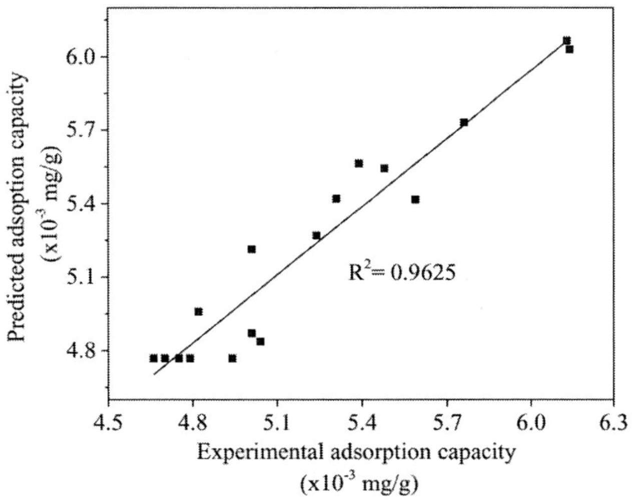

Fig. 1. Experimental and predicted values of NO adsorption capacity onto CuO/EFBC.

For the model terms, Prob. > F less than 0.05 indicates that the model term is significant [15-16] and the largest F-value indicated that the model term has the most significant effect on the response [17]. Based on Table 3, the quadratic model for adsorption capacity has an F-value of 16.24, indicating that this model is significant. In the case of model terms, x_1, x_1x_2, x_2x_3, x_1^2 and x_3^2 can be considered significant with Prob. > F values less than 0.05. However, the rest of the model terms, x_2, x_3, x_1x_3, x_2^2 are insignificant to the adsorption capacity response. The model term which has the most significant effect on the response is x_1^2 with an F-value of 50.64 and the overall effects are in the order of $x_1^2 > x_3^2 > x_1x_2 > x_1 > x_2x_3$.

Based on the results in Table 3, it is also found that the combined effects due to interactions between metal loading and sintering temperature, as well as sintering temperature and sintering time are significant to the adsorption capacity. The effect of these model terms interactions on NO adsorption capacity are shown in graphical form in Figs. 2 (a) and 2(b) using three-dimensional response surface plots.

Fig. 2(a) shows that increasing metal loading from 5 to 20% (at a constant 200°C sintering temperature) enhances the adsorption capacity of NO from 0.00504 to 0.00614 mg/g. However, by keeping metal loading at 20%, an increase in the sintering temperature decreases the adsorption capacity from 0.00614 to 0.00501 mg/g. At the same metal loading of 5%, decreasing the sintering temperature slightly reduces the adsorption capacity from 0.00531 to 0.00504 mg/g. The interaction between high metal loading (20%) and low sintering temperature (200°C) appears to be more effective in enhancing NO adsorption capacity. This result is in agreement with the work done by Ryu et al. [18] which reported that as the metal loading increased, the adsorption capacity increased because the surface sites for reaction increased with the increase in metal loading. Zhu et al. [19] also reported similar finding to this work.

Table 3. ANOVA analysis of NO adsorption capacity by CuO/EFBC.

Source	Sum of squares	Degree of freedom	Mean Square	F-Value	Prob.>F
Model	3.394×10^{-6}	9	3.771×10^{-7}	16.24	0.0006
x_1	4.851×10^{-7}	1	4.851×10^{-7}	20.89	0.0026
x_2	9.461×10^{-8}	1	9.461×10^{-8}	4.07	0.0833
x_3	3.380×10^{-8}	1	3.3800×10^{-9}	1.46	0.2668
x_1x_2	4.900×10^{-7}	1	4.900×10^{-7}	21.10	0.0025
x_1x_3	9.302×10^{-8}	1	9.302×10^{-8}	4.01	0.0854
x_2x_3	3.306×10^{-7}	1	3.306×10^{-7}	14.24	0.0070
x_1^2	1.176×10^{-6}	1	1.176×10^{-6}	50.64	0.0002
x_2^2	2.595×10^{-8}	1	2.595×10^{-8}	1.12	0.3256
x_3^2	5.336×10^{-7}	1	5.336×10^{-7}	22.98	0.0020

As can be observed in Fig. 2(b), at the same sintering time of 2 hrs, decreasing the sintering temperature reduced the adsorption capacity from 0.00539 to 0.00482 mg/g, while increasing sintering time from 2 to 6 hrs (at a constant 200°C sintering temperature) increased the adsorption capacity of NO from 0.00482 to 0.00539 mg/g. For a constant 6 hrs sintering time, higher temperature decreased the adsorption capacity from 0.00539 to 0.00501 mg/g. Therefore, the interaction between high sintering time (6 hrs) and low sintering temperature (200°C) appears to be more effective in enhancing NO adsorption capacity. Prolonged sintering time is expected to allow impregnation of more catalyst on the surface of carbon. The impregnation of catalyst increased the catalytic activity, therefore resulted in higher adsorption capacity [20]. On the other hand, increasing the sintering temperature reduced the adsorption capacity because high sintering temperature destroys the activated carbon porous structures [21].

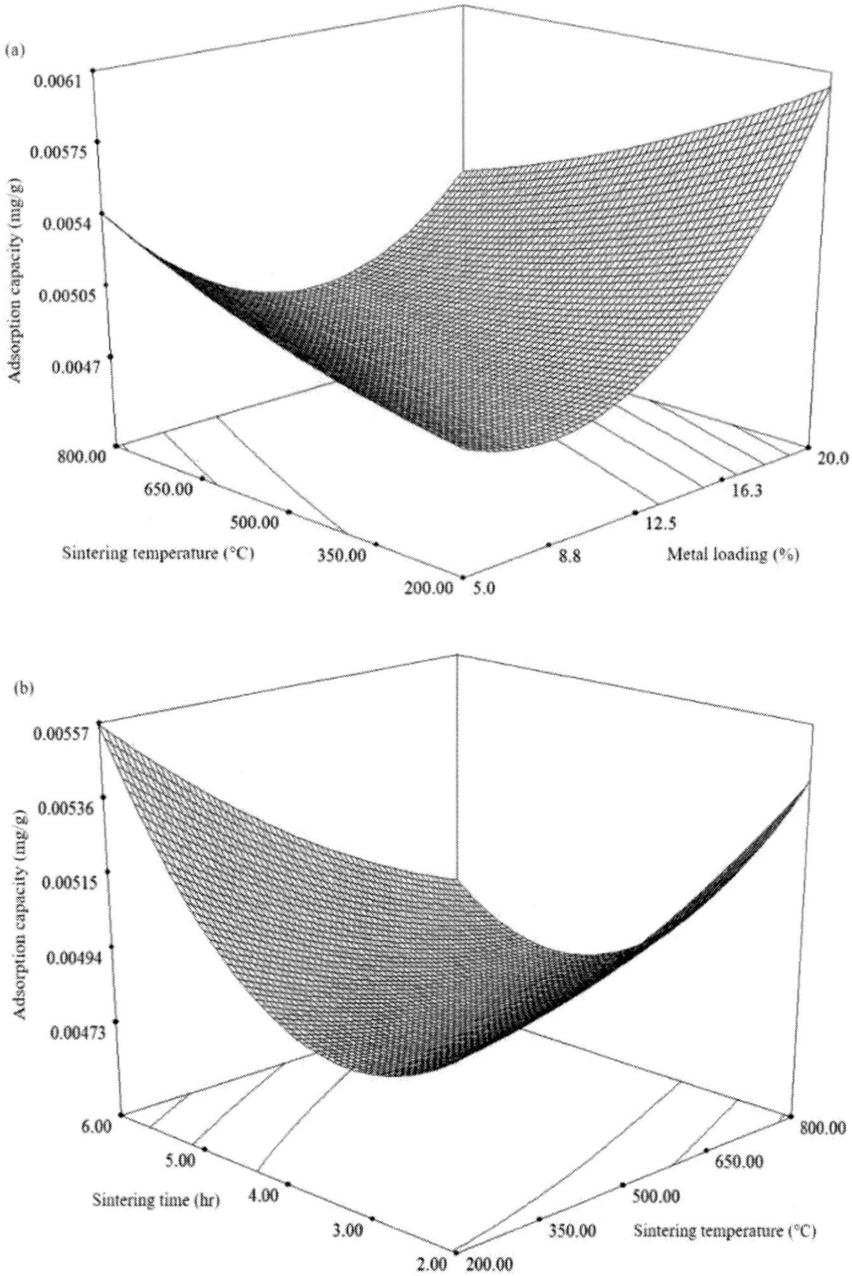

Fig. 2. Three-dimensional response surface plots of NO adsorption capacity by CuO/EFBC due to: (a) interaction between sintering temperature and metal loading, while the sintering time was set constant at 4 hrs; and (b) interaction between sintering time and sintering temperature, while the metal loading was set constant at 12.5%.

3.3 Model optimisation and validation

Model optimisation was carried out over selected range of operating variables for CuO impregnation to obtain CuO/EFBC with optimum NO adsorption capacity. By using Design-Expert software, a set of solution was generated by maximizing NO adsorption capacity and setting the individual variables, x_1, x_2 and x_3. Based on our results and support from literature [20-21], it is known that higher metal loading and lower sintering temperature is significant for NO removal, while prolonged sintering time increases the catalytic active sites of catalyst for the removal of NO. Thus, to achieve maximum NO adsorption capacity. the variables goal of the experimental model were set to be maximum for metal loading whereas the sintering temperature and sintering time were set to be minimum and maximum respectively.

With the optimized variables condition i.e. 20% metal loading, 200°C sintering temperature and 6 hours sintering time, the model predicted an NO adsorption capacity of 0.00694 mg/g, as shown in Table 4. Experimental validation of the NO adsorption capacity predicted by the model results in adsorption capacity of 0.00680 mg/g, with only small error of 2%. Therefore, the model and optimum operating variables developed can be considered valid and applicable for predicting the response.

Table 4. Model validation for activated carbon prepared for NO adsorption.

Carbon supported catalyst	Model desirability	Metal Loading	Sintering temperature	Sintering time	NO adsorption capacity (mg/g)		
					Predicted	Experimental	Error
CuO/EFBC	0.9265	20%	200°C	6 hours	0.00694	0.00680	2%

4 Conclusions

A quadratic model was developed using response surface methodology (RSM) with Box-Behnken Design for optimisation of selected CuO impregnation variables for enhancement of NO removal by CuO/EFBC. The model and variables i.e. metal loading, sintering time and sintering temperature of CuO impregnation was found to be significant in enhancing NO adsorption capacity (NO removal) through analysis of variance (ANOVA). The model term which has the most significant effect on the response is x_1^2 with an F-value of 50.64 and the overall effects are in the order of $x_1^2 > x_3^2 > x_1x_2 > x_1 > x_2x_3$. The interaction between high metal loading and low sintering temperature was noted to be most effective in enhancing NO adsorption capacity, followed by the interaction between high sintering time and low sintering temperature. The optimum operating variables for CuO impregnation onto EFBC at metal loading of 20%, sintering temperature of 200°C and sintering time of 6 hours resulted in predicted NO adsorption capacity of 0.00694 mg/g. Experimental validation of the NO adsorption capacity shows good agreement with the model prediction, with only 2% errors.

The authors would like to acknowledge the support from the Fundamental Research Grant Scheme (FRGS) under a grant number of FRGS/2/2013/SG01/UNIMAP/02/1 from the Ministry of Higher Education Malaysia.

References

1. P. Nowicki, M. Skrzypczak, R. Pietrzak, Chem. Eng. J. **162**, 2, 723 (2010).
2. S. Sumathi, S. Bhatia, K. T. Lee, A. R. Mohamed, Chem. Eng. J. **162**, 1, 51 (2010).
3. X. Li, Z. Dong, J. Dou, J. Yu, A. Tahmasebi, Fuel Process. Technol. **148**, 91 (2016).
4. N. Md-Desa, Z. Ab Ghani, S. Abdul-Talib, C. Tay, Malaysian J. Anal. Sci. **20**, 3, 461 (2016).
5. I. Ghouma, M. Jeguirim, S. Dorge, L. Limousy, and A. Ouederni, C.R. Chim. **18**, 2, 63 (2015).
6. O. A. Bereketidou, N. D. Charisiou, M. A. Goula, *13th International Conference of Environmental Science and Technology*, Athens, Greece (2013).
7. J. M. Rosas, R. Ruiz-Rosas, J. Rodríguez-Mirasol, T. Cordero, Catal. Today **187**, 1, 201 (2012).
8. P. Nowicki, H. Wachowska, R. Pietrzak, J. Hazard. Mater. **181**, 1–3, 1088 (2010).
9. H. H. Tseng, C. Y. Lu, F. Y. Chang, M. Y. Wey, H. T. Cheng, Chem. Eng. J. **169**, 1–3, 135 (2011).
10. N. Ahmad, N. Ibrahim, U. F. M. Ali, S. Y. Yusuf, F. M. Ridwan, Procedia Eng. **148**, 823 (2016).
11. Z. Zhu, Z. Liu, S. Liu, H. Niu, T. Hu, T. Liu, Y. Xie, Appl. Catal. B **26**, 25 (2000).
12. H. Tseng, M. Wey, Y. Liang, K. Chen, Carbon **41**, 1079 (2003).
13. S. A. Pasma, R. Daik, M. Y. Maskat, O. Hassan, Int. J. Polym. Sci. **2013**, 8 (2013).
14. Y. S. Mohammad, S. B. Igboro, A. Giwa, C. A. Okuofu, J. Eng. **2014**, ID 278075 (2014).
15. M. Dutta, P. Ghosh, J. K. Basu, Appl. Sci. Res. **4**, 2, 1053 (2012).
16. J. M. Salman, Arab. J. Chem. **7**, 1, 10 (2014).
17. M. H. Kalavathy, I. Regupathi, M. Ganesa, L. Rose, Colloids Surf. B **70**, 35 (2009).
18. S. K. Ryu, W. K. Lee, S. J. Park, D. D. Edie, Accessed online at url: http://acs.omnibooksonline.com/data/papers/2004_L008.pdf (2004).
19. Z. H. Zhu, L. R. Radovic, G. Q. Lu, Carbon **38**, 3, 451 (2000).
20. G. Yang, Adv. Mater. Res., **428**, 61 (2012).
21. K. H. Chuang, Z. S. Liu, M. Y. Wey, Mater. Sci. Eng. B, **175**, 2, 100 (2010).

 https://doi.org/10.1051/e3sconf/20183402030

Carbon Dioxide Capture by Deep Eutectic Solvent Impregnated Sea Mango Activated Carbon

N.Z. Zulkurnai[1], *U.F.* Md. Ali[1,*], *N.* Ibrahim[1], and *N.S.* Abdul Manan[2]

[1]School of Environmental Engineering, Universiti Malaysia Perlis, Kompleks Pusat Pengajian Jejawi 3, 02600 Arau, Perlis, Malaysia
[2]Department of Chemistry, Faculty of Science, University of Malaya, 50603 Kuala Lumpur, Malaysia

Abstract. The increment amount of the CO_2 emission by years has become a major concern worldwide due to the global warming issue. However, the influence modification of activated carbon (AC) has given a huge revolution in CO_2 adsorption capture compare to the unmodified AC. In the present study, the Deep Eutectic Solvent (DES) modified surface AC was used for Carbon Dioxide (CO_2) capture in the fixed-bed column. The AC underwent pre-carbonization and carbonization processes at 519.8 °C, respectively, with flowing of CO_2 gas and then followed by impregnation with 53.75% phosphoric acid (H_3PO_4) at 1:2 precursor-to-activant ratios. The prepared AC known as sea mango activated carbon (SMAC) was impregnated with DES at 1:2 solid-to-liquid ratio. The DES is composing of choline chloride and urea with ratio 1:2 choline chloride to urea. The optimum adsorption capacity of SMAC was 33.46 mg_{co2}/g_{sol} and 39.40 mg_{co2}/g_{sol} for DES modified AC (DESAC).

1 Introduction

The rising of greenhouse gasses emission into the atmosphere has become a great concern among the public as it is the main contributor to temperature elevation that ultimately causes global warming and climate change. Among the greenhouse gasses, carbon dioxide (CO_2) is regarded as the principal greenhouse gas that approximately contributes more than 60% to global warming [1]. The CO_2 emission is mainly from the anthropogenic sources such as deforestation, transportation, and combustion of carbon emitting fuel by industrial sector and power station.

The continuous increasing of global atmospheric CO_2 concentration will not only bring negatives effect on the environment but also living organisms. According to Hileman [2], the atmospheric CO_2 concentration was estimated to reach about 550 ppm by the year 2035 and leads to a temperature rise of 2°C, which will become a threatened extinction of 15 to 40 species. However, the International Panel on Climate Change (IPPC) had estimated that the global atmospheric CO_2 concentration may increase to 570 ppm by the year 2100 that

* Corresponding author: umifazara@unimap.edu.my

will lead to the average global temperature rise of around 1.9°C and the average sea level rise of 3.8 m. Therefore, immediate and sustained effort should be executed to control and reduce the CO_2 level in the atmosphere.

In order to lessen the CO_2 emission to the atmosphere, several convincing methods of Carbon Capture & Storage (CCS) had been recognized such as cryogenic separation, chemical separation, membrane separation and gas adsorption method. According to Lee [3], adsorption is known as a promising method for CO_2 removal as it is cost effective. Adsorption processes are normally conducted using activated carbon (AC) [4], activated alumina [5], zeolite [6] and polymeric adsorbent [7]. Based on research carried out by Duan [8] Lee [3] and El-Shafey [9], AC has the advantage to capture CO_2 due to its large surface area per unit volume and submicroscopic pores. Besides, AC is also very stable in acid and basic condition and suitable for organic compound removal. In addition, it is also able to regenerate, thus its make AC is cost-effective adsorbent.

In terms of cost-effectiveness in the production of AC, researchers are looking at different precursor from abundant and waste materials such as palm shell, sea mango, cocoa pods shell and rice husks. Research conducted by Siti Noraishah [10] successfully produced rice husks based AC and being utilized to remove CO_2 generated from industrial activities.

Recently, the integration of chemicals into the AC has shown a convincing performance of CO_2 uptakes, thus reducing the amount of CO_2 released into the atmosphere compared to the typical AC without addition of any chemicals. As been measured by Saad [11] and Karousos [12], the impregnation of activated carbon with amine and ionic liquid (IL) has resulted in higher amount CO_2 capture. Although the immobilization of this chemical in the AC assists CO_2 capture, however, this sort of compound is difficult to prepare, toxic and quite expensive thus has caused it rarely used for the commercial usage [13].

Hence, in this research AC prepared from the sea mango will be chemically impregnated with a low-cost green solvent Deep Eutectic Solvent (DES) in order to improve the CO_2 adsorption capacity and on the same time reduce the production cost of AC compare to sample impregnated with amine and IL for CO_2 capture.

2 Procedures

2.1 Preparation of the Sea Mango Activated Carbon

Sea mango (*Cerbera odollam*) was collected around Perlis, Malaysia. In order to obtain fibrous shell, the skin and seeds were removed from the raw material. The samples were then being washed and dried in an oven for 24 h at 105 °C in order to remove the moisture content. Afterward, the samples were crushed and sieved to obtain fine particles measuring 1 mm to 2 mm. The raw material was then subjected to semi-carbonization in a tube furnace for 30 min at 200 °C by flowing of nitrogen gas. Char was subsequently impregnated with 50 wt% H_3PO_4 at 1:2 precursor-to-activant ratio for 4 h and then dehydrated in the oven at 110 °C until dryness.

The char was weighed and placed in a tube furnace until the desired temperature of 519.8 °C was reached. Activation was conducted for 2.28 h. The samples were cooled to room temperature (25°C) and washed few times with hot distilled water to remove excess H_3PO_4.

Sea Mango AC was then filtered and dried in the oven until it is completely dried. The AC samples were denoted as SMAC.

2.2 Preparation of the Deep Eutectic Solvent

Choline chloride and urea (1:2) were added at 80 °C under stirring. The mixture was stirred for approximately 1 h, and the resulting solution was clear and homogeneous.

2.3 Functionalization of Sea Mango Activated Carbon with Deep Eutectic Solvent

The impregnation process of SMAC was conducted with ratio 1:2 AC to DES. The sample was allowed to be impregnated in the vacuum condition for 4 hours with pressure -0.5 Bar. The impregnated sample was allowed to completely dry in an oven at 150^{0}C for 8 hours. The dried DES impregnated SMAC was denoted as DESAC.

2.4 Carbon Dioxide Adsorption

In the adsorption testing, AC was added into the adsorption column. Prior the adsorption process, the sample was subjected to pure nitrogen gas in order to remove CO_2 composition in the sample. This procedure is conducted in order to ensure there is no CO_2 molecule attached to the surface of the sample. 15% of CO_2 gas was subjected to the adsorption column using desired flow rate. The flow rate was controlled by using Aalborg, Stainless Steel Mass Flow Controller. This experiment was conducted at the atmospheric pressure and in the room temperature. The outlet of the column was connected to the Tedlar bag in order to collect exhaust gas for CO_2 measurement. The CO_2 in exhaust gas was measured with Geotech GA 5000 portable gas analyzer. The experimental set-up employed in this study is as shown in Fig. 1.

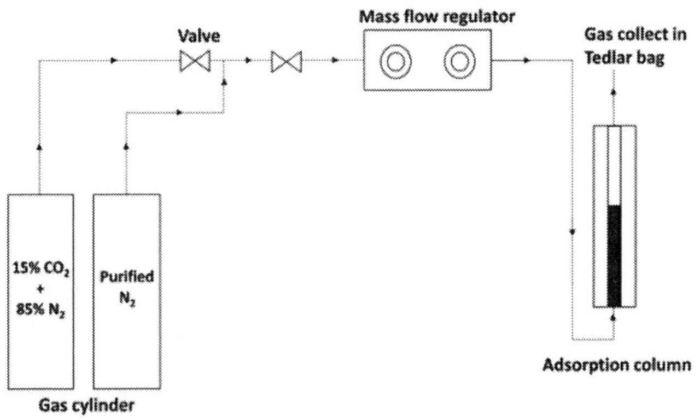

Fig. 1. Experimental setup of continuous CO_2 adsorption system.

2.4.1 Effect of different types of adsorbent on Carbon Dioxide Adsorption

5g of different types of AC was weight and has been subjected to15% of CO_2 in the adsorption column. The flow rate of the gas was fixed at 10ml/min. The outlet gas was collected using the Tedlar bag and measured using Geotech GA 5000 portable gas analyzer. The CO_2 adsorption capacity of the sample was obtained by using the following equation:

$$([v(mL/min) \times \rho (mg/mL) \times c (\%) \times tb (min)])/(w (g)) \quad (1)$$

v = Inlet gas low rate, ρ = CO_2 density at 25°C , c = Concentration of inlet CO_2 , tb = Breakthrough time, w = weight of activated carbon.

2.4.2 Effect of adsorbent dosage on Carbon Dioxide Adsorption

The different adsorbent dosage of DESAC was varied from 1g to 5g and being subjected to the 15% of CO_2 gas in the adsorption column at fixed flow rate 10ml/min. The outlet sample was collected in the Tedlar bag and being tested by using GA 5000 portable gas analyzer. The CO_2 adsorption capacity of each sample was calculated by using Equation 1.

2.5 Characterization of the Activated Carbon

Samples have been characterizing using the standard procedure. In order to determine the total surface area, the BELSORP-Mini machine has been used. Meanwhile, surface morphology and element composition analyses were performed using JEOL JSM-6460LA Analytical Scanning Electron Microscope (SEM).

3 Results and discussions

3.1 Characterization Results

3.1.1 BET Surface Area and Pore Characteristics

Table 1 shows the BET surface area and characteristics for all samples. The surface area after the physiochemical treatment showing a convincing result as it is increasing significantly from 0.7820 m2/g to 857.93 m2/g. The total pore volume of SMAC also higher compares to the raw which is 0.7166 cm^3/g. This result has proved that physiochemical treatment assists in the pore development of SMAC.

After impregnation with DES was conducted, the total surface area and total pore volume of DESAC was reduced to 581.23 m2/g and 0.4977cm^3/g respectively. This is due to the deposition of DES, which blocks a few pores on the AC which had been proved in the SEM test. The average pore diameter for SMAC and DESAC was 3.3363 nm and 3.4255nm respectively. It belongs to the mesopore region according to IUPAC classification [14].

Table 1. The S_{BET} and pore structure characteristic of SMAC and DESAC.

Sample	S_{BET} (m^2/g)	V_{total} (cm^3/g)	V_{meso} (cm^3/g)	V_{micro} (cm^3/g)	D_p (nm)
Raw	0.7920	0.0024	-	-	12.108
SMAC	857.93	0.7166	0.0937	0.6229	3.3363
DESAC	581.23	0.4977	0.0728	0.4249	3.4255

3.1.2 Surface Morphology

Fig. 2 shows the surface morphology of the raw sample, SMAC, and DESAC respectively with same magnification (1000X and 5000X). From the surface morphology analysis, Fig. 2 (a,b) shows the physical structure of a raw sample. Evidently, the pore volume was very limited and not well developed.

 https://doi.org/10.1051/e3sconf/20183402030

However, after physiochemical activation by using H_3PO_4 and carbonization process at 519.8°C, the honey-comb like pore was well developed (Fig. 2 c,d), as confirmed by BET result (Table 1). The development of pore was due to the breakdown of structure in the raw due to the thermal expansion during carbonization process and the addition of the activating agent also assists in the pore development [18].

After the impregnation of the AC with the DES, a few pores in the sample seems had been clogged by the DES which has been shown in Fig 2 e and f. It is proved that the DES was successfully attached to the sample. Consequently, the attachment of DES on the sample has reduced the pore volume in the sample which had been proving by the BET results.

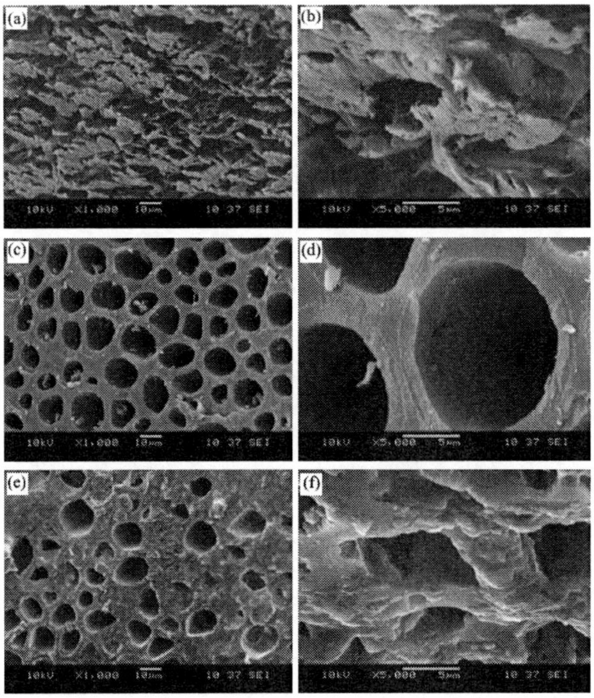

Fig. 2. SEM micrograph of raw precursor (a) 1000×, (b) 5000×; SMAC (c) 1000×, (d) 5000×; and DESAC (e) 1000×, (f) 5000×.

3.1.3 Energy-dispersive X-Ray (EDX) analysis

Elemental analysis revealed that the element Carbon (C), Oxygen (O), Phosphorus (P), Nitrogen (N) and Chlorine (Cl) existed in the Raw, SMAC or DESAC. Based on table 2, the amount of C significantly increased in the SMAC. This finding has proved that the carbonization process assists the C element due to the heat supply during activation which initiates the thermal degradation process. According to the Olugbenga [19], the thermal degradation process will remove the volatile matter in the sample, thus produce a stable carbon element.

The elemental P also had been detected in the SMAC. The elemental P was believed contributed during the impregnation process using H_3PO_4. Meanwhile, the 4.01% of N element detected in the SMAC was believed contributed from the nitrogen purge during physiochemical activation. However, after impregnation process with DES, the amount of N element in the DESAC was drastically increase. It was believed that the N element in the

DESAC was contributed from the urea and choline chloride of the DES. This finding has confirmed that the N element was successfully attached to the surface of DESAC and as a result, it will provide more actives sites for CO_2 adsorption [20]. The elemental Cl also has been detected in the DESAC, as it was contributed from the chlorine chloride of DES during functionalization process.

Table 2. EDS elemental analysis of raw precursor, SMAC, and DESAC.

Sample	Element (wt. %)				
	Carbon	Oxygen	Phosphorus	Nitrogen	Chlorine
Raw	48.49	51.51	0.00	0.00	0.00
SMAC	75.27	14.62	6.10	4.01	0.00
DESAC	48.89	24.45	0.00	18.39	8.27

3.2 Carbon Dioxide Adsorption Performance Test

3.2.1 Different Type of Adsorbent

Fig. 3 shows the breakthrough time curves for SMAC and DESAC at 10mL/min and 5g of adsorbent. Both of the breakthrough curves show a rapid CO_2 uptake in the initial process and increased gradually over contact time and finally reached constant. This is because, at the initial adsorption process, all the adsorption sites of both AC are available for CO_2 capture, however, after a certain time, the available adsorption sites has decreased and finally, all of the adsorption sites are occupied by CO_2 molecule.

For SMAC, the adsorption breakthrough occurred at 60 min, meanwhile, the DESAC breakthrough occurred at 75 min. From the finding, it can say that the DESAC has a better adsorption performance compare to the SMAC as the CO_2 adsorption capacity showing convincing results as tabulated in table 3. Highest breakthrough time obtain by DESAC was due to the N element in the DES which provides more actives site for CO_2 capture and consequently accelerated the CO_2 adsorption [21].

Table 3. Effect of different types of adsorbent dosage on CO_2 adsorption.

Sample	Breakthrough Time (second)	Adsorption Capacity (mg_{co2}/g_{sol})
SMAC	3720	33.46
DESAC	4380	39.40

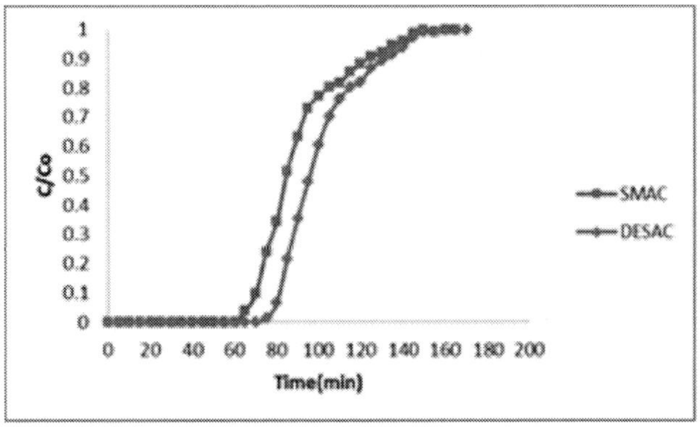

Fig. 3. Effect of different type of adsorbent dosage on CO_2 adsorption.

3.2.2 Difference Adsorbent Dosage

In this study, five different weight of DESAC has been tested in order to study the effect of adsorbent dosage onto the sample. Based on the table. 4, the breakthrough time has increased proportionally with increasing of adsorbent dosage. This result claims that the higher amount of adsorbent dosage will have more available active sites for CO_2 capture. Consequently, the adsorption bed will need a longer time to be saturated.

However, when it's come to the adsorption capacity, increasing in adsorbent dosage has caused the adsorption capacity decreased as been shown in table 4. According to the Allesina [22], increasing adsorbent dosage by using the same adsorption column especially for the low flow rate (10 ml/min) will contribute to the higher resistance and leading to non-uniform of CO_2 gas onto the sample. Consequently, this phenomenon may create channels in the adsorption bed thus reduce the chance active site being exposed to the CO_2 gas which not around the channels area and reduces the adsorption capacity. This finding is in agreement with Nur Hidayah [23], which using AMP impregnated AC for CO_2 capture.

Table 4. Effect of different types of adsorbent dosage on CO_2 adsorption.

Sample (g)	Breakthrough Time (second)	Adsorption Capacity (mg_{co2}/g_{sol})
1	1920	86.35
2	3000	67.46
3	3600	53.97
4	4020	45.20
5	4380	39.40

4 Conclusions

DESAC was successfully utilized as a new modification AC. The CO_2 adsorption capacity of DESAC showing convincing results compares to the SMAC. Even though DESAC obtains lower BET surface area compare to SMAC, the breakthrough time of DESAC showing better results. This is due to the chemisorption action in the actives site provides by N element from DES, which helps in promotion new site for more CO_2 capture.

This study was financially supported by Fundamental Research Grant Scheme (FRGS 9003-00468).

References

1. D. Krishnaiah, A. Bono, S.M. Anisuzzaman, C. Joseph, B.K. Teo, J. Appl. Sci **14**(23), 3142-3148 (2014)
2. B. Hileman, Chem. & Eng. News **84**, 7-8 (2006).
3. C. S. Lee, Y. L. Ong, M. K. Aroua, W.M. A. W. Daud, Chem. Eng. J. **219**, 558–564 (2013)
4. S. Mohammad Saleh, W. D.Wan Mohd Ashri, H. Amirhossein, A. Shamiri, J. Anal. Appl. Pyrolysis. **89**, 143–151(2010)
5. S. S. Tony, K. K. Pant, Sep. Purif. Technol. **36**, 139–147 (2004)
6. K. L. Chung, S. L. Shin, C. J. Lain, C. W. Cheng, S. L Kuen, D. L. Meng, J. Hazard. Mater. **147**, 997–1005 (2007)
7. D. A. Jones, T. P. Lelyveld, S. D. Mavrofidis, S. W. Kingman, N. J. Miles, Res., Conser. and Recycle. **34**, 75–90 (2002)
8. X. Duan, C. Srinivasakannan, J. S. Liang, J. Taiwan Inst. Chem.Eng. **45**, 1618–1627 (2014)

9. E. I. El-Shafey, N. F. A. Syeda, A. Saleh, A. J. Haider, A. Lawati, J. of Environ. Chem. Eng. **4**, 2713–2724 (2016)
10. M. R. Siti Noraishah, J. Saidatul Shima, F. A. S. Noor Asma, I. Nur Aminatulmimi, ARPN J. Eng. and Appl. Sci. **1**, 1- 4 (2016)
11. H. K. Saad, M. K. Aroua, W. M. A. W. Daud, Chem. Eng. J. **183**, 15-20 (2012)
12. D.S. Karousos, O. C. Vangeli, C. P. Athanasekou, A. A. Sapalidis, E. P. Kouvelos, G. E. Romanos, N. K. Kanellopoulos, Chem. Eng. J. **306**, 146–154 (2016)
13. G. García, A. Santiago, U. Ruh, A. Mert, Energy Fuel **29**, 2616-2644 (2015)
14. E.F. Maria, V.N. Gisel, R.B. Pablo, C. Ana Lea, Ind. Crops Proct. **62**, 437–445 (2014)
15. A. Kundu, B.S. Gupta, M.A. Hashim, G. Redzwan, J. Clnr. Prod. **105**, 420-427 (2015)
16. R.H. Hesas, A. Arami-Niya, W.M.A.W. Daud, J.N. Sahu, Bioresources **8(2)**, 2950-2966 (2013)
17. S. Yorgun, D.Yildiz, J. Taiwan Inst. Chem. Eng. **53**, 122-131 (2015)
18. A. Mohd Azmier, A. P. Nur Azreen, S. B. Olugbenga, Wtr. Res. and Ind. **6**, 18–35 (2014)
19. S. B. Olugbenga, A. Mohd Azmier, T. S. Tan, Trends in Appl. Sci. R. **6(**8), 794-812 (2011)
20. M. A. Khalid, M. A. Abdul Hakim, M. Hayyana, A. Shatirah, H. Mohd Ali, Appl. Surf. Sci. **389**, 216–226 (2016)
21. O. Yew Liang, *Master's Thesis,* Universiti Malaya, Kuala Lumpur, Malaysia (2014)
22. G. Allesina, S. Pedrazzi, P. Tartarini. Bioresource Tech. **146**, 704-712 (2013)
23. A. Nur Hidayah, *Master's Thesis,* Universiti Malaysia Perlis, Perlis, Malaysia (2016)
24. X. Wu, J. Xu, J. Tao, New York Sci. J. **6(**11), 33-37 (2013)

The acute toxicity of the metaldehyde on the climbing perch

Syamimi Wahida Mohamad Ismail[1, 2], *Farrah Aini* Dahalan[1,2*], *Ammar* Zakaria[1,3], *Ali Yeon* Mad Shakaff[1,3], *Siti Aqlima* Ahmad[4], *Mohd Yunus* Abd Shukor[4], *Mohd Khalizan* Sabullah[5], *Khalilah* Abdul Khalil[6], and *Mohd Faizal* Ab Jalil[7]

[1] Centre of Excellence of Advance Sensor Technology (CEASTech), Pusat Pengajian Jejawi 2, Taman Muhibbah, Universiti Malaysia Perlis, 02600 Arau, Perlis, Malaysia
[2] Water Research Group, School of Environmental Engineering, Universiti Malaysia Perlis, Pusat Pengajian Jejawi 3, 02600 Arau, Perlis, Malaysia
[3] School of Mechatronic Engineering, Universiti Malaysia Perlis (UniMAP), Malaysia
[4] Department of Biochemistry, Faculty of Biotechnology and Biomolecular Sciences, Universiti Putra Malaysia, 43400 UPM Serdang, Selangor, Malaysia
[5] Faculty of Science and Natural Resources, Universiti Malaysia Sabah, 88400 Kota Kinabalu, Sabah, Malaysia
[6] Department of Biomolecular Sciences, Faculty of Applied Sciences, Universiti Teknologi MARA, 40150 Sec. 2, Shah Alam, Selangor, Malaysia
[7] Department of Environment, Ministry of Natural Resources and Environment, Tingkat 2, Bangunan KWSP Jalan Bukit Lagi 01000 Kangar, Perlis, Malaysia

Abstract. In Asia, Climbing perch (*Anabas testudineus*) is commonly found in paddy fields and irrigation systems. Due to its habitat, Climbing perch is exposed to toxic pesticides used in paddy fields such as metaldehyde which is one of the most widely used molluscicide. This study aims to determine the acute toxicity Lethal Concentration$_{50}$ (LC_{50}) of metaldehyde and its effect on the behaviour and physical changes of the Climbing perch. The fish mortality responses to six different metaldehyde concentrations ranging from 180 to 330 mg/L were investigated. The 96-h LC_{50} values were determined and analysed using three different analysis methods which is arithmetic, logarithmic and probit graphic. The LC_{50} values obtained in this study were 239, 234 and 232 mg/L, respectively. After 96-h of exposure to metaldehyde, the fish showed a series of abnormal behavioural response in all cases: imbalance position, and restlessness of movement. The LC_{50} values show that metaldehyde is moderately toxic to the Climbing perch indicating that metaldehyde is not destructive to Climbing perch. However, long term exposure of aquatic organisms to the metaldehyde means a continuous health risk for the fish population as they are more vulnerable and it is on high risk for human to consume this toxicated fishes.

* Corresponding author: farrahaini@unimap.edu.my

1 Introduction

The excess use of pesticides do not only brought adverse influence on agro ecosystems but also cause alteration in physiological processes of non-target organisms as it can cause serious health or environmental damage [1]. There is growing concern that pesticides applications do not only reach its target organism but also reach non-target organisms. Among aquatic organism, fish are very sensitive to wide range of pesticides, which can cause deleterious effects through bioaccumulation [2]. Accumulation of pesticides in tissues of organism can cause physiological and biochemical changes to the tissues, thus influencing the activities of several enzymes and disturb the metabolic process [3]. The risks of pesticides pose to the surrounding environments depend on the pesticide toxicity level and the duration of exposure to organisms. The effect of pesticides toxicity on fish is alarming and has raised awareness among scientist since the matter is highly related to food and water security issues. In addition, the effect of pesticides toxicity on fish also raised questions about health hazards of fish to consumers as fish is one of the source of protein consumed by human [4].

Pesticides can be categorized based on their specific functional target which is insecticides, herbicides, molluscicide, and rodenticides. This study focuses on metaldehyde which is one of the molluscicide. Metaldehyde (2, 4, 6, 8-tetramethyl-1, 3, 5, 7-tetroxacyclo-octane) is a cyclic tetramer of acetaldehyde. It is the active compound in many snail and slug baits [5].

This molluscicide is used in many regions around the world. In East Asia especially in Thailand, Indonesia, Vietnam and Malaysia, metaldehyde is mainly used to prevent crop damage from golden apple snail (*Pomacea canaliculata*). Golden apple snail is a pest causing major problem in paddy fields [6-7]. According to "Global Invasive Species Database", golden apple snail is known as one of the 100 world's worst invasive species and poses serious threat to many agricultural sectors in the world [8]. Metaldehyde kill snails by causing dehydration and paralysis, it is directly absorbed from the gastrointestinal tract and crosses the blood-brain barrier [9-10]. The toxicity of metaldehyde causes central nervous system (CNS) depression to the organism. According to a study conducted on mice, metaldehyde has been detected in brain, blood, and liver. The study proved that metaldehyde reduced brain concentration of neurotransmitters noradrenaline, serotonin, and gamma amino butyric acid leading to involuntary muscle contraction properties to the mice [11].

This study aims to determine the acute toxicity Lethal Concentration$_{50}$ (LC_{50}) of metaldehyde and behavioural changes of Climbing perch (*Anabas testudineus*) due to the exposure from different metaldehyde concentration. The Climbing perch has been selected for the experiments because Climbing perch is the largest freshwater fish that inhabit paddy fields and irrigation systems and has important commercial value. Moreover, it is known for its high tolerance and survival in extreme environmental changes in paddy field and irrigation system. To date, according to the authors' knowledge, there is no study conducted to investigate the toxicity of metaldehyde to Climbing perch. Therefore, it is important to use Climbing perch as test organism for the toxicity evaluation of metaldehyde.

2 Materials and methods

2.1 Test organism and chemical

Healthy adult fish species *Anabas testudineus* weighing 18 ± 4.4 g and total length 11 ± 0.8 cm were obtained from a commercial hatchery in Arau, Perlis and brought to the laboratory in plastic bags with sufficient air. Fish were acclimated to laboratory conditions for 14 days in 30 litre tank of aquarium with continuous aeration using an electric air pump compressor before being used in the experiment. During this period, the fish were fed with commercial fish food and water quality was regularly maintained at 25 °C, pH 7.0 and dissolved oxygen above 5 mg/L. The water was periodically changed to minimize food waste, urine and faeces contamination based on EPA method [12].

The fish were fed with fish feed pellet during acclimating period but they were not fed during the last 48 h prior to the experiment. This is to allow sufficient time for digestive abstinence. The metaldehyde 99% (CAS no.: 108-62-3) used in this study was purchased from Acros Organics.

2.2 Experimental set up

Before starting the test, all aquaria were cleaned with Decon 90 and filled with 30 L of decholorinated tap water. The experimental water was kept in the tank for 24 h in order to decrease the chorine level before the addition of the metaldehyde. Preliminary screening test was done to determine the appropriate concentration range for the metaldehyde. Seventy fish were randomly transferred to seven aquaria tanks where each tanks consist of ten fish. Once fish were transferred into each aquaria, they were exposed to different concentrations (180, 210, 240, 270, 300 and 300 mg/L) of metaldehyde for 96 hours under semi-static conditions. The experimental tests were conducted as in the following where each group represent different metaldehyde concentration:

Group 1: as a control group with no exposure to metaldehyde
Group 2: received 180 mg/L of metaldehyde
Group 3: received 210 mg/L of metaldehyde
Group 4: received 240 mg/L of metaldehyde
Group 5: received 270 mg/L of metaldehyde
Group 6: received 300 mg/L of metaldehyde
Group 7: received 330 mg/L of metaldehyde

Each concentration was examined by triplicate with 10 individual per test group. A stock solution of metaldehyde was prepared by dissolving it in acetonitrile; the necessary volume of stock solution was mixed to the water in aquaria to achieve the desired final concentrations (solvent = 0.005% per aquaria) [13]. During the experiment, fish were not fed. The number of dead fish were counted every 24 hours and removed from the aquarium as soon as possible. Fish mortality was recorded at the end of the 96 hours. The fish behaviour including their activities, loss of equilibrium, abnormal swimming and colour changes in the skin of the fish were observed. Median lethal concentration (LC_{50}) of metaldehyde was calculated based on arithmetic, logarithmic and Probit transformation of the mortality dose curve analysis by referring to the Karber [14], Bliss [15] and Finney [16] method respectively. Arithmetic and logarithmic analyses were done by using Microsoft Excel software whereas probit analysis was done using Biostat professional version 9. Fig. 1 shows the experimental setup for the toxicity test of metaldehyde.

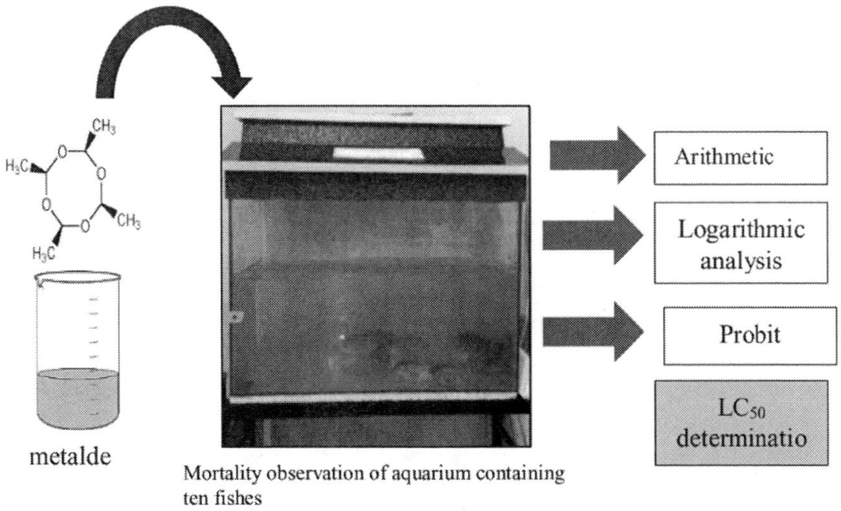

Fig. 1. Experimental setup for toxicity test of metaldehyde and LC_{50} determination methods

3 Results and discussions

The data obtained from the acute toxicity test of metaldehyde for climbing perch revealed that metaldehyde toxicity increased with increasing concentration (Figure 2 and Figure 3). No deaths were recorded after 96 hours in control group (no exposure to metaldehyde) and the lowest metaldehyde concentration of 180 mg/L. On the other hand, the 100% mortality rate was achieved only at 330 mg/L (Fig. 2).

Fish mortality due to exposure of metaldehyde mainly depends on its sensitivity to the toxicants and concentration. At low concentration of metaldehyde approximately less than 220 mg/L, there is high percent of survival (more than 50% survival) or little mortality of Climbing perch. However, as the concentration of Metaldehyde increased, the percentage of survival also decreases. This is due to the toxicant causing severe damage to different systems which leads to the mortality of the fish [17]. The evaluation of LC_{50} value is important step in ecotoxicological studies. There are several methods of calculating the LC_{50}'s of toxicants. In this study the following methods have been used: arithmetic graphic, logarithmic and probit analysis method. From the graph in Figure 2, LC_{50} values using arithmetic graphic method is 239 mg/L, whereas LC_{50} values using logarithmic method is 234 mg/L (Fig. 3).

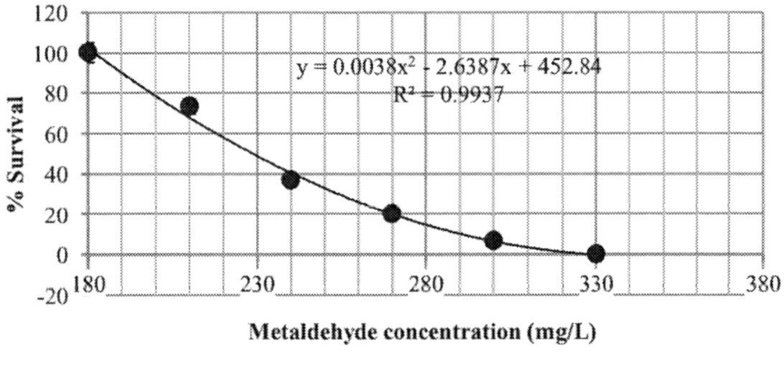

Fig. 2. Arithmetic graphic analysis on percentage (%) survival of Climbing perch (*Anabas testudineus*) after 96 hours incubation with various metaldehyde concentrations

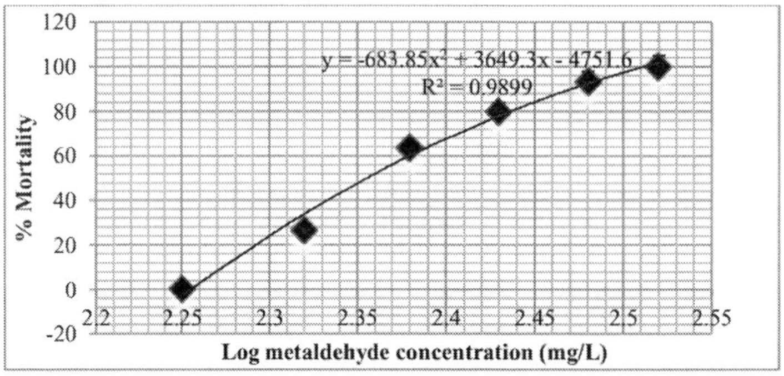

Fig. 3. Logarithmic analysis on percentage (%) Climbing perch (*Anabas testudineus*) 96 hours incubation with various metaldehyde concentrations in logarithmic scale

In additions, the data were also assessed according to probit analysis using Biostat professional version 9. This software was designed for analysis of mortality data from the acute toxicity tests with fish and other aquatic life [18]. Probit analysis minimizes the extremes in the results and maximizes the middle percentage survivals. The graph produced through probit analysis is obtained by excluding the zero and 100% point of survival [19]. Figure 4(a) and 4(b) show the screenshots of probit analysis using Finney method [Lognormal Distribution] and least squares [Normal Distribution] method generated by Biostat Version 9 Software. In Fig. 4(a), the percentage of response is plotted against the log dose whereas Fig. 4(b) shows the percentage of responds plotted against dose in mg/L. Both figures demonstrate an S-shaped idealized concentration response graph, which is the typical of most types of exposure. This "S" shaped curve illustrates that at low doses there is little or no response while at high dose all individuals respond to the effect.

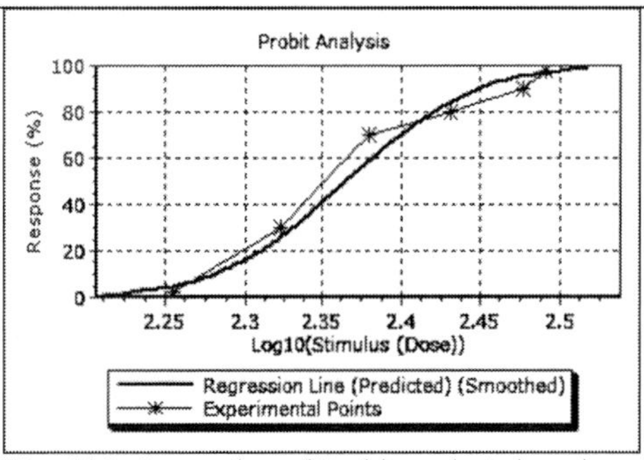

Fig. 4(a). Screenshot of Probit analysis based on Finney method [Lognormal Distribution] generated by Biostat Professional version 9 Software

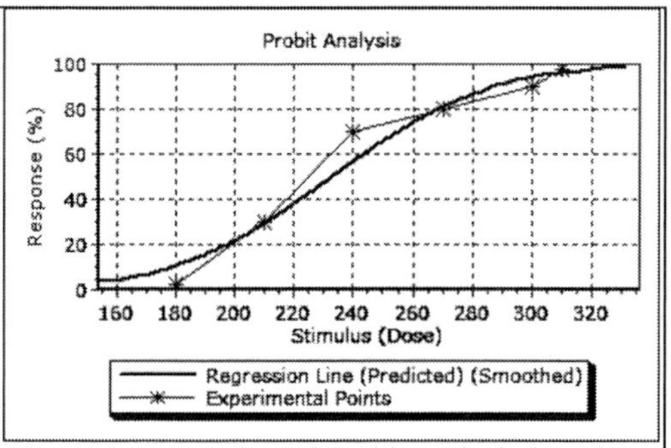

Fig. 4(b). Screenshot of Probit analysis based on least squares [Normal Distribution] method generated by Biostat Professional version 9 Software

Table 1 show the regression statistics of two types of probit analyses generated based on Finney method [Lognormal Distribution] and Least squares [Normal Distribution] method. The two methods of probit analysis gave different values for LC_{50} of metaldehyde. LC_{50} value using Finney method [Lognormal Distribution] is 232 mg/L while LC_{50} value obtained using Least Square [Normal Distribution] method is 233 mg/L. The Alpha value (confidence interval) for both method is 0.05, since p value is greater than 0.05 ($p>0.05$), therefore there is no significant difference.

The determination of LC_{50} value by using the three methods is in good accordance and results revealed that metaldehyde is least toxic to the Climbing perch. According to OECD Test Guidelines scale, the LC_{50} value of the metaldehyde is acute category IV to climbing perch (Table 2) [20].

Table 1. Details of regression statistics for LC_{50} value of metaldehyde concentration determination using Finney method [Lognormal Distribution] and Least Square [Normal Distribution] of probit analysis generated from Biostat Professional version 9 Software

	Finney method	Least square
LC_{50}	**232**	**233**
LC_{50} LCL	214	199
Log_{10} [LC50]	2	0.0236
Beta	15	0.0064
Beta Standard Error	3	178
LC_{50} Standard Error	8	9
LC_{50} UCL	247	266
Intercept	-31	-0.5001

Table 2. The classification of pesticides toxicity on fish.

Toxicity class	Description	LC_{50} (Concentration over 96 hour period for fish) mg/L
Acute I	Extremely high toxicity	$\leq$1 mg/L
Acute II	High toxicity	>1 - $\leq$10 mg/L
Acute III	Moderate toxicity	>10 - $\leq$100 mg/L
Acute IV	Least toxicity	$\geq$ 100 mg/L

Different species correspond differently to the same type of pesticide. Although no research has been done on acute effect of metaldehyde on climbing perch but limited work has been done on acute effect of metaldehyde on other fish species. Reported 96-h LC_{50} values for the metaldehyde on rainbow trout *Oreochromis niloticus, Oreochromis mykiss* and bluegill (Lepomis macrochirus) was 62 mg/L, 75 mg/L and 10 mg/L respectively (Table 3) [21-22]. While the EC_{50} value of metaldehyde for two aquatic invertebrate *D. magna* and *P. corneus* was 90 mg/L and 200 mg/L, respectively. The EC_{50} value for algae, *D. subspicatus* was 200 mg/L [22]. In fact, this study found that the 96-h LC_{50} value of the Climbing perch is higher than other aquatic organisms due to it has high endurance under extreme circumstances.

Climbing perch is highly adapted to life in unfavourable conditions because it possesses an accessory air breathing organ [23-24]. In addition, it also can survive in low oxygen concentrated water and resistant against diseases [25]. [23] found that Climbing perch had high resistance toward carbofuran pesticides, with LC_0 value of 560 ppb and the LC_{100} value of 1560 ppb compared to the marine sheepshead minnow (*Cyprinodon variegatus*) with LC_{50} value of 386 ppb [26].

Besides, Climbing perch as a water-breathing fish takes only less toxicant through gills from the water and it is highly tolerant to toxicants [27]. A study shows that Climbing perch was found to be moderately susceptible to the Epizootic ulcerative syndrome (EUS)

infection causes by aquatic oomycete fungus (*Aphanomyces invadans),* compared to EUS susceptible such as Snakehead (*Channa striata*), tropical snakeskin gourami (*Trichopodus pectoralis*), koi carb (*Cyprinus carpio*), and broadhead catfish (*Clarias macrocaphalus*) [28].

Table 3. Acute toxicity of metaldehyde to fish species.

Species	96 hours LC50 (mg/L)
Rainbow trout (*Oreochromis niloticus*)	62
Rainbow trout (*Oreochromis mykiss*)	75
Bluegill (*Lepomis macrochirus*)	10

Behavioural changes could be the first response of an organism to environmental change. Therefore, in this study, it was observed that Climbing perch individuals exhibited a variety of behavioural changes when subjected to different metaldehyde concentrations ranging from 180, 210, 240, 270, 300 to 330 mg/L. When there is no metaldehyde concentration in the aquarium (control group), there were no deaths, and behavioural and physical appearance were normal. However, the behavioural changes and clinical toxic symptoms were observed as the Climbing perch exposed to various metaldehyde concentrations (180,210, 240, 270, 300 and 330 mg/L). After exposure to metaldehyde concentration, Climbing perch only showed preference for the top area of the aquarium due to escape conduct of the fish and for aerial breathing, in order to avoid contact of the pesticide with gills. This behaviour is similarly reported for *J. multidentata* which showed similar alteration after 96 hours of exposure to the insecticides (chlorpyrifos and cypermethrin) [29].

Furthermore, Climbing perch also continuously secreting excess amount of mucus from the body and thus a thick layer of mucus was deposited in the gills and skin. Massive mucus discharge by the fish is due to the metaldehyde's toxic mode of action causing irreversible damage in the mucous cells of the skin, thus leads to excessive mucus production, and destruction of the mucus cells [30-31]. When the fish died, the mouth and the opercula were wide open and the skin colour of the fish changed from dark to pale. The same appearance was observed for *Etropus suratensis* exposed to different concentration of tannery effluent. [32] observed the changes in appearances such as the fish succumbed to tannery effluent with mouth and operculum wide open. The changes in fish body colour was also noted.

According to [33], at the beginning of the toxic effect of a toxicant, several biochemicals upregulated occurs in maintaining the hemeotasis mechanism such as biochemical action, defensive mechanism and cellular alteration, and at the same time affected protein that is downregulated due to inhibition. Prolong or the increasing concentration of toxicant leads to cellular alteration due to imbalance of protein recovery and toxic effect associated with the induction of program cell death. Hence, death of cell is related to the alteration of fish behaviour and physical appearance [33].

4 Conclusions

The LC_{50} value of molluscicide metaldehyde tested on freshwater fish Climbing perch based on arithmetic, logarithmic and probit analyse revealed the susceptibility of the fish to the toxic stress (LC_{50} 213, 234 and 232 mg/L). Comparison using the LC_{50} data indicates that metaldehyde was less toxic compared to other pesticide due to its robustness and high survivability. However, exposure of metaldehyde may cause alteration in physical and behaviour of Climbing perch at higher metaldehyde concentrations. With the resistant of the Climbing perch in paddy field with high concentration of metaldehyde raised concerns since this fish will be consumed by human as this pesticide is most likely to have some health risk to humans who consume it. Thus, there is a need to abide to regulation and education on proper use of pesticides to preserve the environment.

Authors wish to thank the financial support by the RACE fund (9017-00009), FRGS (9003-00386) by Ministry of Higher Education Malaysia and Incentive Journal funded by UniMAP (9007-00230).

References

1. J. H. Chandra, S. Sridhar, J. Chem. Pharm. Sci. **8**(1), 34–39 (2015)
2. H. Y. Zhang, C. Wang, H. Z. Lu, W. B. Guan, Y. Q. Ma, Ecotoxicol. Environ. Saf., **74**(6), 1653–1658 (2011)
3. K. N. Mukesh, African. J. Environ. Sci. Technol. **7**(7), 725–727 (2013)
4. A. Lazartigues, M. Thomas, D. Banas, J. Brun-bellut, C. Cren-olivé, C. Feidt, Chemosphere. **91**, 530–535 (2013)
5. R. C. Gupta, *Veterinary Toxicology Basic and Clinical Principles* (Elsevier, 2007)
6. A. L. Sanico, S. Peng, R. C. Laza, R. M. Visperas, Crop. Prot. **21**, 137–143 (2002)
7. J. Y. Figueroa, M. L. P. Almazan, F. G. Horgan, Crop Prot. **62**, 23–31 (2014)
8. Global invasive species database. Available: http:// www.issg.org/database/welcome/ [Accessed 31 Dec 2015]
9 L. K. Dolder, Vet. Med. **1**, 213–215 (2003)
10. T. F. Booze, F. W. Oehme, Toxicol. Sci. **6**(3), 440–446 (1986)
11. J. D. Mills, S. E. R. Bailey, C. R. McCrohan, M. A. Wedgwood, *Effect of molluscicides on feeding behavior and neuronal activity, in Symposium organized by the British Crop Protection Council in association with the Malacological Society of London*, 77–83 (1989)
12. United States Environmental Protection Agency, *Ecological Effect Test Guidelines OPPTS 850.1075 Fish Acute Toxicity Test, Freshwater and Marine,* (1996)
13. R. H. Da Cuña, G. Rey Vázquez, M. N. Piol, N. V. Guerrero, M. C. Maggese, F. L. Lo Nostro, Ecotoxicol. Environ. Saf. **74**(4), 1065–73 (2011).
14. G. Karber, **162**, 480–483 (1931)
15. C. L. Bliss, Ann. Appl. Biol., **22**(1), 134–167 (1935)
16. D. J. Finney, "Probit Analysis, 3rd ed.," in Cambridge University Press, XV-333 (1971).
17. C. I. Abraham, "Biodegradation of Dyes," Unpublished Ph.D dissertation. Mahatma Gandhi University (2013)
18. P. B. A. Fai, S. O. Fagade, Ecotoxicol. Environ. Saf. **62**(1), 128–131 (2005)
19. D. L. Reish, P. S. Oshida, *Manual of Methods in Aquatic Environment Research, Part 10: Short-term Static Bioassays.* Food & Agriculture Org. (1987)
20. Organisation for Economic Co-operation and Development, "Hazardous to the aquatic environment," International Labour Organization, 2010. [Online]. Available: http://www.ilo.org/legacy/english/protection/safework/ghs/ghsfinal/ghsc14.pdf. [Accessed: 28-Mar-2016].
21. P. L. Pingali, P. A. Roger, *Impact of Pesticides on Farmer Health and the Rice Environment.* Int. Rice Res. Inst. (1995).
22. E. F. S. Authority, EFSA J. **8**(10), 1–71 (2010)
23. G. Hitchcock, *"Climbing perch (Anabas testudineus) (Perciformes: Anabantidae) on*

Saibai island, Northwest Torres Strait: First Australian Record of this exotic pest fish," in Memoir of the Queensland Museum, 207–211 (2008)
24. G. R. Allen, *Field guide to the freshwater fishes of New Guinea* (1991).
25. M. Khatune-Jannat, M. M. Rahman, M. A. Bashar, M. N. Hasan, F. Ahamed, M. Y. Hossain, Sains Malaysiana **41**(10), 1205–1210 (2012)
26. R. Eisler, U.S. Fish Wildl. Serv. **85**(10), 1–63 (1988)
27. S. S. Jacob, N. B. Nair, and N. K. Balasubramanian, "Toxicity of certain mosquito larvicides to the larvivorous fish Aplocheilus lineatus (Cuv. & Val.) and Macropodus cupanus (Cu. & Val.)," Environ. Pollut. **28**, 7–13 (1982)
28. S. F. Afzali, H. H. M. Daud, I. Sharifpour, M. Afsharnasab, S. Shankar, Vet. World. **8**(9), 1038–1044 (2015)
29. R. I. Bonansea, D. A. Wunderlin, M. V. Amé, Ecotoxicol. Environ. Saf. **129**, 311–319 (2016)
30. K. C. Hallett, A. Atfield, S. Comber, T. H. Hutchinson, Sci. Total Environ. **543**, 37–43 (2016)
31. R. Triebskorn, K. Christensen, I. Heim, J. Molluscan Stud **64**, 467–487 (1998)
32. G. Taju et al., Chemosphere. **87**(1), 55–61 (2012)
33. M. K. Sabullah, S. A. Ahmad, J. Hussin, A. J. Gansau, M. R. Sulaiman, J. Environ. Bioremediation Toxicol. **2**(1), 28–32 (2014)

 https://doi.org/10.1051/e3sconf/20183402032

Prediction of Deepwater FPSO responses using different numerical analysis methods

Matthew Guan[1,*], *Montasir* Osman[1], and *Cheng Yee* Ng[1]

[1]Civil and Environmental Engineering Department, Universiti Teknologi PETRONAS, 32610 Bandar Seri Iskandar, Perak, Malaysia

Abstract. The limitations of existing wave basins present a significant challenge when modelling offshore deepwater systems, particularly due to the basin's relatively shallow depth. Numerical simulation thus becomes valuable in predicting its behaviour during operation at sea. The coupled dynamic analysis is preferred over the traditional quasi-static method, as the former enables the inclusion of damping and added mass properties of the complete mooring line system, which becomes increasingly prominent at greater water depths. This paper investigates the motions and mooring line tensions of a turret moored Floating Production Storage Offloading (FPSO) platform using three numerical models, i.e. a dynamic system, quasi-static system and linear spring system subjected to unidirectional random wave condition. Analysis is carried out using a commercial software AQWA. The first two numerical models utilise a complete system of the same setup and configuration, while the linear spring system substitutes the mooring lines with equivalent linear springs and attempts to match the total mooring line restoring forces with that of the coupled dynamic analysis. The study demonstrates the significance of coupled dynamic analysis on the responses of an FPSO in deepwater. The numerical model of the FPSO is validated against the results of a published work.

1 Introduction

Floating platforms play an important role in the offshore oil and gas industry. Vessels such as the ship-shaped FPSOs are commonly used in the production of oil and gas in offshore deepwater regions. The system is held in place by mooring lines anchored to the seabed during operation. Due to the cost and complexity of manufacturing a floating platform, extensive design and model testing is vital in ensuring the performance and safety of the prototype before installation and operation. As exploration and production progresses into deeper waters, experimental and analytical studies become increasingly crucial to predict the responses and mooring line/riser tensions in these remote and harsh environments.

Physical modelling is traditionally used in predicting platform prototype behaviour and testing new concepts. Ideally a complete system should be tested in order to account for the dynamic effects induced by the mooring lines. Tests with scales up to 1:150 showed the

* Corresponding author: matthewgzh@gmail.com

possibility of achieving good results [1]; however, acute control of environment parameters is necessary for experimental testing on smaller scales, as the model becomes highly sensitive to environment input.

Truncated systems are used for cases where complete system tests are unfeasible. This method can further be divided into passive systems and active systems. For passive systems, the mooring lines are commonly modelled using some mechanical means (e.g. springs) at the cut-off section. This approach ignores the dynamic effects of the mooring line however, and has been proven to yield conservative results. An experiment was conducted by comparing two truncated semisubmersible models with that of a full depth system [2]. The three were designed to have matching static offset curves. It was discovered that the dynamic tensions were underestimated in the truncated systems compared to the full length mooring.

To account for the truncated portion, Stansberg et al. [1] combined physical modeling with numerical simulations to assess the viability of passive hybrid model testing for floating structures [1]. The system was designed to achieve a good reproduction of the total restoring force stiffness and quasi-static offset characteristics of a full depth system. A time domain coupled analysis program RIFLEX-C was applied to simulate the dynamic portion of the system. The results obtained after numerical extension to full depth were promising. Currently hybrid model testing is recommended and extensively used in offshore industry.

In an active truncated system, computer controlled devices at the cutoff portion of the mooring/riser system simulate the effects at the truncated portion in real time. One research project initiated at MARINTEK reexamines the feasibility of a proposed active system HydeMoor [3]. The study used a single mooring line to demonstrate the principles of HydeMoor, in which the effect of the lower portion below the cut-off point was simulated using a numerical modeling run on a computer, and applied to the upper physically modeled portion of the line at the truncation point by means of an actuator. The results showed the possibility of achieving good synchronization of truncation points in the system.

In recent years, numerical simulation has become an attractive option for model testing. These may function as an alternative when inadequate facilities are available for physical testing. Additionally, numerical simulations may reduce the amount of cost as well as time in manufacturing a physical system. Technip has recently modified their design process via the development of a numerical wave basin [4]. This was built on the features provided by the commercial software STAR-CCM+, and coupled with their in-house program to improve certain functionalities that was yet fully developed. Validation of the numerical basin against experimental results showed good agreement after calibration.

In this paper, the motions and line tensions of a turret moored FPSO are studied by applying a coupled dynamic analysis simulation. The results are first validated against a published numerical simulation and its corresponding experiment conducted by Kim et al. [5]. The same model is then later compared with two statically equivalent numerical simulations, a quasi-static analysis and a linear spring system.

2 Numerical Modelling

The general concept of the three systems are described, as well as the setup for the numerical model AQWA used in this study.

2.1 Dynamic system

In analyzing the mooring lines of the dynamic system, the effects of mass, drag and inline elastic tension of the line are considered. The mooring line is discretized along the length of the entire line into a finite number of elements subjected to external environmental forces. Combining the cable dynamics with the hull responses, the following equation of motion is solved through coupled time domain analysis:

$$[M + M_a(\infty)]\ddot{u}_p + \int_0^\infty R(t-\tau)\dot{u}_p d\tau + K_H u_p = F_D + F^{(1)} + F^{(2)} + F_p + F_w + F_c + F_{WD} \quad (1)$$

where M and M_a are the mass and added mass respectively, R = retardation function, K_H = hydrostatic restoring coefficient, F_D = drag force on the hull, $F^{(1)}$ and $F^{(2)}$ = 1st and 2nd wave load on the hull respectively, F_p = transmitted force from the mooring lines to the hull, F_w = wind loading, F_c = current loading on hull, and F_{WD} = wave drift damping. The process is computed iteratively for each time interval.

2.2 Quasi-static system

In the quasi-static system, the multiple segments in each line is defined by its stiffness, length, mass per unit length and equivalent cross-sectional area. The drag and inertia components on the line are neglected, and the environmental forces acting on the system are considered as static forces. The system's initial position is first obtained, and the static analysis of the complete system is carried out. The simulation is solved iteratively at each time interval, similar to the dynamic system.

2.3 Linear spring system

The simplest of the three models, the mooring line is modelled as a massless linear elastic cable. Taking k = stiffness, L = stretched line length, and L_0 = unstretched line length, the line tension is given as:

$$T = k\,(L - L_0) \quad (2)$$

This theory allows tension but not compression forces in the line, hence $T = 0$ if $L < L_0$.

2.4 FPSO model and environmental conditions

The AQWA model attempts to reproduce as close as possible the same model as given in [5], in which the main particulars of the FPSO, its mooring systems as well as its corresponding hydrodynamic coefficients are also referred to. The arrangement of the mooring lines is shown in Fig. 1. All 12 lines are modelled in AQWA.

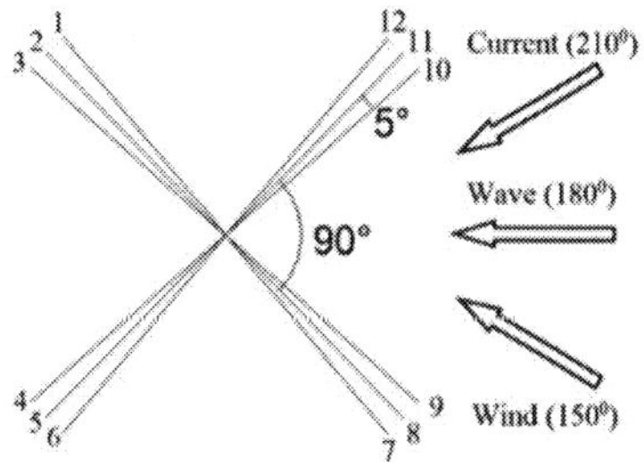

Fig. 1. Mooring arrangement.

In the published work [5], computation of the second order radiation/diffraction for the hull was carried out using Newman's approximation in order to avoid the more computationally expensive QTF method. The former can be reasonably applied when the natural frequency of the floating system is small and the slope of the QTFs near the diagonal is not large. In the case of AQWA, the more accurate full QTF was used for computation.

3 Results and discussions

3.1 Model validation

A 100-year storm condition at the Gulf of Mexico was simulated, the details of which can be found in [5]. For the wind and current force considerations, reference was made to OCIMF [6]. The wind and current coefficients were obtained via linear interpolation in order to obtain the values at 80% loading condition.

The surge static offset was first carried out by displacing the FPSO at varying distances, and plotting the total force in the surge direction, as well as the tensions in the most and least loaded line. The AQWA results were compared with the published work in Fig. 2 and 3.

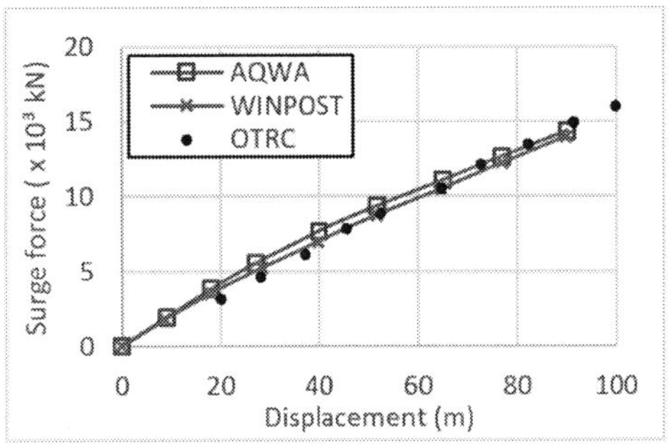

Fig. 2. Static offset results.

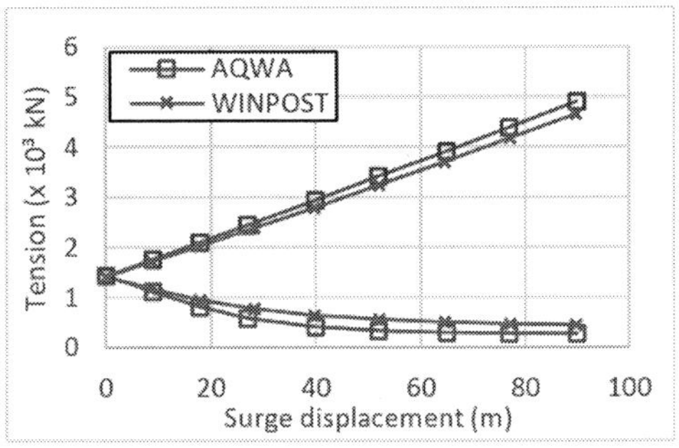

Fig. 3. Tension of most loaded and least loaded line in static offset.

The system's total damping was obtained through free decay tests by displacing the hull at a distance/angle in calm water, and then releasing it to allow it to oscillate freely. The summary comparison is shown in Table 1.

Table 1. Summary of free decay comparison.

Motion	Period (s)			Damping ratio (%)		
	OTRC	WINPOST	AQWA	OTRC	WINPOST	AQWA
Surge	206.8	204.7	192.6	3.0	4.4	12.2
Heave	10.7	10.8	11.1	6.7	11.8	12.5
Roll	12.7	12.7	12.8	3.4	0.7	0.5
Pitch	10.5	10.8	10.1	8.0	10.5	13.5

Fig. 4 to 9 represents the 6 degrees of freedom (DOF) motion responses of the system. The results were first obtained via time domain analysis and later converted to spectral density. It is observed that the surge, sway and yaw are dominated by low frequency (LF) motions, while the heave, roll and pitch are wave frequency (WF) dominated. The surge, heave and pitch show good agreement among the three tests, while the roll spectrum in

AQWA lie between that obtained from WINPOST and OTRC. The discrepancies with WINPOST may be attributed by a couple of factors such as the minor differences in FPSO model design and the calculation of full QTF matrix in AQWA.

Substantial sway and yaw motions are recorded in the AQWA simulation. This is mainly due to the exclusion of hull viscous effects, which are ignored in the 3D potential diffraction/ radiation theory. Additional hull drag damping was manually input in the WINPOST simulation, which was not carried out in AQWA, in order to match that of the OTRC experiment.

The larger surge, sway and roll are recorded to have an effect on the mooring line tensions (Fig. 10-13). In most cases, the line tensions in AQWA are lower than WINPOST, possibly due to the effect of the exclusion of viscous hull drag. It is also noted that the LF components of tension are significantly higher than the WF components.

Disregarding the magnitudes, it is observed that the three tests display similar trends in the respective spectrums. Overall, the AQWA simulation yields satisfactory results when compared to the published literature.

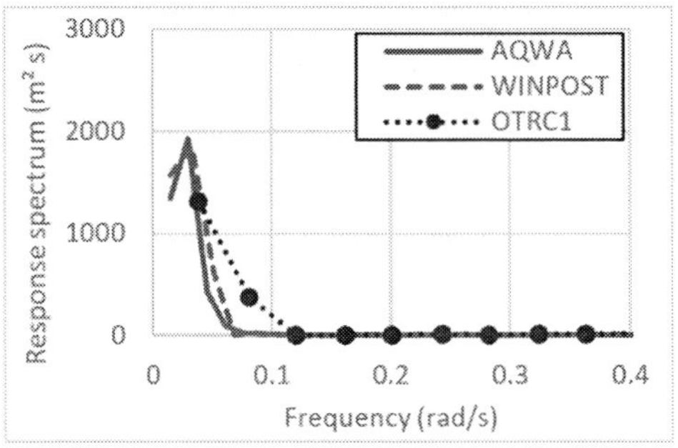

Fig. 4. Surge response.

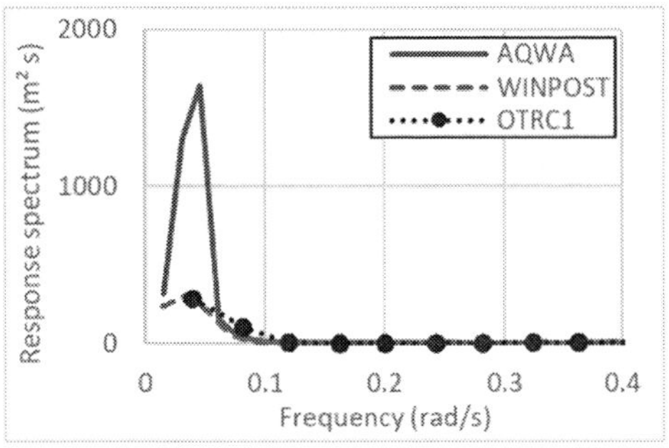

Fig. 5. Sway response.

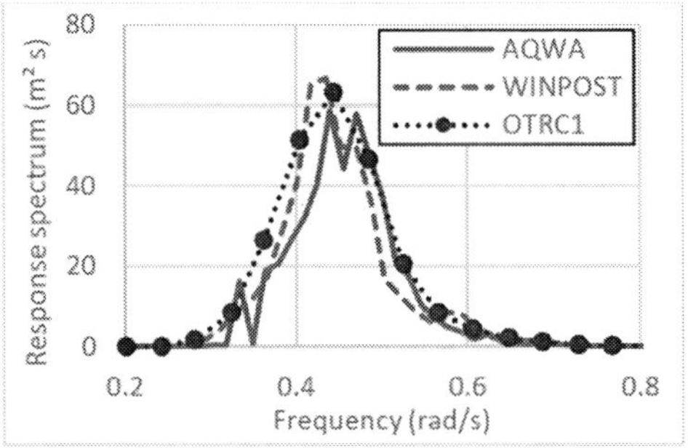

Fig. 6. Heave response.

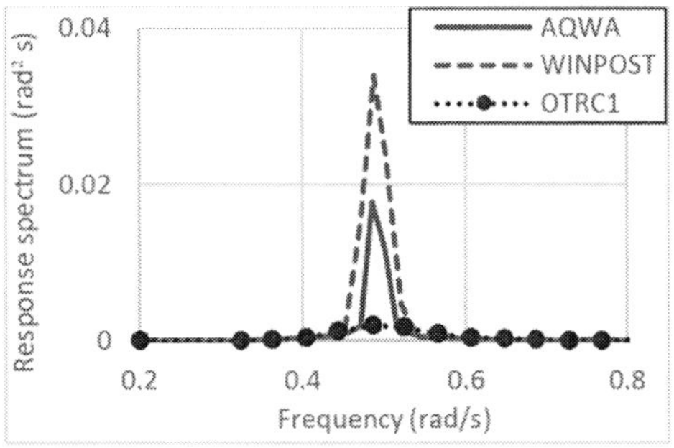

Fig. 7. Roll response.

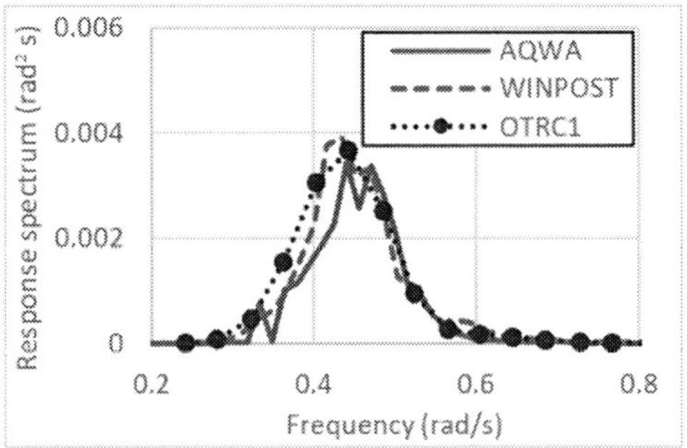

Fig. 8. Pitch response.

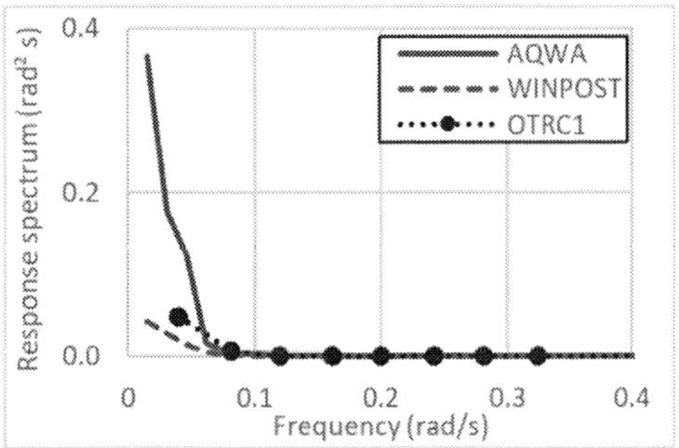

Fig. 9. Yaw response.

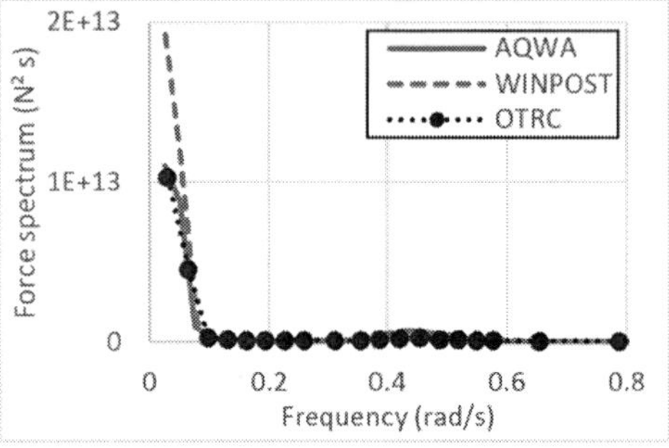

Fig. 10. Combined tensions of lines 1, 2 and 3.

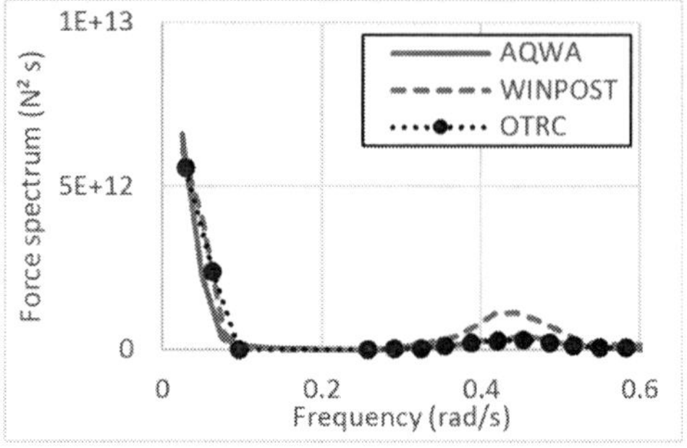

Fig. 11. Combined tensions of lines 4, 5 and 6.

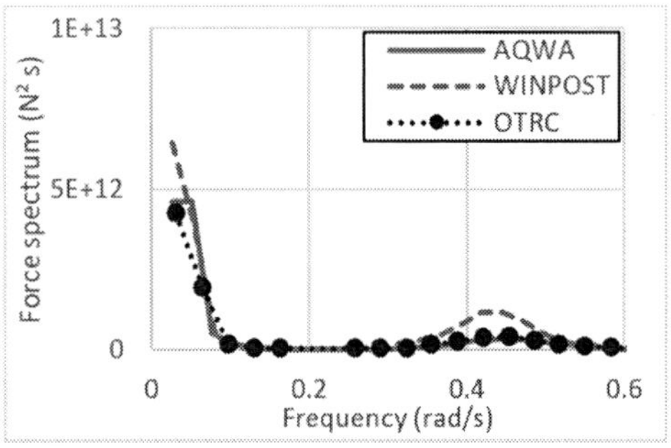

Fig. 12. Combined tensions of lines 7, 8 and 9.

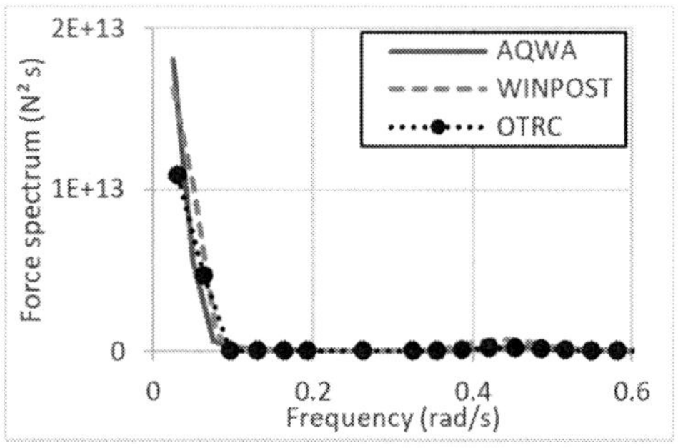

Fig. 13. Combined tensions of lines 10, 11 and 12.

3.2 Dynamic, Quasi-static and Spring Simulation

Unlike in the validation stage, the analyses simulate only the same wave condition in [5], while disregarding the current and wind effects on the FPSO. The previous AQWA model is renamed as the dynamic system and denoted as D. Two additional simulations were modelled: quasi-static (non-dynamic) system and spring system, denoted as ND and SPR respectively. The hull properties remain the same for the three models, however four equivalent linear springs are used in the SPR system instead of the full 12 catenary lines as with D and ND.

The static offset was first conducted, similar to the validation stage (Fig. 14-15). The results of D and ND were found to be the same, and thus placed in the same line. While the resultant surge force for all three cases matches well for the range of surge motions tested, the combined tension of the major loaded lines 7, 8, and 9 for D and ND are significantly higher than its equivalent spring counterpart of the SPR system. However, the stiffness is greater for the SPR, as can be noted in Figure 15 by the higher rate of tension increment against the displacement. Additionally, the static offset of SPR exhibit a linear increase in surge force as opposed to the non-linear weakening behavior of the D and ND system.

The free decay tests were omitted, as matching the damping of the three systems was not considered in this study.

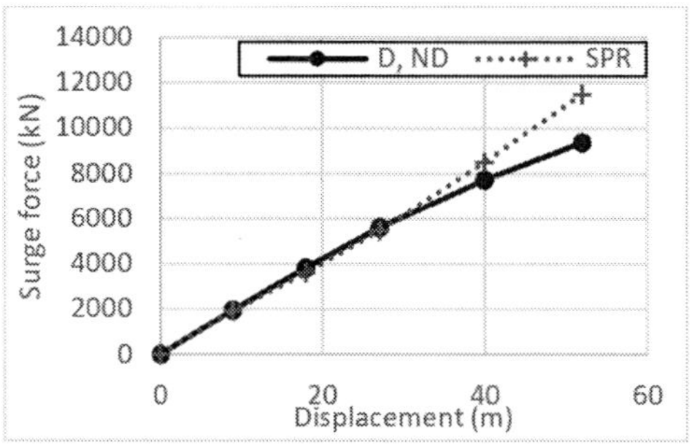

Fig. 14. Surge static offset.

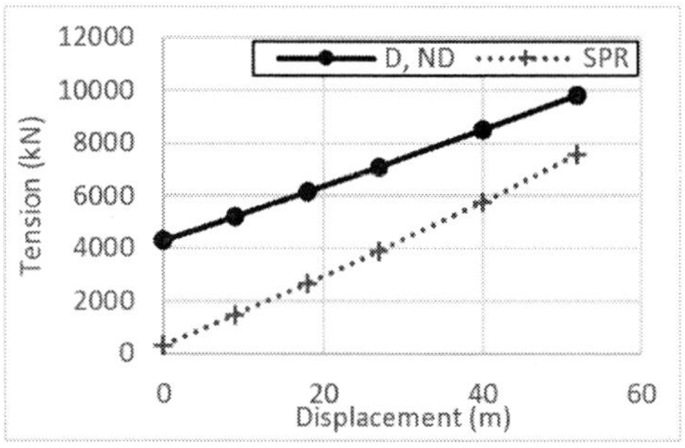

Fig. 15. Combined tension of lines 7,8 and 9.

The station keeping results are shown in Figures 16 to 21. The sway and yaw in unidirectional waves are expected to be insignificant and hence not included, while the combined tensions 6, 7, 8 and 10, 11, 12 is expected to mirror 1, 2, 3 and 7, 8, 9 respectively. It is observed that the WF heave, pitch and roll responses agree well among the three tests, with the exception of the roll for SPR, which overpredicts the motion. However, both ND and SPR overpredict the surge motion. The surge in D is significantly lower as the damping induced from the mooring lines significantly reduce its low frequency motions. For the surge, ND and SPR are almost similar in energy content. This can be expected, as the SPR system can be considered as a type of quasi-static system, but without line mass. Good agreement for ND and SPR is also observed for the combined tensions of the major loaded lines 7, 8, and 9, as surge is the primary contributor to mooring line tension. The combined tensions spectrum of the minor loaded lines 1, 2, and 3 of SPR is smaller than ND due to its shorter line length.

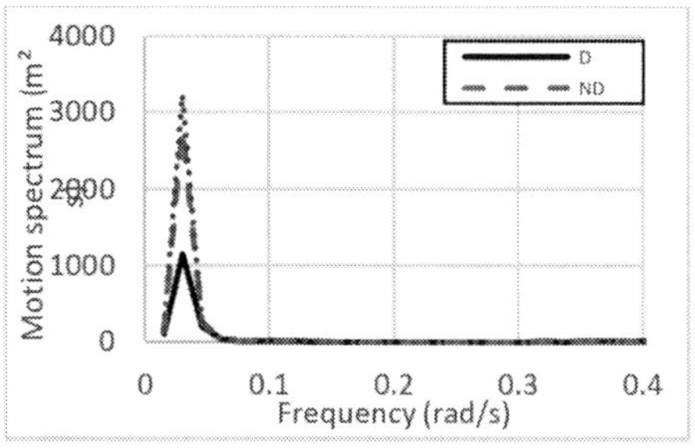

Fig. 16. Surge response.

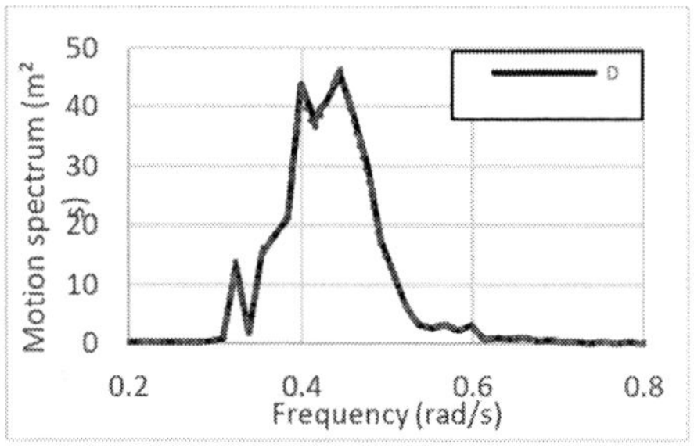

Fig. 17. Heave response.

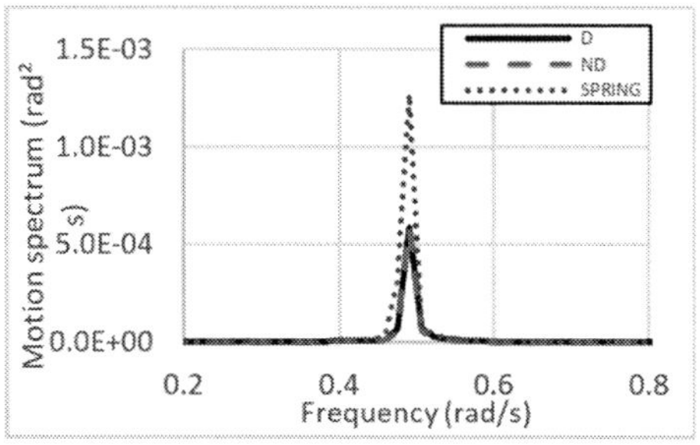

Fig. 18. Roll response.

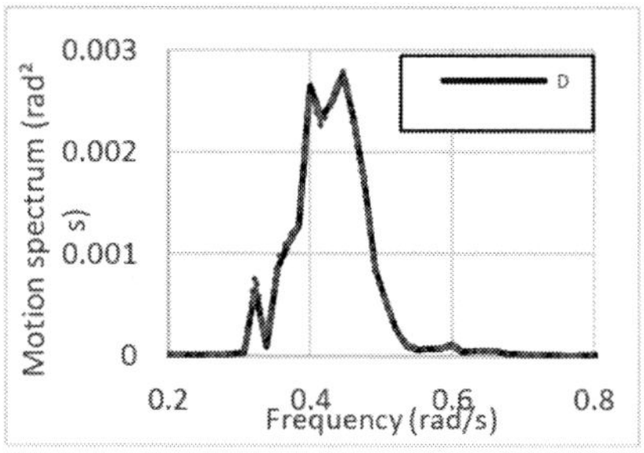

Fig. 19. Pitch response.

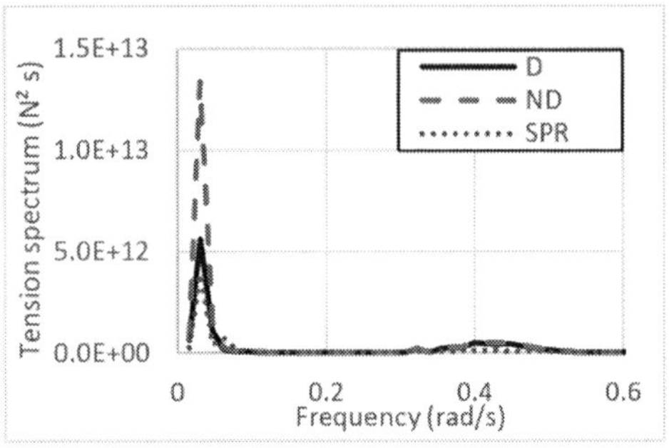

Fig. 20. Combined tensions of lines 1, 2 and 3.

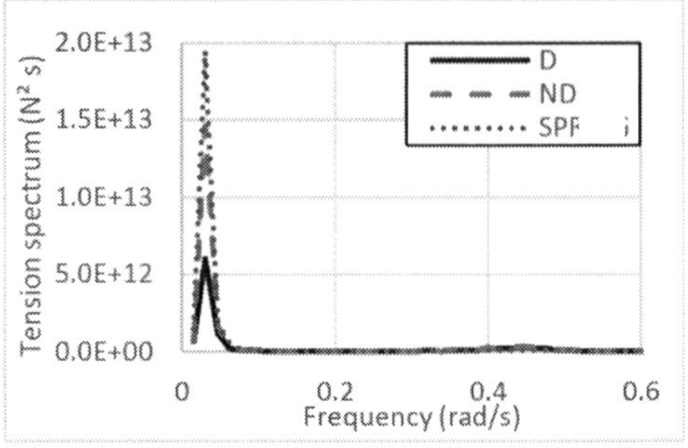

Fig. 21. Combined tensions of lines 7, 8 and 9.

4 Conclusions

The motion responses and line tensions of a turret moored FPSO are obtained using a coupled dynamic analysis program. The results were validated against a published physical test and numerical analysis. The same model was then compared with an equivalent quasi-static system and a linear spring system, under unidirectional wave condition. The heave, roll and pitch are generally in agreement among the three cases, while the surge is overpredicted in the quasi-static and spring system. Consequently, the major loaded line tensions of the two cases are larger than the dynamic system, while the minor loaded line tension for SPR is lower than ND due to its shorter line length. The findings will be used in further research to develop an economical method to improve physical model test results of motions and line tensions in deepwater.

The authors would like to thank the Ministry of Higher Education Malaysia (MOHE), to which the project was funded under the FRGS grant.

References

1. C.T. Stansberg, O. Øritsland, Marintek, G. Kleiven, Norsk Hydro, *Offshore Technology Conference*, 12087 (2000)
2. X. Chen, J. Zhang, P. Johnson, M. Irani, *10th International Offshore and Polar Engineering Conference*, 94-101, 2000
3. Y. Cao, G. Tachiev, *ASME 2013 32nd International Conference on Ocean, Offshore and Arctic Engineering*, OMAE2013-11471, (2013)
4. J.W. Kim, H. Jang, J. Kyoung, A. Baquet, J. O'Sullivan, *Offshore Technology Conference*, 26060 (2015)
5. M.H. Kim, B.J. Koo, R.M. Mercier & E.G. Ward, *Ocean Eng. J.* **32**, 1780-1802 (2005)
6. Oil Companies International Marine Forum, *Prediction of Wind and Current Loads on VLCCs* (Witherby & Co. Ltd., London, 1994)

Treatment of kitchen wastewater using *Eichhornia crassipes*

Rijwana Parwin[1], and Kakoli *Karar* Paul[2,*]

[1]Research Scholar, Department of Civil Engineering, NIT, Rourkela, 769008, India
[2]Assistant Professor, Department of Civil Engineering, NIT, Rourkela, 769008, India

Abstract. The efficiency of *Eichhornia crassipes* for treatment of raw kitchen wastewater was studied in the present research work. An artificial wetland of 30 liter capacity was created for phytoremediation of kitchen wastewater using *Eichhornia crassipes*. Kitchen wastewater samples were collected from hostel of an educational institute in India. Samples were characterized based on physical and chemical parameters such as pH, turbidity, total hardness, nitrate-nitrogen, ammonium-nitrogen, sulphate, dissolved oxygen, total organic carbon and total dissolved solid. The physico-chemical parameter of kitchen wastewater samples were analysed for durations of 0 (initial day), 4 and 8 days. After 8 days of retention period, it was observed that pH value increases from 6.25 to 6.63. However, percentage reduction for turbidity, total hardness, nitrate-nitrogen, ammonium-nitrogen, sulphate, dissolved oxygen, total organic carbon and total dissolved solid were found to be 74.71%, 50%, 78.75%, 60.28%, 25.31%, 33.33%, 15.38% and 69.97%, respectively. Hence water hyacinth *(Eichhornia crassipes)* is found efficient and easy to handle and it can be used for low cost phytoremediation technique.

1 Introduction

Wastewater generated from domestic and industrial activities may be a vital resource if it can be reuse for safe utilization in agriculture [1]. Domestic wastewater contains both black and grey wastewater. Greywater constituted 80% of the total wastewater with maximum contribution (44%) from the kitchen [2]. In developing countries, wastewater from kitchen outlet with high organic material and oil and grease were disposed into rivers and stream without treatment. The characteristics of kitchen wastewater are quite variable among households due to the type of cooking and dietary preference [3]. The presence of less number of pathogenic microorganism and toxic elements in the kitchen wastewater reduces health risk and treatment cost in comparison to other wastewater. The nutrient content of kitchen wastewater can be utilized for reuse in agricultural sector if it can be treated using an ecofriendly, sustainable and low cost technique like phytoremediation. In recent years, the artificial constructed wetland with floating aquatic plant species such as water hyacinth

* Corresponding author: kkpaul@nitrkl.ac.in

 https://doi.org/10.1051/e3sconf/20183402033

[4], water lettuce [5], duck weed [6], vetiver grass [7] have been used for treatment of wastewater. *Eichhornia crassipes* has been considered as problematic aquatic free floating weed because of its uncontrolled growth in water bodies and it is difficult to control and eradicate this plant from water bodies. However, its ability to uptake heavy metals from aquatic ecosystem has been considered as a bioindicator [8].

Limited research has been done on the phytoremediation technique for treatment of kitchen wastewater using *Eichhornia crassipes* [9]. The aim of the present study is to assess the efficiency of water hyacinth (*Eichhornia crassipes*) for the treatment of kitchen wastewater for different retention periods.

2 Materials and methods

2.1 Experimental set up

A bench-scale shallow pond containing water hyacinth was constructed in the premise of the CVR Hall of Residence, NIT Rourkela, India (22.2525° N, 84.9046° E). Wastewater samples were collected from kitchen outlet. The study set up consists of wetland unit (30 L) with water hyacinth (phytoremediation) and another wetland unit (30 L) without water hyacinth to serve as a control and for identifying the pattern of pollution reduction. Healthy young water hyacinth plants collected from nearby pond were chosen for efficient removal of contaminants from kitchen wastewater. The roots of the water hyacinth plants were washed carefully with tap water to remove adhering dirt. Then these plants were kept few weeks in tap water for acclimatization in the stock container. Initially experimental set up was properly cleaned with tap water. The physico-chemical parameters of kitchen wastewater sample were tested at durations of 0 (initial day), 4 and 8 days. Removal efficiency (RF) was calculated using eq. (1). Where C_i and C_e are the influent and effluent concentration respectively.

$$RF = \frac{C_i - C_e}{C_i} \times 100 \quad (1)$$

2.2 Experimental procedure for analysis

The measurement of pH, and temperature were done in the sampling site. Then, collected wastewater samples were taken to laboratory for further detailed analysis of physico-chemical parameters such as turbidity, total hardness (TH), total dissolved solid (TDS), sulphate, dissolved oxygen (DO), nitrate-nitrogen, ammonium-nitrogen and total organic carbon (TOC). The analysis procedures are summarized in Table 1.

 https://doi.org/10.1051/e3sconf/20183402033

Table 1. Experimental procedure for characterization of kitchen wastewater.

Physico-chemical parameters	Standards adopted	Methods / Instrument used
Colour	-	Eye observation
Odour	-	Olfactory sense
pH	-	HQ40D (HACH)
Turbidity (NTU)	-	2100 Q (HACH)
Total hardness (ppm)	IS:3025(Part21):2009	EDTA method
DO (ppm)	IS:3025(Part38):2003	Winkler method
TS, TDS, TSS (ppm)	IS:3025(Part15, 16,17):2009	Gravimetric Method
Sulphate (ppm)	IS:3025(Part24):2003	Turbidity Method
Nitrate-Nitrogen (ppm)	-	DR/890:8039 (HACH)
Ammonium-Nitrogen (ppm)	-	IntelliCAL ammonium ISENH4181 HACH
Total organic carbon (ppm)	-	HACH DR/890:10129

3 Results and discussion

After 8 days of experiment with 4 days interval, physico-chemical parameters were evaluated to check the effectiveness of *Eichhornia crassipes* for kitchen wastewater treatment. As shown in Fig.1, a small increase was observed in pH range from 6.25 to 6.63 (6.08%) by phytoremediation whereas 17.44 % decrease of pH range was observed in control. Study by [10] found an increase of 13% pH value while treating the domestic wastewater using *Eichhornia crassipes*. Study by [11] observed that aeration reduces rate of pH reduction was higher in phytoremediation treatment with *Eichhornia crassipes* in comparison to treatment by *Pistia stratiotes*.

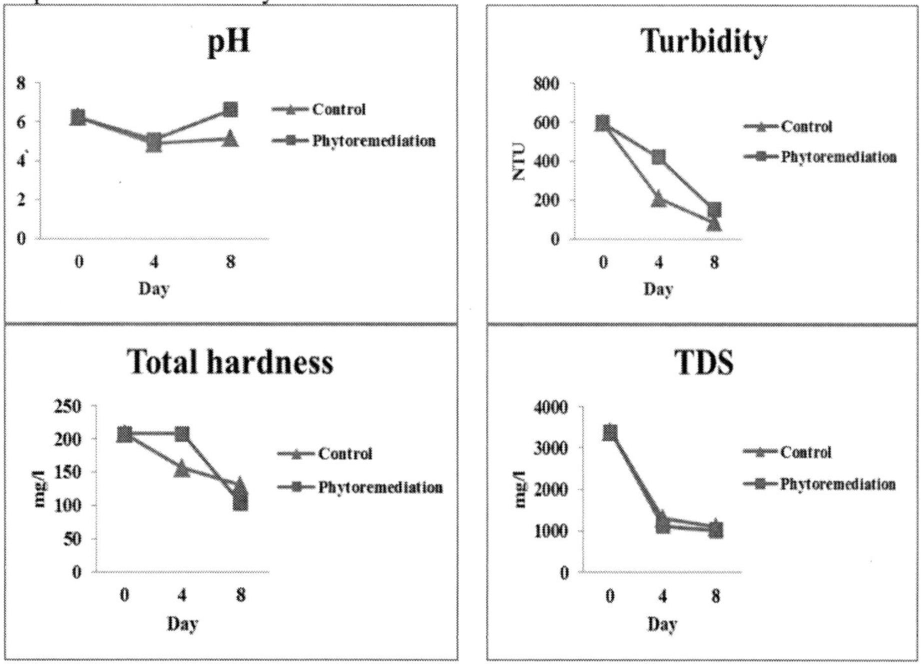

Fig. 1. The removal of pH, turbidity, total hardness and total dissolved solid (TDS) during different time interval.

As demonstrated in Fig. 1 turbidity showed significant reduction from 596.4 NTU to 83.58 NTU (85.98 %) in control whereas 74.71 % reduction efficiency occurred by phytoremediation. The reduction of total hardness was 37.5% for control whereas 50 % reduction for phytoremediation. *Eichhornia crassipes* showed good results over control for total hardness reduction. In this study, the removal efficiency for total dissolved solid was 67.72% in control whereas 69.97% for phytoremediation. About 11% reduction efficiency for TDS was observed by [10] while treating the domestic wastewater using *Eichhornia crassipes*. However, reduction of 26% for TDS was reported by [12] in *Eichhornia crassipes*.

In this study, NH_4^+-N was found to be removed (60.28%) by phytoremediation (Fig. 2) when compared to the control (48.22%). Study by [10] reported 95% reduction efficiency for NH_4^+-N while treating the domestic wastewater using *Eichhornia crassipes*. However in this study, NO_3^--N reduction was 78.75% and 65.62% by phytoremediation and control, respectively (Fig. 2). It was observed that sulphate reduction in control was 69.13% whereas only 25.31% by phytoremediation. Reduction efficiency for TOC was found 13.67% and 15.38% for control and phytoremediation, respectively.

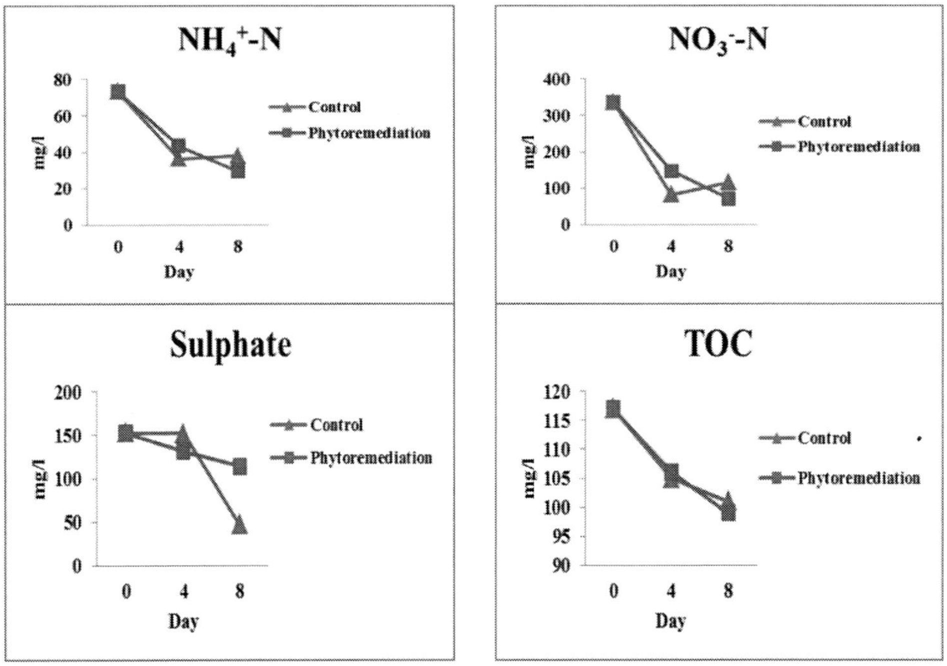

Fig. 2. The removal of ammonium-nitrogen, nitrate-nitrogen, sulphate and total organic carbon (TOC) during different time interval.

In this study, reduction efficiency for DO was same (33.33%) in both control and phytoremediation due to microbial and decomposition activities associated with floating plants (Fig. 3). Aerobic micro-organisms need oxygen for degradation of organic content present in raw kitchen wastewater thus reducing oxygen level in sample on 4th day but on 8th day again oxygen level increases. Dense plant cover on water surface reduces atmospheric oxygen diffusion. The floating plants play vital role in tranferring atmospheric oxygen (30%-40%) to rhizospere through aerenchyma (internal gas space) for aerobic microbial activity [4].

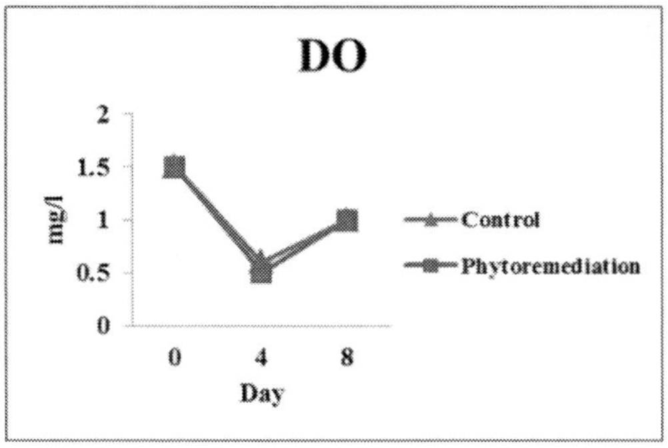

Fig. 3. The removal of dissolved oxygen (DO) during different time interval.

4 Conclusions

In recent years more emphasis has been focused on low cost, eco-friendly, sustainable and green technique like phytoremediation. In present study, an attempt has been made to study on effectiveness of floating aquatic plant, water hyacinth (*Eichhornia crassipes*) for the treatment of kitchen wastewater during different retention periods. After 8 days of retention period, phytoremediation showed good reduction efficiency in total hardness (50%), nitrate-nitrogen (78.75%), total organic carbon (15.38%), ammonium-nitrogen (60.28%) and total dissolved solid (69.97%) whereas, reduction efficiency of turbidity (74.71%) and sulphate (25.31%) were found less in comparison to control condition. However, phytoremediation showed an increase of pH (6.08%). Hence, water hyacinth (*Eichhornia crassipes*) is found efficient in treatment of kitchen wastewater. It can be used for low cost phytoremediation technique.

References

1. M. Raychaudhuri, S. Raychaudhuri, S.K. Jena, A. Kumar, R.C. Srivastava, *DWM Bulletin No. 71*, Directorate of Water Management, Bhubaneswar, India, 43 (2014).
2. K. A. Vakil, M.K. Sharma, A. Bhatia, A.A. Kazmi, S. Sarkar, Separa. and Purifi. Tech. **130**, 160–166. (2014).
3. R.M.S.R. Mohamed, C.M. Chan, H.B. Ghani, M. A. M. Yasin, A.H.M. Kassim, Int. J. of Zero Waste Genera. **1**, 1 (2013).
4. A. Valipour, V.K. Raman, Y.H. Ahn, Water, **7**, 329-347 (2015).
5. Q. Lu, Z.L. He, D.A. Graetz, P.J. Stoffella, X. Yang, Environ. Sci. Pollut. Res. **17**, 84-96 (2010).
6. A. Priya, K. Avishek, G. Pathak, Environ. Monit. Assess. **184**, 4301–4307 (2012).
7. N. Girija, S.S. Pillai, M. Koshy, Ecoscan **1**, 267-273 (2011).
8. E.S. Priya, P.S. Selvan, Arab. J. of Chem. **10**, S3548-S3558 (2017).
9. O.O. Seun, O. N. Adeshina, K. Kolawole, T.S. Adedeji, A. Jamal, Int. J. of Sci. and Engi. Res. **6**, 1, 834 (2015).
10. S. Rezania, M.F.M. Din, S.M. Taib, F.A. Dahalan, A.R. Songip, L. Singh, H. Kamyab, Int. J. of Phytorem. **18**, 7, 679-685 (2016).
11. C.O. Akinbile, M.S. Yusoff, Int. J. of Phytorem. **14**, 3, 201–211 (2012).
12. P. Moyo, L. Chapungu, B. Mudzengi, Adv. Appl. Sci. Res. **4**, 4, 55-62 (2013).

Treatment of hazardous waste landfill leachate using Fenton oxidation process

Pradeep Kumar Singa[1]*, *Mohamed Hasnain* Isa[1], *Yeek-Chia* Ho[1], and *Jun-Wei* Lim[2]

[1] Department of Civil & Environmental Engineering, Universiti Teknologi PETRONAS, 32610 Bandar Seri Iskandar, Malaysia
[2]Department of Fundamental & Applied Sciences, Universiti Teknologi PETRONAS, 32610 Bandar Seri Iskandar, Malaysia

Abstract. The efficiency of Fenton's oxidation was assessed in this study for hazardous waste landfill leachate treatment. The two major reagents, which are generally employed in Fenton's process are H_2O_2 as oxidizing agent and Fe^{2+} as catalyst. Batch experiments were conducted to determine the effect of experimental conditions viz., reaction time, molar ratio, and Fenton reagent dosages, which are significant parameters that influence the degradation efficiencies of Fenton process were examined. It was found that under the favorable experimental conditions, maximum COD removal was 56.49%. The optimum experimental conditions were pH=3, H_2O_2/Fe^{2+} molar ratio = 3 and reaction time = 150 minutes. The optimal amount of hydrogen peroxide and iron were 0.12 mol/L and 0.04 mol/L respectively. High dosages of H_2O_2 and iron resulted in scavenging effects on OH• radicals and lowered degradation efficiency of organic compounds in the hazardous waste landfill leachate.

1 Introduction

Globally, engineered landfill is practiced as the easiest and most economical method of disposing municipal and hazardous solid waste. Generation of leachate has been encountered as a significant issue in landfilling, which contaminates surface and subsurface water leading to the mortality of aquatic flora and fauna. Leachate is a concentrated wastewater generated by passage of rainwater through various layers of landfill. The various factors on which leachate characteristics depends include site hydrology, landfill age, temperature, landfill depth, refuse composition and moisture content [1].

Reverse osmosis, Biological treatment, chemical oxidation, chemical precipitation, activated carbon adsorption, electrochemical oxidation and ion exchange are the various treatment techniques which have been employed for the removal of recalcitrant organic compounds from landfill leachate [2]. Many researchers have reported the chemical treatment techniques, which uses the oxidizing agents viz; Fenton's process, photo-Fenton, UV-Visible light or ozone for leachate treatment. These treatment methods are collectively known as advanced oxidation processes (AOPs), which employ hydroxyl radicals as a strong oxidation agent. Fenton and associated reactions involve reactions of iron with

* Corresponding author: pmsinga@gmail.com

hydrogen peroxide, which form oxygen species that oxidize organic or inorganic compounds. Fenton process was named after H.J.H Fenton in 1894 and later he arrived at the conclusion that hydrogen peroxide could be activated by ferrous as catalyst to oxidize tartaric acid. Past few years, the significance of hydroxyl radical's reactions has been identified and around 1700 rate constants for hydroxyl reactions with inorganic and organic compounds in aqueous solutions have been reported. Recently, Fenton process was vastly used in the treatment of various industrial effluents for the degradation of several hazardous organic compounds [3]. In Fenton's reagents there are two constituents; a catalytic species (Fe^{2+}) and a strong oxidant (H_2O_2) [3]. The Fenton oxidation process has been commonly employed to degrade refractory organic compounds from leachate as a pre and/or post treatment method.

The definition of Fenton's process can be stated as the catalytic production of hydroxyl radicals (OH•) arising from a series of reactions between Fe^{2+} ions and H_2O_2.

Hydroxyl radicals are produced by catalysis of hydrogen peroxide by ferrous ions [4]:

$$Fe^{2+} + H_2O_2 \rightarrow Fe^{3+} + OH^- + OH\bullet \quad (1)$$

Fe^{3+} is again converted to Fe^{2+}

$$H_2O_2 + Fe^{3+} \rightarrow Fe^{2+} + HO + H^+ \quad (2)$$

Additionally, hydroxyl radicals rapidly destroy ferrous ions.

$$Fe^{2+} + OH\bullet \rightarrow Fe^{3+} + OH^- \quad (3)$$

Thereafter the organic matter would be degraded by hydroxyl radicals generated in the landfill leachate:

$$RH + OH\bullet \rightarrow H_2O + R\bullet \quad (4)$$

$$R\bullet + Fe^{3+} \rightarrow R^+ + Fe^{2+} \quad (5)$$

Hence, sufficient of ferrous ion is required to retain the moderate hydroxyl radicals' reduction. But this leads to production of high volume of ferric hydroxide sludge in neutralization step of Fenton process, which has to be separated and disposed. In the Fenton's reaction each of the factors are optimized to boost the reaction between the hydroxyl radicals and the pollutants [5]. Fenton's reaction has some advantages as compared to other AOPs. The advantages of this process include easy implementation and operation, low cost of reagents, high efficiency, lower production of harmful by-products and no requirement of energy for H_2O_2 activation. Since Fenton's process has the above mentioned advantages, it has been used in treatment of various wastewaters and also for treatment of landfill leachate. In order to make the biological treatment economical Fenton's process was used as a post-treatment instead of pre-treatment in the treatment of landfill leachate. By combining the biological and chemical treatment methods COD removal efficiency can be enhanced [6].

Nevertheless, this process also has a few disadvantages, which include fast consumption of Fe^{2+} relative to its rate of regeneration, generation of sludge that requires post treatment, narrow pH range to operate, the complexity of few iron species and the possible waste of oxidants. H_2O_2 is eco-friendly, as it gradually breaks up into O_2 and H_2O. Fe^{2+} is nontoxic in nature, highly available and can be easily removed from water, which makes it the most frequently used transition metal for Fenton's oxidation process. The main objective of this experiment is to determine the effect of significant operating conditions viz., reaction time, molar ratio and Fenton's reagent dosages on COD removal from hazardous waste landfill leachate using Fenton's oxidation process [7]. The outcomes of this research might be useful for increasing COD removal efficiency using Fenton oxidation in the field.

2 Materials and methods

2.1 Sampling

Hazardous waste landfill leachate samples were obtained from Kualiti Alam, Malaysia. Approximately 100 tons of hazardous waste is received by this facility per day and an average of 150 m^3 of leachate is generated per day. Table 1 presents the characteristics of the hazardous waste leachate. The samples were kept in a cold room at 4°C until use. All parameters were measured according to the Standard Methods for the Examination of Water and Wastewater (APHA, AWWA, WEF, 1992).

Table 1. Leachate characteristics.

Parameter	pH	BOD_5	COD	Suspended solids	Nitrates	NH_3-N
Value (mg/L)*	9.53	960	3715	1168	23	1668

**All the values are in mg/L except pH*

2.2 Fenton Experiments

Batch experiments were conducted on Fenton oxidation of hazardous waste landfill leachate using beakers of 500 ml capacity. H_2SO_4 was used for adjusting pH to 3. In every run, desired amounts of $FeSO_4.7H_2O$ and H_2O_2 (30%) were added to determine the optimum dosages of the reagents. All experiments were carried out at atmospheric pressure and room temperature. Fenton oxidation has better efficiency in acidic medium hence initial pH of leachate was reduced to 3 using 1 M sulphuric acid. A magnetic stirrer was used to stir the leachate thoroughly. After completion of each experiment, the pH was increased above 10 by addition of NaOH to suppress the reaction. The degradation efficiency of organics via oxidation ceases at high pH values [8]. The sample was kept undisturbed for half an hour at the end of the experiment to allow the sludge to settle down and the supernatant was collected for COD measurement.

2.3 Analytical methods

COD analyses were carried out by using a DR/5000 spectrophotometer. All solutions were prepared with distilled water. Hach pH meter was made use of to measure the pH. In order to get the accuracy in the results the experiments were conducted in triplicate. Percentage removal was determined by using the equation 6.

$$R = \left(\frac{Co - C}{Co}\right) * 100 \tag{6}$$

Where Co = COD before treatment and C = COD after treatment.

3 Results and discussions

3.1 Optimum molar ratio

H_2O_2 and Fe^{2+} are two important reagents in Fenton process that govern the efficiency and cost of operation. It is essential to find the optimum H_2O_2/Fe^{2+} molar ratio to enhance the efficiency of the process. H_2O_2 and Fe^{2+} play a vital role in the degradation of organic compounds by Fenton process; understanding the phenomenon helps in determination of the optimum dosages of Fenton reagents, as they are utilized to generate the hydroxyl radicals required to degrade organic compounds (Equation 7).

$$\text{Organic substances} + OH\bullet \rightarrow \text{Oxidation Products} \quad (7)$$

It is required to pay additional attention on Fe^{2+} and H_2O_2 dosages, as high dosages of these reagents lead to generation of undesired OH• radicals which has scavenging reactions leading to decrease in the process efficiency [9]. To determine the optimum H_2O_2/Fe^{2+} molar ratio, various H_2O_2/Fe^{2+} molar ratios i.e., 1, 2, 3, 4, 5, 6 and 7 were tested as shown in Fig. 1. The maximum removal of COD was 29.95% at H_2O_2/Fe^{2+} molar ratio of 3 and Fe^{2+} dose of 0.01mol/L.

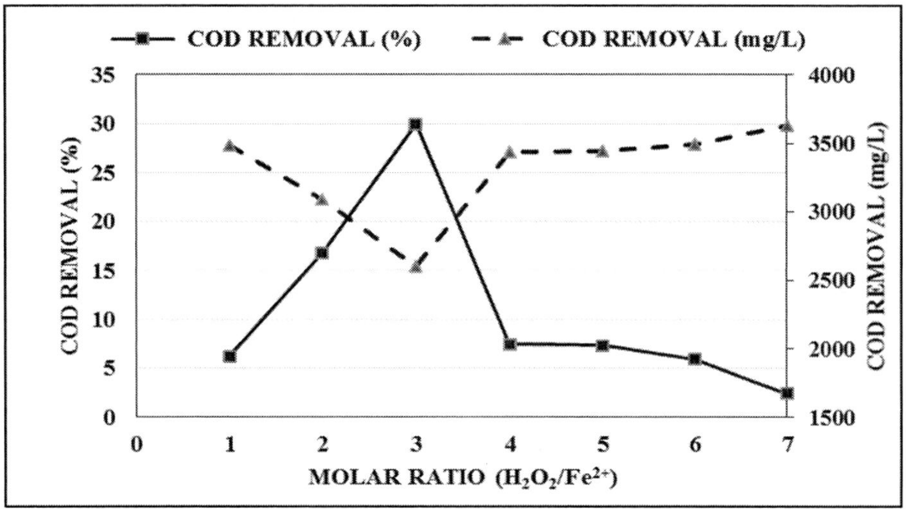

Fig. 1. Molar ratio vs COD removal (Fe^{2+} = 0.01 mol L^{-1}, pH = 3, time = 60 min).

It seems that higher hydrogen peroxide doses result in scavenging effect on hydroxyl radicals (Equation 8).

$$H_2O_2 + OH^{\bullet} \rightarrow H_2O + HO^{\bullet}_2 \quad (8)$$

The produced hydro-peroxyl radical has weak oxidizing potential in comparison with hydroxyl radicals. In addition, the high dosages of H_2O_2 leads to its auto-decomposition forming oxygen and water, and also when the hydroxyl radicals recombine, results in the reduction of concentration of hydroxyl radicals and in addition reduces compound removal efficiency. COD removal efficiency was reduced as a result of scavenging effect of excess Fe^{2+} when the molar ratio was lower than 3 as shown in equation 3. Hence when a high amount of Fe^{2+} is available, Fe^{2+} and organic compounds compete to react with hydroxyl radicals. Thus, Fe^{2+} exhausts hydroxyl radical, finally leading to reduced COD removal efficiency as shown in Equation 3 [10].

Consequently, the excess Fe^{2+} dosage and the less quantities of hydroxyl radicals are present to oxidize organic matters. Additional of Fe^{3+} produced react with hydrogen peroxide to produce Fe^{2+} and hydroperoxyl radicals. Hence, molar ratio 3 was selected as the optimum molar ratio. On the other hand, the concentrations of Fenton reagents that can be found in literature for treatment of leachate are vastly different. As an example Lopez et al. [6] reported optimum molar ratio of H_2O_2 to Fe^{2+} of 20:1, whereas optimum molar ratio was 1.5 as reported by Zhang et al. [11]. It may be due to the variations in the characteristics of leachate, which lead to the vast variations in optimal ratios of H_2O_2 to Fe^{2+}.

3.2. Optimum reaction time

The impact of reaction time on Fenton oxidation was investigated to find the optimum duration of reaction. The reaction time was starting from 30 minutes and varied up to 240 minutes with an interval of 30 minutes each. The initial pH was adjusted to 3 and the dosage of H_2O_2 and Fe^{2+} were kept constant at 0.03mol L^{-1} and 0.01mol L^{-1}, respectively. Fig. 2 shows that COD removal reached equilibrium at 150 min with COD removal of 40.98%. There was no substantial change in COD removal efficiency with further increase in the reaction time.

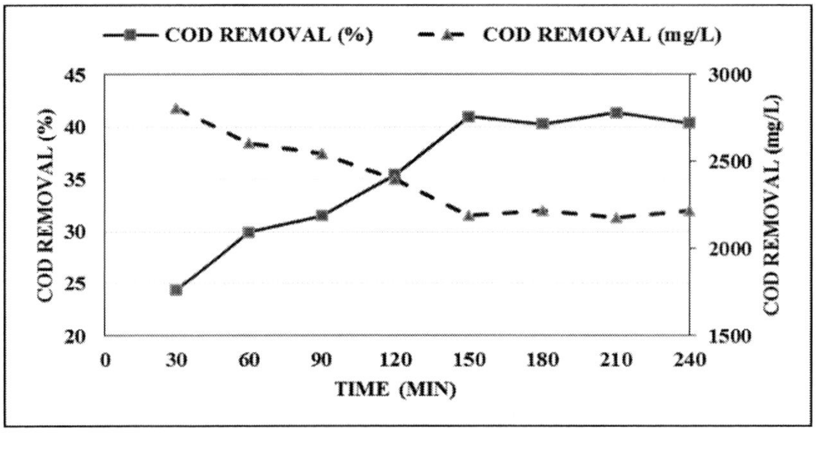

Fig. 2. Reaction time vs COD removal (pH = 3, H_2O_2/Fe^{2+} = 0.03 mol L^{-1}/0.01 mol L^{-1}).

3.3 Optimum Fenton reagents dosages

To evaluate the correlation between the reagents dosage and COD removal efficiencies, a series of experiments were conducted at H_2O_2/Fe^{2+} molar ratio and reaction time of 3 and 150 minutes respectively, and with varying dosages of H_2O_2 and iron.

Fig. 3 shows that there is gradual increase in COD removal percentage as Fenton reagent dosages are increased. Highest COD removal percentage was 56.49% at optimum dosage of H_2O_2/Fe^{2+} = 0.12 mol L^{-1}/0.04 mol L^{-1}. Further increase in the dosages lead to decrease in the COD removal efficiency. This can be explained by taking in to account that the hydroxyl radicals might have been scavenged by the reaction with higher dosages of H_2O_2 and Fe^{2+} as shown in Equations 3 & 8. From the results it can be concluded that the highest COD removal can be attained when both H_2O_2 and Fe^{2+} are dosed appropriately.

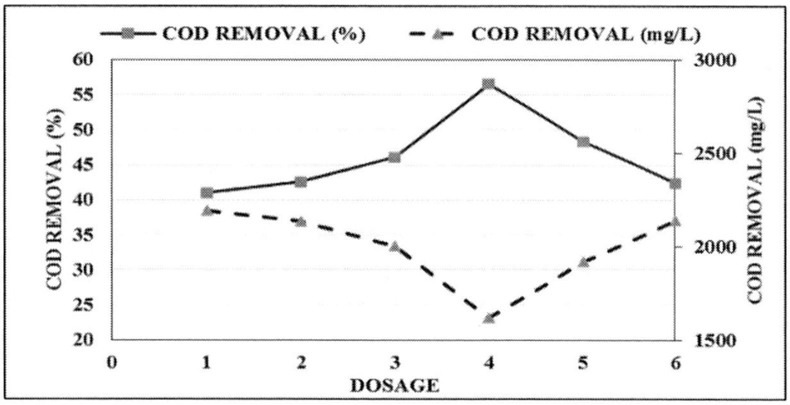

Fig. 3. Dosage of H_2O_2/Fe^{2+} vs COD removal (pH = 3, Reaction time = 150 min).

Table 2. Optimum dosage.

Dosage	1	2	3	4	5	6
H_2O_2	0.03	0.06	0.09	0.12	0.15	0.18
Fe^{2+}	0.01	0.02	0.03	0.04	0.05	0.06

4 Conclusions

Fenton oxidation process seems to be a suitable treatment technique for degradation of organic compounds from hazardous waste landfill leachate. Highest COD removal of 56.49% was obtained under optimum operating conditions. The optimum operating conditions were molar ratio (H_2O_2/Fe^{2+}) of 3.0, pH 3 and reaction time of 150 minutes. The optimum dosage of hydrogen peroxide and ferrous sulphate were 0.12 mol L^{-1} and 0.04 mol L^{-1} respectively.

References

1. S. Renou, J. G. Givaudan, S. Poulain, F. Dirassouyan, P. Moulin, J. Hazard. Mat. **150**(3), 468–493, (2008)
2. Y. Deng and J. D. Englehardt, A. Babuponnusami, K. Muthukumar, J. Environ. Chem. Eng., **2**(1), 557–572, (2014)
3. N. Wang, T. Zheng, G. Zhang, P. Wang, Biochem. Pharmacol. **4**(1), 762–787, (2016)
4. A. R. Ribeiro, O. C. Nunes, M. F. R. Pereira, A. M. T. Silva, Environ. Int., **75**, 33–51, (2015).
5. A. Lopez, M. Pagano, A. Volpe, A. Claudio, D. Pinto, Chemosphere **54,** 1005–1010, (2004)
6. F.J. Beltrán, M. González, F. J. Rivas, P. Alvarez, Water, Air, and Soil Pollution **105**, 685–700 (1998)
7. S. S. A. Amr, H. A. Aziz, Waste Manag. **32**(9), 1693–1698, (2012)
8. S. K. Singh, W. Z. Tang, G. Tachiev, Waste Manag. **33**(10), 2116–2122, (2013)
9. S.Mohajari, H.Aziz, M. Isa, M.Bashir, L.Mohajari, M. Adlan, Journal of Environ Sci and Health, (2010)
10. H. Zhang, H. Jin, C. Huang, Journal of Hazard. Mat. **125**(1-3), 166–174, (2005).

Removal of organic pollutants from produced water using Fenton oxidation

Talia Afzal[1]*, *Mohamed Hasnain* Isa[1], and *Muhammad* Raza ul Mustafa[1]

[1]Department of Civil and Environmental Engineering, Universiti Teknologi PETRONAS, 32610 Seri Iskandar, Malaysia

Abstract. Produced water (PW) is the largest stream of wastewater from oil and gas exploration. It is highly polluted and requires proper treatment before disposal. The main objective of this study was to investigate the effectiveness of Fenton oxidation in degradation of organic matter in PW. The role of operating factors viz., H_2O_2 concentration (0.12×10^{-3} moles/L to 3 moles/L), $[H_2O_2]/[Fe^{2+}]$ molar ratio (2 to 75), and reaction time (30 to 200 minutes), on COD removal was determined through a series of batch experiments conducted in acidic environment at room temperature. The experiments were conducted with 500 mL PW samples in 1L glass beakers covered on the outside with aluminum foil to protect them from sunlight. Pre-decided amounts of ferrous sulfate heptahydrate ($FeSO_4.7H_2O$) and hydrogen peroxide (H_2O_2) were added to initiate the Fenton reaction. An increase in COD removal was observed with increase in reaction time and $[H_2O_2]/[Fe^{2+}]$ molar ratio. COD removal also increased with H_2O_2 concentration up to 0.01 moles/L; further increase in H_2O_2 concentration decreased the COD removal efficiency. Over 90% COD removal was achieved under optimum reaction conditions. The study indicates that Fenton oxidation is effective for remediation of PW in terms of organic matter removal.

1 Introduction

In recent years, an increasing concern about monitoring water quality has been reflected in many studies. The amount of freshwater on Earth is limited and its quality is being constantly threatened due to various types of pollutions. Hence there is a demand for the protection of water resources, in order to prevent their contamination by toxic compounds and pathogenic agents [1].

Organic compounds especially persistent organic pollutants (POPs) may enter the environment by either natural or anthropogenic sources. The former includes volcanic eruptions and forest fires. However, the largest fraction is produced by the latter, namely, by incomplete combustion of fossil fuels, petrochemical processing, automobile exhausts, and tobacco smoke [2-4]. These compounds provoke adverse effects in the ecosystems, even at low concentrations (ng/L-μg/L). They are toxic and persistent, some of them show bioaccumulation effects [5], and are endocrine disrupting as well as tumorigenic substances

* Corresponding author: talia.afzal@gmail.com

[6]. In addition, a class of organic compounds i.e. aromatic hydrocarbons with four or more rings are carcinogenic and mutagenic as a result of their ability to undergo metabolic transformations [7].

Like many production activities, a large volume of liquid waste is produced from oil and gas exploration. This liquid waste is termed Produced water (PW). Produced water is the water that exists in subsurface formations and is brought to the surface during oil and gas production. One of the causes of the harmful characteristics of produced water is the presence of toxic organic compounds. The most abundant organic pollutants in produced water are water-soluble low molecular weight organic acids, monocyclic aromatic hydrocarbons, polycyclic aromatic hydrocarbons (PAHs), and dissolved hydrocarbons like BTEX and high molecular weight alkyl phenols [8,9].

In recent years, research and development of enhanced options to more effectively treat produced water and particularly to remove toxic organic compounds have received growing interests in order to meet the increasingly stringent regulations.

The removal of organic compounds from wastewater can be performed through physical, chemical and biological processes. Physical processes such as volatilization and adsorption [10] can greatly reduce the amount of persistent organic compounds in the wastewater, but they do not solve the problem of pollution due to their inability of contaminant degradation [11]. Conventional phase separation techniques will not remove the water-soluble organics from the aqueous phase. Thus, various biological [12, 13] and chemical [14] oxidation methods have been used to treat produced water. By far, biological degradation has been considered as the most cost-effective treatment method [15,16] However, conventional biological systems exhibited limited ability to remove aromatic hydrocarbons from PW [10,17] due to the bio-recalcitrance, toxicity and high chemical stability of the fused benzene rings [18]. Since some organic compounds are quite resistant to the first steps of biological oxidation [19], chemical oxidative treatment, such as advanced oxidation processes (AOPs), can be employed as an alternative approach for their remediation. AOPs are oxidative technologies by which the hydroxyl radicals are generated to degrade organic contaminants. Among the AOPs, the Fenton method is one of the most promising treatments, due to its high performance, technological simplicity, and moderate cost [20–23].

Commonly, it takes place at temperatures between 20 to 40 Celsius [24] and no energy input is necessary to activate hydrogen peroxide [25]. In Fenton initiation reaction, Fe^{+2} catalyses the decomposition of hydrogen peroxide to generate the hydroxyl radical (OH•) (Equations 1 and 2), which is a strong and relatively non-specific oxidant (with redox potential E of 2.8 V) [26]. It is the main oxidizing specie that can degrade most refractory organics including persistent organic pollutants e.g. PAHs [27]. The reaction between OH• and hydrocarbons is very fast, and the rate constants can be as high as 10 mol /L s [24].

$$H_2O_2 + Fe^{2+} \rightarrow HO^{\bullet} + OH^{-} + Fe^{3+} \quad (1)$$

$$HO^{\bullet} + RH \rightarrow Oxidation\ products \quad (2)$$

Literature reveals that Fenton's reagent is one of the most efficient compositions based on hydrogen peroxide. This method can be effectively used to degrade various contaminants owing to the activity of hydroxyl radical which allows it to oxidize nearly all organic pollutants [28], e.g. phenol ketone, alcohol, benzene, nitrobenzene, perchloroethylene, toluene, p-toluene, aniline, p-nitrophenol, humus compounds etc. Fenton's reagent was also shown to be useful for destabilizing oil emulsions and decomposition of the oxidizing ethyl-t-butyl ether (ETBE) in water petrol extracts [29].

The present work evaluates the feasibility of organic matter degradation in produced water by applying Fenton's reagent at room temperature. The effect of variables (initial concentrations of hydrogen peroxide, reaction time and $[H_2O_2]/[Fe^{2+}]$ molar ratio) that influence the Fenton degradation of organic matter will be determined.

2 Materials and methods

2.1 Produced water and reagents

Produced water was obtained from a local Crude Oil Terminal and used without further treatment before experiments. Table 1 shows the characteristics of PW. Hydrogen peroxide (30%) and $FeSO_4.7H_2O$ were purchased from R & M chemicals Ltd. The pH of the produced water was adjusted by the addition of 98% H_2SO_4 and 98% NaOH solution.

Table 1. Characteristics of PW.

Parameters	Values
pH	7.81
COD	3200 mg/L
TOC	891 mg/L

2.2 Method

To understand the role of different factors influencing Fenton oxidation in COD removal, batch experiments were carried out at room temperature and acidic environment. The experiments were conducted in a 1L glass beaker serving as a reactor. The outside of the reactor was covered with aluminium foil to protect it from light. Homogeneous mixing was provided using a magnetic stirrer. In each of the experimental runs, 500 ml of produced water was stirred in the glass reactor at room temperature. After the adjustments of pH up to 3 using H_2SO_4, a calculated amount of ferrous sulfate heptahydrate ($FeSO_4.7H_2O$) was added as the source of Fe^{2+}. After $FeSO_4.7H_2O$ was completely dissolved in PW, necessary amount of H_2O_2 was added cautiously into the reactor to start the reaction. Time zero coincided with the H_2O_2 addition. After the pre-decided reaction times, which ranged from 30 to 200 minutes, Fenton reaction was seized by increasing the pH of the reacting solution up to 9 by adding NaOH solution. Aliquots were taken from the reactor at measured time intervals and immediately analyzed for COD [30].

2.3 Analysis

The COD concentrations of treated produced water samples were determined by using HACH DR 2800 spectrophotometer under Program 430 using the USEPA Method 8000. TOC measurement of the PW sample was carried out using Shimadzu TOC-VCSH- TOC analyser following the standard operating procedure for the analysis of dissolved and Total organic carbon CCAL 20A.0.

3 Results and discussions

All the experiments were carried out at room temperature. The initial pH of each sample was adjusted at 3.5. Effects of concentration of H_2O_2, $[H_2O_2]/[Fe^{2+}]$ molar ratio and reaction time on COD removed were studied during the experimentation. Details of these experiments are discussed below.

3.1 Effect of H_2O_2 concentration

The concentration of H_2O_2 for Fenton's treatment of produced water was varied in the range of 0.12 mmol/L to 3 mol/L to determine an experimental condition for further research. Results from the experiments are outlines in Fig.1. In all experiments, it was observed that the maximum COD removal was achieved after 120 minutes. On the other hand, the increase in hydrogen peroxide concentration from 1.2 mmol/L to 0.01 mol/L yielded rising removal efficiencies and the percentage of COD removal increased from 76% to 91.6% (Fig. 1). The increase of H_2O_2 dosage leads to an increase in the generation of hydroxyl radical species responsible for the degradation of organic matter, thus increasing the removal efficiency of COD. However, higher concentrations of H_2O_2 lead to lower degradation rates due to the fact that in the presence of excess amount of hydrogen peroxide, chemical oxidation of organics can be inhibited due to the formation of oxygen bubbles instead of hydroxyl radicals [31].

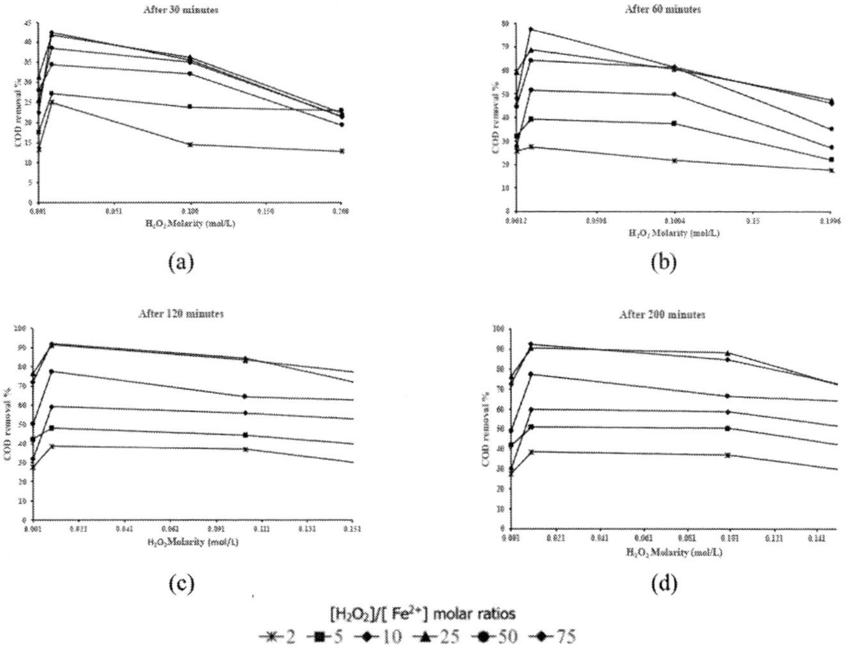

Fig. 1. COD removal with respect to varying H_2O_2 concentrations at different $[H_2O_2]/[Fe^{2+}]$ molar ratios.

3.2 Effect of $[H_2O_2]/[Fe^{2+}]$ molar ratio

To study the role of Fe^{2+} concentration on Fenton treatment of produced water, a series of experiments were conducted with different Fe^{2+} concentrations by keeping the dosage of H_2O_2 at 0.01 mol/L, reaction time 120 minutes and pH was adjusted to the value of 3.5. As by applying these experimental conditions, COD removal efficiency achieved was the highest under the present study. Based on six different $[H_2O_2]/[Fe^{2+}]$ molar ratios, the results obtained are presented in Fig. 2.

During Fenton oxidation, $OH^{\bullet}$ and Fe^{2+} ions both enter a series of reactions that lead to the oxidation of organics. The $[H_2O_2]/[Fe^{2+}]$ molar ratio of 1:5 to 1:25 has been reported in literature [32]. According to the results, it is evident that the maximum COD removal was achieved between $[H_2O_2]/[Fe^{2+}]$ molar ratios of 10 to 25. In Fenton oxidation, Fe^{2+} species act as catalyst which are accountable for speeding up the formation of $OH^{\bullet}$ radicals. In experiments where the amount of Fe^{2+} is in low concentrations, the degradation of organic compounds due to Fenton oxidation is slower due to the unavailability of sufficient amount of catalyst (Fe^{2+}) for the oxidant. Though, under the present study, the removal efficiency of organic compounds decreased by increasing Fe^{2+} concentration due to the scavenging effect exhibited by high concentration of Fe^{2+}. Moreover, if the amount of Fe^{2+} used is too high, there would be production of large quantities of ferric based sludge, which further muddles the oxidation process due to the necessity of its handling and disposal [33].

Bearing in mind the potential generation of sludge and its handling and disposal, the molar ratio of 10 was thought to be suitable to achieve effective reduction of COD in present experimental conditions.

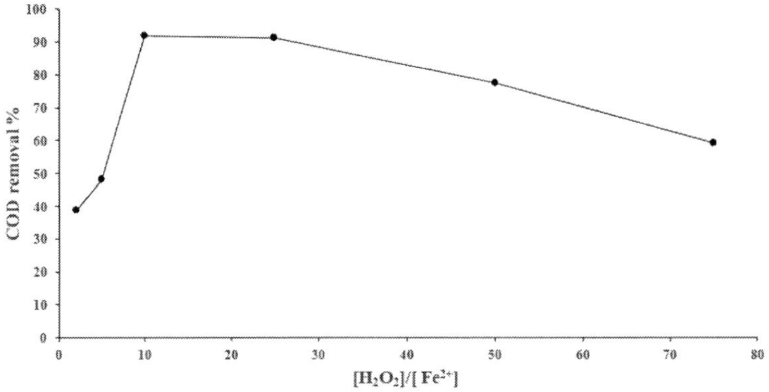

Fig. 2. COD removal at different $[H_2O_2]/[Fe^{2+}]$ molar ratios by keeping H_2O_2 at 0.01 mol/L and reaction time 120 minutes.

3.3 Effect of reaction time

The reaction time of Fenton's reagent with produced water was varied in the range of 30-180 minutes. The highest COD removal of almost 90% was observed at 120 minutes as shown in Figure 3, with initial pH of produced water adjusted at 3.5, dosage of H_2O_2 at 0.01 mol/L and $[H_2O_2]/[Fe^{2+}]$ molar ratio at 10. Reaction time beyond 120 minutes did not give any further COD removal. Hence 120 minutes was considered to be ideal with the afore-mentioned conditions.

It can be seen from the graph that the COD reduction of PW was very rapid in the first 60 minutes of Fenton oxidation, after which the COD reduction became gradual up to 120 minutes. After 120 minutes there was only a slight increase in COD reduction which became steady after 180 minutes. Even though the reaction was seized after 200 minutes, but no further COD removal was achieved after 180 minutes. Since in Fenton oxidation $OH^{\bullet}$ is the key oxidizing species and its production is very quick and vigorous. Thus, in the present study, Fenton reaction of organics initiated immediately as soon as H_2O_2 was mixed with Fe^{2+}.

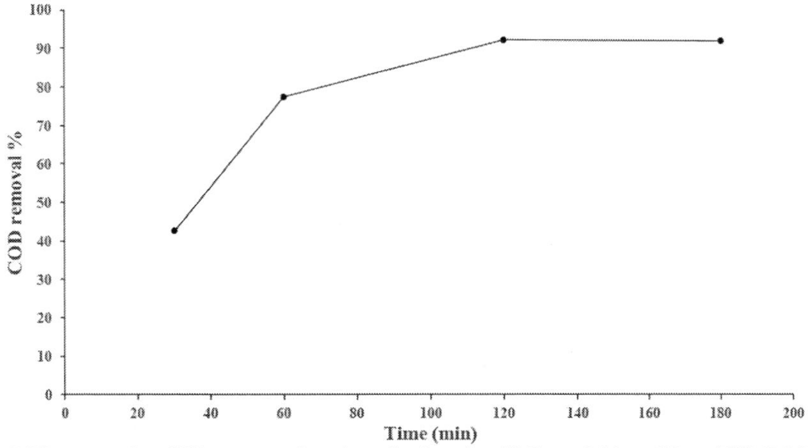

Fig. 3. COD removal at different reaction times by keeping H_2O_2 at 0.01 mol/L and $[H_2O_2]/[Fe^{2+}]$ molar ratio at 10.

4 Conclusions

This paper studied the degradation of organic compounds in produced water by Fenton's reagent at room temperature to make the treatment process more economical. Maximum COD removal efficiency of up to 91% was achieved. Results showed that chemical degradation of organics was highly dependent on Fe^{2+} and H_2O_2 dosage as well as the reaction time. The hydrogen peroxide played an important role during the oxidation as changing hydrogen peroxide concentration changes the $[H_2O_2]/[Fe^{2+}]$ molar ratio , although the degradation of organics increased with the increasing H_2O_2 concentration, at high concentrations of the oxidant COD removal was reduced. With an initial concentration of 0.01 mol/L, up to 91% COD removal was achieved while with 0.1 mol/L only 50% COD removal was observed. The same effect was observed for the concentration of Fe^{2+} ions.

This study demonstrates that Fenton oxidation process is effective for the remediation of produced water in terms of COD removal without increasing the temperature of PW. COD removal percentage may have been higher than the present results if the temperature of PW was increased, but the goal of the study was to make Fenton oxidation efficient as well as economical at the same time, so no heating was provided. It could be considered a promising technology for treatment of produced water.

References

1. Y. Liu, Y.Hashi, J.M. Lin, Analytica Chimica Acta **585** (2), 294–299 (2007)
2. N. Ratola, S. Lacorte, D. Barcel´o, A. Alves, Talanta **77** (3),1120–1128 (2009)
3. S.K. Samanta, O.V. Singh, R.K. Jain, Trends in Biotechnology **20**(6), 243–248 (2002)
4. N. Nadarajah, J. Van Hamme, J. Pannu, A. Singh, Applied Microbiology and Biotechnology **59**(4-5), 540–544 (2002)
5. J. H. Sun, G. L. Wang, Y. Chai, G. Zhang, J. Li, J. Feng, Ecotoxicology and Environmental Safety **72**(5), 1614-1624 (2008)

6. L. Li-Bin, L. Yan, L. Jin-Ming, T. Ning, H. Kazuichi, M. Tsuneaki, Journal of Environmental Science and Health **19**, 1–11 (2007)
7. X. Luo, B. Mai, Q. Yang, J. Fu, G. Sheng, Z. Wang, Marine Pollution Bulletin **48**, 1102–1115 (2004)
8. M. Neff, K. Lee, Produced Water, 3-54 (2011)
9. A. Yaqub, M. Isa , M.S. Kutty, H. Ajab, Electrochemistry **82**(11), 979–984 (2014)
10. R. Crisafully, M.A.L. Milhome, R.M, Cavalcante, E.R. Silveira, D. De Keukeleire, R.F. Nascimento, Bioresource Technology **99**(10), 4515-4519 (2008)
11. A. Rubio-Clemente, R.A. Torres-Palma, G.A. Peñuela, Science of The Total Environment **478**, 201 225 (2014)
12. G.T. Tellez, N. Nirmalakhandan, J.L. Gardea-Torresdey, Adv Environ Res **6**, 455–470 (2002)
13. D.D.C. Freire, M.C. Cammarota, G.L. Sant'Anna, Environ Technol **22**, 1125–1135 (2011)
14. Z.G. Yang, N.S. Zhang, J Xi'an Shiyou Univ Nat Sci **20**, 50–53 (2005)
15. S.M. Bamforth, I. Singleton, Journal of Chemical Technology and Biotechnology **80**(7), 723 736 (2005)
16. T. Ebihara, P. Bishop, Journal of Environmental Engineering **128**(6), 505-513 (2002)
17. F. Fatone, S. Di Fabio, D. Bolzonella, F. Cecchi, Water Research **45**(1), 93-104 (2011)
18. S. Guha, P.R. Jaffe, Environmental Science & Technology **30**(2), 605-611(1996)
19. A.L. Juhasz, R. Naidu, International Biodeterioration & Biodegradation **45**(1-2), 57-88 (2000)
20. V. Flotron, C. Delteil, Y. Padellec, V. Camel, Chemosphere **59**(10),1427–1437, (2005)
21. R. Oliveira, M. F. Almeida, L. Santos, L. M. Madeira, Industrial and Engineering Chemistry Research **45**(4), 1266–1276, (2006)
22. H. Lee, M. Shoda, Journal of Hazardous Materials **153**(3),1314–1319 (2008)
23. S. Wang, Dyes and Pigments **76**(3) ,714–720 (2008)
24. R.J. Bigda, Chemical Engineering Progress **91**(12), 62-66 (1995)
25. M.L. Rodriguez, V.I. Timokhin, S. Contreras, E. Chamarro, S. Esplugas, Advances in Environmental Research **7**(2), 583-595 (2003)
26. J. Nakrst, M. Bistan, T. Tisler, J. Zagorc-Koncan, J. Derco, A.Z. Gotvajn, Water Science and Technology **63**(10), 2131-2137 (2011)
27. K. Nam, W. Rodriguez, J.J. Kukor, Chemosphere **45**(1), 11-20 (2001)
28. J.D. Rodgers, N.J. Bunce, Water Research **35**(9), 2101-2111 (2001)
29. W.R. Haag, C.C.D. Yao, Environmental Science & Technology **26**(5), 1005-1013 (1992)
30. E. Ebrahiem , N. Al-Maghrabi, R. Mobarki , Arabian journal of Chemistry **10**, S1674-S1679 (2017)
31. T.P. Gu, F. Guo, H. Yang, H.P. Wu, G.F. Xu, W. Liu, Z.G. Xie, L. Shi, X. He, S.G. Ji, Nature **477**, 606-610 (2011)
32. K.V. Padoley, S.N. Mudliar, S.K. Banerjee, S.C. Deshmukh, R.A. Pandey, Chemical Engineering Journal **166**(1), 1-9 (2011)
33. Y. Ma, B. Chen, B. Zhang, J. Zheng, B. Liu, H. Zhang, *International Environmental Specialty Conference*, Canada (2014)

Electricity Generation in Microbial Fuel Cell (MFC) by Bacterium Isolated from Rice Paddy Field Soil

Fakhriah Fakhirruddin[1], *Azura* Amid[1], *Wan Wardatul Amani* Wan Salim[1], and *Azlin Suhaida* Azmi[1*]

[1]Department of Biotechnology Engineering, Faculty of Engineering, International Islamic University Malaysia (IIUM), P.O. Box 10, 50728 Kuala Lumpur, Malaysia

Abstract. Microbial fuel cell (MFC) is an alternative approach in generating renewable energy by utilising bacteria that will oxidize organic or inorganic substrates, producing electrons yielded as electrical energy. Different species of exoelectrogenic bacteria capable of generating significant amount of electricity in MFC has been identified, using various organic compounds for fuel. Soil sample taken from rice paddy field is proven to contain exoelectrogenic bacteria, thus electricity generation using mixed culture originally found in the soil, and pure culture isolated from the soil is studied. This research will isolate the exoelectrogenic bacterial species in the rice paddy field soil responsible for energy generation. Growth of bacteria isolated from the MFC is observed by measuring the optical density (OD), cell density weight (CDW) and viable cell count. Mixed bacterial species found in paddy field soil generates maximum power of 77.62 μW and 0.70 mA of current. In addition, the research also shows that the pure bacterium in rice paddy field soil can produce maximum power and current at 51.32 μW and 0.28 mA respectively.

1 Introduction

Microbial Fuel Cell (MFC) is a device, which enables generation of electricity by exploiting bacterial action in metabolizing organic or inorganic substrate under anaerobic condition. The breakdown of the process occurring is that in the anodic chamber, substrate is oxidized by bacteria, producing electrons that are transferred to anode electrode. Transfer of electrons is mediated either by membrane-associated components or soluble electron shuttle. The electrons will then be further transported to the cathode via resistor. Subsequently, current is generated by the positive terminal and flows to the anode, in contrast to the direction of electron flow [1, 2]. Energy generated from MFC is typically evaluated in the form of maximum power density and is renewable and safe for environment. There are several designs of MFC currently being used, the single chamber MFC, dual or two chambers MFC and triple chambers MFC [3]. Single chamber MFC

* Corresponding author: azlinsu76@iium.edu.my

contains only one vessel residing both anode and cathode electrodes. Substrate is filled into the vessel, sandwiched between the two electrodes.

Bacteria responsible in powering up the MFCs are specifically termed exoelectrogens, microorganisms capable of generating and transferring electrons extracellularly [4-6]. Various substrates/ fuels are utilised by exoelectrogens in an MFC, ranging from simple molecules to complex substrates containing different types of organic matter such as wastes and soil [7]. Several exoelectrogenic bacteria species have been successfully isolated from paddy field soil as presented by research from [8]. Denaturing gradient gel electrophoresis (DGGE) of polymerase chain reaction (PCR)-amplified gene fragments was employed for identification of bacterial species residing on the anode and cathode of the MFC. Through database search of the bacteria profiles, several species of bacteria were identified and among them, one of the strains was a close relative to *Natronocella acetinitrilica* (accession number EF103127, 96% identical in nucleotide sequence). Other than that, strains closely related to uncultured *Beijerinckiaceae* bacterium clone GASP-38KB-490-B06 (EU044137, 97%), *Rhizobiales* bacterium A48 (AB081581, 96%), and *Rhodobacter gluconicum* (DQ363135, 99%) were also identified. These bacterial strains were responsible in electricity generation of MFC fed with rice paddy field soil [8].

In another research [9], bacterial species closely related with *Acetobacterium sp.* HAAP 1 (98% identical) from the phylum *Firmicutes* was the dominant species identified an air-cathode MFC. The MFC utilised rice paddy field soil as substrate cum inoculum and the identification was made through 16S rRNA identification method. The dominancy was then followed by a relative of *Firmicutes* bacterium BV9-3a (94% identical) and strain related to *Holophage foetida* (89% identical) [9]. Further, close relative of bacterial strain *Geobacter sulfurreducens* PCA (AE017180) with 99.53% identity (named active strain D-8) was also identified in research [10]. *Geobacter sulfurreducens* is a known exoelectrogenic species with high efficiency in generating electricity and has been isolated from various environment such as freshwater and lotus field mud [11]. Relative of *G. sulfurreducens* isolated from paddy field soil can generate electricity better (i.e. 1.5 times higher current density) than the predecessor (i.e. *G. sulfurreducens* PCA). Strain D-8 could generate up to $1088 \pm 7.7 \mu A/cm^2$ whilst *G. sulfurreducens* PCA was only capable of producing current density at highest $722 \pm 2.6 \mu A/cm^2$ [10].

Bacteria associated with the rhizosphere of paddy plant can also produce electricity. Kouzuma et al. [12] analysed the microbiomes present at anodes of the MFCs set-up in paddy field through pyrotag sequencing and 16S rRNA method. Additionally, two MFC set-ups with paddy field soil as substrate were fed with acetate and glucose and a comparative analysis was made between the three (i.e. rhizosphere MFC, acetate-fed MFC, glucose-fed MFC). For control, bulk soil was also sampled for microbial identification. Bacteria from genus *Geobacter* were identified and were more prominently discovered in the anodes of MFCs rather than in bulk soil. The findings confirmed that *Geobacter* was responsible for the electricity production in MFC. Different strains were in abundance in different MFCs. *G. psychropillus* was identified as the main exoelectrogen present in glucose-fed MFC and rhizophere MFC, whereas *G. sulfurreducens* was more abundant in acetate-fed MFC [12].

Rice is a staple food in a lot of South Asian countries including Malaysia and Japan has been doing multiple researches on MFC with rice paddy field soil as substrate. Mirroring the effort by the Japanese, and with the abundance of rice paddy field soil in Malaysia, research combining MFC and rice paddy field soil seems to be promising. Currently, there are no researches on MFC utilising rice paddy field soil in Malaysia and the absence of similar researches creates a unique opportunity in creating novel research in this country. Additionally, the optimum temperature for all growth stages of rice is in between 18-40°C, a temperature fit of Malaysian weather for the execution of the research [13]. In [14], five samples of paddy field soil with different chemical properties were collected to study the effect of soil condition in electricity generation from MFC. Upon investigation, soil with higher dissolved organic compound (DOC) and ammonium (NH_4^+) content could generate more electricity than the other soil samples [14].

In the present study, a single chamber MFC was prepared with rice paddy field soil as both inoculum and substrate to study the electricity generation by the microbiota present in the soil. A species of exoelectrogenic bacterium was isolated from the mixed microbiota and its capability in generating electricity was observed. The isolated bacterium species is to be identified for further researches on the mechanism of electron transfer and substrate uptake of the bacteria. The results suggested that mixed culture produced more electricity as compared to pure culture inoculated in the paddy field soil.

2 Materials and methods

2.1 MFC configuration

Single chambered MudWatt MFC purchased from KeegoTech, USA was used throughout the research. The anode and cathode were made up of graphite fibre felt with thickness of 0.5 cm and diameter of 8 cm and 1 cm thickness with diameter of 8.5 cm, respectively. Both anode and cathode electrodes were connected to a titanium wire of 22.9 cm and 19.1 cm respectively. The MFC vessel was made from plastic materials, and could accommodate up to 400 mL (equivalent to 400 cm^3) of water/liquid feed. Blinker board with 10 μF capacitor and light emitting diode (LED) was linked with the end of titanium wires connecting to anode and cathode electrodes. The temperature was fixed to room temperature (27^{0}C) at all time, by placing the MFC at designated place in the laboratory of International Islamic University Malaysia (IIUM).

2.2 Feed and inoculum

Rice paddy field soil was taken from paddy field in Sungai Limau, Kedah Malaysia. The MFC vessel was filled with 342.9 g of the paddy field soil that acted as the substrate and also the inoculum from where the bacteria were isolated. The paddy field soil was packed in a very saturated manner to prevent air bubbles formation, especially at anode for no oxygen should be present in this area. Presence of oxygen at anode was undesirable and precautions were taken to prevent oxygen from disturbing with normal MFC function.

2.3 Experimental designs and operation

2.3.1 Generation of electricity and application of resistor

As the LED blinking stabilised (i.e. blinking at similar intervals consistently), the electricity generated by the MFC was measured using two methods. The first method is using Mudwatt explorer application provided by KeegoTech and the second method using resistor of values 47, 100, 220, 470, 1000, 2200 and 4700 Ω.

2.3.2 Isolation of bacteria from soil sample

The bacterial species present at anode electrode and responsible for powering up the MFC were isolated in petri dish. LED that was connected to the blinker board of the MFC must be blinking at the time of isolation, to ensure the presence of exoelectrogenic bacteria. The MFC was disassembled to reach for the anode, and the anode graphite felt was rinsed with sterile distilled water. Sterilised cotton bud was used to swap the surface of anode graphite felt and diluted in 10μL of sterile distilled water. Subsequently, the cotton bud was spread on pre-prepared nutrient agar. Single colonies of bacterial species formed after 24 hours of incubation time on the agar and a colony was transferred to a different agar plate. The steps were repeated until pure culture/colony was obtained. The isolation was performed under sterile condition to avoid contamination of other bacteria aside from the exoelectrogenic bacteria.

Nutrient broth was also prepared as medium for growth of the unidentified bacterial species from rice paddy field soil. When only a pure colony of a single species of bacteria was obtained on agar, the bacterium was then inoculated in 7 mL of nutrient broth and was left overnight. The overnight culture (containing the desired bacteria) was then mixed with 28 mL of nutrient broth medium. The inoculation was done with 20% cell to 80% nutrient broth percentage, and the experiment was done in triplicates.

2.3.3 Optical Density (OD) measurement of growth of bacteria isolated from anode of MFC

Growth of the bacteria in broth was observed and measured hourly (until 8th hour) and daily for seven days using spectrophotometer at 600 nm wavelength. Immediately after inoculation, 1.5 mL of the bacteria cultured in nutrient broth was transferred into a 3 mL cuvette and had the OD_{600} read, and continued at every 30 minutes interval for eight hours. Nutrient broth was used as blank. Based on the OD readings, the doubling time (t_d) for the growth of bacteria was calculated.

2.3.4 Cell Dry Weight (CDW) measurement

The CDW was measured to obtain biomass reading of the bacterial culture. Three aluminium weighing boat were prepared and dried overnight in the oven. The initial weight of the three aluminium boats were recorded upon overnight drying (M1). 1 mL of bacterial culture was transferred into microcentrifuge tube and centrifuged for 30 minutes at 5000 rpm. After 30 minutes, pellet formation was observed on the microcentrifuge tubes. If pellet formed, the supernatant was to be discarded and 1 mL of distilled water was used to resuspend the pellet before the liquid was poured into aluminium weighing boat. If no pellet was formed, the supernatant was poured directly into the aluminium weighing boat. The weighing boats were left overnight in the oven for drying, and the final weight was recorded (M2). The DCW (in g/mL) was calculated using equation (1):

$$Dry\ Cell\ Weight\ \left(\frac{g}{mL}\right) = \frac{[M2(g) - M1(g)]}{1\ mL} \qquad (1)$$

2.3.5 Viable cell count using Hemocytometer

2.3.5.1 Bacterial suspension preparation

Tube containing bacterial culture was swirled to ensure even distribution of bacteria concentration across the tube. 100 μL of suspension from the tube was transferred to a sterilised microcentrifuge tube using sterile pipette. 100 μL of trypan blue solution was then immediately transferred to the same microcentrifuge tube containing bacterial culture. Mixing was done by pipetting the suspension in the tube up and down.

2.3.5.2 Viewing with compound light microscope

After proper mixing of the suspension with trypan blue solution, sufficient amount of suspension was transferred to hemocytometer and let to seep through under the cover slip via capillary action. The suspension was viewed under compound light microscope at 40x magnification. Total number of cells in five squares of the hemocytometer was counted and using equation (2), the viable cell count was calculated;

$$= \frac{\Sigma\, number\ of\ cells\ in\ five\ squares\ \times 1000\ \times dilution\ factor\ (ie: 2)}{number\ of\ squares\ (ie: 5) \times\ 0.004\, mm^3} \quad (2)$$

2.3.6 Inoculation of exoelectrogenic bacterium species into soil

Soil sample was sterilised at 121°C for 30 minutes, twice in 48 hours interval [15] and the MFC chamber was exposed to UV light for the purpose of killing all bacteria present. Pure bacterial culture initially prepared in nutrient broth was mixed with the soil after 8-16 hours growth in the broth. The relevant behind this action is that the growth of bacteria is potentially at exponential phase at period 8-16 hours. After 8-16 hours, 8.6 mL of pre-prepared culture was mixed evenly with sterilised soil sample using spatula [16].

3 Results and discussions

3.1 Isolation of bacteria in nutrient agar

3.1.1 Isolation and growth in petri dish

As a result of isolation of bacteria residing on anode, two major colonies were observed (Fig.1). The two colonies differ in sizes with one colony (labelled as 1 in Fig.1) was significantly smaller than the colony labelled 2. A single colony of 1 and 2 were further streaked, and colony 2 did not grow well on nutrient agar. Growth of colony 2 on nutrient agar plate halted at 8th hour after streaking, hence colony 2 was not considered for inoculation in pure culture MFC. A single colony from colony 1 was also further streaked (Fig.2) and the morphology matched with the original small single colony isolated from anode from Fig

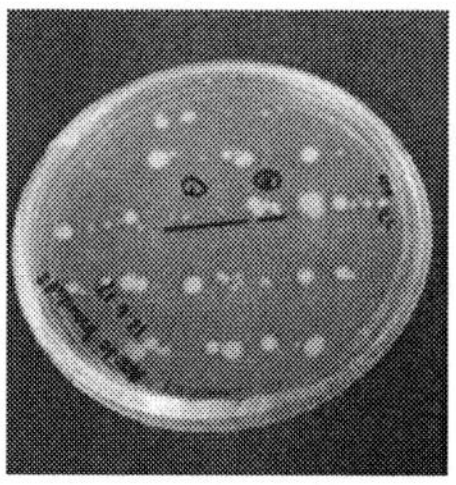

Fig. 1: Isolation of bacteria at anode.

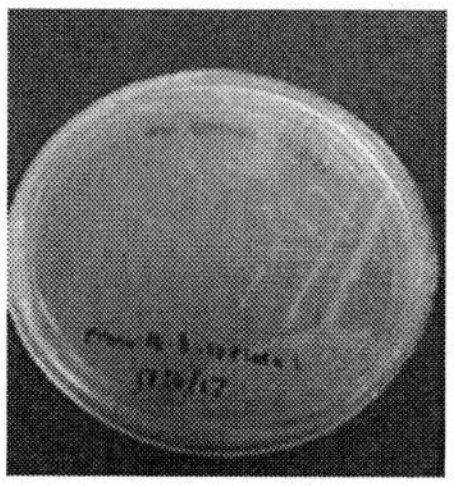

Fig. 2: Pure culture streak.

3.2 Growth of pure/isolated bacteria in nutrient medium

3.2.1 OD reading

Growth of bacteria isolated from anode of MFC was observed through OD reading at 600 nm wavelength for seven days, and at every 30 minutes interval for the first eight hour. The exact species of the bacteria isolated were unknown, hence 600 nm wavelength was chosen based on literature provided in [17]. The experiment was done and the average growth (depicted as ln(OD)) from triplicates of the bacteria was shown in Fig. 3. The OD_{600} reading increased rapidly within the 8 hours of inoculation, and steadily on day 2 onwards, and declined on day 7. The growth pattern of the isolated microorganism was similar to the typical growth pattern of bacteria, confirming the microorganism isolated to be bacteria. Since the aim of the study was to investigate electricity generation by pure monoculture isolated from rice paddy field soil, the characterisation of the bacterial species present was essential. Therefore, the isolated bacteria were grown outside the soil environment to observe the behaviour and characteristics (i.e. morphology of the cell, etc) for comparison and validation with the behaviour of bacterial cells in the soil.

From graph plotting, value of μ was found out to be 0.004 min^{-1}, and the doubling time was calculated to be 173.3 minutes or equivalent to 2.9 hours (Fig. 3). Growth of bacteria was at exponential phase on the first day of inoculation, therefore for pure culture MFC set-up, the bacteria were inoculated in nutrient broth for five hours (300 minutes) prior to being inoculated into the sterilised paddy field soil. Since the doubling time is at the 173th minutes, pure bacteria were inoculated for five hours to ensure decent bacteria concentration exist to power up the MFC when the bacteria were inoculated into the soil. The highest OD_{600} reading (average of three replicates) recorded was on day 6, and started to decline on day 7. The decline in OD_{600} reading indicated death phase for the bacteria.

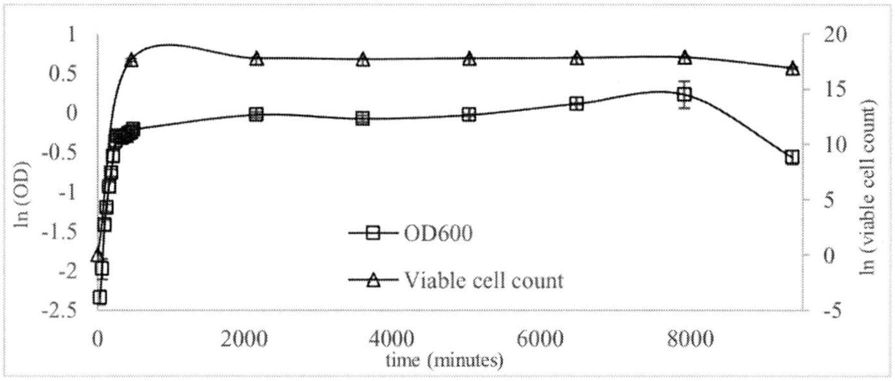

Fig. 3. Growth of bacteria observed as ln (OD_{600}) and ln (viable cell count).

3.2.2 Viable cell count

Total viable cells were calculated every day (i.e. 7 days) based on the protocol mentioned in 2.3.5.2. Total viable cell counts were the highest on day 6, based on the average taken from triplicates (Fig. 3). Data collected from total viable cell count protocol matched with the OD_{600} data shown in section 3.2.1, further confirming that the bacteria will enter death phase on day 7 and is at exponential growth on the first eight hours of day 1, hence inoculation of isolated bacteria into soil for pure culture MFC needed to be done before the growth curve declined. Further, the viable cell counts data ruled out the relevancy of carrying cell dry weight protocol on the isolated bacteria. CDW is informative only if the cell count of microorganism is more or equal to 10^9 [18]. Since the viable cell counts from the triplicates are all lesser than 10^9 (i.e. maximum cell counts are at 10^7) CDW values are not relevant.

3.3 Electricity generation by mixed and pure culture

In this experiment, power generated by mixed culture of exoelectrogenic bacteria and pure exoelectrogenic bacterium species were compared over a period of 30 days. The unknown microbiota residing in anode of the MFC produced up to 27 μW of power on day 14. Power generated from mixed culture MFC was relatively stable until 24th day. However, a decline was observed from day 24 onwards, presumably due to reduction of substrate available (Fig.4). Similarly, pure culture inoculated in the soil generated highest power output on day 13, at 24 μW before decreasing and stabilising at 22 μW from day 14 onwards. Decrease of power generated was observed starting on day 27 onwards (Fig. 4). Limitation of substrate available may be the cause of decline in power output. There were two major colonies formed on nutrient agar upon isolation of bacteria from anode, and only one colony was inoculated into the soil sample for study of electricity generation by pure culture, due to the survival of the colony on petri dish for more than two days.

Both pure culture and mixed culture have been successfully inoculated inside MFC for successful production of current. Despite, the amount of electricity generated when mixed culture was utilised significantly differ as opposed to pure culture. Mixed culture generally produced more electricity than pure culture, due to complex syntrophic relationship between the microorganism species in the MFC. Individual bacterial species in mixed culture was assumed to have specific mechanism in transferring electrons to produce high electricity generation, but the complex relationship has yet to be explained scientifically [1, 19]. Hence, in this experiment mixed culture MFC was able to produce slightly higher power output compared to pure culture MFC, agreeing claims from literature.

Some of pure bacterial species typically studied in MFC research were *Geobacter sp.*, *Shewanella sp.*, and *Rhodoferax sp.*. Pure exoelectrogenic species require longer time for growth in the chamber, a characteristic attributed to the anaerobic condition of the anode chamber. On the other hand, mixed culture usually grows faster, and is able to produce higher power output for the MFCs. Mixed culture is preferred as inoculum in MFC, but using mixed culture increase the startup time and longer time is required to reach stability of the power generated by the fuel cell. Furthermore, handling mixed culture is more challenging than pure culture. The difference in startup time of MFC with mixed and pure culture could be seen from Fig. 4, at which MFC with mixed culture light up after four days of inoculation, while pure culture MFC only took two days for the LED to be blinked. Additionally, mixed culture MFC required longer time (i.e. 17 days) to reach steady power output as opposed to pure culture MFC (i.e. 14 days).

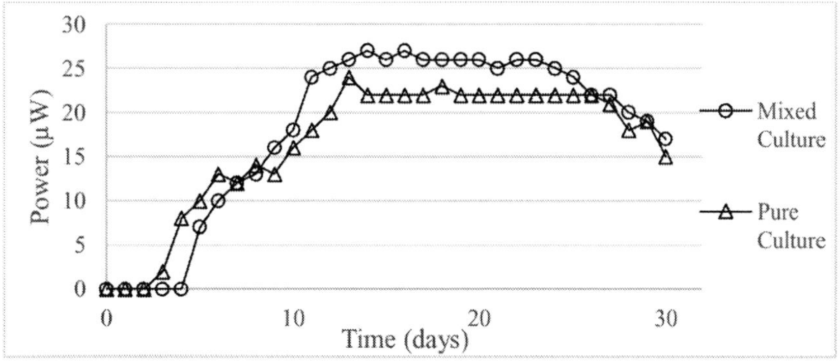

Fig. 4. Power generation in MFC by mixed and pure culture for 30 days.

3.4 Maximum power and current output

The production of current (I) and voltage(V) from both MFC supplied with mixed and pure culture, linked with resistors of different values, and the maximum power and current output for mixed bacterial species were 77.62 μW and 0.70 mA, respectively. While for pure culture, the maximum power and current output were 51.32 μW and 0.28 mA, respectively.

4 Conclusions

In this research, the ability of bacteria isolated from rice paddy field soil to produce electricity calculated in power and current production was studied. Single chamber MFC provided with rice paddy field soil and the microbial consortia found in the soil generates 77.62 μW of power and 0.70 mA of current. Pure exoelectrogenic bacteria isolated from the same microbiota generated maximum power of 51.32 μW and 0.28 mA. The maximum power and current could be achieved only when external resistance was equal to the internal resistance, and each MFCs (i.e. mixed culture and pure culture MFC) gave out different resistance values, with mixed culture MFC at 470 Ω and pure culture MFC at 2200 Ω.

Growth of unknown bacteria was observed, and the bacteria reached the exponential phase of growth on the first day of inoculation, and the doubling time was calculated to be at 2.9 hours (173.3 minutes). Knowledge on the growth phase of the bacteria helped in deciding the exact time for inoculation of bacteria into rice paddy field soil for electricity generation by the pure culture MFC. Though the research was done in small scale, but this research act as a basis for the implementation of MFC device in actual rice paddy field in the future.

References

1. K. Rabaey, W. Verstraete. Trends Biotechnol **23**, 291-8 (2005)
2. B.E. Logan, B. Hamelers, R.A. Rozendal, U. Schrorder, J. Keller, S. Freguia, et al., ES & T **40**, 5181-92 (2006)
3. M. Rahimnejad, A. Adhami, S. Darvari, A. Zirepour, S-E Oh, AEJ **54**, 745-56 (2015)
4. D.R Lovley, Nature Rev. Microbiol. **4**, 497-508 (2006)
5. I.S. Chang, H. Moon, O. Bretschger, J.K Jang, H.I Park, K.H Nealson, et al., J. Microbiol. Biotech. **16**, 163-77 (2006)
6. B.E. Rittmann, R. Krajmalnik-Brown, R.U. Halden, Nature Rev. Microbiol. **6**, 604-12 (2008)
7. B.E. Logan, Nature Rev. Microbiol. **7**, 375-81 (2009)
8. N. Kaku, N. Yonezawa, Y. Kodama, K. Watanabe, Appl. Microbiol. Biotechnol. **79**, 43-9 (2008)
9. S. Yamamoto, K. Suzuki, Y. Araki, H. Mochihara, T. Hosokawa, H. Kubota, et al., JSME **29**, 145-53 (2014)
10. D.D. Deng, Y.C. Zhang, Y. Liu, RSC Adv. **5**, 43978-89 (2015)
11. S. Viulu, K. Nakamura, A. Kojima, Y. Yoshiyasu, S. Saitou, K. Takamizawa, J. Gen. Appl. Microbiol. **59**, 325-34 (2013)
12. A. Kouzuma, T. Kasai, G. Nakagawa, A. Yamamuro, T. Abe, K. Watanabe, Plos One, **8** (2013)
13. D. Datta, *Principles and practices of rice production,* (Int. Rice Res., 1981)
14. N. Wang, Z. Chen, H.B Li, J.Q. Su, F. Zhao, Y.G. Zhu, J. Soils Sediments **15**, 926-36 (2015)

15. R. Shuab, R. Lone, J. Naidu, V. Sharma, S. Imtiyaz, K.K. Koul, JAES **14**, 527-35 (2014)
16. U.C. Brinch, F. Ekelund, C.S. Jacobsen, Appl. Environ. Microbiol. **68**, 1808-16 (2002)
17. P.M. Steubing, *Tested studies for laboratory teaching,* **14,** 81-114 (1993)
18. J.A. Myers, B.S. Curtis, W.R. Curtis, BMC Biophys. **6** (2013)
19. Y. Zuo, D.F. Xing, J.M. Regan, B.E. Logan, Appl. Environ. Microbiol. **74**, 3130-7 (2008)

Potential of Micro Hydroelectric Generator Embedded at 30,000 PE Effluent Discharge of Sewerage Treatment Plant

M.A. Che Munaaim[1*], *N.* Razali[1], *A.* Ayob[1], *N.* Hamidin[1], and *M.A.* Othuman Mydin[2]

[1]School of Environmental Engineering, Universiti Malaysia Perlis, Arau Perlis, Malaysia
[2]School of Housing, Building and Planning, Universiti Sains Malaysia, Pulau Pinang, Malaysia

Abstract. A micro hydroelectric generator is an energy conversion approach to generate electricity from potential (motion) energy to an electrical energy. In this research, it is desired to be implemented by using a micro hydroelectric generator which is desired to be embedded at the continuous flow of effluent discharge point of domestic sewerage treatment plant (STP). This research evaluates the potential of electricity generation from micro hydroelectric generator attached to 30,000 PE sewerage treatment plant. The power output obtained from calculation of electrical power conversion is used to identify the possibility of this system and its ability to provide electrical energy, which can minimize the cost of electric bill especially for the pumping system. The overview of this system on the practical application with the consideration of payback period is summarized. The ultimate aim of the whole application is to have a self-ecosystem electrical power generated for the internal use of STP by using its own flowing water in supporting the sustainable engineering towards renewable energy and energy efficient approach. The results shows that the output power obtained is lower than expected output power (12 kW) and fall beyond of the range of a micro hydro power (5kW - 100kW) since it is only generating 1.58 kW energy by calculation. It is also observed that the estimated payback period is longer which i.e 7 years to recoup the return of investment. A range of head from 4.5 m and above for the case where the flow shall at least have maintained at 0.05 m^3/s in the selected plant in order to achieved a feasible power output. In conclusion, wastewater treatment process involves the flowing water (potential energy) especially at the effluent discharge point of STP is possibly harvested for electricity generation by embedding the micro hydroelectric generator. However, the selection of STP needs to have minimum 4.5 meter head with 0.05 m^3/s of continuously flowing water to make it feasible to harvest.

* Corresponding author: arkam@engineer.com

1 Introduction

Recently, energy crisis related problem such as depletion of fossil fuel, climate change and further increasing demand of electricity supply is continuously raised world-wide [1]. One of the popular alternatives is using the renewable energy resources that avoid from bringing impact to the environment while resolving the problem of energy crisis. The renewable energy sources include wind, solar and hydroelectric power plant is normally used [2, 3]. The application of micro hydro for electrical generation is something promising for future especially for small scale user.

Micro hydroelectric power is considered as a non-polluting, renewable and environmental-friendly source of energy [4]. According to [5], it is utilized as a source of natural energy since water is the only source that is required by the system and does not cause impact to the environment. Besides, neither a large dam nor a land flooded is required for the application of a micro hydropower system [5]. However, it is capable of consistent in supplying stable electrical energy by only using pressure of treated wastewater from the discharge of sewerage treatment plant. Recently, this system is becoming more popular as it is reliable and involves minimum maintenance and care.

In sewerage treatment plant, the solid waste or the treated sludge which suitable for disposal of reuse are usually processed as fertilizers. On the other hands, the treated water in fact does not exploit in any other usage but only discharge into the river or drain. Hence, there are actually possessed of a high potential of hydropower application in this sector. The treated water can be utilized for generating electricity by a micro hydro power plant before releasing it back to the environment. [6] stated that treated water of sewerage at a high pressure or flowing with a high velocity possess of a high potential energy can be used to run turbine or water wheel attached to generator and therefore of electrical power can be generated.

Micro hydro, as one of the small hydropower systems with a power rating of 100kW or less and a 100kW system will produce 100 standard units of electricity in one hour on ideal case [7]. This paper promotes small-scale hydro schemes that desired to generate up to 12 kilowatts as targeted power output. There is approximately 50% to 65% of the net power demand for a typical activated sludge wastewater treatment plant is consumed by the aeration system. Hence, if the implementation of micro hydropower system is successful achieved, it can be contributing to a significant figure in term of cost saving for the electrical bill since the system can facilitate in cutting down the expenses of electrical generation of the plant operation especially the consumption from aeration system.

This paper targeted to explain the potential of generating electricity through the micro hydro generator at 30,000PE sewerage treatment plant by assessing the available parameters on actual site in Perlis, Malaysia that been conducted in 2016.

2 Materials and methods

The aim of this research is to identify the potential of micro hydroelectric generator embedded at 30,000 PE effluent discharge of sewerage treatment plant. The plant that has been chosen for this research is located in the Rural Sewerage Treatment Plant (RSTP) at Wang Bintong, KNR059, Jalan Raja Syed Saffi, 01000 Kangar, Perlis Indera Kayangan. Population equivalent that served by this STP facility is given as maximum serving value of 30,000 PE and the normal service value is maintain at 21,000 PE, which is recorded at a percentage of 70% as compare to the maximum state. The effluent from this STP facility is referring to Standard B (Category II) and direct discharge into river. Total head of difference between effluent discharge point and the connected discharge channel is 3.5 meter.

2.1 Preliminary data

Table 1 indicated the preliminary data obtained from RSTP Wang Bintong, KNR 059 and used for computation of potential power harvested by the micro hydro generator and its feasible payback period.

Table 1. Preliminary data from RSTP Wang Bintong, KNR 059.

Preliminary data	Values
Population equivalent, PE	30,000
Average flow rate of effluent, Q	0.054 m^3/s
Head of effluent, H	3.5 m
Gravitational acceleration, g	9.81 m^2/s
Density of effluent, ρ	1000 kg/m^3
Efficiency of turbine, η	0.85
Average total electrical energy consumption per month	53,004 kWh
Average total electrical bill per month	RM 22,569.50
Average rates of electrical bill per kWh usage	RM 0.426/kWh

2.2 Calculation of the potential electrical power output

Based on head, flow, and constants related to the water density, gravitational acceleration, and efficiency, the electrical power that can be generated from a turbine at the RSTP Wang Bintong was determined using Equation:

$$P_e = P_h \times \eta = Q \times H \times g \times \rho_w \times \eta \quad (1)$$

Where,

P_e is electrical power (kW)
P_h is hydraulic power (kW)
η is the efficiency of the turbine (unitless)
ρw is the density of the water (kg/m^3)
g is the acceleration due to gravity (m/s^2)
Q is the flow of water through the generator (m^3/s)
H is the head (m)

Electrical power that can be generated is depend on the water flow rate and head besides the gravity and density of the water as constant.

2.3 Estimation of feasible payback period

The payback period is obtained by initially identify estimated power generation by the generator and also its contribution to the saving in electrical bill of plant facility based on annual operation. The monthly contribution in term of saving in electrical bill is then compared to the total electrical bill for the plant operation in RSTP Wang Bintong, KNR 059 in order to attain the feasible payback period of the implementation of the generator.

2.4 Viability of hypothetical micro hydropower scenarios

In order to define the viability of installing micro hydroelectric power generator for the hypothetical cases of head and flow that were created, three steps were taken. First, potential power generation (in kWh) was calculated for all possibilities of flow and head combinations within the ranges. These values were calculated based using equation of potential power output. Second, the potential power generation values were converted into savings, using average rates of RM 0.426 per kWh usage in electrical energy consumption. Yearly savings (rounded to the nearest Ringgit Malaysia) for each combination of artificial head and flow cases computed. Finally, the yearly savings were compared to an estimated capital of investment of RM 150,000. Payback periods were calculated for each of the cases. The hypothetical cases and their payback periods are then listed in tables.

3 Results and discussions

3.1 Potential power output

From this research, the output power obtained, 1.58 kW is found to be categorized as pico hydro power which is outside the scope of micro hydro power as shown in Table 1.1. As a result, this plant may not be recommended to implement the system as the outcome may not be satisfied and some modification need to be done before practical implementation. In the meanwhile, other bigger capacity plant shall be proposed in order to suit the criteria for the implementation of the system.

Table 2. Power output calculated from the equation-Equation (1).

Power output	Values
The amount of hydraulic power available in the system, P_h	1.85 kW
Conversion of electrical power from hydraulic power, P_e	1.58 kW

On the other hands, in order to attain the targeted output power for implementing the system at RSTP Wang Bintong, the minimum flow to coordinate is derived. Rearrange the equation (1) previously of electrical power, the required flow rate is found to be,

$$Q_{req}=P_{req}/(H\times g\times\rho\times\eta) \qquad (2)$$

Table 3. The comparison of required flow based on the difference in electrical power required.

Electrical Power Required, P_{req} (kW)	Head, H (m)	Gravitational Constant, g (m^2/s)	Turbine Efficiency, η	Flow Required, Q_{req} (m^3/s)
5.0	3.5	9.81	0.85	0.17
12.0	3.5	9.81	0.85	0.41

As shown in the table above, it is observed that a minimum flow rate of 0.41 m^3/s shall be required at the effluent discharge point to achieve 12kW under consideration of the other input data remain as usual. The required flow rate, Q_{req} is resulted to be around eight-fold compared to the actual flow rate of effluent, Q if the head is assumed to be maintained as the same, H = 3.5 meter. According to [6] earlier, a micro hydro power plant usually generated electrical power with a minimum value of 5 kW. Thus, it is found that the required flow for effluent discharge shall maintained at an average of 0.17 m^3/s in order to fulfil the minimum electrical power generation of micro hydro power if other input data is remained at constant. This value is differing about 3-fold compared to the actual flow at the effluent discharge at RSTP Wang Bintong, KNR 059.

Table 4. The comparison of power output based on different head and flow rate in varies locations of facilities.

Facility	Gross Head (m)	Flow Rate (m^3/s)	Power output (kW)	References
Emmerich, Germany	3.6 - 3.8	0.40	13	[8]
Puan Hydro, Korea	19.6	1.18	200	[8]
Poggio Cuculo, Italy	28.0	0.38	44	[9]
Armary, Switzerland	105.0	0.09	68	[9]
Marchfeldkanal, Austria	2.0	6.00	70	[9]
Llysy Fran, Scotland	25.0	0.16	29	[9]
Sangüesa, Spain	11.0	1.16	75	[9]
RSTP Wang Bintong -KNR059	3.5	0.054	1.58	This research

The flow rate obtained after conversion is 0.054 m^3/s, which is significantly smaller as it is comparing to the 400 l/s which is converted as 0.4 m^3/s. It was mentioned that for the micro hydro scheme at a sewerage treatment plant in Emmerich of Germany that able to generate an output power up to 13 kW [8]. On other hand, according to [6], the estimated power output of 12 kW can be resulted from a maximum head of up to 30 m of height and a low water flow rate of 60 l/s or 0.06m^3/s. It is therefore the possible output power obtained in the STP at Wang Bintong in this research could not be achieved up to the targeted value of 12 kW but only 1.58 kW. This is only 12.75% of the expected outcome. Furthermore, [9] reporting that other facilities in Poggio Cuculo of Italy, Armary of Switzerland, Llysy Fran of Scotland, Sangüesa of Spain and Marchfeldkanal of Austria are able to offer a reasonable output power under range of below 100 kW (micro hydro power) generated by either high in gross head and/or high in the nominal discharge as shown in Table 4. Thus, the factors of obtaining a smaller output power than targeted value is contributed by a small effluent flow rate and/or low head in the sewerage treatment plant for this research.

3.2 Feasible payback period

Table 5 shows the estimated power generation and also its contribution to the saving in electricity bill according to the monthly and annual operation. This can be obtained by using Equation (1) and the electrical power P_e that can be generated is 1.58kW for the particular RSTP Wang Bintong, KNR059 in Perlis.

Table 5. Estimated output for the practical application of micro hydroelectric generator in RSTP Wang Bintong, KNR059.

Category of operation	Total Estimated Power Generated (kWh)	Contribution to the Saving in Electricity Bill*
Monthly	4,550.40	RM 1,938.60
Annual	54,604.80	RM 23,261.64

**Currency exchange at RM 4.2 for 1 USD*

The estimation of power generated is based on the 24 hours operation and current power utility provider electricity tariff is used to estimate the electricity charges as shown in Table 5 above. It showed in both average monthly and average yearly.

The monthly contribution in term of saving in electricity bill is then compared to the total electrical bill for the plant operation in RSTP Wang Bintong, KNR 059. It is discovered the percentage of monthly savings for electricity bill of the plant operation is obtained at 8.6%. Nonetheless, the obtaining output power may be differing from the expected and not an effective outcome, the estimated power output generating by the micro hydropower system still can be contributed to the internal use of STP and eventually facilitated on saving in monthly electricity bill with around RM 1,938.60.

As for the applying of micro hydroelectric generator in the STP at Wang Bintong, with theoretical annual savings of approximately RM 23,261.64, and a total system installation cost is estimated at approximately RM 150,000.00, the payback period for this facility would be around 7 years. [10] defined that staff members at multiple wastewater treatment facilities in Massachusetts were interviewed, whom a plant manager at Deer Island in Boston of MA, O'Brien claims that a reasonable payback period should be between 0 and 10 years. It is thus deemed this project potentially feasible according to the payback period.

3.3 Viability of hypothetical micro hydropower scenarios

Cases that had a payback period of five years or less were deemed viable, while cases with payback periods greater than five years were determined to be potentially viable with some modification, or not viable. Table 6 illustrates the viability of the different combinations of hypothetical flow and head cases. A total of 144 hypothetical cases were analysed by previous researchers, with 79.86% of the cases being viable, 4.86% being potentially viable with some modification, and 15.28% being not viable. As displayed in the table, 100% of flow cases greater than or equal to $0.30m^3/s$ would be viable for the implementation of this micro hydropower project, as long as the available head is within the range of 1.0 meter and above. For cases with average flows less than $0.30m^3/s$, 16 out of the 45 cases were viable (35.56%) and 7 were potentially feasible with modification (15.56%). More than half (48.89%) of the cases with flows less than 0.30 cubic meter per seconds were deemed not viable.

4 Conclusions

On the basis of result, the output power obtained is fairly lower than expected output power, 12 kW. The power output estimated is fall beyond of the range of a micro hydro power (5kW - 100kW) but the computational calculation shows that it is only generating 1.58 kW of electrical energy. On the other hand, it is also obtained that the estimated payback period is slightly longer whereby the investor need to be waited for around 7 years to have the return of investment or obtaining the profit. The primary factors are due to a lower flow rate, head and population equivalent, PE numbers are available from the selected sewerage treatment plant. Thus, research on other sewerage treatment plant that is available with a higher PE, flow rate and pressure head will contribute to achieve on obtaining the targeted output power. Besides that, it is also discovered that the total saving in monthly electricity bill is estimated at RM 1,938.60 per month or 8.6% of the average monthly electricity bill. Nevertheless, the obtaining output power may not be considered as an expected and effective outcome, this research has found that the estimated power output generating by the micro hydropower system still can be contributed to the internal use of STP and eventually facilitated on saving in monthly electricity bill. In conclusion, wastewater treatment process involves the flowing water (potential energy) especially at the effluent discharge point of sewerage treatment plant is possibly harvested for electricity generation by embedding the micro hydroelectric generator.

https://doi.org/10.1051/e3sconf/20183402037

Table 6. The viability of the different combinations of hypothetical flow and head cases.

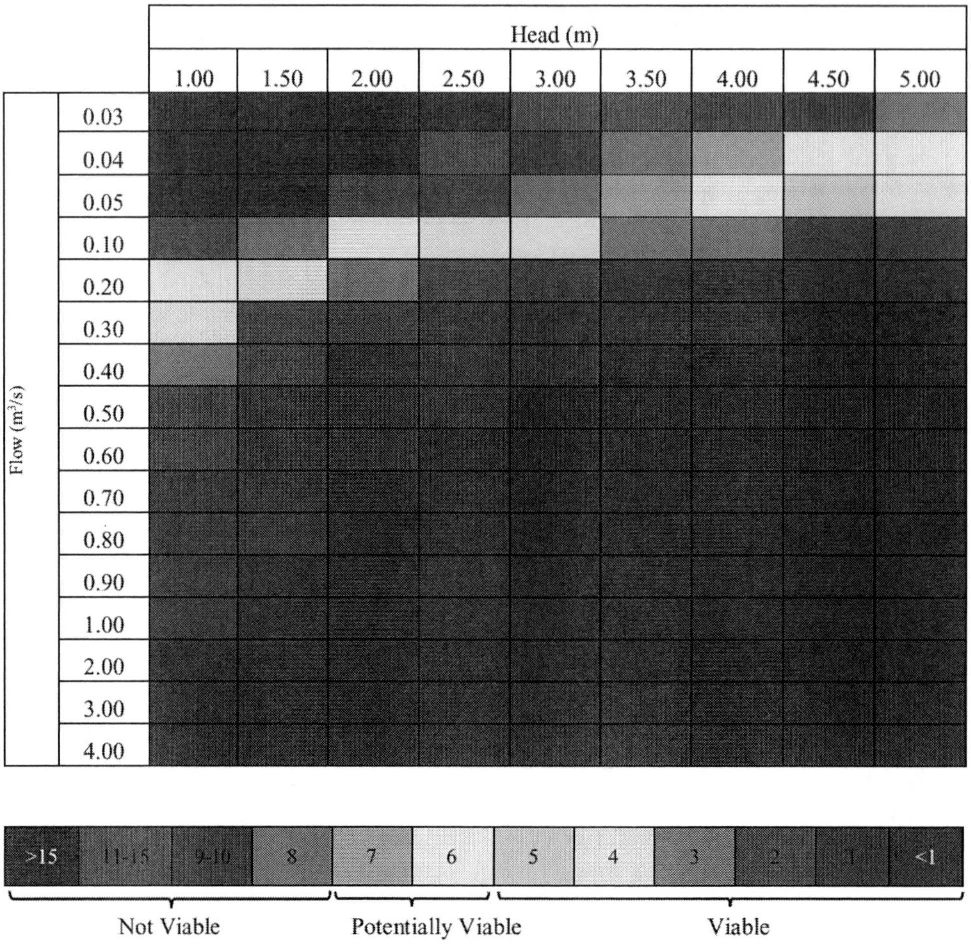

5 Recommendations

[10] reported that the sewerage treatment plant at Deer Island practices a combination of renewable energy technologies, and these technologies have the potential to be applied at other facilities. Furthermore, research on the application of micro hydroelectric generator in other sectors such as domestic use, commercial use and public use are suggested in order to provide the alternative way to reduce the energy usage, recycle, reuse and reproduce the energy. It is hoped that this research can be further explore using actual empirical data collection using actual generator embedded to confirm the theorical findings. Next, future research work is proposed for micro, pico or even nano hydro power generation from overhead water tanks of the bathrooms and kitchen to charge the mobiles and small chargeable facilities. This sustainable approach and design is eventually facilitated in preventing the energy deficiency problem and thus benefit the entire social, country and even the future generation.

6 Commercialization

From this research, the potential for building a business around harvesting energy in wastewater effluent streams through implementing a micro hydroelectric generator is considered as a feasible approach. This business could easily grow into other markets, notably the potable public water supply in which pressure reduction valves are used to reduce the pressure by dissipating energy rather than capturing it. Industrial processes that are heavy users of water are also one of the possible candidates. The factors of focusing the commercialization of the micro hydroelectric generator system on wastewater is due to sewerage treatment plants (STPs) are large consumers of electrical energy as well as a stable and continuous energy source available, thereby creating a ready use for any energy captured from the effluent stream. Besides that, sewerage treatments plants are famished to diminish their energy consumption, making them are more likely to try the new technology with an acceptable risk mitigation plan.

The authors would like to express our gratitude to Mega Jati Consult Sdn Bhd, Suria Infiniti Sdn Bhd and D'Bumiwater Sdn Bhd for their consultations and to Indah Water Konsortium Malaysia (IWK) for allowing and approved this research at the treatment plant facility.

References

1. B. A. Nasir, Int. J. Eng. Adv. Techn. **2**(5) 39-47 (2013)
2. M. A. C. Munaaim, N. Hamidin, A. Ayob, Jurnal Teknologi (Sciences & Engineering), **78**(5) (2016)
3. W. J. Klunne, Microhydro web portal (2017). {Assessed online via http://www.microhydropower.net/ on 14th August 2017}
4. M. Pauzi M. Kassim, Karam M. Al-Obaidi, M. A. C. Munaaim and Abd. Mokhti Salleh, American Journal of Engineering and Applied Sciences **8**(2), 210-222 (2015)
5. S. Eslamian, *Urban Water Reuse Handbook*- 1st edition (CRC Press, USA, 2016)
6. T. Archana, S. K. Pandey, S. C. Dubey **10**(2) 179-183 (2015)
7. Microhydro Power Basics (2017), Home Power web portal {Assessed online via https://www.homepower.com/articles/microhydro-power/basics/what-microhydro-power on 14th August 2017}
8. C. Bousquet, Cécile; Samora, Irene Manso, Pedro, In Renewable Energy **113**, 64-73 (2017).
9. C. Aline, D. Vincent, P. Petras, Integration of Small Hydro Turbines into Existing Water Infrastructures, Hydropower-Practice and Application, INTech {Assessed online via https:// www.intechopen.com/books/ hydropower-practice-and-application/ integration – of – small – turbines – into – water - infrastructure on 14th August 2017}
10. C. Marissa, D. Jessica, H. Christopher, *Reclamation of Power in Wastewater Treatment Facilities,* Bachelor Degree Desssertation (Faculty Of Worcester Polytechnic Institute, United States, (2014)

Preliminary screening oxidative degradation methyl orange using ozone/ persulfate

Nur Aqilah Razali[1], *Che Zulzikrami* Azner Abidin [1*], *Ong* Soon An[1], *Fahmi* Muhammad Ridwan[1], *Abdul Haqi* Ibrahim[1], *Siti Nasuha* Sabri[1], and *Su Huan* Kow[1]

[1]Universiti Malaysia Perlis, School of Environmental Engineering, Kompleks Pusat Pengajian Jejawi 3, 02600 Arau, Perlis, Malaysia.

Abstract. The present study focusing on the performances of advanced oxidation process by using ozonation method towards Methyl Orange based on the efficiency of colour removal and Chemical Oxygen Demand (COD) removal. Factorial design with response surface methodology (RSM) was used to evaluate the interaction between operational conditions, such as pH, initial concentration, contact time and persulfate dosage to obtain the optimum range conditions using a semi-batch reactor. The range of independent variables investigated were pH (3-11), initial concentration (100-500mg/L), contact time (10-50min) and persulfate dosage (20-100mM) while the response variables were colour removal and COD removal of Methyl Orange. The experimental results and statistical analysis showed all the parameters were significant. Thus, from this findings, optimization of operational conditions that had been suggested from the ozone/persulfate RSM analysis were (pH 3, 100 mg/L, 50min, 60mM) that would be produced 99% Colour Removal and 80% COD Removal and help in promoting an efficient ozonation process. The effect list data that showed the most contributed effects to increase the percentages of colour removal were pH and persulfate dosage whereas the contact time and initial concentration had the highest positive effects on the COD removal. Other than that, the interaction between pH, contact time and persulfate dosage were found to be the most influencing interaction. Therefore the least influencing interaction was interaction between persulfate dosage and pH. In this study, the correlation coefficient value R^2 for colour removal and COD removal of Methyl Orange were R^2= 0.9976 and R^2= 0.9924 which suggested a good fit of the first-order regression model with the experimental data.

1 Introduction

Main traits to determine the textile wastewater based on their colour intensity and chemical oxygen demand (COD) content, large amount of suspended solids, condition of pH and bio toxicity which can cause harmful towards water environment [1]. Current environmental issues either encountered by developed and under developed countries over the world

* Corresponding author: zulzikrami@unimap.edu.my

associated with residual dye content or residual colour in treated textile effluents are always a concern for each textile operations because they are directly discharged from sewage treatment works and commercial textile operations.

Textile industries discharged large volumes of residual dye which contributed a significant effect towards environmental pollution [2]. Due to its complex aromatic structures, it can stand as refractory in nature and resistant to the degradation environmental activities, consequently threaten the environment especially towards human and aquatic organisms. Azo dyes containing wastewater must be treated well before discharged into the environment due to the fact that about 15% of azo dyes of the total world production are from the textile wastewater [3]. Hence, it makes degradation of azo dyes in the industry has been appealed around the world [4].

Over the years, many conventional methods are used in order to get an achievement of the treatment of textile wastewater. Those methods are mainly the combinations of physical, chemical and biological processes which interferences could occur in a large scale operation. The drawbacks include secondary pollution would occur when some chemical oxidants are added into the water and their unsatisfied yields are formed due to the recalcitrant characteristics and non-biodegradable by the conventional activated sludge process [5].

Advanced oxidation process (AOP) is a process that able to degrade numerous pollutants such as hazardous compounds into mineral end-product for example CO_2 and inorganic ions. The AOP chemistry consists of hydroxyl radicals ($HO\cdot$), which is a non-selective reactive species [6] and has a high standard potential which is 2.8V [7].Ozonation process is an effective and environment-friendly method that has been proven performed well in the colour removal problem in textile wastewater among all the AOPs [8]. Ozonation does not influence by the addition wastewater volume, sludge, and removed colour and their organic contents will not produce by-products that need to be retreated back. In order to get a plausible overall cost, effluent treatment with addition of catalyst or combination with other methods were suggested due to ozonation process consumed a high cost [9]. Abu Amr and co-workers reported that ozonation process when combined with the usage of persulfate yields higher removal efficiencies for COD, colour and ammoniacal nitrogen (NH_3-N) as compared to ozone only [10].

There are various available oxidants that can be used in ozonation process, and persulfate is the one of latest and least studied, where still need an improvement [11]. The advantages that make persulfate a promising choice among others are it has a relative stability in subsurface, strong oxidation potential, high reactivity, high aqueous solubility, low cost and favourable end products that treating aromatic organic compounds [12]. Persulfate also has a strong, reactive sulfate radical with a high aqueous solubility and high stability oxidizing agent at room temperature, SO_4^-(E^o =2.6V) that often obtained from peroxydisulfate (PS, E^o=2.1V), compared to hydrogen peroxide (H_2O_2), E^o=1.77V) [13].

There are many factors that influences the oxidative degradation process which are pH of solution, initial concentration of solution, contact time and persulfate dosage. Thus, the process to study each factor will may risk towards time consuming and slow processes [14].Therefore, 2^4 level factorial experimental designs can be adopted to minimize all the possibilities of the difficulties by screening all the affecting parameters collectively at a time. In this research, the oxidative degradation of methyl orange using 2^4 factorial experimental designs was performed. The design consists of four factors such as initial concentration, pH, persulfate dosage and contact time with two responses, in which the effects and interaction of factors and significances of the model will be analysed based on the responses.

2 Materials and methods

2.1 Reagent

Analytical grade methyl orange (MO) was obtained from Sigma-Aldrich. MO was selected as model due to its common application in industries. Its molecular weight and molecular formula were 327.33 g/mol and $C_{14}H_{14}N_3NaO_3S$ respectively. The appropriate weight of dye will be dissolved into ultrapure water for the preparation of stock solution. Fig. 1 represented the chemical structure of the studied dye.

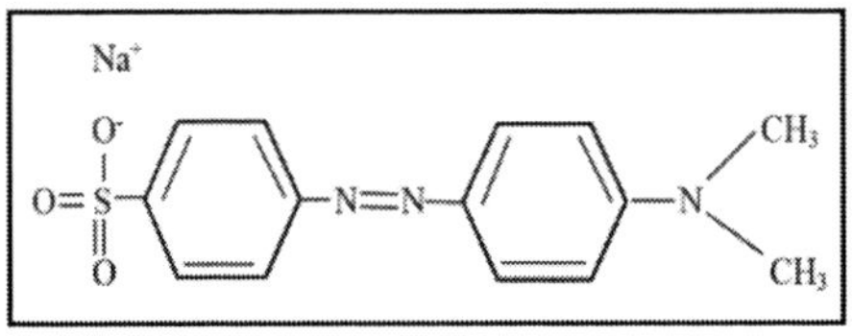

Fig. 1. Chemical structure of methyl orange.

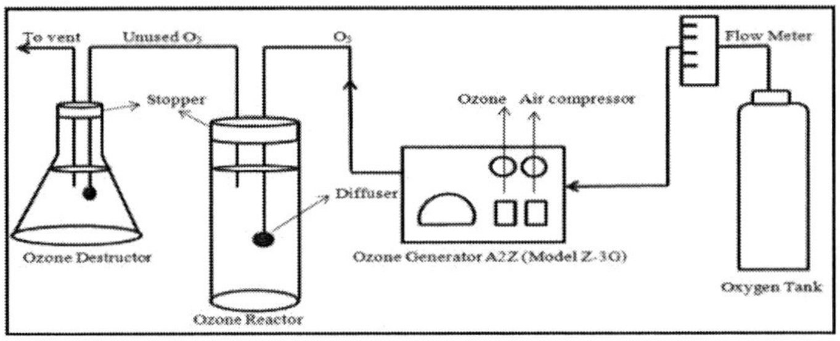

Fig. 2. Schematic diagram of experimental set-up.

2.2 Experimental procedures

In this experiment, ozone gas was generated and constantly flowed through a diffuser to produce fine bubbles into 2L dye sample by using A2Z (model Z-3G) Ozone Generator which continuously supplied with purified O_2. It was equipped with 2L of cylindrical glass reactor connected to 2% of potassium iodide trap which acted as O_3 destructor. Then, 2L sample was prepared with a certain desired condition of initial concentration, pH and persulfate dosage before O_3 entered the reactor based on run that constructed in the Table 2. 2 L/min O_2 flow rate was applied to generate O_3 throughout the experiments. Different ranges of persulfate dosage (20, 60, 100mM), pH (3, 7, 11), initial concentrations (100, 300, 500 mg/L) were conducted at room temperature with different conditions and undergoing various contact time (10, 30, 60 min). Finally, the samples were withdrawn at definite intervals for analyses.

2.3 Analysis methods

UV-Vis spectrophotometer at 461nm was used to analyse the absorbance colour change and concentration of the dye samples. The pH value before and after the treatment were tested by using pH meter (Hannah Instruments H1223). The value of COD was tested with the standard APHA method. HACH DR200 COD reactor and DR2800 spectrophotometer were used to react and COD readings were taken in terms of mg/L, respectively. The removal efficiency of COD and colour were determined by using Eq. 1.

$$\text{Removal (\%)} = [C_i - C_f/C_i] \times 100 \quad (1)$$

where C_i and C_f were respectively referred to the initial and final colours and COD concentration (mg/L).

2.4 Experimental design

Factorial design of Design-Expert® (Version 7.1.5) Stat-Ease, Inc., USA) was utilized for the statistical analysis and interaction of the experiments. Experimental factors and responses were applied to investigate the relationships between the three independent factors, as presented in Table 1.

For this research, factorial design for experimental data was chosen, because the design allowed the determination of factors with the highest impact on the process. Generally, the design achieved best result in screening and optimization processes [15]. The main advantage was able to analyse the effects of each factor towards the response that varies with different levels and interaction effects of different factors could be verified using design of experiment only compared to one-factor-at a time experiment [16]. Thus, the software will reduced the number of experiments, time, and overall processes costs in order to obtain a good response.

Full factorial design of 2^k runs, where k was the number of variables, was selected for the screening design. The full factorial screening design involved runs at every possible combination at the defined upper and lower limit for each variable, see Table 1.

Table 1. Experimental ranges and levels of independent variables used in the screening.

Level of value	pH (A)	Initial concentration (mg/L) (B)	Contact time (min) (C)	Persulfate dosage (mM) (D)
-1	3	100	10	20
0	7	300	30	60
1	11	500	50	100

3 Results and discussions

3.1 Multiple linear regression and Analysis of Variance (ANOVA)

Table 2 showed that Run 6 (pH 3, 100mg/L, 50min, 100mM) and Run 14 (pH 3, 100mg/L, 50min, 20mM) gave the highest percentages of colour and COD removal, which were 98.59% and 98.52% respectively. However, Run 2 (pH 3,100mg/L, 10min, 100mM) and Run 17 (pH 11, 100mg/L, 50min, 100mM) gave the lowest data for both responses, which were 89.26% and 69.49% for colour removal and 36.84% and 63.13% for COD removal.
ANOVA was performed to determine the significant effect of the main factors and their interactions, which influencing towards both responses. Statistically, a factor that significant for 95% confidence level, the p-value should be less than or equal to 0.05 [17]. Both colour and COD removal MO responses gave the best data of significant p-value. However, COD Removal response showed a non-significant curvature in this study. This was due to the technical problem where the factor B gave a disturbance towards COD removal. All factors were significant where R^2 values were approximately to 1, which were 0.9976 and 0.9924 as stated in Table 3. The fit of the model was further checked by the

coefficient of determination R^2. R^2 value should be always between 0 and 1. The closer the R^2 value to 1.0, the better the model predicts the response. A high R^2 value described a good agreement of results and shows that a desirable and reasonable agreement with the adjusted [18]. Fig. 3(a) and (b) showed the Pareto charts for colour and COD removal of Methyl Orange. Pareto chart indicated the relative importance of each independent model's parameter and their interactions towards both responses. Table 3 showed the contribution percentage of each factor and interaction between factors.

These screening analyses were necessary to know specifically the percentage of each factor's effect on the colour removal and COD removal. The findings showed that all factors (A, B, C, D) as stated in Table 1 previously acted as influencer in this experiment. Interaction between factors ABCD was the highest amongst all, in which the percentages of contribution for colour and COD removal of MO were 14.4% and 0.87%, respectively, while for factor AB and ABC referred to the best percentages of contribution which were 7.73% and 10.30%. However, factor C and AD gave the least percentage of contribution. Based on Table 3, the model and curvature that had been formed were significant. There indicated that all parameters were significant to be optimized at certain range condition based on the value of ABCD factors for colour and COD removal were 14.44% and 0.87% respectively. As a result, one of the factors must be excluded to be proceeding in optimization stage study to enhance the result for both responses.

3.2 Effect of parameters studied on 2D contour plot on colour and COD removal of methyl orange

3.2.1 Effect of pH on colour and COD removal of MO

Fig.5 a) and c) depicted a 2D contour plot of significant factors for COD removal of MO. Based on the figures, pH had the greatest influence in both colour and COD removal of MO. This was because at pH 3-11, conditions in this experiment were still able to get the highest percentage of colour and COD removal. Colour removal of MO decreased as the pH increased from pH 3 to pH 11 but the COD removal increased as the pH increased from acidic to alkaline condition as shown in Fig.4 a) and b).

Generally, the percentages of colour removal MO in alkaline condition was lower than in acidic condition, due to the fact that under neutral or basic pH conditions, sulfate anion radical can react with H_2O to produce OH^- and $HO^\cdot$. Therefore, it was likely that more HO^- would be scavenged by sulfate ions at alkali rather than acidic or neutral pH that resulting in the decreased of colour removal .While in an acidic condition, further $SO_4^{\cdot-}$ could be formed due to acid-catalyzation. Under alkaline condition, the formation of bicarbonate and carbonate ions may inhibit oxidation due to carbon dioxide that was formed in the process. Thus, sulfate and hydroxyl radicals decay can be accelerated by reason of their reaction with hydroxyl ions. Under acidic conditions, breakdown of persulfate into sulfate free radicals can be further undergoes acid catalyzed as stated in Eq.2 and Eq.3 [19].

https://doi.org/10.1051/e3sconf/20183402038

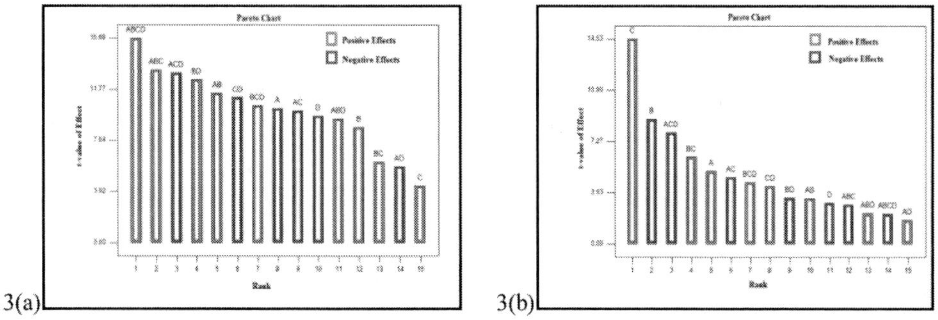

Fig.3. Pareto chart methyl orange (a) colour removal (b) COD removal.

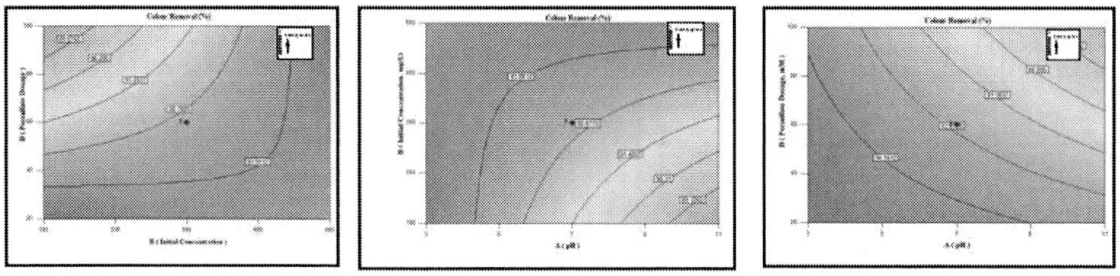

a) b) c)

Fig. 4. 2D contour plot of significant factors accordingly (Highest, average, lowest) towards color removal a) BD (9.20%) b) AB (7.73%) c) AD (2.01%).

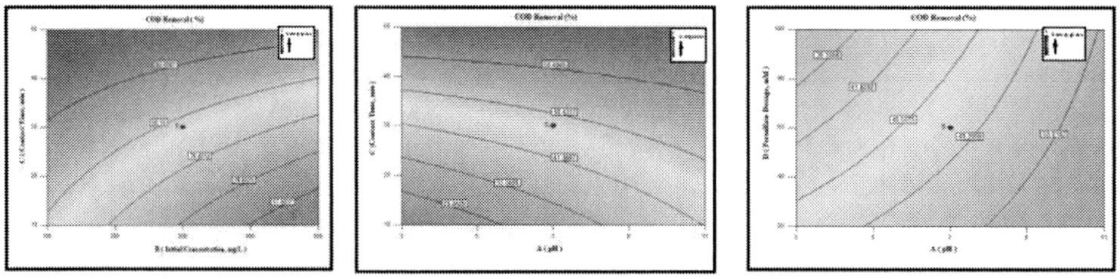

a) b) c)

Fig.5. 2D contour plot of significant factors accordingly (Highest, Average, Lowest) towards COD removal a) BC (7.20%) b) AC (4.26 %) c) AD (0.56 %).

Sulfate anion radicals when react with hydroxide anion, or water, will formed hydroxyl radical as referred to Eq.4 and Eq.5 [20]. An increased of pH will significantly enhanced COD removal due to alkaline pH that accelerated COD removal by generating OH^- radicals, which contained electrochemical oxidation potential (EOP) that was stronger than the ozone molecule (2.8/2.08V) [21].

$$SO_2^{8-} + H^+ + 2e^- \rightarrow HS_2O_8^- \quad (2)$$
$$HS_2O_8^- \rightarrow H^+ + SO_4^{2-} + SO_4^{\cdot -} \quad (3)$$
$$SO_4^{\cdot -} + OH^- \rightarrow HO^{\cdot} + SO_4^{2-} \quad (4)$$
$$SO_4^{\cdot -} + H_2O \rightarrow H^+ + HO^{\cdot} + SO_4^{2-} \quad (5)$$

Table 2. Percentage of colour and COD removal.

Run	A	B	C	D	Colour removal (%)	COD removal (%)
1	3	500	50	100	94.73	74.72
2	**3**	**100**	**10**	**100**	**89.26**	**36.84**
3	11	500	10	100	93.06	28.21
4	11	500	50	100	98.19	60.13
5	3	500	10	100	93.36	-54.42
6	**3**	**100**	**50**	**100**	**98.59**	**80.35**
7	7	300	30	60	95.27	39.86
8	3	500	10	20	92.58	24.63
9	11	500	10	20	91.6	30.2
10	7	300	30	60	93.86	51.21
11	7	300	30	60	94.44	51.05
12	11	100	50	20	98.18	80.12
13	7	300	30	60	94.93	47.37
14	**3**	**100**	**50**	**20**	**98.52**	**62.99**
15	11	100	10	20	89.97	45.8
16	11	100	10	100	94.13	62.07
17	**11**	**100**	**50**	**100**	**69.49**	**63.13**
18	7	300	30	60	93.78	40.56
19	3	500	50	20	96.09	51.52
20	11	500	50	20	95.43	70.85
21	3	100	10	20	93.81	49.67

Table 3. Analysis of Variance (ANOVA).

4FI	% Colour removal contribution	% COD removal contribution
A	6.24	5.05
B	4.60	14.75
C	**1.13**	**39.86**
D	5.55	1.62
AB	**7.73**	1.99
AC	6.01	4.26
AD	**2.01**	**0.56**
BC	2.27	7.20
BD	9.20	2.03
CD	7.29	3.17
ABC	**10.30**	1.49
ABD	**5.33**	**0.91**
ACD	**9.99**	**11.78**
BCD	6.47	3.61
ABCD	**14.44**	**0.87**
Model	Significant p-value :0.0002	Significant p-value:0.0018
Curvature	Significant p-value :0.0104	Not Significant p-value:0.5352
R^2 value	0.9976	0.9924

3.2.2 Effect of initial concentration on colour and COD removal of MO

Strong evidence of initial concentration factor was found when it gave significant effect towards contact time and persulfate dosage factors as shown in Fig.4a) and Fig 5 a). This can be supported with a hypothesis, that the higher initial concentration of MO, the longer their contact time to decolourize it. However, due to the least percentage of colour removal produced at 100mM, the range of persulfate dosage from 20-100mM would give some disturbance if the usage were higher than 60mM. Basically in AOPs, increasing initial concentration of dyes would cause formation of intermediates during the degradation of the primary dye and consequently, may interfere the desirable oxidation [22]. Besides, the contact time for colour removal would be longer at higher initial dye concentration in constant operational conditions in order to get a higher percentage for both responses.

3.2.3 Effect of persulfate dosage on colour and COD removal of MO

The results indicated that at acidic condition than basic condition with medium persulfate dosage at 60mM achieved the highest colour removal of MO as stated in Fig.4(c) and Fig. 5(c). This disturbance would gave a negative effect to both colour and COD removal efficiencies at a persulfate dosage that higher than 60mM. Based on [23] indicated that excess SO_4^- would generated at higher persulfate dosage and these excessive may work as scavengers for $S_2O_8^{2-}$ .While based on Fig.4. a), persulfate dosage clearly can enhance the efficiency of colour removal for all range of initial concentration.

3.2.4 Effect of contact time on colour and COD removal of MO

As referring to contact time factor, it can be concluded that the longer of time treatment, the higher the percentage of colour removal MO. The result showed that process can enhanced more than 90% colour removal while COD removal will be increased about 80% as well as contact time will be increased. Based on [24] some part of organic products were not oxidized completely in the presence of ozone only, due to carboxylic acids directly attacked by ozone from some dyes. Thus, some of the organic compounds were not completely degraded in this process. As a result, persulfate dosage which higher than 60mM gave least percentage of response and less affected towards contact time factor.

Overall, the response of colour removal was faster compared to removal of COD in oxidative degradation of methyl orange by ozone/persulfate. This was because the colour of MO can be removed once the chromophore in MO molecular, azo bonds (-N=N-) were destructed, that made the speed of removal of chemical oxygen demand slower than the colour removal efficiency [25].

4 Conclusions

The important finding to emerge in this study is preliminary screening oxidative degradation MO using ozone/persulfate by 2^4 level factorial designs allowed us to determine the importance and suitable range of operational conditions for a better degradation performance. According to this study, the main parameters that influenced the ozonation process were pH and persulfate dosage for colour removal MO while contact time and initial concentration for COD removal response. The interaction between pH, initial concentration, persulfate dosage and contact times is the most important interaction as it has the highest percentage contribution. The excellent correlation between predicted and observed degradation efficiency, R^2 for colour removal and COD removal of Methyl Orange were R^2=0.9976 and R^2=0.9924 which ensured to give a good accordance between the model and experimental data which confirmed the validity and practicability of the model.

The authors would like to express their gratitude to the Ministry of Higher Education (MOHE), Malaysia for economic support through Fundamental Research Grant Scheme (FRGS 9003-00513).

References

1. A. S. Koparal, Y. Yavuz, C. Gurel, U. B. Ogutveren, J. Hazard. Mater **145**, 100-108 (2007)
2. F. Torrades, J. García-Montano, Dyes Pigments **100**, 184-189 (2014)
3. H. N. Liu, G. T. Li, J. H. Qu, H. J. Liu, J. Hazard. Mater **144**, 180–186 (2007)
4. Y. Li, F. S. Zhang, J. Chem. Eng **158**,148–153 (2010)
5. H. S. El-Desoky, M. M. Ghoneim, R. El-Sheikh, N. M. Zidan, J .Hazard. Mater **175**, 858–865 (2010)
6. A.L. Estrada, Y. Y Li, A. Wang, J. Hazard.Mat **41**, 227-228 (2012)
7. J. Garcıa-Montano, L. Perez-Estrada, I. Oller, M. I. Maldonado, F. Torrades, J. Peral, J. Photochem. Photobiol., A **195**, 205–214 (2008)
8. L. W. Lackey, R. O. Mines Jr, P. T. Mc. Creanor, J. Hazard. Mater **138**, 357–362 (2006)
9. S. M. de Souza, K. A. Bonilla, A. A. de Souza. J. Hazard. Mater **179**, 35 (2010)
10. S. S. Abu Amr, H. A. Aziz, M. J. K. Bashir, S. Q. Aziz, T. M. Alsaibi, J. Eng. Res. & Technol **2**, 122–130 (2015)
11. B. Roshani, N. J. Karpel Vel Leitner, J. Hazard. Mat **190**, 403–408 (2011)

12. H. Lin, J. Wu, H. Zhang, J. Sep. Pur. Technol **117**, 18–23 (2013)
13. H. Liu, J. Environ. Sci. & Technol **48**, 10330-10336 (2014)
14. A. Elhalil, H. Tounsadi, R. Elmoubarki, F. Z. Mahjoubi, M. Farnane, Water Resour. Ind **15**, 41-48 (2016)
15. L. Brasil, C. Martins, R. R. Ev, J. Dupont, S. L. P. Dias, J. A. A. Sales, C. Airoldi, E. C. Lima, Int. J. Environ. Anal. Chem **15**, 475–491 (2005)
16. I. Erper, M. S. Odabas, M. Turkkan, J. Agric **98**, 195–204 (2011)
17. D. C. Montgomery, Design and Analysis of Experiments (John Wiley and Sons, 1997)
18. M. Y. Nordin, V. C. Venkatesh, S. Sharif, S. Elting, A. Abdullah, J. Mater. Process. Technol **145**, 46–58 (2004)
19. Y. H. Guan, J. Ma, X. C. Li, J. Y. Fang, L. W. Chen, J. Environ. Sci. Technol **45**, 9308–9314 (2011)
20. P. Drzewicz, L. Perez-Estrada, A. Alpatova, J. W. Martin, M. Gamal El-Din, J. Environ. Sci, Technol **46**, 8984–8991 (2012)
21. K. Turhan, Z. Turgut, Desalination **242**, 256-263 (2009)
22. A. Ghauch, A. M. Tuqan, N. Kibbi, S. Geryes, J. Chem. Eng **213**, 259–71 (2012)
23. C. Wang, & C. Liang, J. Chem. Eng **254**, 472–478 (2014)
24. M. Constapel, M. Schellentriager, J. M. Marzinkowski, S. Gab, Water Res **43**, 733-743 (2009).
25. A. Shimizu, Y. Takuma, S. Kato, A. Yamasaki, T. Kojima, K. Urasaki, S. Satokowa, J. Fac. Sci. Technol **50**, 1–4 (2013)

The performance and decolourization kinetics of O_3/H_2O_2 oxidation of reactive green 19 dye in wastewater

S. N. Sabri[1], *C. Z. A.* Abidin[1,*], *Fahmi*[1], *S.H.* Kow[1], and *N. A.* Razali[1]

[1]School of Environmental Engineering, Universiti Malaysia Perlis (UniMAP), Kompleks Pusat Pengajian Jejawi 3, 02600 Arau, Perlis, Malaysia.

Abstract. The degradations characteristic of azo dye Reactive Green 19 (RG19) was investigated using advanced oxidation process (AOPs). It was evaluated based on colour and chemical oxygen demand (COD) removal. The effect of operational parameters such as initial dye concentration, initial dosage of hydrogen peroxide (H_2O_2), contact time, and pH was also being studied. The samples were treated by ozonation (O_3) and peroxone O_3/H_2O_2 process. Advanced oxidation processes (AOPs) involve two stages of oxidation; firstly is the formation of strong oxidant and secondly the reaction of organic contaminants in water. In addition, the term advanced oxidation is referring to the processes in which oxidation of organic contaminants occurs primarily through reactions with hydroxyl radicals. There are several analyses that use to determine the efficiency of the treatment process, which are UV-Vis absorption spectra, COD, Fourier Transform Infrared (FT-IR), and pH. The results demonstrated that the ozone oxidation was efficient in decolourization and good in mineralization, based on the reduction of colour and COD. Additionally, results indicate that H_2O_2 is able to perform better than ozonation in order to decolourize the dye wastewater with 0.5 mL H_2O_2/L dye dosage of H_2O_2 at different initial concentration, initial pH, with contact time.

1 Introduction

The industrial wastewater pollution is a serious problem that impacts negatively on the environment especially water resources. The biggest industrial wastewater comes from textile, leather, paper and other dying industries. The treatment of industrial wastewater poses significant problems in wastewater treatment industry. However, the AOPs have become the most promising procedures to treat dye-containing in terms of effective decolourization and degradation of the refractory pollutants [1]. Nearly all AOPs are based on the generation of reactive species, hydroxyl radicals which degrade a broad range of organic pollutants quickly and non-selectively. The generation commonly accelerated by a

Corresponding author: zulzikrami@unimap.edu.my

combination of oxidants (ozone (O_3) or H_2O_2), irradiation (UV light), and catalysts (metal ions or photocatalysts). The oxidation potential of hydroxyl radicals (2.80V) exceeds that of O_3 (2.08V).

The textile industry is one of the largest industries in the world in terms of its output or production and employment [2]. The dye is a colorant which can be used to dye textile fibers, which means that it has an affinity to the material to which it is applied. Dye in wastewater of textile industry mainly generated from the wet-processing originates from washing and bleaching [3]. Dyes are widely used in the textile industries due to favourable characteristics, such as intense colour, water fastness, and uncomplicated application techniques. In the industries, the azo are among the most common synthetic dyes both in terms of the number and amount produced with azo bond groups [4]. The azo dye wastewaters are high biological oxygen demand (BOD) and COD content that impact visible colour, obstruct light penetration and oxygen transfer into water bodies. Besides, azo dyes account for approximately 60-70% of all dyes used in food and textile manufacture [5]. O_3 reacts directly with pollutants via ozonolysis and through radical reaction [6]. Typically, ozonation rarely produces complete mineralization to carbon dioxide (CO_2) and water (H_2O), but leads to partial oxidation products such as organic acid, aldehydes, and ketones where oxygen is introduced into carboneous sites within the product molecules [7]. The peroxone (O_3 with H_2O_2 oxidation) is used basically for the contaminants which conventional oxidation is difficult and consumes large amounts of oxidant [8]. The reaction comprises of O_3 dissolution and H_2O_2 addition (Equation 1).

$$H_2O_2 + 2O_3 \rightarrow HO\cdot + HO\cdot + 3O_2 \quad (1)$$

The H_2O_2 is relatively inexpensive and readily available chemical oxidant [9], and the O_3/H_2O_2 can be easily be automated. The independence of the O_3/H_2O_2 from any light source (UV irradiation) gives a specific benefit to this process, that can be used in high turbidity or dark wastewater [8]. The O_3/H_2O_2 of synthetic dye highly depended on the pH. The O_3 decomposed rate was increased with pH, as more H_2O_2 was dissociated into HO_2^-. For this reason pH 11 was considered to be the most suitable pH for the O_3/H_2O_2 treatment [10]. Moreover, complete discoloration achieved for vinyl sulphone dyes of Reactive Blue 220 and Reactive Yellow 15 in 90 min, and experience more difficulty in decolourization [11]. The COD values obtained were also below the allowed limit, but exceeded the allowed limit value of the other dyes. Therefore, it is obvious that performance of O_3/H_2O_2 depends on the type of dyes. Therefore, in order to get the most suitable and economical process, it is important that appropriate contact time, H_2O_2 dosage, and pH are being determined.

2 Materials and methods

2.1 Materials

The analytical grade diazo RG19 was selected as a model due to its common application in industries obtained from Sigma-Aldrich. Its molecular weight and the molecular formula are 1,418.94g and $C_{40}H_{23}C_{12}N_{15}Na_6O_{19}S_6$, respectively. The appropriate weight of dye will be dissolved in ultrapure water for the preparation of the stock solution. In order to prepare 5,000 mg/L concentration of the stock solution, 5 g RG19 will be dissolved in 1 L of ultrapure water. Figure 1 shows the molecular structure of RG19.

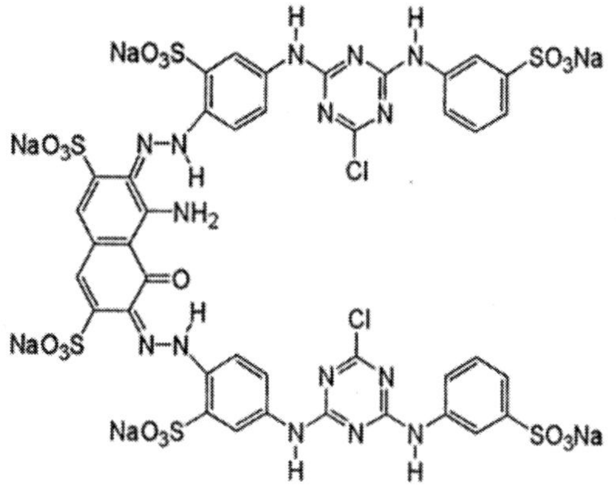

Fig. 1. Molecular structure of RG19.

2.1 Experimental Setup

The O_3 was generated in a laboratory scale cylindrical glass reactor from purified O_2 with a total volume of 2 L, equipped with O_3 destructor. The O_2 will continue introduced into ANA-13 A2ZZ-3G Ozone Generator to produce O_3 gas into the reactor. Thus, the gas continues flow through a diffuser and produce fine bubbles at constant flow rate supplying to 2 L dye sample. Predetermined H_2O_2 dosage will be added to the dye sample just before O_3 entered the reactor. The 2% of potassium iodide (KI) will trap the leftover O_3. Figure 2 shows the schematic diagram of experimental set-up.

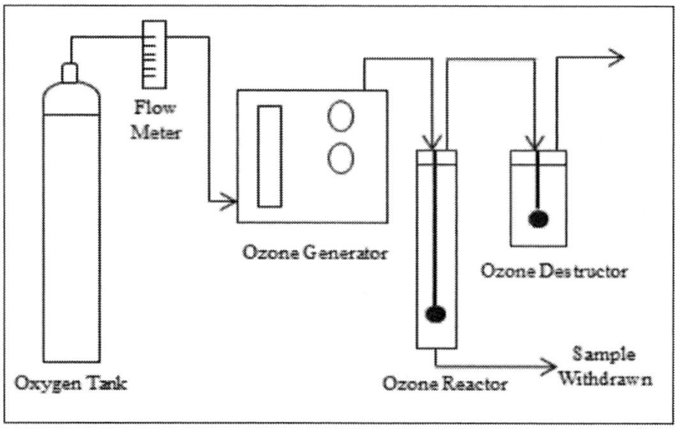

Fig. 2. Schematic diagram of experimental set-up for O_3 and O_3/H_2O_2.

2.3 Experimental Procedure

Constant O_3 flow rate has been applied, with different dosage of 0.25, 0.50, 0.75 and 1.00 mL H_2O_2/L dye at pH 7. Moreover, the O_3 and optimum H_2O_2 dosage were applied to the study of the effects of pH (3, 5, 7, 9, 11), together with the initial concentration of 100, 300, 500 mg/L. The experiments were conducted at room temperature with different contact time of 0, 1, 2, 3, 4, 5, 10, 15, and 20 minutes. Finally, the samples were withdrawn at definite intervals for analyses.

 https://doi.org/10.1051/e3sconf/20183402039

2.4 Analytical Method

The KI titration will use to determine the O_3 flow rate dye samples. UV-Vis spectrophotometer will be used to analyse the colour and dye concentration of the dye samples. The value of COD will be tested before and after treatment by O_3/H_2O_2 oxidation with the standard APHA method. HACH DR200 COD reactor and DR2800 spectrophotometer will be used to react and take the COD reading in terms of mg/L, respectively. The FT-IR (Spectrum 400 FT-IR/FT-Nir Perkin Elmar) spectroscopy is used to identify and compare the basic functional group of the dye samples before and after O_3/H_2O_2. The pH of the samples will be adjusted and measured by using Hanna Instrument HI 2211 pH meter.

3 Results and discussions

3.1 Percentage of Colour Removal

Fig. 3 shows the percentage removal for RG19 that has been treated with H_2O_2 in the dosage 0.5 mL H_2O_2/L dye with ozonation for different contact time. For the comparison between the two experiment result O_3 and O_3/H_2O_2, the graph below shows that the decolorization of RG19 is faster with the presence of H_2O_2. This is because, when H_2O_2 is added into the ozonation system, the decomposition of O_3 into hydroxyl radical is accelerated the process and shift the process entirely to an AOP [12]. Therefore, with the more number of hydroxyl radicals, they can attack the azo bond faster than breaking down the sulphonate group and naphthalene ring in dye [13]. Fig. 4 shows the colour removal result for RG19 treated with O_3/H_2O_2 with different dosage of H_2O_2 in the different contact time. According to the graph dosage, 0.5 mL H_2O_2/L dye is the suitable dosage for due to the faster decolourization of RG19 compared to other. Thus, 0.5 mL H_2O_2/L dye was selected as the optimum dosage and the dosage has been used to treat RG19 for higher concentration (300 and 500 mg/L).

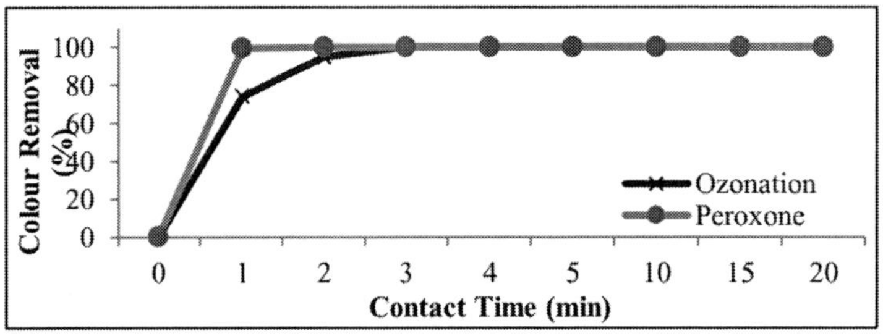

Fig. 3. Colour Removal for RG19 Treated with Ozonation and O_3/H_2O_2 of 100 ppm of RG19.

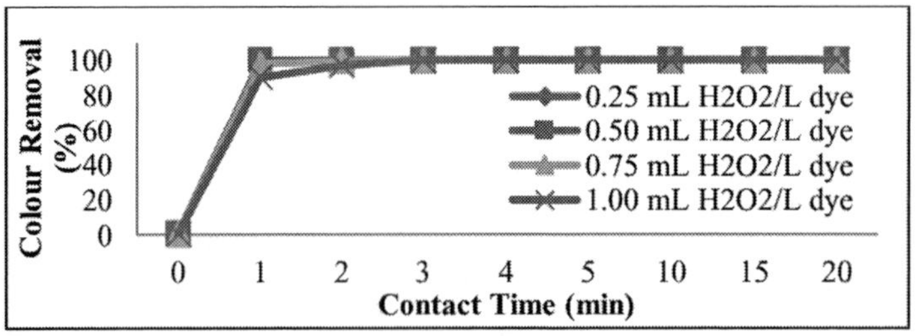

Fig. 4. Colour Removal for 100 ppm of RG19 for Peroxone with Different Dosage of H_2O_2.

3.2 COD Results

Fig. 5 shows the COD result for different dye treatment in the contact time 0, 1, 2, 3, 4, 5, 10, 15, and 20 minutes. According to the graph shown, the treatment with O_3/H_2O_2 has higher COD reading compared to treatment with O_3 only. This is because that H_2O_2 can interfere with COD analysis by consuming oxidation reagent, for example, $K_2Cr_2O_7$. Thus it is lead to the overestimating of COD measurements [14]. Generally, the addition of H_2O_2 during ozonation may enhance the rate of COD degradation compared to treatment without the interference of H_2O_2. Fig. 6 shows the COD result for RG19 treated with O_3/H_2O_2 in a different dosage of H_2O_2 with different contact time. According to the graph below, higher dosage of H_2O_2 can give a higher reading of COD value. Therefore, the results show that H_2O_2 is not suitable to remove COD compared to ozonation. This is because of its radical scavenging effect [14].

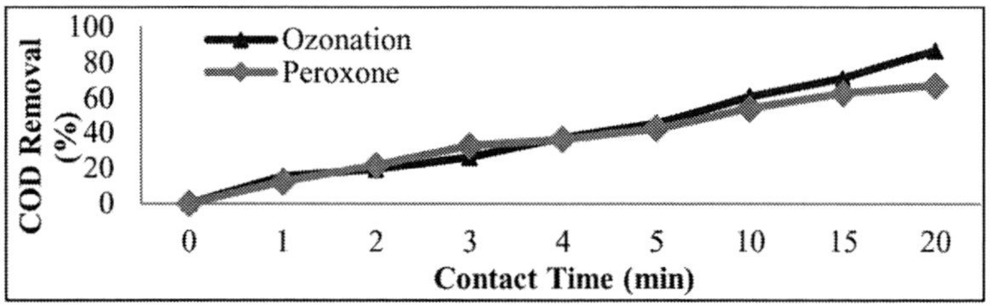

Fig. 5. COD for Treated RG19 with O_3 and O_3/H_2O_2.

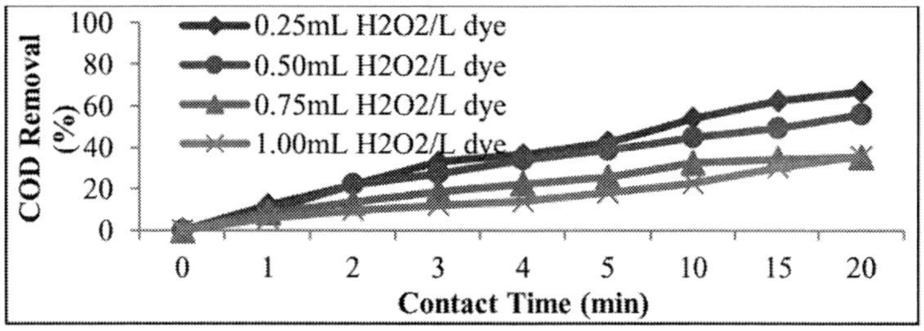

Fig. 6. COD for Treated RG19 with O_3/H_2O_2 with Different Dosage of H_2O_2.

Fig. 7 shows the COD result for RG19 treated with O_3/H_2O_2 in optimum dosage with different concentration of dye. According to the graph shown below, the higher concentration of RG19 can give the given value of COD. In addition, it is also shown that longer treated contact time give the lower value of COD. This can explain that longer period of treatment gives higher ability to treat dye. Thus, less residual COD found as most of them have been oxidized to H_2O and CO_2 [15].

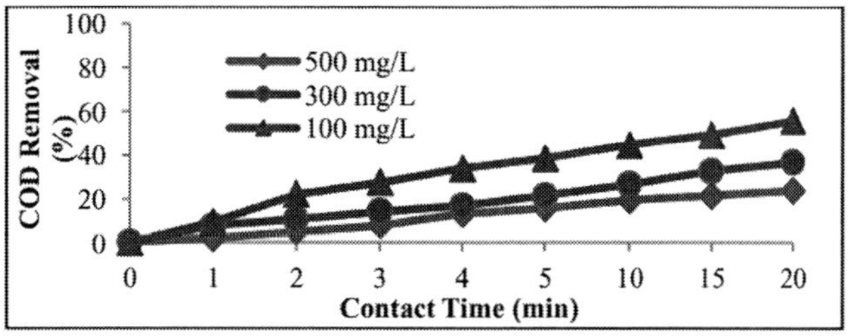

Fig. 7. COD Removal for Treated RG19 with O_3/H_2O_2 in Optimum Dosage of H_2O_2 with Different Concentration of RG19.

3.3 FT-IR Spectra

Fig. 8 (a) below shows the FT-IR result for RG19 for ozonation in different contact times 0, 10, and 20 minutes in form of split view. We can observe that some of the absorption peaks will be shifted slightly or disappeared after the ozonation for some period of time. By observing 0 min treatment process, peak at 1,500 cm^{-1} can be seen, which is indicates the presence of azo bond in the RG19. But, it is shown that as the time of treatment increase the azo bond has been shifted and disappeared. Besides, benzene ring that represents in peak 1,600 cm^{-1} to 1,400 cm^{-1} has been decreasing with the treatment period of ozonation. Benzene ring has been oxidized by hydroxyl ion. Around peak 1,180 cm^{-1} that represent sulphonate group, it is decreased after ozonation from 0 to 20 minutes. this shows that it has been mineralized. 3,400 cm^{-1} to 3,000cm^{-1} show a very weak peak that represents carboxylic acid. Fig. 8 (b) below shows the FTIR result for RG19 treated with O_3/H_2O_2 in different contact time in form of split view. For O_3/H_2O_2 treatment, it is almost similar to ozonation for the peak absorption peak. However, the peak for representing each functional groups are stronger than the ozonation process. This happens due to the addition of H_2O_2 in the process.

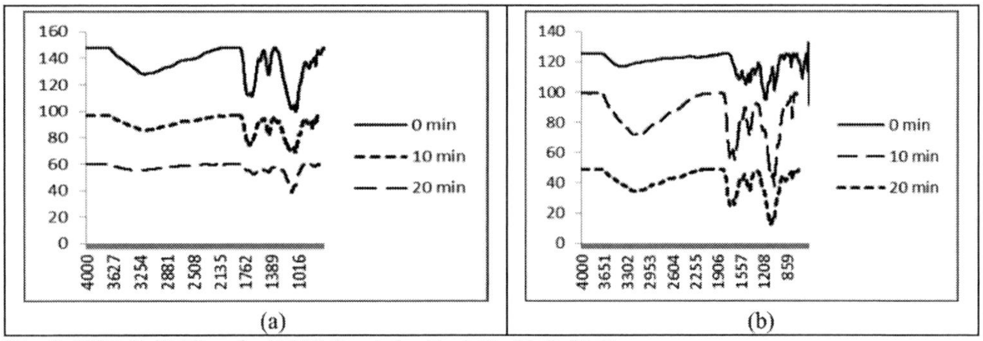

Fig. 8. FTIR Split View for RG19 Treated with a) O_3, b) O_3/H_2O_2.

3.4 Effect of pH

Fig. 9 shows the pH of treated sample of RG19 for the different dosage of H_2O_2 treated with peroxone in the contact time 0, 1, 2, 3, 4, 5, 10, 15 and 20 minutes. Based on the graph, 1.0 mL H_2O_2/L dye give the lowest pH and 1.5 mL H_2O_2/L dye give the highest pH of treated sample compared to other. The O_3/H_2O_2 of synthetic dye highly depended on the pH. The O_3 decomposed rate was increased with pH, as more H_2O_2 was dissociated into HO_2- [8]. Therefore, in order to get the most suitable and economical process, it is important that appropriate contact time, H_2O_2 dosage, and pH are being determined. Figure 10 shows the COD value against dosage of H_2O_2 with different initial pH. According to the graph below initial pH, 11 give the lower COD value and initial pH 3 give the higher COD value. Therefore, initial pH 11 proves that pH 11 is the most suitable initial pH of RG19 to be treated compared to others.

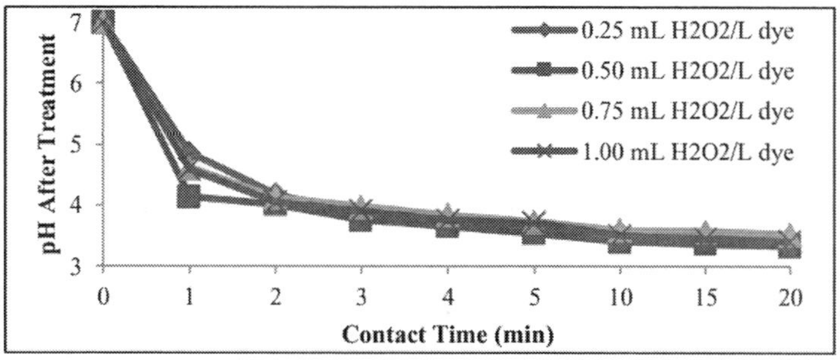

Fig. 9. pH After Treatment versus Contact Time with Different Dosage of H_2O_2.

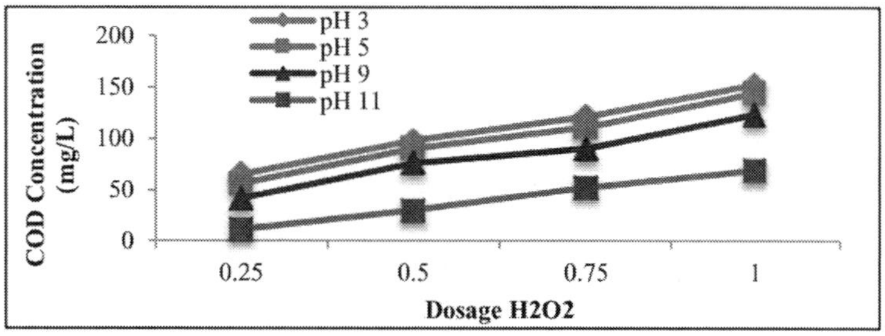

Fig. 10. COD Concentration versus Dosage of H_2O_2 from Different Initial pH.

4 Conclusions

In conclusion, ozonation and AOPs are able to remove colour from dye wastewater. It is decolourised azo dyes by breaking the azo bond and degrades the organic matter to become small fragments of an organic compound. In addition, by comparing ozonation and AOP, the addition of H_2O_2 in the treatment process can enhance the efficiency of colour removal and degradation process. This is explained that H_2O_2 able to enhance the decomposition of O_3 into hydroxyl radicals. In addition, from the UV-Vis result shows that 0.5 mL H_2O_2/L dye is the optimum dosage of H_2O_2 to treat RG19. This is because it gives faster time to decolourized RG19; give the lower value of COD; and able to decolourized the high concentration of dye. The degradation of dye can be seen from the wavelength scan. It is shown that AOPs has faster degradation of wavelength compared to ozonation. For COD value, AOPs has the higher value compared to ozonation due to the presence of H_2O_2 in the treatment. Besides, initial pH 11 gives the lower COD value compared to others. The same trend happens on FTIR results, the absorption peak for AOPs is stronger than ozonation. However, AOPs presence lower peak azo bond compared to ozonation indicates that azo bond has been breaking more during AOPs treatment process. Therefore, in order to get the most suitable and economical process, it is important that appropriate contact time, H_2O_2 dosage, and pH are being determined.

This work was carried out with the support of Ministry of Higher Education (MOHE), Malaysia for financial support under the grant of FRGS 9003-00513.

References

1. H. J. Hsing, P. C. Chiang, E. E. Chang, M. Y. Chen, J. Hazard. Mater. **141** 8-16 (2007)
2. C. Lakshami, *Approach Towards Eco-friendly Textiles.* Bangalore University (2012)
3. V. Smita, A. R. Quaff, N. D. Pandey, Desalin. Water Treat. **57** 4135-4145 (2014)
4. Q. Muqing, S. Jianxin, R. Peiqi, J. Kaihui, J. Chem. Pharm. Res., **7** 2046–2051 (2014)
5. Ashraf, M. A. Rauf, S. Salman, *Dyes and Pigments : New Research*. (Nova Science Publishers, 2009)
6. K. Sarayu, K.. Swaminathan, S. Sandhya, Dyes and Pigments, **75** 362–368 (2007)
7. R. Gerald, B. Rupert, H. Gunter, Chemosphere **28**,1447–1454 (1994)
8. T. Yonar, In Advanced in treating textile effluent 1-26 (2011)
9. R. Munter, Proc. Estonian Acad. Sci. Chem **50** 59–80 (2001)
10. I. Arslan, I. A. Balcioglu, T. Tuhkanen, Environ. Technol., **20**, 921–931 (1999)
11. T. Kurbus, Dye. Pigment. **58**, 245–252 (2003)
12. O. Griffini, P. Iozzelli, Ozone Sci. Eng. **18**, 117 (1996)
13. K. Pachhade, S. Sandhaya, K. Swaminathan, J. Hazard. Mater. **167** 313-318 (2009)
14. E. Lee, H. Lee, Y. K. Kim, K. Sohn, K. Lee, Int. J. Environ. Sci. Tech. **8** 381–388 (2010)
15. I. Talinli, G. K. Anderson, Water Res **26** 107–110 (1992)

Distribution and Analysis of Heavy Metals Contamination in Soil, Perlis, Malaysia

Ain Nihla Kamarudzaman[1,*], *Yee Shan* Woo[1], and *Mohd Faizal* Ab Jalil[2]

[1]School of Environmental Engineering, Universiti Malaysia Perlis, Kompleks Pusat Pengajian Jejawi 3, 02600 Arau, Perlis, Malaysia
[2]Perlis State Department of Environment, 2nd Floor, KWSP Building, Jalan Bukit Lagi, 01000 Kangar, Perlis, Malaysia

Abstract. The concentration of six heavy metals such as Cu, Cr, Ni, Cd, Zn and Mn were studied in the soils around Perlis. The aim of the study is to assess the heavy metals contamination distribution due to industrialisation and agricultural activities. Soil samples were collected at depth of 0 – 15 cm in five stations around Perlis. The soil samples are subjected to soil extraction and the concentration of heavy metals was determined via ICP - OES. Overall concentrations of Cr, Cu, Zn, Ni, Cd and Mn in the soil samples ranged from 0.003 - 0.235 mg/L, 0.08 - 41.187 mg/L, 0.065 - 45.395 mg/L, 0.031 - 2.198 mg/L, 0.01 - 0.174 mg/L and 0.165 - 63.789 mg/L respectively. The concentration of heavy metals in the soil showed the following decreasing trend, Mn > Zn > Cu > Ni > Cr > Cd. From the result, the level of heavy metals in the soil near centralised Chuping industrial areas gives maximum value compared to other locations in Perlis. As a conclusion, increasing anthropogenic activities have influenced the environment, especially in increasing the pollution loading.

1 Introduction

Soil is very important fundamental of natural resource for human survival [1]. Plot of land is important for recycling and redistribution of nutritious elements. Heavy elements and trace elements in the ground system is increasingly becoming a global concern, mainly because the soil is an important component of urban and rural areas [2]. In general, heavy metals deposition on the soil surface is due to natural formation on earth and also human activities such as industrialisation and agricultural activities. Heavy metals can be accumulated in topsoil through atmospheric deposition by sedimentation, impaction and interception [3]. Although some of the trace elements are important in the soil, however, the concentration of heavy metals in the soil at high quantities can be harmful to the environment [2].

In recent years, the heavy metals contamination in soil is a serious threat to the environment, ecosystems and human health. The atmospheric dust and aerosol deposition, automobile exhaust emissions, and various industrial activities are important sources of

* Corresponding author: ainnihla@unimap.edu.my

heavy metals contamination of soil [2]. Increasing anthropography activities will influences on the environment, particularly the burden of pollution, adverse changes in natural ecosystems and biodiversity decline. Besides, heavy metals pollution in agricultural soils also increased due to reuse of untreated industrial wastewater and the use of fertilizers and pesticides in agriculture activities [4]. As a result, excessive accumulation of heavy metals through agriculture activities can cause pollution not only to the soil but also to the quality and safety of the food. The dangerous possibility may occur such as contamination of the crop and influence the plant growth.

Heavy metals pollution in soil can cause harm and damage to the human and ecology through the digestive directly or contact with contaminated soil, food chain, drinking groundwater contaminated, increasing the problem of land and land use rights in agricultural production leading to food security [5]. Many studies have shown that exposure to toxic contaminants is influenced by a variety of factors, including exposure, absorption, metabolism and distribution in the human body [2]. Therefore, the objective of the study is to assess the heavy metals contamination distribution in the soil due to industrialisation and agricultural activities.

2 Methodology

2.1 Study Area

Perlis is located northern part of the west coast of Peninsular Malaysia (6° 30'N, 100° 15'E). This area undergoes various anthropogenic activities such as cement, chemical, quarry, power plant, transportation and agriculture activities that might contribute to heavy metals contamination. Soil samples were collected at 5 stations around Perlis such as Jejawi, Kangar, Chuping, Kuala Perlis and Beseri. Soil samples at depth 0 - 15 cm were collected using hand auger and stored in plastic bag. All soil samples were properly labelled before transported to the laboratory. Global Positioning System (GPS) records the coordination of sampling points. Table 1 show the location coordinates for soil sampling in Perlis.

Table 1. Location of sampling point.

Location	Latitude (N)	Longitude (E)
Jejawi	6.26.265	100.13.420
Chuping	6.30.465	100.19.097
Kangar	6.26.262	100.11.541
Kuala Perlis	6.24.210	100.09.169
Beseri	6.32.320	100.18.519

2.2 Soil Preparation

After sampling, the grass and unwanted materials were removed from the soil samples and the soil samples were stored in polyethylene bottles. The soil samples were dried in an oven at 105°C for 2 days. After drying, the soil samples were crushed using mortar and pestle. Subsequently, the soil samples were sieved through a 2 mm sieve and stored in polyethylene bottles under room temperature.

2.3 Analytical Method

In order to identify the elemental composition in the soil samples, the X-ray fluorescence (XRF) test was conducted. Prior to the XRF analysis, the soil samples were oven dried at 105°C overnight, grinded and sieved to obtain soil samples size of less than 75 μm and pressed into pallet by using hydraulic Pellet Press model PP 25. Meanwhile, in order to assess the concentration of heavy metals in the soil samples, the Inductively Coupled Plasma Optical Emission Spectrometry (ICP - OES) was conducted. The soil extraction was carried out using EDTA method. 0.05 mol of EDTA solution was prepared by dissolving 14.61 g of EDTA powder into 1 L of distilled water. The mixture was stirred until fully dissolved. Next, about 5 g of soil sample was mixed with 20 ml of 0.05 mol of EDTA solution and shaken at 100 rpm for 60 minutes on a rotary shaker. The extraction solution was filtered through Whatman No. 42 paper. All heavy metals analysis performed by Inductively Coupled Plasma Optical Emission Spectrometry (ICP - OES).

3 Results and discussions

3.1 Elemental Composition in Soil

The elemental composition analysis in the soil samples using XRF includes major and minor elements such as S, Si, Ca, Fe, K (major) and Cd, Ti, V, Mn, Cr, Cu, Ni, Pb, Sr, and Zn (minor). Table 2 and Table 3 showed that the elemental composition of the soils collected from five locations in Perlis. From Table 2, the percentages of Si, Ca, Fe and Al were higher than other elements. The percentage of calcium was the highest (4- 89%) in the XRF result. The highest percentage of calcium due to the soil is collected nearby to the cement factory at Beseri. In addition, the XRF result also shows that the percentage of Si is around 18-59 % in Perlis, which is the second higher compared with other elements in the soil samples. Si is the second most abundant element in the crust of the earth after oxygen [6].

Table 2. Major element compositions in the soil samples.

Element	Percentage of elements at five location in unit %				
	Beseri	Chuping	Jejawi	Kangar	Kuala Perlis
S	-	0.93	0.44	0.14	0.48
Si	3.67	54.2	33.6	58.9	18.1
Ca	89.07	17.3	30.8	4.35	63.65
Fe	3.41	14.1	16.07	15.22	7.85
K	-	2.11	4.18	3.25	2.92
Al	2.80	7.39	12.10	13.60	5.29

3.2 Distribution of Heavy metals

Five different location of soils were analysed by ICP- OES for six elements which are Cr, Cu, Zn, Ni, Cd and Mn. The concentrations of each heavy metals are shown in Table 4. Obviously, the concentration of Mn in soil was detected at highest concentration in all location. The highest Mn concentration in soil, which is 63.79 mg/L was found at the centre of Chuping. On the other hand, the maximum concentration of Ni, Zn and Cu in the soil samples are 2.198 mg/L, 45.39 mg/L and 41.187 mg/L respectively. These heavy metals were also detected at the same location in Chuping.

Table 3. Minor element compositions in the soil samples.

Element	Percentage of elements at five location in unit %				
	Beseri	Chuping	Jejawi	Kangar	Kuala Perlis
Ti	2.80	7.39	12.1	13.60	5.29
V	0.03	0.03	0.06	0.05	0.03
Cr	0.02	0.05	0.04	0.05	0.03
Mn	0.20	0.16	0.20	0.38	0.24
Cu	0.06	0.17	0.11	0.11	0.09
Zn	0.04	0.10	0.16	0.05	0.04

Table 4. The concentration of heavy metals by ICP – OES.

Element	Concentration of heavy metals at five location in mg/L				
	Beseri	Chuping	Jejawi	Kangar	Kuala Perlis
Cr	0.147	0.235	0.056	0.004	0.003
Cu	0.404	41.187	7.412	0.086	0.080
Zn	24.077	45.390	24.439	0.140	0.065
Ni	0.116	2.198	0.26	0.031	1.840
Cd	0.098	0.174	0.010	0.047	0.048
Mn	32.159	63.789	37.00	0.165	2.823

The result showed that all six elements such as Cu, Cr, Ni, Cd, Zn and Mn were highly distributed in Chuping. The sampling location for this soil was located near the Chuping industrial area and near to the main road, which loads with heavy traffic. Such industrial activities include cement production, chemical plant processing, quarry, and sugar plant have operated in the Chuping industrial area. Therefore, these industrial activities indirectly contributed to the heavy metal pollution in the soil. A study by [7] reported that Cu, Zn and Cr mainly derived from industrial activities and traffic emissions. Heavy vehicles also give large contribution in depositing of Cu, Cr and Ni in soil through traffic emission.

3.3 Descriptive Statistics

Table 5 presents the descriptive statistics of the heavy metals concentration such as mean, standard deviation, maximum and minimum for soils of Perlis state. The mean values of each heavy metals concentration in soils showed that the values were below the guidelines limit as shown in Table 6. The total mean concentrations of heavy metals in five different locations are Cr (0.089 mg/L), Cu (9.834 mg/L), Zn (18.822 mg/L), Ni (0.889 mg/L), Cd (0.075 mg/L) and Mn (27.187 mg/L). The values of the mean concentrations of heavy metals on soil decreased in the order Mn > Zn > Cu > Ni > Cr > Cd. The maximum concentrations of Mn, Zn and Cu give high values, however, it still not exceeding the permissible limit [8].

Table 5. Descriptive statistics of heavy metals concentration in the soil samples.

Variables	Mean	Standard Deviation	Minimum	Maximum
Cr	0.089	0.100	0.003	0.235
Cu	9.834	17.826	0.003	41.187
Zn	18.822	19.143	0.065	45.390
Ni	0.889	1.0424	0.031	2.198
Cd	0.075	0.0638	0.010	0.174
Mn	27.187	26.384	0.165	63.789

Table 6. The guidelines limit of heavy metals in soil [8].

	Cr	Cu	Zn	Ni	Mn
This study	0.089	9.802	18.822	0.889	27.187
World soil average levels	59.5	38.9	70	29	488

4 Conclusions

The soil samples were collected from five locations in Perlis includes Jejawi, Kangar, Chuping, Kuala Perlis and Beseri, and undergo the analysis concentration heavy metals in soil. In general, the results showed the concentration of certain elements are higher in the soil samples. From this result, it was found that the level of heavy metals in the soil near to the centralized Chuping industrial areas gives maximum values compared with other location in Perlis. The result showed that the concentration of Mn has the highest value, whereas Cd has the least and the order observed in this study is Mn >Zn>Cu>Ni>Cr>Cd. As a conclusion, increasing anthropogenic activities such as industrial activities and traffic emission have influenced the pollution of heavy metals in the soil.

We are grateful for the university resources provided by Universiti Malaysia Perlis (UniMAP), Malaysia. We also thank to School of Environmental Engineering, UniMAP for providing us with laboratory facilities.

References

1. Liu, C. Lu, L. Huang, T., Huang, Y., Ding, L., Zhao, W. Environmental research and public health (2016)
2. Rahman, S. H. Applied Sciences **2**, 586-601 (2012)
3. Li , Xue, Poon, Chao., Liu, Pin, Applied Geochemistry **16**, 1361-1368 (2001).
4. Hezbullah, M., Sultana, S., Chakraborty, S., Patwary, M., Journal of Toxicology and Environmental Health Sciences **8**, 1-5 (2016)
5. Wuana, R. A., Okieimen, F. E. ISRN Ecology, 1-20 (2011)
6. Tubaña, B. S., Heckman, J. R., Silicon and plant diseases 7-35 (2015)
7. Mat Ripin, S. N., Hasan, S., Kamal, M. L., Mohd Hashim, N., Malaysian Journal of Analytical Sciences **18**, 155-161 (2014)
8. Hasan, S., Kamal, M. L., Mat Ripin, S. N., Mohd Hashim, N., Malaysian Journal of Analytical Sciences **18**, 155-161 (2014)

Environmental Flow for Sungai Johor Estuary

A.Kadir Adilah[1,*], *Yusop* Zulkifli[2] , *Z. Noor* Zainura[1], and *Baharim* N.Bakhiah[3]

[1] Faculty of Chemical and Energy Engineering, Universiti Teknologi Malaysia
[2] Resource Sustainability Research Alliance, Universiti Teknologi Malaysia
[3]School of Marine and Environmental Sciences, Universiti Malaysia Terengganu

Abstract. Sungai Johor estuary is a vital water body in the south of Johor and greatly affects the water quality in the Johor Straits. In the development of the hydrodynamic and water quality models for Sungai Johor estuary, the Environmental Fluid Dynamics Code (EFDC) model was selected. In this application, the EFDC hydrodynamic model was configured to simulate time varying surface elevation, velocity, salinity, and water temperature. The EFDC water quality model was configured to simulate dissolved oxygen (DO), dissolved organic carbon (DOC), chemical oxygen demand (COD), ammoniacal nitrogen (NH_3-N), nitrate nitrogen (NO_3-N), phosphate (PO_4), and Chlorophyll a. The hydrodynamic and water quality model calibration was performed utilizing a set of site specific data acquired in January 2008. The simulated water temperature, salinity and DO showed good and fairly good agreement with observations. The calculated correlation coefficients between computed and observed temperature and salinity were lower compared with the water level. Sensitivity analysis was performed on hydrodynamic and water quality models input parameters to quantify their impact on modeling results such as water surface elevation, salinity and dissolved oxygen concentration. It is anticipated and recommended that the development of this model be continued to synthesize additional field data into the modeling process.

1 Introduction

Estuaries face a host of common challenges. As more people flock to the shore, we are upsetting the natural balance of estuaries and treating their health. We endanger our estuaries by polluting the water and building on the lands surrounding them. These activities can contribute to unsafe drinking water and beach, closing of shellfish bed, harmful algal blooms, declines in fisheries, loss of habitat, fish kills, and a host of other human health and natural resource problems.

Sungai Johor has a total length of about 122.7 km with catchment area of 2636 km2. The river originates from Gunung Belumut and Bukit Gemuruh in the north and flows to the southeastern part of Johor and finally into the Straits of Johor. The major tributaries of Sungai Johor are Sungai Sayong, Sungai Linggiu, Sungai Semanggar, Sungai Lebam, Sungai Seluyut and Sungai Tiram. A great amount of pollutants from various sources, such

* Corresponding author: adilahabdulkadir@gmail.com

as the sewerage network of the Johor Bahru, Pasir Gudang, Ulu Tiram, and Kota Tinggi cities, Industrial wastewaters from many industries in the surroundings, and agricultural wastewater containing fertilizers and pesticides are discharged into the Sungai Johor.

The primary purpose of the study is to apply hydrodynamic model and water quality model to Sungai Johor estuarine system using EFDC. The objectives of the study are listed as follows:

i. To calibrate the model utilizing historical data and field data collection.

ii. To do sensitivity analysis of input parameters of hydrodynamic and water quality model.

Depending on the characteristics of a system, one may choose an appropriate hydrodynamic model. For a large coastal system where both horizontal and vertical gradient are significant, one needs to apply a three dimensional hydrodynamic model. Therefore the Environmental Fluid Dynamic Code, EFDC [1], is selected for the model build up. An intensive field observation surveys is conducted by the UTM research group during January- June 2015. Field data are required to specify the initial conditions, boundary conditions, and forcing function distributions in time and space for the calibration, verification and application of a three-dimensional hydrodynamic and water quality model of the study area.

The river quality water in lestuaries, indicating multiple inputs of pollutants along the river systems [2]. As shown in Fig. 5 and 6, stations 3JH19 and 3JH20 are categorized as clean or Class II except in June 1997 and the year 2000, which have been categorized as polluted river and Moderate River, respectively. Station 3JH20 is located near Kota Tinggi bridge and station 3JH19 is approximately 7 kilometers upward from station 3JH20. Stations 3JH13 and 3JH15 are categorized as clean river or Class II. Station 3JH13 is located near Kampung Rantau Panjang, Kota Tinggi and station 3JH15 is located approximately 5 kilometers downward of station 3JH13.

2 Results

All model coefficients should be consistent between the calibration period and the verification period. The method used in determining the values for the model coefficients is essentially one of trial and error. An intensive survey conducted in January 14-30, 2008 was used as calibration data set. The order in which the hydrodynamic model is calibrated is performed to address issues such as bathymetry, friction, tidal volume, cross-sectional area, and heat budget before salinity is calibrated. Salinity is the predominant signal in the model to ensure that mass is being moved horizontally and vertically with the appropriate timing and direction.

During the calibration process, Bottom roughness values are adjusted until the predicted results reasonably matched the observed data. The model results were compared against observed data in the study period (January 14-30, 2008). The temporal profiles of observed and predicted tide level are compared in for the calibration period. As shown in Fig. 1, the predicted tide levels reasonably matched the observed data at two sampling stations: Teluk Sengat and Tanjung Surat. These figures indicate that the model reasonably simulated the tide range and phase at a number of locations throughout Sungai Johor estuary. The simulated water level showed a good and fairly good agreement with observations, indicated by the low values of root mean square as shown in Fig. 2 and 3 respectively. Bottom roughness is very sensitive to water surface elevation at upper Sungai Johor and insensitive to lower Sungai Johor.

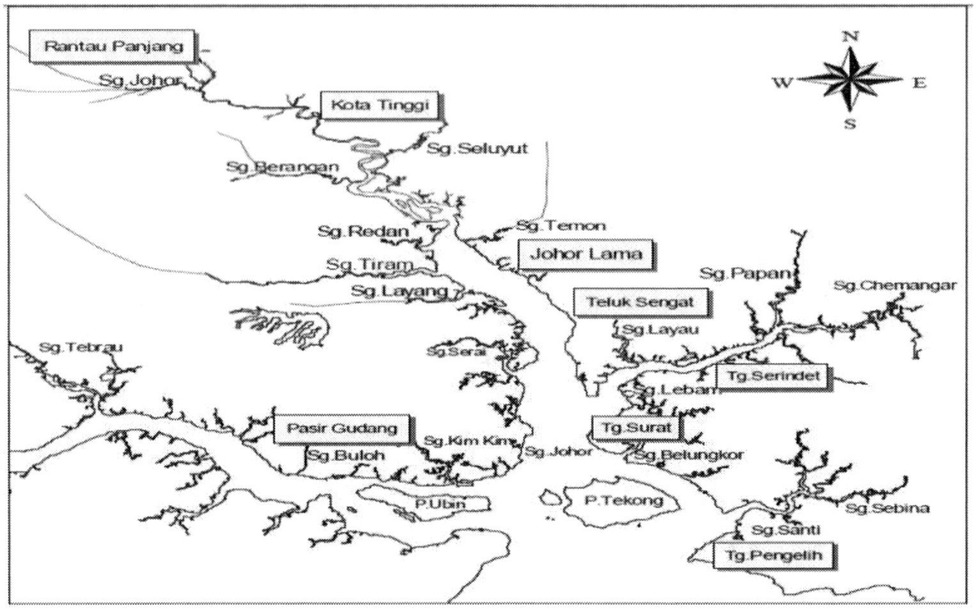

Fig. 1. Map of Sungai Johor study area.

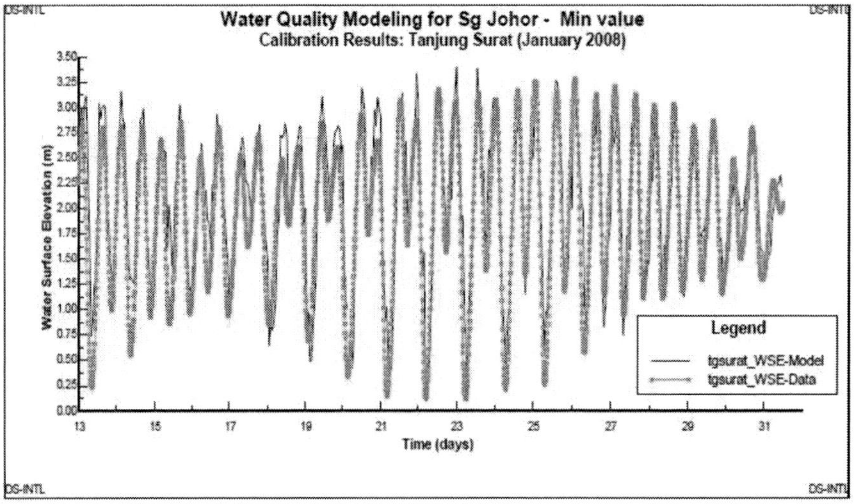

Fig. 2. Observed and simulated water surface elevations at Station Teluk Sengat.

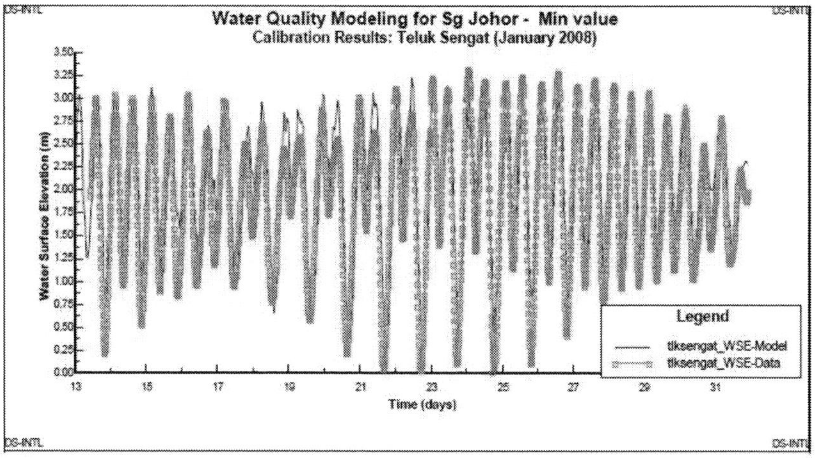

Fig. 3. Observed and simulated water surface elevations at Station Teluk Sengat.

3 Conclusions

The mechanisms that appear to be of significant include the absorption of phosphate to sediment particles and subsequent settling, transportation of sediment in response to high freshwater flow and the release of sediment phosphate. It was not because of the model but because of the quality and quantity of the field data used for the current calibration.

References

1. A. Elshafie, N. Ali, A.O. Karim, O. Jaafar, European Journal of Scientific research **28**(3), 422-435 (2009)
2. R.B. Ambrose, Journal of Environmental Engineering **113**(4), 703-721 (1987)
3. R.B. Ambrose and J.L. Martins, *Technical Guidance Manual for Performing Waste Load Allocations, Book III: Estuaries, Part I, Estuaries and Waste Load Allocations*, (U.S. Environmental Protection Agency, Washington, D.C., 1990)
4. R.B. Ambrose, T.A. Wool, J.L. Martin, *Centre for Exposure and Assesment Modelling*, (Athens, GA, 1993)
5. J.M. Beman, K.R. Arrigo, P.A. Matson, Nature **434,** 211-214 (2005)
6. A. Blumberg and G.L. Mellor, *Three dimensional coastal models*, (Americn Geophysical union, Washington D.C., 1987)

Characterization of Limestone as Raw Material to Hydrated Lime

Rajeb Salem Hwidi[1], *Tengku Nuraiti* Tengku Izhar[1,*], and Farah *Naemah* Mohd Saad[1]

[1]School of Environmental Engineering, Universiti Malaysia Perlis, Kompleks Pusat Pengajian Jejawi 3, Arau 02600, Perlis, Malaysia.

Abstract. In Malaysia, limestone is essentially important for the economic growth as raw materials in the industry sector. Nevertheless, a little attention was paid to the physical, chemical, mineralogical, and morphological properties of the limestone using X-ray fluorescence (X-RF), X-ray diffraction (X-RD), Fourier transform infrared spectroscopy (FTIR), and Scanning electron microscopy / energy dispersive x-ray spectroscopy (SEM-EDS) respectively. Raw materials (limestone rocks) were collected from Bukit Keteri area, Chuping, Kangar, Perlis, Malaysia. Lab crusher and lab sieved were utilized to prepare five different size of ground limestone at (75 µm, 150 µm, 225 µm, 300, and 425 µm) respectively. It is found that the main chemical composition of bulk limestone was Calcium oxide (CaO) at 97.58 wt.% and trace amount of MnO, Al_2O_3, and Fe_2O_3 at 0.02%, 0.35%, and 0.396% respectively. XRD diffractograms showed characteristic peaks of calcite and quartz. Furthermore, main FTIR absorption bands at 1,419, 874.08 and 712.20 cm^{-1} indicated the presence of calcite. The micrographs showed clearly the difference of samples particle size. Furthermore, EDS peaks of Ca, O, and C elements confirmed the presence of $CaCO_3$ in the samples.

1 Introduction

In Malaysia, limestone is characterized by wide karstification with the development of a complex and large network of caves [1]. Limestone hills naturally with sharp sides and rising up to many hundred meters above the flat alluvial plains are a well-known feature of the Malaysian landscape. Most of these alluvial plains are underlain at moderately shallow depths by the same limestone that forms, the prominent hills. The detection of rich tin deposits in the alluvium has led to the improvement of several urban centers adjacent to and later increasing on to the former mining regions [1].

Limestone is mainly composed of calcium carbonate mineral $CaCO_3$. This compound is one of the most common materials among the chemically precipitated sedimentary rocks [2]. Biological and also biochemical methods are the main processes in the carbonate sediments formation. Nevertheless, the inorganic precipitation of calcium carbonate from seawater can also happen [3]. After $CaCO_3$ formation, physical and chemical processes of diagenesis can significantly change the characteristics of $CaCO_3$.

* Corresponding author: nuraiti@unimap.edu.my

https://doi.org/10.1051/e3sconf/20183402042

Limestone is widely used in architectural applications for walls, decorative trim and veneer. It is less frequently used as a sculptural material, because of its porosity and softness. However, it is a common base material. It may be found in both bearing (structural) and veneer applications. In Malaysia, limestone is recently used in removal of heavy metals (Pb, Cd, Cu, Ni, Zn, and Cr(III)) from water [4]. It's found that limestone was able to remove more than 90% of heavy metals from polluted water [4].

Another important use of limestone (after converting to hydrated lime) is mitigation of poisonous and dangerous gases emission. Lime treatment reduces odors, particularly hydrogen sulfide, which is not only a nuance odor but also can be very dangerous if localized high concentrations build up [5]. In addition to high pH, lime provides free calcium ions, which react and form complexes with odorous sulfur species such as hydrogen sulfide and organic mercaptans [5, 6].

In the current work, the characterization of limestone as a raw material of hydrated lime was conducted using wide range of analysis to investigate the physical, chemical, mineralogical, and morphological properties of limestone.

2 Materials and methods

2.1 Sample location

The main material used in this work is the limestone, which was obtained from Bukit Keteri area, Chuping, 02450 Kangar, Perlis, Malaysia, 6.5035795,100.26139 GPS coordination.

2.2 Sample preparation

In order to prepare the samples, limestone were crashed using lab crasher. The ground limestone samples were sieved into five different sizes at 75 μm, 150 μm, 225 μm, 300, and 425 μm as shown in Figure 1.

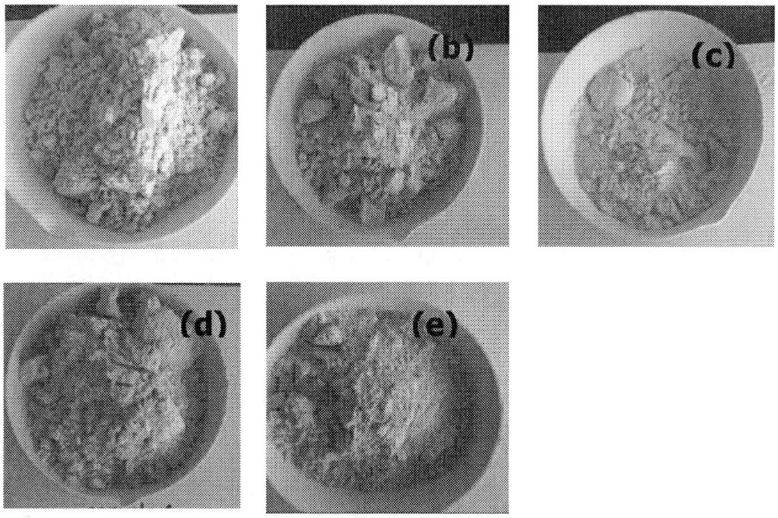

Fig. 1. Optical images of limestone samples at (a) 75 μm, (b) 150 μm, (c) 250 μm, (d) 300, and (e) 425 μm.

2.3 Sample characterization

After preparation of limestone samples, the mineralogical, chemical, physical and morphological properties of the hydrated lime samples were characterized and evaluated using X-ray fluorescence (X-RF) spectrometry, X-ray diffraction (X-RD), Fourier Transform Infrared Spectroscopy (FTIR), and Scanning Electronic Microscopy (SEM) attached with Energy Dispersive X-ray analysis (EDS) respectively.

2.3.1 X-ray fluorescence (X-RF)

X-rays fluorescent (XRF) was conducted using XRF spectrometer (Model MiniPAL 4 Brand: PANanalytical PW4030). X-RF analysis was carried out to determine elemental composition of the materials, which is based on the principle of individual atoms. When the sample is irradiated by X-rays, it measures the individual component wavelengths of the fluorescent emission produced by the sample.

2.3.2 X-ray diffraction (X-RD)

First, the specimen was pressed in stainless steel holder and then the Quantitative mineralogical evaluation was conducted by X-ray diffraction (X-RD) using CuKα radiation, 0.02° step size, 3 s counting time, 10° b 2θ b 80° range and Rietveld refinement method (X'Pert MPD - PANalytical X-ray B.V.). This method is designed to obtain a high quality diffraction data, associated with ease of use and flexibility to quickly switch to different applications.

2.3.3 Fourier transform infrared spectroscopy (FTIR)

Transmission FTIR spectra of composite films were recorded from thin KBr disc of the samples with Perkin Elmer 2000 FTIR spectrophotometer at room temperature. The samples were scanned from 4000 to 400 cm^{-1} with resolution of 0.4 cm^{-1}. By using FTIR, the functional groups in a molecule can be identified.

2.3.4 Scanning electron microscopy / energy dispersive x-ray spectroscopy (SEM-EDS)

The morphology and elemental chemical composition of the limes were determined respectively by scanning electronic microscopy (SEM) – using secondary electrons detector – and energy dispersive X-ray analysis (EDS) (LEO Stereoscan 440). The specimens were coated by an extremely thin layer of gold (1.5 - 3 nm) using sputter coater machine. The purpose of specimens coating is to avoid the poor resolution of the image and also to prevent the electrostatic charging during test. The test was conducted at 1000x magnification.

3 Results and discussions

The chemical composition of bulk limestone is listed in Table 1. The high contents of Calcium oxide (CaO) (97.58 wt.%) in the sample indicated that the sample was high limestone purity [7], but silica constituted the common impurity (0.90 wt.%). The high percentage of the main element (CaO) in the specimen showed high enrichment factor thus concerning cement as the main contributing source to airborne particulate matter in these factories and environs. This has a clear environmental implication, which should be of interest to the environmental protection agency and also to the government. The occurrence of faint magnesium oxide in the sample bears witness to the presence of trace amounts of smectite [7, 8]. It is also found that a trace amount of Al_2O_3, Fe_2O_3, and SrO elements are existed in the sample.

Table 1. Chemical composition of bulk limestone sample (wt.%).

Chemical Compounds	Al Al_2O_3	Si SiO_2	Ca CaO	Ti TiO_2	Mn MnO	Mg MgO	Fe Fe_2O_3	Cu CuO	Sr SrO	Ru RuO_2	Eu Eu_2O_3	Total
Weight %	0.35	0.90	97.58	0.13	0.02	0.01	0.395	0.065	0.092	0.37	0.08	100

Fig. 2 shows the XRD diffractograms of the representative limestone samples, indicating the presence of characteristic peaks of calcite as identified by the distinctive reflections at 3.85–3.86 A° (102), 3.03 A° (100), 2.84 A° (006), 2.49 A° (110), 2.28 A° (113), 2.09 A° (202), 1.97 A° (108), 1.87 A° (116) and 1.60 A° (212). Furthermore, the sample showed additional peaks prevailing at 3.33–3.34 A° (101), 1.54 A°, (211), 1.37 A° (203) and 1.28 A° (104) showed the presence of quartz [7]. Detailed clay mineralogy was investigated and identified by the characteristic reflections according to Moore and Reynolds [9].

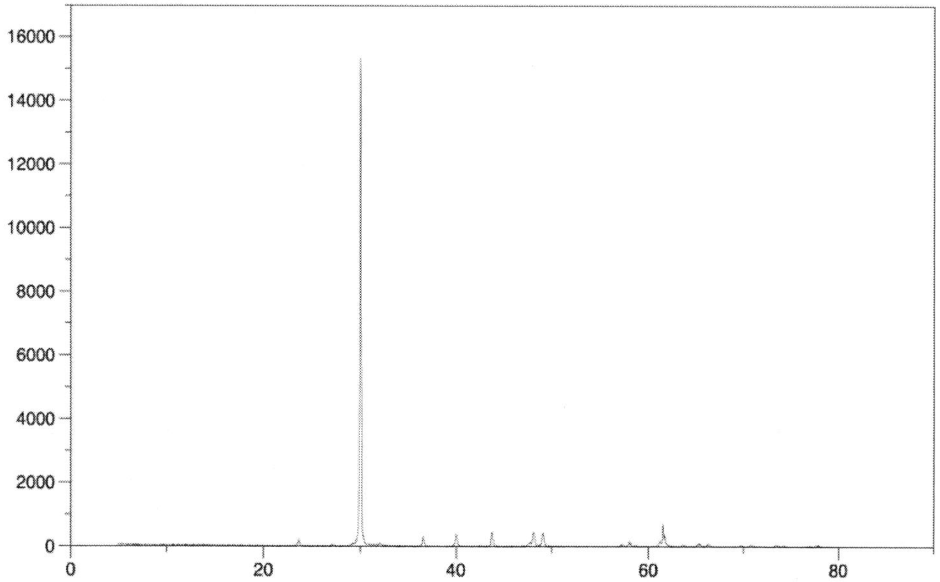

Fig. 2. X-RD of the limestone sample.

Fig. 3 shows the FTIR spectrum of the limestone sample. It shows the characteristic bands of calcite at 1,419, 874.08 and 712.20 cm^{-1}. The spectrum peaks appearing at 1,799 and 2,513.04 cm^{-1} are also an indication of the presence of calcite [10-13]. These data indicate that the studied limestone is mainly composed of calcium in the form of calcite as identified by its main absorption bands [13]. The reference bands observed at 1,419, 874.08 and 712.20 cm^{-1} can be assigned to the asymmetric stretching, out-of-plane bending and in-plane bending modes of CO_3^{2-}, respectively [14]. Gunasekaran and Anbalagan mentioned in their research that the observed out-of-plane bending mode occurs at 877 cm^{-1} for ^{12}C.

This band shifts to lower wave numbers for other isotopes of carbon (^{13}C and ^{14}C). However, limestone sample has a splitting band at 874.71 cm^{-1} [12]. This clearly indicates that there is no isotopic shift. According to the author in his research, the bands observed at 1,799 and 2,513 cm^{-1} are attributed to the $v_1 + v_4$ combination mode. Moreover, the stretching vibrations of the surface hydroxyl groups (Si-Si-OH or Al-Al-OH) were found at 3,544.52 and 3,619.73 cm^{-1} [15]. Infrared techniques have been frequently used for the identification of clay minerals [10, 11] as well as the natural calcite minerals [12].

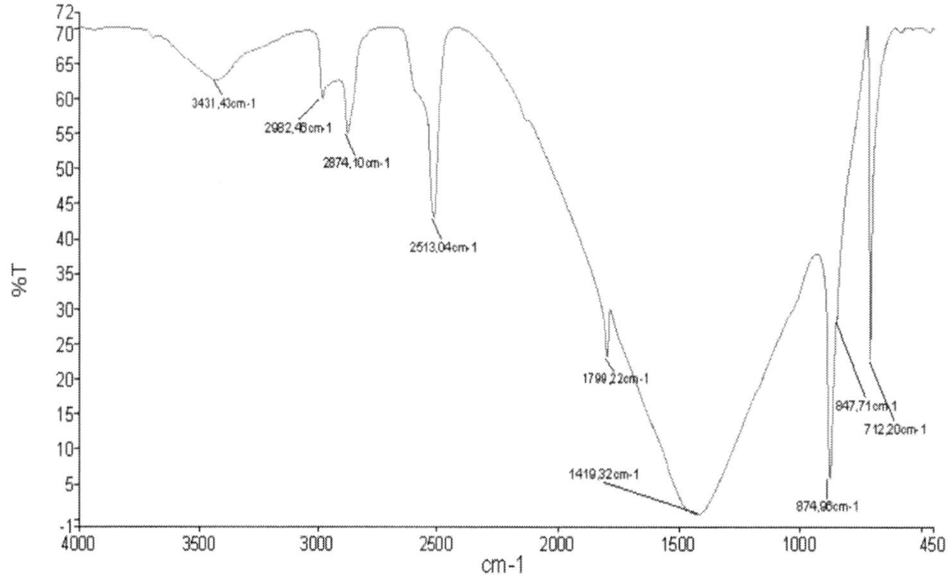

Fig. 3. FTIR spectrum of bulk limestone sample.

The observation of the morphology using the scanning electron microscope of the limestone, with three different sizes (fine, med, and coarse size) is reported in Fig. 4 a, b, and c respectively. It's clearly observed that the samples have different sizes due to different grinding and sieving processing applied on the samples. The samples show compact morphology, so the samples have less porous and the carbonates represent the main phase with calcite peaks [16]. Furthermore, the samples have morphology with angular grains of quartz size, covered with fine particles and the grains of calcite appear clearly. This analysis of the morphology by SEM is associated by EDS analysis of the elements exist on choised line on area with maximum amount of informations as it is showing in Fig. 4 and Table 2 respectively. The presence of Ca, O, and C, as major element confirms the importance of carbonate phases, $CaCO_3$ compound in the samples (Table 2).

https://doi.org/10.1051/e3sconf/20183402042

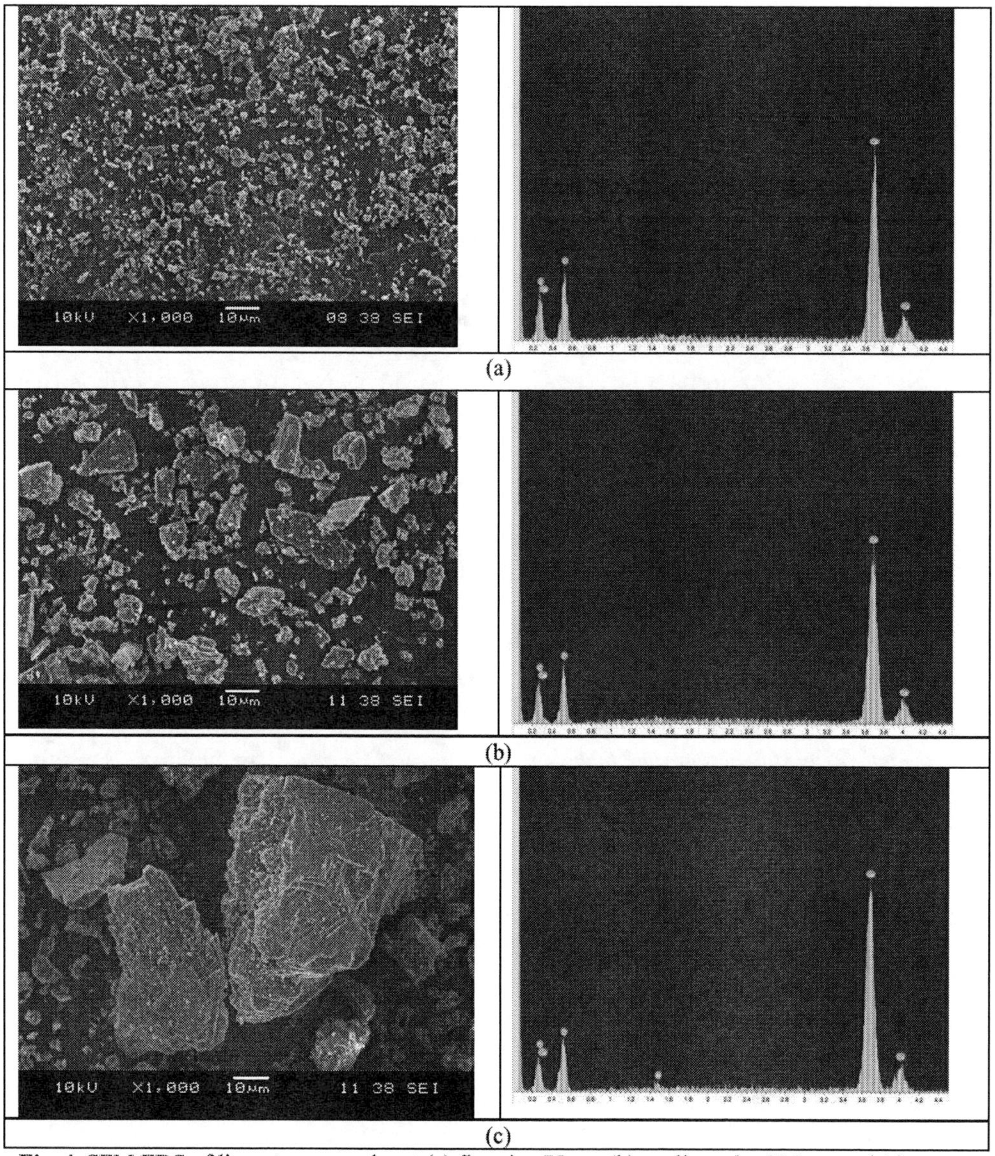

Fig. 4. SEM-EDS of lime stone samples at (a) fine size 75 μm (b) medium size 225 μm and (c) coarse size 425 μm.

Table 2. Elemental analysis of limestone samples.

Element	Weight%	Atomic %
C K	17.33	27.63
O K	45.69	54.70
Ca K	36.98	17.67
Total = 100		

4 Conclusions

In this work, it can be concluded that the limestone samples obtained from Bukit Keteri area had high limestone purity due to high calcium oxide and low silica percentage in the samples. Calcite mineral was identified by X-ray diffractograms and FTIR spectrum. Furthermore, the Ca, O, C elements proved the presence of $CaCO_3$ in the samples (as shown in SEM-EDS micrographs).

References

1. B. K. Tan, Bull. Eng. Geol. Environ. **35**, 57 (1987)
2. S. Cornelius, Jr. Hurlbut, K. Cornelis, *Dana's Manual of Mineralogy*, Wiley: New York, (1977)
3. M.E. Tucker, *Sedimentary Petrology: An Introduction to the Origin of Sedimentary Rocks*, Blackwell Scientific Publishing: Malden, (2001)
4. H.A. Aziz, M.N. Adlan, K.S. Ariffin, Bioresour. Technol. **99**, 1578 (2008)
5. R.J. Fairweather, M.A. Barlaz, J. Environ. Eng. **124**, 353 (1998)
6. R. Anderson, J.R. Jambeck, G.P. McCarron, Environmental Research and Education Foundation (2010)
7. A. Sdiri, T. Higashi, T. Hatta, F. Jamoussi, N. Tase, Environmental Earth Sciences **61**, 1275 (2010)
8. M. Felhi, A. Tlili, M.E. Gaied, M. Montacer, Appl Clay Sci **39**, 208 (2008)
9. D.M. Moore, R.C. Reynolds, *Quantitative analysis. In: Moore DM, Reynolds RC Jr (eds) X-Ray diffraction and the identification and analysis of clay minerals*, Oxford Univ Press: Oxford, 272, (1989)
10. Y.S. Al-Degs, M.I. El-Barghouthi, A.A. Issa, M.A. Khraisheh, G.M. Walker, Water Res **40**, 2645 (2006)
11. S.N. Preeti, B.K. Singh, Bull Mater Sci **30**, 235 (2007)
12. S. Gunasekaran, G. Anbalagan, Spectrochim. Acta, Part A **68**, 656 (2007)
13. S. Gunasekaran, G. Anbalagan, S. Pandi, J Raman Spectrosc **37**, 892 (2006)
14. N.V. Vagenas, A. Gatsouli, C.G. Kontoyannis, Talanta **59**, 831 (2003)
15. M. Hajjaji, S. Kacim, A. Alami, A. El Bouadili, M. El Mountassir, Appl Clay Sci **20**, 1 (2001)
16. K. Besnard, D.E.A. Hydrologie, G. Hydrogéologie, G.F. Hydrologie, Hydrogéologie, Géostatistique et Géochimie, Université Paris-Sud. (2000)

Identification of Trihalomethanes (THMs) Levels in Water Supply: A Case Study in Perlis, Malaysia

Mohd Faizal Ab Jalil[1*], *Nasrul* Hamidin[2], *Ahmad Anas* Nagoor Gunny[3], and *Ain Nihla* Kamarudzaman[2]

[1]Perlis State Department of Environment, 2nd Floor, KWSP Building, Jalan Bukit Lagi, 01000 Kangar, Perlis, Malaysia.
[2]School of Environmental Engineering, Universiti Malaysia Perlis, Kompleks Pusat Pengajian Jejawi 3, 02600 Arau, Perlis, Malaysia.
[3]Department of Chemical Engineering Technology, Faculty of Engineering Technology, Universiti Malaysia Perlis, Kampus UniCITI Alam, Sungai Chuchuh, 02100 Padang Besar, Perlis, Malaysia.

Abstract. In Malaysia, chlorination is used for drinking water disinfection at water treatment plants due to its cost-effectiveness and efficiency. However, the use of chlorine poses potential health risks due to the formation of disinfection by-products such as trihalomethanes (THMs). THMs are formed due to the reaction between chlorine and some natural organic matter. The objective of the study is to analyze the level of THMs in the water supply in Perlis, Malaysia. The water samples were collected from end-user tap water near the water treatment plant (WTP) located in Perlis, including Timah Tasoh WTP, Kampung Sungai Baru WTP, Arau Phase I, II, III, and IV WTPs. The THMs were analyzed using a Gas Chromatography-Mass Spectrometry (GC/MS). The results showed that the water supply from Timah Tasoh WTP generates the most THMs, whereas Kuala Sungai Baru shows the fewest amounts of total THMs. In conclusion, the presence of THMs in tap water has caused great concern since these components can cause cancer in humans. Therefore, the identification of THM formation is crucial in order to make sure that the tap water quality remains at acceptable safety levels.

1 Introduction

Chlorine is usually used in water treatment as a disinfectant agent and is very effective in reducing and eliminating waterborne diseases such as typhoid, cholera, hepatitis, dysentery, and gastrointestinal illness. [1, 2, 3, 5]. Chlorine reacts with organic matter in water to form Disinfected By-Products (DBPs) such as trihalomethanes (THMs), haloacetic acids (HAAs), and other compounds that are hazardous to human health [5, 6, 8, 13, 14, 15]. DBPs have established a large family of compounds and produce various stages of toxicological effects. THMs and HAAs are the major groups of DBP in drinking water.

* Corresponding author: faj@doe.gov.my

Their establishment should be known and their level of compliance enforced in many countries (especially THMs and HAAs) [4, 9, 11].

The formation, speciation, and concentration of CBPs in drinking water vary according to several factors related to water source characteristics, as well as operational parameters during treatment and drinking water distribution system characteristics [6, 10]. Hence, the aim of this paper is to identify the presence of DBP in the water distribution system in Malaysia, particularly in Perlis state. For this reason, the study is carried out on the following procedures. Firstly, water samples are collected from end-user tap water near several water treatment plants (WTP) located in Perlis such as Timah Tasoh WTP, Kampung Sungai Baru WTP, and Arau Phase I, II, III, and IV WTPs. Second, the samples are analyzed to detect the presence of several types and concentration of THMs using Gas Chromatography-Mass Spectrometry.

2 Methodology

Water samples were collected from four different water treatment centers in Perlis which are Timah Tasoh, Kampung Sungai Baru, Arau Phase I, II, and III, and Arau Phase IV water treatment plants as shown in Fig. 1, Fig. 2 and Fig. 3.

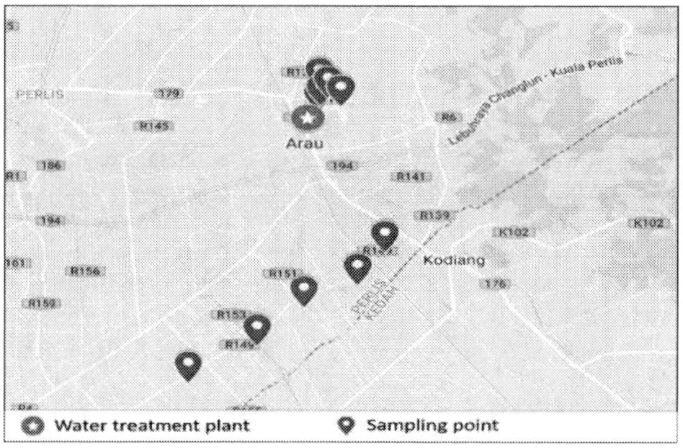

Fig. 1. Location of Arau Phase I, II, III & IV WTC and sampling point for the following location. Taken from Map Data ©2017 Google.

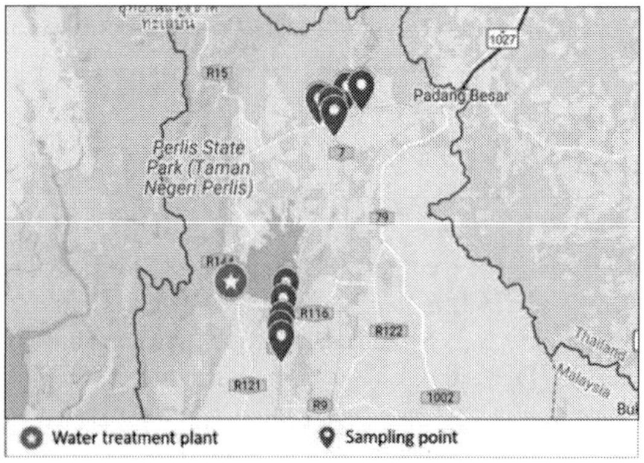

Fig. 2. Location of Timah Tasoh WTC and sampling point for the following location. Taken from Map Data ©2017 Google.

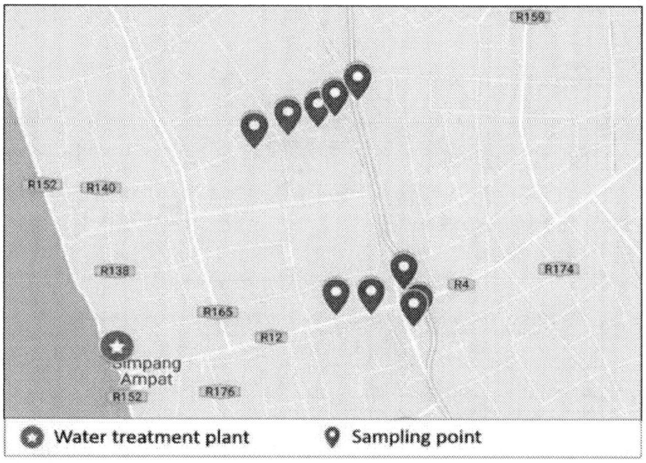

Fig. 3. Location of Kuala Sungai Baru WTC and sampling point for the following location. Taken from Map Data ©2017 Google.

However, since the water supplies from Arau Phase I, II, and III are combined with the Arau Phase IV water treatment plant inside a storage tank before being sent to the users, these treatment plants are considered to be one plant instead of two. After the sampling process was done, it was analyzed in the lab within 14 days of the extraction time, as suggested in USEPA 551.1. Standard quality controls also are carried out according to USEPA to check for instrument and laboratory capability. The water samples were analyzed by using Gas Chromatography-Mass Spectrometry with the following condition: the types of column are HP-5, the length of the column is 30m, and the Injector and Detector Temperature are 200°C and 260°C respectively.

3 Results and discussions

3.1 THMs exposure levels

The total number of THMs was found to be different in every location. The location is indicated by the number of samples, as shown in following; a) Sample 1 to 30: WTP Timah Tasoh, b) Sample 31 to 60: WTP Kuala Sungai Baru and c) Sample 61 to 90: WTP Arau Phase 1, 2, 3 & 4. Chloroform was found to be higher than other elements, especially in the Timah Tasoh area where some of it has violated the limit provided by WHO and USEPA as shown in Fig. 4.

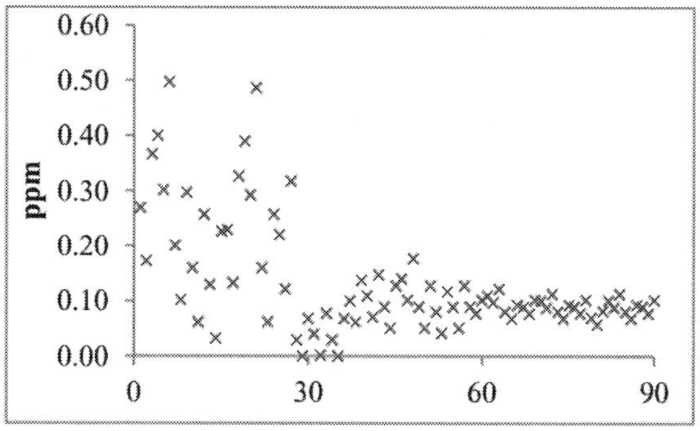

Fig. 4. Chloroform of end-user water samples.

 https://doi.org/10.1051/e3sconf/20183402043

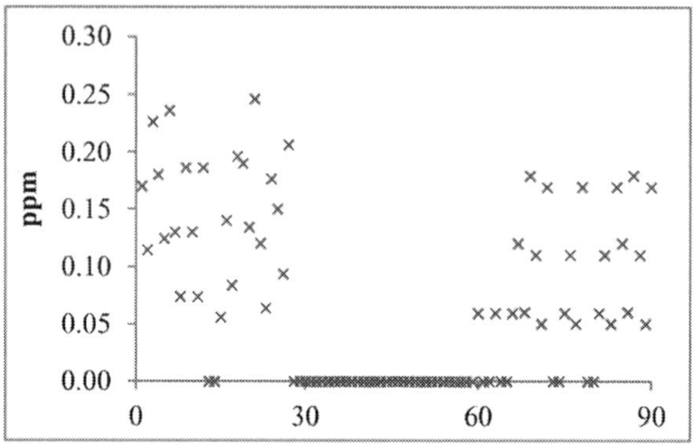

Fig. 5. Bromodichloromethane at end-user water samples.

Even though the pH around the area is slightly higher as compared to other locations, the chloroform formation seems to be contrasted to the report by Singer & Shi (2008) [12] where they claim the formation of chloroform is supposed to be lower in higher pH. However, this does not count other factors such as the present of ammonia in the raw water and the time contact between the chlorine and water. It is known that the pipeline between WTP to the household is the longest at Timah Tasoh, and due to the fact that it is a large area and with the current earth surface condition, it requires a long pipeline. The longer the pipeline system is, the higher the amount of organic matter buildup, especially when the storage and delivery tank are involved. So, all these factors will affect the formation of chlorine despite the fact that it has a higher pH.

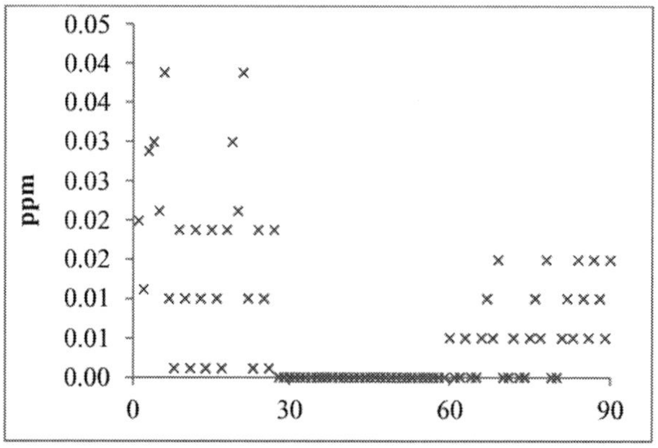

Fig. 6. Dibromochloromethane at end-user water samples.

The area that receives a water supply from Kuala Sungai Baru shows the best results in this study. This is because its location is close to the residential area, thus required a shorter piping system. Since there is no storage tank is being used in this area, contributing factors such as organic buildup were automatically eliminated. That explains the unavailability of certain THMs in the water samples.

Table 1. WHO guidelines [15].

Parameter	WHO Guidelines value (mg/L)
Chloroform	0.20
Dichlorobromomethane	0.06
Dibromochloromethane	0.10
Bromoform	0.10
Total THM	<0.001

Table 2. Variation of THMs level in the tap water received by end-user in 3 WTP areas.

Location	TCM	DCBM	DBCM	TBM	Unit
Timah Tasoh					
Max	0.50	0.25	0.04	<0.03	mg/L
Min	0.00	<0.07	0.00	<0.03	
Mean	0.22	0.12	0.01	<0.03	
Kuala Sungai Baru					
Max	0.18	<0.07	0.01	<0.03	mg/L
Min	0.00	<0.07	0.00	<0.03	
Mean	0.09	<0.07	0.00	<0.03	
Arau Phase I, II, III & IV					
Max	0.12	0.18	0.02	<0.03	mg/L
Min	0.06	<0.07	0.00	<0.03	
Mean	0.09	0.08	0.01	<0.03	

4 Conclusions

This paper shows that THM formation is the highest in the tap water supplied by WTP Timah Tasoh, and the lowest amount of THMs is in tap water supplied by WTP Kuala Sungai Baru. Some of it even violated the guidelines provided by WHO [15]. The range of THM concentrations in all the areas is between 0.00mg/L to 0.50mg/L for all compounds. Chloroform ($CHCl_3$) dominated the results, followed by bromodichloromethane ($CHCl_2Br$). The other two almost give no results in this study, especially bromoform ($CHBr_3$). The formation of THMs in the water received by the end-user has caused great concern since these substances can cause cancer in humans. This paper describes a few parameters that need to be taken consideration when conducting a standard testing. This is to reduce the possible error that can distort the results of the analysis.

The authors would like to thank School of Environmental Engineering, UniMAP and Universiti Teknologi MARA (Arau) for providing the instruments used in the analysis throughout the study.

Reference

1. R.L., Calderon, Food Chem. Toxicol. **38** (12) (2000)
2. W., Gan, W., Guo, J., Mo, Y., He, Y., Liu, W., Liu, Y., Liang, X., Yang, Sci. of the Total Environ, **447**, 108 – 115 (2013)
3. S.K., Golfinopoulos, A.D., Nikolaou, Desalination **176**, 13 – 24 (2005)
4. S., Guilherme, M.J., Rodriguez M.J., Chemosphere **117**, 425–432 (2014)
5. N., Hamidin, Q.J., Yu, D.W., Connell, Water Res. **42**, 3263 – 3274 (2008)

6. C. Legay, M.J., Rodriguez, R., Sadiq, J.B., Serodes, P., Levallois, F., Proulx, J. of Environ. Man. **92**, 892 – 901 (2011).
7. M.Y. Perlis, Google Maps. Google. Retrieved from https://www.google.com.my/maps/place/Perlis/@6.2895831,100.3726868,10z/data=!4m5!3m4!1s0x304c7916f2d4bdf9:0x675ec9fc75fcfb6f!8m2!3d6.4449129!4d100.2047691?hl=en
8. S.D., Richardson, Trends Anal. Chem. **22** (10), 666 – 884 (2003)
9. S.D., Richardson, *Encyclopedia of Environmental Health* (Elsevier, Burlington, 2011)
10. M.J., Rodriguez, J.B., Serodes, P., Levallois, Water Res. **38** (20), 4367 – 4382 (2004)
11. P.C., Singer, Water Sci. Technol. Water Supply **2** (5–6), 487–492 (2002)
12. P.C., Singer, J., Shi, *Control of Haloacetic Acid Formation in North Carolina Drinking Water* (2008)
13. V., Uyak, K., Ozdemir, I., Toroz, Sci. of the Total Environ. **390**, 417 – 424 (2008)
14. J., Wei, B., Ye, W., Wang, L., Yang, J., Tao, Z., Hang, Sci. of the Total Environ. **408**, 4600 – 4606 (2010)
15. WHO, *Guidelines for drinking-water quality 3rd Ed.* (World Health Organization, Geneva, 2004)

Study on Molasses Concentration from Sugarcanne Bagasse for Biohydrogen Production using Enriched Granular Activated Carbon (GAC) Immobilised Cells by Repeated Batch Cultivation

Norfatiha Idris[1], *Nabilah Aminah* Lutpi[1*], *Che Mohd Ruhaizul* Che Ridzuan[1], *Wong Yee* Shian[1], and *Tengku Nuraiti* Tengku Izhar[1]

[1]School of Environmental Engineering, Universiti Malaysia Perlis, Arau, Perlis, 02600, Malaysia

Abstract. Repeated batch cultivation is known as most attractive method in improving hydrogen productivity, due to the facts that this approach could minimize the reuse of the cell and the inoculum preparation. In addition, with the combination of attach growth system during the fermentation processes to produce biohydrogen, the density of cells will be increased and the cell washout could be avoided. Therefore, this study aimed to examine the effectiveness of repeated batch cultivation for enrichment of anaerobic mixed culture onto granular activated carbon (GAC) and investigate the effect of molasses concentration during immobilization of mixed culture onto the GAC. The molasses concentration using 50 %, 40 %, 30 %, 20 % and 10 % of diluted molasses were used as feedstock in the fermentation process. The maximum hydrogen production of 60 ml was obtained at 30 % of molasses concentration with 831 ppm of hydrogen concentration. Thus, the kinetic parameter obtained from the batch profiling based on modified Gompertz equation are, H_m= 58 ml for the maximum hydrogen production and R_m= 2.02 ml/h representing the hydrogen production rate.

1 Introduction

Nowadays, the utilization of fossil fuels was rapidly increase as energy supply especially in transportation and electricity. About 75% of the increasing CO_2 concentration in the atmosphere contributes from burning of fossil fuels. Burning fossil fuels resources could increase four times the present value of the CO_2 concentration in the atmosphere which of about 350 to 1500 ppmv (parts per million by volume) [1]. Momirlan et. al stated that in the next 30 years the consumption of petroleum based is predicted to increase 80% [2]. Therefore, there is a need to replace the fossil fuels with a new renewable energy sources such as hydrogen. Hydrogen has high energy yield which is 122 kJ/g compared with other fuels [3]. There is increased and developing interest in biological hydrogen production by

* Corresponding author: nabilahaminah@yahoo.com

utilizing the use of microorganism. In biological production, there are few types of processes, which are light-dependent processes (e.g. photo-fermentation) and light-independent processes (e.g. dark fermentation) [4]. However, dark fermentation is more attractive due to the ability of anaerobic microorganism to produce hydrogen gas.

The used of simple sugars (e.g. glucose and sucrose) as substrate or feedstock for dark fermentation in biohydrogen production are widely been research but these are considered as expensive [5]. Therefore, reuse of waste such as molasses wastewater as substrate in fermentative biohydrogen production is best choice because it's high composition of sugar, easy to obtain and cut off the overall project costing. Repeated batch cultivation is most attractive method in improving the hydrogen production due to reuse of bacteria cell, minimizing fermentation preparation and cut the overall costing. Repeated batch cultivation is applied for process to attach the bacteria cell on the surface of support carrier [6].

In this study, mesophilic fermentative hydrogen production was carried out by immobilization of the anaerobic sludge through surface attachment on granular activated carbon (GAC). As a support carrier, GAC functioned to attach growth cells of molasses that operate under mesophilic temperature at suitable pH and nutrient needed [6]. This study will focus on to examine the effectiveness of repeated batch cultivation for enrichment of anaerobic mixed culture onto GAC and to optimize the concentration of molasses as the substrate use for the biohydrogen fermentation.

2 Material and methods

2.1 Sources of seed sludge, feedstock and support carrier for fermentative hydrogen production

The origin of seed sludge and molasses as feedstock in the study were from sugarcane bagasse that was obtained from Fempro, Chuping, Perlis. The bacterial sludge used in this study was the continuity from previous study that had operated for two months in a 14 L anaerobic reactor by feeding with molasses. The effluent samples of the reactor were collected every week and stored at 4°C as stock culture for further use as inoculum. The anaerobic sludge will undergo heat shock by heating at 80°C for 60 minutes prior to be used for experiment. This is to inactivate methanogenic bacteria and non-hydrogen producing bacteria prior to batch fermentation. Granular activated carbon (GAC) in the range of 2 – 3 mm size was used as support carrier to attach the hydrogen producing bacteria in this study.

2.2 Preparation of molasses at different dilution percentage for batch fermentation

First step prior to prepare the molasses concentration at different dilution percentage is to determine the total carbohydrates content of the raw molasses. After that, the suitable weight of glucose will be added to the sample according to equation 1 and 2 so that each of the molasses from different percentage of dilution with have the same concentration of 5g/l of total carbohydrate.

$$\text{Total carbohydrate still needed (g/L)} = 5\text{g/l} - \text{Total carbohydrate obtained (g/l)} \quad (1)$$

$$\text{Amount addition glucose (g)} = \text{Total carbohydrate still needed (g/l)} \times \text{Volume of sample (L)} \quad (2)$$

Experiment was conducted in 100 ml serum bottle with different percentage (50%, 40%, 30%, 20% and 10%) of molasses. Different glucose weight in (g) were added to the different percentage of molasses to equalised all of the sample to 5g/L of total carbohydrate (shown in Table 1). The batch experiments were carried out under mesophilic condition (37°C) in the incubator.

Table 1. Fermentation set-up parameter for batch experiment.

Serum bottle	Molasses concentration				
	50%	40%	30%	20%	10%
Total volume of serum bottle (ml)	100	100	100	100	100
Working volume (ml)	55	55	55	55	55
Ratio of molasses:distilled water (for 50 ml of working volume)	1: 1 (25:25)	1: 1.5 (20:30)	1: 2.33 (15:35)	1: 4 (10:40)	1: 9 (5:45)
Inoculum/sludge (mm)	5	5	5	5	5
Support carrier, GAC (g)	5	5	5	5	5
Glucose added (g)	0.82	0.86	0.92	0.94	0.97

2.3 Repeated batch cultivation

Repeated batch cultivation was conducted according to the set-up parameter in Table 1. Fermentation with GAC was conducted at various molasses concentration (50 %, 40 %, 30 %, 20 %, and 10 %) and similar experiments were conducted in parallel without GAC as control. Each of the serum bottle was then purged with nitrogen for a few minutes to provide an anaerobic conditions. Finally, the serum bottle was incubated at mesophilic temperature, 37 °C for 24 hours in incubator. This method was repeated for several successive batch until the performance of hydrogen gas is stable in order to compare the hydrogen production at different molasses concentration.

2.3.1 Biogas production

The total gases collected were measured using Gas Analyser (biogas 5000 standard) Model GA 5000. The sulphuric acid with pH 2-3 was prepared in serum bottle to store the biogas prior to analysis. The biogas was analysed for the composition of hydrogen, carbon dioxide, oxygen and methane (if presence).

2.3.2 Kinetic analysis using Modified Gompertz equation for batch profiling

Kinetic analysis using Modified Gompertz equation for the optimal molasses concentration was applied whereby batch profiling for 24 hours with interval of 3 hours sample collection were observed. Theoretically, the modified Gompertz equation is as in equation 3:

$$H_t = H_m \exp\left\{-\exp\left[\frac{R_m e}{H_m}(\lambda - t) + 1\right]\right\} \tag{3}$$

where H_t is the cumulative hydrogen production (ml), H_m is the maximum hydrogen production (ml), R_m is the maximum hydrogen production rate (ml h^{-1}), e is Euler's number (e =2.73), λ is the lag phase time (h), and t is the fermentation time (h).

2.4 Analysis procedure

2.4.1 Total carbohydrate

The total carbohydrate present in the seed sludge, molasses, and fermentation effluent were tested by using phenol sulphuric acid method. Each of samples was prepared with volume of 0.5 ml in a vial. Then, 2.5 ml of sulphuric acid and 0.5 ml of phenol will be added to the vial. The samples were mixed evenly and leave for 30 minutes in dark place before being tested using spectrophotometer. The equation to determine the percentage of sugar consumption was as in equation 4.

$$\frac{Total\ carbohydrate\ before\text{-}Total\ carbohydrate\ after}{Total\ carbohydrate\ before} X\ 100\% \quad (4)$$

2.4.2 Alkalinity

The dilution of 2.7 ml of sulphuric acid, (0.1N) in 1000 ml of volumetric flask and 50 ml of sample was prepared in a beaker. The diluted sulphuric acid was filled in 50 ml of burette and titration process is started. The titration stops when the samples reach pH 4.5. To determine the total alkalinity, the equation 5 is used.

$$\text{Total alkalinity, mgCaCO}_3\text{/l} = \frac{S\ x\ N\ x\ 50000}{volume\ of\ sample,ml} \quad (5)$$

3 Results and discussions

3.1 Characterization of seed sludge and molasses

The characterization of seed sludge as inoculum to immobilized on GAC, and molasses as feedstock were listed as follow:

Table 2. Characteristics of seed sludge and molasses.

Parameter	Value	
	Molasses	Seed sludge
Chemical oxygen demand (COD), (g/L)	28.00	6.27
Total suspended solid (TSS), (g/L)	0.05	0.01
Initial pH	7.538	7.808
Initial temperature, (°C)	26	26

3.2 Batch fermentation at various molasses concentration with the presence of GAC

The hydrogen production from the repeated batch fermentation for 5 different molasses concentration (50%, 40%, 30%, 20% and 10%) were collected for 10 repeated batch fermentation cycles (days) with the presence of GAC as support carrier.

https://doi.org/10.1051/e3sconf/20183402044

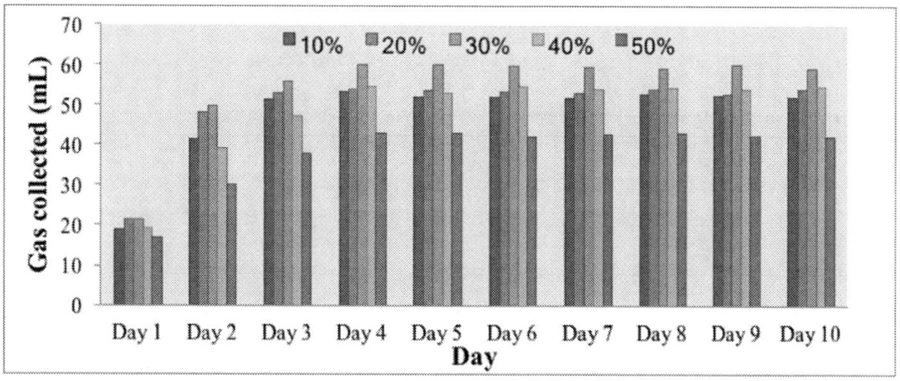

Fig. 1. Graph of hydrogen gas collection for different molasses concentration in repeated batch fermentation of 10 days.

Based on Fig. 1, the hydrogen produce by the mixed culture at different molasses concentration on the first cycle (day) was the lowest compared to other cycles. The hydrogen collected was increased from 19 ml to 21.3 ml when the molasses concentration is increase from 10% to 20% and 30%. However when further increased in substrate concentration to 40%, the hydrogen collected is decrease to 19 ml. The decreased trend was continued when the molasses concentration is increased to 50 % with 17.7 ml. The usage of 50% of molasses concentration was the lowest hydrogen production in all cycles (days) of repeated batch fermentation. It seem that the mixed culture bacteria were difficult to adjust to the environment with higher concentration of molasses. This might be due to the existence of fermentation inhibitor in the sugarcane bagasse molasses such as furfural. According to [7], pentose sugar such as in molasses could degrade to furfural. The inhibitory effects of furfural depend on its concentration and yet, no clear effect either positive or negative on the cell growth for molasses from sugarcane bagasse was reported. Further study and experimental need to be conduct for further justification on this matter. When the mixed culture bacteria growing in a rich medium, however were inoculated together with the presence of the inhibitor in the medium, the bacteria will take time to synthesising the necessary proteins, co-enzymes and vitamins needed for their growth.

Overall, the hydrogen production in all molasses concentration were rapidly evolved from day 1 to day 4 and start to stabilised until the day 10 of fermentation. The hydrogen producing bacteria might have increased their metabolic activity and also strengten their adhesion onto GAC during the repeated batch cycle of fermentation. The repeated batch fermentation seems to caused the exploitation of growth medium at the maximal rate and growth of culture reaches the maximum rate. The used of repeated batch cultivation was successfull due to increasing in hydrogen performance. To sum up, the molasses concentration of 30 % depicted the highest hydrogen performance with 59 ml from day 4 of fermentation and the performance was sustained until day 10 of repeated batch fermentation.

3.3 Hydrogen production with GAC and without GAC (control) at various initial substrate concentration

The hydrogen production for 50%, 40%, 30%, 20% and 10% of molasses concentration with the presence of GAC were compared with the performance without GAC. The result obtained was shown in Fig. 2.

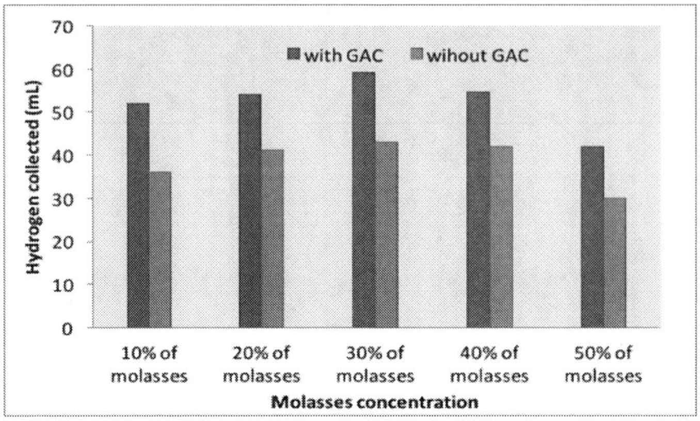

Fig. 2. Hydrogen production with GAC and without GAC at various molasses concentration.

The effect of initial substrate concentration at temperature 37°C with GAC and without GAC were studied in the range of 10% to 50% of molasses concentration. Based on Fig. 2, with the presence of GAC attached biofilm, the hydrogen production is increased from 52 ml to 54 ml when the initial substrate concentration was increased from 10% of molasses to 20% of molasses. However, a further increased in the initial substrate concentration to 30% of molasses was found to be the highest hydrogen collected with 59 ml. However, the hydrogen production collected is decreased when the initial substrate increased from 40% of molasses to 50% of molasses. This observation is similar to that of [8], who also reported that the hydrogen production is increase from 2.82 mmol $H_2L^{-1}h^{-1}$ to 4.3 mmol $H_2L^{-1}h^{-1}$ in fermentation with GAC attached biofilm.

On other hand, the production of hydrogen trend without the presence of GAC were the same as with GAC attached biofilm when the concentration of molasses were increased. The biohydrogen collected is increase from 36 ml to 41 ml when the substrate concentration increased from 10 % to 20% of molasses. At 40 % to 50 % of molasses, the hydrogen collected was declined from 43 ml to 30 ml. The highest production of hydrogen without GAC is at 30% of molasses concentrtaion with 45 ml of hydrogen produced. However, the overall performance without the presences of GAC were lower than the performance of hydrogen with GAC immobilised cells.

3.4 Kinetic analysis of the optimal molasses concentration with GAC immobilised cells

The kinetic parameter of the optimal molasses concentration (30%) with GAC immobilised cells was obtained from modified Gompertz model as in equation 3. The batch profiling of cumulative hydrogen production between the data from experimental and predicted modified Gompertz model is shown in Fig. 3. Based on the predicted modified Gompertz model, the cumulative hydrogen data fitted with maximum hydrogen production, H_m =58 ml, hydrogen production rate, R_m = 2.02 ml/h and lag phase time, λ = 6.13 h.

Based on Fig. 3, at the first 6 hours of fermentation process, the hydrogen production was starting to increase. This phase is known as the lag phase whereby the mixed culture bacteria have just started to adapt to new environment and medium (molasses). Mixed culture activity is accelerated due to its cellular metabolism and increasing in their cells size. The bacteria start to synthesize the necessary proteins, co-enzyme and vitamins needed for their growth rate hence it will take time to produce high hydrogen production.

At 9 hours to 24 hours of fermentation process, the hydrogen collected was increased rapidly from 7 ml to 35 ml and this is known as the exponential or logarithmic phase. At this phase, bacteria are rapidly growing and dividing state occur due to their metabolic activity. The medium (molasses) is exploited at the higher rate and reaches their higher growth rate to produce higher hydrogen performance. The maximum of hydrogen concentration for this batch profiling is 831 ppm.

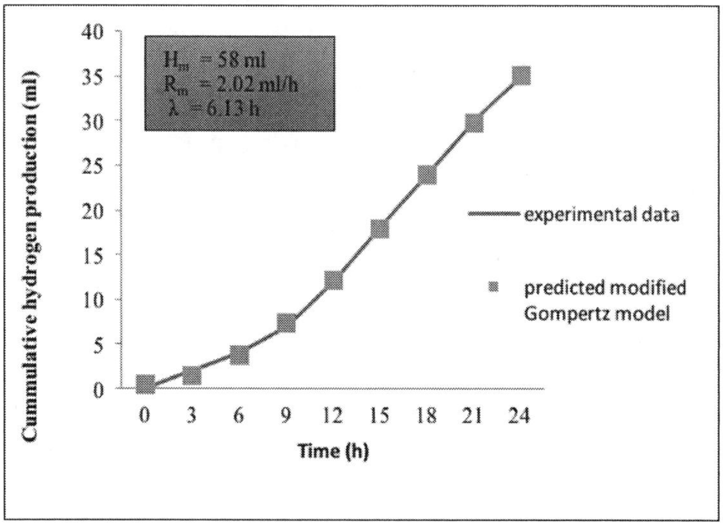

Fig. 3. Graph of experimental data and predicted Modified Gompertz Model for optimal molasses concentration (30 %).

3.5 Sugar consumption and total alkalinity of optimal molasses concentration with GAC immobilised cells

Fig. 4 shows the sugar consumption for the optimal molasses concentration (30 %) from the batch profiling. The results decreased over time of fermentation, which is from 100 % at first 3 hour to 36 % at 24 hour. This result was inline with the results obtained in Fig. 3, whereby the higher the sugar is consumed by the hydrogen producing bacteria, the higher the conversion of sugar into hydrogen is produced. Fig. 4 also depicted the total alkalinity, which represent the volatile fatty acid (VFAs) produced. VFAs and solvents, collectively known as soluble microbial products (SMPs) are considered as beneficial indicators to observe the hydrogen production performance. The result of total alkalinity was increased over time with 54 mg $CaCO_3$/l at first 3 hours and increased 3 fold to 94 mg $CaCO_3$/l at 24 hour of fermentation. The increment of total alkalinity was concurrent with the evolution of hydrogen production, which reflect the norm of anaerobic biohydrogen process.

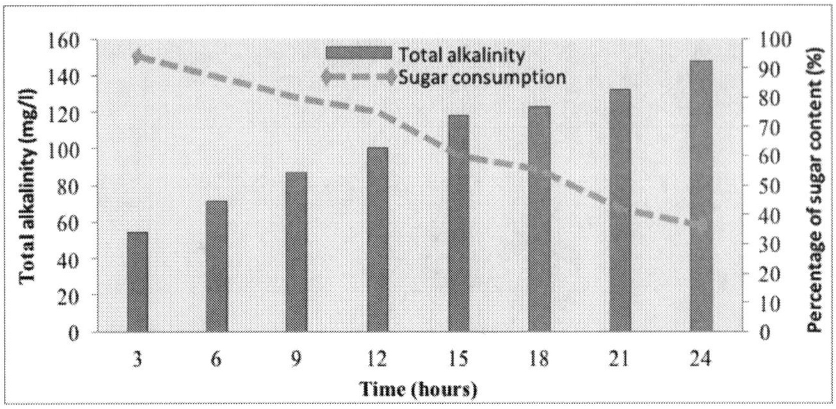

Fig. 3. Total alkalinity and sugar consumption of batch profiling fermentation process for optimal molasses concentration (30 %).

4 Conclusions

Ten cycles of repeated batch cultivation provide positive result in hydrogen production performance. The dark fermentation using GAC immobilised cell through repeated batch cultivation depict stable hydrogen production performance. The maximum biohydrogen obtained was optimal using 30% of molasses concentration. Batch profiling was done for the optimal molasses concentration, and the data was fitted with predicted modified Gompertz equation. The kinetic parameters were obtained with maximum hydrogen production, H_m= 58 ml and hydrogen production rate, R_m= 2.02 ml/h, with hydrogen concentration of 831 ppm. The performance of substrate concentration (molasses) during immobilisation of mixed culture bacteria onto granular activated carbon (GAC) has been proved by comparison with control sample (without GAC).

This research is fully supported by FRGS grant (FRGS No: 9003-00511). The authors fully acknowledged Ministry of Higher Education (MOHE) and Universiti Malaysia Perlis (UniMAP) for the approved fund, which makes this important research viable and effective.

References

1. R. R. Judkins, W. Fulkerson, M. K. Sanghvi, Energy Fuels. **7,** 14-22 (1993)
2. M. Momirlan, T. N. Veziroglu, Int. J. Hydrogen Energy. **30,** 795-802 (2005)
3. R. Chamousis, *Hydrogen: Fuel of the future.* (The Scientific Research Society, 2000)
4. J. J. Lay, Y. J. Lee, T. Noike, Water Res. **33,** 2579-2586 (1999)
5. W. Han, B. Wang, Y. Zhou, D. X. Wang, Y. Wang, L. R. Yue, N. Q. Ren, Bioresour. Technol. **110,** 219-223 (2012)
6. N. A. Lutpi, J. M. Jahim, T. Mumtaz, S. Harun, P. M. Abdul, Process Biochem. **51,** 297-307 (2016)
7. S. I. Mussatto, I. C. Roberto. Bioresour Technol. **93,** 1-10 (2004)
8. W. Q. Guo, N. Q. Ren, X. J. Wang, W. S. Xiang, Z. H. Meng, J. Ding, L. S. Zhang, Int. J. Hydrogen Energy. **33,** 4981-4988 (2008)

Enhancement of Biohydrogen Production via pH Variation using Molasses as Feedstock in an Attached Growth System

C.N.S. Che Zuhar[1], *N.A.* Lutpi[1*], *N.* Idris[1], *Y.S.* Wong [1], and *T.N.* Tengku Izhar[1]

[1]School of Environmental Engineering, Universiti Malaysia Perlis, Arau, Perlis, 02600, Malaysia

Abstract. In this study, mesophilic biohydrogen production by a mixed culture, obtained from a continuous anaerobic reactor treating molasses effluent from sugarcane bagasse, was improved by using granular activated carbon (GAC) as the carrier material. A series of batch fermentation were performed at 37°C by feeding the anaerobic sludge bacteria with molasses to determine the effect of initial pH in the range of 5.5 to 7.5, and the effect of repeated batch cultivation on biohydrogen production. The enrichment of granular activated carbon (GAC) immobilised cells from the repeated batch cultivation were used as immobilised seed culture to obtain the optimal initial pH. The cumulative hydrogen production results from the optimal pH were fitted into modified Gompertz equation in order to obtained the batch profile of biohydrogen production. The optimal hydrogen production was obtained at an initial pH of 5.5 with the maximum hydrogen production (H_m) was found to be 84.14 ml, and maximum hydrogen production rate (R_m) was 3.63 mL/h with hydrogen concentration of 759 ppm. The results showed that the granular activated carbon was successfully enhanced the biohydrogen production by stabilizing the pH and therefore could be used as a carrier material for fermentative hydrogen production using industrial effluent.

1 Introduction

Today, energy demand and resources has become a crucial aspect in sustainable development due to the growing concerns over the environmental impact and the inevitable depletion of fossil fuels. Due to that reason, the development of alternative energy sources has become matter of urgency. According to Ni et al. [1], two important issues which are energy crisis and environmental sustainability were the major concern in order to achieve global sustainable development. Since mostly the focus on fossil fuels, it leads to climatological changes and rapid exhaustion of natural energy resources.

Hydrogen is well known as clean energy carrier since it is renewable energy source with carbon free [2]. Hydrogen also contributes substantially to the reduction of greenhouse gas (GHG) emissions as well as its might be considered as ultimate clean and climate neutral energy system [3]. If compared to fossil fuels, hydrogen yield obtained higher energy. Hydrogen gas guaranteeing other possibility to fossil fuels, otherwise it suggests the possibility of generating a valuable energy carrier that is renewable and carbon unbiased.

* Corresponding author: nabilah@unimap.edu.my

Besides, hydrogen need an helter - skelter vitality yield about 122 kJ/g, which will be 2.75 times more terrific over hydrocarbon fills [4]. Various processes are being used in order to produce hydrogen gas, including photo fermentation, dark fermentation, photolysis and electrolysis of water. Among all, the most suitable processes are to be dark fermentation [5].

Dark fermentation is an anaerobic digestion of organic substances and it is considered the simplest processes of obtaining biohydrogen. Different factors have been investigated to improved biohydrogen production either as a suspended cells system or immobilised cells system using dark fermentation processes in this study. Many factors that affecting dark hydrogen fermentation such as inhibits of undissociated acid and changes of pH, hydrogen partial pressure and metal ions.

In terms of feedstock, real wastewater have demonstrated strong evidence of hydrogen production. Sugarcane, Saccharum officinarum L sp. is carbon - rich as well as the biomass residue [6]. There are various products that can be obtained from sugarcane, and one of it is molasses that are widely used and known to be practical feedstock in terms of availability, cost, and biodegradability for fermentative hydrogen production.

Therefore, this study will focus on the enhancement of batch biohydrogen production using immobilised cells system by feeding molasses from sugarcane bagasse as feedstock. This study will emphasis on the effect of initial pH on fermentative hydrogen production by varying the pH, which is the key environmental parameter in biohydrogen process.

2 Material and methods

2.1 Sample collection

The origin of seed sludge and molasses used in the study were from sugarcane bagasse that was obtained from Fempro, Chuping, Perlis, an industry producing ethanol. The bacterial sludge used in this study was the continuity from previous study that had operated for two months in a 14 L anaerobic reactor by feeding with molasses. The operation was done in continuous mode at 37°C. The effluent samples of the reactor were collected every week and stored at 4°C as stock culture for further use as inoculum.

2.2 Mixed culture and carrier material

The anaerobic sludge underwent heat shock by heating at 80°C for 60 minutes prior to be used for experiment. This is to inactivate methanogenic bacteria prior to batch fermentation. Granular activated carbon (GAC) in the range of 2 – 3 mm size was used as carrier material to attach the hydrogen - producing bacteria in this study.

2.3 Repeated batch cultivation

Repeated batch cultivation was conducted in 100 mL serum bottle with 50 mL working volume filled with 10 % of anaerobic sludge, and 90 % of molasses. This repeated batch of fermentation were conducted into different range of initial pH from 5.5 to 7.5 with 0.5 of increments (pH 5.5, 6.0, 6.5, 7.0, and 7.5). The immobilization of the anaerobic sludge was carried out by adding GAC in a 1 : 1 ratio of anaerobic sludge volume (mL) to GAC weight (g) in the serum bottle, while fermentation without GAC was conducted in parallel as control. Each of the serum bottle was then purged with nitrogen for a few minutes to provide an anaerobic conditions. Finally, the serum bottle was incubated at mesophilic temperature, 37 °C for 24 hours in incubator. This method was repeated for several successive batch until the performance of hydrogen gas is stable in order to compare the hydrogen production at different initial pH values.

2.3.1 Biogas collection

Sampling of biogas from the serum bottles were done after 24 hours for each cycle of fermentation. The collected biogas were kept in 30 mL serum bottle containing acid at pH 2.0 before the biogas were analysed. The production of hydrogen was measured by using Gas Analyser (biogas 5000 standard) Model GA 5000. The samples were also analysed for other biogas composition such as carbon dioxide, oxygen and methane (if presence).

2.3.2 Batch profile fermentation

Generally, the experimental procedure on batch profile was same as in the Section 2.3. However, this batch profile was only employed for the initial pH with optimal hydrogen production that was obtained from Section 2.3. The data were recorded and the sampling were done every 3 hours of interval for 24 hours of incubation time.

2.3.3 Calculation of hydrogen production

The cumulative hydrogen production in the batch experiments was determined according to a modified Gompertz equation using Sigma Plot 10.0 (Systat Software Inc., USA). Theoretically, the modified Gompertz equation is as follows.

$$H_t = H_m \exp\left\{ -\exp\left[\frac{R_m e}{H_m}(\lambda - t) + 1\right]\right\} \tag{1}$$

where H_t is the cumulative hydrogen production (mL), H_m is the maximum hydrogen production (mL), R_m is the maximum hydrogen production rate (mL h^{-1}), e is Euler's number (e =2.73), λ is the lag phase time (h), and t is the incubation time (h).

2.4 Analysis procedure

2.4.1 Total carbohydrate

The Phenol - Sulphuric acid reaction is feasible method used to determine the amounts of sugar for carbohydrate analysis of the fermentation broth and effluent. This method was widely used to determine sugar compositions due to its simplicity and sensitivity. Simple sugar such as glucose are stained orange - yellow when treated with phenol and concentrated sulphuric acid. This quality can be used to perform quantitative analyses of sugars using colometric methods such as spectrophotometer. The absorption of each sample is then measured by using spectrophotometer at 490 nm.

2.4.2 Alkalinity

The titration was used to determine the alkalinity of the effluent sample after fermentation to obtained the total value of volatile fatty acids. 50 mL of the sample was taken and then titrated with 0.1 N of sulphuric acid solutions. The electrode of pH meter was rinsed and the placed into the sample. The sample was mixed thoroughly until constant reading is obtained. The sample was titrated to pH 4.5 to an end point value for alkalinity. Then, the value of alkalinity of the sample were calculated by using the equation 2:

$$Alkalinity,\ mg\ CaCo_3/\ l = \frac{S\ x\ N\ x\ 50000}{Volume\ of\ sample,\ ml} \qquad (2)$$

where S = ml of titrant to achieve pH 4.5 as end point and N = equivalents H_2SO_4 per liter titrant.

2.4.3 Chemical Oxygen Demand (COD)

In this study, 2.5mL of the mixed culture samples were added in vial tubes. Then, the vial tubes were added with 1.5mL potassium dichromate solutions and 3.5mL of sulphuric acid solutions. Once it was done, the vials were placed in preheated DRB200 Reactor for two hours. After that, the vials were placed at room temperature for 30 minutes. The COD value were measured by using spectrophotometer.

3 Results and discussions

3.1 Characterization of seed sludge and molasses

Before conducting the batch fermentation, the sample of seed sludge and molasses were characterised. The results were presented in Table 1.

Table 1. Characteristics of anaerobic sludge and molasses.

Parameter	g/L (except pH)	
	Anaerobic sludge	Molasses
pH	7.583 ± 0.008	7.808 ± 0.004
TSS	0.0480	0.014
COD	28.053 ± 1.083	6.270 ± 0.673
Total Carbohydrate	3.46	2.60

3.2 Batch fermentation at various initial pHs

3.2.1 Effect of repeated batch cultivation at various initial pHs

pH can be the main control parameter for the bacteria activities in order to enhance the hydrogen performance. Therefore, batch fermentation were conducted in series to determine the effects of different initial pH during fermentative hydrogen production in this study. The fermentation broth consist of effluent sludge and molasses were conducted triplicate each for different initial pH of 5.5, 6.0, 6.5, 7.0 and 7.5. The result shown in Fig.1 was the data of hydrogen performance at different initial pH values for ten days of repeated batch cultivation. On the first day, the highest collected gas was at pH 6.0 with 44 ml. However, at pH 7.5, no gas was produced and this might be due to the bacteria still adjusting to their new environment. The performance of gas seems to be evolved for all pHs and start to stabilised starting from day 5.

At initial pH 5.5, the hydrogen produced only 37 ml on the first day, however starting the second day, the gas increased to 40 ml. Then it continue to increase up to day five which

 https://doi.org/10.1051/e3sconf/20183402045

reached to 70 ml. It can be stated that the hydrogen gas production was two folds. Start from day five, the gas from initial pH 5.5 reached its stability and it sustain until day ten. Meanwhile, for initial pH 6.0, the biogas produce almost 50 ml of hydrogen gas on the first day. Then, on the second day until third day the gas produced were increase continuously up to 70 ml. However, after day five, the hydrogen gas production at initial pH 6.0 had depicted its stability and could only produce a maximum of 60 ml of hydrogen gas. On the other hand, initial pH 7.5 did not show any production of gas on the first day. However, it start to produce gas on the second day until day ten. The maximum of gas that can be produced at initial pH 7.5 is only 30 ml of hydrogen.

From the graph, it shown that pH between 5.5 to 6.0 produce hydrogen gas the most while decrease when pH are set between 6.5 to 7.5. Our studies revealed that between those pH, the fermentation process do not favour the production of hydrogen. This result was found to be likely same as previous studies by Chen et al. (2009) [7]. Normally, pH values between 5.5 and 6.0 are reported to be appropriate for better microbial activity as well as produce better hydrogen. Meanwhile, pH between 6.5 to 7.5 that operates under natural condition reduced the hydrogen production. As previous reported, hydrogenase activity is suppressed at low pH during fermentation.

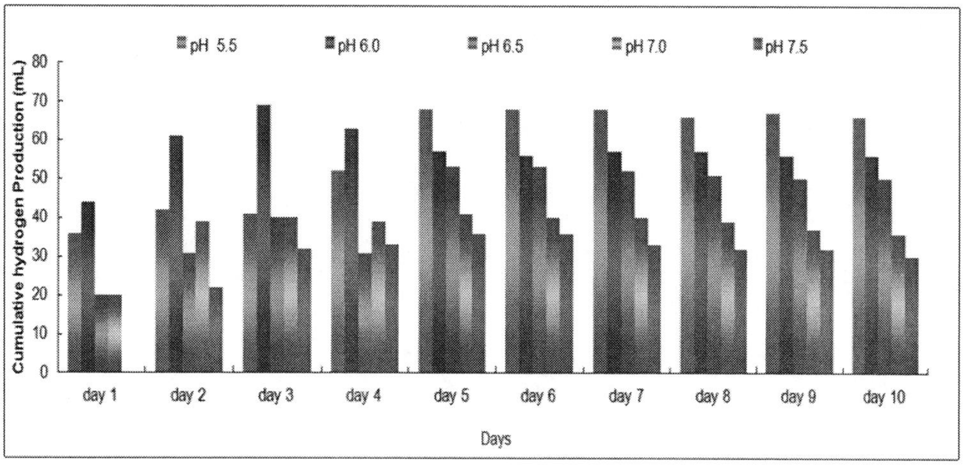

Fig. 1. Production of hydrogen with GAC immobilized cells at various initial pH.

3.2.2 Production of Hydrogen with GAC and without GAC (control) at various initial pH

Fig.2 show the comparison of hydrogen production between different initial pH values with immobilised cells and control (without the presence of GAC). For all pH values, the samples which contains GAC were more efficient in producing biohydrogen compared to the sample without GAC. The study shows that pH 5.5 produce the highest hydrogen if compare to others pH values. This finding is similar to previous study by Wang & Wan (2009) [8] which state that the fermentative anaerobic bacteria were mostly favour to produced hydrogen at optimal pH of 5.5 and 6.0.

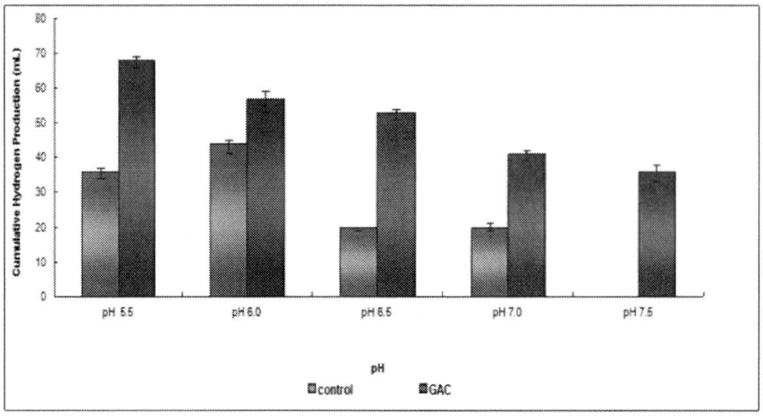

Fig. 2. Hydrogen production with GAC and without GAC (control) at 37°C.

3.2.3 Kinetic analysis of batch fermentation using GAC immobilised cells

The kinetic analysis of batch fermentation using GAC immobilised cells were performed for only pH 5.5 due to the highest hydrogen performance in section 3.2.2. Fig.3 depicts the experimental data and the modified Gompertz model as in equation (1) of the cumulative hydrogen production of the GAC attached biofilm at pH 5.5. Table 2 presented the kinetic parameters obtained for hydrogen production based on data fitted by the modified Gompertz model, whereby H_m = 84.14 ml, R_m = 3.63 ml/h and λ = 2.95 h. The pH of 5.5 is considered weakly acidic, which urges microorganisms to release express protons from cytoplasm for hydrogen production evolution and for the recommencement of cell growth [9]. This study had revealed the greater effects of pH on biohydrogen production that influence the hydrogenase activity.

Table 2. Modified Gompertz equation parameter values for batch profile at optimal performance of biohydrogen.

pH		Modified Gompertz Equation Parameter values for hydrogen production			Hydrogen concentration GAC – cells (ppm)
Initial	Final	H_m (ml)	R_m (ml/h)	λ (h)	
5.5	4.1 ± 0.03	84.14	3.63	2.95	759

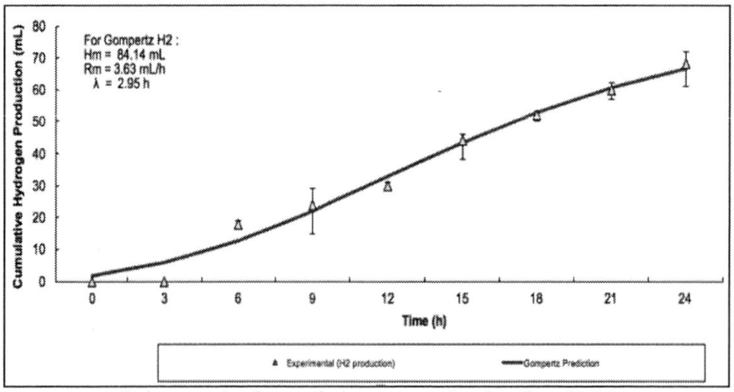

Fig. 3. Gompertz curve- fitting graph of cumulative gas production for pH 5.5.

3.2.4 Total alkalinity and sugar consumption

The result in Fig.4 represented the total alkalinity and sugar consumption of the batch profile for initial pH 5.5, which also reflected the hydrogen production in Fig.3. The total alkalinity increased as the time increases in profile fermentation. The analysis of alkalinity as in equation (2) in this study represent the volatile fatty acids produced as by product besides the production of biohydrogen as the main product. Meanwhile, sugar consumption was decreased with an increase in fermentation time. At time 0 hour, the percentage of sugar is 100%, however after 24 hours, the sugar percentage start to reduce until 60 %. This reduction happened due to rapid growth and dividing state of bacteria resulting from the increasing of their metabolic activity. Thus, bacteria had consumed the sugar and start to adapt with the environment under mesophilic temperature.

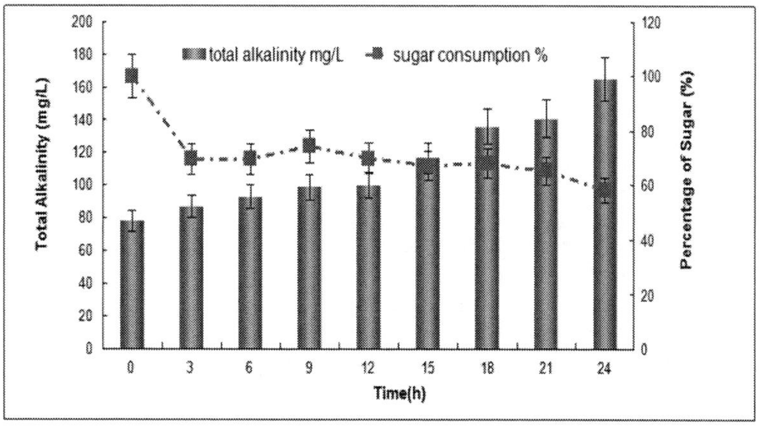

Fig. 4. Graph of total alkalinity and sugar consumption at pH 5.5.

3.2.5 Chemical oxygen demand (COD) analysis

Fig.5 illustrated the Chemical Oxygen Demand (COD) that were obtained after batch profile fermentation for pH 5.5 under mesophilic temperature. The result shows that there is reduction in COD, whereby this finding reveals that the process of fermentative biohydrogen production has simultaneous advantage by not only producing clean renewable energy (hydrogen), but also treating the wastewater used.

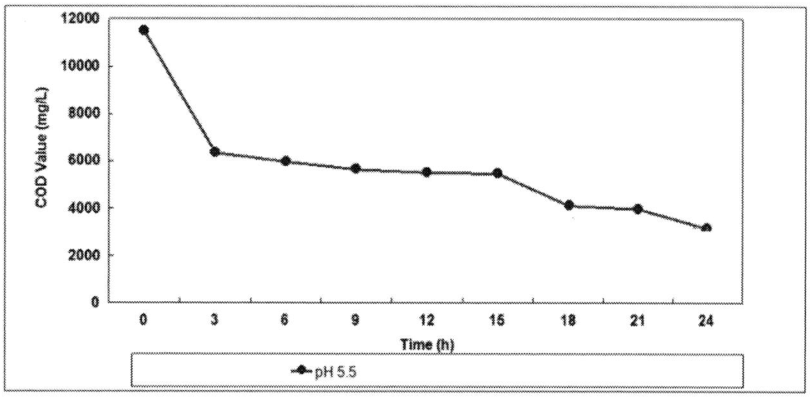

Fig. 5. COD results for pH 5.5 under mesophilic temperature.

4 Conclusions

The biohydrogen production performance of effluent sludge and molasses with GAC immobilized cells were investigated successfully. The optimal conditions for producing hydrogen using GAC immobilised cells at 37°C were achieved at initial pH of 5.5. Repeated batch cultivation have shown the enhancement of biohydrogen production and the durability of the enriched GAC immobilised cells were stable even for several cycles of repeated batch fermentation. The maximum cumulative hydrogen production from batch profiling of initial pH 5.5 was 84.14 ml and the maximum hydrogen concentration obtained was 759 ppm. Overall, the formation of total volatile fatty acid was increased parallel to the increased of the hydrogen production. Therefore, the hydrogen fermentation performed by mixed culture is very much dependent on the initial pH parameter to achieve better performance of hydrogen.

This research is fully supported by FRGS grant (FRGS No: 9003-00511). The authors fully acknowledged Ministry of Higher Education (MOHE) and Universiti Malaysia Perlis (UniMAP) for the approved fund, which makes this important research viable and effective.

References

1. M. Ni, D. Y. C. Leung, M. K. H. Leung, K. Sumathy, Fuel Process. Technol. **87,** 461-72 (2006)
2. S. Dunn, Int. J. Hydrogen Energy. **27,** 235–264 (2002)
3. S. M. Kotay, D. Das, Int. J. Hydrogen Energy. **33,** 258–63 (2008)
4. I. K. Kapdan, F. Kargi, Enzyme Microb. Technol. **38,** 569-582 (2006)
5. P. K. Poddar, O. Sahu, Appl. Water Sci. **7,** 461-468 (2017)
6. C. Bryant, W.Y. Yassumoto, Int. Sugar J. **111,** 696-700 (2009)
7. S.D. Chen, K.S. Lee, J.F. Wu, C.Y. Lin, Int. J. Hydrogen Energy. **34,** 799–811 (2009)
8. J. Wang, W. Wan, Int. J. Hydrogen Energy. **34,**799-811 (2009)
9. P.S. Chong, J.M. Jahim, S. Harun, S.S. Lim, S.A. Mutalib, O. Hassan, M.T.M. Nor. Int. J. Hydrogen Energy **38,** 9592–9599 (2013)

Role of UASBs in River Water Quality Conservation in India

Veeresh Gali[1*], *Manisha* Thakur[2], *Ashok* Kumar Gupta[1], and *Rajiv* Ganguly[1]
[1]JUIT Waknaghat, Civil Department, Teh. Kandaghat, Dist. Solan,173234 H.P, India
[2]Chandigarh Group of Colleges Jhanjeri, Civil Department, Dist. Mohali, Punjab, India

Abstract. Appropriate low-cost treatment technologies are a prerequisite for sound management of natural water resources against pollution in developing countries. Among the existing technologies available, UASB is found to be economically viable for India when considering all factors including operation and maintenance cost and treatment efficiency. However, this technology suffers setbacks in meeting the effluent guidelines prescribed by the government of India. Post treatment is supplemental to this process to meet the effluent standards in terms of removal of organic matter, suspended solids, pathogens and nutrients. Recent stringent effluent guidelines notified by the Ministry of Environment, Forests and Climate Change, Government of India has further reduced the limits of BOD by 3 times, COD and TSS by 5 times, NH4-N and total Nitrogen by 10 times as compared to the previous guidelines. Fe_{cal} Coliforms has been specified as <100MPN/100mL. In this paper, the present scenario of UASB based STPs and their role in river conservation is reviewed against the backdrop of stringent effluent guidelines. The minimum removal rates of BOD, COD and TSS in these plants are around 42 – 44% and the average removal rates are reported to be 66%, 61% and 65% respectively. The enhanced removal of BOD (97%), COD (98%) and TSS has been reported in STPs in conjunction with post treatment facilities such as facultative aerated lagoons, aeration tanks and polishing ponds.

1 Introduction

India is the second most populous country in the world with an estimated population of over 1.3 billion during 2017. There are 29 states and 7 Union Territories(UTs) in the country. Central Pollution Control Board (CPCB), is an apex body under the Ministry of Environment, Forest and Climate Change (MOEF & CC), Government of India (GOI) for control and abatement of pollution in India. According to CPCB, the estimated sewage generation in the country during 2015 was 61,948Million Litres per Day (MLD) against available treatment capacity of 23,277 MLD. There are 816 Sewage Treatment Plants (STPs) in the country out of which 522 are operational, 79 are non-operational, 145 are under construction and 70 are under planning stage. While the existing installed treatment capacity corresponds to around 37.6%, the operational capacity of the STPs is only 30.5%

* Corresponding author: veeresh_ghali@yahoo.com

of the total sewage generated. The remaining untreated sewage is discharged into nearby water bodies. Domestic sewage is considered responsible for about 80% of the water pollution in India due to inadequate sewage treatment facilities. The National River Conservation Directorate (NRCD) which functions under MOEF & CC provides assistance to various State Governments for implementing the River Action Plans in 29 identified grossly polluted stretches of various rivers of India under the National River Conservation Plan (NRCP). The objective of NRCP is to improve the water quality of these rivers, which are the major fresh water sources in the country, through the implementation of pollution abatement schemes. MOEF & CC in its draft notification released on 24 November 2015 under the Environment (Protection) Act, 1986 has set up new standards for sewage treatment plants along with time frame for implementation as shown in Table 1. Achievement of standards for existing STPs is within 05 years from date of notification.

Table 1. Standards for Sewage Treatment Plants.

Parameters	Standards for New STPs	Existing general effluent standards
pH	6.5 – 9.0	5.5 – 9.0
BOD	10	30
COD	50	250
TSS	20	100
NH4-N	5	50
N-Total	10	100
Fe_{Cal} Coliforms (MPN/100mL)	<100	Not specified earlier

Note:
(i) All values in mg/L except pH and Coliform.
(ii) These standards will be applicable for discharge in water resources as well as for land disposal. The standards for Fe_{Cal} Coliform may not be applied for use of treated sewage in industrial purposes.

2 Overview of Sewage Generation in India

There are 53 urban agglomerations in India with a population of 1 million or more as of 2011 against 35 in 2001. The rapid rate of growth of urbanization from 10.7% in 1971 to 37.7% in 2011 has put lot of stress on urban infrastructure including potable water supply and safe disposal of wastewater. According to CPCB report on inventorization of STPs in India published in March 2015, the states/UTs generating various amounts of sewage are shown in Fig. 1.

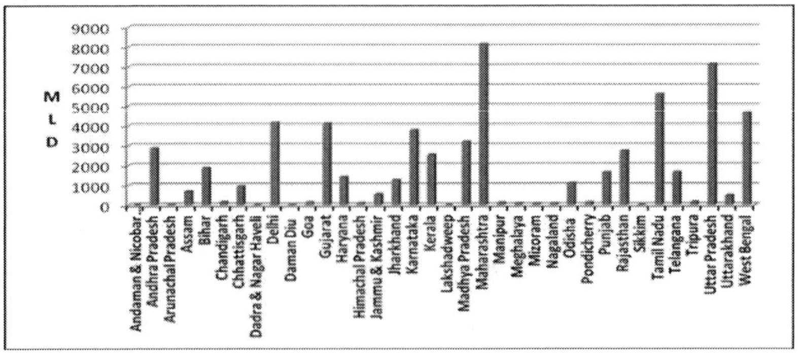

Fig. 1. Sewage Generation of Urban Population in 2015 (calculated @ 148 Lpcd) [5].

There are three states which contribute to more than 5000MLD of sewage whereas Lakshadweep is a UT contributing to 10MLD as shown in Table 2. The five states i.e., Maharashtra (13%), Gujarat (7%), Uttar Pradesh (12%), Delhi (7%) and Tamil Nadu (9%) contributed to around 50% of the sewage generated in India whereas 67% of the total sewage treatment capacity is installed in these states/UTs.

Table 2. Sewage Generation Classification.

Sewage Generation, MLD	No. of States/UTs	State/UT
Less than 10	1	Lakshadweep, Andaman & Nicobar Islands, Arunachal Pradesh, Dadra
10 to 50	5	and Nagar Haveli, Daman Diu and Sikkim
50 - 100	3	Meghalaya, Mizoram and Nagaland
100 - 500	7	Chandigarh, Goa, Himachal Pradesh, Manipur, Pondicherry, Tripura and Uttarakhand
500 - 1000	3	Assam, Chhattisgarh and Jammu & Kashmir
1000 - 2000	6	Bihar, Haryana, Jharkhand, Odisha, Punjab and Telangana
2000 - 5000	8	Andhra Pradesh, Delhi, Gujarat, Karnataka, Kerala, Madhya Pradesh, Rajasthan, West Bengal
>5000	3	Maharashtra, Tamil Nadu and Uttar Pradesh
Total	36	

The overall BOD load discharged into surface water bodies has been assessed as 14352.7 TPD out of which less than 1% is contributed by industries [6]. The raw sewage characteristics with respect to BOD, COD and TSS are reported to range from 50 – 250mg/L (185.5mg/L), 100 – 700mg/L (481 mg/L) and 100 – 500 mg/L (328mg/L) with average values in parenthesis. Average COD to BOD ratio is found to be around 2.6 [3].

2.1 Status of Sewage Treatment in India

During 2015, the installed treatment capacity of STPs was 37.58% of the total sewage generated by Urban Agglomerations out of which 81.12% of the sewage was under operational capacity of installed STPs while 5.32% was non-operational, 10.86% of the sewage would be treated in STPs under construction and 2.70% of the sewage was under proposed capacity as shown in Fig. 2.

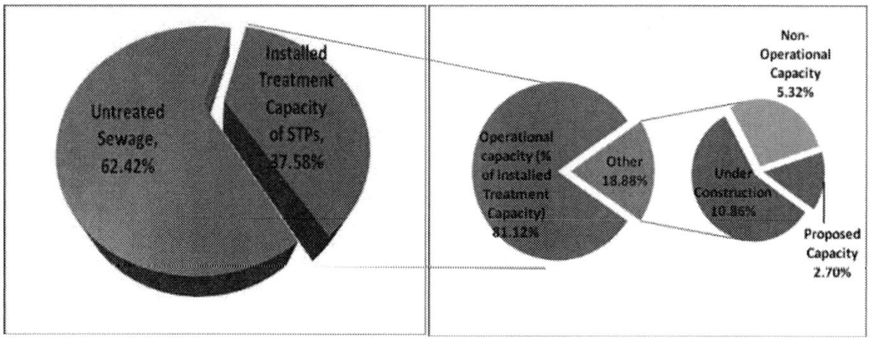

Fig. 2. Sewage treatment potential in India.

 https://doi.org/10.1051/e3sconf/20183402046

Out of 816 STPs, 522 were in operation whereas 79 STPs were non-operational and 145 STPs were under construction as shown in Fig. 3. The number of STPs proposed are 70 with treatment capacity of 628.64 MLD.

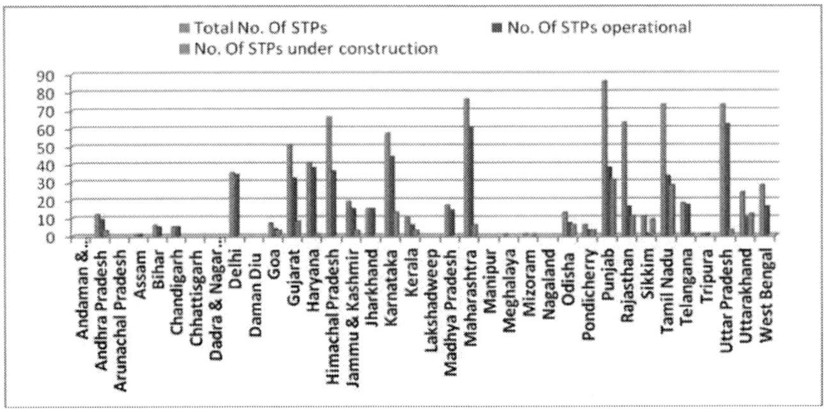

Fig. 3. Status of STPs in India.

There are different technologies which are adopted for the treatment of sewage in India, which includes Activated Sludge Process (ASP), Waste Stabilization Ponds(WSP), Aerated Lagoons (AL), Trickling Filters (TF), Fluidizied Aerated Bed (FAB), Upflow Anaerobic Sludge Blanket (UASB), Oxidation Pond (OP) and advanced technologies like Sequencing Batch Reactor (SBR), Membrane bioreactor (MBR) [5]. The STPs prevailing in India are shown in Fig. 4.

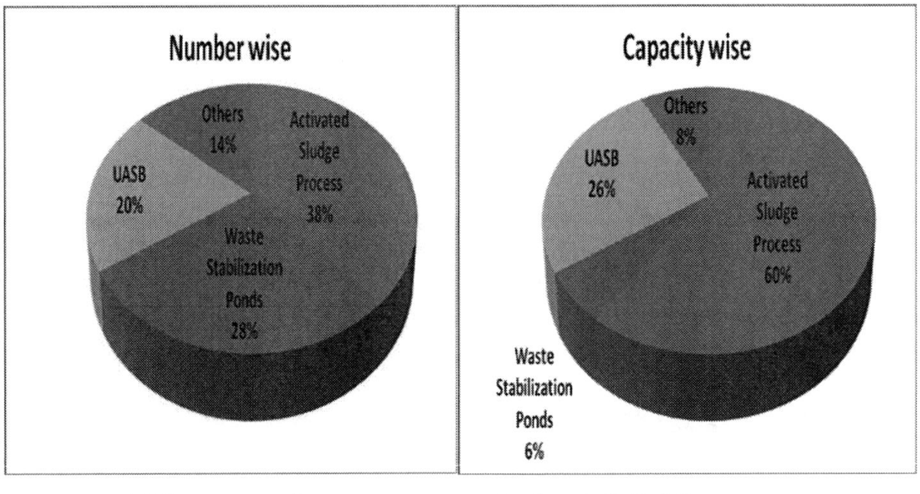

Fig. 4. Status of different types of sewage treatment technology in India [1].

In recent years, UASB technology has been extensively employed for treatment of domestic sewage in India. It is claimed that 80% of total UASB reactors installed globally for domestic wastewater treatment are in India [10]. 48 UASB based STPs are in operation accounting for around 72% of the plants and remaining are either non-operational or under construction and commissioning phase [5] as shown in Fig. 5.

 https://doi.org/10.1051/e3sconf/20183402046

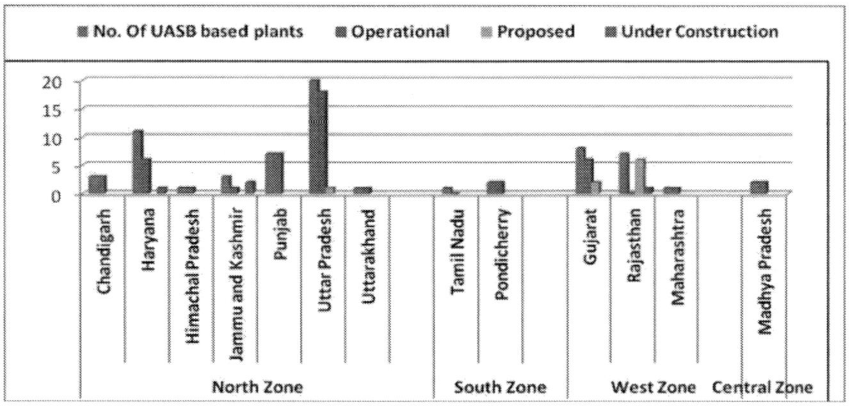

Fig. 5. UASB based STPs in India, Zone-wise [5].

2.2 Performance of UASB based STPs in India

The Government of India initiated the Yamuna Action Plan (YAP) in 1993 for conservation of river Yamuna under which 16 UASB STPs were commissioned [10] and 19 UASB plants under Ganga Action Plan-Phase I (GAP-I) [5]. The average BOD, COD and TSS removal efficiency of UASB based plants in India is reported to be 66%, 61% and 65% respectively. The maximum BOD and COD removal efficiency has been found in UASB based STPs in Hyderabad. 97% BOD removal efficiency is reported from 30MLD Nallacheruvu plant with 9hrs HRT in UASB reactor followed by 1 day HRT in facultative aerated lagoons. Similarly, 98% COD reduction is obtained in 339 MLD STP in Amberpet, Hyderabad. The HRT of UASB reactor is 8.88 hrs and that of Aeration tanks and Polishing Ponds are 1day and 12 hrs respectively. Minimum BOD removal efficiency of 44% is reported from 14MLD STP in Jaganpur, Agra with UASB technology followed by Polishing Ponds for post treatment whereas 56MLD STP in Ghaziabad has a minimum COD removal efficiency of 42%. The designed capacity of this plant is 56 MLD whereas actual treatment capacity is 58 MLD [4]. HRT of UASB is 10.7hr and that of Final Polishing Unit is 1.5day [13]. The average TSS removal efficiency of UASB reactors treating sewage is 65%. Maximum removal efficiency is found in 78 MLD STP in Kabt Khedi, Indore, Madhya Pradesh and minimum TSS removal efficiency is of 40 MLD STP in Karnal which is about 42%. UASB alone does not meet discharge standards; therefore, various post-treatments were used in combination with UASB reactor to achieve desired BOD and SS reduction [10]. The best operating full scale UASB reactor has achieved 67, 70 and 75% of removal efficiencies for COD, BOD and TSS in which temperature varies between 17 and 27°C [15].

3 State of Art Post-Treatment

The prime objective of installing post-treatment in combination with UASB reactor is to meet the effluent standards set by NRCD for public health, safety, water usage and environmental considerations [13]. There are various types of post-treatment configurations based on different combinations such as UASB + Activated Sludge, UASB + Biofilm Aerobic Reactor System, UASB + Polishing Ponds etc. UASB with ASP is considered most appropriate combination for treatment in warm conditions than the conventional AS system alone [11]. Polishing ponds are very interesting alternative post-treatment option because they require less investment costs and simple design and operation [13]. Utilization of UASB reactor and polishing ponds can reduce the problems related to odours in plants [2]. Polishing pond technology has been used commonly in many developing countries because the polishing pond effluent can be used for agriculture purposes as it aims at the removal of pathogens present in sewage [7]. UASB/SAB can maintain stable operational conditions without being affected by influent load variations [7]. TF can be adopted as post-treatment for treating UASB effluent for low hydraulic and organic rates under mesophilic conditions in tropical countries [11].

4 UASB Operations and Maintenance

The overall efficiency of STPs depends on the operation and maintenance of reactor. The UASB technology should be constructed, operated and maintained by professionals. UASB reactor operation requires various activities which are necessary for better performance of a reactor. Activities like process performance monitoring, collection of data, collection of influent and effluent samples for testing, collection of sludge samples, etc. Operator of the plant is fully responsible for the operation and maintenance of plant and a skilled and experienced operator is required to monitor the reactor and repair the part. Plant operation also includes start-up process and shut down of the plant. UASB plant requires a long start-up period as it takes several months for sludge to form and to adapt to the characteristics of the wastewater. During starting phase, the granulation process accelerates by high organic loading rates in connection with lower organic loading rates [14]. Monitoring of sludge and its profile inside the reactor is one of the major operational activities related to UASB plants [10]. Maintenance of UASB plant includes removal of floating layers inside the gas dome at least once in six months, cleaning and repairing of gutters, V- notch weir plates, baffles and feeding boxes, checking of the level of the overflow weirs yearly, removal of floating scum/material at the top of the water surface of the reactor once a day and cleaning of chocked feeding pipes [12]. Consequences of operational failure include deterioration of effluent quality [8] and impose a barrier for energy recovery by blocking the natural passage of gas [9]. Operation and maintenance cost of UASB plants is less than 1% of the capital cost per annum [10].

5 Conclusions

It is observed that all UASB based STPs are not capable of producing effluent that follows the discharge standards in terms of BOD, COD and TSS removal. Minimum removal rates of BOD, COD and TSS of 42-44% have been found in some plants despite of having post treatment technologies. This may be because of poor operation and maintenance of STPs such as improper cleaning of screens at inlet and ponds, excess sludge accumulation in reactors and ponds and hiring of persons with less knowledge of plant operations.

The underperformance of UASB based STPs in achieving the regulatory standards set forth by the MOEF & CC has been attributed to some of the following reasons by various researchers and reports:

- Actual Capacity of the plants exceeds the Design Capacity which would affect the HLR and HRT of the reactors.
- Mixing of Industrial Wastewater along with sewage affects the influent characteristics and design flow for which the STPs are designed. In turn, the STPs may be subjected to shock organic and hydraulic loading. The toxic and inhibitory compounds present in industrial wastewater can affect the biological processes.
- Leakage of raising mains/pipelines.

Improper operation and Maintenance involves non-functioning of bar screens, pumps, flow measuring devices, Diesel Generator (DG) sets during power failure, gas collection system etc.

6 Recommendations based on Literature, Reports and personal opinion

- The estimation of sewage generation is based on 80% of the fresh water supplied by the Urban Local Bodies (ULBs) / Municipal Corporation (MC). However, the discharge may vary given the individual household supplies through private wells and other sources.
- The performance evaluation of UASB based STPs are made on grab sampling done one time. However, data based on composite sampling would be more realistic. Studies may be carried out under different seasons
- Need for trained and experienced workers to analyse the treatment performance at defined time intervals
- UASBs are vulnerable to shock loads, shock temperature, organic loading rate, hydraulic loading rate etc. Equalization tanks may be installed to overcome such shock loads.
- Operational guidelines should be strictly followed
- Periodic Sludge Profiling of UASB reactor as a Sludge Management tool can enhance UASB performance
- Upgradation of STP facilities and physical improvement of FPUs by adding an aerator and extending retention time or re-evaluation of FPUs by considering alternate technologies such as ASP, TF and RBC. UASB with ASP is considered the most suitable combination as it can achieve 85-95% removal rates
- Consideration of Down Hanging Sponge System (DHS) as a novel technology in post treatment
- Since the STPs vary in both size and effluent quality, a parameter such as Effluent Quality Index (EQI) may be generated to evaluate the STPs on a common platform.

To optimize the performance of STPs, the UASB reactors should be maintained and operated properly. Maximum removal rates are found in some plants which are in combination with post treatment facilities such as facultative aerated lagoons, aeration tanks and polishing ponds. While most of the STPs are evaluated for previous discharge standards, they should be re-evaluated for current standards and necessary measures be taken.

References

1. A. Majumder, [online] Available: http://www.cseindia.org/userfiles/arunabha.pdf
2. P. F. F. Cavalcanti, A.Van Haandel, G. Lettinga, Water Sci. and Technol **44** (4), 237-245 (2001)
3. Central Pollution Control Board(CPCB), [Online] Available: http://cpcb.nic.in/newitems/12.pdf (2005, November)
4. Central Pollution Control Board(CPCB) (Ministry of Environment and Forests, Govt. of India), [online] Available: http://cpcb.nic.in/upload/NewItems/NewItem_195_STP_REPORT.pdf (2013)
5. Central Pollution Control Board(CPCB) (Ministry of Environment and Forests, Govt. of India) , [online]

Available:http://nrcd.nic.in/writereaddata/FileUpload/NewItem_210_Inventorization_of_ Sewage-Treatment_Plant.pdf (2015, March)

6. Central Pollution Control Board(CPCB), Bulletin, **1** [online], Available: http://cpcb.nic.in/upload/Latest/Latest_123_SUMMARY_BOOK_FS.pdf (2016, July)
7. C. D. L. Chernicharo, Reviews in Environ. Sci. and Bio/Technol **5** (1), 73-92 (2006)
8. C. D. L. Chernicharo, P.G.S. Almeida, *Sustainable Solutions for Small Water and Wastewater Treatment Systems, Spain,* (2010)
9. C. D. L. Chernicharo, J. B. van Lier, A. Noyola, T.B. Ribeiro, Reviews in Environ. Sci. and Bio/Technol **14** (4), 649-679 (2015)
10. N. Khalil, R. Sinha, A.K. Raghav,A.K. Mittal, IWTC **12**, 1411-1427 (2008)
11. A. A. Khan, R. Z. Gaur, V. K. Tayagi, A. Kursheed, B. Lew, I. Mehrotra, A. A. Kazmi, *Resources, Conservation and Recycling* **55**, 1232-1251 (2011)
12. Operation and Maintenance Manual for 19.35MLD Sewage Treatment Plant at Lalpani, Shimla, MWH India Pvt. Ltd.
13. N. Sato, T. Okubo, T. Onodera, A. Ohashi and H. Harada, J. Envir. Mgmt. **80**, 198-207 (2006)
14. L. Sasse, Bremen, Bremen Overseas Reasearch and Development Association (BORDA) (1998)
15. J. B. Van Lier, A. Vashi, J. Van Der Lubbe, B. Heffernan, H. Fang, *Anaerobic sewage treatment using UASB reactors: engineering and operational aspect* (Imperial College Press, London, UK, 2010)

Temporal Characterisation of Ground-level Ozone Concentration in Klang Valley

Nur Izzah Mohamad Hashim[1], *Norazian* Mohamed Noor[1*] and *Sara Yasina* Yusof[1]

[1]School of Environmental Engineering, Universiti Malaysia Perlis (UniMAP), Kompleks Pusat Pengajian Jejawi 3, 02600 Arau, Perlis, Malaysia.

Abstract. In Malaysia, ground-level ozone (O_3) is one of the most significant air pollutants due to the increasing sources of ozone precursors. Hence, the surface O_3 concentration should have received substantial attention because of its negative effects to human health, vegetation and the environment. In this study, hourly air pollutants dataset (i.e O_3, Carbon monoxide (CO), Nitrogen dioxide (NO_2), Particulate matter (PM_{10}), Non-methane hydrocarbon (NmHC), Sulphur dioxide (SO_2)) and weather parameters (i.e. wind speed (WS), wind direction (WD), temperature (T), ultraviolet B (UVB)) for ten years period (2003-2012) in Klang Valley were selected for analysis in this study. Two monitoring stations were selected that are Petaling Jaya and Shah Alam. The aim of the study is to determine the diurnal variations of O_3 concentrations according to the seasonal monsoon and the correlation between the ground-level O_3 concentration and others parameter. A high concentration of ground-level O_3 was observed during the first transition (April to May) for both of the stations. While at a low surface, O_3 concentration was found out during the southwest monsoon within June to September. Pearson correlation was used to find the correlation between the O_3 concentration and all other pollutants and weather parameters. Most of the relationship between O_3concentrationswas positively correlated with NO_2 and negative relationship was found out with NMHC. These results were expected since these pollutants are known as the O_3 precursors. Besides that, O_3 concentration and its precursors show a positive significant correlation with all meteorological factors except for relative humidity.

1 Introduction

Tropospheric ozone (O_3) is an important atmospheric constituent because it plays a main role as both an oxidant and a greenhouse gas [1, 2]. Most of the O_3 concentrations occupy in ozone layer; which is located in the middle of the stratosphere and acts as a protecting shield that protects Earth from any harmful radiation. However, a small concentration of O_3 also exists in the troposphere.

At the surface or ground level, ozone (O_3) is a secondary air pollutant that created by chemical reaction that occurs in atmosphere which give harmful effects on human health,

* Corresponding author: norazian@unimap.edu.my

crop production, material quality, and the ecosystem [3, 4]. This is the primary cause why the presence of O_3 is becoming a severe environmental problem because it is hard to control and predict [5]. Besides that, ozone is unique among pollutants because it is not directly emitted by combustion processes or any other sources of air pollutants but it is formed photochemically from the sunlight-initiated oxidation of volatile organic compounds (VOCs) in the presence of nitrogen oxides (NOx) in the atmosphere [6]. Collectively, NOx and VOCs are referred to as ozone precursors [7]. In addition, ozone concentrations are strongly correlated to meteorological conditions. Therefore, to predict the concentration of surface ozone, the researchers must have good understanding not only on ozone itself but also the situation that make a contribution to its formation. In Malaysia, ground level ozone (O_3) becomes one of the most significant air pollutants due to the increasing sources of ozone precursors [8]. It was proved to negatively impact the air quality status around the world due to the growth in population and rapid development. It arises from natural sources (e.g., vegetation), mobile sources (e.g., motor vehicles), and stationary sources (power plants, industrial facilities, residential and commercial establishments).

Therefore, the study of surface ozone is important due to its long-term and short-term impacts on human health, well-being, vegetation and the environment when its exposure to the high ozone concentration [9, 10, 11]. Thus, to predict the ozone concentration, it is important to apply a model that describes the relationships within the ozone concentrations and many variables that cause its production. It is necessary and quite urgent to gain a good understanding of the characteristics of O_3 pollution. Hence, the modeling of ozone's fluctuations provides the good prediction model. The outcomes from the analysis can be used to the next analysis such as principle component analysis (PCA) and developing the air pollution model that can be used to give early warning to the public in the future in Malaysia.

2 Materials and methods

In this research, the measured records of hourly air pollutants (i.e. O_3, CO, NO_2, PM_{10}, NMHC, and SO_2) and meteorological parameters (i.e. WS, WD, T, and UVB) in Klang Valley, Malaysia was chosen. In this research, two sites were selected namely Petaling Jaya and Shah Alam. All the hourly O_3 concentration and meteorological data from 2003 to 2012 were provided by Department of Environment (DoE) Malaysia. These data were regularly subjected to quality control due to the high level of O_3 concentration were released or emitted in atmosphere during that particular time. All of these data were sorted by locations, years, month, and day for analysis purposes.

2.1 Study area and data collection

Air quality data were obtained from two stations located at the Klang Valley. Petaling Jaya and Shah Alam monitoring stations were chosen due to the consistency of showing high frequency of higher O_3 concentration if compared to the DOE standard, which are 0.06 ppm (8-h average) and 0.1 ppm (1-h average). Besides that, these two areas were suggested to be selected due to the heavily industrialized area with high pollution, with a population of more than half a million, and high traffic density. Ten years of ground level ozone concentration from 2003 to 2012 were obtained from air quality monitoring in these stations. The data were owned by Department of Environment (DOE) Malaysia, however are managed by Alam Sekitar Malaysia Sdn. Bhd.

The monitoring station for Shah Alam is located at Sekolah Kebangsaan Taman Tun Dr Ismail Jaya, with the coordinate of N 3.1066° and E 101.5573°. This station is located in residential areas and close to the industrial zone. While, the monitoring station in Petaling Jaya is located in Sekolah Menengah Kebangsaan Bandar Utama (N 3.1315, E 101.6082). This station is located in a housing area and beside the Jalan Utara (road) and light industrial area. This station was surrounded by residential, commercial, light industrial and green areas.

2.2 Data analysis

The aim of the study was to determine the diurnal variations of O_3 concentrations according to the monsoon and also the correlation between ground level O_3 concentration with others parameter. Hence, all of these data were sorted by locations, years, month, and day for the analysis. After that, the hourly and diurnal data of ground level ozone concentration were examined visually using diurnal plot and Pearson correlation.

Correlation is a measure of association between two variables. Pearson correlation is one of the correlation is reported by 'r' and it is used to determine the strength of a linear relationship between two variables. Thus, the relationship between ozone, pollutant and meteorological parameter are being investigated by using this correlation.In this study, the Pearson correlation was conducted to determine the relationship between different air pollutants at each station as well as the connection between O_3 concentration and meteorological factors, such as temperature, wind speed and humidity.

3 Results and discussions

3.1 Monthly diurnal plot

Fig.1 shows the monthly diurnal plot of surface O_3 concentration in Petaling Jaya and Shah Alam. The surface O_3 concentration were plotted according to the four (4) dominant monsoon throughout a year. Malaysia annually experiences four distinct monsoonal changes, there are the Northeast Monsoon (NEM), the First Transition (1st transition), the Southwest Monsoon (SWM) and the Second Transition (2ndtransition). NEM occurs during early November and ends in March, while 1st transition occurred from April to May. SWM occurred during Jun to September and 2nd transition are in October. The northeast monsoon brings heavy rainfall to the country between November to January; while February to March usually associated with hot weather. The southwest monsoons mainly bring hot and dry weather from June to September. However, during the two transitions monsoon, the winds are generally light and variable [12].

Overall, from the plot, the highest concentrations of ozone were observed during the transition monsoon (2ndintermonsoon) due to the high penetration of light and variable wind speed. However, the highest O_3 concentration was recorded during the transition monsoon, followed by the northern-east monsoon, and the southern-west monsoon respectively.From this result, it also shows the average diurnal profiles of O_3 concentration for both dry and wet season shows almost the same pattern that suggested the O_3 concentrations had lower concentration compared to the transition monsoon. Normally in the inter-monsoon (wet season) it was supposed to have lower pollutant concentration due to the precipitation of rain that will carry the pollutants to the earth and can reduce the level of pollutants in the atmosphere [13].

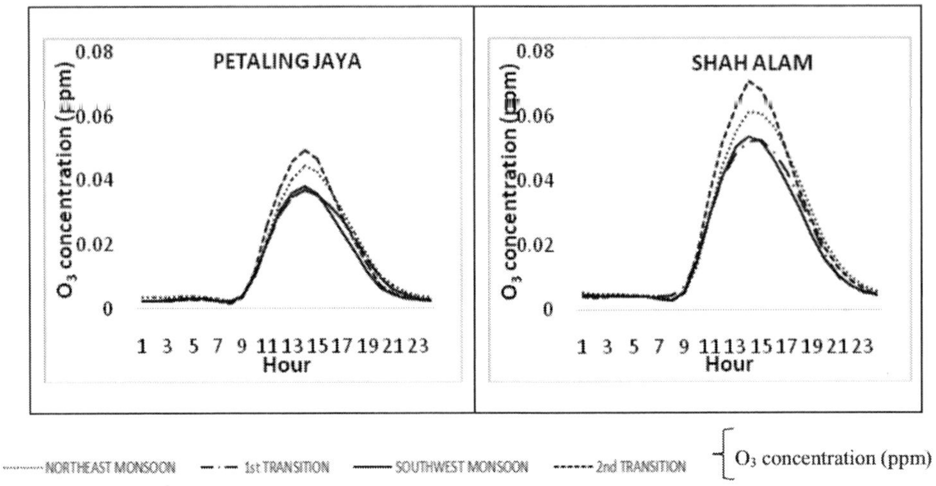

Fig. 1. Line graph of hourly average of ozone concentrations by monsoon.

3.2 Correlations of air pollution and meteorological parameters

Table 1 shows the Pearson correlation matrix of different variables for Petaling Jaya and Shah Alam during the peak concentration of O_3 (1300-1600 pm). While, Fig. 2 shows the scatter plots for the highest correlated variables (from the results of Table 1) in these two locations. The relationships between pollutant-pollutant and pollutant-meteorological parameter are being investigated by using these correlations. These tables show that the correlation between O_3 concentration with the wind speed (WS), wind direction (WD), temperature (T), ultraviolet (UVB), humidity (H), non-methane hydrocarbons (NMHC), sulphur dioxide (SO_2), nitrogen dioxide (NO_2), carbon monoxide (CO), and particulate matter (PM_{10}). These values show the correlation is significant at the 0.01 levels.

Mostly, the positive correlation was found out to be between O_3concentrationsand NO_2, whereas, negative relationship was observed between NMHC and O_3 concentration. This result was expected since these pollutants are known as precursors of O_3. It indicates that high O_3 concentration is associated with the low level of all these pollutants. Besides that, O_3concentration shows a positive significant correlation with all meteorological factors except for relative humidity. The precursors also have significantly negative relationships with all the meteorological factors except for relative humidity.

Fig. 2 shows that the highest correlated values for Petaling Jaya and Shah Alam. For Petaling Jaya, a positive correlation ($p<0.01$) between SO_2-CO and NO_2 which were valuated at ($r=0.611$, $r^2=0.3735$) and ($r=0.731$, $r^2=0.535$). While, the correlations are negative between Temperature (T)-Humidity (H), which was($r=-0.750$, $r^2=0.0.5631$). While, for Shah Alam, a significant negative correlation was observed between T-H, which was ($r=-0.860$, $r^2=0.7403$) and positive correlation for PM_{10}-CO ($r=0.637$, $r^2=0.4058$).

Table 1: Pearson correlation matrix of different variables for Petaling Jaya and Shah Alam.

PJ	WS	WD	T	UVB	H	NMHC	SO_2	NO_2	CO	PM_{10}	O_3
WS	1	0.126	0.094		-0.239		-0.214	-0.388	-0.361	-0.119	-0.050
WD		1	0.138		0.033		0.215	0.261	0.237	0.098	-0.108
T			1		-0.750		-0.073	-0.088	-0.081	0.091	0.071
H					1		0.123	0.188	0.164	-0.080	-0.137
SO_2							1	0.575	0.611	0.351	
NO_2								1	0.731	0.406	
CO									1	0.544	
PM_{10}										1	0.091
O_3											1
SA	**WS**	**WD**	**T**	**UVB**	**H**	**NMHC**	**SO_2**	**NO_2**	**CO**	**PM_{10}**	**O_3**
WS	1	0.132	0.047	0.139	-0.058	-0.176		-0.310	-0.162		-0.057
WD		1	-0.058	0.076	0.147	-0.153	0.087	-0.273	-0.153	-0.080	0.047
T			1	0.592	-0.860	-0.130		-0.227		0.135	0.130
UVB				1	-0.531	-0.265	-0.209	-0.461	-0.288	-0.242	0.152
H					1	0.166		0.211		-0.142	-0.161
NMHC						1	0.193	0.531	0.387	0.322	
SO_2							1	0.330	0.090	0.180	
NO_2								1	0.439	0.279	
CO									1	0.637	0.091
PM_{10}										1	0.147
O_3											1

*PJ- Petaling Jaya, SA- Shah Alam

4 Conclusions

In conclusion, from the hourly diurnal plot of O_3 concentration in Petaling Jaya and Shah Alam, it can be summarized that during daytime (8 am – 7 pm), the peak started to develop at 8.00 a.m. and reached a peak at 1 p.m. - 4 p.m. and decreased after 4 p.m. However, the concentration of O_3 during night-time was far lower than the daytime concentration due to the absence of photochemical reactions. This is because during night-time O_3 concentration had lower concentration due to no titration and also the deposition [1]. For the correlation between ground level O_3 concentration with others parameters, Pearson correlation was used in this study. Most of the relationship between O_3 concentration with NO_2 and NMHC were negatively correlated since these pollutions are known as precursors to O_3 concentration, which indicates that they originated from vehicle exhaust. Besides that, O_3 and its precursors show a positive significant correlation with all meteorological factors except for relative humidity.

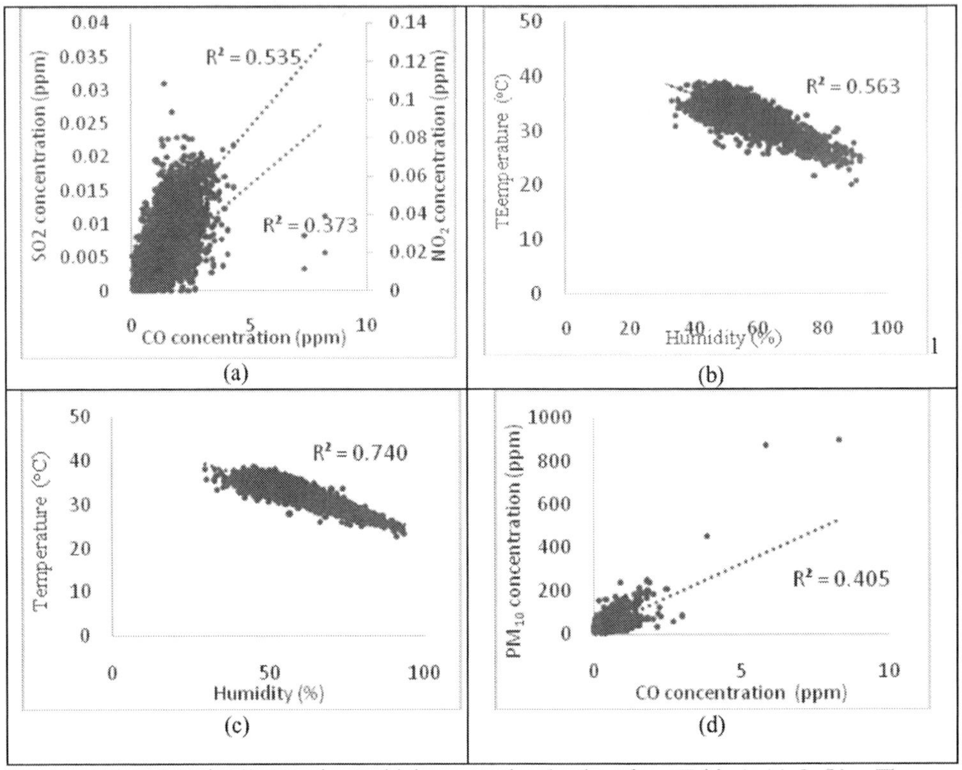

Fig. 2. Scatter plot of pollutants for the highest correlated values from Table 1. (a) & (b) = The most significant parameters in Petaling Jaya; (c) & (d) = The most significant parameters in Shah Alam.

The authors would like to thank the Ministry of Higher Education Malaysia (MOHE) for the Fundamental Research Grant Scheme (FRGS 9003 00508) and the Department of Environmental (DOE) Malaysia for the air pollutants data.

References

1. N. R. Awang, N. A. Ramli, A. S. Yahaya, M. Elbayoumi, Aerosol Air **Qual. Res**. **15**(4), 1357–1366 (2015)
2. Y. Ying, S. Fook, R. V. Glasow, Atmos. Environ. **70**, 435–446 (2013)
3. T. Wang, L. Xue, P. Brimblecombe, Y. Fat, L. Li, L. Zhang, Sci. Total Environ. 575, 1582–1596 (2017)
4. K. C. Tan, H. San Lim, M. Z. M. Jafri, Atmos. Poll. Res. 1–14, (2016)
5. S. A. Abdul-Wahab, C. S .Bakheit, S. M. Al-Alawi, Environ. Modell. Softw. **20**, 1263-1271 (2005)
6. N. R.Awang, A.Ramli, N. I. Mohammed, A. S. Yahaya, International Journal of Engineering and Technology **3**(3), 390–394, (2013)
7. S.A.Abdul-Wahab, C.S. Bakheit, S.M. Al-Alawi, Environ. Modell. Softw. **20**, 1263-1271 (2005)
8. Air Quality Expert Group (AQEG). *Ozone in the UK, the fifth report produced by air quality expert group,* (Department for the Environment, Food and Rural Affairs. DEFRA publication London, 2009)
9. N. I. Mohammed, N. A.Ramli, A. S.Yahya, Atmos. Environ. **68**, 343–349 (2013)

10. B. R. Gurjar, T. M. Butler, M. G. Lawrence, J. Lelieveld, Atmos. Environ. **42**, 1593–1606 (2008)
11. Ozden, T. Dogeroglu, S. Kara. Environ. Int. **34**, 678–687 (2008)
12. S. A .Rahman, M. S.Hamzah, M. S. Elias, N. A. A. Salim, A. Hashim, S. Shukor, A. K. Wood. Aerosol Air Qual. Res. **15**(6), 2291–2304 (2015)
13. Z. Syazwani,*Application of GIS in Air Quality Monitoring of a Road Construction Project.* (Universiti Teknologi Malaysia, Malaysia, 2010)

Spatial Interpolation of Historical Seasonal Rainfall Indices over Peninsular Malaysia

Zulkarnain Hassan[1,*], *Ahmad* Haidir[1], *Farah Naemah* Mohd Saad[1], *Afizah* Ayob[1], *Mustaqqim* Abdul Rahim[1], and *Zuhayr* Md. Ghazaly[1]

[1]School of Environmental Engineering, Universiti Malaysia Perlis, Kompleks Pusat Pengajian Jejawi 3, 02600 Arau, Perlis, Malaysia.

Abstract. The inconsistency in inter-seasonal rainfall due to climate change will cause a different pattern in the rainfall characteristics and distribution. Peninsular Malaysia is not an exception for this inconsistency, in which it is resulting extreme events such as flood and water scarcity. This study evaluates the seasonal patterns in rainfall indices such as total amount of rainfall, the frequency of wet days, rainfall intensity, extreme frequency, and extreme intensity in Peninsular Malaysia. 40 years (1975-2015) data records have been interpolated using Inverse Distance Weighted method. The results show that the formation of rainfall characteristics are significance during the Northeast monsoon (NEM), as compared to Southwest monsoon (SWM). Also, there is a high rainfall intensity and frequency related to extreme over eastern coasts of Peninsula during the NEM season.

1 Introduction

Evaluation of rainfall behaviour has drawn an attention throughout the world [1]. The relationship between rainfall and climate has become an intensive study nowadays since the climate change is likely to cause changing the magnitudes, frequencies, and intensities of rainfall as well its seasonal variability and distributions [2,3]. Therefore, the effect of these changes is needed to focus on the areas that are already vulnerable to rainfall characteristics due to geographical position and prominent diurnal variability in rainfall such as the coastal regions of southeast Asia [4] including Malaysia.

Many recent studies have evaluated historical rainfall and rainfall-related extremes in Malaysia such as Deni et al. [5]; Suhaila et al. [1]; Mayowa et al. [3]; and Syafrina et al. [6]. All these studies are analysed rainfall pattern and trend over Peninsular Malaysia based on only a few stations. However, the data is not up-to-date and limited to some stations, which do not reflect the current pattern of rainfall characteristics over the Peninsula.

This study aims to spatial interpolate of historical seasonal rainfall indices based on recent 40 years (1975-2015) daily rainfall data series collected from 37 stations located over Peninsular Malaysia. Rainfall indices including rainfall-related extremes such as total

* Corresponding author: zulkarnainh@unimap.edu.my

 https://doi.org/10.1051/e3sconf/20183402048

amount of rainfall, frequency of wet days and rainfall intensity, extreme frequency, and extreme intensity are used in this study.

2 Materials and methods

Peninsular Malaysia is known as West Malaysia, is a part of Malaysia which lies on the Malay Peninsula and surrounding islands. With an area of 130,598 square kilometres, it shares a land border with Thailand in the north and Singapore in the south. The climate of Peninsular Malaysia is described by four seasons, namely two monsoon seasons and two inter-monsoon seasons. In this study, the Southwest monsoon (SWM) occurs from May to August while the Northeast monsoon (NEM) from November to February were used as the main seasons for the rainfall indices evaluation.

2.1 Rainfall stations

The rainfall stations in each of the region are chosen according to the homogenous of data produce in the rainfall stations and the completeness of data series based on the study period of 1975 to 2015. The details of 38 rainfall stations and their locations are shown in Table 1 and Fig. 1, respectively.

2.2 Rainfall indices

The rainfall indices are designed to explain the changes of the variety aspects in the rainfall distribution. The rainfall indices used in this study to explain the rainfall distribution: total amount of rainfall (TAR) is the total amount of rainfall value on each season, frequency wet days (FWD) is the frequency of wet days that are defined as rainfall day with at least 1mm, rainfall intensity (RI) is the average amount of rainfall annually, extreme frequency (XF) is the frequency of wet days exceeding the 95th percentile, and extreme intensity (XI) is the rainfall intensity exceeding the 95th percentile.

2.3 Inverse Distance Weighted interpolation method

Inverse Distance Weighted (IDW) interpolation method is used for the mapping of the spatial pattern of rainfall indices. Quantum GIS (QGIS) under spatial analysis is used for this purpose. IDW is a point based interpolation method, in which widely used for surface interpolation and has a point influence that decreases with each point distance [7]. The details of the IDW method can be referred to Harman et al. [8].

3 Results

3.1 Total Amount of Rainfall

The detail of spatial distribution of the total amount of rainfall (TAR) on each season is shown in Fig. 2. It can be seen that the large of TAR is happened on NEM season (Fig. 2a), as compared to SWM (Fig. 2b). During the NEM season, the large value of the TAR can be seen in the area of eastern regions, with the range 1174 mm to 1972mm. The largest of the TAR value is recorded at Station E5, with the value of 1972 mm. On the SWM season, it can be seen that a small area at southern region are given a large of the TAR value, with the largest value is 967 mm at Station S3 (Pintu Kawalan Tg. Agas). From the record, the lowest of TAR value is happened at Londah, Negeri Seremban (Station S8), with the value of 228 mm and 143 mm for NEM and SWM seasons, respectively.

Table 1. Rainfall stations.

States	Station Code	Station Details	Station ID.	Lat. (° N)	Long. (° E)
Kedah	N1	Kg. Jelutong	6005044	6.0458	100.5042
	N2	Komplek Rumah Muda	6108001	6.1056	100.8472
	N3	Jabatan Penjara	5604002	5.6292	100.4861
	N4	Pulai	5609074	5.6750	100.9150
	N5	Rumah Penjaga JPS	5105106	5.1278	100.5069
Kelantan	E8	Gua Musang	4819027	6.4458	100.1875
	E9	Ladang Kuala Nal	5521050	6.4306	100.2708
	E10	Station Pertanian Melor	5922001	6.6569	100.3097
Melaka	S1	Hospital Jasin	2324033	5.3917	100.2125
	S2	Jambatan Sungai Simpang Empat	2421003	5.3756	100.4756
Pahang	E1	Rumah Rehat Rompin	2834181	4.8792	101.9694
	E2	Permatang Pauh	3433105	5.5708	102.1639
	E3	Paya Besar	3732020	5.9639	102.2917
	E4	Peri Jerantut	3924071	2.8111	103.4944
	W10	Ladang Teh Cameron Highlands	4514032	3.4694	103.3833
Perak	W1	Rumah JPS Sungai Meng	4011139	3.7722	103.2806
	W2	JPS Telok Sena	4209093	3.9625	102.4278
	W3	Ladang Kuda Keb. Ulu Kinta	4611001	5.0708	103.0139
	W4	Pekan lenggong	5109070	4.2319	103.4222
Perlis	N6	Padang Katong	6401002	5.3181	103.1333
	N7	Arau	6402007	4.5167	101.4167
	N8	Padang Besar	6603002	4.0056	101.1028
Penang	N9	Kolam Air Bukit Berapit	5304045	4.6806	101.1694
Selangor	W5	Pejabat JPS Sungai Manggis	2815001	5.1194	100.9833
	W6	Ladang West Pulau Carey	2913121	2.8264	101.5417
	W7	Stor JPS Tg. Karang	3411017	2.9111	101.3583
	W8	Rumah Pam JPS Bagan Terap	3710006	3.4236	101.1733
Terengganu	E5	Hospital Kuala Berang	5030039	3.7286	101.0831
	E6	JPS Kemaman	4234109	3.1514	101.6847
	E7	Stor JPS Kuala Terengganu	5331048	2.3083	102.4319
Kuala Lumpur	W9	JPS Wilayah Persekutuan	3116003	2.4389	102.1861
Johor	S3	Pintu Kawalan Tg. Agas	2025001	1.4708	103.7528
	S4	Stor JPS Johor Bharu	1437116	1.6264	103.2028
	S5	Pintu Kawalan Tampok	1632096	2.0514	102.5778
Negeri Sembilan	S6	Hospital Port Dickson	2517033	2.5306	101.7972
	S7	Kg. Bahru Pantai	2720041	2.6389	102.5944
	S8	Londah	2625084	2.7850	101.9950

3.2 Frequency wet days

In general, more than 70% of Peninsular Malaysia is facing a high number of frequency wet days (FWD) for both monsoons, as shown in Fig. 3. It can be seen that the eastern and western coasts of the Peninsular are facing a high number of FWD, with the range between 61 days to 70 days as compared with other regions. During the NEM season (Fig. 3a), several areas of the northern and southern regions are recorded with a lower number of FWD, with the lowest value of 26 days at Kg. Jelutong (Station N1). This pattern is also displayed at Londah (Station S8), in which gives a low value of FWD on the NEM and SWM seasons with the value of 31 days and 26 days, respectively.

3.3 Rainfall intensity

Rainfall intensity (RI) or mean rainfall amount on wet days for the Peninsular Malaysia is displayed in Fig. 4. It shows that the high value of RI can be seen on the east coast of Peninsular, including Kelantan, Pahang, and Terengganu with the range of 30-36 mm/day during the NEM season (Fig. 4a). The high of RI during the NEM season is also recorded near to Tg. Agas at Johor (Stations 5) with the range between 25 mm/day to 29 mm/day. During the SWM season, the high of the RI values are seen over southern coast, especially at Melaka (Station S1) and Johor (Station S5) with the value of 20 mm/day to 27 mm/day. The lowest of RI value is recorded at Ladang Teh Cameron Highlands (Station W10), Pahang and Bagan Terap (Station W8), Selangor during the NEM and SWM seasons, respectively with the value of 11 mm/day.

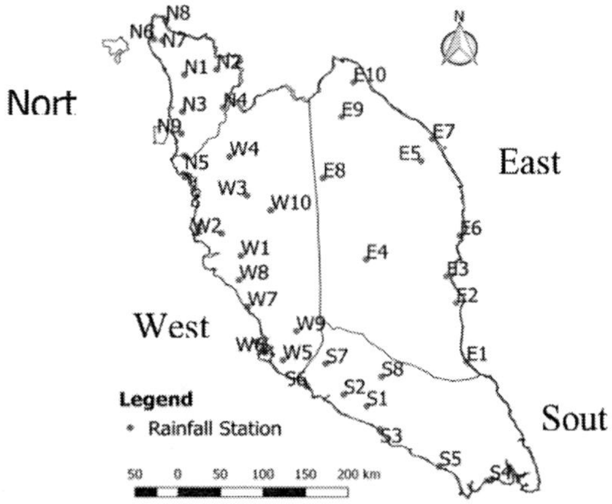

Fig. 1. Locations of rainfall stations.

3.4 Extreme frequency of wet days

Fig. 5 shows the extreme frequency (XF) of the Peninsular Malaysia obtained from the 40 years data (1975-2015). Result reveals that XF is recorded more than 3 days/year during both monsoons for over 80% of the Peninsular. The highest XF is obtained along eastern coasts during the NEM season, in which the highest value of XF is 5 days/year.

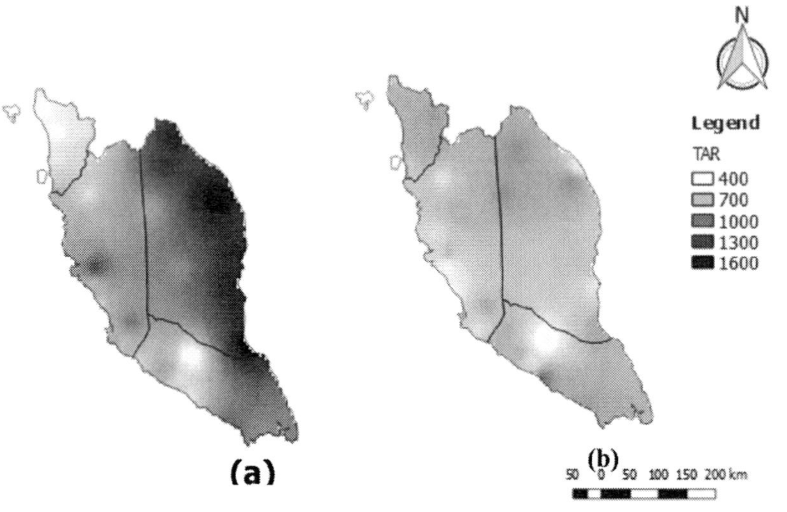

Fig. 2. Spatial pattern in TAR value on **a)** NEM and **b)** SWM.

3.5 Extreme intensity

Extreme intensity (XI) can be defined as a rainfall intensity which exceeds 95th percentile. As shown in Fig. 6, it can be seen that the XI is high in the NEM season as compared with the SWM. During the NEM, the high of XI are recorded at the eastern coasts, with the range between 119 and 151 mm/day. The lowest of XI value for both seasons is recorded at Ladang Teh Cameron Highlands (Station W10), Pahang with the value of 44 mm/day.

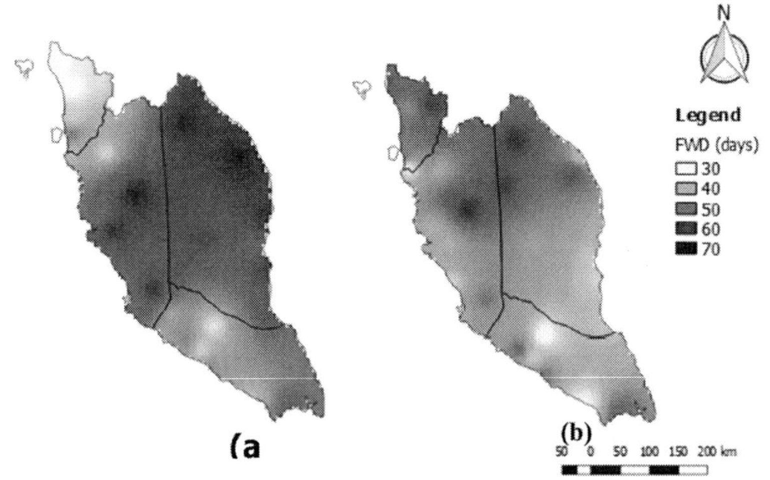

Fig. 3. Spatial pattern in FWD value on **a)** NEM and **b)** SWM.

 https://doi.org/10.1051/e3sconf/20183402048

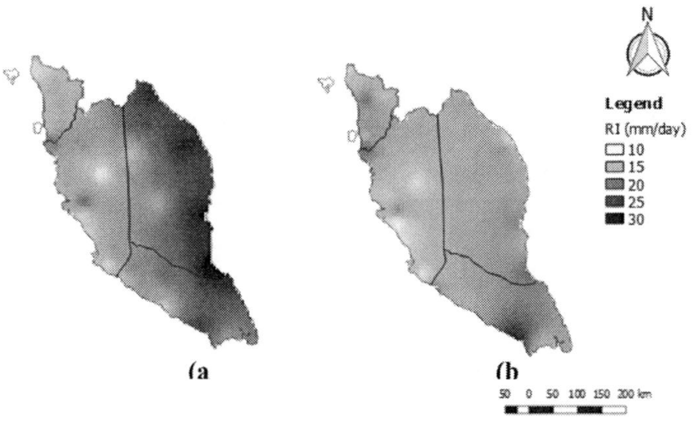

Fig. 4. Spatial pattern in RI value on **a)** NEM and **b)** SWM.

4 Discussions

Based on the results of the rainfall indices, it can be said that the Peninsular Malaysia, especially along eastern coasts is facing a high rain intensity and duration during the NEM season as compared to the SWM since all rainfall indices tend to be higher than those during another season. This situation occurs due to interaction with the land since the northeasterly winds bring heavy rainfall when passage over the South China Sea [9-10]. The result is similar with Suhaila et al. [1], which identified that some areas at the eastern region have received a high intensity of rainfall amount as compared to other regions in the Peninsula.

Conversely, less rainfall during the SWM is happenned because the southwesterly winds which come to Peninsula from Sumatra, Indonesia are brought low of the rainfall intensity and the wind speed is literally slow due to the high mountain ranges over Sumatra that create rain sheltering effects for Peninsular Malaysia [11]. During this season, the eastern coasts of the Peninsular is also recorded less rainfall as compared to the western coast since the presence of Titiwangsa range which limits the in-land penetration of the southwesterly winds [12].

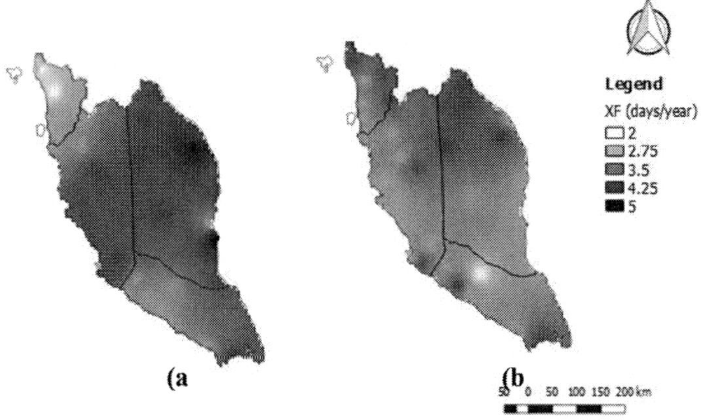

Fig. 5. Spatial pattern in XF value on **a)** NEM and **b)** SWM.

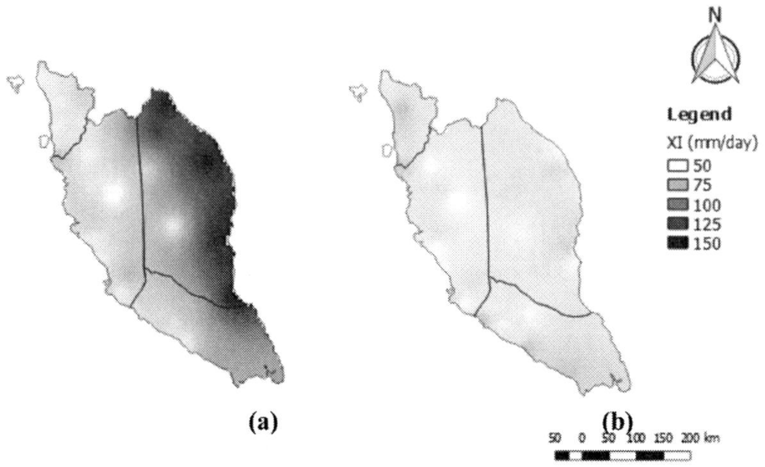

Fig. 6. Spatial pattern in XI value on **a)** NEM and **b)** SWM.

This study also found that Londah (Station S8) at southern region is the driest station on the Peninsula. Although it is found that northern region is a dry region as illustrated in the result and claimed as the driest region in Peninsular Malaysia by Suhaila et al. [1], still the area at Londah give the driest record. This difference occurs probably due to the effect of climate change toward the current climate in which increase the drying period at the area of southern region.

In term of extreme rainfall in Peninsular Malaysia, the study found that several areas especially east coastal like Kelantan, Pahang and Terengganu coasts had to face a long heavy rainfall in which high value of XF. Those records indicated that the areas had to be facing a possible flood due to high rainfall duration and intensity. Study from Leman et al. [13] also found that the areas of the east coast of the Peninsular are occurred the major flood event for the sixties of the last century due to abnormally heavy rainfall intensity especially during the NEM season.

5 Conclusions

Seasonal patterns have been interpolated by using five rainfall indices on four regions in Peninsular Malaysia for the period of 1975-2015. It can be concluded that the highest mean for each indices are on the eastern coasts of the Peninsula during the NEM season. During the SWM, it has less significance in characterizing the rainfall pattern over Peninsula and greater impact in the southern and western regions. The driest part is likely to be located in the southern region of Peninsula. In term of extreme rainfall, both seasons give a high rainfall frequency and less than 30% of the regions have faced the high rainfall intensity.

Results of this study will be aid guidance to the understanding the pattern of rainfall characteristics in Peninsular Malaysia. It also will help to increase effectiveness in risk (such as crop, irrigation, and flood) planning purpose and policy. Nevertheless, further studies should be conducted to consider more rainfall indices and further analyses such as trend and re-sampling, in which should display the significant trends to each station are recommended.

The authors thank the Department of Irrigation and Drainage Malaysia (DID), Malaysia for providing the data.

 https://doi.org/10.1051/e3sconf/20183402048

References

1. J. Suhaila, S. Mohd Deni, W.Z. Wan Zin, A.Z. Jemain, Sains Malaysiana **39**, 533-542 (2010)
2. X.J. Wang, J.Y. Zhang, S. Shahid, E.H. Guan, Y.X. Wu, J. Gao, R.M. He, R. M. Mitigation & Adapt. Strategies for Global Change **21**(1), 81-99 (2016)
3. O.O. Mayowa, S.H. Pour, S, Shahid, M. Mohsenipour, S, Harun, A. Heryansyah, T. Ismail, J. Earth Syst. Sci. **124**(8), 1609–1622 (2015)
4. H. Chang, W.T. Kwon, Environ. Res. Lett. **2**, 045012 (2007)
5. S.M. Deni, J. Suhaila, W.Z. Wan Zin, A.A. Jemain, Theor. Appl. Climatol. **99**, 357–371 (2010)
6. A.H. Syafrina, M.D. Zalina, A. Norzaida, A. J. of Appl. Sci. **14**(3), 392-405 (2017)
7. P.M. Bartier, C.P. Keller, Comp. & Geosci. **22**(7), 195-799 (1996).
8. B.I. Harman, H. Koseoglu, C.O. Yigit, Appl. Acous. **112**, 147-157 (2016)
9. C.P. Chang, P.A. Harr, H.J. Chen, Mon. Weather Rev. **113**, 489-503 (2005)
10. E. Salimum, F.T. Tangang, L. Juneng, Meteorol. Atmos. Phys. **107**, 33-49 (2010)
11. Z.A. Mohtar, A.S. Yahaya, F. Ahmad, S. Suri, M.H. Halim, J. of Civil Eng. Res. **4**(3A), 222-227 (2014)
12. C.L. Wong, J. Liew, Z. Yusop, T. Ismail, R. Venneker, S. Uhlenbrook, Water **8**, 500 (2016)
13. A.M. Leman, K.A. Rahman, M,N.M. Salleh, D. Feriyanto, S.N.H. Muhammad, The Soc. Sci. **12**(5), 781-786 (2017)

Activated alumina preparation and characterization: The review on recent advancement

A. R. Rabia[1], *A. H.* Ibrahim[1*], and *N. N.* Zulkepli[2]

[1]School of Environmental Engineering, Universiti Malaysia Perlis, Kompleks Pusat Pengajian Jejawi 3, 02600 Arau, Perlis, Malaysia
[2]Faculty of Engineering Technology, Kampus Uniciti Alam Sg. Chuchuh, 02100 Padang Besar, Universiti Malaysia Perlis, Perlis, Malaysia

Abstract. Aluminum and aluminum based material are significant industrial materials synthesis because of their abandonment, low weight and high-quality corrosion resistance. The most advances in aluminum processing are the ability to synthesize it's under suitable chemical composition and conditions, a porous structure can be formed on the surface. Activated alumina particles (AAP) synthesized by the electrochemically process from aluminum have gained serious attention, inexpensive material that can be employed for water filtration due to its active surface. Thus, the paper present a review study based on recent progress and advances in synthesizing activated alumina, various techniques currently being used in preparing activated alumina and its characteristics are studied and summarized

1 Introduction

The level of heavy metals in drinking water is getting very high in both villages and cities due the activities such as mining and industrial waste disposal which will bring clean environment and sanitation [1]. Therefore, it becoming important physicochemical needed to be factor, which needed be considered when assessing water quality for both human, animal and plant consumption according to [1-3] this will enhance the condition of wellbeing animal and human. It is well known medically that heavy metal excess causes health problems to the human beings as stated by [2] which lead to several health problems such bladder blocking in order hard urinary track blockage etc [3]. The consumption of drinking water with a heavy causes prevailing health problem in the population, and the constant consumption of water containing heavy metals provokes serious high concern such as urine tract blockage and infection and equally problem that is associated with serious bone abnormalities and recently discovered problem as mentioned in [1-5]. The concentration of heavy metal and other impurities in water for human consumption may be reduced through the following methods: such as ion exchange on polymeric resins,

* Corresponding author: abdulhaqi@unimap.edu.my

adsorption, reverse osmosis, and electro-dialysis which is illustrated in [3, 5]]. However, the most effective method proposed is the adsorption on activated alumina as clarify by [6]. Activated alumina is very available in sustainable quantity, which can be harness with little effort and is an inorganic porous material, semi-crystalline and essentially comprised by aluminum oxide that can be easily produce with high quality in reasonable amount as stated by [1, 5-10]. The material shows amphoteric character that allows the material to behaves like a base in acidic solutions, and as an acid in basic solutions, which easily pre-treated for desired goal [6]. Moreover, it has be reported unwanted material such as heavy metals and other materials in water has been a major challenge and the pollution in the water is quite heavy today due to the release of toxic component from various sources; this has caused various health disorders. There are several attempts to treat or pollutant degradation and purifying various products such as water [7, 10].

According to [11] among these approaches are biological, chemical and bio-electrochemical dealt with chemical treatments this needing to consuming time and costly. However, adsorptive removal pollutant using different adsorbents is more promising and economical which could used in many applications [12-15]. Among various adsorbents used, activated alumina is well known for its high adsorption capacity due to large surface area and pore volume can be engineered and utilised in various application [1]. The water treatment technology via aluminum has gained a great importance in recent time competing with conventional technologies in various industrial fields in particular for chemical and petrochemical industries that material have shown capability to treat water long time in history and equally some other inorganic and organic such as phenol etc. which appeared in [12]. Currently, Aluminum play a special role in the field of energy and electric power as one of the elemental parts of conducting material due to availability and cheapness that material has gain a wider application, it has be use through the with help addition component such membrane, use in separation process technology [14]. Since the beginning of 1960's, aluminum was used as coagulant in local water purification. At the beginning of 1970's, the first industrial installation of membrane using aluminum was initiated and using this platform, the water purification related activities rapidly grown and developed large-scale commercial. With this initiative, it began to insert significant improvements in membrane selectivity / permeability and process design which could be used in many applications and processs [4]. The main goal of improving the membranes permeability is to achieve constant quality, high performance membrane and systems for desired applications with enhanced troubleshooting capabilities, which could be used in many applications and process [4].

There are several methods to treat heavy metals arsenic at household level; oxidation of As (III) can be done through chemical like bleaching powder and chemical such as potassium permanganate as stated in [2]. Unfortunately, the bleaching based powder are usually bring about an unpleasant taste in the treated water where in most cases, this material equally cause serious lung diseases moreover, these chemical are prone to cause discomfort by formation of methanes and methane related chemical species which could be used in many applications and processs [5].

Thus, is a very dangerous chemical to use for daily treatment of the house hold water and the quality of bleaching powder available in the open market of varies widely, it is very difficult to ensure the proper dosing of bleaching powder which could be used in many applications and processs [15]. On the other hand, potassium permanganate is a stable chemical and easily available locally to be used in industrial level or local level [17]. However, it produces a color in the treated water making it less attractive for drinking [18]. If the problem of color can be solved, potassium permanganate is preferable as an oxidizing agent, which could be used in many applications and processs [3].

2 Activated Alumina

Activated alumina is a high-surface-area, highly porous form of aluminium oxide which recently attracted the attention of various researchers [19] . It can absorb gases and liquids without changing its form. It works as a desiccant through adsorption and other impurities such as in water or other product; it gives clean water due to ability to attract impurities [12]. As water passes through the alumina, the polluted in the water sticks it and becomes trapped and water that passes through an activated alumina filter out [17]. Activated alumina is materials, which utilizes the principle of adsorption to prevent environmental pollution and can be regenerated merely burning it with high temperature between300 to 400^0C [18]. It has been used for recently in industrials as adsorbent to remove impurities from gaseous or liquids and to polish effluent for meeting stringent discharge standard; as catalyst to improve process efficiency; and as gases container to store pure gas and other product. The adsorption performance of alumna is determined by available surface area for physisorption and available adsorption site for chemisorption. Many researches had studied on methods for alumina regeneration due to the fact that regeneration is cheaper than replacement [20]. Furthermore, regeneration and reuse of alumina is also a practical way to reduce secondary pollution and conserve natural resources [19]. Generally, activated alumina behaves in much way like activated coal, which is a familiar adsorbent. The adsorption process depends on the force field that exists at the surface of a solid. Activated alumina possesses several uniqueness such as high adsorption capacity, high surface area, wide spectrum of functional group and variety of porosity sizes which made it suitable to be applied for the treatment of wastewater treatment plant effluent throughout the world [22, 23].

It has been reported by [23] that the activated alumina was used in wastewater treatment plant and they claimed that its can be used to reduce the concentration of heavy metal in the wastewater effluent for the compliment of Malaysian regulation of Environmental Quality (Industrial Effluent) Regulations 2009 for permissible discharge levels [3] There are many research of heavy metals adsorption using the activated. Normally activated alumina is used for adsorption since activated alumina has a wide variety of target pollutants such as organic pollutant, dyes and even heavy [19]. According to [23]. The Inhibitory effects of phenol compound on microbial activity have been reported for concentrations larger than 500 ppm [24]. Therefore high loaded effluents require additional solutions to conventional biological treatment. Among possible techniques, adsorption on activated alumina is extensively applied, as it is most effective for removing organic compounds such as surfactants, pesticides, dyes, and aromatic molecules [24]. Economic viability of processes using adsorption is thus directly related to their efficiency in regenerating and recycling the activated alumina [25].

A wide spread contamination of shallow ground waters of African has been reported and over 25% of this water is producing arsenic contaminated water. This is a serious public health concern as the shallow water is the major source of drinking water for almost all of the population [23]. Thousands of people are reported to have already been suffered from arsenicosis and other heavy metal by drinking arsenic contaminated water and millions are at risk of arsenic toxicity due to exposure to arsenic contaminated water. The most important measure needed to combat the arsenic problem is to provide arsenic safe drinking water to the exposed people [25]. Unless suitable alternative drinking water sources are made available, arsenic removal from the shallow water through simple and low cost technologies appears to be an immediate and short-term solution of this problem. In this perspective, an activated alumina is proposed [26]. Activated alumina an effective adsorbent, has been widely proposed in tertiary treatment due to its large specific surface area and pore structure. There have been many researches on activated alumina adsorption for organic pollutant removal with the effects of temperature, solution pH, ionic strength, etc. on the adsorption isotherms [20].

Moreover, another important characteristics are the adsorption and regeneration is two processes in which a substance (adsorbate), in gas or liquid phase, accumulates on a solid surface and later regenerated to clean away the adsorbed particle [4]. It is based on the capability of porous materials with large surfaces to selectively retain compounds on the surface of the solid (adsorbent). There are two types of adsorption; physical and chemical adsorptions [27].

2.1 Physical Adsorption mechanism

The material have been reported by many research communities for its capability showing physical adsorption and physical adsorption is achieved by Van der Waals forces, dipole interactions, and Hydrogen binding and other ions which give additional priority [15]. There is no electron exchange between adsorbent and adsorbate. Because there is no activation energy required for physical adsorption, the time needed to reach equilibrium is very short. Physical adsorption is a non–specific and a reversible process and allow regenration[16]. The Fig. 1, show the composite indicating high intensity by activated aluminium

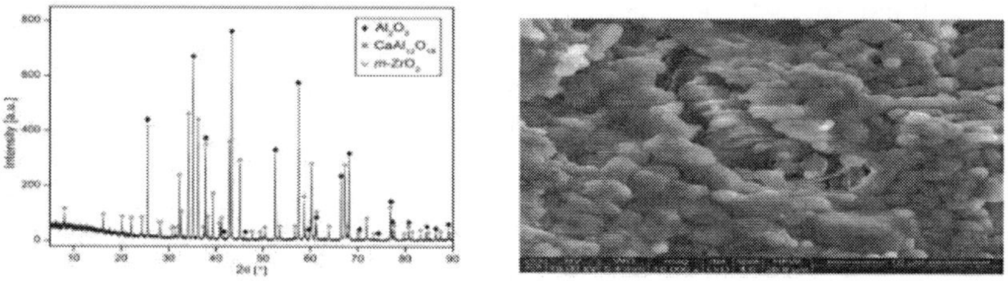

(a) (b)

Fig. 1. (a) The XRD pattern of the Al_2O_3–$CaAl_{12}O_{19}$–ZrO_2 ceramic composite material (b) SEM photomicrograph of the fractured section of the Al_2O_3–$CaAl_{12}O_{19}$–ZrO_2 ceramic composite material fabricated with homemade method [16].

As can be observed Fig. 2 shown below an indicate the existence of chemical adsorption results from the chemical link between adsorbent and adsorbate molecule, therefore it is specific as well as irreversible and chemical as well as electronic properties of adsorbent are changed. Binding between adsorbent and adsorbate by covalent bond is called weak chemical adsorption, and that by ionic bonds is called strong chemical adsorption this typical example hydrogen radical for phenol [17].

2.2 Chemical Adsorption mechanism

Several researches have reported on chemical mechanism, according [17] the alumina chemical reaction capacity is found to be between 3 mg/g in alkaline water and 20 mgF/g in acidic water. At pH 7 in their study, they equally claimed that, the de fluoridation capacity was found to be 5.6 mg/g. The capacity decreases with increased bicarbonate concentration but is found to be independent of the presence of other anions like chloride and sulphate [19]. The saturated medium could be regenerated by 2 % hydrochloric acid, 2 % sodium hydroxide or 1 % sulphuric acid. The fluoride removal obeyed Langmuir's adsorption isotherm indicating that the forces of adsorption are governed by chime-sorption. According to AFP group technologies the chemical mechanism is directly related to surface phenomena (These phenomena are surface reactions, which control the reaction rate. This is the reason why specific surface area is an important feature for designing adsorbents in dehydration and other applications) as shown in table 1. Moreover, due to the afore mentioned phenomena the activated alumina have a high adsorption capacity at low partial pressures. Adsorption capacity decreases with increasing temperatures.

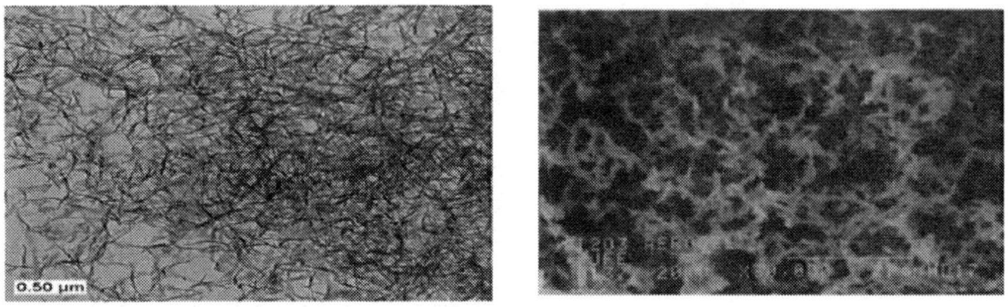

(a) (b)

Fig. 2. (a) TEM image branch like structure (b) SEM showing high porosity [12].

The adsorption process of pollutants by activated alumina includes the following four steps, its begin with transport of absorbates from the bulk fluid to the fluid-solid interface external surface of adsorbent particle, the material have shown capabilities and potentials for treatment industrial product and is equally employed in water and water related applications, like wise it used in gas absorptions etc. This step is also called film diffusion because of the passing through the stationary film of the surrounding of activated alumina before reaching the external surface treatment industrial product and is equally employed in water and water related applications, like wise it used in gas absorptions [26]. Moreover, intra particle transport of absorbates into the activated alumina particle, also called pore diffusion. Where, the adsorptions of absorbates at interior sites of the activated alumina particle and desorption of absorbates from interior sites and transport into the bulk fluid. The specific surface area is another important property that determines adsorbent usage and its capacity which shown in treatment industrial product and is equally employed in water and water related applications, like wise it used in gas absorptions [28]. The total surface area of activated alumina, ranging from 500 to 2000 m2. g−1, quantifies adsorption sites for molecules to attach treatment industrial product and is equally employed in water and water related applications, like wise it used in gas absorptions [29]. Mesopores, macropores and the nonporous surface of sample represent the external surface. Despite most of the adsorption takes place in the micropores, the meso- and macropores play also an important role in any adsorption process because they serve as passage for the adsorbate to reach micropores.

3 Preparation and Characteristic of Alumina

Alumina preparation was proposed by many researchers and especially; the Nano sized gamma alumina the material was proposed [30]. Their approach solely depend on precipitation by aluminium nitrate and sodium carbonate, it was very simple reaction where reaction between aluminium nitrate and sodium carbonate precipitation was done [31-35]. Their results showed that a stable gamma alumina can be prepared in high yield and with high specific surface area with controlled pH 8.5 value, 70°C reaction temperature. Their study equally proposed the aging process at 70°C temperature. Eliassi and Ranjbar 2014 [25-36] proposed the impurities removal to remove the impurities by washing and then centrifuging rather than using filter paper Fig. 3 shows the gamma alumina calcinated in

 https://doi.org/10.1051/e3sconf/20183402049

500, (a) SEM image for Nano gamma alumina produced by calcinations in 550°C and (b) the XRD Nano gamma alumina produced by calcinations in 550°C [37-40].

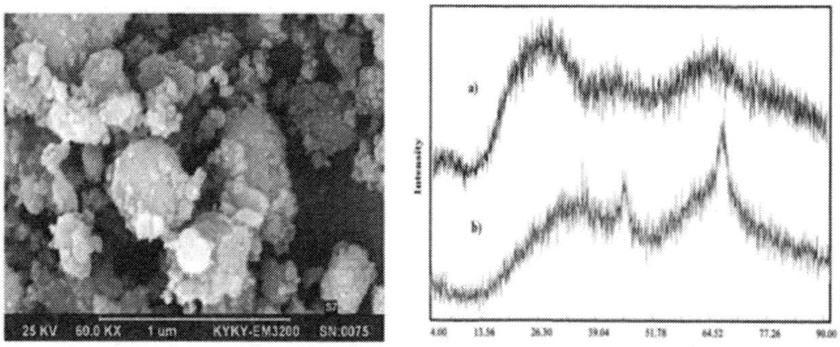

(a) (b)

Fig. 3. (a) gamma alumina calcinated in 500, (b) nano gamma alumina produced by calcinations in 550°C [14].

Similar study was conducted by [25] which stated that they used Sol–gel transition with the aid of pyrolysis of the alumina-based gels prepared via alumatrane precursor. According to [16] the alumina gels were prepared by the sol–gel method via alumatrane precursor the materials were synthesised directly from the reaction of readily available and inexpensive compounds, the compound include aluminium hydroxide and triisopropanolamine were mixed in careful and in a step wise manner. With controlled Sol–gel process parameters, these include time gel time, pH, and temperature and hydrolysis ratio. Bussarin and his team were able to monitor the sol–gel transition of alcoholic alumatrane solution via multiple waveform rheological measurements to characterize the materials [37-40].

4 Conclusions

The study summarized the recent development in synthesis and characterization of the activated alumina. The recent techniques in both electrochemical and anodization process employed in synthesizing activated. Different sizes such as Nano sized gamma alumina preparation were equally reviewed. Very good results were observed and summarized various optimum conditions for preparing activated alumina leading to high yield and high specific surface area were summarized, this included degree of pH, reaction temperature etc. The systematic characterization based on SEM, TEM, FTIR, X-ray diffraction and EDX were equally summarized.

References

1. Ahmed S., Rasul M.G., Martens W.N., Brown R., Hashib M.A., Desalination **261**, 3–18 (2010)
2. Musmarra D., D. Karatza, A. Lancia, M. Prisciandaro, G.M. di Celso, Chem. Eng. Trans. **32**, 547–552 (2013)
3. Anirudhan T., S. Sreekumari, J. Environ. Sci. **23**, 1989–1998
4. Guo Y., E. Du, Energy Procedia **17**, 444–449 (2012)
5. Salvador F., N. Martin-Sanchez, R. Sanchez-Hernandez, M.J. Sanchez-Montero, C. Izquierdo, Microporous Mesoporous Mater. **202**, 259–276 (2015)

6. F. Salvador, N. Martin-Sanchez, R. Sanchez-Hernandez, M.J. Sanchez-Montero, Microporous Mesoporous Mater. **202**, 277–296 (2015)
7. P.-J. Lu, H.-C. Lin, W.-T. Yu, J.-M. Chern, J. Taiwan Inst. Chem. Eng. **42** 305–311 (2011)
8. Ahmed S., Rasul M.G., Brownb R., Hashib M.A., Journal of Environmental Management **92**, 311-330. (2011)
9. Beker U., Ganbold B., Dertli H., Duranog D., Gulbayir I. Energy Conversion and Management **51**, 235–240 (2010)
10. Cadoret L, Rossignol C, Dexpert-Ghys J, et al. Materials Science and Engineering B-Advanced Functional Solid- State Materials **170** (1-3), 41-50 (2010)
11. Darko S. A., Maxwell E., Park S. Thin Solid Films **519** (1). 174-177 (2010)
12. Lam S-M., Sin J-C., Mohamed A-R.. Korean J. Chem. Eng. **27**(4), 1109-1116 (2010)
13. Saka, C. Journal of Analytical and Applied Pyrolysis **95**, 21–24 (2012)
14. Shah, I. K., Pre, P. & Alappat, B., Journal of the Taiwan Institute of Chemical Engineers **45**, 1733–1738 (2014)
15. Xin-hui, Chemical Engineering and Processing: Process Intensification **53**, 53– 62 (2012)
16. Barakat, M.,. New trends in removing heavy metals from industrial wastewater. Arabian Journal of Chemistry **4**, 361–377 (2011)
17. Berenguer, R. et al., ScienceDirect **48**, 2010, 2734-2745.
18. Eliassi A M. Ranjbar, Int. J. Nanosci. Nanotechnol. **10**(1), 13-26 (2014)
19. Bussarin Ksapabutr , Erdogan Gulari, Sujitra Wongkasemjit, Colloids and Surfaces A: Physicochem. Eng. Aspects **233**, 145–153 (2004)
20. Michael A. Karakassides, Dimitrios Gournis, Athanasios B. Bourlinos, Pantelis N. Trikalitis and Thomas BakascMagneti, J. Mater. Chem. **13**, 871–876 (2016)
21. Ramana Reddy P., Ajith K. M. and N. K. Udayashankar, Adv. Mater. Lett. **7**(5), 398-401 (2016)
22. Poco J .F. J. H. Satcher Jr. L. W. Hrubesh, Journal of nano-crystaline solids **285**, 57-63 (2001)
23. Karthikeyan G, B. V. Apparao, S Meenakshi, *2nd international workshop on fluorosis prevention and defluoridation of water*, 79-82 (2014)
24. Abdul Rashid, J. I., Abdullah, J., Yusof, N. A., Hajian, R., Abdul Rashid, J. I., Abdullah, J., … Hajian, R. The Development of Silicon Nanowire as Sensing Material and Its Applications. Journal of Nanomaterials, 1–16 (2013)
25. Ahmed, R. A., Fekry, A. M. International Journal of Electrochemical Science **8**(5), 6692–6708 (2013)
26. Boken, J., Kumar, D. Nanoparticles **4**(4), 303–308 (2014)
27. Chen, K. I., Li, B. R., Chen, Y. T. Nano Today. Elsevier Ltd. https://doi.org/10.1016/j.nantod.2011.02.001 (2011)
28. Chua, J. H., Chee, R. E., Agarwal, A., She, M. W., Zhang, G. J. Analytical Chemistry **81**(15), 6266–6271(2009)
29. Cui, L., Wu, J., Ju, H. Electrochemical sensing of heavy metal ions with inorganic, organic and bio-materials. Biosensors and Bioelectronics **63**, 276–286 (2015)
30. Cui, Y., Wei, Q., Park, H., & Lieber, C. Science **293** (5533), 1289–1292 (2001)
31. Cui, Y., Zhong, Z., Wang, D., Wang, W. U., Lieber, C. M. Nano Letters **3**(2), 149–152 (2003)
32. Dhahi, T. S., Hashim, U., Ahmed, N. M. Science of Advanced Materials **3**(2), 233–238 (2011)
33. Dielacher, B. K. A Combined Electrical , Plasmonic and Fluidic Measurement System for Metal Nanostructure based Ion Sensing, 22493 (2015)

34. Eshkeiti, A., Narakathu, B. B., Reddy, A. S. G., Rebrosova, E., Rebros, M., Joyce, M., Atashbar, M. Z. Nano chemistry, 0–3 (2012)
35. Guo, Z., Seol, M., Gao, C., Kim, M., Ahn, J., Choi, Y., Huang, X. Electrochimica Acta **211**, 998–1005 (2016)
36. Harrison, B. S. Applications of Nanotechnology. In Principles of Regenerative Medicine (pp. 554–561). https://doi.org/10.1016/B978-012369410-2.50033-4 (2008)
37. Joyner, J. J., V., D. K. International Journal Of Science & Technoledge **3**(4), 80–90 (2015)
38. Korent Urek, Š., Frančič, N., Turel, M., Lobnik, A. Journal of Nanomaterials, https://doi.org/10.1155/2013/501320 (2013)
39. Li, Z., Chen, J., Guo, H., Fan, X., Wen, Z., Yeh, M. H., … Wang, Z. L. Adv Mater, 2983–2991(2016)
40. Long, F., Zhu, A., & Shi, H. Sensors **13**(10), 13928–13948 (2013)

The Effects of Saltwater Intrusion to Flood Mitigation Project

Azinoor Azida Abu Bakar[1,*], and *Muhammad Khairudin* Khalil[2,]

[1]Faculty of Civil Engineering, Universiti Teknologi MARA, Johor Branch, Pasir Gudang Campus, Bandar Seri Alam, 81750 Masai, Johor, Malaysia.
[2]Department of Irrigation and Drainage, Bangunan Dato' Muhamad Ibrahim Munsyi, Pusat Pentadbiran Kota Iskandar, 79626 Nusajaya, Johor, Malaysia.

Abstract. The objective of this study is to determine the effects of saltwater intrusion to flood mitigation project located in the flood plains in the district of Muar, Johor. Based on the studies and designs carried out, one of the effective flood mitigation options identified is the Kampung Tanjung Olak bypass and Kampung Belemang bypass at the lower reaches of Sungai Muar. But, the construction of the Kampung Belemang and Tanjung Olak bypass, while speeding up flood discharges, may also increase saltwater intrusion during drought low flows. Establishing the dynamics of flooding, including replicating the existing situation and the performance with prospective flood mitigation interventions, is most effectively accomplished using computer-based modelling tools. The finding of this study shows that to overcome the problem, a barrage should be constructed at Sungai Muar to solve the saltwater intrusion and low yield problem of the river.

1 Introduction

Sg. Muar is being extracted for water supply to the States of Johor, Malacca and Negeri Sembilan. There have been reported incidences of saltwater intrusion during drought into the raw water intake at Panchor, 68km upstream from Sg. Muar river mouth, indicating either that the station was not located sufficiently upstream to avoid the saltwater intrusion or the raw water extraction from Sg. Muar has increased over the years and caused the saltwater intrusion to move further upstream during drought when fresh water discharge moving downstream is very low.

The Sungai Muar river basin was affected by two major flood events in December 2006 and January 2007, which caused massive floods with devastating damages within the river basin. Based on the original studies and designs carried out and completed in 2011 [1] the Kg. Tg. Olak Bypass and the Kg. Belemang Bypass have been identified as one of the effective flood mitigation components that will reduce future flooding and consequential hazards within the Sg. Muar river basin. There is concern that construction of the Kg. Belemang and Tg. Olak Bypasses, while speeding up flood discharges, may also increase

* Corresponding author: khairudinkhalil@yahoo.co.uk

saltwater intrusion during drought low flows since the 2 new bypasses will reduce the river distance between the river mouth to the raw water intake station at Panchor [1].

A variety of different measures are commonly used to control seawater intrusion and protect groundwater resources, although the target of most is to increase the flow of fresh groundwater and/or reduce the flow of saltwater. As such there are a number of methods with which to prevent saltwater contamination of groundwater, including subsurface barriers and artificial recharge [1]. Various reports reported [2, 3, 4], the insufficient yield of Sg. Muar to cater for the daily water supply demands of households in Johor, Malacca and Negeri Sembilan during drought season, which will be further aggravated if saltwater intrusion problem increases.

Contributory factors to flooding problems includes development in the catchment, leading to reductions in the time of concentration and increased flood discharges, human settlement and encroachment on the floodplain, the occurrence of high tides, exacerbated by low pressure and/or wind/wave set-up, siltation of waterways, due to erosion of soil from the catchment (exacerbated by poor sediment control practices at development sites), existence of rock outcrop (e.g. at Batu Liang) in the river channel and trees falling into Sg. Muar due to bank erosion [2, 5].

2 Methodology

2.1 Site Area

The project is situated at the lower reaches of the Sg. Muar basin, which is located in the north-western part of Johor State as shown in Fig. 1.

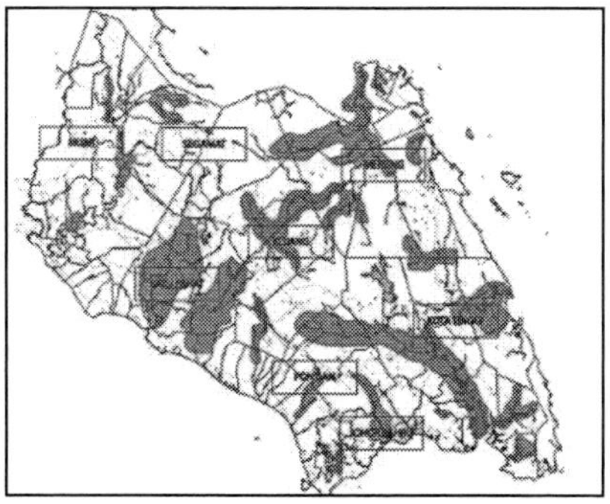

Fig. 1. Project location plan.

With a total catchment area of 6,595 km^2, part of the Sg Muar catchment boundary is within the states of Negeri Sembilan and Pahang. Sg. Muar rises at an elevation of 350m at Kg Lubok Kadam (District of Kuala Pilah, Negeri Sembilan) and flows in a predominantly easterly direction through Gemas before turning near Segamat to flow south and turning west to reach the Straits of Melaka at Muar. In its lower reaches, Sg Muar is characterised by numerous large meanders. The main channel of Sg Muar is about 310km long and the average annual discharge at its estuary is about 140m^3/s. The two main population centres within the river basin are Segamat (on Sg Segamat) and Muar at the Sg Muar river mouth. The total population of the river basin is made up of 250,000 at Segamat District (2,825 km^2) and 450,000 at Muar District (2,346 km^2) [5].

2.2 Climate

The study area experiences typical humid equatorial climate with high temperature and humidity all year round. The average monthly low temperature is 23°C - 24°C and the high temperature is 31°C - 33°C. The highest air temperature usually occurs from April to May and the lowest is from December to January. The mean annual humidity is above 80%. The annual rainfall in the project area ranges from 1,300mm to 2,400mm. Heaviest rainfall is from the Northeast monsoon in November to December. From May to July, the Southwest monsoon brings moderate rainfall to the project area.

2.3 Soil and Geological Characteristics

The study area being within the lower reaches of Sg. Muar, is generally flat and low lying with ground level of RL 1 to 2.5m. Oil palms and fruit trees dominate the landscape, with mangroves along the riverbanks. The geological formation of the project area basically consists of Quaternary Alluvium, which consists of very soft marine and fluviatile clays about 20m thick underlain by sands and silts. No bedrocks are encountered down to 60m below ground level. Groundwater is at 0 to 3m below ground level, and fluctuates with wet / dry seasons and tides. The very soft marine clays are expected to pose very serious stability and settlement problems for the proposed works both during and after construction. Sand lenses in between and below the soft clays can also cause serious piping and boiling conditions during and after construction of the works.

2.4 Recent Flood Events

The two major flood events in recent years affecting the Sg Muar River Basin occurred in December 2006 (with a 48 hour rainfall of over 300mm) and January 2007. The extreme rainfalls from 17th -20th December 2006 caused Sg. Muar and its tributaries to overflow their banks and resulted in massive and widespread floods which were remarkable in extent, duration and depth and with devastating damages. The January 2007 flood event was of a lesser magnitude, but was nevertheless significant enough to cause widespread damages and disruption [2].

The township of Gerisek, Panchor, Pagoh, Kundang Ulu, Lena and Bukit Kepong were badly affected by the Dec 2006 and Jan 2007 flooding when Sungai Muar and Sungai Segamat overtopped their banks and flooded a large extent of the low-lying populated areas. Some 25,000 people in the Sg Muar River Basin left their homes to seek shelter in relief centres. Lives were lost as were crops and livestock. Millions of Ringgit worth of damages was caused to infrastructure such as roads, utilities, as well as personal properties and belongings. In some cases, road and rail embankments were eroded by floodwater and have necessitated expensive and time consuming reconstruction works. Many roads including the PLUS Expressway around KM 152 were inundated and made impassable. Some 397 km^2 of the Sg. Muar river basin was inundated with floodwater during the Dec 2006 and Jan 2007 floods [11]. Based on previous studies [2, 6], the Dec 2006 and Jan 2007 floods have been assessed to be of average recurrence interval (ARI) of between 70 to 100 years. The estimated flood map of the Lower Muar River Basin for 100 year ARI is shown in Fig. 2.

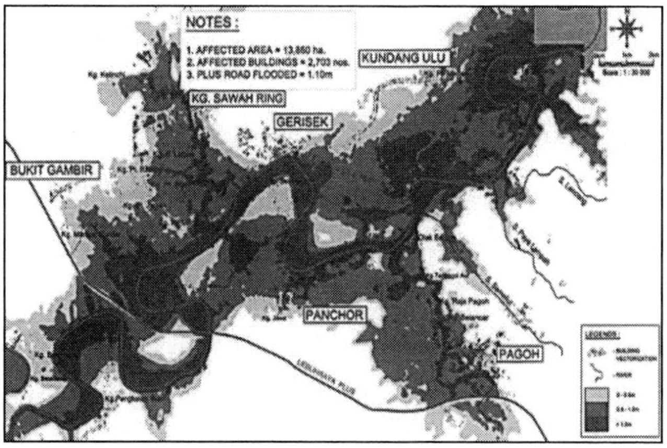

Fig. 2. 100 years ARI flood map for lower Sg Muar River basin.

2.5 Data Collection and Model Calibration

The Intensity-Duration-Frequency (IDF) analysis for six rainfall stations in Sg Muar basin was carried out by in 2003. Flood Frequency Analysis (FFA) for four water level gauging stations was conducted. A NAM hydrological model was established to simulate the rainfall runoff process for the entire Sg Muar basin.

The rainfall temporal patterns and Intensity-Duration-Frequency curves for six rainfall stations located within Sg Muar basin with different storm duration (15 min, 30 min, 60 min, 120 min, 12 hours, 24 hours and 72 hours) and return period (2, 5, 10, 20, 50 and 100 years) has been derived. These patterns were then used as the rainfall input in the hydrological model to produce the design flood.

The study area, which involves a total of 6,200 km^2, was divided into 21 sub-catchments, and was modelled using the NAM hydrological model. The model was calibrated with the flood data from two major flood events taking place in December 2006 and January 2007. In addition, the model was also calibrated based on Stream flow data obtained for the Segamat, Buloh Kasap and Bt. 57 stations.

2.6 Flood Mitigation Proposal

Although the events of Dec 2006 and Jan 2007 may have been unusual, it is intended to identify and carry out such mitigation works that may reduce future flooding and consequential hazards within the Sg Muar river basin. Based on the studies and designs carried out and completed in the year 2011 [1] one of the effective flood mitigation option identified are the Kg. Tg. Olak Bypass and Kg. Belemang Byass at the lower reaches of Sg Muar. The proposed Tg Olak and Kg Belemang Bypasses are located in the flood plains of Lower Muar at Tg Olak and Kg Belemang respectively in the district of Muar, Johor. The main function of these bypasses is to cut through 2 meanders to reduce the flow distance from 27km to 6km to effectively increase the hydraulic gradient and hasten the evacuation of flood waters to the sea thereby alleviating flooding and mitigating floods in the town of Gerisek, Panchor, Pagoh and Lenga for the 100 year ARI event.

Hydraulic profile of the river flood water level before and after the construction of the bypasses is shown in Fig. 3. It is to be noted that flooding along Sg. Muar will still occur after completion of Kg. Belemang and Tg. Olak Bypasses as these are just part of the whole of Sg. Muar flood mitigation works.

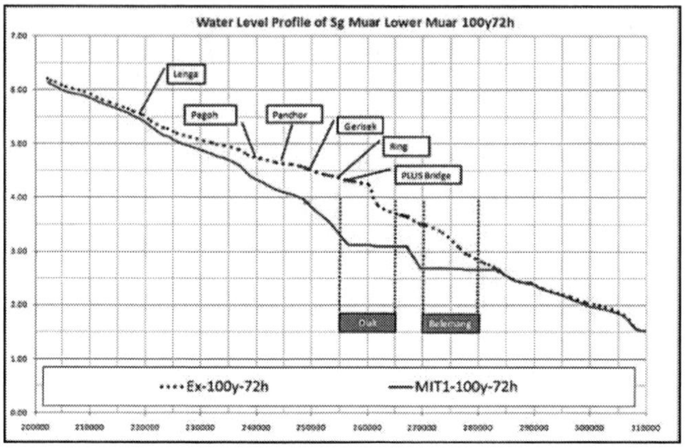

Fig. 3. Profile of water level at lower Muar with bypasses.

3 Results and discussions

3.1 Sungai Muar Flood Mitigation

The Intensity-Duration-Frequency (IDF) analysis for six rainfall stations in Sg Muar basin was carried out [3]. Flood Frequency Analysis (FFA) for four water level gauging stations was conducted. A NAM hydrological model was established to simulate the rainfall runoff process for the entire Sg Muar basin. Table 1 shows the 100 year ARI design runoff values obtained from Flood Frequency Analysis and Simulation from the Hydrological model at location upstream of Tg Olak.

Table 1. 100 year ARI design runoff estimation from Iktisas (2003) for specific gauge locations.

Name	Station No	FFA 100 year ARI (m^3/s)	Modelled 100 year ARI (m^3/s)
Bt 57	2625412	660	1690
Buloh Kasap	2527411	500	1500
Segamat	2528414	1,350	Not Available

The rainfall temporal patterns and Intensity-Duration-Frequency curves for six rainfall stations located within Sg Muar basin with different storm duration (15 min, 30 min, 60 min, 120 min, 12 hours, 24 hours and 72 hours) and return period (2, 5, 10, 20, 50 and 100 years) has been derived. These patterns were then used as the rainfall input in the hydrological model to produce the design flood.

The study area, which involves a total of 6,200 km^2, was divided into 21 sub-catchments, and was modelled using the NAM hydrological model. The model was calibrated with the flood data from two major flood events taking place in December 2006 and January 2007. In addition, the model was also calibrated based on Stream flow data obtained for the Segamat, Buloh Kasap and Bt. 57 stations.

3.2 Reviewed and Updated the Hydrological Procedures

In the recent years, Malaysia has been experiencing extreme weather conditions of frequent floods. The utilisation of larger volume and longer record of available rainfall data ensured more accurate rainfall intensity estimation particularly at high return period [7]. It also provides the estimation of design rainstorm/rainfall intensity and IDF relationship at ungauged site such as Tg. Olak and others. A new Areal Reduction Factor (ARF) for catchment rainfall was developed in the study. Equation in the revised HP1 (2010) [7],

$$i = \frac{\lambda T^{\kappa}}{(d+\theta)^{\eta}} \tag{1}$$

The derived parameters including λ, κ, θ and η was generalised for the construction of ungauged IDF relationship where λ is the mean number of peaks per year included in the partial duration series ; κ is the important shape parameter ; T is the return period ; d is duration ; and θ and η are non-negative coefficients. IDF curves for a control point have been developed using HP1 [7].

The Town and Country Planning Department reviewed the Draf Rancangan Tempatan Daerah Muar in 2010 [8]. The review indicates that the increase in development area including Residential, Industrial, Trading and Services, and Institutions and Community Facilities is up to 4307.94 hectare. On the other hand, there is a decrease of 814.9 hectare of land in the Open Space and Recreation area. These imply that the impervious area within the catchment area has increased accordingly, which contribute to greater runoff during storm event

3.3 Hydrological Modelling

The hydrological modelling was carried out using DHI's NAM. NAM a parametric lumped hydrologic model that simulates the land phase of the hydrologic cycle. Meteorological inputs to the model are rainfall and potential evapotranspiration. Input rainfall data used in the hydrological modelling are rainfall records from JPS stations (various) and the design rainfall for various return periods and storm durations.

Rainfall-runoff outputs are catchment runoff (split into surface flow, interflow and groundwater or baseflow), groundwater levels and other information such as moisture content and groundwater recharge.

3.4 Future IDF curves

HP1 [7] allows IDF curves to be developed for an ungauged site, for example, at Tg. Olak (102° 40' E, 2° 10' N), where no JPS hydrological stations are available. Derived parameters for Tg Olak are extracted using GIS and the corresponding values are shown in Fig. 4 and future IDF curve for Tg. Olak are shown in Fig. 5 and salt water intrusion as shown in Fig. 6 and Fig. 7 respectively. Based on the findings, the bypass will be designed for discharge of 2170 m^3/s with 1m freeboard, as with the old river meander open and by making use of the 1m freeboard the whole system can still cater for discharge of 2170 m^3/s multiplied by Climate Change Factor of 1.58. From the calculations at design flow, about 15% of the flow is through the old river meanders and about 85% through the bypasses. The channel side slopes of 1 in 5 with 6 m wide berms at 6 m vertical intervals, are dictated by geotechnics. Manning's Formula with coefficient of 0.035 is used to compute channel discharge capacity.

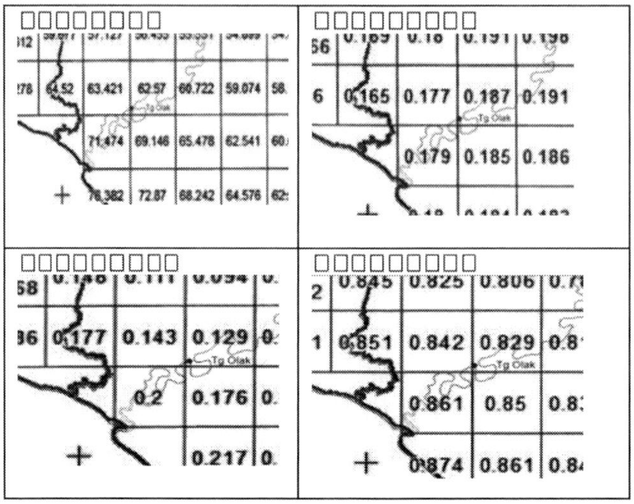

Fig. 4. The derived parameters from for Tg Olak site.

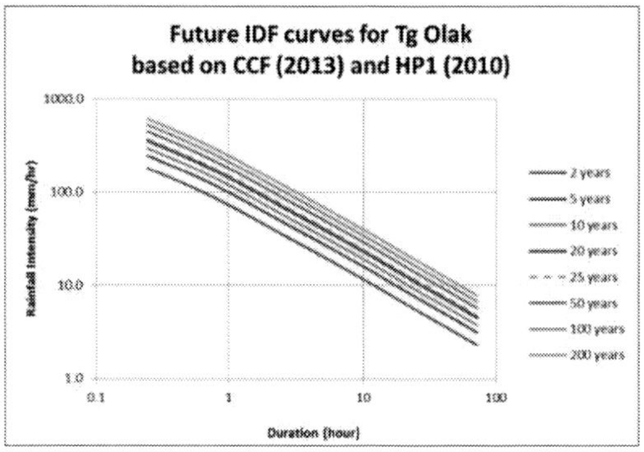

Fig. 5. The IDF curves for Tg Olak based on HP1 [7] and CCF [9].

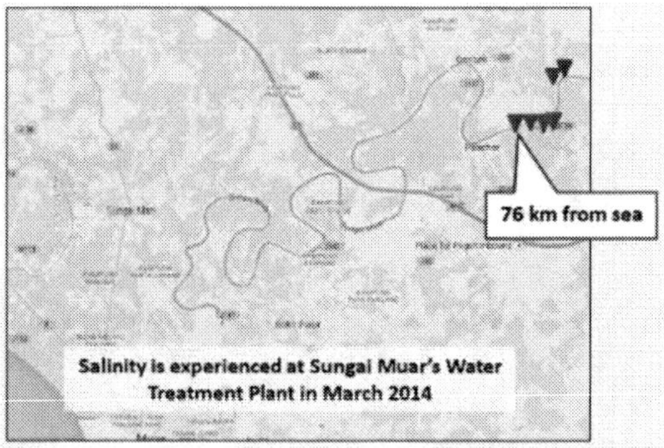

Fig. 6. Salt water in Sg Muar.

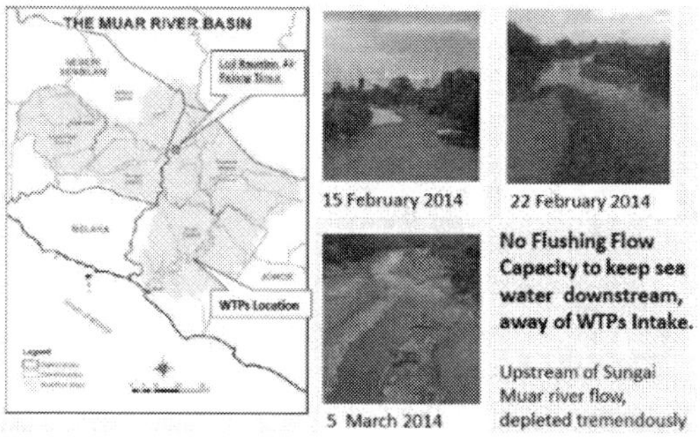

Fig. 7. Incidents on salt water intrusion.

The channel base width has been reduced from 50 m to 30 m, side slopes from 1 in 6 to 1 in 5 and this has resulted in substantial earth excavation and cost savings. Fom European experience that straightening rivers by removing meanders in flood mitigation works can have negative impacts on riparian flora and fauna and river life [11], it is a wise to keep the old river meanders at both bypasses open to river flow. This not only enables barges and other bigger boats to use Sg. Muar for navigation on a status quo basis, but will also reduce any negative environmental impacts. The hydraulic calculations to determine the division of flow at the bifurcation where the river divides into the new bypass and the old meander is based on the method [12]. From the calculations it can be seen that at design flow, about 15% of the flow is through the old river meanders and about 85% through the bypasses.

3.5 Success Criteria

The criteria that considered in the evaluation of flood mitigation in the Muar basin are the Peak Water Level and Extent of Inundation. These criteria can be measured by significant consequence to the householder and flood mitigation works. For example for a property with a floor level of 3.5 m, if the peak flood level exceeds 3.5 m, then it is of significant consequence to the householder. If then, mitigation can reduce the likelihood in any given year of such a flood height being achieved from the 50 year to the 100 year flood level, then this is of consequence. Whereas flood mitigation works which reduce the extent of inundation for a given event are generally considered ideal, if the area of inundation is reduced whilst the total flood volume is not, then it is likely that the areas that remain inundated are experiencing higher velocities, higher discharges and higher levels of hazard.

4 Conclusions

Sg. Muar is being extracted for water supply to the States of Johor, Malacca and Negeri Sembilan. There have been incidences of salt-water intrusion into the raw water intakes, especially during the low flow event. Meanwhile, construction of the Kg. Belemang and Tg. Olak Bypasses, while speeding up flood discharges, may also increase salt-water intrusion during low flows. However, there are plans and studies being carried out now to address this problem like the construction of weir or barrage to resolve this problem. By constructing the barrage at the study area, the saltwater intrusion can be controlled. But a detailed study should be conducted in near future in order to obtain more precise information on other potential environmental and ecological effects to the area.

The authors acknowledges support from the Ministry of Higher Education Malaysia through a Fundamental Research Grant Scheme (FRGS) (Grant 600-RMI/ST/FRGS 5/3/(18/2015). The authors also would like to thank FKA UiTM, IRMI UiTM and JPS for their support to this project.

References

1. HSS Integrated (HSSI), *Cadangan Kerja-kerja Tebatan Banjir Pintasan Tanjung Olak dan Belemang di Hilir Sungai Muar* (2011)
2. Research Division Malaysian Meteorological Dept, Ministry Of Science, Technology And Innovation, *Report on Heavy Rainfall That Caused Floods in Johor, Melaka, Negeri Sembilan and Pahang During the Period 17th – 20th December 2006* (2006)
3. Iktisas Sdn Bhd. *Sg Muar – Flood Mitigation Masterplan* (2003)
4. HSS Integrated (HSSI), *Hydrology Report* (2009)
5. Shafie, *Extreme Flood Event: A case study on Floods of 2006 and 2007 in Johor, Malaysia* (2009)
6. Jabatan Pengairan dan Saliran Malaysia (JPS), *HP4 – Magnitude and Frequency of Floods in Peninsular Malaysia (Revised and Updated)* (1987)
7. HP1 -National Hydraulic Research Institute of Malaysia (NAHRIM), *Reviewed and Updated the Hydrological Procedure No.1 (Estimation of Design Rainstorm in Peninsular Malaysia)* (2010)
8. Town and Country Planning Department, Draf Rancangan Tempatan Daerah Muar 2002-2015 (Pengubahan) (2010)
9. National Hydraulic Research Institute of Malaysia (NAHRIM), *Estimation of Future Design Rainstorm under the Climate Change Scenario in Peninsular Malaysia,* Nahrim Technical Guide No.1 (2013)
10. A. Khomine, S. János and K. Balázs, Adv. in the Res. of Aquatic Env., 251-257 (2011)
11. V. T. Chow, *Open-Channel Hydraulics.* McGraw-Hill Book Company ((1973)
12. Y. L. Wan, *National Seminar on Socio-Economic Impacts of Extreme Weather and Climate Changes* (2007)

Biomass Morphology Subjected to Different Chemical Treatment

Norsuzailina Mohamed Sutan[1,*], *Siti Masjida* Mazlan[1], *Siti Noor Linda* Taib[1] ,*Delsye Teo Ching* Lee[1] , *Alsidqi* Hassan[1] , *Siti Kudnie* Sahari[2] , *Khairul Anwar* Mohamad Said[3], and *Habibur Rahman* Sobuz[4]

[1]Civil Engineering Department, Universiti Malaysia Sarawak, 94300 Kota Samarahan, Sarawak, Malaysia
[2]Electrical and Electronic Engineering Department, Universiti Malaysia Sarawak, 94300 Kota Samarahan, Sarawak, Malaysia
[3]Chemical Engineering and Energy Sustainability Department, Universiti Malaysia Sarawak, 94300 Kota Samarahan, Sarawak
[4]Department of Building Engineering and Construction Management, Khulna University of Engineering and Technology (KUET), Khulna, Bangladesh

Abstract. A growing interest of sugarcane bagasse fibre composite has been observed in recent years due to its attractiveness properties such as low specific weight, renewable source and producible with low investment at low cost. However, these materials have a low interfacial adhesion between fibre and matrix which lead to reduction in certain mechanical properties of the composite. To overcome this problem, studies show that certain chemical treatments on the surface of the fibres are some alternatives that significantly increase the adhesion reinforcement/matrix, in some cases improving its mechanical properties. The objective of this study was to evaluate the effect of different type of chemical treatment which are alkali and acid treatment on sugarcane bagasse fibre surface morphology. Seeking to improve the adhesion fibre matrix, the fibre has been treated with 5% of NaOH and 5% of HCL solution with added of bagasse fibre used in the range of 0% to 3% of cement weight respectively. Through SEM investigation, it is confirmed that chemical treatment helps to remove hemicelluloses from raw bagasse fiber as well as improved fibre matrix adhesion.

1 Introduction

Nowadays, composites have become an essential part and used by us almost every day. In the sense of being high performance materials and costs effective, lead composite to be used in broader volume in various engineering structures with ultra difficult applications such as spacecraft. Owing to the fact that in the aeronautics industry, the main goal of using composite material is the 'reduction of weight combined with strength' or 'the reduction of weight and saving precious fuel'. The uses of nature have been applied for million years by ancient society in order to change heterogeneous materials combining with the best aspects

* Corresponding author: msnorsuzailina@unimas.my

of different constituents.

Despite its exorbitant price, fibre reinforced composites materials have gained popularity as it has capability to achieve high strength [1]. It has been used broadly in rehabilitation and repair of damaged masonry and concrete structures due to its high strength to weight ratio, corrosion resistance and ease of application [2]. Generally, fibres divided into two groups such as man-made fibre and natural fibre. Natural fibre is subdivided based on their origins like plants, animals or minerals. As for man-made fibres, it classified by regenerate fibres in such a way that natural materials brought into fibres form by a chemical reaction or a physical action, fibres made from synthetic polymers (synthetic fibres) and inorganic fibres [3].

In recent years, the natural fibres composites have attracted the attention of scientists due to its substantial importance as competitive and greatly potential structure of materials. Fibrous material is an advanced composite material embedded in a resin matrix, generally laminated with fibers oriented in alternating directions to give material strength and stiffness. It is originating from annual plants and has been used in broad variety of purposes. The most common natural fibres product is bagasse fibres. Bagasse is bio product of sugar cane fibres which is left after the juice has been extracted from the sugar cane. From the analysis of bagasse composition indicates that the main constituent are cellulose, hemicelluloses, lignin, ash and wax [4]. Thus, these compositions of sugarcane bagasse possess distinct physical and chemical properties make it an ideal candidate to be considered for value-added applications such as reinforcement in plastic composites for the purposes of creating new materials. Besides that, the utilization of bagasse fibres promotes low production cost with such a high quality of green products.

It cannot be denied that the merger of natural fibres into synthetic polymers helps to enhance mechanical properties. However, the major shortcomings of using natural fibres are their high degree of moisture absorption and poor dimensional. This statement is true when subjecting the use of bagasse fibres. In fact, bagasse fibres have not been subjected to any chemical surface treatments and thus contribute to weak interface between the bagasse and the resin matrix. According to a investigation by [5] in [6] research noted that the high content of lignin (21% on average) in sugarcane bagasse is the reason why this happened. The author also highlight that high lignin content is considered to prevent a good surface wettability between polymer matrix material and natural fibres [7] although it increases the resistance to chemical and microbial attack [8]. Therefore, a good bond between the polymeric matrix and the fibres are vital as it is responsible to transfer the load to the stiff fibres through shear stresses at the interface. Modifying the fibres surface by using chemical treatment can enhance bond strength between fibres and matrix. Chemical modifications are considered to optimize the adhesion between the fibres surface and the polymer matrix may not only modify the fibres surface but also increase fibres strength. Besides that, water absorption of composites will be reduced and enhanced their mechanical properties.

In the present study is focused on the effect of applying the different type of chemical modifications on bagasse fibres for use in natural fibres reinforced composite. Chemical treatments include alkali and acid are discussed and compare with untreated fibres. Apart from that, the benefits of using the surface treatment process and how it can improve and change the properties of fibres and composites are also discussed. Then changes in cement composite properties investigated were then related to varying the percentage of bagasse fibres proportions subjected to different chemical treatments.

2 Materials

Sugarcane bagasse (Saccharum Officinarum) is the fibrous solid residue left over from the extraction of sugarcane juice. It was collected from the juice shop after they were crushed by using a hand crushing machine. It is white in colour with a length of 100- 300 mm and free from hard, thick outer layer. After that, the collected bagasse is allowed to dry under sunlight for 2 to 3 days and spread on a waterproof sheet in order to remove the wetness that present in it. The bundled fibres were peeled out and separated manually by means of scissors. For the untreated fibres, the fibres are cut into dimension of approximately 10 mm.

Fig. 1. Raw Bagasse Fibre.

Fig. 2. Bagasse fibres that has been peeled out.

Table 1. Average Bagasse composition [6].

Items	Percentage (%)
Moisture	49
Soluble solids	2.3
Fiber	*48.7*
Cellulose	41.8
Hemicellulose	28
Lignin	21.8

3 Methods

3.1 Fibre treatment

Once the fibres has been prepared then it has undergone the surface treatments. The treatment of bagasse fibre was performed by alkaline treatment and acid treatment with similar concentration solution. For alkaline treatment, the dry fibres were treated with 5% by mass of NaOH solution while the acid treatment was carried out in a solution of 5% by mass of HCL. The bagasse fibres were soaked for 1 hour and it has been taken out and washed several times with fresh water to make sure all the water soluble chemicals and by products were removed. Next, the bagasse fibres washed again with distilled water until final pH of 7 was maintained. The fibres were then dried at room temperature for 48 hours followed by oven drying at 100°C for 6 hours. After pretreatment reaction, the fibres were cut into the dimensions with length of approximately 10 mm to be mixed with cement. This was done for acid, alkaline and untreated fibres. Then it was stored in sealed plastic bag to prevent it from moisture sorption from surrounding air and moist. Fig. 3 shows the soaking fibres into alkaline and acid solution.

3.2 Production of cement composite

The amount of bagasse fibres was the major parameter of this study. These fibres were in used in the range of 0% to 3% of cement weight respectively as shown in Table 2. Composite cements were designed and made with w/c ratio of 0.42 which is the suitable content used in the cement mixture. Before the mixing procedure is conducted, bagasse fibres with various percentage of 1%,2%,3% (by weight relative to cement weight) of short fibres of 10 mm in length, with random fibres orientation, were prepared from the untreated and treated fibres. In order to produce a cement paste, particular amount of cement weight and water were mixed together. Then, particular amount of bagasse fibres is added into the cement paste and mixed well with a spatula to obtain homogenous paste. The mixture was then poured into a cylinder moulds in order to be introduced to make paste samples. Crushing of the cubes was carried out at 7,14, 28 and 56 days respectively to determine the compressive strength of the composite.

3.3 Morphological and physical characteristics of the fibres

3.3.1. Scanning Electron Microscopy (SEM)

To characterize the morphology of untreated and treated fibres surfaces and the mode of the material removal, the samples were observed under a scanning electron microscopy (SEM). The SEM used in this experiment is JEOL JSM-6390 microscope which is located at Mechanical and Manufacturing laboratory, UNIMAS. In this research, SEM will be conducted before and after fibre treated and when the fibre mixed with cement paste. Right after the sample failed undergone compressive strength; small piece of sample with the size of 1cm x 1cm is polished, coated with thin layer of gold then stored in a sealed plastic bag before it is taken to conduct SEM. The samples are magnified up to 500 times in order to allow the submicron scale features of the samples to be seen.

4 Results and discussions

4.1 Effect of fibre treatment on physical characteristics

4.1.1 Colour of Bagasse Fibre

As mentioned previously, two different procedures for surface modifications were applied to bagasse fibres which are alkali treatment and acid treatment. As a result of applying chemical treatment on bagasse fibres can be seen clearly by the colour changes after treatment. The colour of treated bagasse fibres is visible in Figure 3 below. It can be seen that the appearances of the bagasse fibre changes after the chemical treatment take place. After treated with NaOH solution in 1 hour, it has been observed that the sugarcane bagasse fibres turns into yellowish in colour. At this point, the pH increase as high as pH 9~10, which means the fibre is alkalizing and turns the colour to become even more yellowish as illustrated in Figure 3(a). This may stem from the removal of the impurities that exist on the surface of the fibre. Furthermore, the surface of the fibre becomes smoother and a bit sticky when it soaked into NaOH solutions compared to untreated and acid bagasse fibre.

Basically, raw bagasse fibre has impurities that accumulated on the surface of the fibre during harvest, transportation and storage after the extraction of the juice. They may contain dust, soil and other debris. These impurities however can be removed from the surface by fibre treatment. How the way it removes can be seen clearly based on the colour

 https://doi.org/10.1051/e3sconf/20183402051

change. In this case, the finding provides evidence that there is a significant change of colour that leads to removal of impurities on the surface of the fibre become more yellowish. This is supported by [9] in his study on the effect of alkali treatment on properties of kenaf fibre. The author found that the changing of the colour of fibre treated with alkali to yellowish may be due to the removal of waxy layer and impurities from the fibre surface. This finding is similar to bagasse fibre as it turns to yellowish. Similarly, the author found that after kenaf fibre treated with alkali treatment, the surface become rather smoother as compared to that of untreated kenaf fibre.

In addition, according to [9], it was found those higher pHs are more effective to modify the fibre structure to obtain better mechanical properties. The efficiency in removing lignin, wax and oils that cover the external surfaces of the cell wall is higher when treated with alkali treatment. Based on the physical observation, the surface of the fibre becomes smoother, clean and a bit sticky when it soaked into NaOH solutions compared to untreated and acid bagasse fibre. Under the circumstances, the author reported that there is a possibility of interfacial interaction between the cement matrix and the reinforcing fibre. Next, the increase of effective surface area causes the possibility of contact between the fibre and cement matrix to occur. Plus, it also indicates that the possibility of loading transfer between the fibre and cement matrix. Hence, the surface texture would increase the number of possible interactions thereby improve mechanical fibre matrix adhesion. Meanwhile, for the case of acid treatment, no color change is identified for HCL treated fibre. This is because acid solution is not substantive to cellulosic fibres.

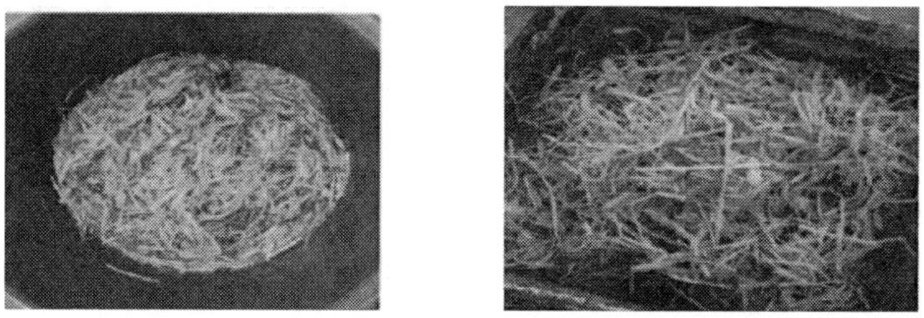

(a) (b)

Fig 3. Bagasse fibre after treated in solution after 1 hour (a) 5% NaOH treated, (b) 5% HCL treated.

4.1.2 Thickness dimension

The effect of chemical treatments of bagasse fibre on the thickness dimension was examined using treated fibre composites. The discussion of the results begins with alkali treatment after being heated in oven for 6 hours in 100 °C. As seen in Fig. 4, the greatest decrease in diameter of the bagasse fibre was observed for 5% NaOH treated fibre. This happened because of the removal of hemicelluloses and lignin that cause the surface of the bagasse fibre become smooth. Thus it leads to a decrease in diameter and fibrillation of the fibre bundles into small fibre. Even though it decreases the thickness dimension but it promotes the crystallinity of the fibre [10]. However for acid treatment not much different in terms of thickness dimension are observed. The dimension still the same like the untreated but acid fibre have shining effect and it does not fibrils into small fibre just like alkali fibre.

4.2 Surface Morphology of fibre

4.2.1 Alkali Treatment

To confirm the effect of alkali treatment using 5% NaOH solution on morphological characteristics of bagasse fibres, the samples were studied using SEM. Fig. 4 shows that differences are apparent in surface morphology between untreated and alkali treated fibres. Fig. 4 (a) show the SEM images of untreated bagasse fibre. The results indicate that the superficial layer of bagasse fibre is evident and the filaments in the untreated fibre were still packed together. In other words, the micrograph of untreated fibre looks very rigid and compact. These findings reveal that, on the surface of the fibre, the cellulose microfibrils were still intact with a layer of fatty tissue and sugarcane lignin [10].

The alkali treatment has been investigated by many researchers as one of the most effective pre treatment methods which predominantly eliminate lignin and hemicelluloses on fibres surface. A significant change in the morphological structure of the bagasse fibres occurred after treated with 5% NaOH for 1 hour is illustrated in Fig. 4 (b). The fibrils were starting to separate from each other and breaking down the untreated fibres into smaller parts This happened due to the dissolution of hemicelluloses that plays a major role in removing bagasse surface from impurities, wax and hemicelluloses structure. As a result, the diameter of fibres diameter decreases with a highly fibrillated surface or an increase of the surface area.

The most striking result to emerge from this finding is that the effect alkaline solution manages to removed lignin or other cementing components. As a consequence, surface area becomes increase and helps to improve fibre-matrix adhesion in composite [10-11]. In another study [12-14] research, emphasize on the effect of NaOH pretreatment of lignocellulosic materials. The author observed that this treatment cause swelling and thus leading to an increase in internal surface area and a decrease in the degree of polymerization and crystallinity. As highlighted by the author, NaOH treatment is more effective to enhance the enzymatic digestibility compared to acid treatment due to the fact that alkali has a stronger ability in de-lignification. On the other hand, if prolong the duration of the treatment, the rough and degraded surface on fibres skeleton reflected the partial removal of crystalline cellulose material.

4.2.2 Acid Treatment

Researchers have studied the effects of acid treatments on vegetable fibres. It was found that it varies with the nature and concentration of the acid and temperature of the experiment [15]. In this kind of treatment, the hemicelluloses are easily hydrolyzed by acids and later form of their correspondent monomer elements. Fig. 4(c) shows the effects of an acid attack on bagasse fibres. From the SEM observation, it is apparent that the skeleton of the fibres and its surrounding layer are partly destroyed. Therefore, the fibre become weaker.

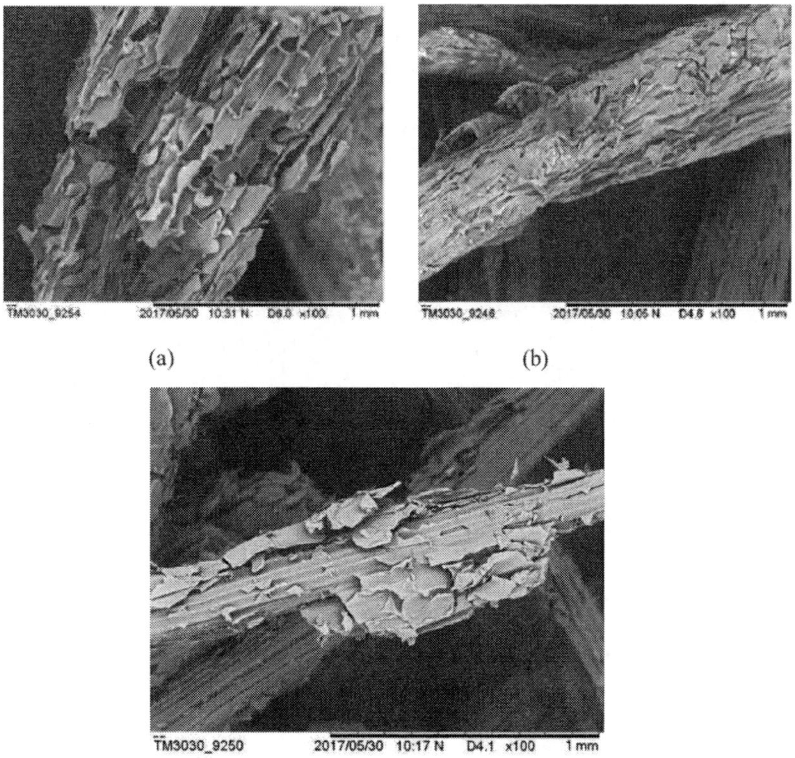

(a) (b) (c)

Fig. 4. SEM micrographs of bagasse fibre (a) Untreated fibre, (b) Alkali treated bagasse fibre, (c) Acid treated bagasse fibre.

5 Conclusions

Based on this study natural fibre can potentially provides satisfactory solution in terms of economical and environmental concerns regarding waste management. By modifying the fibre surface by using chemical treatment can enhance bond strength between fibre and matrix. Alkali treatment is more effective in removing hemicelluloses, lignin and wax from fibre surface by making it rougher as compared to acid attack. This was confirmed by SEM micrograph analysis. The treatment promotes better fibre matrix interface and improved mechanical properties of cement paste.

The authors wish to acknowledge Ministry of Higher Education and University Malaysia Sarawak for supporting this work under FRGS/TK04(02)/1138/2014(05)

References

1. Agarwal, B. D., Broutman, L. J., & Chandrashekhara, K., *Analysis and Performance of Fiber Composites* (Wiley, 2006)
2. Natali, A., Manzi, S., & Bignozzi, M. C., Procedia Engineering **21**, 1124–113 (2011)
3. Bartl, A., Mihalyi, B., & Marini, I., Chemical and Biochemical Engineering **18**(1), 21–28 (2004).

4. Loh, Y. R., Sujan, D., Rahman, M. E., & Das, C. A., Resources, Conservation & Recycling **75**, 14–22 (2013)
5. Cadenas, G. A., Mitrani, R. B., Pena, C. G., Munilla, M. H., and Correa, J. L., In: Silverio, H. N. (ed.) *Manual de los Derivados de la Cana de Azucar, 2nd Ed. (in Spanish). GEPLACEA, Mexico City, Mexico*, 47 (1990)
6. Acharya Samir Kumar, Mishra Punyapriya, & Mehar, S. K., BioResources **6**(3), 3155–3165 (2011)
7. Navarro, R. F., Medeiros, J., Mariz, T. F., and Maia, D. F. *Determination of cellulose and lignin content from luffa cylindrical fibers (in Portuguese), Proceedings of the 1st Brazilian Polymer Conference, Sao Paulo, Brazil* **2**, 679-681 (1991)
8. Godfried, L. M. Unsaturated Polyester Resin. *Its Production and Reinforcement with Natural Organic Fibers*, (Fokker-VFW B.V. Report, The Netherlands, 1975)
9. Lam, T. F., & Yatim, J. M., *IJRET: International Journal of Research in Engineering and Technology*, 37-40 (2015)
10. A., N. R. Juliana, *Alkali treatment on sugarcane bagasse to improve properties of green composites of sugarcane bagasse. Ceramic Engineering and Science Proceedings* **35**(2), (2014)
11. A. E Milena, C. G. Antonio, J. E. B. Freitas, P. B. A. Fechine, A. S. B. Sombra, E. Corradini, G. Mele, A. Maffezzoli, S. E. Mazzetto, Bioresources **5**(4), 2478-2501 (2010)
12. Y. Sun, J. Cheng, Bio-resource Technol **83**, 1-11 (2002)
13. Y. Sudiayani, S. Alawiyah, K. Sembiring, Menara Perkebuan **78**(2), 70-74 (2010)
14. N.M. Sutan, S. Hamdan, H.R. Sobuz, V.Laja, M.S. Islam, Concrete Research Letters **2**(4), 326-335 (2012)
15. Arsène, M.-A., Bilba, K., Savastano Jr, H., Ghavami, K, Materials Research **16**(4), 903–923. (2013)

 https://doi.org/10.1051/e3sconf/20183402052

Preparation and Characterization of Activated Alumina

A R. Rabia[1*], *A.H.* Ibrahim[1], and *N.N.* Zulkepli [2]

[1]School of Environmental Engineering, Universiti Malaysia Perlis, Kompleks Pusat Pengajian Jejawi 3, 02600 Arau, Perlis, Malaysia
[2]Faculty of Engineering Technology, Kampus Uniciti Alam Sg. Chuchuh, 02100 Padang Besar, Universiti Malaysia Perlis (UniMAP)

Abstract. Activated alumina is a high surface area and highly porous form of aluminum oxide that can be employed for contaminant species adsorb from ether gases or liquids without changing its form. The research in getting this material has generated huge interested. Thus, this paper presented preparation of activated alumina from chemical process. Pure aluminum (99.9% pure) reacted at room temperature with an aqueous NaOH in a reactor to produce a solution of sodium aluminate ($NaAlO_2$). This solution was passed through filter paper and the clear filtrate was neutralized with H_2SO_4, to pH 6, 7 or 8, resulting in the precipitation of a white gel, Al(OH)3·XH2O. The washed gel for sulfate ions were dried at 80 ◦C for 6 h, a 60 mesh sieve was to separate and sort them into different sizes. The samples were then calcined (burn) for 3h in a muffle furnace, in air, at a heating rate of 2 ◦C min^{-1}. The prepared activated alumina was further characterized for better understanding of its physical properties in order to predict its chemical mechanism.

1 Introduction

The water treatment technology via aluminum has gained a great importance in recent time competing with conventional technologies in various industrial fields in particular for chemical and petrochemical industries that material have shown capability to treat water long time in history [1]. Currently, Aluminum play a special role in the field of energy and electric power as one of the elemental parts of conducting material due to availability and cheapness that material has gain a wider application, it has be use through the with help addition component such membrane, use in separation process technology which appeared [2-5]. Since the beginning of 1960's, aluminum was used as coagulant in local water purification. At the beginning of 1970's, the first industrial installation of membrane using aluminum was initiated and using this platform, the water purification related activities rapidly grown and developed large-scale commercial [6]. With this initiative, it began to insert significant improvements in membrane selectivity / permeability and process design [3].

*Corresponding author: elmahdy1999@yahoo.com

The main goal of improving the membranes permeability is to achieve constant quality, high performance membrane and systems for desired applications with enhanced troubleshooting capabilities [7]. Moreover, based on this membrane development, the removal of contaminate from in water is taking great interest in the last decades from the environmental and energy perspective, respectively [8]. From the environmental view point, heavy metals considered one of the major element which causes a pollution in the water and the environment as well as being a main reason of the health issue because its increased concentrations of the this element due to the wide use of the industrial process as a main source of heavy metal [9]. While from the health perspective, metal heavy is one of the most pollutants in the feeding of natural environment and therefore must be removed because it subsequently lead to reduction of the bring health problem [10].

The alumina based technology in addition to what has been mentioned, its can equally play a major role in making water purification processes economically feasible. Today, treatment with aluminum are a promising, effective and reliable solution when compared to the traditional process in terms of effective cost, preserving environmental and capability to remove contaminants such as carbon dioxide and hydrogen sulfide across wide range of operating conditions [11]. The process of forming good treatment depends mainly on the selection of appropriate materials meaning making aluminum active and is one of the key factors to produce a good treatment and the success of the performance process. In general, material prepared which consists of polymer, solvent and non-solvent using are the major techniques to produce asymmetric membrane structural [12].

For the membrane fabrication technique, the mechanism to invert phase is introduced by called phase inversion and subsequent immersion precipitate process is a proposed and well-known approach for making polymeric based membranes device [13]. However, this is not well promising like activated alumina. By the activated aluminum with the immersion of a designed substrate in pre-coagulation bath and solvent in the solution is transferred with a non-solvent in the precipitating media and phase thus, phase inversion occurs [14]. Good number of research communities has tried to attempt to give the explanation on the membrane-forming and development mechanism during and after the phase inversion process [14-18].

For instance, in membrane fabrication, addition of non-solvent additive into activated alumina formulation might be useful due to its influence to physical properties of membrane such as pore formation, pore distribution by suppression of macro voids or micro voids evolution [19-23].

2 Materials and Methods

The samples of aluminum powder was obtained in Meridian World Sdn. Bhd, which is one of the environmental service providing company at Sungai Petani, Kedah Darul Aman, Malaysia. Pure aluminum (99.9% pure), reacted, at room temperature, with a aqueous NaOH in a reactor to produce a solution of sodium aluminate ($NaAlO_2$). This solution was passed through filter paper and the clear filtrate was neutralized with H_2SO_4, to pH 6, 7 or 8, resulting in the precipitation of a white gel, $Al(OH)_3{\cdot}XH_2O$. The gel was washed until no more sulfate ions were detected in the washings. Finally, they were dried at 80 ◦C for 6 h, then ground and passed through a 60 mesh sieve .The samples were then calcined (burn) for 3h in a muffle furnace, in air, at a heating rate of 2 ◦C min^{-1}. Finally activated alumina was obtained.

Fig.1. Show the prepared alumina (a) Bubbles (b) Bubbles settled (c) Slake alumina (d) final process for the alumina preparation.

The prepared alumina via the pure aluminum (99.9% pure) reaction was observed during proves Fig.1 shown the bubbles raising during the reaction, from the Fig. 1b , the bubbles settle down at room temperature, with a aqueous NaOH in a reactor to produce a solution of sodium aluminate ($NaAlO_2$). This solution was passed through filter paper and the clear filtrate was neutralized with H_2SO_4, to pH 6, 7 or 8, resulting in the precipitation of a white gel, $Al(OH)_3 \cdot XH_2O$. The gel was washed until no more sulfate ions were detected in the washings, which were observed in Fig. 3 and 4. Finally, they were dried at 80 ◦C for 6 h, then ground and passed through a 60 mesh sieve. The samples were then steadily calcined (burn) to remove other unwanted for 3h in a muffle furnace, in air, at a heating rate of 2 ◦C min^{-1} as shown in Fig. 1c and 1d.

3 Results and discussions

The Fig. 2 shown the spectrum of alumina, however, EDX spectrum shows that the existence of oxygen (oxide) and chloride. This is happening due to the diffusion between the oxide layers, which is located underneath the alumina. When the diffusion takes place, the oxygen may exist. An investigated was equally conducted on the present of prepared alumina structures based on EDX. Fig. 2 shows the of highest peak of the alumina structures. The alumina exhibits a homogeneous size over the whole substrate, uniform distribution of the might be expected on structures since show only 1 peak as can be observed.

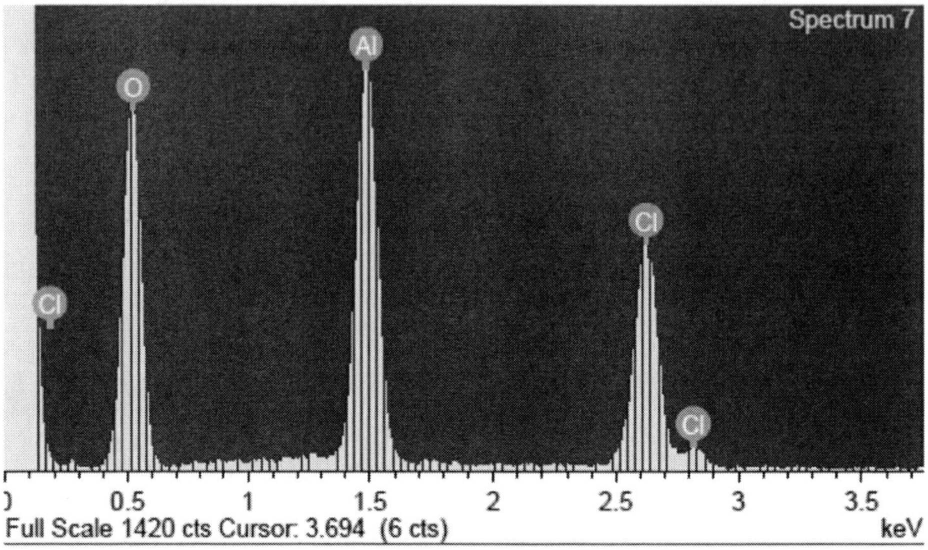

Fig. 2. Cross section of alumina highlighted with X35000.

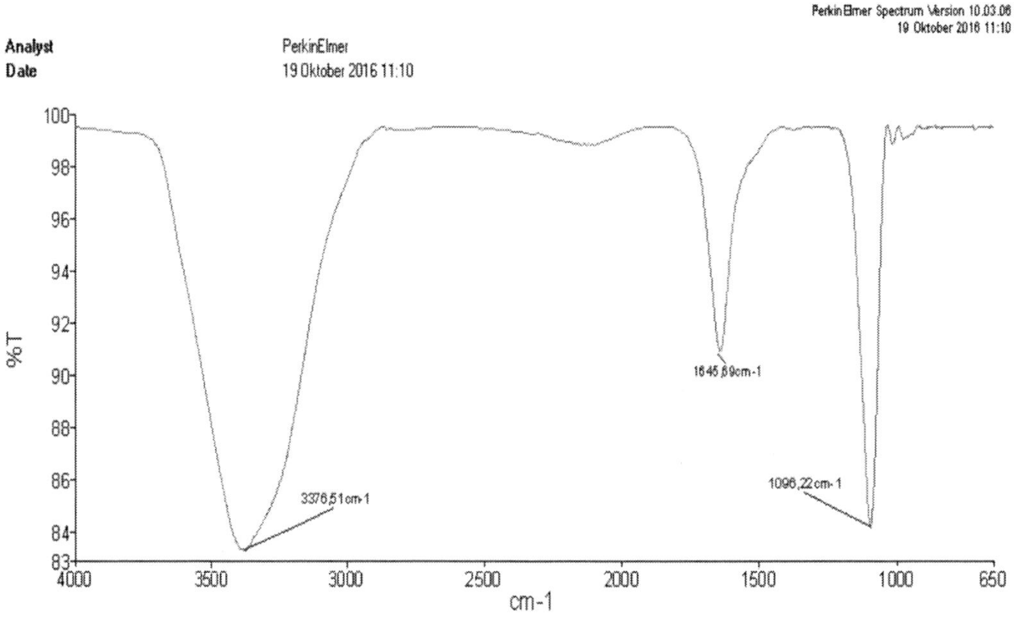

Fig. 3. FTIR spectra of prepared alumina before washing sulfate ion.

The Fourier transport infra-red spectra (FTIR) shown from Fig. 3, the typical peak in the FTIR spectra were obtained after observed in the region comprised of bands assigned to the pure region to alumina. The peaks are shown 3376.5cm^{-1}, 1645cm^{-1}, 1096.22cm^{-1} stretching of alumina-alumina bond. The ion stretching region is particularly useful for elucidating alumina-bonding patterns. At wave numbers between 1000 and 3500 cm-1 the intensities decreased due to the washing. This peak has been assigned to the asymmetrical stretching

 https://doi.org/10.1051/e3sconf/20183402052

of alumina. As the washing of sulfate ions was done, the Fourier transport infra-red spectra (FTIR) in shown in Fig. 4, the typical peak in the FTIR spectra were obtained after observed in the region comprised of bands assigned to the pure region to alumina. The peaks are shown 3376.5cm^{-1}, 1645cm^{-1}, 1096.22cm^{-1}, 1422 cm^{-1} and 1030 cm^{-1} this indicate that alumina getting pure and stretching of alumina-alumina bond exist. The ion stretching region is particularly useful for elucidating alumina-bonding patterns. The typical peak for pure alumina is expected between 1000 and 3500 cm-1 the intensities decreased due to the washing. This peak has been assigned to the asymmetrical stretching of alumina.

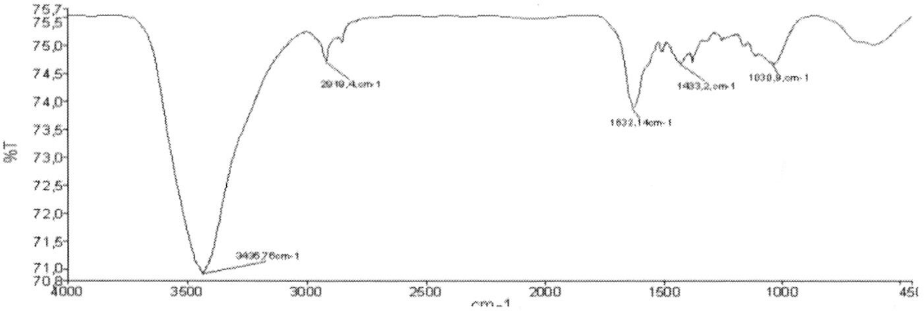

Fig.4. FTIR spectra of prepared alumina after washing sulfate ion.

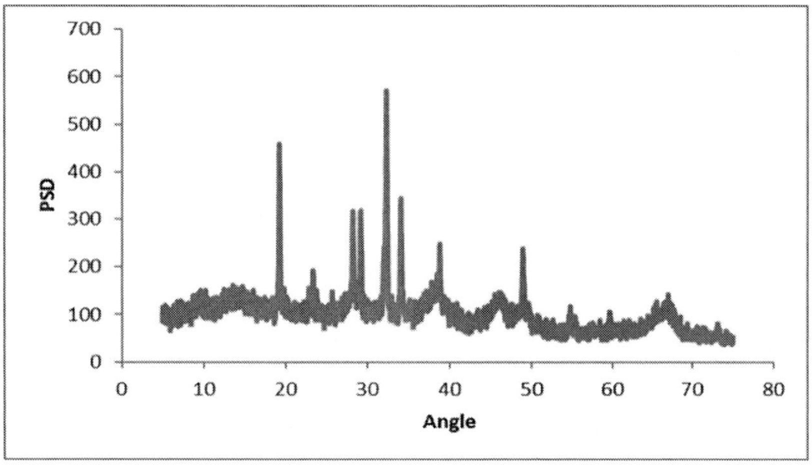

Fig.5. XRD patterns of alumina particles.

Fig. 5 shows the XRD patterns of the ZnO after washed different water flow rate. Generally can be observed that main peaks observed in the sample which locate at 2θ = 18.76, 23.51 and 33.885 which can be attributed to the active alumina bond at xyz, (100) and (200) crystal planes of the alumina active phase of alumina. It is important to note that the X-ray diffraction patterns of alumina particle over the entire purification range show no additional peaks exist for corresponding to secondary impurities such as sulfate ions and also other closely related cation-ordered orthorhombic structures of other elements such radicals. The intensity of the 35 peak becomes more intense and sharp, with the increase wash rate, the diffraction peaks monotonously shifts towards small angles as can be observed indicating that thc lattice constant increases which in other attributed for being getting pure.

4 Conclusions

The study demonstrated the preparation and characterization of activated aluminum via pure aluminum (99.9% pure), which subsequently reacted at room temperature, with a aqueous NaOH in a reactor to produce a solution of sodium aluminate ($NaAlO_2$). This solution was passed through filter paper and the clear filtrate was neutralized with H_2SO_4, to pH 6, 7 or 8, resulting in the precipitation of a white gel, $Al(OH)_3 \cdot XH_2O$. The washed gel for sulfate ions were dried at 80 °C for 6 h, a 60 mesh sieve was to separate and sort them into different sizes. The samples were then calcined (burn) for 3h in a muffle furnace, in air, at a heating rate of 2 °C min^{-1}. The characterization of the activated alumina revealed that the activated alumina is obtained. The approach is simple yet could produce pure activated alumina.

References

1. S. Ahmed, M.G. Rasul, W.N. Martens, R. Brown, M.A. Hashib. Desalination **261**, 3–18 (2010)
2. D. Musmarra, D. Karatza, A. Lancia, M. Prisciandaro, G.M. di Celso, Chem. Eng. Trans. **32,** 547–552 (2013)
3. T. Anirudhan, S. Sreekumari, J. Environ. Sci. **23**, 1989–1998 (2011)
4. Y. Guo, E. Du, Energy Procedia. **17**, 444–449 (2012)
5. F. Salvador, N. Martin-Sanchez, R. Sanchez-Hernandez, M.J. Sanchez-Montero, C. Izquierdo, Microporous Mesoporous Mater. **202**, 259–276 (2015)
6. F. Salvador, N. Martin-Sanchez, R. Sanchez-Hernandez, M.J. Sanchez-Montero, Microporous Mesoporous Mater. **202**, 277–296 (2015)
7. S. Ahmed, M.G. Rasul, R. Brownb, M.A. Hashib. Journal of Environmental Management **92**, 311-330 (2011)
8. U. Beker, B. Ganbold, H. Dertli, D. Duranog, I. Gulbayir. Energy Conversion and Management **51**, 235–240 (2010)
9. L. Cadoret, C. Rossignol, J. Dexpert-Ghys, et al. Materials Science and Engineering B-Advanced Functional Solid- State Materials **170** (1-3), 41-50 (2010)
10. S.A. Darko, E. Maxwell, S. Park, Thin Solid Films **519**(1), 174-177 (2010)
11. S.M. Lam, J.C. Sin, A.R. Mohamed, Korean J. Chem. Eng. **27**(4), 1109-1116 (2010)
12. C. Saka, Journal of Analytical and Applied Pyrolysis **95**, 21–24 (2012)
13. I.K. Shah, P. Pre, B. Alappat, Journal of the Taiwan Institute of Chemical Engineers, **45**, 1733–1738 (2014)
14. Xin-hui, Process Intensification **53**, 53– 62 (2012)
15. M. Barakat, Arabian Journal of Chemistry **4**, 361–377 (2011)
16. R. Berenguer, J.P. Marco-Lozar, C. Quijada, D. Cazorla-Amorós, E. Morallón, Carbon **48**, 2734-2745 (2010)
17. A. Eliassi, M. Ranjbar, Int. J. Nanosci. Nanotechnol. **10**, 1, 13-26 (2014).
18. B. Ksapabutr, E. Gulari, S. Wongkasemjit, Physicochem. Eng. Aspects **233** 145–153 (2004)
19. M.A. Karakassides, D. Gournis, A.B. Bourlinos, P.N. Trikalitis, T.B. Magneti, J. Mater. Chem. **13**, 871–876 (2016)
20. P.R. Reddy, K.M. Ajith, N.K. Udayashankar, Adv. Mater. Lett. **7**(5), 398-401(2016)
21. G. Karthikeyan', B.V. Apparao, S. Meenakshi, *Defluoridation properties of activated alumina* (2nd international workshop on fluorosis prevention and defluoridation of water, 79-82, 2014)

Influence of Courtyard Ventilation on Thermal Performance of Office Building in Hot-Humid Climate: A Case Study

Esra'a Sh. Abbaas[1]*, Ala'eddin A. Saif[2], MAC Munaaim[1], and Md. Azree Othuman Mydin[2]

[1]School of Environmental Engineering, Universiti Malaysia Perlis, 02600 Arau, Perlis, Malaysia
[2]School of Microelectronic Engineering, Universiti Malaysia Perlis, 02600 Arau, Perlis, Malaysia
[3]School of Housing, Building and Planning, Universiti Sains Malaysia, Pulau Pinang. Malaysia

Abstract. The influence of courtyard on the thermal performance of Development Department office building in University Malaysia Perlis (UniMAP, Pauh Putra campus) is investigated through simulation study for the effect of ventilation on indoor air temperature and relative humidity of the building. The study is carried out using EnergyPlus simulator interface within OpenStudio and SketchUp plug in software to measure both of air temperature and relative humidity hourly on 21 April 2017 as a design day. The results show that the ventilation through the windows facing the courtyard has sufficient effect on reducing the air temperature compared to the ventilation through external windows since natural ventilation is highly effective on driving the indoor warm air out to courtyard. In addition, the relative humidity is reduced due to ventilation since the courtyard has high ability to remove or dilute indoor airborne pollutants coming from indoor sources. This indicates that the presence of courtyard is highly influential on thermal performance of the building.

1 Introduction

The increment in human activities and industrial revolution leads to sharp changes in the ecosystem and accordingly in climate due to the incredibly increase in earth temperature, which in turn causes global warming phenomena. Building sector consumes about 40% of total energy in the world, moreover, the office type of buildings required 50% of total energy consumption of non-residential buildings since many persons are existed in limited zone in the same time [1]. The continued increases in global warming and energy consumption required to apply sustainable concepts in order to reduce their effect in daily life. Applying sustainable concepts in buildings considered one of main approaches to maintain energy consumption [2]. Many solutions have been proposed to improve the thermal comfort of buildings in hot humid areas. Among of them, introducing courtyard to buildings enhances the thermal performance for the building's interior due to its impact on the passive cooling [3].

There are many reports reveal that using internal courtyard helps the air to flow within the interior spaces of the building and it also helps getting rid of the hot air inside the building [4].

* Corresponding author:esraabbas91@gmail.com

The courtyard building architecture form enhances the thermal performance of the building by natural ventilation due to shade creating [5]. However, constructing appropriate openings in both sides of the adjacent spaces for courtyard can enhance the natural ventilation, which in turn improves the thermal performance [6-7].

[8] studied the effect of natural ventilation in office building in hot-humid climate and reported that the thermal condition of the spaces increases, and the occupants feel comfort when the wind speed higher than 0.1 m/s, so to get better results, one has to use the natural ventilation when the outdoor temperature is lower than indoor temperature. Courtyard had been used as a part of buildings in different countries located in hot-humid areas such as China and India [9], however, it still not common in Malaysia. In this work, a simulation study for the effect of natural ventilation due to internal courtyard in an existence office building in Malaysia on its thermal performance.

2 Case Study Descriptions

In this work, Development Department building in University Malaysia Perlis (UniMAP) that is located in Pauh Putra campus has been taken as a case study. It has central rectangular courtyard with a dimension of 14.9×26.1 m^2 and all offices and meeting rooms surrounding it. Fig. 1 shows the two dimensional AutoCAD ground floor plan drawing for the building, which has been divided into four zones, i.e. North, South, East and West zone with respected to the central courtyard with an area of 333.18 m^2, 373.93 m^2, 338.86 m^2 and 415.48 m^2, respectively. Regarding to the weather conditions, Malaysia is considered a high humidity region and the humidity slightly varies along the country, the average value of humidity is within the range of 70% - 90%. On the other hand, the average temperature is within the range of 29 °C to 34 °C [5].

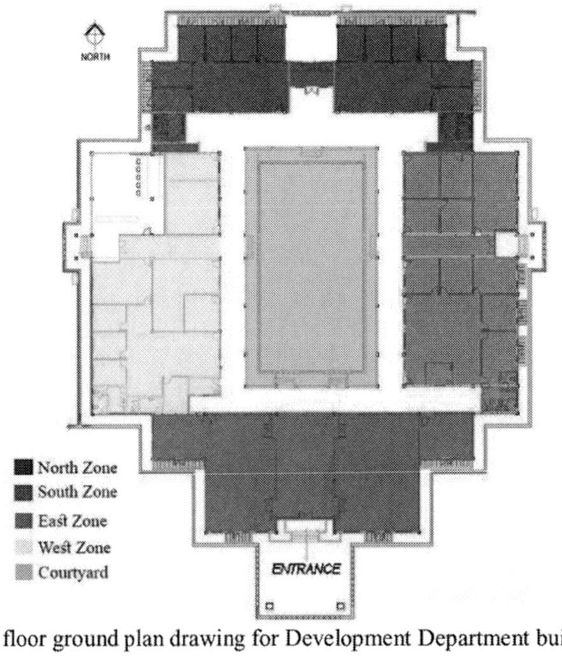

Fig. 1. 2D AutoCAD floor ground plan drawing for Development Department building in UniMAP with the zones.

3 Methodology

The Development Department building has rectangular courtyard with offices surrounding it which are set as four thermal zones as shown in Fig. 1. The two dimensional plan for the building is drawn using AutoCAD, then the three dimensional model for the building has been built using SketchUp with OpenStudio software. The model construction is set based on the materials for the exiting building, and then the internal heat loads for building including people, lights and electric equipment are scheduled, moreover, the ventilation system is activated. Before the model is simulated using EnergyPlus simulator, the climate file for Malaysia that was obtained from US DOE website in EPW format was imported to the software in order to set all environmental conditions, then the design day from DDY file was also imported, where 21 April 2017 is selected to be the design day in this study. After that, the physical output quantities, particularly the air temperature and relative humidity are specified and set to be measured hourly.

4 Results and discussions

Fig. 2 shows the hourly changing in indoor air temperature for the adjacent zones to the courtyard on 21 April 2017 design day with and without ventilation. The results show that for all cases opening the windows after midnight until early morning, particularly, from 12:00 am to 8:00 am has unnoticeable effect on the interior air temperature since the outdoor air temperature is very close to the indoor air temperature. However, starting from 9:00 am until 11:00 pm, the ventilation through windows is more effective. In all cases the nature ventilation through windows reduces the air temperature for the interior zones, indicating that the thermal performance of the building is enhanced. From Fig. 2, it can be observed that the ventilation through the external windows slightly reduces the air temperature for the interior spaces of the building compared to the case where all windows are closed. For instance, the maximum air temperature difference between the case of ventilation through external windows and the case of no ventilation for the east zone is recorded at 10:00 am and it is 0.49 °C.

A lower interior air temperature has been obtained due to ventilation through the windows that are facing the courtyard compared to the case of ventilation through external windows. For example, the maximum air temperature difference between the case of ventilation through the windows on courtyard and the case of no ventilation for the east zone is also recorded at 10:00 am and it is 1.56 °C, which is triple the air temperature difference obtained in case of external windows ventilation. The better performance of ventilation through windows on courtyard can be explained as follows: courtyard acts as an air tube that let out the internal air into sky due to the deference of air density between indoor and outdoor and this mechanism is called the stack driven ventilation [10].

The lowest indoor air temperature is obtained for the case when all windows of the building are opened. For instance, the maximum air temperature difference between the case of ventilation through all windows and the case of no ventilation for the east zone is also recorded at 10:00 am which is 1.71 °C. The large air temperature difference for ventilation through all windows is due to the existence of opposite sides window, which creates path for the air through the building, hence, better natural ventilation circulation that minimize indoor air temperature and enhancing thermal performance of the building. This ventilation mechanism is called cross driven ventilation [10].

It is interesting to mention here that ventilation through the windows facing the courtyard has very close results to the cross ventilation case. This implies that the presence of courtyard is highly effective on thermal performance of the building resultant due to nature ventilation. In addition, it is important to clarify here that due to non-presence of windows on the courtyard for south zone in the building, no stack and cross ventilation values for the indoor air temperature are recorded in this zone.

[11] studied the performance of courtyard of a single story building in hot-humid area and they found that creating openings for the building on the courtyard effects on airflow, hence, improves of the thermal behavior of the interior zone, which is in line with our results. Another study by [10] reported the effect of natural ventilation strategies on thermal behavior of existing school building with a courtyard in Italy and they concluded that all ventilation strategies enhance the thermal comfort for the indoor zones of building. Additionally, the cross ventilation gives better results compared to stacks driven ventilation strategy which is similar to our findings for hot-humid climate.

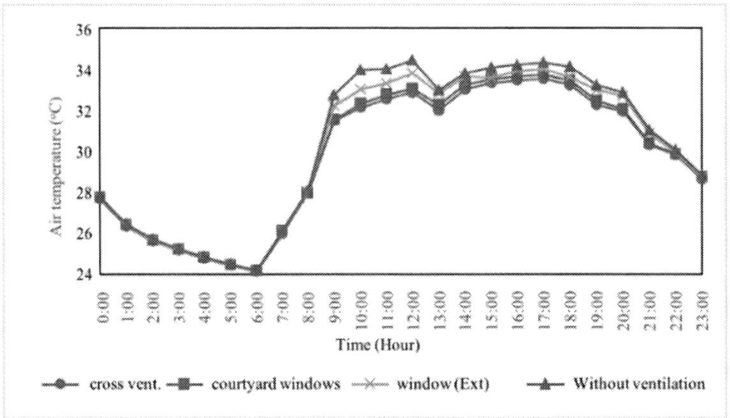

(a)

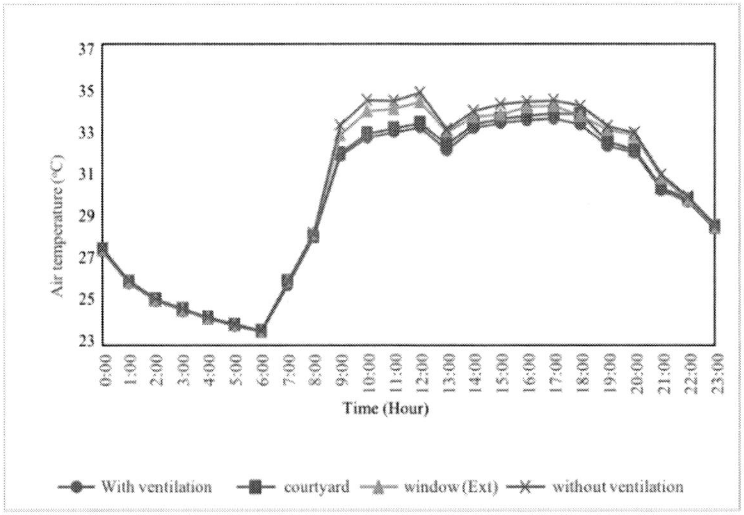

(b)

https://doi.org/10.1051/e3sconf/20183402053

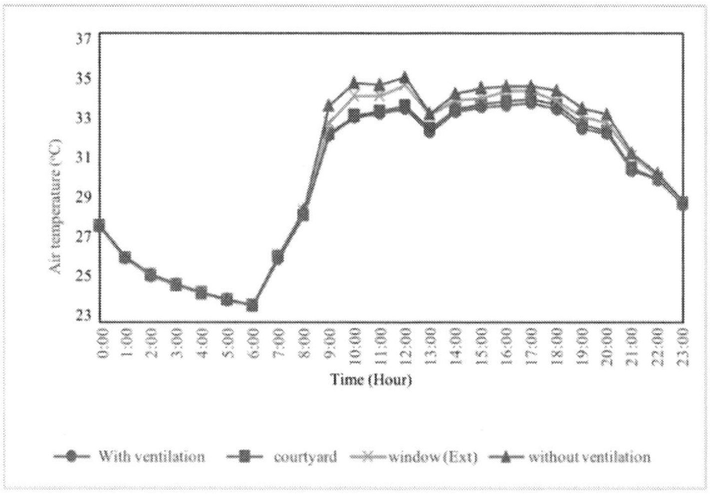

(c)

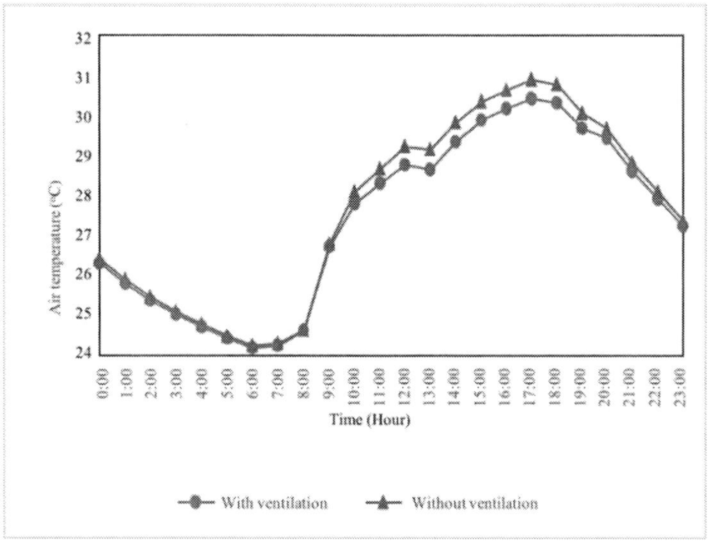

(d)

Fig. 2. Hourly indoor air temperature variation for Development Department building (Jabatan Pembangunan) at University Malaysia Perlis (UniMAP, Pauh Putra campus) on 21 April 2017 for: (a) North façade, (b) East façade, (c) West façade and (d) South façade.

Fig. 3 illustrates the variation of indoor relative humidity for the zones surrounded to the courtyard on 21 April 2017 for the cases with and without ventilation. The results show that the relative humidity for the building is quite high and reaches to 100% at most of the measurement time for the case without ventilation. This is attributed to presence of the electrical equipment loads and light loads in addition the occupants who are using the building as set earlier. The existence of all these type of loads, especially human factor, increases the amount of moisture for the indoor air that in turn increases the relative humidity.

It is also observed that the relative humidity for the case without ventilation of the north zone during the period between 2:00 am and 7:00 am has lower values compared to the other zones within this duration of time. This is due the higher air temperature measured for this zone compared to the other zones at these times as shown in Figure 2, which it is inversely proportional to the relative humidity [12].

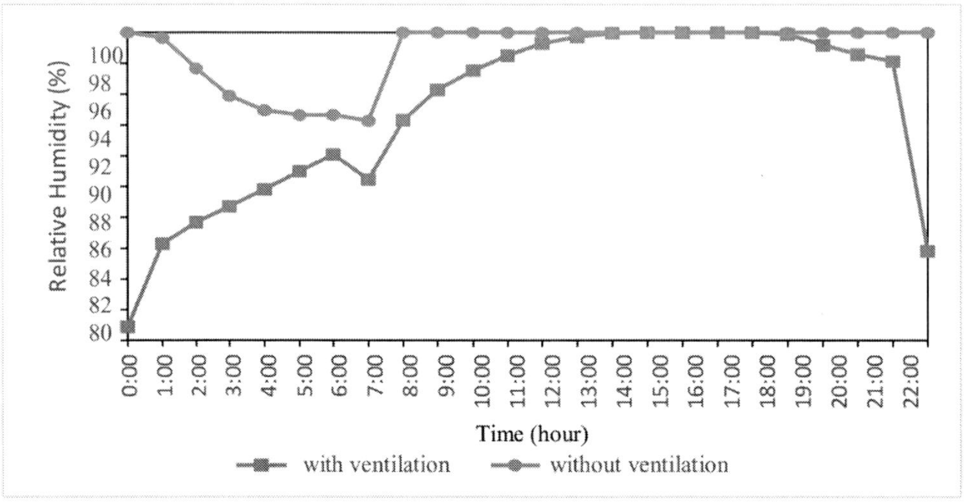

(a)

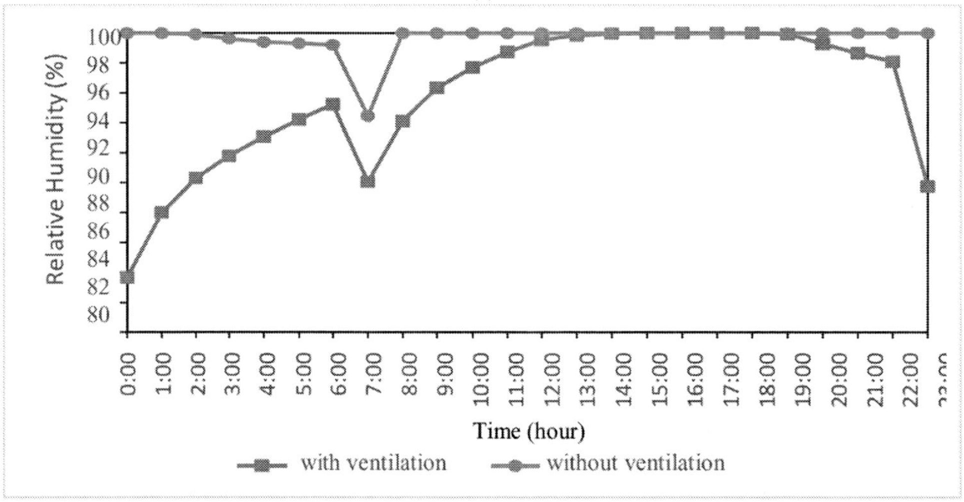

(b)

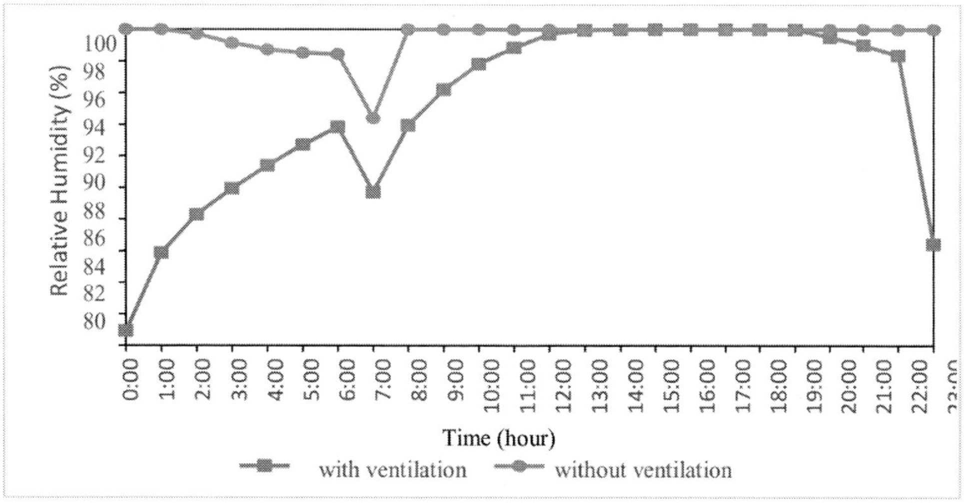

(c)

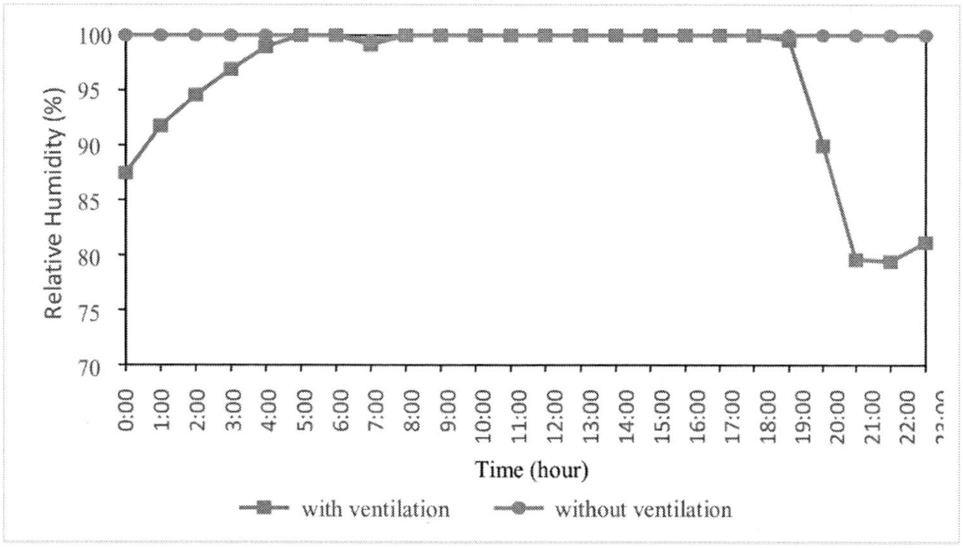

(d)

Fig. 3. Hourly indoor relative humidity variation for Development Department building (Jabatan Pembangunan) at University Malaysia Perlis (UniMAP, Pauh Putra campus) on 21 April 2017for: (a) North façade, (b) East façade, (c) West façade and (d) South façade.

In all zones, the nature ventilation through windows reduces the relative humidity for the interior zones, indicating that the thermal comfort of the building is enhanced. For instance, the relative humidity for the north zone drop from 100% to 81% at 12:00 am due to the natural ventilation then it increases gradually with the time and reaches to 100% again at 3:00 pm. The drop in humidity values at night. Hence, it flows into the adjacent interior areas and drives the warm air out of these areas, which in turn intends to remove or dilute indoor airborne pollutants concentration that are coming from indoor sources. This reduces the level of contaminants and improves indoor air quality [13]. Same drop in relative humidity values is also noticed after 7:00 pm for all zones.

It is important to mention here that in this experiment stack driven ventilation is considered for north, east and west zones since its high efficiency compared to the other ventilation types as discussed in Fig. 2, while ventilation through external windows for south zone is considered since it is the only ventilation possible for this zone. Therefore, higher drop difference in relative humidity is observed for north, east and west zones compared to the one obtained for the south zone as illustrated in Fig. 3(d).

5 Conclusions

This paper introduces a simulation study for the influence of natural ventilation due to the central courtyard at Development Department building in University Malaysia Perlis (UniMAP, Pauh Putra campus) using EnergyPlus software. The interior air temperature and relative humidity for the building have been measured hourly for the design day of 21 April 2017. It has been found that the presence of courtyard is highly effective on reducing both of indoor air temperature and humidity due to the natural ventilation through the windows that are facing it. Where ventilation through windows allows the warm indoor air to be released at the adjacent zones at the courtyard, which in other way reduces the air temperature. Additionally, it helps to remove or dilute indoor air pollutants that are originate from indoor sources.

References

1. L. Perez-Lombard, J. Ortiz, C.H. Pout, Energy. Build. **40**, 3 (2008).
2. S. Cho, N. Mohammadazadeh, IBPSA (2013).
3. D. H. Ch. Toe, Doctoral Dissertation, (2013).
4. N. Moosavi, N. Mahyuddin, N. Abd Ghafar, M. A. Ismail, Ren. Sus. Energy. Rev. 34 (2014).
5. M. A. Zakaria, and L.H.Ismail, IConCEES (2011).
6. N. Sadafi, E. Salleh, L. Ch. Haw, Z. Jaafer, J. App. Sci. **8**, 15 (2008).
7. T. Tabesh, B. Sertyesilisik , Amer. Trans. Eng. App. Sci. **3**, 2 (2014)
8. R. Daghigh, K. Sopian, J. Mostagh, Eur. J. Sci. Res. **26**, 2 (2009).
9. C. Chan, Y. Xiong, Arch. Conf. Proc. Pres. 62 (2005).
10. L. Lion, A. Belleri, R. Lollini, D. Zardi, L. Giovannini, 33rd AIVC- 2nd TightVent. Con. (2012).
11. Rajapaksha, H. Nagai, M. Okumiya , Ren. Energy. 28 (2003).
12. H. Djamila, Ch. Chu, S. Kumaresan, J. Build. Const. Plan. Res. 2 (2014).
13. K. Ponsoni, M. S. G. Raddi, Braz. Arch. Biol. Tech. **53**, 1 (2010).

https://doi.org/10.1051/e3sconf/20183402054

Acclimatization Study for Biohydrogen Production from Palm Oil Mill Effluent (POME) in Continuous-flow System

N. Idris[1], *N.A.* Lutpi[1,*], *Y. S.* Wong [1], and *T.N.* Tengku Izhar[1]

[1]School of Environmental Engineering, Universiti Malaysia Perlis, 02600, Arau, Perlis, Malaysia

Abstract. This research aims to study the acclimatization phase for biohydrogen production from palm oil mill effluent (POME) by adapting the microorganism to the new environment in continuous-flow system of thermophilic bioreactor. The thermophilic fermentation was continuously loaded with 0.4 L/day of raw POME for 35 days to acclimatize the microorganism until a steady state of biohydrogen production was obtained. The significance effect of acclimatization phase on parameter such as pH, microbial growth, chemical oxygen demand (COD), and alkalinity were also studied besides the production of biogas. This study had found that the thermophilic bioreactor reach its steady state with 1960 mL/d of biogas produced, which consist of 894 ppm of hydrogen composition.

1 Introduction

Dependency on fossil fuel to fulfil energy demands has become a big issue for developing countries. Due to that reason, the development of alternative energy source has become matter of urgency. Hydrogen is a one of the promising energy supply to replace the dependency on fossil fuels. According to Ni et al., (2006) [1] energy crisis and environmental sustainability are two important issues that need to take into consideration for a global sustainable development. Hydrogen is well known as clean energy carrier since it is renewable energy source with carbon free [2]. No residue that would adverse the environment or human life when hydrogen energy is burned [3]. Hydrogen also contributes substantially to the devaluation of greenhouse gas emissions as well as its might be considered as clean and climate neutral energy system [4]. Besides, hydrogen need a helter-skelter vitality yield about 122 kJ/g, which will be 2.75 times greater over hydrocarbon fills [5]. Various processes are being used to produce hydrogen gas, including photo fermentation, dark fermentation, photolysis and electrolysis of water. Among all, the most feasible processes are to be dark fermentation [6].

Dark fermentation is the fermentative conversion of organic substrate to hydrogen during adaptation of organic substances into volatile fatty acids and mainly involving acidogenic bacteria [7]. Microorganism in fermentation process hydrolyze organic substance into monomers and reduce molecular weight of organic acids by acidogenic

* Corresponding author: nabilah@unimap.edu.my

bacteria [8]. Application of wastewater as a potential substrate for biohydrogen production has been major interest in recent years especially in the dark fermentation process. As for this study, palm oil mill effluent has been chosen as the wastewater source to produce biohydrogen.

Malaysia is the largest producer and exporter of crude palm oil (CPO). The production of crude palm oil reached 29 million tonnes in the year 2015 from 28 million tonnes in the year 2014 [9]. Palm oil is produced using huge amount of water in mills where oil is extracted from the palm fruits. During the eradication of crude palm oil from the fruits, 50% of the water results in palm oil mill effluent (POME) [10]. This huge quantity of POME that contains high concentration of chemical oxygen demand (COD), biochemical oxygen demand (BOD), total nitrogen, total phosphorus, oil and grease, and total solids were produced and accompanied with the production of crude palm oil [11]. Therefore, this study will focus on the characteristic study of POME itself and the acclimatization phase of dark fermentation for biohydrogen production in thermophilic bioreactor.

2 Materials and method

2.1 Wastewater source and characterization

This research project was conducted by using palm oil mill effluent (POME) that was obtained from Malpom Industries Sdn Bhd, Sungai Bakap, Pulau Pinang. The POME wastewater was collected from the cooling pond whereby it takes place before it was discharge for further treatment. The characteristics of raw POME and anaerobic sludge were also being studied based on various parameters such as; pH, temperature, total solid (TS), total volatile solid (TVS), mixed liquor suspended solid (MLSS), mixed liquor volatile suspended solid (MLVSS), chemical oxygen demand (COD), biochemical oxygen demand (BOD), alkalinity, total carbohydrate (TC) and total volatile fatty acid (TVFA) [12].

2.2 Acclimatization study

In this study, substrate that was used for fermentation in bioreactor start-up was taken from anaerobic pond treatment plant. About 14 liters of the anaerobic digested wastewater was used for the start-up and acclimatization. This acclimatization phase was continued for about 35 days to allow microorganisms present in the mixed liquor perfectly acclimatized to the new environment and reached a steady state condition before proceed to the next phase of research studies. The pH of the wastewater was adjusted to the range of pH 6 ± 0.5 to promote the production of biohydrogen. The schematic diagram of the thermophilic bioreactor is shown in Fig. 1.

2.3 Analysis procedure

The treated effluents were analyzed with the same parameter as in characteristic study, while the biogas composition that were collected from tedlar bag will be analyzed after every 24 hours of fermentation. The production of hydrogen, carbon dioxide, oxygen and methane will be measured using Gas Analyzer (biogas 5000 standard) Model GA 5000.

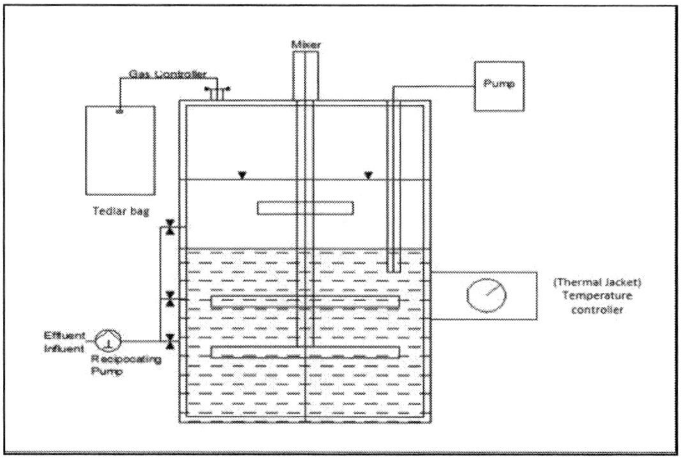

Fig. 1. Schematic diagram of experimental set-up of thermophilic bioreactor.

3. Results and discussions

3.1 Wastewater characteristic

Table 1 indicates that the BOD and COD of both raw POME and anaerobic sludge are extremely high. When POME is released into water resources, it can pollute the water supply. According to Cheng et al., (2015) [13] COD concentration is within the range of 45,000 mg/L to 65,000 mg/L while BOD_5 is about 18,000 mg/L to 48,000 mg/L. Furthermore, high value of suspended solids will lead to nasty odor and may cause epidemic. POME has high value of turbidity due to many organic compounds that not dissolved before the treatment.

Table 1. Wastewater characteristic of raw POME and anaerobic sludge.

	Raw POME	Anaerobic sludge
pH	4.42	4.91
Temperature	56 °C	32 °C
Total solid (TS)	33 930 mg/L	29 015 mg/L
Total volatile solid (TVS)	27 260 mg/L	24 615 mg/L
Mixed liquor suspended solid (MLSS)	56 800 mg/L	50 650 mg/L
Mixed liquor volatile suspended solid (MLVSS)	54 750 mg/L	48 600 mg/L
Chemical oxygen demand (COD)	86 333 mg/L	82 066 mg/L
Biochemical oxygen demand (BOD)	34 300 mg/L	34 000 mg/L
Alkalinity	-	2569 mg/L
Total carbohydrate (TC)	12.74 g/L	6.40 g/L
Volatile acidity (TVFA)	11542.86 mg CH_3COOH/L	7257.14 mg CH_3COOH/L

3.2 Analysis procedure

3.2.1 pH variation during acclimatization phase

The pH variations of this fermentation during acclimatization phase are presented in Fig. 2. The initial pH is 5.71 with 0.4 L/day of raw POME loaded every day for 35 days. According to Gonzalez et al., (2001) [14] in acclimatization phase, loading rate should be low to avoid drawback on the system. As shown Fig. 2, pH decrease from 5.71 to 5.59 between day 1 to day 7. This is mostly due to increase of fatty acid concentration that clearly explained acidogenesis phase is occurred. Then, from day 8 to day 16, pH reading fluctuated within range of 5.61 to 5.63. But from day 17 until day 35, pH variation reached its stability of the anaerobic digestion process within range of 5.67 to 5.65.

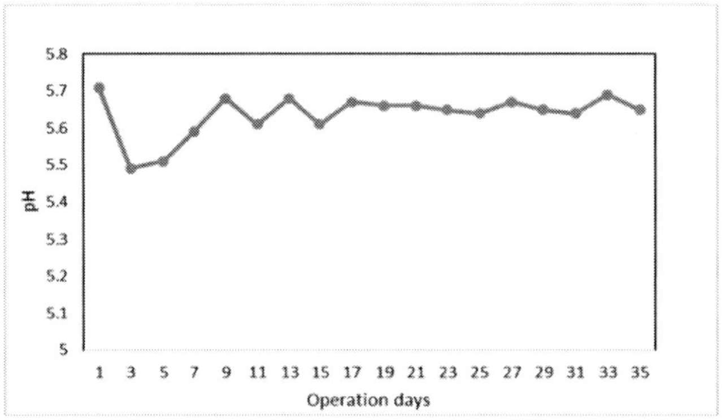

Fig. 2. pH variation during acclimatization phase.

3.2.2 Microbial growth variations during acclimatization phase

During acclimatization phase, microbial growth is referred as mixed liquor volatile suspended solid (MLVSS). Mixed liquor suspended solid (MLSS) was also measured as fraction ratio between MLVSS and MLSS. Commonly, the MLVSS used as indicator of the amount of organic matter contained in the sludge. Therefore, the desolation of MLVSS in sludge stabilization process can be used to determine the effectiveness of sludge stabilization process to the organic component in the sludge [15]. The microbial growth variations during acclimatization phase are shown in Fig. 3. At the beginning of acclimatization phase, the MLVSS started to increase gradually from 2550 mg/L to 8750 mg/L from day 1 until day 6. Then, the MLVSS fluctuated between 8500 mg/L and 12800 mg/L for the next 12 days. At this phase, the colony of acidogenic bacteria was more than methanogenic bacteria to convert raw waste into organic acids, carbon dioxide and hydrogen gas [16]. Finally, biomass regained its stability from day 19 to day 35 with MLVSS reading is within 13350 mg/L and 13900 mg/L.

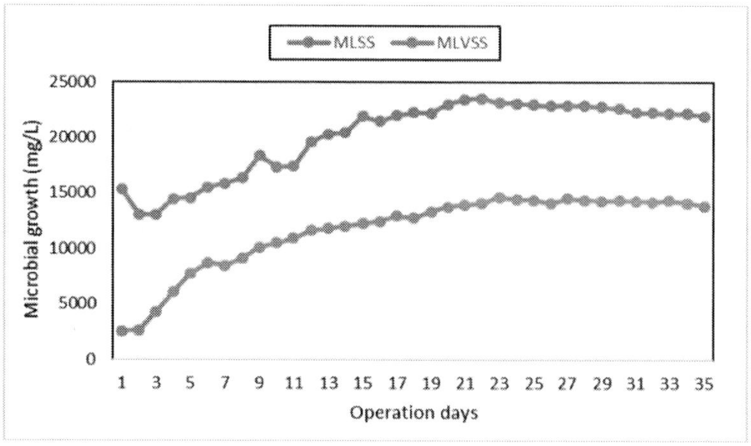

Fig. 3. Microbial growth during acclimatization phase.

3.2.3 COD variations during acclimatization phase

Fig. 4 shows the performance of bioreactor for Total COD (TCOD) and Soluble COD (SCOD) during acclimatization phase. TCOD represent the total COD in the bioreactor while SCOD represent the soluble (suspended of effluent) in the bioreactor. COD variations is important to evaluate microorganism activities toward environmental changes during acclimatization phase. Removal efficiency of TCOD and SCOD were calculated as the difference between initial and final concentrations inside the bioreactor. Removal efficiency for TCOD is 14.8% while for SCOD is 22.6%. This is proved that this bioreactor not only focusing on producing biogas especially hydrogen, but also decrease the COD value itself. Based on Figure 4, TCOD reading fluctuated from day 1 until day 15 which is from 70900 mg/L to 78600 mg/L. After that, from day 16 until day 35, TCOD reading reached its stability as microbial growth in the bioreactor are adapting to the new environment. The pattern of SCOD reading almost same with TCOD as its reach their stability on day 16. COD reading for both soluble and total reach stability as the bacteria already acclimatized themselves to the new environment through increasing the number of active microorganism [17].

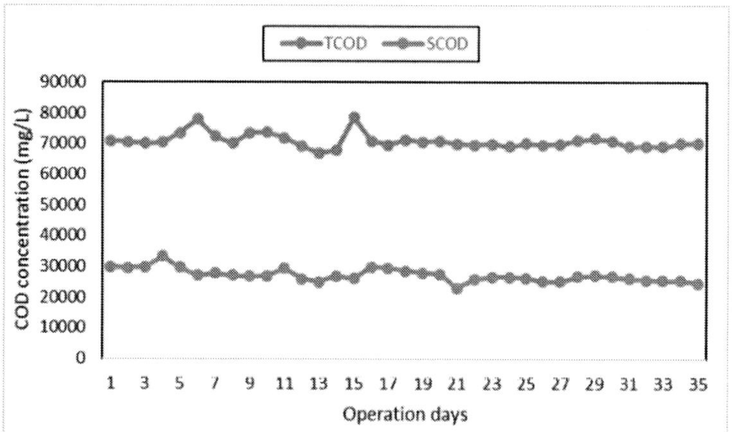

Fig. 4. TCOD and SCOD concentration of effluent during acclimatization phase.

3.2.4 Alkalinity variations during acclimatization phase

Fig. 5 describes the alkalinity concentration of mixed effluent during acclimatization phase. Initial concentration of alkalinity dropped from day 1 to day 3 around 5790 mg/L to 2170 mg/L. After that, from day 4 to day 11, it started increasing slightly from 3140 mg/L to 5470 mg/L for 8 days. This occurred due to the increasing of buffering capacity of the bioreactor impact from volatile fatty acid (VFA) that started to accumulate due to slow conversion of VFA that caused of the population of biomass increased [18]. In addition, the alkalinity of wastewater will increase as increasing in biomass concentration in wastewater [19]. For the next 24 days, the alkalinity remains constant around 5980 mg/L to 6010 mg/L due to it reached the steady state condition since the biomass growth remains constant.

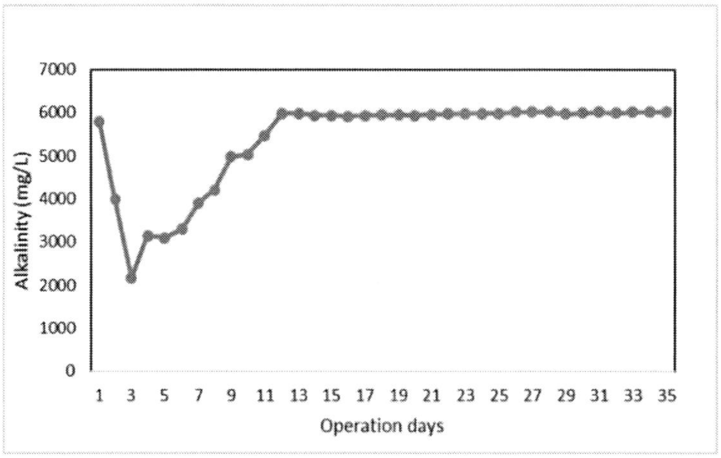

Fig. 5. Alkalinity variations during acclimatization phase.

3.2.5 Biogas variations during acclimatization phase

Biogas performance was also monitored every day at this acclimatization phase. Based on Fig. 6, the initial volume of biogas in bioreactor is 900 mL for day 1. However, the biogas started to increase gradually from 900 mL to 1910 mL in day 11. Then, it sustained its stability from day 12 to day 35 which from 1900 mL to 1950 mL. Once this acclimatization phase had achieved its stability, the biogas production is around 1960 mL/day. Hydrogen gas composition by using Gas Analyzer (biogas 5000 standard) Model GA 5000 was around 827 to 894 ppm once it reaches its steady state stating from day 11. Hydrogen gas reading from gas analyzer is shown in Fig. 7 from day 1 until day 35 of acclimatization process.

 https://doi.org/10.1051/e3sconf/20183402054

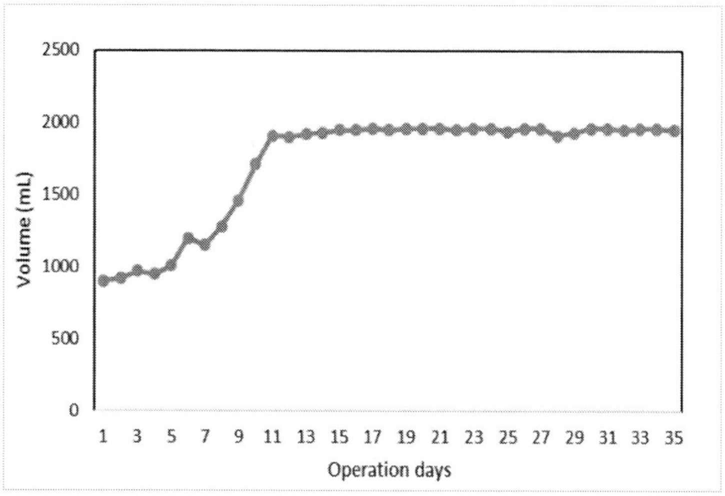

Fig. 6. Biogas variations during acclimatization phase.

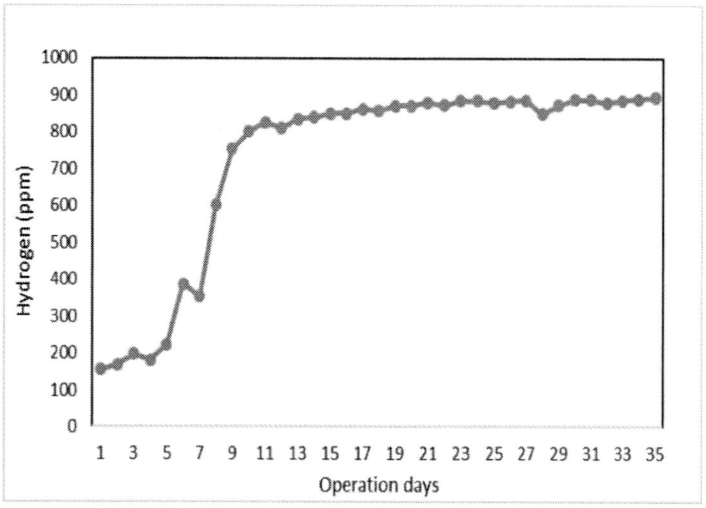

Fig. 7. Production of hydrogen gas reading based on daily intake.

4. Conclusions

In this research, the acclimatization phase of biohydrogen production through dark fermentation has been studied and achieved its stability in continuous-flow system. From the liquid analysis and biogas variations, it showed that these treated effluents are adapting to the new environment in the bioreactor. There is no report from previous study that clarify how long is required to acclimatize the microorganism to adapt to the new environment. However, in this study, 35 days is needed to reach its stability, with a maximum of 1960 mL/d of biogas is produced and 894 ppm of hydrogen composition.

This research is fully supported by FRGS grant (FRGS No: 9003-00511). The authors fully acknowledged Ministry of Higher Education (MOHE) and Universiti Malaysia Perlis (UniMAP) for the approved fund, which makes this important research viable and effective.

References

1. M. Ni, D. Y. C. Leung, M. K. H. Leung, K. Sumathy, Fuel Process. Technol. **87,** 461-72 (2006)
2. S. Dunn, Int. J. Hydrogen Energy. **27,** 235–264 (2002)
3. S. Apak, E. Atay, G. Tuncer, Int. J. Hydrogen Energy. **42,** 2446-2452 (2017)
4. S. M. Kotay, D. Das, Int. J. Hydrogen Energy. **33,** 258–63 (2008)
5. I. K. Kapdan, F. Kargi, Enzyme Microb. Technol. **38,** 569-582 (2006)
6. K. Nath, D. Das, Int. J. Hydrogen Energy. **34,** 7497-7501 (2009)
7. C. Chen, M. Yang, K. Yeh, C. Liu, J. Chang, Int. J. Hydrogen Energy. **33,** 4755-4762 (2008)
8. C. Cote, D. I. Masse, S. Quessy, Bioresour Technol. **97,** 686-691 (2006)
9. Malaysia Palm Oil Board, http://www.mpob.gov.my/ accessed on 17th November 2017.
10. Okwute, L. Ojonoma, Isu, R. Nnennaya, African Journal of Agricultural Research. **2,** 656-662 (2007)
11. S. Dong, Y. Pi, Q. Li, L. Hu, Y. Li, X. Han, J. Wang, J. Sun, J. Alloys Compd. **663,** 1-9 (2016)
12. APHA. Standard methods for the Examination of Water and Wastewater. Washington DC (1992)
13. C. K. Cheng, M. R. Derahman, M. R. Khan, J. Env. Chem. Eng. **3,** 261-270 (2015)
14. A. C. Gonzalez, M. E. Poy, C. Duran-de-Bazua, Anaerobe Biotech. J. **7,** 143-149 (2001)
15. C. Arnaiz, J. C. Gutierrez, J. Lebrato, Bioresour Technol. **97,** 1179-1184 (2006)
16. M. Chong, V. Sabaratnam, Y. Shirai, M. A. Hassan, Int. J. Hydrogen Energy. **34,** 3277-3287 (2006)
17. G. Gonzalez, H. Urrutia, M. Roeckel, E. Aspe, Chem. Technol. Biotechno. J. **80,** 151-157 (2015)
18. S. Yacob, Y. Shirai, M. A. Hassan, M. Wakisaka, S. Subash, Process Biochem. **41,** 962-964 (2006)
19. Metcalf & Eddy, F. L. Burton, H. D. Stensel, G. Tchobanoglous, G. (2003)
20. Wastewater engineering: treatment and reuse. McGraw Hill

Endosulfan Toxicity to *Anabas testudineus* and Histopathological Changes on Vital Organs

I.L. Nordin[1], *N.* Ibrahim[1,*], *S.A.* Ahmad[2], *N.I* Hamidin[1], *F.A.* Dahalan[1], and *M.Y.* Abd. Shukor[2]

[1]School of Environmental Engineering, Kompleks Pusat Pengajian Jejawi 3, Universiti Malaysia Perlis, 02600 Arau, Perlis, Malaysia.
[2]Department of Biochemistry, Faculty of Biotechnology and Biomolecular Sciences, Universiti Putra Malaysia, UPM 43400 Serdang, Selangor, Malaysia.

Abstract. The toxicity of endosulfan, an organochlorine type insecticide to a commonly consumed freshwater fish species, *A. testudineus* (40.68±9.03 g; 13.49±0.99 cm), was investigated under static conditions. The nominal endosulfan concentrations ranging from 10 to 80 µg/L subjected to the fish population results in 96-hour median lethal concentration, LC_{50}, of 35.2±3.99 µg/L. The toxicity is a function of both endosulfan concentration and exposure time ($p>0.05$). Histopathological analysis on vital organs exposed to sublethal concentrations indicates that structural changes started at sublethal dose and the effects aggravated with increasing endosulfan concentration. Gill was found to experience aneurism, hyperplasia in lamellar and autolysis of mast cell. Pyknotic nuclei and necrosis were observed in liver cell, while the lumen of renal tubule was found to narrow and haemorrhage was observed in cytoplasm cell. High LC_{50} compared to other fishes indicates that *A. testudineus* has high tolerant to endosulfan, however, endosulfan slowly alters the fish biochemistry and is potentially transferable to human

1 Introduction

Anabas testudineus, commonly known as climbing perch is a freshwater fish species widely populating irrigation canal and rice field of Southeast Asia countries. It is known as hardy fish as it can survive in mud and shallow water and for hours on land. The fish ability to endure harsh condition could lead to an increase in bioaccumulation of toxicants it is exposed to. On the other hand, endosulfan is widely known as toxic insecticide acting as a contact poison for variety of insects and mites. Various negative effects have been described as a result of fish exposure to endosulfan. Chronic effects on fish include oxidative damage [1], genotoxicity [2], damage to testes [3], changes in circulating thyroid hormones [4] and alteration of acetylcholinesterase activity in the brain [5]. Depressed levels of testosterone and estradiol have also been found in fish with elevated residues of endosulfan [6]. Other effects are reduced feeding behaviour [7], alterations in development

* Corresponding author: naimah@unimap.edu.my

[8], sexual and escape behaviour, and reproductive physiology [9-10]. Studies on endosulfan toxicity also showed that it is acutely toxic to most fishes with LC_{50} ranging from 0.1 to 41 µg/L [5, 11–22]. Although currently banned in many countries due to its adverse effects on human health and the environment, endosulfan is still continuously employed in several developing countries in Asia for high commercial value crops. In Malaysia, its residue was recently detected in a stream and tap water at a concentration of 0.112 µg/L endosulfan sulphate and 0.072 µg/L endosulfan II [23]. Bioaccumulation of endosulfan in the commonly consumed *A. testudineus* may lead to biomagnification of the toxicant from the non-targeted organism to consumer. A risk assessment on infants in Punjab, India [24] reported mean values of 90.69 ng g^{-1} β-endosulfan and 14.00 ng g^{-1} endosulfan sulphate excretion in breastmilk. Due to the high toxicity of endosulfan, this study aims to assess acute toxicity of endosulfan to juvenile *A. testudineus* when exposed to endosulfan-containing water samples. High median lethal concentration (LC_{50}) would be evidence for high survival. The toxic effects of endosulfan to the fish cellular was also investigated by histopathological analysis on several vital organs i.e. gill, liver and kidney.

2 Experimental

2.1 Acute toxicity and determination of LC_{50}

Juvenile *A. testudineus* of 40.68 g (±9.03 g) weight and 13.49 cm (±0.99 cm) length were obtained from local fish supplier in Perlis, Malaysia. Acclimatisation was done for 2 weeks in 200 gallons of dechlorinated tap water. The fishes were fed twice daily with commercial pellets during acclimatisation, however, food was not given for 48 h before and during the test to empty their guts. Endosulfan powder (GmbH, 99.5%, 64-67% α-endosulfan; 29-32% β-endosulfan) was mixed with 40 L water to produce desired stock solutions of 0.4 g/L. Acetone (Merck, 99.5%) was used at 0.005% (v/v) to aid in pesticide dissolution.

Five glass aquarium tanks were each filled with 40 L chlorine-free tap water and kept for 24 h prior to endosulfan addition. Water quality parameters in the aquarium tanks were measured daily. Average value of DO is 70±2%, temperature is 25±2°C, total alkalinity is 80±3 mg/L $CaCO_3$, pH 6.9±0.2 and total hardness is 23±3 mg/L $CaCO_3$. Air compressor was used to maintain DO level in water. Static acute toxicity experiment was carried out based on Standard Methods [25]. Samples containing different nominal endosulfan concentrations (10, 20, 40, 60 and 80 µg/L) and two control samples (tap water) were used in the test. One of the control samples was added with acetone (0.005% v/v). Ten fishes were transferred in random into three replicate sample tanks to obtain a total of 30 fish population. Fish mortality was observed and dead fishes were immediately removed from the tank. None of the control fishes died throughout acclimatisation period and toxicity test. The LC_{50} was estimated with Probit Analysis, using Biostat Profesional Version 5.

2.2 Histopathology

Gill, liver and kidney were dissected and fixated in 10% neutral buffered formalin for 24 h at room temperature. Tissue processing started with dehydrating gill, liver and kidney samples through graded alcohol series, cleared in alcohol and embedded in paraffin. Paraffin sections of 5-10 µm thickness were trimmed using Jung Multicut RM2045 microtome and stained with haematoxylene-eosin. The tissues were then examined under light inverted microscope and photographed using Nikon photomicroscope.

3 Results and discussions

3.1 Acute toxicity and determination of LC_{50}

Fig. 1(a) and 1(b) show dose-response curves of fish mortality due to exposure to endosulfan, generated by Probit Analysis based on Finney Method (lognormal distribution) and Least Squares Method (normal distribution) respectively. The 96-h LC_{50} values are 32.39±4.32 µg/L (Finney) and 38.03±5.67µg/L (Least Squares) with no significance error ($p>0.05$). The average 96-h LC_{50} is 35.21±4.99 µg/L based on the two methods. This value is very high compared to those reported in studies on other fish species as shown in Table 1, except for European eel [22] at 41 µg/L, indicating its high survival.

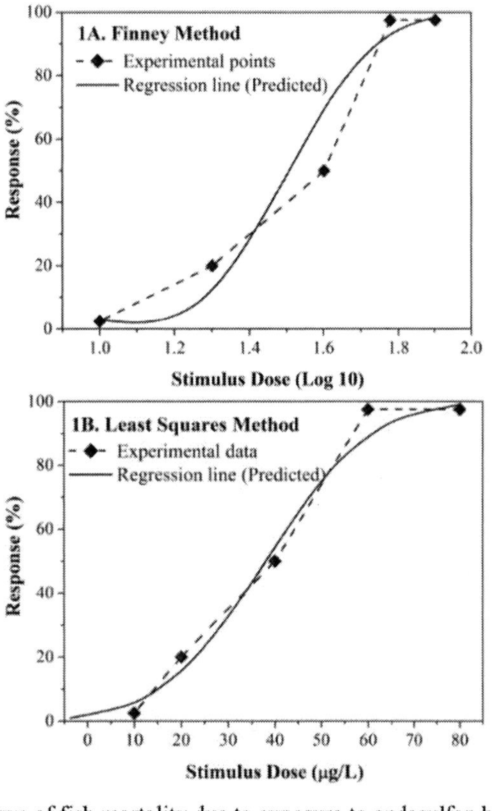

Fig. 1. Dose-response curve of fish mortality due to exposure to endosulfan by Probit Analysis based on (a) Finney Method (lognormal distribution) and (b) Least Squares Method (normal distribution).

Table 1. LC_{50} values for various fish species exposed to endosulfan.

Fish species	Weight/ Length	Loading	Endosulfan purity	Water quality	Test type	LC_{50} (96 h)	Ref.
Climbing perch (Puyu), *Anabas testudineus*	40.68±9.03 g/ 13.49±0.99 cm	10.17 g/L	99.5%	T: 25±2°C; pH: 6.9±0.2; DO: 70±2%; A: 80+3 mg/L $CaCO_3$; H: 23±3 mg/L $CaCO_3$	Static	35.2	This study
Florida flagfish, *Jordanella*	Larval fish (2-3 days)	10 fish/ 200 mL	99%	T: 25°C	Semi static	4.35	[11]

jordanella							
Spotted murrel, *Channa punctatus*	35.6±0.7 g	7.12 g/L	99.0%	T: 26-28°C; pH: 7.2-7.5; DO: 6.5-7.1 mg/L; A: 48-58 mg/L; H: 48-60 mg/L	Static	24.3	[12]
Teleostei, Perciformes, *Cichlasoma dimerus*	21.2±1.5 g/ 7.5±0.2 cm	3.18 g/L	94.9%	T: 25±1°C; pH: 7.3	Semi static	2.6	[13]
Tilapia, *Oreochromis mossambicus*	46.78 g	4.68 g/L	99%	T: 26.4-28.8°C	Static	3.609	[14]
Tilapia, *Oreochromis niloticus*	10.26±2.84 g/ 8.36±0.78 cm	10.26 g/L	99%	T: 24±1°C; pH: 7.5	Semi static	10.20	[15]
Nile tilapia, *Oreochromis niloticus*	150 g	37.5 g/L	35%	T: 28±2°C	Static	12.775	[16]
Tilapia, *Oreochromis mossambicus*	Fingerling	-	35%	-	-	1.42	[17]
European carp, *Cyprinus carpio*	5months/ 6-13 cm	5 fishes/ 20 L	-	T: 25±1°C; pH: 6.0-7.3	Static	2	[18]
Asian swamp eel, *Monopterus albus, Zuiew*	150-180 g/ 30-40 cm	49.5 g/L	33%	pH: 7.2±1°C; A: 80±1 mg/L $CaCO_3$	Static	0.42	[19]
Rainbow trout, *Onchorhynchus mykiss*	10.61 ± 1.69 g/ 10.0±0.4 cm	4.24 g/L	32.9%	T: 12.1±0.6°C; pH: 7.84±0.8; A: 16.5±0.8 mg/L $CaCO_3$	Semi static	1.75	[20]
Silver perch, *Bidyanus bidyanus*	50 mm	-	-	-	Semi Static	7.75	[21]
European carp, *Cyprinus carpio*	50 mm	-	96.2%	T: 25°C	Semi static	0.1	
Mosquito fish, *Gambusia affinis*	Age: 21 d/ 40 mm	-	96.2%	T: 25°C	Static	2.3	
Golden Perch, *Macquaria ambigua*	Age: 21 d/ 40 mm	-	96.2%	T: 25°C	Semi static	0.5	
Eastern rainbowfish, *Melanotaenia duboulayi*	Age: 48 d/ 40 mm	-	96.2%	T: 25°C	Static	11.4	
Bony bream, *Nematolosa erebi*	Age: 22 d/ 50 mm	-	96.2%	T: 26°C	Semi Static	0.2	
Rainbow trout, *Oncorhynchus mykiss*	Age: 24 d/ 55 mm	-	96.2%	T: 4°C	Static	1.6	
Harlequin fish, *Rasbora sp.*	30 mm	-	96.2%	T: 25°C	Flow through	0.2	
European eel, *Anguilla anguilla*	20-30 g/ 16-20 cm	-	96%	T: 20±2°C; pH: 7.9±0.2; A: 4.0±0.5	Static	41	[22]

	mmol/L; H: 240±10 ppm as $CaCO_3$	

Note: T = temperature, DO = dissolved oxygen, A = alkalinity, H = hardness

However, it is important to note that LC_{50} is also dependent on various factors such as the type of acute toxicity test and the size of fish used in the toxicity tests. Sunderam et al. [21] and Capkin et al. [20] for example, obtained LC_{50} values of 1.6 and 1.75 µg/L for Rainbow trout using static and static renewal test respectively. On the other hand, Kumar et al. [14] and Kegley et al. [17] reported LC_{50} values of 3.609 and 1.42 µg/L for Tilapia of different sizes; i.e. 46.78-g fish and fingerlings respectively. Previous preliminary work on *A. testudineus* of bigger sizes [26] than the present study also produced higher LC_{50} (41.70 µg/L). Environmental factors including temperature, pH, alkalinity and turbidity were also found to correlate with endosulfan toxicity [27]. For instance, a pH of less than 5 would accelerate hydrolyzation of endosulfan to endosulfan sulphate, which is more toxic [28]. Capkin et al. [20] showed that LC_{50} of Rainbow trout did not only depend on fish size, but also the temperature and water quality including pH, alkalinity and hardness.

3.2 Histopathological Changes

Endosulfan toxicity at sublethal concentrations is also important as the fish biochemical reactions may be altered while being alive, leading to poisonous effect on the genotype and bioaccumulation. Therefore, histology analysis was done to investigate its effect on vital organs; gill, liver and kidney at sublethal phases. Fig. 2 shows gill cross sections of control fish (A) and fish exposed to 8.8 (B), 17.6 (C) and 35.2 (D) µg/L endosulfan under light microscope. Fig. 2A shows regular filament arrangement of primary lamellae with central cartilages, afferent and efferent arterioles in control fish. About 5 secondary lamellae or strata are attached to primary lamellae on basement membranes supported by pillar cells. Erythrocytes can be seen in spaces between lacunae and pillar cells.

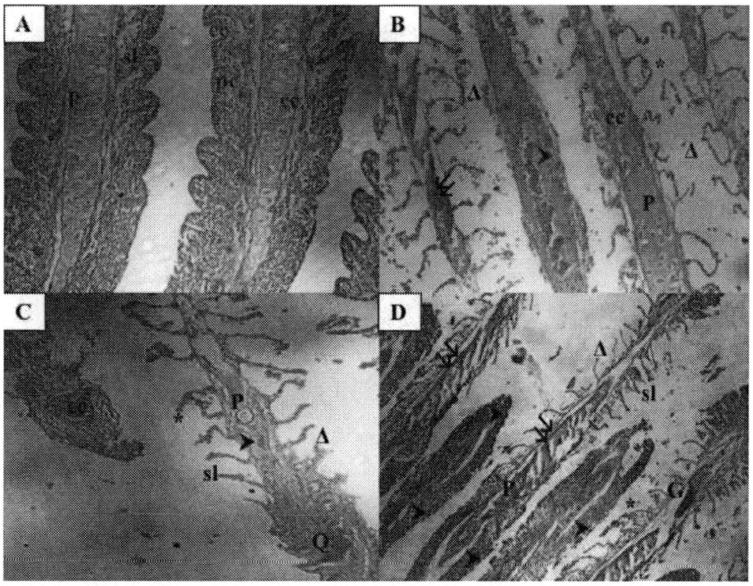

Fig. 2. Gill cross section of (A) control fish and fish exposed to (B) 8.8, (C) 17.6 and (D) 35.2 µg/L endosulfan under light microscope (H&E stain, x200 for ABC and x100 for D). Symbol: primary lamella (P); secondary lamella (sl); pillar cell (pc); erythrocytes (ec); chloride cell (cc); hyperplasia of

interlamellar epithelium (➤); epithelial lifting (Δ); fusion of secondary lamella (*); aneurism in primary lamella (↓); desquamation (Q); autolysis (G).

Fig. 2B shows that endosulfan causes hyperplasia of interlamellar epithelium, epithelial lifting, fusion of the secondary lamellae, aneurism in primary lamella and chloride cell lysis. Fig. 2C indicates that higher endosulfan concentration causes aneurism on secondary lamella and desquamation of the gill. In Fig. 2D, autolysis of mast cell and chloride cell on primary lamellar can be observed. Autolysis can lead to cellular necrosis, which may be indicative of cell injury or death. In previous studies, hyperplasia of the interlamellar epithelium was also observed on *Cichlasoma dimerus* [13], while Rainbow trout was reported to have lifting of lamellar epithelium, distal hyperplasia of lamellas and combination of lamellas [29] due to exposure to endosulfan.

Fig. 3 shows liver cross sections of control fish and fish exposed to 8.8, 17.6 and 35.2 μg/L endosulfan solutions under light microscope. Fig. 3A shows that liver tissue of control *A. testudineus* is composed of polyhedric hepatocytes with normal sinusoidal spaces. Kupffer cells can also be observed. On the other hand, Fig. 3B shows that cytoplasm of liver cell is high in vacuolated aspect and there is an increase in Kupffer cells. The structure of polygonal hepatocytes cannot be identified due to wide vacuolation, and the cell appears to become pyknotic, which may lead to fragmentation of nucleus in cytoplasm. Fig. 3C shows that the liver cell has swelled and ruptured indicating the presence of apoptosis, a biomarker of toxic exposure. The liver also experiences hyalinization of hepatocytes. In Fig. 3D, abnormal sinusoids can be observed with obstruction, while the accumulation of erythrocytes can be seen in the sinusoids probably due to inflammation in the blood vein or artery branch. Hepatocytes also become denser in the vacuolated cytoplasm of liver. Similar liver alteration such as cellular swelling, large vacuolation of cytoplasm and pyknotic nuclei of hepatocytes due to exposure to endosulfan were also reported by previous studies [30–31]. Comparative vacuolar distrophy and hypertrophy of hepatocytes were observed in Rainbow trout [29], while Onesided livebearer was found to experience hydropic degeneration, sinusoids dilation and necrosis in liver [32].

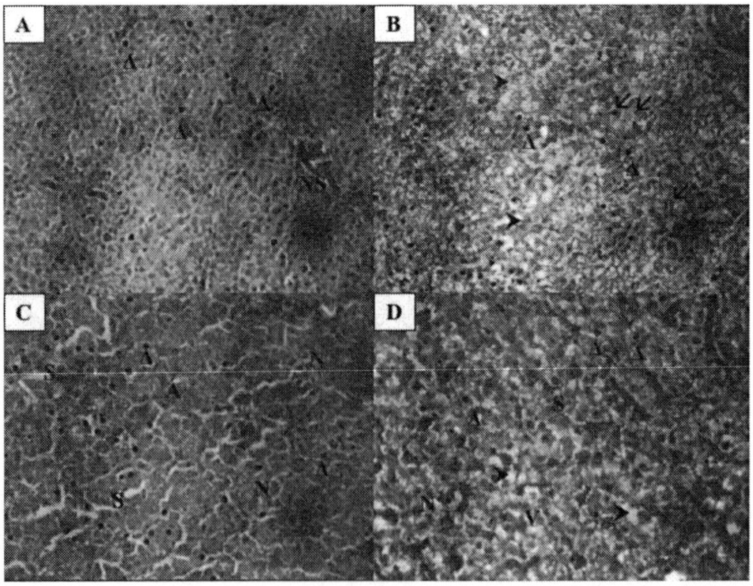

Fig. 3. Liver cross section of (A) control fish and fish exposed to (B) 8.8, (C) 17.6 and (D) 35.2 μg/L endosulfan under light microscope (H&E stain, x40). Symbol: normal sinusoids (NS); Kupffer cell (Δ); vacuolation of cytoplasm (➤); pyknotic nuclei (↓); sinusoid (S); necrosis (N); apoptosis (A).

Fig. 4 shows kidney cross sections of control fish and fish exposed to 8.8, 17.6 and 35.2 μg/L endosulfan under light microscope. The control fish in Fig. 4A is composed of distal convulated tubule, proximal convulated tubule, glomerulus and hematopoietic tissues, but the Bowman's space cannot be clearly observed. The peritubular capillaries can be observed with heavy nuclei of epithelial cells. Exposure to endosulfan causes inflammation and interstitial fluid mixing with blood as shown in Fig. 4B. Higher endosulfan concentration worsens the kidney condition, induces heavy bleeding and causes the lumen of the tubules to degenerate and become dilated (Figs. 4C and 4D). Other than that, intracytoplasmic vacuoles can be seen in epithelial cells of distal tubule, while haemorrhage occurs in hemopoietic cells. Previously, significant decrease in the dimension of Bowman's capsule and destruction of glomerulus shape due to precipitation of cytoplasm and karyolysis were also reported in *Channa punctatus* treated with endosulfan [33].

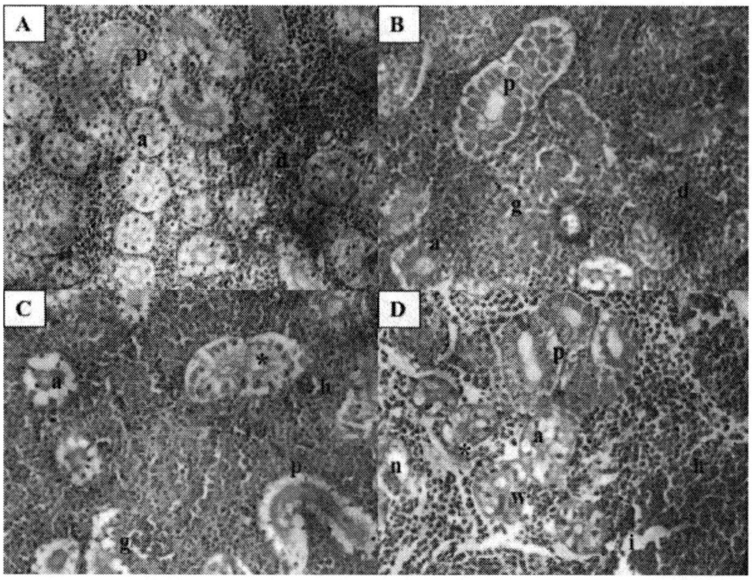

Fig. 4. Kidney cross section of (A) control fish and fish exposed to (B) 8.8, (C) 17.6 and (D) 35.2 μg/L under light microscope (H&E stain, x400). Symbol: renal tubule (a); hematopoietic tissue (d); proximal convoluted tubules (p); glomerulus (g); necrosis (n); haemorrhage (h); renal tubule degeneration (*), pyknotic (p); (D) intracytoplasmic (i) vacuoles; narrowing (w) of tubular lumen.

4 Conclusions

This work investigates the toxicity of endosulfan to juvenile *A. testudineus* using dose-response relationship and histopathological analysis. The toxicity was found to increase with endosulfan concentration. The high LC_{50} value obtained indicates that *A. testudineus* has high survival in contaminated surroundings. However, various structural changes were already induced on the morphology of the vital organs, i.e. gill, liver and kidney even with exposure to low, sublethal endosulfan concentration.

The authors acknowledge funding provided by Ministry of Higher Education Malaysia through Research Acculturation Collaborative Effort Grant (9017-00023).

References

1. M.L. Ballesteros, D.A. Wunderlin, M.A. Bistoni, Ecotoxicol. Environ. Saf. **72**, 199 (2009)
2. T. Neuparth, J.W. Bickham, C.W. Theodorakis, F.O. Costa, M.H. Costa, Bull. Environ. Contam. Toxicol. **76**, 242 (2006)
3. H.M. Dutta, D. Misquitta, S. Khan, Arch. Environ. Contam. Toxicol. **51**, 149 (2006)
4. A.M. Coimbra, M.A. Reis-Henriques, V.M. Darras, Comp. Biochem. Physiol. - C Toxicol. Pharmacol. **141**, 8 (2005)
5. H.M. Dutta, D.A. Arends, Environ. Res. **91**, 157 (2003)
6. P.B. Singh, V. Singh, P.K. Nayak, Food Chem. Toxicol. **46**, 2533 (2008)
7. G. Giusi, R.M. Facciolo, R. Alò, A. Carelli, M. Madeo, P. Brandmayr, M. Canonaco, Environ. Health Perspect. **113**, 1522 (2005)
8. J.B. Willey, P.H. Krone, Aquat. Toxicol. **54**, 113 (2001)
9. A. Balasubramani, T.J. Pandian, Curr. Sci. **94**, 883 (2008)
10. K.L. Gormley, K.L. Teather, Ecotoxicol. Environ. Saf. **54**, 330 (2003)
11. L. Beyger, R. Orrego, J. Guchardi, D. Holdway, Ecotoxicol. Environ. Saf. **76**, 71 (2012)
12. K. Sarma, A.K. Pal, K. Baruah, Turkish J. Fish. Aquat. Sci. **11**, 93 (2011).
13. R.H. Da Cuña, G. Rey Vázquez, M.N. Piol, N.V. Guerrero, M.C. Maggese, F.L. Lo Nostro, Ecotoxicol. Environ. Saf. **74**, 1065 (2011)
14. N. Kumar, P. Antony Jesu Prabhu, A.K. Pal, S. Remya, M. Aklakur, R.S. Rana, S. Gupta et al., Pestic. Biochem. Physiol. **99**, 45 (2011)
15. W. Kenneth, S. Willem, Open Environ. Pollut. Toxicol. J. **2**, 21 (2010)
16. M.C. Tellez-Bañuelos, A. Santerre, J. Casas-Solis, A. Bravo-Cuellar, G. Zaitseva, Fish Shellfish Immunol. **27**, 105 (2009)
17. S.E. Kegley, B.R. Hill, A.H.C. Orme, PAN Pesticide Database, Pestic. Action Network, North Am. (San Fr. CA). (2009). http://www.pesticideinfo.org/.
18. L.M. Salvo, I.L. Sinhorini, B.E. Malucelli, C. Klemz, D.C.O. Sanchez, L. Nicaretta, M.I.C. Malucelli et al. Brazilian J. Vet. Res. Anim. Sci. **45**, 87 (2008)
19. Y.S. Hii, M.Y. Lee, T.S. Chuah, Pestic. Biochem. Physiol. **89**, 46 (2007)
20. E. Capkin, I. Altinok, S. Karahan, Chemosphere. **64**, 1793 (2006)
21. R.I.M. Sunderam, D.M.H. Cheng, G.B. Thompson, Environ. Toxicol. Chem. **11**, 1469 (1992)
22. M.D. Ferrando, E. Sancho, E. Andreumoliner, J. Environ. Sci. Heal. Part B-Pesticides Food Contam. Agric. Wastes. **26,** 491 (1991)
23. M.P. Abdullah, Y.F.A. Aziz, M.R. Othman, W.M.A.W.M. Khalik, Iran J. Energy Environ. **6**, 141 (2015)
24. J.S. Bedi, J.P.S. Gill, R.S. Aulakh, P. Kaur, A. Sharma, P.A. Pooni, Sci. Total Environ. **463–464**, 720 (2013)
25. APHA. *Standard Methods for the Examination of Water and Wastewater*. 14th ed. Washington, DC. (American Public Health Association, 1998)
26. I.L. Nordin, N. Ibrahim, N. Hamidin, A.L. Othman Lutifi, N. Abdul Ghani, Adv. Environ. Biol. **9**, 341 (2005)
27. B.R. Pillai, K.D. Mahapatra, R.W. Ponzoni, L. Sahoo, P.L. Lalrinsanga, N.H. Nguyen, S. Mohanty et al. Aquaculture **319** (3–4) 347 (2011)
28. S.E. Lee, J.S. Kim, I.R. Kennedy, J.W. Park, G.S. Kwon, S.C. Koh, J.E. Kim, J. Agric. Food Chem. **51** (5) 1336 (2003)
29. I. Altinok, E. Capkin, Toxicol. Pathol. **35** (3) 405 (2007)
30. T. Geerinckx, M. Brunain, D. Adriaens, J. Morphol. **268**, 254 (2007)
31. A.L. Miranda, H. Roche, M.A.F. Randi, M.L. Menezes, C.A.O. Ribeiro, Environ. Int. **34** (7) 939 (2008)

32. M.L. Ballesteros, G.E. Bianchi, M. Carranza, M.A. Bistoni, J. Environ. Sci. Health. B. **42**, 351 (2007)
33. S. Banerjee, S. Bhattacharya, Ecotoxicol. Environ. Saf. **29** (3) 265 (1994)

AUTHOR INDEX

AUTHOR INDEX

AUTHOR INDEX

AUTHOR INDEX